W0043950

POST-TRANSCRIPTIONAL CONTROL OF GENE EXPRESSION IN PLANTS

Post-Transcriptional Control of Gene Expression in Plants

Edited by

Witold Filipowicz and Thomas Hahn

Reprinted from *Plant Molecular Biology*, Vol. 32(1, 2), 1996

KLUWER ACADEMIC PUBLISHERS

DORDRECHT / BOSTON / LONDON

Library of Congress Cataloging-in-Publication Data

ISBN-13: 978-94-010-6644-0   e-ISBN-13: 978-94-009-0353-1
DOI: 10.1007/978-94-009-0353-1

Published by Kluwer Academic Publishers,
P.O. Box 17, 3300 AH Dordrecht, The Netherlands

Kluwer Academic Publishers incorporates
the publishing programmes of
D. Reidel, Martinus Nijhoff, Dr. W. Junk and MTP Press.

Sold and distributed in the U.S.A. and Canada
by Kluwer Academic Publishers,
101 Philip Drive, Norwell, MA 02061, U.S.A.

In all other countries, sold and distributed
by Kluwer Academic Publishers Group,
P.O. Box 322, 3300 AH Dordrecht, The Netherlands.

Printed on acid-free paper

All Rights Reserved
©1996 Kluwer Academic Publishers
Softcover reprint of the hardcover 1st edition 1996

No part of the material protected by this copyright notice may be reproduced
or utilized in any form or by any means, electronic or mechanical,
including photocopying, recording or by any information storage and
retrieval system, without written permission from the copyright owner.

CONTENTS

Preface i–ii
Dedication iii
List of Contributors iv–viii

RNA processing and stability

Splicing of precursors to messenger RNA in higher plants: mechanism, regulation and sub-nuclear organisation of the spliceosomal machinery
G.G. Simpson, W. Filipowicz 1–41
Plant mRNA 3′-end formation
H.M. Rothnie 43–61
Control of mRNA stability in higher plants
M.L. Abler, P.J. Green 63–78
RNA as a target and an initiator of post-transcriptional gene silencing in transgenic plants
D.C. Baulcombe 79–88
RNA structure and regulation of gene expression
P. Klaff, D. Riesner, G. Steger 89–106

Translation

The plant translational apparatus
K.S. Browning 107–144
Translational control of cellular and viral mRNAs
D.R. Gallie 145–158
Translation in plant: rules and exceptions
J. Fütterer, T. Hohn 159–189

Fate of translation products

Molecular chaperones and protein folding in plants
R.S. Boston, P. Viitanen, E. Vierling 191–222
Transport of proteins in eukaryotic cells: more questions ahead
M. Bar-Peled, D.C. Bassham, N.V. Raikhel 223–249
Plasmodesmal cell-to-cell transport of proteins and nucleic acids
L.A. Mezitt, W.J. Lucas 251–273
Proteolysis in plants: mechanisms and functions
R.D. Vierstra 275–302

Organelles

Regulation of gene expression in plant mitochondria
S. Binder, A. Marchfelder, A. Brennicke 303–314
Regulation of gene expression in chloroplasts of higher plants
M. Sugita, M. Sugiura 315–326

Post-transcriptional regulation of chloroplast gene expression in *Chlamydomonas reinhardtii*
J.-D. Rochaix   327–341
RNA editing in plant mitochondria and chloroplasts
R.M. Maier, P. Zeltz, H. Kössel, G. Bonnard, J.M. Gualberto, J.M. Grienenberger   343–365

Special topics

Gene expression from viral RNA genomes
I.G. Maia, K. Séron, A.-L. Haenni, F. Bernardi   367–391
Optimizing expression of transgenes with an emphasis on post-transcriptional events
M.G. Koziel, N.B. Carozzi, N. Desai   393–405

Subject index   407–414

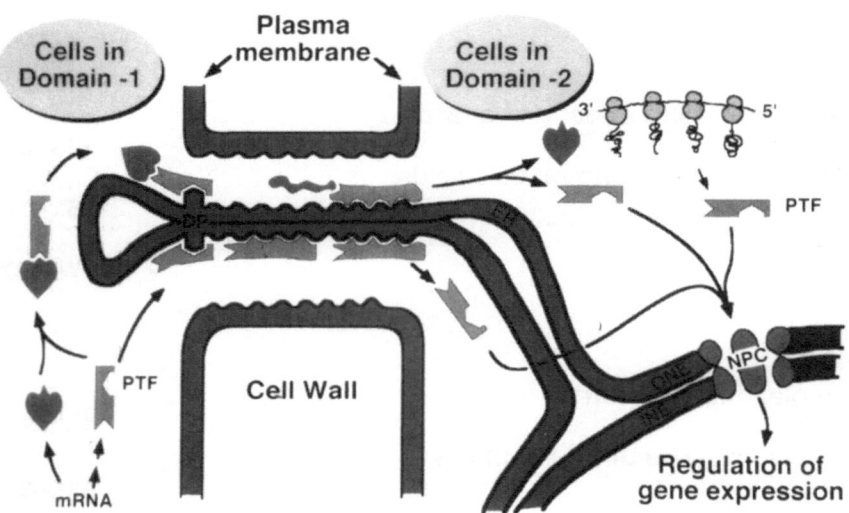

For details see Mezitt and Lucas, pp. 251–273

Plant Molecular Biology **32:** i–ii, 1996.

Preface

A recent volume in this series (Signals and Signal Transduction Pathways in Plants (K. Palme, ed.) Plant Mol. Biol. **26**, 2137–1679) described the relay races by which signals are transported in plants from the sites of stimuli to the gene expression machinery of the cell. Part of this machinery, the transcription apparatus, has been well studied in the last two decades, and many important mechanisms controlling gene expression at the transcriptional level have been elucidated. However, control of gene expression is by no means complete, once the RNA has been produced. Important regulatory devices determine the maturation and usage of mRNA and the fate of its translation product. Posttranscriptional regulation is especially important for generating a fast response to environmental and intracellular signals.

This book opens with chapters describing the processing of mRNAs, i.e. splicing (Simpson and Filipowicz) and polyadenylation (Rothnie). Usually, neither the splicing nor the polyadenylation signals of animals and fungi are recognized efficiently and correctly in plants, indicating that they are not strictly conserved among the eukaryotic kingdoms. In contrast to animal introns, the processing of plant introns depends on the presence of AU- and U-rich clusters. The known plant polyadenylation sites do not include a defined element downstream of the AATAAA motif as found in animals, the AAUAAA motif is not strictly conserved, and upstream elements are important. More and more examples of alternative mRNA splicing and polyadenylation sites are being identified in plants.

Degradation of RNA might be as important as RNA production for the well-being of cells. Some mRNAs encoding regulatory proteins must have a high turnover to avoid undesirable overproduction. Short-lived mRNAs have been found to contain *cis* elements that label the RNA for degradation (Abler and Green). An especially interesting mechanism of mRNA disposal occurs during posttranscriptional silencing (Baulcombe). In this case, RNA that had apparently been produced originally at high excess is degraded specifically. The mechanism of posttranscriptional silencing is basically unknown, difficult to explain in mechanistic terms and might involve events described by chaos theory.

Protein synthesis in eukaryotic cells is much more complex than originally thought. More and more factors governing pretranslational processes (initiation complex formation and scanning), the initiation itself, and elongation are becoming known (Browning). Sequences both on the leader and the trailer of mRNAs play important roles in the efficiency and regulation of translation (Gallie). Furthermore, not all eukaryotic mRNAs are strictly monocistronic. Plant viral RNAs, in particular, are frequently provided with complex leaders containing short open reading frames which control translation, and viral RNAs are often polycistronic (Fütterer and Hohn). In general, strategies of gene expression from viral genomes are unique and frequently also very complex (Maia et al.). For all the properties and functions of viral and cellular mRNAs, their secondary and tertiary structure is very important (Klaff et al.). An increasing number of RNA-binding proteins which specifically recognize different structural elements in RNA are being characterized.

The correct folding of many proteins has been found to depend on the action of chaperonins (Boston et al.). While some proteins remain in the cytoplams, others are transported to various subcellular localizations or are secreted (Bar-Peled et al.). A unique property of plant cells is that some proteins, and also nucleic acids, can be actively transported between cells using gates in the cell wall, the plasmodesmata. This novel macromolecular trafficking pathway may play a role in the programming of plant development (Mezitt and Lucas). Like mRNA, turnover, that of proteins is also regulated. Protein degradation in plants involves many proteolytic pathways operating in different cellular compartments (Vierstra).

Due to their probably origin from prokaryotes, plant organelle gene expression uses mechanisms more related to those of prokaryotes than those used by the cell for nuclear encoded genes. Regulation of organelle gene expression occurs mainly at the posttranscriptional level, i.e. RNA *cis* and *trans* splicing, processing of polycistronic transcripts (Rochaix; Sugita and Sugiura; Binder et al.) and finally, RNA editing (Maier et al.). The latter process occurs both in mitochondria and chloroplasts but its biochemistry remains largely unknown.

From the many examples of posttranscriptional regulation described in this book it is obvious that optimizing gene expression in transgenic plants cannot stop with the design of optimal promoters but must also consider the stability and translatability of the transcripts as well as certain structural properties of the protein products themselves (Koziel et al.).

We would like to dedicate this volume to the memory of Hans Kössel, our colleague and friend, and also coauthor of one of the chapters (Meier et al.) of this book, who passed away suddenly in December 1995. For many years Hans Kössel was a Professor of Molecular Biology and Genetics at the University of Freiburg, Germany. His research focused on various aspects of the chemistry and biology of nucleic acids. His more recent work on the structure and function of chloroplast genomes culminated in the discovery of RNA editing in these organelles.

Witold Filipowicz and
Thomas Hohn
Guest Editors

Hans Kössel (1934–1995)

List of Contributors

RNA processing and stability

Splicing of precursors to messenger RNA in higher plants: mechanism, regulation and sub-nuclear organisation of the spliceosomal machinery

G.G. Simpson
Friedrich Miescher Institute
P.O. Box 2543
CH-4002 Basel
Switzerland
Phone: 41-61-697-4128
Fax: 41-61-697-3976
E-mail: simpson@fmi.ch

W. Filipowicz
Friedrich Miescher Institute
P.O. Box 2543
CH-4002 Basel
Switzerland
Phone: 41-61-697-6993
Fax: 41-61-697-3976
E-mail: filipowi@fmi.ch

Plant mRNA 3′-end formation

H.M. Rothnie
Friedrich Miescher Institute
P.O. Box 2543
CH-4002 Basel
Switzerland
Phone: 41-61-697-6684
Fax: 41-61-697-3976
E-mail: rothnie@fmi.ch

Control of mRNA stability in higher plants

M.L. Abler
Plant Research Laboratory
Michigan State University
East Lansing, MI 48824-1312
USA
Phone: 1-517-353-4838
Fax: 1-517-353-9168

P.J. Green
Plant Research Laboratory
Michigan State University
East Lansing, MI 48824-1312
USA
Phone: 1-517-353-4838
Fax: 1-517-353-9168
E-mail: 22313pgi@ibm.cl.msu.edu

RNA as a target and an initiator of post-transcriptional gene silencing in transgenic plants

D.C. Baulcombe
The Sainsbury Laboratory
Norwich Research Park
Colney
Norwich NR4 7UH
UK
Phone: 44-1603-452571
Fax: 44-1603-250024
E-mail: baulcombe@bbsrc.ac.uk

RNA structure and regulation of gene expression

P. Klaff
Heinrich-Heine-Universität Düsseldorf
Institut für Physikalische Biologie
Universitätsstrasse 1
D-40225 Düsseldorf 1
Germany
Phone: 49-211-8115153
Fax: 49-211-8115167

D. Riesner
Heinrich-Heine-Universität Düsseldorf
Institut für Physikalische Biologie
Universitätsstrasse 1
D-40225 Düsseldorf 1
Germany
Phone: 49-211-8115153
Fax: 49-211-8115167

G. Steger
Heinrich-Heine-Universität Düsseldorf
Institut für Physikalische Biologie
Universitätsstrasse 1
D-40225 Düsseldorf 1
Germany
Phone: 49-211-8114840
Fax: 49-211-8115167

Translation

The plant translational apparatus

K.S. Browning
Department of Chemistry
University of Texas
Austin, TX 78712-1096
USA
Phone: 1-512-471-3973 or 4562
Fax: 1-512-471-8696
E-mail: kbronwing@mail.utexas.edu

Translation control of cellular and viral mRNAs

D.R. Gallie
Department of Biochemistry
University of California
Riverside, CA 92521
USA
Phone: 1-909-787-7298
Fax: 1-909-787-3590

Translation in plant-rules and exceptions

J. Fütterer
Institute of Plant Sciences
ETZ
Universitätsstrasse 2
CH-8092 Zürich
Switzerland
Fax: 41-1-6321044

T. Hohn
Friedrich Miescher Institute
P.O. Box 2543
CH-4002 Basel
Switzerland
Phone: 41-61-6977266
Fax: 41-61-6973976
E-mail: hohn@fmi.ch

Fate of translation products

Molecular chaperones and protein folding in plants

R.S. Boston
Department of Botany
North Carolina State University
Raleigh, NC 27695
USA

P. Viitanen
Dupond de Nemours Co.
Central Research and Development
Bldg. 402
Wilmington, DE 19880-0402
USA
Phone: 1-302-695-7032
Fax: 1-302-695-4296

E. Vierling
Department of Biochemistry
University of Arizona
Tucson, AZ 85721
USA
Phone: 1-520-621-1601 or 3977
Fax: 1-520-621-3709
E-mail: eliz@biosci.arizona.edu

Transport of proteins in eukaryotic cells: more questions ahead

M. Bar-Peled
Plant Research Laboratory
Michigan State University
East Lansing, MI 48824-1312
USA
Phone: 517-353-3518
Fax: 517-353-9168

D.C. Bassham
Plant Research Laboratory
Michigan State University
East Lansing, MI 48824-1312
USA
Phone: 517-353-3518
Fax: 517-353-9168

N.V. Raikhel
Plant Research Laboratory
Michigan State University
East Lansing, MI 48824-1312
USA
Phone: 517-353-3518
Fax: 517-353-9168

Plasmodesmal cell-to-cell transport of proteins and nucleic acids

L.A. Mezitt
University of California
Division of Biological Sciences
Section of Plant Biology
Davis, CA 95616
USA
Phone: 916-752-0617
Fax: 916-752-5410

W.J. Lucas
University of California
Division of Biological Sciences
Section of Plant Biology
Davis, CA 95616
USA
Phone: 916-752-0617
Fax: 916-752-5410
E-mail: wjlucas@ucdavis.edu

Proteolysis in plants: mechanisms and functions

R.D. Vierstra
University of Wisconsin-Madison
Department of Horticulture
1575 Linden Drive
Madison, WI 53706-1590
USA
Phone: 608-262-1490
Fax: 608-262-4743
E-mail: vierstra@macc.wisc.edu

Organelles

Regulation of gene expression in plant mitochondria

S. Binder
Universität Ulm
Abt. Allgemeine Botanik (Biol. II)
Albert-Einstein-Allee 11
D-89081 Ulm
Germany
Phone: 49-731-502-2610/2611
Fax: 49-731-502-2626

A. Marchfelder
Universität Ulm
Abt. Allgemeine Botanik (Biol. II)
Albert-Einstein-Allee 11
D-89081 Ulm
Germany

A. Brennicke
Universität Ulm
Abt. Allgemeine Botanik (Biol. II)
Albert-Einstein-Allee 11
D-89081 Ulm
Germany
Phone: 49-731-502-2610/2611
Fax: 49-731-502-2626

Regulation of gene expression in chloroplasts of higher plants

M. Sugita and M. Sugiura
Center for Gene Research
Nagoya University
Chikusa
Nagoya 464-01
Japan
Phone/Fax: 81-52-789-3081
E-mail: h44979a@nucc.cc.nagoya-u.ac.jp

Post-transcriptional regulation of chloroplast gene expression in *Chlamydomonas reinhardtii*

J.-D. Rochaix
Department of Molecular Biology
University of Geneva
30, Quai E. Ansermet
CH-1211 Genève
Switzeland
Phone: 41-22-7026111
Fax: 41-22-7026868

RNA editing in plant mitochondria and chloroplasts

R.M. Maier
Institut für Biologie III
Universität Freiburg
Schänzlestrasse 1
D-79104 Freiburg
Germany

P. Zeltz
Institut für Biologie III
Universität Freiburg
Schänzlestrasse 1
D-79104 Freiburg
Germany

G. Bonnard
Institut de Biologie Moléculaire des Plantes
12, rue du General Zimmer
F67084 Strasbourg Cédex
France
Phone: 33-88-417238
Fax: 33-88-614442

J.M. Gualberto
Institut de Biologie Moléculaire des Plantes
12, rue du General Zimmer
F67084 Strasbourg Cédex
France
Phone: 33-88-417238
Fax: 33-88-614442

J.M. Grienenberger
Institut de Biologie Moléculaire des Plantes
12, rue du General Zimmer
F67084 Strasbourg Cédex
France
Phone: 33-88-417238
Fax: 33-88-614442
E-mail: grienen@mito.u-strasbg.fr

Special topics

Gene expression from viral RNA genomes

I.G. Maia
Institut Jacques Monod
Université Paris 7/Tour 43
2 Place Jussieu
F-75251 Paris Cédex 05
France

K. Séron
Institut Jacques Monod
Université Paris 7/Tour 43
2 Place Jussieu
F-75251 Paris Cédex 05
France
Phone: 33-1-44274036
Fax: 33-1-44273580

A.L. Haenni
Institut Jacques Monod
Université Paris 7/Tour 43
2 Place Jussieu
F-75251 Paris Cédex 05
France
Phone: 33-1-44274036
Fax: 33-1-44273580

F. Bernardi
Institut Jacques Monod
Université Paris 7/Tour 43
2 Place Jussieu
F-75251 Paris Cédex 05
France
Phone: 33-1-44274036
Fax: 33-1-44273580

Optimizing expression of transgenes with emphasis on post-transcriptional events

M.G. Koziel
Ciba-Biotechnology Research
P.O. Box 12257
Research Triangle Park, NC 27709-2257
USA
Phone: 1-919-541-8593 or 8527
Fax: 1-919-541-8585

N.B. Carozzi
Ciba-Biotechnology Research
P.O. Box 12257
Research Triangle Park, NC 27709-2257
USA
Phone: 1-919-541-8593 or 8527
Fax: 1-919-541-8585

N. Desai
Ciba-Biotechnology Research
P.O. Box 12257
Research Triangle Park, NC 27709-2257
USA
Phone: 1-919-541-8593 or 8527
Fax: 1-919-541-8585

Plant Molecular Biology **32:** 1–41, 1996.
© 1996 *Kluwer Academic Publishers.*

1

Splicing of precursors to mRNA in higher plants: mechanism, regulation and sub-nuclear organisation of the spliceosomal machinery

G.G. Simpson[1] & W. Filipowicz
Friedrich Miescher-Institut, P.O. Box 2543,CH-4002 Basel, Switzerland
[1]*Present address: John Innes Centre, Colney, Norwich, NR4 7UH, UK*

Key words: pre-mRNA splicing, U snRNPs, alternative splicing, intron enhancement, coiled bodies, interchromatin, plant gene expression, *Arabidopsis thaliana*

Contents

Abstract	2
Introduction	2
The mechanism of pre-mRNA splicing	3
The reaction	3
Spliceosome composition	4
The spliceosome cycle	5
A dynamic RNA skeleton at the heart of the spliceosome	6
Conservation of the RNA skeleton in higher plants	8
Heterologous introns are often not processed in higher plants	8
A monocot-dicot difference in the processing of heterologous introns	8
Species differences in intron processing	8
Structural features of higher-plant introns	9
Intron size	9
5′ splice site	9
3′ splice site	11
Branch site	12
Higher-plant pre-mRNA introns are AU-rich	13
3′ splice site-proximal polypyrimidine tract	13
The higher-plant splicing machinery	14
U snRNAs	14
U snRNPs	14
SR proteins	15

The mechanism of pre-mRNA splicing in higher plants	15
5′ splice site selection	15
3′ splice site selection	16
Splice site pairing	17
Finding the splice sites: the importance of exons	17
Breaking down the contribution of AU-sequence: the importance of U	17
The role of AU-rich sequence	19
A mosaic of cis-acting elements determines the accuracy and efficiency of pre-mRNA splicing	19
Pre-mRNA secondary structure can affect pre-mRNA splicing	20
Alternative splicing	21
Introns affect the pattern of plant gene expression	24
Intron enhancement of gene expression	24
Effects of stress on pre-mRNA splicing	26
The sub-nuclear organisation of spliceosomal components	26
Mammalian nuclei: perichromatin fibrils, interchromatin granule clusters and coiled bodies	26
Higher-plant spliceosomal components exhibit a discrete sub-nuclear organisation	27

Higher-plant spliceosomal
  components localise to an
  interchromatin network   27
Higher-plant spliceosomal components
  concentrate in structures that
  resemble mammalian coiled bodies   28
Sliceosomal components in the plant
  nucleolar vacuole   29
The localisation of plant spliceosomal
  components is dynamic   29
Ribonucleoprotein domains in plant
  nuclei: unification of old and new
  observations   30

A classification of nucleolar associated
  bodies and dense bodies   30
Nucleolus-associated
  bodies/foci/coiled bodies   30
Dense bodies   30
Multiple small foci in late telophase   31
Resolution of previous ultrastructural
  analyses with recent
  immunofluorescence studies   32
Perspective   32
Acknowledgements   33
References   33

Abstract

The removal of introns from pre-mRNA transcripts and the concomitant ligation of exons is known as pre-mRNA splicing. It is a fundamental aspect of constitutive eukaryotic gene expression and an important level at which gene expression is regulated. The process is governed by multiple *cis*-acting elements of limited sequence content and particular spatial constraints, and is executed by a dynamic ribonucleoprotein complex termed the spliceosome. The mechanism and regulation of pre-mRNA splicing, and the sub-nuclear organisation of the spliceosomal machinery in higher plants is reviewed here.

Heterologous introns are often not processed in higher plants indicating that, although highly conserved, the process of pre-mRNA splicing in plants exhibits significant differences that distinguish it from splicing in yeast and mammals. A fundamental distinguishing feature is the presence of and requirement for AU or U-rich intron sequence in higher-plant pre-mRNA splicing. In this review we document the properties of higher-plant introns and *trans*-acting spliceosomal components and discuss the means by which these elements combine to determine the accuracy and efficiency of pre-mRNA processing. We also detail examples of how introns can effect regulated gene expression by affecting the nature and abundance of mRNA in plants and list the effects of environmental stresses on splicing.

Spliceosomal components exhibit a distinct pattern of organisation in higher-plant nuclei. Effective probes that reveal this pattern have only recently become available, but the domains in which spliceosomal components concentrate were identified in plant nuclei as enigmatic structures some sixty years ago. The organisation of spliceosomal components in plant nuclei is reviewed and these recent observations are unified with previous cytochemical and ultrastructural studies of plant ribonucleoprotein domains.

Introduction

Precursor-mRNA splicing is the process by which intervening sequences (introns) are excised from pre-mRNA transcripts and exons are concomitantly ligated. It is a fundamental feature of constitutive gene expression in eukaryotes and a key step at which gene expression is regulated. It is closely linked to multiple processing events which effect the maturation of pre-mRNA; for example, there is mounting evidence that splicing and 3′ end formation, and splicing and nuclear mRNA export are coupled processes [reviewed in 108 and 18 respectively].

Pre-mRNA splicing can, but need not necessarily, occur co-transcriptionally [14, 292, 304]. It takes place within a large ribonucleoprotein complex termed the spliceosome, which assembles on substrate pre-mRNA in a step-wise manner governed by multiple *cis*-acting elements of limited sequence content and particular spatial constraints. It is a dynamic complex that exhibits multiple conformational and compositional changes as it progresses through a cycle of assembly, disassembly and reassembly [reviewed in 148, 171, 235].

Interest in pre-mRNA splicing in higher plants increased when it was realised that heterologous introns are usually not processed in transformed plant cells (see below). This indicated that splicing in plants has some unique requirements that distinguish it from splicing in mammals and yeast. A basic understanding of plant gene expression therefore requires a dissection of those aspects which distinguish pre-mRNA splicing in higher plants from that in other organisms.

Besides this fundamental question, pre-mRNA splicing in higher plants has attracted attention because of its impact on diverse aspects of plant cell biology and pathology. Pre-mRNA splicing underpins, for example, regulated gene expression through alternative splicing [63, 272], mutant phenotypes of maize through cryptic splicing of transposable elements [reviewed in 143, 279], mutant phenotypes of barley, such as the *Hooded* mutation, through duplication of intron sequence [173] and plant pathogenesis: certain plant viruses [73, 111, 174, 229] and even prokaryotic pathogens of plants such as *Agrobacterium rhizogenes* [149] possess introns that must be spliced for pathogenicity. Recently, higher-plant spliceosomal components have been localised to discrete sub-nuclear compartments and have thus provided compositional markers for, and a means to access the function of, previously enigmatic nuclear structures first identified in plant nuclei almost 60 years ago [13, 38, 52, 57].

In addition, pre-mRNA splicing in higher plants has been exploited in the optimisation of transgene expression. The presence of introns in transcription units can dramatically enhance transgene expression in higher plants [reviewed in 143, 245, Koziel *et al.*, this volume] and, as introns themselves can code for functional RNAs (many yeast and vertebrate snoRNAs are expressed, after processing, from introns of RNA Pol II transcribed mRNA genes [157]), they offer a novel route for transgene expression (of ribozymes, for example). Recent studies with transgenic plants exhibiting post-transcriptional transgene cosuppression have raised the possibility that pre-mRNA splicing may be perturbed in such plants [36, see Baulcombe, this volume].

It is therefore clear that the analysis of pre-mRNA splicing in higher plants is essential to the understanding of fundamental aspects of plant gene expression and diverse aspects of plant cell biology and pathology as well as the derivation of expression strategies for transgenes.

The mechanism of pre-mRNA splicing

Pre-mRNA splicing can be studied *in vitro* in extracts prepared from yeast cells (unless otherwise stated, reference to yeast in this review corresponds to *Saccharomyces cerevisiae*) or HeLa nuclei. This facility, coupled with the rigorous genetic analysis of RNA processing *in vivo* in yeast, has meant that, overwhelmingly, the study of pre-mRNA splicing has been the study of the process in these organisms or experimental systems. In contrast, no plant-derived extract capable of faithfully processing plant pre-mRNAs *in vitro* has yet been developed and, as a result, the mechanistic analysis of pre-mRNA processing in higher plants has been severely hampered. In order to introduce the process of pre-mRNA splicing here, we have therefore drawn primarily from knowledge obtained from the study of the process in yeast and HeLa cells.

The reaction

Pre-mRNA splicing proceeds in two steps through successive in-line *trans*-esterification reactions [170]. The first step involves nucleophilic attack at the 5' splice site phosphate by the 2'hydroxyl of the branchpoint adenosine (Fig. 1). The reaction products are the lariat intermediate and free 5' exon. The lariat is formed by the joining of the 5' end of the intron to the adenosine of the branch site via an unusual 2'–5' phosphodiester bond. The second step involves nucleophilic attack at the 3' splice site phosphate by the 3' hydroxyl of the free 5' exon liberated in the first step. The resultant products are ligated exons and released lariat intron. The spliceosome probably shifts between two active sites (or remodels a single site) for catalysis of the two *trans*-esterifications [170].

The many mechanistic similarities between nuclear pre-mRNA splicing and group II self-splicing and the analogy of the successive in-line *trans*-esterification

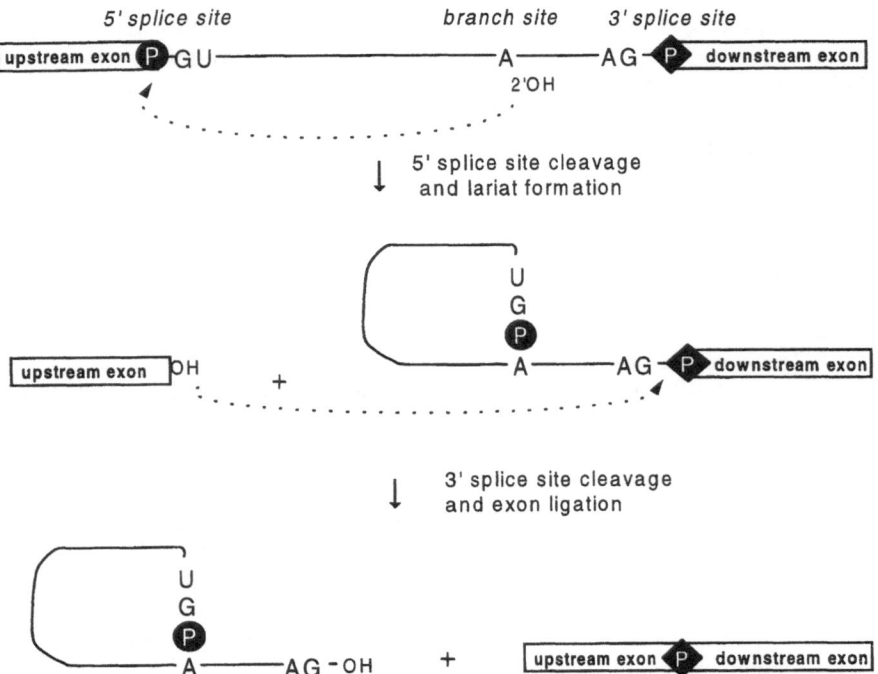

Figure 1. Pre-mRNA splicing is a two-step reaction. Conserved nucleotides and the phosphates at the splice sites are shown. The dashed arrows represent nucleophilic attack of the hydroxyl groups on the splice junctions. The exons are shown as boxes and the intron as a line. (After [117].)

reactions to catalysis in group I self-splicing has led to a widespread expectation that spliceosomal catalysis will be RNA-mediated [see 148, 181, 183, 271, 280, for recent discussions]. The active spliceosome is assembled in a stepwise manner and at its heart lies a dynamic skeleton of multiple RNA-RNA interactions that are stabilised and destabilised by proteins that facilitate their rearrangement as the reaction proceeds. The mechanistic dissection of spliceosomal catalysis is therefore based on an analysis of spliceosomal composition and assembly, and the functional mapping of the dynamic RNA skeleton.

Spliceosome composition

The major subunits of the spliceosome are the uridylate-rich small nuclear ribonucleoprotein particles (U snRNPs), U1, U2, U4/U6 and U5. Each U snRNP contains one U snRNA, except U4/U6 which contains U4 and U6 snRNAs complexed through base-pairing interactions. Each U snRNA (except U6) possesses a tri-methyl guanosine cap, and a short, single-stranded conserved sequence (consensus: $RAU_{3-6}GR$) flanked by stems, known as the Sm-antigen binding site (or Sm site). This sequence is the site of assembly of

eight core proteins (B, B', D1, D2, D3, E, F and G) common to each U snRNP. The core proteins share a common structural motif which can be used to identify them [232]. In addition, all U snRNPs contain characteristic proteins which are specifically associated with them. For example, the human U1 snRNP has three specific proteins stably associated with it, U1A, U1C and U1-70K, while the U5 snRNP has eight [see 117].

In addition to the spliceosomal U snRNPs, a large number of other non-U snRNP proteins are essential for constitutive and regulated splicing. Reed and co-workers have revealed the presence of 30–50 different spliceosome associated proteins in purified mammalian spliceosomes and spliceosome assembly intermediates [11, 86], while genetic analyses in *S. cerevisiae* have identified more than 30 genes encoding factors necessary for pre-mRNA splicing (genes identified as mutants defective in pre-mRNA splicing typically possess the prefix *PRP* for precursor RNA processing). The complex protein composition of mammalian and yeast spliceosomes has recently been reviewed [117 and 9 respectively], but two classes of protein warrant additional attention: the SR proteins and the RNA-dependent ATPases or helicases.

SR proteins are characterised by the presence of one or more copies of an RNA-binding domain called the RNP motif (also known as the RNA-binding domain [RBD] or RNA recognition motif [RRM]) and the presence of phosphorylated domains highly enriched in alternating serine/arginine repeats (hence the name) [70, 221, 299]. They participate in both constitutive [71, 116] and regulated pre-mRNA splicing [reviewed in 272], acting early to facilitate commitment of pre-mRNAs to splicing [48, 69] and, later, to escort U snRNP addition [219]. They remain associated with the spliceosome through both steps of the reaction and some remain associated with either the excised intron lariat or the spliced exon products [17]. They are numerous, abundant and exhibit tissue-dependent differences in expression [300]. They concentrate in sub-nuclear domains called inter-chromatin granule clusters (see below), the targeting to which is facilitated by the SR domain itself [133]. The SR domain is a likely target of phosphorylation-dependent regulation of splicing [163]. A family of SR proteins (the 104 family) was first purified by a simple $(NH_4)_2SO_4$ fractionation, $MgCl_2$ precipitation procedure [299] and could be specifically detected by a monoclonal antibody, mAb 104, which recognises a phosphoepitope in the SR domain. More recently, the development of an alternative purification procedure, the Mg^{2+} dependent precipitation/high salt protocol, has revealed the existence of many more SR proteins (at least 25) in HeLa nuclei, not all of which react with mAb 104 [17]. Despite their widespread role in metazoan pre-mRNA splicing and their identification in plants (see below) no yeast analogues have yet been described.

RNA-dependent ATPases and/or helicases can be identified by a set of core conserved motifs that include the DEAD (Asp-Glu-Ala-Asp) box, which gives the broader family its name [reviewed in 72]. Two yeast genes essential for pre-mRNA splicing: *PRP5* and *PRP28*, encode DEAD-box proteins. Three other essential genes, *PRP2*, *PRP16* and *PRP22*, encode proteins which possess a DEAH rather than a DEAD-box motif and which, in addition, possess conserved sequence distinctions in some of the other core motifs [44, 230]. None of these proteins have been shown to possess RNA helicase activity, but they do possess RNA-dependent ATPase activity [109, 230]. They play critical roles in modulating RNA-RNA interactions in the spliceosome and in switching spliceosome conformation [231]. By determining the timing of such conformational switches, in a manner coupled to ATP hydrolysis, they can affect splicing fidelity [26, 27].

The conformational switches themselves contribute to the timed progression of the spliceosome through the splicing reaction. This progression can be blocked by the expression of dominant negative mutants of these proteins; for example, the overexpression of a dominant negative mutant of PRP2 in yeast cells leads to the accumulation of stalled splicing complexes [206].

The spliceosome cycle

U snRNPs assemble on the pre-mRNA via an ordered pathway that requires the transient association of diverse non-U snRNP proteins. The defined pathway of assembly and disassembly is called the spliceosome cycle.

Nascent pre-mRNA transcripts first associate with multiple distinct heterogenous nuclear ribonucleoprotein particle (hnRNP) proteins to form hnRNP complexes [56]. A similar set of proteins bind to exogenously added pre-mRNA *in vitro*, in HeLa extract, generating the so-called H complex [29, 113]. This complex is not a functional intermediate in spliceosome assembly, nor a complex specific to the splicing reaction, but it is the substrate recognised by the splicing machinery. Moreover, hnRNP proteins exhibit a transcript-dependent association with pre-mRNAs [10] and they can influence subsequent processing events, such as splice site choice [134, 158]. Most hnRNP proteins dissociate from the pre-mRNA upon assembly of the active spliceosome, but some, like hnRNP C, may be important at later stages [42], while others, like hnRNP F, are present in regulated spliceosomal complexes [165]. No yeast hnRNP proteins that are obvious homologues of any mammalian hnRNP protein have yet been identified (yeast hnRNP proteins have been reviewed by Swanson [258]).

In yeast and mammals, the first splicing-specific complex formed on pre-mRNA is the ATP-independent commitment complex, or E complex [reviewed in 97]. The term commitment is operationally defined by the preferential splicing of committed pre-mRNA when excess competitor pre-mRNA is subsequently added. It is the rate-limiting step of pre-mRNA splicing *in vitro*. In mammals, the complex is composed (at least) of U1 snRNP bound at the 5' splice site, the U2 snRNP auxilliary factor (U2AF: a heterodimer composed of subunits U2AF[65] and U2AF[35]) bound to the 3' splice site-proximal polyypyrimidine tract and SR proteins that stabilise the binding of U1 snRNP [112] and bridge interactions between it and U2AF [291]. Individual SR proteins exhibit differing capacities to commit pre-mRNAs to splicing [69] and purified SR proteins can commit pre-mRNAs to splicing in the absence of U1 snRNP [48, 263, 264]. U2AF is required for the subsequent addition of the U2 snRNP. A functionally similar complex can be identified in *S. cerevisiae*, but the details are somewhat different. Splicing of pre-mRNAs in *S. cerevisiae* does not require a 3' splice site-proximal polypyrimidine tract, but a strictly conserved branch site sequence is needed. A yeast protein, MUD2, which resembles U2AF^{65} and which requires an intact branch site for its association, is present in yeast commitment complexes [1]. MUD2 interacts with proteins of the U2 snRNP and directly or indirectly with the U1 snRNP, indicating that it may function in a related manner to U2AF [1].

The next step in spliceosome assembly, in both yeast and mammals, is the ATP-dependent addition of the U2 snRNP, leading to the formation of the prespliceosome or A complex. In mammals, A complex can be formed in the absence of the 5' splice site but requires a 3' splice site and a proximal polypyrimidine tract. In yeast, however, the 5' splice site and branch site are required, but sequences downstream of the branch site are dispensable. Proteins associated with the U2 snRNP bind to an anchoring site upstream of the branch site and associate with one another in a manner that predicts they wrap the RNA into a particular three-dimensional structure [87]. U2 snRNP proteins involved in this binding [87, 282] and the addition of U2 snRNP [see 97, 117] are conserved in mammals and yeast.

Next, U4/6 and U5 snRNPs enter the prespliceosome as a tri-U snRNP particle (U4/U6.U5) resulting in the formation of splicing complex B, which is converted to the active splicing complex C after a conformational rearrangement. There is a change in the proteins that bind around the branch site from those seen in A complex [87, 146]. The *trans*-esterification reactions proceed and spliced RNA is released. The excised intron lariat remains associated with U2, U4, U5 and U6 in a post-splicing complex prior to being debranched and degraded. The unusual 2'-5' phosphodiester bond of the intron lariat is debranched by a specifc enzyme that has been purified from mammals [224] while the gene encoding it, *PRP26*, has been cloned from yeast [40]. Disassembly of the spliceosome is an active ATP-dependent process [44, 228].

A dynamic RNA skeleton at the heart of the spliceosome

The stepwise assembly of the spliceosome mediates a dynamic array of RNA-RNA interactions at the heart of the catalytically active complex. The documentation of this RNA skeleton is outlined below and illustrated in Fig. 2.

The 5' splice site (AG/GURAGU) is initially recognised by U1 snRNP through Watson-Crick base-pairing interactions with the 5' sequence of U1 snRNA (5'-GpppAU<u>ACUUACCU</u>-3') [306]. This interaction is not sufficient to specify the exact location of 5' splice site cleavage but contributes to the commitment of the pre-mRNA to splicing [234]. Subsequently, U2 snRNP binds at the branch site [201, 290, 305]. The duplex formed between the U2 snRNA sequence 5'-GUAGUA-3' and the branch site sequence 5'-UACUA<u>A</u>C-3' (or its variant) shifts the branchpoint adenosine (<u>A</u>) into a bulged position (it has no base opposite it), thereby contributing the 2'OH group for the first step of splicing [210]. U4/U6.U5 enters the spliceosome and there is then a major rearrangement of the spliceosome involving all five U snRNAs prior to the first catalytic step. The precise order of events regarding these rearrangements is not certain. The invariant loop I of U5 snRNA (5'-GCCUUUUAC-3') associates with non-conserved nucleotides close to the 5' splice site in exon 1 [182, 250, 293]. The base-pairing between U1 and the 5' splice site is destabilised and replaced by a new interaction in which part of an invariant U6 sequence (<u>ACAGAG</u> in *S. cerevisiae*, <u>ACAGAG</u> in mammals) base-pairs with a subset of the same intron nucleotides [106, 132, 278]. The extensive base-pairing between U4/U6 is dissolved and replaced with two new helicies; an intermolecular helix between U2 and U6 called helix I, which has two parts: helix Ia and Ib, and a U6 intramolecular helix. This network of RNA interactions between U6 and the 5' splice site, U6 and U2 (helix I), and between U2 and the branchpoint, establishes a tight physical link between the reaction components of the first step: the 5' splice site and the branchpoint. Sun and Manley [256] failed to find evidence for helix Ib in HeLa cells, and instead, described a U2-U6 helix III that involves bases extending 3' from the branch site recognition sequence in U2 and 5' from an evolutionarily conserved U6 sequence which would also function in bringing the 5' splice site and branch site together for the first catalytic step. The interaction that signals catalytic activation is not known, but the first *trans*-

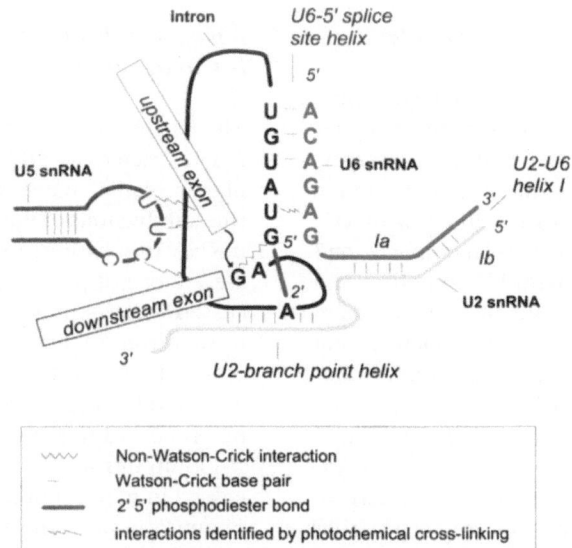

Figure 2. A network of RNA-RNA interactions in the spliceosome. A model for the catalytic core of the spliceosome poised for the second *trans*-esterification reaction. Interactions that have been identified by either genetic analyses or photochemical *in vitro* cross-linking are shown. Represented in this way, the RNA-RNA interactions exhibit similarities with group II self-splicing introns. (Derived from [148, 181, 289]; for further details, see [271].)

esterification proceeds, yielding the lariat intermediate and free 5′ exon. The transition to the second step of splicing is accompanied by another major conformational change within the spliceosome [231]. After the first cleavage reaction, U5 remains associated with the free 5′ exon and establishes a new interaction with the 3′ exon [182, 250]. U5 may therefore serve to tether the free 5′ exon and facilitate alignment of the nucleophile (the 3′ hydroxyl of the free exon) for the second step. The position of U6 also shifts: the penultimate nucleotide of the sequence ACAGAG becomes closely apposed to the conserved U of the GU dinucleotide at the 5′ splice site [250] and the adjacent G of the sequence ACAGAG makes a tertiary interaction with the bulged nucleotides in the U2-U6 helix I [147]. Mutational analyses in yeast has revealed that both the terminal A and G residues in the sequence ACAGAG can influence cleavage at the 3′ splice [132, 147]. A non-Watson-Crick interaction between the two terminal guanosines of the intron (/GU; AG/) is required for the second step of splicing to proceed [200] and this interaction indicates a means by which U6, associated with the 5′ splice site through the sequence ACAGAG, is brought into the proximity of the 3′ splice site. The association of U5 with the 3′ exon is destabilised after the second step and it associates with sequence in the lariat intron [278].

The U2-U6 helix I is widely conjectured to represent the catalytic core of the spliceosome. Its positional rearrangement with respect to the 5′ splice site and 3′ splice site in the course of the reaction is consistent with the spliceosome having two active sites, or re-modelling a single site, for the two *trans*-esterification reactions. The evidence supporting this conjecture and a model framing the corresponding data has been reviewed by Madhani and Guthrie [148]. A representation of the RNA skeleton based on known interactions is illustrated in Fig. 2. A degree of consensus has supported this structure, which, represented in this way, displays structural similarities to the arrangement of self-splicing group II introns [148, 181, 183, 289]. However, the definitive dissection of the dynamic RNA skelton is far from complete. A U2-U6 helix II has been reported to be required for splicing in mammals, but not in yeast, while U2-U6 helix Ib is required in yeast but not mammals. In addition, Sun and Manley [256] have suggested that incorporation of the U2-U6 helix III structure enables one to draw parallels with the structure of a hairpin ribozyme. Field and Friesen [61] have revealed that U2-U6 helix II does form in yeast, but it is essential only if there are mutations disrupting U2 helix I and the U2-U6 helix Ib. That neither U2-U6 helix Ib nor II is essential, but that both are involved in splicing, suggests they are both part of a larger structure that may act cooperatively [61].

Conservation of the RNA skeleton in higher plants

Each of the key RNA sequences that form the dynamic skeleton of the spliceosome are conserved in higher plants: the 5′ sequence of U1 snRNA that interacts with the 5′ splice site; the U2 sequences that interact with the branch site and that form helix I with U6; the U5 loop I sequence that makes exon contacts; and the U6 sequences that interact with U2 in helix I and that interact with the 5′ splice site, are all invariant among plant U snRNAs [84, 252]. The sequences that comprise the U2-U6 helix III [256] are highly conserved but not invariant in plant U snRNAs. The *cis*-acting elements are similarly conserved. The 5′ and 3′ splice site dinucleotides (/GU; AG/) are almost invariant in higher-plant introns and the consensus derived for the extended splice site sequences are very similar to those derived for *S. cerevisiae* and mammals (see below). In addition, functional branch sites have been mapped to adenosine residues within wider sequences that closely fit the degenerate branch site consensus of mammals [136]. These multiple lines of relatedness in the spliceosomal RNA skeleton enable us to draw a simple conclusion: the fundamental mechanism of pre-mRNA splicing in higher plants, in all likelihood, closely resembles that of yeast and mammals. The most direct evidence that this is indeed the case, comes from the demonstration that pre-mRNA splicing in plants proceeds through a branched lariat intermediate [136].

Heterologous introns are often not processed in higher plants

Despite the similarities in the sequences that act at the core of the spliceosome, the splicing of pre-mRNAs in heterologous systems is not without restrictions. For example, pre-mRNAs possessing mammalian introns or hybrids of plant and mammalian introns are generally not spliced or are incorrectly processed when expressed in plant cells [6, 83, 202, 226, 275, 285]. One exception is the small t intron of Simian Virus 40 which is efficiently and accurately processed in tobacco [101]. When tested reciprocally, certain plant introns are either not processed, or are inefficiently or incorrectly processed in transfected mammalian cells or HeLa cell *in vitro* splicing extracts [24, 226, 285] while others are efficiently and accurately processed [24, 92, 226, 285].

A monocot-dicot difference in the processing of heterologous introns

The barriers to processing of heterologous introns are not restricted to differences between mammals and higher plants, but extend to define a monocot/dicot (the subdivsions of the angiosperms) distinction in pre-mRNA processing. Several introns of monocot origin are either not processed or are only inefficiently processed when expressed in dicot cells [83, 107]. In contrast, introns of dicot origin appear to be accurately and efficiently processed when expressed in monocots [83, 205, 261]. In fact, introns of mammalian origin can be processed in monocot plants. For example, human β-globin intron 1 is correctly, albeit inefficiently, processed in maize, but is not processed at all in *N. plumbaginifolia* [83]. In general terms, it appears that pre-mRNA splicing in monocots is more permissive than in dicots. Intron secondary structure has a more profound inhibitory effect on pre-mRNA splicing in dicots than in monocots (see below) [83] underlining the fact that pre-mRNA processing in the different subdivisions of the angiosperms is not equivalent.

Species differences in intron processing

In addition to the monocot/dicot distinction in processing pre-mRNA described above, different dicot species may also exhibit differences in processing activity, although this has not been well studied. The first indication that this is the case comes from the characterisation of the maize *En/Spm* transposon expressed in different dicot species. The *En-1*-encoded genes, *tnpA* and *tnpD*, are generated by alternative splicing. In maize, the natural host, the ratio of *tnpA* transcripts to *tnpD* transcripts is 100:1 and a similar ratio is observed upon expression in *Arabidopsis* [30, 76, 152]. However, when expressed in potato, this ratio is overturned and *tnpD* transcripts predominate. In addition, northern analyses reveal the presence of additional, aberrant *tnpA* and *tnpD* transcripts in tobacco and potato, that indicate more extensive cryptic splicing events in these plants than in *Arabidopsis* [30, 64, 102]. Another example of such a difference comes from the study of splicing of a transposable gene trap derived from the maize transposable element, *Activator*, in *Arabidopsis* and tobacco. Nussaume *et al.* [189] incorporated RT-PCR, cloning and sequence analysis of spliced products and revealed qualitative and quantitative differences in processing in these two dicot plants. In comparison to tobacco, processing in *Ara-*

bidopsis was at reduced efficiency and with the selection of fewer cryptic 3′ splice sites [189]. The molecular bases underpinning these subtle differences are not known.

Structural features of higher-plant introns

The distinctions in the processing of heterologous introns outlined in the preceding sections reveal that other *cis*-acting elements contribute to pre-mRNA splicing in addition to the splice sites and branch site sequences and/or that the relative importance of particular *cis*-acting elements differs in different species. In order to determine the molecular basis that underpins these distinctions, it is first necessary to consider the structural features of plant introns and define the functional role they play.

Most higher plant mRNA genes contain introns and the majority contain multiple introns. Transcription units can be very complex; the gene encoding the 140 kDa subunit of RNA Polymerase (Pol) II in *Arabidopsis* contains 24 introns [126] while a gene encoding an *Arabidopsis* acetyl-coenzyme A carboxylase contains 31 introns [217]. Genes encoding highly conserved eukaryotic proteins such as histones which lack introns in mammals, can contain introns in higher plants [37], but introns are not a universal feature of plant pre-mRNAs; maize zein genes, for example, lack introns.

Intron size

Higher-plant pre-mRNA introns vary widely in length, but around two-thirds are shorter than 150 nt and most fall within a size range of 80–139 nt [63, 239, 240]. Higher-plant introns are therefore typically much shorter than most mammalian introns. A minimal functional length of 70 nt has been determined for the efficient processing of synthetic introns in higher plants [82] and consistent with this, naturally occurring introns shorter than 70 nt are rare [143, 239]. Very few plant introns are longer than 2–3 kb and the longest identified plant intron, as far as we are aware, at 7 kb, is encoded by the maize *Pericarp* gene [88].

5′ splice site

The 5′ splice site dinucleotide /GU is conserved in virtually all naturally occurring plant introns (Table 1). However, in rare instances (ca. 0.5% of *Arabidopsis* introns) the 5′ splice site dinucleotide is /GC [115,

239] (Table 1). Interestingly, a /GC 5′ splice site is conserved in seven myrosinase genes characterised from three different plant species [295]. Mutation of /GU in plant introns typically results in abolition of splicing to this site [31, 111] and, if available, the activation of cryptic splice sites upstream [160]. However, unusual progression of splicing of a synthetic pre-mRNA, called Syn7 (described in detail below), bearing mutations in the 5′ splice site, have been observed [136]. Mutation of the 5′ splice site /$\underline{G}$U → A$\underline{U}$ in yeast, mammals and higher plants results in an inhibition of the second step of splicing (see above) and an accumulation of the intermediates [39, 136, 200, 222]. The suppression of this defect by mutations at the terminal A$\underline{G}$/ has indicated that a nonWatson-Crick interaction between these bases is required for the second step of splicing (see above) [200]. However, the /$\underline{G}$U → A$\underline{U}$ mutation in derivatives of Syn7 allows some splicing through to AG/ in addition to a cryptic AU/ 3′ splice site [136]. A /G$\underline{U}$ → G$\underline{A}$ mutation in the same intron has only a minor inhibitory effect on RNA processing [136] and yet results in a strong inhibition of splicing, with accumulation of intron-exon 2 lariat intermediate, in mammals [2] and yeast [222]. It may be that employing a pre-mRNA substrate devoid of cryptic 5′ splice sites is essential to visualise splicing to the GA 5′ splice site. In the context of a natural pea rubisco gene intron, the /G$\underline{U}$ → G$\underline{A}$ mutation inhibits utilisation of the wild-type site and activates cryptic sites upstream [160].

The wider 5′ splice site consensus in higher plants, AG/GUAAGU, matches that derived for mammals and closely resembles that derived for *S. cerevisiae*, /GUAUGU (Table 1). However, differences in detail are apparent: the consensus nucleotide A_{+4} is less strictly conserved in higher-plant introns than in mammals and yeast, but plant introns exhibit a clear bias against G at this position. Similarly, G_{+5} is less strictly conserved in plant 5′ splice sites than in either yeast or mammals. G_{+5} plays an important role in splice site fidelity in *S. cerevisiae* [199], but in mammals, mutation of this residue only causes a decrease in the efficiency of splicing when the degree of complementarity to U1 snRNA is low [179]. Mutation of this residue in pea *rbcs3A1* intron 1 weakens the competitive strength of the site it is in, but results in only a slight reduction of splicing efficiency [160]. This indicates that, as in mammals, this residue does not play as critical a role in pre-mRNA splicing in higher plants as it does in *S. cerevisiae*.

Table 1. Nucleotide frequencies and consensus sequences at the 5′ and 3′ splice sites. Frequencies for vertebrate sequences are taken from Goodall *et al.* [84]. The *S. cerevisiae* consensus is from Rymond and Rosbash [225]. The monocot and dicot sequences are from Simpson *et al.* [239] and the *Arabidopsis* data from Korning *et al.* [115]. The 5′ and 3′ consensus nucleotides are in bold. Enrichment in T or T + A nucleotides is not highlighted since it continues throughout the intron sequence. T residues in the −20 to −6 region of the 3′ splice site are on average 8 and 6% more frequent than further upstream (−50/−21 region) in the dicot and monocot plants, respectively. This contrasts with the 25% enrichment in T + C residues in the −15/−5 region, corresponding to the polypyrimidine tract, in vertebrate introns.

		5′ Splice site									
		exon				intron					
		−3	−2	−1	↓	1	2	3	4	5	6
Vertebrate	%G	16	12	**79**	↓	**100**	0	30	12	**80**	21
	%A	33	**63**	9	↓	0	0	63	73	9	15
	%T	11	13	6	↓	0	**100**	3	7	5	**49**
	%C	40	12	6	↓	0	0	3	9	5	15
Consensus			**A**	**G**	↓	**G**	**T**	**A**	**A**	**G**	**T**
Dicot plant	%G	17	8	**79**	↓	**100**	0	11	3	**50**	10
	%A	36	**62**	9	↓	0	0	69	58	23	22
	%T	15	20	9	↓	0	**99**	15	25	18	**53**
	%C	31	10	2	↓	0	1	4	14	9	16
Consensus			**A**	**G**	↓	**G**	**T**	**A**	**A**	**G**	**T**
Monocot plant	%G	19	7	**78**	↓	**100**	0	19	7	**59**	10
	%A	40	**65**	7	↓	0	0	64	44	17	17
	%T	7	14	7	↓	0	**99**	9	22	13	**53**
	%C	34	14	8	↓	0	1	8	26	10	20
Consensus			**A**	**G**	↓	**G**	**T**	**A**	**A**	**G**	**T**
Arabidopsis	%G	16	9	**79**	↓	**100**	0	11	4	**52**	11
	%A	36	**62**	9	↓	0	0	67	57	20	23
	%T	12	16	8	↓	0	**99.6**	17	25	20	**51**
	%C	37	13	4	↓	0	0.4	5	14	7	15
Consensus			**A**	**G**	↓	**G**	**T**	**A**	**A**	**G**	**T**
Yeast consensus					↓	G_{100}	T_{98}	A_{96}	T_{89}	G_{100}	T_{94}

The consensus derived for 5′ splice sites is complementary to the 5′ end of U1 snRNA, which recognises it early in spliceosome assembly (see above). The 5′ splice site is probably recognised at least twice, however. Crispino and Sharp [49] have shown that the full set of consensus sequences at the 5′ splice site is recognised during splicing of pre-mRNA in HeLa extracts depleted of U1 snRNP (as mentioned above, splicing in the absence of U1 snRNP proceeds upon the addition of excess SR proteins). The sequential recognition of the 5′ splice site involves U6 snRNA (see above) and additional unidentified factor(s) [49]. The degeneracy of the 5′ splice site consensus sequence is such, that like mammalian sites, relatively few plant 5′ splice sites actually possess a perfect match to the consensus sequence [143, 245]. Mutation of the wider 5′ splice site region to either improve or reduce the match to the consensus can affect both the efficiency and accuracy of splicing as well as the abundance of detectable transcript [160]. For example, introduction of a consensus match to the 5′ splice site in maize *Adh1* intron 1 improved the efficiency of its processing in *N. benthamiana* by 2.5-fold [138]. Similarly, mutation of the maize *bronze*-2 intron 5′ splice site to improve its match to the consensus improved the efficiency of its processing in maize [31]. In contrast, weakening the consensus match in the pea *rbcs3A1* intron 5′ splice site resulted in a reduction in its use and activation of cryptic competing 5′ splice sites upstream [160].

Table 1. Continued.

3' Splice site																	exon		
intron																			
	−15	−14	−13	−12	−11	−10	−9	−8	−7	−6	−5	−4	−3	−2	−1	↓	1	2	3
Vertebrate																			
%G	14	16	10	12	8	6	10	9	6	5	7	25	0	0	100	↓	55	25	26
%A	10	10	9	8	6	10	8	9	10	5	8	26	4	100	0	↓	21	22	21
%T	45	45	48	54	59	46	47	43	44	46	47	22	19	0	0	↓	9	34	26
%C	31	29	33	26	28	37	34	39	39	44	38	27	77	0	0	↓	15	19	27
Consensus	Y	Y	Y	Y	Y	Y	Y	Y	Y	Y	N	C	A	G		↓	G		
Dicot plant																			
%G	14	17	15	15	14	19	17	14	18	17	13	42	1	0	100	↓	57	20	26
%A	21	21	20	24	23	23	24	26	21	23	16	30	4	100	0	↓	21	22	32
%T	52	51	54	50	51	44	48	48	49	48	64	21	32	0	0	↓	11	44	28
%C	13	11	11	11	13	14	11	11	12	12	7	7	63	0	0	↓	11	14	14
Consensus	T	T	T	T	T	T	T	T	T	T	G	Y	A	G		↓	G	T	
Monocot plant																			
%G	16	20	19	16	18	20	17	20	21	22	11	51	1	0	100	↓	60	23	23
%A	20	20	18	16	19	22	17	17	15	19	10	15	3	100	0	↓	15	17	20
%T	44	42	42	48	43	37	45	46	41	41	68	21	15	0	0	↓	11	40	31
%C	20	19	20	19	20	21	22	16	22	18	11	13	80	0	0	↓	13	20	24
Consensus	T	T	T	T	T	T	T	T	T	T	G	C	A	G		↓	G	T	
Arabidopsis																			
%G	20	17	18	14	18	22	17	18	19	17	10	38	1	0	100	↓	57	18	22
%A	20	17	17	17	15	20	20	19	19	19	16	30	4	100	0	↓	23	24	30
%T	49	52	53	57	55	44	51	52	51	53	63	26	29	0	0	↓	12	43	32
%C	12	14	13	12	11	14	12	11	12	12	11	6	66	0	0	↓	8	16	·17
Consensus	T	T	T	T	T	T	T	T	T	T	N	C	A	G		↓	G	T	
Yeast consensus												Y_{98}	A_{100}	G_{100}		↓			

3' splice site

The 3' splice site dinucleotide, AG/, is invariant in naturally occuring higher-plant introns (see Table 1). Mutation of the terminal AG/ by either deletion [138], substitution [111] or point mutation of the terminal G [31, 241] results in abolition of its use, with, if available, the selection of cryptic 3' splice sites downstream at reduced efficiency [138, 241]. The inhibition of splicing by mutation of the terminal G in each of these cases is consistent with its requirement in the second step of splicing [200]. Point mutation of the 3' dinucleotide AG → UG in the maize *bronze-2* intron abolished processing [31]. However, point mutation at the 3' dinucleotide AG → CG in a cauliflower mosaic virus (CaMV) 35S RNA intron only partially inhibited processing [111]. The corresponding mutation in actin intron reduced splicing efficiency to only 4% of wild type in *S. cerevisiae* [200]. This might indicate that either a wider consensus in the 3' splice site (see below) or other *cis*-acting elements can compensate for such a mutation in higher plants.

The wider consensus derived for plant 3' splice sites resembles, but is more extensive than that derived for mammals (UGYAG/GU compared to YAG/G, respectively) (Table 1). Mutations in the consensus can affect the processing of plant introns. Improving the match of the 3' splice site of β-globin intron 1 and maize *waxy* intron 9 to the 3' splice site consensus improves the efficiency of their processing in maize [83], while a consensus 3' splice site improves the efficiency of processing of synthetic introns in *N. plumbaginifolia* [79]. The 3' splice site may be recognised at least twice in the course of the splicing reaction [135 and references therein] and in plants at least, the contribution of the nucleotides in the site appears to be different each time [135]. Since the recognition of the 3′ splice site is poorly understood [271], the molecular basis that underpins the distinctions in the 3′ splice site or how they affect splicing is unknown.

Branch site

A highly conserved branch site sequence, UACUAAC, is required for efficient pre-mRNA splicing in *S. cerevisiae* [225], but is neither strictly conserved in, nor required for the processing of, mammalian or higher-plant introns [81, 83, 178]. Instead, the functionally defined mammalian branch site consensus is the degenerate CURAY [178]. A close match to this degenerate sequence can be found in the 3′ region of most plant introns [24, 241, 245, 284], but no independent consensus for plant sequences has been defined. This might be possible, however; a more sophisticated analysis of *Drosophila* intron sequences has identified a branch site consensus, related to, but distinct from, that derived for mammals [172].

The processing of pre-mRNA *in vivo* is rapid and intermediates do not accumulate. The absence of a plant-derived extract capable of faithfully processing plant pre-mRNAs *in vitro*, therefore means that mapping of functional branch sites in plant pre-mRNA is a far from trivial task. In order to overcome this, Liu and Filipowicz [136] exploited the fact that a G_{+1} → A mutation blocks the second, but not the first, step of splicing and therefore causes accumulation of the branched lariat intermediate (see above). Using an *Arabidopsis* mutant (*rca*) which bears such a mutation in intron 3 of a Rubisco activase gene [194] to map the functional branch site, they identified a stop upon primer extension analyses of rubisco activase mRNA in total RNA isolated from mutant *rca* plants which could be specifically suppressed by incubation with purified HeLa debranching enzyme (see above). The functional branch site maps to the sequence UUGAU, 32 nt from the 3′ splice site. The introduction of the same mutation in the 5′ splice site of the synthetic intron Syn7 [81] enabled the mapping of the functional branchpoint in this intron to the sequence CUAAC, 31 nt from the 3′ splice site (but in this case the primer extension stop could not be debranched) [136]. Both of the mapped branch site sequences resemble the mammalian branch site consensus, CURAY.

Branch sites and 3′ splice sites exhibit a spatial constraint with respect to one another. In *S. cerevisiae*, the branch site is typically found 10–50 nt upstream of the 3′ splice site, while in mammals, it is typically 18–40 nt upstream of the 3′ splice site. The average position of the first match to the mammalian branch site consensus in plant introns is 27 nt upstream of the 3′ splice site (C.G. Simpson, personal communication) and thus similar to the position of the functionally mapped branch sites mentioned above [136].

Recognition of the branch site is an early step in the assembly of splicing complexes and the identity of the branch nucleotide is important for both *trans*-esterification reactions. However, mutation of a mammalian consensus branch site in an efficiently spliced synthetic intron, Syn7 (see below), had no effect on either the accuracy or efficiency of its processing in monocot or dicot plants [81, 83]. Moreover, introns devoid of A residues downstream of position +13 could also be processed, indicating that residues other than A can be used as the branch acceptor [81]. This had indicated that the requirement for a consensus branch site was certainly not as strict as in *S. cerevisiae* and that the sequence that could serve as a branch site in higher plants was relatively permissive. More recently however, point mutation of putative branch site adenosines in synthetic introns and natural introns of either monocot or dicot origin were found to reduce the efficiency of their processing when expressed in *N. tabacum* [241]. Strikingly, point mutation of the putative branch site adenosine of a pea legumin intron, which had previously been shown to serve as the branch site when tested in the heterologous HeLa *in vitro* splicing system [25], completely abolished its processing in *N. tabacum* [241]. This was particularly surprising since other sequences matching the branch site consensus were available close to the mutated site. An explanation for this effect comes from recent work in mammals, which has revealed that the presence of a consensus branch site that cannot present a reactive nucleophile suppresses splicing and the use of cryptic branch sites elsewhere [210]. The mutated pea legumin branch site may represent such a site, since mutation of the putative branch site A → U, results in an intron sequence in proximity to the branch site that has the potential to form a helix of nine consecutive basepairs with the branch site interacting sequence of U2 snRNA. Thus, despite the presence of adjacent candidate branch sites, U2 snRNA can apparently be committed to a consensus sequence that results in a dead-end complex [210, 241]. However, if the preferred branch site is not so clear cut, a shift to alternative branch sites can occur [210]. Such a shift may [241] or may not

Table 2. Mean nucleotide base composition in the regions extending 50 nucleotides upstream and downstream of the 5′ and 3′ splice sites. The mean percentage of base composition was calculated for 271 dicot and 146 monocot plant introns surveyed by Goodall *et al.* [84].

Base	5′ site region (exon | intron)	3′ site region (intron | exon)
Dicots		
T	27.1 | 41.7	42.7 | 25.4
A	27.3 | 30.1	28.9 | 29.7
G	21.6 | 13.3	16.6 | 25.1
C	24.0 | 15.0	11.7 | 19.9
Monocots		
T	20.6 | 35.0	34.9 | 20.2
A	23.7 | 22.4	23.2 | 23.5
G	25.9 | 18.5	21.0 | 31.5
C	29.4 | 24.1	20.9 | 25.1

[81, 83] result in a decrease in splicing efficiency and it seems likely that this distinction will be mediated by sequence context or the relative contribution of other *cis*-acting elements in governing efficiency of splicing of a particular pre-mRNA.

Higher-plant pre-mRNA introns are AU-rich

Higher-plant introns, and in particular dicot plant introns, are characteristically AU-rich. Introns of dicot origin average 70% AU and very few (2%) have an AU content of less than 59% [50, 81, 239]. In monocots, the average AU content of introns is considerably lower than in dicots, but the difference in AU content between introns and flanking exon sequences is similar (Table 2). When compared to flanking exon sequences, the U content of both monocot and dicot introns increases markedly, but the content of A changes little [63, 143, 245] (Table 2) and therefore reference to the U-richness of plant introns may be more appropriate. However, the enrichment in U residues in introns is at the specific expense of G and C residues, but not A. Furthermore, some dicot introns (ca. 10% [81]) are more A-rich than U-rich. We have therefore retained the conventional 'AU-rich' terminology here.

The requirement for AU-rich sequence in efficient processing of dicot pre-mRNAs was first demonstrated using a synthetic gene called *Syn7* [81]. *Syn7* consists of an 85 nt intron (close to the modal length of plant introns) with higher-plant consensus 5′ and 3′

splice sites and a branch site matched to the mammalian consensus. The remainder of the intron is composed of arbitrary sequence reflecting the AU compositional bias of dicot plant introns. It is efficiently processed *in vivo*, but gradual replacement of AU-rich intronic sequence with GC sequence drastically reduces splicing efficiency. However, reinsertion of AU-rich sequence into such defective introns rescues their processing [81]. AU-rich sequence is spread throughout the entire length of plant introns and extended stretches of AU-rich sequence exist [240] that Schuler and coworkers first termed 'islands' (arbitrarily defined as four or more A or U residues in a row [138]).

Pre-mRNA splicing in monocots does not exhibit the same requirement for AU-rich sequence. Introns of monocot origin average 60% AU, but many (38%) have an AU content of less than 59%, with some being as low as 30–35% [81, 83, 239]. In contrast to dicots, GC-rich introns can be efficiently processed in monocot plants [83]. For example, introns 9 and 10 of the maize *waxy* gene, which are only 40% and 42% AU respectively, are processed at 95% efficiency in transfected protoplasts of maize, but are not processed at all in protoplasts of *N. plumbaginifolia* [83]. Although not absolutely required for splicing, AU-rich sequence can still promote the efficient processing of certain introns in maize [83, 142]. For example, the efficiency of processing of synthetic introns with weak matches to consensus 5′ and 3′ splice sites can be improved if the introns are AU-rich, in line with observations that monocot introns with poor matches to splice site consensus sequences are often AU-rich [83].

The significance of these findings is that not only do they distinguish pre-mRNA splicing in higher plants from *S. cerevisiae* and mammals, which do not possess distinctly AU-rich introns, but that they provide a unifying explanation for restricted processing of heterologous introns in higher plants [83]. In every documented case, the absence of processing or inefficient processing of mammalian or monocot introns in dicot plants can be explained by an AU content of less than 50%. Conversely, those heterologous introns that are processed, like the small t intron of SV40, are AU-rich [101].

3′ splice site-proximal polypyrimidine tract

A distinguishing element of metazoan intron structure is the 3′ splice site-proximal 10-15 nt long polypyrimidine tract. It is the site of binding of U2AF[65] and is required for commitment complex and prespliceosome complex assembly (see above). The tract is subsequently recognised by PSF (a mammalian splicing factor required for constitutive splicing) in B and C complex [86] and is a major target for regulated alternative splicing [244, 272]. However, a compilation of intron sequences from a diverse range of organisms indicates that it is not a universal element [50]; lower eukaryotes do not exhibit the same degree of pyrimidine enrichment close to the 3′ splice site that mammals show and this is true of higher plants [285]. Moreover, such a sequence is not required for accurate and efficient splicing of a synthetic intron in either monocots or dicots [81, 83]. However, there is a small enrichment of U residues (but not C) close to the 3′ splice site in both monocot and dicot introns. Indeed the consensus nucleotide at each position from −5 to −14 upstream of the 3′ splice site in higher plant introns is U, but as U is so abundant throughout plant introns, this enrichment although significant, is not dramatic [63] (Table 1). Some higher-plant introns do (perhaps coincidentally), possess prominent 3′ splice site-proximal polypyrimidine tracts (which probably explains why some plant introns, but not others, are accurately and efficiently processed in transfected mammalian cells or HeLa cell *in vitro* splicing extracts [see above]).

S. cerevisiae introns exhibit an enrichment in pyrimidine residues close to the 3′ splice site. The yeast tract is less conserved and less prominent than in mammalian introns, being restricted to mostly U residues. In contrast to the situation in mammals, the yeast pyrimidine tract is not required for commitment complex formation, but contributes to 3′ splice site selection in the second step of splicing. A similar role of the mammalian polypyrimidine tract is often experimentally obscured by its importance in commitment complex formation for the first step [discussed in 271]. U-rich sequence upstream of the 3′ splice site can influence splice site selection in plant introns [139, 142], but whether this sequence functions in commitment complex formation in a manner akin to mammals has not been tested.

Interestingly, higher plants (C. Domon and W. Filipowicz, unpublished results) and *Schizosaccharomyces pombe* [207] (which has characteristically short introns that also lack prominent 3′ splice site-proximal polypyrimidine tracts) possess highly conserved analogues of U2AF^{65}, and the *S. pombe* protein, at least, is required for viability [207]. This raises the question of whether U2AF functions in these organisms only in the processing of introns that possess distinct polypyrimidine tracts and not constitutively, or

whether the enrichment in U residues close to the 3′ splice site together with other signals (e.g. the 3′ splice site sequence itself, or other elements, see below) is sufficient to allow U2AF^{65} function, or even, whether U2AF^{65} binds U-rich sequence elsewhere in the intron, but still facilitates commitment. It should be noted that mammalian introns that lack prominent polypyrimidine tracts can still be processed via U2AF action; in one documented example, the binding of U2AF^{65} to the intron is facilitated by the binding of U1 snRNP at a downstream 5′ splice site which is communicated through a complex spanning the downstream exon [98]. Notably, the only reported test of 3′ splice site-proximal polypyrimidine tract requirement in higher plants was performed with synthetic introns in constructs that also possessed a natural intron (flanked by its natural exons) positioned downstream of the experimental intron [81, 83]. It is therefore possible that under some experimental conditions, a requirement for a 3′ splice site proximal polypyrimidine tract in natural plant introns may be identified.

The higher-plant splicing machinery

U snRNAs

Genes encoding each of the known spliceosomal U snRNAs, U1, U2, U4, U5 and U6 have been cloned from several different plant species. They are highly conserved with their mammalian counterparts in terms of size, sequence and secondary structure, and they possess key sequence elements that have been shown to be essential for pre-mRNA splicing in yeast and mammals. Higher-plant U snRNAs have been comprehensively reviewed [84, 252] and are not considered in detail here. The most striking feature to have emerged from their characterisation is the existence of an extremely large number of U snRNA sequence variants. Notably, U snRNA variants exhibit developmental differences in their pattern of expression [91]. While this raises the possibility that compositionally or functionally distinct U snRNPs act to effect developmental gene expression patterns, there is, as yet, no direct evidence to support this.

U snRNPs

The conservation of the spliceosomal U snRNAs in higher plants predicts that U snRNP composition will be conserved too. Consistent with this, several examples of the antigenic relatedness of higher-plant, yeast and mammalian U snRNPs have been described, identifying probable analogues of U snRNP core proteins and the extremely large U5 snRNP-specific protein, PRP8 [118, 198, 240]. In addition, the application of the common structural motif of the U snRNP core proteins as a probe in searches of plant gene sequences reveals the existence of conserved U snRNP core proteins in higher plants [232]. The first cDNA encoding a plant spliceosomal protein, the U2-specific protein U2B″ of potato, was cloned by Simpson *et al.* [242]. Full-length cDNAs encoding the related U1-specific protein, U1A, have been cloned from *Arabidopsis* and potato [243], but no other full-length cDNAs encoding higher-plant U snRNP proteins have yet been published. U1A binds U1 snRNA stem-loop II and U2B″ binds U2 snRNA stem-loop IV [243]. Both the sequence of the target RNA binding sites and the sequence of the proteins themselves are highly conserved between higher plants and vertebrates (although neither are well conserved in *S. cerevisiae*). The functional relatedness of these vertebrate and plant U snRNP proteins is exemplified by the ability of a second human U2 snRNP-specific protein, U2A′, to facilitate sequence-specific binding of potato U2B″ to U2 snRNA of either plant or vertebrate origin [243]. This degree of conservation indicates that higher-plant U snRNPs will exhibit further and more general homologies with their mammalian counterparts (many related proteins can already be identified in the Expressed Sequence Tag database).

Notably, the binding sites of U1A and U2B″ are variant in higher-plant U1 and U2 snRNAs; this variation is restricted to the terminal residues of the loops they bind. However, sequence distinctions in these positions do not perturb the binding of either U1A or U2B″ [243], probably because this part of the RNA is not recognised upon RNA binding [195, 269]. Therefore the sequence variants of U1 and U2 snRNA bearing distinctions in these positions are probably not functionally significant.

SR proteins

SR proteins are present in higher plants [15, 128, 137]. Like metazoan SR proteins, they can be purified by the $(NH_4)_2SO_4$, $MgCl_2$ precipitation procedure and are recognised by the SR domain-specific antibody, mAb 104 (see above) [128, 137]. SR proteins have been prepared in this way from *Arabidopsis*, tobacco and carrot [128, 137], and when added to HeLa S100 extracts are able to complement the deficiency in constitutive pre-mRNA splicing indicating that they are splicing factors [137]. The size distribution of SR proteins prepared in this way and revealed by mAb 104 probing is not conserved in different plant species [137].

An *Arabidopsis* cDNA encoding an SR protein, SR1, has been cloned and characterised [128]. SR1 is highly related to the human protein SF2/ASF, but may not necessarily be its functional homologue. It can promote a switch in 5′ splice site selection when added to HeLa cell *in vitro* splicing extracts but does not complement a splicing-defective HeLa cell S100 extract.

The U1-specific protein U1-70K and U2AF^{65} both possess SR domains [266, 302]. A partial cDNA from *Arabidopsis* encoding a protein strikingly similar to human U1-70K has been described [213] and full-length cDNAs encoding likely homologues of U2AF^{65} have been isolated from *N. plumbaginifolia* (C. Domon and W. Filipowicz, unpublished results).

The mechanism of pre-mRNA splicing in higher plants

In the preceding sections we have outlined the structural and functional properties of higher-plant introns, and components of the higher-plant spliceosome. It is therefore apparent that in many ways the process of pre-mRNA splicing is conserved. The fundamental feature that distinguishes pre-mRNA splicing in plants from splicing in vertebrates and *S. cerevisiae* is the presence in plants of AU-rich intron sequence and its requirement in efficient processing. This feature, and the absence of both an absolutely conserved branch site (required in *S. cerevisiae*) or universal 3′ splice site-proximal polypyrimidine tract (required in mammals) explain the barriers to heterologous splicing between the species. The key mechanistic issues in higher-plant pre-mRNA splicing then, are splice site selection, relating, in particular, to selection of the 3′ splice site and the function of AU-rich sequence.

5′ splice site selection

Recognition of the 5′ splice site by the U1 snRNP is a key early feature of 5′ splice site selection. While the degree of complementarity to the 5′ end of U1 snRNA is important in determining splice site strength, it is not sufficient to specify choice; sequence context is important too. For example, 5′ splice sites sequestered in secondary structure are not selected in *N. plumbaginifolia* [135]. Since introns are markedly more AU-rich than flanking exons, a dramatic AU/GC transition in sequence composition surrounds both 5′ and 3′ splice sites. Therefore, a more general consideration has been the possibility that the AU/GC transition might contribute to splice site selection.

McCullough *et al.* [160] analysed the pattern of 5′ splice site selection in the pea *rbcs3A1* intron after weakening the consensus match of the wild-type site which lies at an AU/GC transition, and strengthening upstream or downstream cryptic sites. Weakening of the wild-type 5′ splice site did not prevent its selection. However, when the weakened wild-type site was coupled with mutations to an upstream cryptic site that strengthened it to a perfect match with the consensus, a preferential switch in 5′ splice site usage to the upstream site was observed. In contrast, when a downstream site buried in AU-rich intron sequence was strengthened in a similar manner, it was rarely selected. However, when the AU-rich sequence between the wild-type site and the downstream site was replaced with unrelated exon sequence, then the downstream site was selected. Further mutation of this region illustrated that the first 5′ splice site upstream of AU-rich sequence could be preferentially selected, even if it presented the weakest match to the 5′ splice site consensus in the transcript. Thus, while splice site strength contributes to splice site selection, it is not sufficient to govern splice site choice and, in plants, a polarity in splice site selection involving selection of 5′ splice sites upstream and not downstream of AU-rich sequence exists.

3′ splice site selection

Schuler and co-workers have investigated whether the AU/GC transition contributes to 3′ splice site selection. Lou *et al.* [139] progressively deleted the 3′ end of maize *Adh1* intron 3 and analysed processing of the mutant RNAs in *N. benthamiana*. In contrast to the situation in mammalian introns, where internal cryptic 3′ splice sites are typically not selected, efficient activation of previously unused, internal cryptic 3′ splice sites brought into proximity of the AU/GC transition occurred. When the order of the cryptic sites was rearranged, so that normally internal sites were swapped with the wild type site, then the outermost site was always selected, regardless of its previous location [139]. Mutation of individual AU-rich regions or 'islands' in proximity to the 3′ splice site promoted the

same switch. In addition, the mutation of more than one AU island in concert resulted in the same switch, but to a more pronounced extent, indicating that the AU islands may act cooperatively. This result, together with the fact that mutation of an AU island considerably further upstream of the 3′ splice site strongly promoted the same splice site switch, indicated that a continuum of AU-rich sequence may be important for function.

A model based on 3′ splice site choice requiring a continuum of AU-rich sequence and selection of the splice site at AU/GC transitions explains the non-selection of internal AGs by their burial in AU-rich sequence [139]. The processing of several other introns as studied by other workers is consistent with this model [81, 135, 142]. There are however, several examples of preferential usage of internal 3′ splice sites over consensus sites located at the AU/GC transition [135, 139, 142, 189, 192, 241, 261] indicating that the transition alone may not be sufficient to specify 3′ splice site choice.

In mammals, 3′ splice site selection displays properties of both a scanning process and competition between AGs based on immediate sequence context. Scanning proceeds in a 5′ → 3′ direction from the branch point and the first AG encountered downstream is usually selected as the 3′ splice site [247, 248]. Several documented examples of 3′ splice site selection are consistent with a scanning process taking place in plants. Firstly, insertion of an AG 6 nt upstream of the wild type AG in intron 1 of castor bean catalase gene, results in its exclusive use [261]; secondly, point mutation of wild type 3′ splice sites results in activation of downstream cryptic site in plants [241] as it does in mammals [2]; thirdly, insertion of a branch site consensus sequence between two adjacent AGs results in activation of the downstream site [241]; and finally, AG dinucleotides are rarely found in the region upstream of the 3′ splice site in plant introns (unpublished observations).

The implication for the existence of a branch point-3′ splice site scanning process in plants is that selection of the branch site can contribute to 3′ splice site selection; the selection of a downstream 3′ splice site, following insertion of a branch site between two AGs in an experimental transcript expressed in *N. tabacum* demonstrates this directly [241]. Therefore, reference to branch site selection needs to be incorporated within the broader model of 3′ splice site selection in higher plants. Simpson *et al.* [241] recently summarised the available data stating, essentially, that in higher plants, 3' splice site recognition is probably based on recognition of the AU/GC transition, the 3' splice site sequence itself and AU-rich sequences upstream of the splice site with subsequent selection of a proximal consensus-type branch site sequence from which the AG used as an acceptor is later selected via a scanning mechanism. Reflecting the lack of complete understanding of 3' splice site selection in general, however, there are several examples of 3' splice site selection in nature which are not consistent with a scanning process [discussed in 271] and diffusion collision probably explains selection in these instances.

Splice site pairing

The analysis of inter-species hybrid intron constructs bearing either mammalian or plant 5' exon/introns sequences fused to the plant or mammalian 3' intron/exon sequences often results in cryptic splicing events [285, 275]. Waigmann and Barta [275] have provided evidence that in plants authentic or cryptic splice site choice at either the 5' or 3' splice site is influenced by the identity of the other site. In other words, for some pre-mRNAs there is a close co-operation between splice site regions that exerts a mutual influence on splice site selection. Lamond et al. [125] reached a similar conclusion in considering mammalian spliceosome assembly. The 5' and 3' splice sites are functionally associated as early as E complex [164] although fine tuning of selection occurs throughout the splicing reaction. The mechanism that promotes genotypic splice site pairing however, has not been defined.

Finding the splice sites: the importance of exons

Berget and co-workers described a model that predicts that the initial unit of definition in pre-mRNA splicing is the exon [216, reviewed in 12]. Exon definition would occur by virtue of interactions between factors bound at a 3' splice site and a downstream 5' splice site with a network of interactions spanning the exon and communicating the two sites. After this initial selection, a rearrangement at the intron must take place before the chemistry of the reaction can proceed. In vertebrates, introns can be many kb in length, but exons rarely exceed 300 nt. Exon definition may facilitate splice site selection in transcripts with extremely long introns by virtue of the fact that much less primary transcript has to be cleared of hnRNPs before splice sites are identified and communicated [254]. There is

considerable experimental support for the model [12, 187, 254], but it does not explain processing of either the first or last introns within a transcript. In the case of the first intron, interactions between the 5' cap-binding complex and the spliceosome may be important for its processing [103, 191], whereas for the last intron, an interaction communicating splicing and polyadenylation that thereby defines the last exon may be important [184, 185, 186].

Several pre-mRNAs contain an exonic element required for splicing of the upstream intron. These elements, designated exon enhancers, can potently stimulate processing of pre-mRNAs with poorly conserved splice sites and are important elements in regulated alternative splicing [127, 254, 257, 267 and references therein]. Members of the SR family have been shown to bind different exon enhancers [127, 257, 267] and be essential for enhancer function [257, 267] underlining their importance in exon-mediated definition in pre-mRNA processing.

Does exon definition take place in plants? There is, as yet, no definitive data to answer this question. However, since higher-plant introns are typically much shorter than vertebrate introns and possess AU-rich sequence required for splicing throughout their entire length, exon definition may not be necessary, but rather, the intron, as others have speculated [143], may be the unit of definition [discussed in 12]. If so, it could point to a fundamental difference between plants and mammals. However, the many lines of relatedness between higher-plant and mammalian pre-mRNA splicing means that exon definition should not be quickly dismissed. Exon sequences certainly influence pre-mRNA splicing in plants; mutation of exon sequence flanking the maize bronze-2 intron can promote an increase in the efficiency of its processing in maize [31] and the selection of splice sites at the AU/GC border by exon based definition is as plausible an interpretation of available data as is outward scanning from internal intron sequence [143]. Since the model makes testable predictions, it will be interesting to study exon definition in higher plants experimentally.

Breaking down the contribution of AU-sequence: the importance of U

The demonstration that insertion of segments of AU-rich sequence into GC-rich synthetic introns improves the efficiency with which they are processed provided the first indication that AU sequence acts as a positive element that may serve as a target site for RNA-binding protein(s) that facilitate processing [81, see also above]. By implication, it should be possible to define a recognition element within the vagueness of 'AU-richness' that protein(s) might bind. The analysis of compiled plant intron sequences has failed to identify a universally conserved element [284]. Since AU-rich synthetic introns of arbitrary sequence [81] and plant introns expressed in either sense or antisense orientations [238] are spliced efficiently, any signal that might exist must be degenerate. In addition to degeneracy in the sequence of the signal, there is also redundancy in its position. AU-rich sequence is spread throughout the entire length of plant introns and extensive internal deletions of AU-rich introns can be made that do not perturb either the efficiency or accuracy of splicing [141, 142]. Schuler and co-workers have studied the contribution of individual AU-rich islands in plant introns to pre-mRNA processing (see above) [138, 139]. Significantly, minor mutations that disrupt the continuity of AU sequence in single islands, but which have a negligible effect on overall AU composition, can affect 3' splice site selection [139]. This result clearly illustrates that AU sequence can act as a specific recognition element.

In an attempt to define the minimal properties of such a recognition element, Gniadkowski et al. [79] investigated the ability of different sequence insertions to rescue the efficiency of splicing of a synthetic GC-rich intron. This work uncovered four significant findings: firstly, the insertion of either single or multiple copies of a U-rich sequence, UUUUUAU, rescued splicing, but the insertion of A-rich sequence (AUAAAAA) did not, indicating that it is the U, rather than the A residues in AU-rich sequence that play a key role in the promotion of efficient splicing; secondly, limited mutational analysis of a single centrally located U-island indicated that repeats of only two consecutive U residues function effectively in rescue, so long as the Us are not flanked by G residues; thirdly, multiple U-rich sequences inserted in proximity to either the 5' splice site, 3' splice site or the mid-region of the intron each rescued splicing, indicating that U island rescue is position independent and fourthly, the degree of rescue promoted by multiple insertions at either a single site or multiple sites is additive [79].

The unequal contribution of U and A residues in AU-rich islands is consistent with the fact that introns are strikingly enriched with U residues and is an effect that has been noted before by other workers. As mentioned above, mutations in AU-rich islands close to the 3' splice site of Adh1 intron 3 inhibit selection of the wild-type 3' splice site. Lou et al. [139] showed that U → A mutations also prevented efficient utilisation of the proximal downstream 3' splice site and thus provided the first demonstration that the contribution of A and U residues to pre-mRNA splicing in plants is not equivalent. In addition, Carle-Urioste et al. [31] have identified a centrally located U-rich island (an 11 nt long sequence, in which 8 positions are occupied by U) within the maize bronze-2 intron (an otherwise AU-poor intron). Mutation of each of the U residues (in concert) in this island to G residues results in the abolition of detectable splicing, while improving the quality of the island by mutating each residue to U, results in a more than 3-fold enhancement of splicing efficiency over wild type. Consistent with the documented functional importance of U residues, UU is the most frequently occuring dinucleotide in monocot and dicot introns (this reflects the abundance of U residues rather than a statistically significant clustering of the UU dinucleotide above what can be expected by chance alone) [284].

The importance of U within the degenerate signal indicates that RNA-binding proteins that make critical contacts to U residues will be candidate mediators of the requirement for AU-rich sequence. Proteins have been identified in nuclear extracts of N. plumbaginifolia that can be cross-linked to higher-plant pre-mRNA introns by treatment with UV light and whose binding can be specifically competed by poly(U) [79]. Particularly prominent cross-linking species that migrated at 50 and 54 kDa upon SDS-PAGE were identified. The overall complexity in the pattern of proteins that were cross-linked to pre-mRNA introns in N. plumbaginifolia was much more simple than the pattern observed with extracts of HeLa nuclei. Since these binding experiments were performed under nonfunctional conditions ie. the extracts were not pre-mRNA splicing competent, they must be interpreted with caution. Indeed, in the absence of a functional in vitro assay for higher-plant pre-mRNA splicing, the identification of, and provison of functional evidence for the role of, U-rich RNA-binding proteins in higher-plant pre-mRNA splicing will be a difficult task. In terms of protein binding, the degeneracy of the minimal element that can function is such, that specific recognition by an RNA-binding protein seems improbable in the absence of co-operative interactions either with itself or with factors recognising other cis elements.

The position independence of U-island rescue described by Gniadkowski et al. [79] provides little indication as to the manner in which rescue is effected, but it is consistent with the fact that AU-rich sequence is distributed throughout the entire length of dicot introns. However, since the molecular basis upon which rescue is effected is not known, it may not necessarily result from a functionally equivalent process in each case. A notable feature of the results of Gniadkowski *et al.* [79] is that insertion of U-rich sequence in the 5′ region of the intron promoted rescue of splicing; no precedent for a constitutively acting *cis* element located in this position has been described for any other pre-mRNA splicing system. Perhaps the most related *cis*-acting pre-mRNA splicing element in nature is the anchoring site, the region upstream of the branch site to which many U2 snRNP associated proteins bind upon pre-spliceosome complex formation [87]. However, binding of proteins to this region in mammalian introns is independent of sequence composition [87]. Two other studies have also noted the particular importance of AU-rich sequence located between the 5′ splice site and the probable branch site [81, 241].

The role of AU-rich sequence

Although synthetic GC-rich introns are inefficiently processed in dicot plants, they are processed accurately and without the accumulation of intermediates [81, our unpublished results]. Therefore, AU-rich sequence is not required for either the first or second step of splicing, but functions at an early event prior to catalysis; that is, at the potential plant equivalents of either H-complex, commitment complex or pre-spliceosome complex formation. As described above, although yeast and mammals exhibit close similarities in spliceosome catalysis, the assembly of these early complexes are distinct and involve non-conserved *trans*-acting factors and different *cis*-acting elements. Thus, although catalysis may well be conserved, the pathway of assembly to catalysis exhibits distinctions. It seems likely that plant proteins that facilitate the association of U1 or U2 snRNP with the substrate pre-mRNA by promoting addition and/or stabilisation of the interaction will be involved in the AU-mediated effect. Alternatively, the role of AU-binding proteins may be more passive. They may, instead, function like metazoan hnRNP proteins, binding RNA with a low degree of sequence specificty, but exhibiting a preference for U-rich RNA and act as RNA chaperones, maintaining an open conformation of the intronic RNA and facilitating the active identification of other *cis*-acting elements, like the splice sites, for example, by spliceosomal components. Since sequential recognition of

the same element is a recurring theme in pre-mRNA splicing, the involvement of AU-rich sequence at more than one of these steps is also possible.

A mosaic of cis-acting elements determine the accuracy and efficiency of pre-mRNA splicing

The efficiency of pre-mRNA splicing is a result of the combined contribution of multiple *cis*-acting elements. Mutations that decrease AU composition, the quality of specific AU- or U-rich islands, or the match of either the wider 5′ splice site or 3′ splice site to the consensus sequence all reduce the efficiency of splicing. In contrast, the converse 'up' mutation of these elements promotes efficient splicing. Mutations that alter putative branch sites can result in a reduction in splicing efficiency and so, by implication, a consensus or optimal branch site sequence likely promotes efficient splicing in plants (in mammals, the sequence UACUAAC, the strictly conserved branch site in *S. cerevisiae* introns, serves as an optimal site, even though, the branch site can be highly degenerate [214, 307]).

The efficiency of splicing is not determined by these *cis*-acting intronic elements alone; exon sequences [31] and context in the complete transcript also contribute. Both the accuracy and efficiency of splicing of *Adh* introns can be perturbed if expressed individually as opposed to within the context of their native transcript [138] and likewise, the maize *bronze-2* intron is processed at 100% efficiency when transiently expressed in maize protoplasts in the context of the native gene, but processing is reduced to 25–35% when the intron is placed into a reporter gene context [31]. It is well established in mammals, that upstream introns can influence the processing of downstream introns and vice versa [177, 180], and that splicing and polyadenylation interact to influence the efficiency of both processes [see 108 for review]. These interactions occur in plants too. Modifications that improve the efficiency of processing of a synthetic intron result in an improved processing of a downstream intron present in the same transcript [81]. In addition, we have preliminary evidence that in plants, upstream introns promote an increase in polyadenylation efficiency (our unpublished results).

There are two important points relating to the mosaic nature of *cis*-acting signals. The first is that multiple means of recognition and biases in the significance of particular recognition events must exist. These different aspects of processing must take place within the same transcript. The processing of some introns may be mediated by exon definition, others by intron definition, while the first and terminal introns via spliceosomal interactions with the 5' cap-binding complex and polyadenylation machinery respectively. There is then, co-operativity between multiple *cis* elements that combine to determine processing. It is also clear that the processing of some introns is more dramatically affected by mutation of particular *cis*-acting elements than others, indicating that the compensatory effect of other elements is greater. In other words, there are differences in the relative importance of certain processing *cis* elements in different introns. This means that processing is complex, compound and dynamic, and multiple routes to commitment complex assembly exist.

The second important point about such a complex presentation of processing signals in higher plants is how much it reveals the particular relatedness of plant and metazoan pre-mRNA splicing as opposed to processing in *S. cerevisiae*. In yeast, relatively few (2–5%) genes contain introns. Those that do usually contain only one and this is located close to the 5' end of the transcript. In addition, the splice site and branch site consensus sequences of yeast are strictly conserved while the same sequences in plants and metazoans can be highly degenerate [9]. These distinctions in transcript organisation extend into differences in the *trans*-acting factors; all of the higher-plant spliceosomal U snRNAs and U snRNP proteins cloned to date exhibit much greater relatedness with their functional analogues in metazoans than in yeast. In addition, while likely functional analogues of key proteins that facilitate commitment complex formation and alternative splicing in metazoans: U2AF^{65} and SR proteins, exist in higher plants, no homologues of these proteins have been identified in *S. cerevisiae*.

Pre-mRNA secondary structure can affect pre-mRNA splicing

An RNA molecule has an inherent capability to form secondary and tertiary structures through multiple intra-molecular interactions. The propensity of RNA molecules to form such structures has been extensively documented (see Klaff *et al.*, this volume) and affects RNA function, ribonucleoprotein complex function and specific recognition of RNA by RNA binding proteins. The management of RNA structure by chaperones is therefore a fundamental feature of the processing of RNA *in vivo* [94].

Pre-mRNA secondary structure can perturb or influence processing events. In yeast, sequestration of the 5' splice site or the branch site into secondary structure (double-stranded stems) inhibits pre-mRNA splicing [296, 80]. The effects in mammals are generally less dramatic [58, 251] and are more pronounced *in vitro* than *in vivo*, indicating that the chaperone effect of, most likely, hnRNP proteins binding to the emerging nascent transcript *in vivo* can prevent extensive secondary structure formation [58, 94].

Pre-mRNA structure can have profound effects on splicing in higher plants too. Sequestration of a 5' splice site into a double-stranded stem of either 18 or 24 bp results in a strong inhibition of its use *in vivo* in *N. plumbaginifolia* [135]. In addition, double-stranded stems of 18–24 bp placed in the middle of short synthetic introns (derivatives of Syn7, see above) inhibit pre-mRNA splicing [135]. However, sequestration of 3' splice sites into secondary structure does not necessarily inhibit processing. In fact, Liu *et al.* [135] revealed that, under certain conditions, inclusion of the 3' splice site into a double-stranded stem was actually necessary for its selection. Usage of a sequestered 3' splice site required a single-stranded region between the 5' splice site and the base of the stem of at least 45 nt and the presence of a second 'helper' 3' splice site downstream. The experiments performed with *N. plumbaginifolia* suggest that the spliceosome or spliceosomal components have the potential to unfold secondary structure in the downstream portion of an intron and that this probably occurs at a late stage in spliceosome assembly, most likely during scanning for the AG acceptor (see above).

The detrimental effect of secondary structure(s) placed in the middle of synthetic introns observed in *N. plumbaginifolia* is much less pronounced when processing of the same introns is examined in maize [83]. This distinction illustrates and underlines that the processing of pre-mRNA in monocots and dicots is not identical (see above).

Since intronic secondary structure can perturb pre-mRNA splicing, particularly in dicots, it is possible that one reason dicot introns are AU-rich is that it reduces the potential for stable secondary structures to form (a G-C base-pair is stronger than an A-U base-pair, necessitating an elevated temperature to melt it). The ability of monocots to process GC-rich introns, which are more prone to forming secondary structures, may be related to their ability to either prevent secondary structure formation or unwind such structures upon processing and thereby tolerate secondary structure in the processing of pre-mRNA [81]. Although the decreased potential of AU-rich sequence to form secondary structures may be important, the indications outlined in preceding sections that U-rich sequence acts as a recognition element that facilitates processing may provide a more compelling rationale for the existence of and requirement for AU-rich sequence in plant introns and this may be the major function of such sequence in introns. Indeed it should be noted that AU-rich sequences can form secondary structures and that AU-rich stem-loop structures can inhibit pre-mRNA splicing in dicot plants [81, 135].

Alternative splicing

Alternative splicing is a common mechanism of gene regulation in animals [246]. Primary transcripts of many genes can be differentially spliced to produce multiple mRNAs giving rise to proteins with different functions: as many as twenty different proteins can be produced in this way from a single transcription unit. The process is often subject to developmental or tissue-specific regulation. In a few cases the mechanism underlying alternative splicing events in vertebrates and insects has been elucidated and *cis* elements and *trans*-acting regulators characterised [reviewed in 171, 272].

Relatively few examples of alternative pre-mRNA splicing have been identified in plants and in most of the described cases it is not yet known whether they are biologically significant. However, more recent reports indicate that this mechanism of regulation of gene expression may be exploited more often than originally anticipated. Known examples of alternative splicing are listed in Table 3. They are subdivided into different categories such as utilisation of multiple 5′ or 3′ splice sites and alternative splicing patterns resulting from differential initiation of transcription or alternative polyadenylation events. The accumulation of unspliced or partially spliced RNAs (splicing/no splicing category) is quite common in plants; in some instances as much as 50% of a specific gene transcript may exist in unspliced form [175]. This phenomenon probably reflects low efficiency of normal splicing [81, 83, 159, 175] rather than a regulated process. This contrasts with the situation in mammalian cells which usually do not accumulate unspliced RNAs. Hirose *et al.* [96] investigated whether unspliced RNA transcribed from the tobacco *RGP-1c* gene undergoes translation but found no evidence of association of the transcript with polysomes.

In some cases, unspliced RNA can be exported from the nucleus. A fraction of the polycistronic 35S RNA transcribed from the circular genome of plant pararetroviruses undergoes splicing. This process is required for expression of some viral proteins and is essential for the life cycle of the virus [111, 73; see also Table 3]. Some animal viruses possess special mechanisms that enable the accumulation of both spliced and unspliced forms of RNA in the cytoplasm [reviewed in 102] and such mechanisms may also be part of plant pararetroviral gene expression.

Many different alternatively processed mRNAs are observed among transcripts of some plant genes encoding proteins with important regulatory roles. For example, the *ZEMa* gene, which encodes a MADS box-type transcription factor in maize, yields at least five different RNAs, some produced by alternative usage of the 5′ splice sites while others by exon skipping or alternative polyadenylation [166]. Alternative splicing could further increase the repertoire of MADS box factors which often exist as heterodimers. Differentially spliced transcripts of the *RGP-1c* gene (which encodes an RNA binding protein) in tobacco were found in association with polysomes indicating that they probably give rise to variant protein products [96].

Studies of different alleles (or haplotypes) of the *S* locus which controls self-incompatibility in *Brassica* revealed many examples of alternatively processed transcripts originating from both *SLG* and *SRK* genes. *SLG* (S locus glycoprotein) genes encode a (usually) secreted glycoprotein, while *SRK* genes code for the S receptor kinase which consists of an intracellular kinase domain, a transmembrane domain, and an extracellular domain that shares similarity with the SLG protein. The *S2* haplotype *SLG* gene (SLG_2) produces two transcripts of 1.8 and 1.6 kb in stigmas and anthers [262]. The 1.8 kb transcript represents the RNA from which the single SLG_2 intron (833 nt) is excised, while the 1.6 kb transcript corresponds to RNA which is not spliced but is polyadenylated approximately 150 nt downstream of the 5′ splice site. Since the downstream exon of SLG_2 codes for a transmembrane domain similar to that found in the SLK_2 gene, the spliced 1.8 kb mRNA produces a membrane anchored form of SLG_2 while the 1.6 kb mRNA produces a secreted protein. The formation of both protein isoforms has been demonstrated in *Brassica* and in transgenic tobacco. Interestingly, there are differences in the ratio of the 1.6 and 1.8 kb transcripts in different organs.

Table 3. Examples of alternative splicing in plants.

Gene/intron/organism	Biological function	Comments	Refs.
Alternative 5′ splice sites			
Rubisco activase ivs6 (soybean, spinach *Arabidopsis*, barley)	?	Two expected protein forms identified but physiological role unknown.	[223, 281]
RNA binding protein RGP-1a,-b,-c (tobacco)	?	RNAs found in association with polysomes. RNAs with retained introns also identified. Organ-specific differences.	[96]
Chorismate synthase, ivs3 (tomato)	?	Organ-specific differences.	[85]
MADS box *ZEMa* gene ivs1 (maize)	?	Three alternative 5′ss; also exon skipping and polyadenylation within introns. Tissue-specific differences.	[166]
Hydroxy pyruvate reductase pumpkin	?	Alternative splicing may produce proteins with and without putative C-terminal microbody targeting signal.	[93]
Alternative 3′ splice sites			
P gene ivs2 (maize)	?	Alternatively spliced RNAs are polyadenylated at different sites	[88]
En/Spm transposable element (maize)	+	Other rare splicing events, including alternative usage of 5′ss, also occur (see text).	[76, 152]
Glycine decarboxylase H subunit (*Flaveria trinervia*)	?		[114]
Splicing/no splicing[a]			
Monopartite geminiviruses (MSV, MDV, DSW)[b]	+	Splicing is required to produce the protein that is essential for replication. Significance of unspliced RNAs is not known.	[2, 129, 130, 229]
Pararetroviruses		Partial splicing of the 35S RNA is essential for	[111]
CaMV[b]	+	viral life cycle. Several alternative 5′ss identified.	
RTBV[b]	+/?	Processing of the 6.3 kb ivs may be required for the expression of internal ORF IV.	[73]
RNA polymerase II, ivs in 3′ UTR (*Arabidopsis*, soybean)	?		[55]
IBP2 transcription factor (maize)	?	Unspliced RNA would encode truncated myb-like protein, devoid of DNA binding domain; organ-specific differences.	[144]
Waxy, ivs1 (rice)	+/?	Efficiency of intron excision controls endosperm amylose content in different cultivars.	[276]
MuDR transposable element (maize)	?	Two *mudrA* introns fail to splice ca. 20% of the time. Inefficient removal of cryptic ivs in *mudrB*.	[95]
Differential transcription initiation/splicing			
Pyruvate, orthophosphate dikinase (maize)	+	Transcription from two promoters combined with alternative splicing. Protein products targeted to different cellular compartments.	[237]

Table 3. Continued.

Gene/intron/organism	Biological function	Comments	Refs
Homeobox *OSH45* (rice)	+/?	Two promoters, combined with alternative splicing Protein products differ in transactivation activity. In addition, usage of three alternative 3′ss in ivs6.	[260]
Splicing/polyadenylation S-locus *SRK3, SLG2*; (*Brassica*)	+/?	Protein products identified (see text). Alternative processing of other S-locus genes, including cryptic ivs excision in *SRK29* [119] also noted.	[77, 262]

[a] mRNA-like transcripts which retain introns are frequently found in plant cells. In most instances, they probably reflect low efficiency of normal splicing rather than a regulated process [62,143,175]. Only few recent examples are listed in Table 3; for others, see [175]. Transcripts of some genes (e.g., *SRK* and *SLG* genes, *RPG*-1) listed in other sections of Table 3 also include unspliced RNAs.

[b] CaMV, cauliflower mosaic virus; RTBV, rice tungro bacilliformis virus; MSV, maize streak virus; MDV, wheat dwarf virus; DTV, digitaria streak virus.

Alternative RNA processing is also responsible for the formation of the truncated form of the SRK receptor kinase identified in the *S3 Brassica* haplotype. *SRK* genes contain one long 5′-proximal and five short downstream introns and the correctly spliced mRNA produces a full-length receptor kinase. Of several *SRK₃* gene-specific transcripts analysed by Giranton *et al.* [77], mRNAs polyadenylated in intron 1 and unspliced RNA would encode the truncated protein containing only a soluble extracellular domain of the receptor. Similar alternatively processed transcripts have also been observed for other haplotype *SRK* genes [119, 255].

Some alternative splicing events may be subject to tissue-specific [85, 96, 166, 262; see Table 3] or developmental [223] regulation but the significance of these observations is unknown. No attempts have yet been made to investigate the underlying mechanisms of any alternative splicing event detected in plants. Inspection of differentially utilized splice sites or intron sequences does not obviously reveal why these and not other RNAs (or introns) undergo differential processing. It is possible that the relatively frequent occurence of alternative polyadenylation within introns (*SRK* and *SLG* genes, *ZEMa* gene, *P* gene; see Table 3) results from the competition between splicing and polyadenylation reactions. In plant pre-mRNAs, both the introns and 3′ UTRs are abundant in U and A nucleotides, and AU- or U-rich motifs may serve as positive signals in both reactions [141].

The best documented and also biologically relevant case of differential splicing in plants is the processing of RNA transcribed from the transposable element *Em/Spm* [76, 152]. 5′-proximal introns of the 8.3 kb long *Em/Spm* transcript undergo alternative processing to yield mRNAs encoding TnpA and TnpD, the two proteins required for transposition. In maize, a natural host of the element, a major product of the *Em/Spm* splicing is *tnpA* mRNA in which the 5′-proximal 4.4 kb of the transcript is removed as an intron. Notably, this same long intron contains two open reading frames, ORF1 and ORF2, and both of them are included in the much less abundant *tnpD* RNA. In *tnpD*, a short 5′-terminal exon and ORF1 and ORF2 alternative exons are spliced together and joined to the downstream portion of RNA which consists of another ten exons, shared by *tnpA* and *tnpD* RNAs. RT-PCR analysis also identifies several additional *Em/Spm* splicing variants, the function of which (if any) remains to be established. Some of them contain ORF2 or its fragments spliced out. Others are generated by alternative usage of two different 5′ splice sites of intron 1. The latter event can generate mRNAs containing different 5′-untranslated regions which could be of importance for the efficiency of translation and regulation of transposition [152]. *En/Spm* transcripts are also processed into functional mRNAs in heterologous hosts. However, while in *Arabidopsis* the pattern of *En/Spm* mRNAs is similar to that in maize (ratio of *tnpA* to *tnpB* being ca. 100), in potato, the *tnpD* RNA is much more abundant than *tnpA* [30, 64]. These observations indicate that splicing requirements differ not only between dicots and monocots but also among different dicot plants (see also above).

Transposable elements inserted into host genes at different locations are frequently excised as introns from the mosaic pre-mRNA transcripts. This process often follows complex alternative patterns using natural and cryptic splice sites in the element and the target gene [273, 283; for reviews, see 63, 143, 279]. In addition, some of these events may be tissue-specific [273]. In many instances, splicing of transposon insertions permits the formation of considerable levels of functional mRNA. Detailed analyses of transposon splicing in maize provided some insight into the mechanism of pre-mRNA processing in plants and also suggested that some contemporary introns may originate from transposable element insertions [208, 279]. It is possible that the frequent occurrence of transposable element excision in maize is a consequence of the relatively relaxed requirements for splicing in this organism (see above). Since many of the transposable elements are relatively AU-rich [176, 283] making them similar to plant introns, the presence of cryptic splice sites at the element termini or in the flanking gene sequences may be the only requirement for splicing of transposon introns.

Introns affect the pattern of plant gene expression

With the exception of one gene in *Arabidopsis*, all higher-plant sucrose synthase genes cloned to date possess a very large intron conserved in position within the 5′ UTR. In an interesting series of papers, Park and co-workers have cloned and characterised two classes of sucrose synthase genes from potato, *Sus3* and *Sus4* and analysed the effect that this conserved intron has on expression in stably transformed transgenic plants [66, 67, 68]. Removal of the 5′ UTR intron from *Sus4*-derived constructs results in an eight-fold reduction of expression in tubers and a four-fold reduction in expression in roots. Strikingly, removal of the intron results in changes in the pattern of expression: while expression normally takes place in the root cap and apical meristem, removal of the intron results in a switch in expression from these cells to the procambium. It seems that the regulatory effect exerted through this intron is influenced by regulatory elements in the promoter and the 3′ UTR: the alterations in tissue-dependent expression observed upon removal of the intron are dependent on the presence of promoter and

3′ UTR sequences [67]. Similar analyses performed with the *Sus3* gene [68] also result in changes in the level and pattern of expression upon removal of the 5′ UTR intron. A reduction in expression in vascular tissue of tobacco anthers at later stages of development was observed, but strikingly, a more than 100-fold increase in expression was detected in pollen. This intron therefore confers positive and negative tissue-specific regulated expression.

Notably, the 5′ UTR intron of the maize sucrose synthase gene, *Sh1*, is also extremely important for *Sh1* gene expression in maize and can confer a dramatic enhancement of gene expression to heterologous genes (see below). The conservation of large introns in the 5′ UTR of plant genes is not a phenomenon restricted to sucrose synthase genes, however. Higher-plant polyubiquitin genes (from *Arabidopsis*, maize and sunflower) also possess a conserved large intron in their 5′ UTR [188]. It will be interesting to determine whether this intron also contributes to the regulation of the pattern of expression and whether such introns play a more general or widespread role in the regulation of plant gene expression. Furthermore, it will be interesting to determine whether these regulatory effects are indeed post-transcriptional.

Intron enhancement of gene expression

The presence of introns in transcription units can dramatically enhance gene expression through an increase in the steady-state level of mRNA. This is true in mammals and in higher plants. The characterisation of intron enhancement of gene expression in higher plants stems historically from studies of expression of genes that lack their native introns. Phaseolin is expressed equally well with or without introns [41], but the removal of introns from maize *Adh1* severely reduced its expression [28]. While in the case of *Adh1*, introns are required for expression of the host gene, there are numerous examples where the insertion of introns into heterologous genes can promote a significant enhancement in their normal level of expression (see below). This capacity has obvious implications for the strategic optimisation of transgene expression in higher plants [see Koziel *et al.*, this volume]. To date, this phenomenon has been most extensively documented in monocots. For example, maize *Adh1* and *Sh1* introns can increase expression of reporter genes in maize 100-fold. However, intron enhancement of gene expression occurs in dicots too, but with enhancements in the range of only 2–5-fold, the effects reported so far, are much less dramatic [45, 53, 159, 188].

The extent of intron enhancement of expression is dependent on many factors, including the intron used, its location in the transcript, the nature of other sequences in the construct (like the promoter and exon sequences), the cell-type in which expression takes place and the physiological status of those cells. Examples describing the influence of each of these effects are documented below.

A wide range of introns can promote enhancement of gene expression [reviewed in 143, 245, Koziel *et al.*, this volume). Introns from one species can promote enhancement of gene expression in another [261, 274] and even synthetic introns can exert this effect [45]. However, the degree of enhancement effected by different introns is variable. For example, under otherwise identical conditions, the maize *Sh1* intron 1 increased expression of a reporter gene in maize more than 40-fold, whereas the *Adh1* intron 1 increased expression only 4-fold [274]. Indeed, not all introns promote this effect, for example, the maize *Hsp81* intron 1 did not enhance the expression of reporter genes that were enhanced by numerous other introns [245] and although maize *Adh1* introns 2 and 6 mediated enhancement of reporter gene activity in maize, *Adh1* intron 9 did not [151].

The degree of enhancement observed can be affected by the promoter used to drive test transcripts [28, 274]. For example, the enhancement of gene expression effected by *Adh1* intron 1 is consistently more dramatic if expression is driven by the native *Adh1* promoter than if the CaMV 35S promoter is used [28]. The degree of enhancement may be inversely proportional to the basal activity of the promoter employed [28, 151].

Exon sequences too can influence the degree of enhancement. An exon size of 60–90 nt from the 5′ cap is required for optimal enhancement of gene expression by *Hsp82* intron 1 in maize cells, but very short 3′ exons eliminate the enhancement effect [245]. Likewise, deletion of exon sequences adjacent to maize *Adh1* intron 2 abolish its ability to enhance reporter gene activity [151]. The maize *Sh1* intron 1 can enhance gene expression 100-fold, but this increases to 1000-fold if its native upstream exon is included in the test construct (the exon contains a transcriptional enhancer element, which itself, can increase gene expression 10-fold) [145].

In terms of position, introns located in the 5′ region of transcripts generally promote enhancement, where-as insertions in the 3′ UTR either do no not, or promote negligible enhancement [28, 151].

Finally, the degree to which an intron enhances gene expression can vary with cell type and the physiological status of the cells at the time of the test. For example, *Adh1* intron 1 increases reporter gene expression in maize 44.1-fold in aleurone protoplasts, 16.5-fold in endosperm protoplasts, but only 3.4-fold in suspension cells [75], while the enhancement mediated by a castor bean catalase gene intron was greater in transformed rice calli than in rice protoplasts [261].

The mechanism that mediates intron enhancement of gene expression is unknown and despite the dramatic nature of the effect and its potential commercial importance, no advance in understanding the mechanism has been made. Indeed, an important criticism of research in this field has been the widespread (but not universal) tendency to measure reporter gene activity enzymatically in the absence of a detailed analysis of RNA processing events. It is highly probable that alterations in exon or intron sequence effect cryptic splicing events that perturb translation of active reporter enzymes, but such phenomena are largely overlooked. This might explain some of the variability in the enhancement effect seen with the same intron in different experimental conditions.

In order to exert an enhancing effect, the intron must reside within the transcribed sequence of the gene [28, 145, 151] and since introns inserted in anti-sense orientations do not provide enhancement and extensive internal deletions to intron sequence can be made that do not prevent enhancement [28, 145], the effect cannot be explained by conventional transcription enhancer sequences residing in the intron: the effect must therefore be post-transcriptional. Although splicing seems to be required for the enhancement effect, it alone is not sufficient to explain the mechanism in view of the variability in the effects seen. Notably, while intron enhancement typically promotes an increase in the steady-state level of mRNA, Mascarenhas *et al.* [151] reported that this increase (4-fold) was not sufficient to account for the accompanying increase in reporter gene activity (12-fold) they observed and suggested that pre-mRNA splicing must somehow improve the quality, as well as quantity of mRNA. Subtle distinctions in the specificity of association of hnRNP or spliceosomal proteins with the pre-mRNA plus the inter-play between splicing and other RNA processing events like polyadenylation, mRNA turnover and nucleo-cytoplasmic transport [18, 108, 156] must account for the variability that is seen, but in a complex way, we do not yet understand. While it is possible that spliceosome association promotes the stabilisation of mRNA, the results of Mascarenhas *et al.* [151] argue against this as a universal explanation for intron enhancement in plant cells. In one case in mammals, intron enhancement has been shown to result from an increase in the efficiency of 3' end processing [100] and we have preliminary evidence that the same effect occurs in plants (our unpublished results).

Effects of stress on pre-mRNA splicing

The reported effects of heat shock on pre-mRNA splicing in plants are variable. While, metazoan heat shock genes are generally intron-less, many higher-plant heat shock genes contain introns, indicating that splicing may persist under heat stress. Consistent with this, exposure of plants to heat shock generally does not block splicing of heat shock pre-mRNAs [197, 288] although in some cases processing is inhibited [99, 259]. The effect of heat shock on the splicing of other transcripts is also variable. While the processing of *bronze-2* in maize is unaffected by heat shock [175], maize polyubiquitin pre-mRNAs accumulate, indicating that splicing has been partially inhibited [43]. In addition, heat shock promotes a marked rearrangement of spliceosomal components in pea nuclei [13]. This degree of variability in the effect of heat shock probably reflects the differential response of different species and ecotypes to heat stress, as well as the fact that different cell-types have been studied. What is clear, however, is that when affected, pre-mRNA splicing in plants is only inhibited and not completely blocked by heat shock and this distinguishes plants from metazoans. This may reflect the sessile nature of plant life, in the natural environment a plant may be exposed to extreme temperature fluctuations, but is unable to move to avoid temperature shock.

Other environmental stresses have been shown to inhibit pre-mRNA splicing in plants. For example, unspliced heat shock pre-mRNAs were detected in plant tissues exposed to heavy metals [51, 288] and splicing was also reported to be affected under anaerobic conditions [196].

The sub-nuclear organisation of spliceosomal components

One of the most striking features of nuclear mRNA processing activity is the highly organised distribution of factors that mediate such processing events within the nucleus. The increased documentation of specific marker probes for sub-nuclear structures and the revelation of their dynamic nature has catalysed a dramatic new interest in the relationship between nuclear structure and the execution of diverse RNA processing events.

Mammalian nuclei: perichromatin fibrils, interchromatin granule clusters and coiled bodies

Factors involved in the processing of pre-mRNA exhibit a distinct and dynamic pattern of nuclear organisation [reviewed 123, 155, 218, 253]. Spliceosomal components can be detected in three distinct sub-nuclear compartments: perichromatin fibrils [59], interchromatin granule clusters [253] and coiled bodies [19, 22, 124]. The functional inter-relationships between these structures are not entirely clear, but the major sites of pre-mRNA splicing activity likely correspond to perichromatin fibrils rather than interchromatin granule clusters or coiled bodies. Perichromatin fibrils (which can only be identified at the ultrastructural level) are typically located adjacent to chromatin and on the surface of and between interchromatin granule clusters. They are rapidly labelled by [3H] uridine and co-localise with a population of poly(A)+ RNA which rapidly disappears following inhibition of RNA Pol II and, therefore, probably represent nascent pre-mRNA transcripts. In contrast, neither interchromatin granule clusters nor coiled bodies co-localise with endogenous nascent transcripts [59, 104, 277] and more specifically, they do not directly co-localise with sites of active pre-mRNA splicing [304]. Rather, active pre-mRNA splicing occurs at or close to the sites of transcription [304].

The precise role of interchromatin granule clusters and coiled bodies is unclear, but they represent functionally distinct compartments as evidenced by their differential composition (see below), the activity of a kinase, SRPK1, which regulates splicing factor association with interchromatin granule clusters but not coiled bodies [89] and by the contrasting patterns of spliceosomal component association with them upon changes in nuclear RNA processing activity: interchromatin granule clusters increase in size and content of splicing factors following microinjection into living cells of oligonucleotides or antibodies which inhibit splicing *in vitro* [193] and when transcription is inhibited by α-amanitin or actinomycin D [34]. In contrast, when RNA Pol II transcription is inhibited by α-amanitin or actinomycin D, U snRNPs no longer concentrate in coiled bodies. Coiled bodies are more prominent in rapidly dividing, metabolically active cells and tumours [5, 22, 34]. Such distinctions may indicate that interchromatin granule clusters represent sites of splicing factor storage, or more simply, sites where excess splicing factors accumulate as indicated in a study of *Drosophila* polytene nuclei [298] (it has been shown in yeast that splicing factors are present in large functional excess [234]). However, such a storage role for interchromatin granule clusters is controversial; multiple SR proteins required for splicing concentrate in these regions and Lawrence and co-workers have provided evidence that at least some gene transcripts may be processed at or near such domains [294]. The dynamics of U snRNP association with coiled bodies in response to inhibition of transcription does not fit well with a storage function; they may play a role in the transport, assembly, modification or recycling of spliceosomal components; functions that would be consistent with the relative abundance of coiled bodies in transcriptionally active cells (the exception to this indication is that in hibernating doormice, coiled bodies are prominent [150]). A sub-set of coiled bodies co-localise with a sub-set of histone, U1 and U2 gene loci [65, 249] indicating some function related to their expression.

Although both interchromatin granule clusters and coiled bodies contain spliceosomal U snRNAs and multiple U snRNP proteins, they are structurally and compositionally distinct. When imaged by immuno or *in situ* fluoresence microscopy, interchromatin granule clusters typically show a punctate or irregular speckled pattern (20–50 speckles per nucleus) throughout the nucleoplasm whereas labelling of U snRNPs in coiled bodies shows fewer, more uniformly round and intensely stained foci 0.1–1 μM in diameter, typically, but not exclusively, located close to nucleoli. While certain antibodies recognise both interchromatin granule clusters and coiled bodies, antibodies which recognise SR proteins label interchromatin granule clusters, but not coiled bodies [253] as do antibodies which recognise the inactive, hypophosphorylated form of the large sub-unit of RNA Pol II [23]. In contrast, antibodies against both U2AF subunits label coiled bodies much more prominently than interchromatin granule clusters [301, 303]. Several non-splicing

factor probes label coiled bodies but not interchromatin granule clusters, the most notable of which, are antibodies against p80 coilin, which (despite labelling interchromatin associated domains [209]) serve as a working marker for coiled bodies in immunofluorescent microscopy [4]. In addition to splicing U snRNPs, mammalian coiled bodies contain U7 snRNA (a small RNA involved in histone mRNA 3' end formation) and many nucleolar components, like fibrillarin, Nopp 140, NAP57 and the ribosomal protein S6 [162, 211, 212]. Ribosomal RNA is not present in coiled bodies but the presence of snoRNAs is not so clear: U3 snoRNA, for example, has been localised to coiled bodies by some workers [105] but not others [33], but this may reflect differences in the technical approaches taken. Other snoRNAs apparently do not concentrate in coiled bodies [153].

Higher-plant spliceosomal components exhibit a discrete sub-nuclear organisation

Three different studies of higher-plant spliceosomal component distribution, with diverse probes and different plant species reveal a related pattern: spliceosomal components localise to an interchromatin network and concentrate in intensely stained foci [13, 38, 78]. This pattern is illustrated in the pea nuclei depicted in Fig. 3. The most detailed and comprehensive of these studies is that by Beven *et al.* [13], who also observed the localisation of individual spliceosomal U snRNAs in the nucleolar vacuole.

Higher-plant spliceosomal components localise to an interchromatin network

Spliceosomal components localise to a fibrous interchromatin network in plant nuclei (Fig. 3) [13, 38, 78]. This network may include perichromatin fibrils and interchromatin granule clusters, but the large, irregular-shaped speckled pattern which characterises prominent splicing factor association with interchromatin granule clusters in mammalian nuclei is not apparent [13, 38]. This distinction cannot be due to the nature of the probes employed, since both the anti-U2B''-specific mAb 4G3 employed by Beven *et al.* [13] and the anti-Sm mAb Y12 employed by Chamberland and Lafontaine [38] have been shown to label prominent interchromatin granule clusters in mammalian cells (in addition to coiled bodies and a diffuse interchromatin network) [35, 253]. Although this

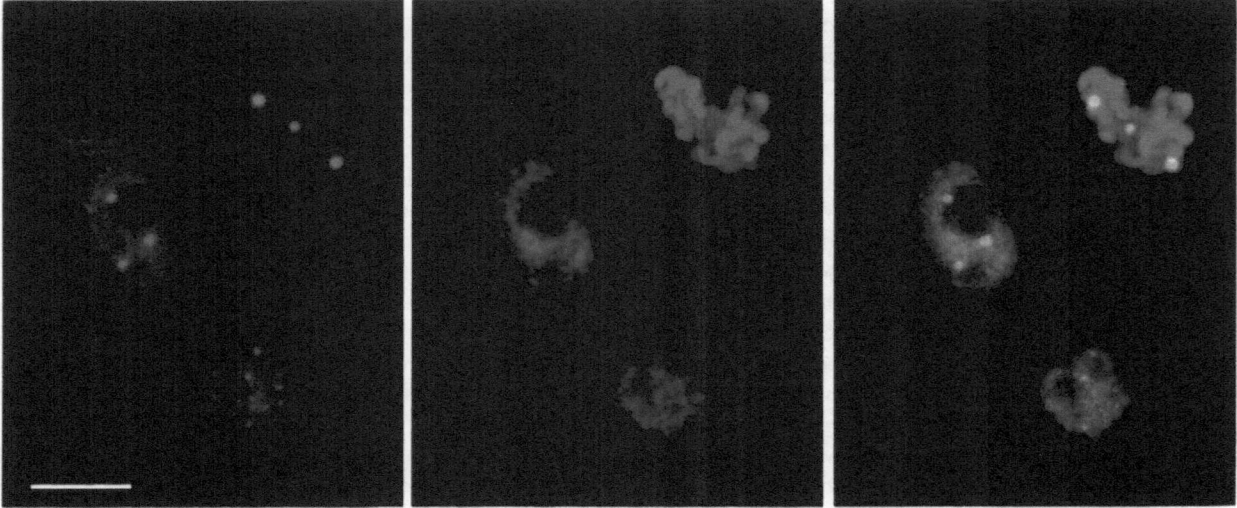

Figure 3. Sub-nuclear localisation of spliceosomal components in higher-plant nuclei. Single confocal optical sections of three different pea root cells labelled with anti-U2B″ mAb 4G3 (green, left panel) and counterstained with 7AAD to reveal chromatin (red, centre panel). The antibody strongly labels the nuclei and specifcally labels round foci and a network located around and between the regions of dense chromatin labelling as revealed by the double labelling (right panel). The cells displayed are in different stages of the cell cycle: clockwise from the top, metaphase, very early G1 and interphase respectively, and illustrate, respectively, that foci persist in mitosis, that in early G1 numerous small bodies are seen and that in late interphase fewer (usually one or two) larger foci are observed which are typically, but not exclusively, located close to nucleoli. Bar = 10 μm (Courtesy of P.J. Shaw.)

might indicate a fundamental difference in spliceosomal distribution in plant cells, such a conclusion cannot be made yet. Ultrastructural and cytochemical analyses reveal the presence of interchromatin granules in higher plants (20–25 nm in diameter), interconnected by fibrils and which resemble the corresponding mammalian structures [161]. However, interchromatin granules in plants do not cluster [161]. Therefore, immunofluorescent labelling of spliceosomal components associated with them in a functionally similar manner to those in mammalian nuclei would not exhibit a prominent speckled pattern.

While the ultrastructural data of Medina *et al.* [161] provides an appropriate explanation for the absence of the prominent speckled pattern of immunofluorescent labelling of spliceosomal components in plant nuclei, additional explanations are possible. Firstly, it may be that such structures do form, but do not survive the fixing conditions that have been applied. Secondly, their absence may reflect the cell types that have been studied to date. Since the association of U snRNPs with interchromatin granule clusters is dynamic and responsive to the transcriptional and splicing activity of the nucleus, the prominence of interchromatin granule clusters can vary even in mammalian cells and varies in different cell types, in different strains

of the same cell type, and in the same cells grown under different conditions [35, 154]. Chamberland and Lafontaine [38], and Beven *et al.* [13] studied root-tip meristem and suspension culture cells. Since such proliferative cells are particularly transcriptionally active, one might not expect prominent spliceosomal association with interchromatin granule clusters. In order to resolve this issue, an examination of alternative cell types, the effects of transcription inhibitors and the use of specific marker probes is required. Probes which recognise SR proteins and RNA Pol II have served as specific markers for interchromatin granule clusters in mammals [16, 23, 253, 268] and since SR proteins are present in higher plants (see above) and plant interchromatin granules possess a high content of phosphorylated proteins [161], they represent good candidates for effective probes in plant cells.

Higher-plant spliceosomal components concentrate in structures that resemble mammalian coiled bodies

The concentration of spliceosomal components in nuclear foci is illustrated in Figs. 3 and 4. The appearance, composition, structure and distribution of the strongly labelled foci closely resembles the properties of mammalian coiled bodies. Like mamalian coiled bodies, they are significantly enriched in U2B″ and U2 and U6 snRNAs, while U1 snRNA, although not excluded, does not concentrate in them [13]. They are labelled by an antibody which recognises p80 coilin and like mammalian coiled bodies they are labelled by antibodies against fibrillarin (P. Shaw, pers. comm.). They are often, but not always, located close to nucleoli, are quite uniformly round in shape and range in size from less than 0.5 μm to more than 2 μm in diameter (somewhat larger than mammalian coiled bodies).

Since the function of coiled bodies is not known, it is not possible to formally state that the foci in plant cells are functionally equivalent. However, the compositional similarities indicate that they are, at least, related structures. The characterisation of these structures in higher plants provides a valuable evolutionarily distinct system for analysing coiled body function. *Xenopus* oocytes contain distinct sub-nuclear structures enriched in spliceosomal components, termed spheres, which are comprised of different substructures called B and C snurposomes [74, 220], but the functional relationship of these structures, in these somewhat specialised cells, to the structures seen in the nuclei of mammalian somatic cells is not clear. Although they are related, there are differences between the structures [discussed in 19, 220]. For example, although fibrillarin is present in the plant foci and mammalian coiled bodies, it has not been detected in *Xenopus* spheres. In contrast, U7 snRNA, which functions in histone pre-mRNA 3′ end formation, has been detected in mammalian coiled bodies and *Xenopus* spheres [220], but is unlikely to be present in plant foci, since plant histone pre-mRNAs are cleaved and polyadenylated [37] and plants therefore presumably lack U7 snRNA. Another possible difference is that while mammalian coiled bodies may contain U3 snRNA, the foci in plants apparently do not [13]. Notably, U3 snRNA in plants is expressed from an RNA Pol III promoter, while in mammals, it is expressed by RNA Pol II [110]. Plant U3 snRNA can be expressed from an RNA Pol II promoter, but when it is, it does not associate with higher-order nucleolar complexes [110].

Lamond and co-workers have recently proposed that the coiled body may function in RNA modification [19]. Both rRNA and U snRNAs undergo extensive post-transcriptional modification on base and sugar residues. It is believed that modifications on rRNA take place in the nucleolus; if processing of U snRNAs takes place in the coiled body, the conservation of the processing machinery may explain the shared antigens

between coiled bodies and the nucleolus. Since coiled bodies possess components that shuttle between the nucleus and cytoplasm (Nopp 140, NAP57) some role connected to U snRNP transport or assembly cannot be ruled out either. Indeed, the only way to resolve the divergent observations on these structures may be to invoke multiple functions.

Spliceosomal components in the plant nucleolar vacuole

The nucleolus primarily consists of three well characterised structural compartments; the dense fibrillar component, the granular component and the fibrillar centre material. In addition, there are various nucleoplasmic inclusions that are generally called nucleolar vacuoles, about which little is known [236]. Notably, Beven *et al.* [13] reported that U2B″ and spliceosomal U snRNAs could be detected in the nucleolar vacuole. The significance of this localisation is not yet clear.

The localisation of plant spliceosomal components is dynamic

The pattern of U snRNP distribution in plant nuclei undergoes a marked rearrangement in the course of the cell cycle [13]. Pea interphase nuclei are characterised by U snRNP association with the interchromatin network and concentration in usually one or two nucleolus associated foci. On entry into mitosis, the interchromatin labelling disappears but the foci persist. The redistribution of the foci into daughter cells is apparently random and while some are closely associated with the condensed chromosomes, others are located far from them, indicating that they do not associate with any specific loci. As soon as chromosome decondensation begins in telophase, the interchromatin network reappears and multiple small nuclear bodies can be seen throughout the nucleoplasm. These small bodies are labelled by probes against U2 snRNA, U6 snRNA and U2B″ but poorly, if at all, by p80 coilin antibodies (Fig. 3).

The association of spliceosomal components with foci is affected by heat shock. Upon heat shock, which inhibits the transcription of most RNA Pol II genes (except for a specific set of heat shock genes) and, as mentioned above, can disrupt pre-mRNA splicing in animals and to some extent also in plants [20, 43, 297], U snRNPs no longer concentrate in foci in plant cells [13] and likewise do not associate with coiled bodies in HeLa cells [34]. However, this rearrangement is reversible and upon recovery, U snRNPs reassociate with coiled bodies in plants and mammals.

Ribonucleoprotein domains in plant nuclei: unification of old and new observations

A classification of nucleolar associated bodies and dense bodies

Light microscope, ultrastructural and cytochemical analyses have long since revealed multiple nuclear bodies in the nuclei of higher plants which, although distinct, are not membrane bound, do not contain DNA, do not incorporate [^{3}H] uridine and are composed of ribonucleoprotein [286]. Their classification is uncertain because no single criterion adequately distinguishes them and morphological variation between species or cell types complicates attempts to define them. Furthermore, attempts to classify such structures may be misleading since there is no clear evidence to suggest that the different nuclear bodies have any functional relationship to one another. Bearing such fundamental reservations in mind, we re-outline here the classification of these plant nuclear bodies, based on their appearance in the electron microscope, as proposed by Williams et al. [286], in order to prepare a foundation for comparing such observations with more recent studies that employed probes against spliceosomal components.

Although nuclear bodies in plants exhibit a range in size and ultrastructure, in order to frame a classification, Williams et al. [286] considered only extreme forms. Two classes of ribonucleoprotein containing nuclear bodies were defined: the nucleolus associated body and the dense body. The nucleolus associated body is typically located close to nucleoli, but is occasionally free in the nucleoplasm. There are usually less than 2 (1–3) nucleolus-associated bodies per nucleus and they are approximately 0.4–0.7 μm in diameter. Nucleolus associated bodies are more common in differentiating cells and more common late in the cell cycle. In contrast, the dense body is typically free in the nucleoplasm, being only occasionally associated with nucleoli. They are more abundant than nucleolar associated bodies (5–20 per nucleus) and smaller (0.2–0.4 μm in diameter). Dense bodies are more common in mitotic cells and early in the cell cycle. Both nucleolus associated bodies and dense bodies are composed of 5 nm fibrils, but while these are closely packed in a homogenous electron dense structure in dense bodies, nucleolus associated bodies possess additional major fibrils (10–45 nm in diameter) with spaces between them resulting in a structure more 'open' in appearance (the presence of the major fibrils and the spaces between them is subject to variation between species). There are some indications too, that nucleolus associated bodies are more prominent in plant species with low nuclear DNA content, while dense bodies are present in species with a high DNA content, but both can co-exist [8, 286].

Nucleolus-associated bodies/foci/coiled bodies

Nucleolus associated bodies correspond to the intensely stained foci observed by Chamberland and Lafontaine [38] and Beven et al. [13] as demonstrated by Chamberland and Lafontaine [38] and which are highly related to mammalian coiled bodies (see above; see Fig. 4). Moreover, they correspond to structures first reported in plant nuclei about 60 years ago [52, 57, 120, 203, 286] and which have been variously referred to as spherules, spherical bodies, karyosomes or loose bodies. Previously performed ultracytochemical and ultrastructural analyses of these nucleolus associated bodies indicated that they possess an RNP similar in structure and composition to the fibrillar component of the nucleolus [140], that they often appear to exhibit some continuity with this portion of the nucleolus [286], are intimately associated with intranucleolar chromatin [122] and exhibit changes in abundance upon changes in ploidy [121]. In many ways, these properties underscore the relatedness of these structures to mammalian coiled bodies [see 19, 22, 124] and the intimate interrelationship between the coiled body and the nucleolus.

Dense bodies

In contrast to the consensus that surrounds the identification of the nucleolar associated body and its relatedness to the coiled body, confusion surrounds reference to the dense body, pre-nucleolar bodies and structures identified and studied by Moreno Díaz de la Espina and co-workers [167, 168, 215], which, to add to the confusion they called coiled bodies. This uncertainty stems from the inadequacy of both the classification and the data available on these nuclear RNP structures. The observations on the coiled bodies studied by Moreno Díaz de la Espina and co-workers [167, 168, 215] were reinterpreted by Williams et al. [286] who assigned them to the dense body class; an inter-

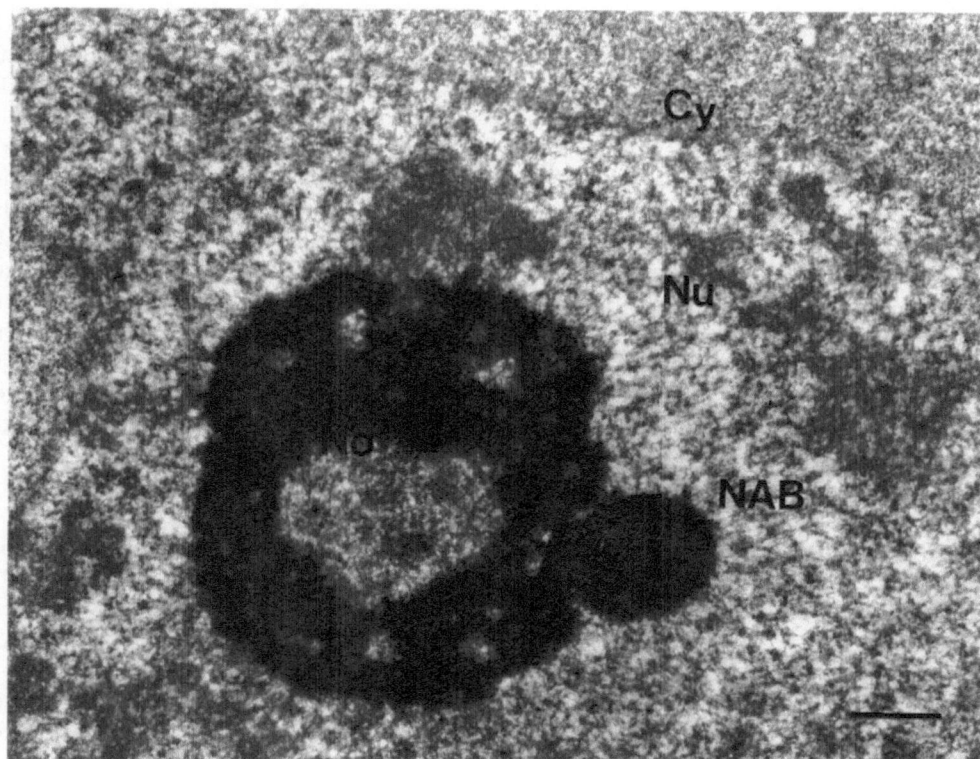

Figure 4. Electron micrograph of a *Pisum sativum* nucleus following incubation with anti-tri-methyl guanosine cap mAb, K121. Immunogold particles concentrate over nucleolus associated bodies. Cy, cytoplasm; Nu, nucleus; No, nucleolus; NAB, nucleolus-associated body. Bar = 0.5 μm (Courtesy of J-G. Lafontaine.)

pretation that has been accepted by Moreno Díaz de la Espina *et al.* [169]. To further underline the confusion, dense bodies have also been interpreted as corresponding to pre-nucleolar bodies which appear in late telophase/early G1 [8, 227], but Moreno Díaz de la Espina *et al.* [169] have argued that the ultrastructural appearance and timing of pre-nucleolar bodies is quite different from the dense bodies (they first called coiled bodies) they have identified.

Ultrastructural analyses have identified both nucleolus associated bodies and dense bodies in pea root-tip nuclei [8, 287]. There is a negative correlation in their appearance: in early interphase, the smaller, more numerous dense bodies are prominent, while in late interphase, one or two nucleolus associated bodies are prominent [8]. The analysis of spliceosomal component distribution in pea nuclei by Beven *et al.* [13] exhibits striking parallels. Late in interphase/G2, only one or two intensely labelled foci, typically closely associated with the nucleolus are seen, while in early interphase, multiple smaller foci located throughout the nucleoplasm are detected [13]. The relationship between the intensely stained foci and the nucleolus associated bodies of late interphase has been established [see above, 38]. It is clearly possible that the more numerous smaller foci located throughout pea nucleoplasm in early interphase correspond to dense bodies. While dense bodies and nucleolus-associated bodies may be related structures, the extent or nature of this relationship is not yet clear.

Multiple small foci in late telophase

Beven *et al.* [13] identified the appearance of a large number of very small bodies in late telophase/early G1 that contain U2B''. Similarly, ultrastructural analyses have identified numerous small ribonucleoprotein bodies arising at late telophase/early G1 and it is probably these structures around which so much of the coiled body/dense body/pre-nucleolar body classification confusion mentioned above has arisen [8, 169, 227]. The coincidence in the timing of appearance of these RNP-containing structures seen at the ultrastructural level and the U2B″ containing bodies detected by immunofluoresence raises the question as to whether they may be corresponding structures.

Resolution of previous ultrastructural analyses with recent immunofluorescence studies

There is clearly a limit to which previous data can be credibly re-interpreted, but the connections outlined above prompt further study and while size, distribution and timing of appearance indicate a relatedness between these structures, in terms of composition, specific marker probes and combined ultrastructural/confocal data are required to clarify this.

By analogy with pre-mRNA splicing in mammals, it seems unlikely that pre-mRNA splicing itself will be localised within these many different RNP domains. However, such structures likely play a key role in the spliceosome cycle in addition to defining major plant nuclear structures. The striking regularity with which such structures are found in plant nuclei and the concentration of U snRNPs within at least some of them is indicative of their fundamental functional importance. Since multiple bodies of diverse morphology and of unknown relationship to one another exist, an essential pre-requistite to advancing their study is the identification of specific compositional markers. This is clearly a field of fundamental plant research awaiting clarification and functional classification.

Perspective

The study of pre-mRNA splicing has thrown up two recent surprises. The first is that a sub-class of introns (AT-AC) are spliced within novel spliceosomes that do not contain U1, U2, U4 or U6 snRNA, but instead contain U5, U11 and U12 snRNA [90, 265]. The second is that introns themselves can code for functional RNAs, dispelling the concept of introns as 'junk DNA'. In fact, in some genes, HUG for example, the exons are non-coding and apparently serve only to provide a splicing substrate for the expression of snoRNAs encoded within the introns that separate them; in other words, the introns rather than the exons specify the functional products of the gene [270]. The degree of conservation of the novel U11, U12 snRNA containing spliceosome in nature is presently unknown, but intron encoded snoRNAs are widespread. Despite this, no intron-encoded snoRNAs have yet been definitively

described in higher plants [see 131]. It will therefore be interesting to determine whether such intron-encoded snoRNAs exist in plants and, if so, to determine the manner in which they are processed.

As outlined in this review, the feature of plant pre-mRNA splicing that most markedly contrasts it with the process in *S. cerevisiae* and mammals is the presence of and functional requirement for AU-rich intron sequence. AU-rich introns are not restricted to plants, but are a feature of intron composition that is widespread in nature [81, 50]. Although the functional requirement for such sequence has been most extensively detailed in dicot plants, the functional requirement for AU-rich sequence has also been demonstrated in nematode *cis* and *trans* splicing [46, 47] illustrating the importance and generality of AU-rich intron sequence in nature.

Clearly then, the fundamental mechanistic issue in higher-plant pre-mRNA splicing is the means by which the requirement for AU-rich intron sequence is mediated. The resolution of this question would define the distinctions between splicing in plants from that in mammals and yeast and provide a unifying description of the evolution of differential spliceosome assembly. The most efficient and direct means to investigate this issue requires a plant extract capable of faithfully processing higher-plant introns *in vitro*. However, repeated attempts to produce such an extract by different laboratories around the world have only met with failure. The absence of a competent *in vitro* splicing extract will restrict the mechanistic dissection of alternative splicing and the analysis of the role the newly discovered plant SR proteins are likely to play in this process. A likely development to come however, is the characterisation of *Arabidopsis* mutants (since so many are under examination) defective in pre-mRNA processing events affecting developmental or stress responses that will thereby provide the genetic route into the analysis of *trans*-acting RNA processing factors in plants, that has so far, been absent.

What will also be interesting and likely within our grasp, is the determination of the mechanisms by which certain intron disruptions promote dramatic phenotypic changes in plants. The duplication of a 291 bp segment of the barley *Knox3* gene intron 4 results in a phenotype that involves the production of an extra flower of inverse polarity [173], while the insertion of the Tam3 transposon into the large intron of *ple* in the sense orientation results in a homeotic conversion of sex organs in flowers of *Antirrhinum majus* (the so-called plena phenotype) [21]. Insertion of Tam3 in the opposite orientation in the same intron results in a complementary phenotype, in which sex organs replace sterile organs (known as the ovulata phenotype) [21]. *Knox3* is a homeobox gene and *ple* a MADS box gene and thus the intron disruptions result in profound visible developmental phenotypes. However, in neither case has a mechanistic analysis of the means by which the intron insertions result in such changes been undertaken.

The ever increasing documentation of alternatively spliced transcripts, the need to exploit introns for optimised gene expression, the impact of post-transcriptional mediated transgene cosuppression and the integrated nature of multiple fundamental RNA processing events will ensure continued attention on pre-mRNA splicing. In addition, the localisation of spliceosomal components to sub-nuclear domains previously identified through cytochemical and ultrastructural analyses has provided compositional markers that should facilitate rapid progress in their characterisation and clarify their confused and inadequate classification.

Acknowledgements

We thank A. Barta, J.W.S. Brown, A. Leitch, C.G. Simpson and R.D. Thompson for communicating results prior to publication. We are grateful to J-G. Lafontaine and P. Shaw for kindly providing figures and to S. Brunak for supplying the splice site sequences of *Arabidopsis* introns. We thank J.W.S. Brown, F. Dragon, M. Lambermon, S. Moreno Díaz de la Espina, and C.G. Simpson for helpful discussions. G.G.S. was supported by the Royal Society (London).

References

1. Abovich N, Liao XC, Rosbash M: The yeast MUD2 protein: an interaction with PRP11 defines a bridge between commitment complexes and U2snRNP addition. Genes Devel 8: 843–854 (1994).
2. Aebi M, Hornig H, Padgett RA, Reiser J, Weissmann C: Sequence requirements for splicing of higher eukaryotic nuclear pre-mRNA. Cell 47: 555–565 (1986).
3. Accotto GP, Donson J, Mullineaux PM: Mapping of Digitaria streak virus transcripts reveals different RNA species from the same transcription unit. EMBO J 8: 1033–1039 (1989).
4. Andrade LE, Chan EK, Raska I, Peebles CL, Roos G, Tan EM: Human autoantibody to a novel protein of the nuclear coiled body: immunological characterisation and cDNA cloning of p80 coilin. J Exp Med 173: 1407–1409 (1991).
5. Andrade LE, Tan EM, Chan EK: Immunocytochemical analysis of the coiled body in the cell cycle and during cell proliferation. Proc Natl Acad Sci USA 90: 1947–1951 (1993).
6. Barta A, Sommergruber K, Thompson D, Hartmuth K, Matzke MA, Matzke AJM: The expression of a nopaline synthase-human growth hormone chimaeric gene in transformed tabacco and sunflower callus tissue. Plant Mol Biol 6: 347–357 (1986).
7. Barlow PW: Nucleolus-associated bodies (karyosomes) in dividing and differentiating plant cells. Protoplasma 115: 1–10 (1983).
8. Barlow PW: Changes in the frequency of two types of nuclear body during the interphase of meristematic plant cells. Protoplasma 118: 104–113 (1983).
9. Beggs JD: Yeast splicing factors and genetic strategies for their analysis. In: Lamond A (ed) Pre-mRNA Processing. R.G. Landes Publishers, Georgetown, TX (1995).
10. Bennet M, Pinol-Roma S, Staknis D, Dreyfuss G, Reed R: Differential binding of heterogenous nuclear ribonucleoproteins to mRNA precursors prior to spliceosome assembly in vitro. Mol Cell Biol 12: 3165–3175 (1992).
11. Bennet M, Michaud S, Kingston J, Reed, R: Protein components specifically associated with prespliceosome and spliceosome complexes. Genes Devel 6: 1986–2000 (1992).
12. Berget SM: Exon recognition in vertebrate splicing. J Biol Chem 270: 2411–2414 (1995).
13. Beven AF, Simpson GG, Brown JWS, Shaw PJ: The organization of spliceosomal components in the nuclei of higher plants. J Cell Sci 108: 509–518 (1995).
14. Beyer AL, Osheim YN: Splice site selection, rate of splicing and alternative splicing on nascent transcripts. Genes Devel 2: 754–765 (1988).
15. Birney E, Kumar S, Krainer AR: Analysis of the RNA-recognition motif and RS and RGG domains: conservation in metazoan pre-mRNA splicing factors. Nucl Acids Res 21: 5803–5816 (1993).
16. Blencowe BJ, Nickerson JA, Issner R, Penman S, Sharp PA: Association of nuclear matrix antigens with exon-containing splicing complexes. J Cell Biol 127: 593–607 (1994).
17. Blencowe BJ, Issner R, Kim J, McCaw P, Sharp PA: New proteins related to the Ser-Arg family of splicing factors. RNA 1: 852–865 (1995).
18. Boelens WC, Dargemont C, Mattaj IW: Export of mRNA through the nuclear pore complex. In Lamond A (ed) Pre-mRNA Processing, pp. 173–186. R.G. Landes Publishers, Georgetown, TX (1995).
19. Bohmann K, Ferrreira J, Santama N, Weis K, Lamond AI: Molecular analysis of the coiled body. J Cell Science (Suppl) 19: 107–113 (1995).
20. Bond U: Heat shock but not other stress inducers leads to the disruption of a subset of snRNPs and inhibition of in vitro splicing in HeLa cells. EMBO J 7: 3509–3518 (1988).
21. Bradley D, Carpenter R, Sommer H, Hartley N, Coen E: Complementary floral homeotic phenotypes result from opposite orientations of a transposon at the *plena* locus of *Antirrhinum*. Cell 72: 85–95 (1993)
22. Brasch K, Ochs RL: Nuclear bodies (NB): a newly 'rediscovered organelle' Exp Cell Res 202: 211–223 (1992).
23. Bregman DB, Du van der Zee S, Warren SL: Transcription dependent redistribution of the large subunit of RNA polymerase II to discrete nuclear domains. J Cell Biol 129: 287–298 (1995).

34

24. Brown JWS: A catalogue of splice junction and putative branchpoint sequences from plant introns. Nucl Acid Res, 14: 9549–9559 (1986).

25. Brown JWS, Feix G, Frendewey D: Accurate in vitro splicing of two pre-mRNA plant introns in a HeLa cell nuclear extract. EMBO J 5: 2749–2758 (1986).

26. Burgess SM, Guthrie CA: Beat the clock-paradigms for NTPases in the maintenance of biological fidelity. Trends Biochem Sci 18: 381–384 (1993).

27. Burgess SM, Guthrie CA: Mechanism to enhance mRNA splicing fidelity:the RNA dependent ATPase Prp16 governs usage of a discard pathway for aberrant lariat intermediates. Cell 73: 1377–1391 (1993).

28. Callis J, Fromm M, Walbot V: Introns increase gene expression in cultured maize cells. Genes Devel 1: 1183–1200 (1987).

29. Calvio C, Neubaue, G, Mann M, Lamond AI: Identification of hnRNP P2 as TLS/FUS using electrospray mass spectrometry. RNA 1: 724–733 (1995).

30. Cardon GH, Frey M, Saedler H, Gierl A: Mobility of the maize transposable element En/Spm in *Arabidopsis thaliana*. Plant J 3: 773–784 (1993).

31. Carle-Urioste JC, Ko CH, Benito M-I, Walbot V: *In vivo* analysis of intron processing using splicing-dependent reporter gene assays. Plant Mol Biol 26: 1785–1795 (1994).

32. Carmo-Fonseca M, Pepperkok R, Sproat BS, Ansorge W, Swanson MS, Lamond AI: In vivo detection of snRNP-rich organelles in the nuclei of mammalian cells. EMBO J 10: 1863–1873 (1991).

33. Carmo-Fonseca M, Tollervy D, Barabino SML, Merdes A, Brunner C, Lamond AI: Mammalian nuclei contain foci which are highly enriched in components of the pre-mRNA splicing machinery. EMBO J 10: 195–206 (1991).

34. Carmo-Fonseca M, Pepperkok R, Carvalho MT, Lamond AI: Transcription-dependent colocalization of the U1, U2, U4/6, and U5 snRNPs in coiled bodies. J Cell Biol 117: 1–14 (1992).

35. Caromo-Fonseca M, Ferreira J, Lamond AI: Assembly of snRNP-containing coiled bodies is regulated in interphase and mitosis: evidence that the coiled body is a kinetic nuclear structure. J Cell Biol 120: 841–852 (1993).

36. de Carvalho Niebel F, Frendo P, Van Montagu M, Cornelissen M: Post-transcriptional cosuppression of β-1,3-glucanase genes does not affect accumulation of transgene nuclear mRNA. Plant Cell 7: 347–358 (1995).

37. Chaboute ME, Chaubert N, Gigot C, Philips G: Histones and histone genes in higher plants: structure and genomic organization. Biochemie 75: 523–531 (1993).

38. Chamberland H, Lafontaine JG: Localization of snRNP antigens in nucleolus associated bodies: study of plant interphasic nuclei by confocal and electron microscopy. Chromosoma 102: 220–226 (1993).

39. Chanfreau G, Legrain P, Dujon B, Jacquier A: Interaction between the first and last nucleotides of pre-mRNA introns is a determinant of 3′ splice site selection in *S. cerevisiae*. Nucl Acids Res 22: 1981–1987 (1994).

40. Chapman KB, Boeke JD: Isolation and characterization of the gene encoding yeast debranching enzyme. Cell 65: 483–492 (1991).

41. Chee PP, Klassy RC, Slighton JL: Expression of a bean storage protein 'phaseolin minigene' in foreign plant tissues. Gene 41: 47–57 (1986).

42. Choi YD, Grabowski PJ, Sharp PA, Dreyfuss, G: Heterogeneous nuclear ribonucleoporteins: role in RNA splicing. Science 231: 1534–1539 (1986).

43. Christensen AH, Sharrock RA, Quail PH: Maize polyubiquitin genes: structure, thermal perturbation of expression and transcript splicing and promoter activity following transfer to protoplast by electroporation. Plant Mol Biol 18: 675–689 (1992).

44. Company, Arenas J, Abelson J: Requirement of the RNA helicase-like protein PRP22 for release of messenger RNA from spliceosomes. Nature 349: 487–493 (1991).

45. Connelly SC, Filipowicz W: Activity of chimeric U small nuclear RNAs (UsnRNA)/mRNA genes in transfected protoplasts of *Nicotiana plumbaginifolia*: UsnRNA 3′ end formation and transcription initiation can occur independently in plants. Mol Cell Biol 13: 6403–6415 (1993).

46. Conrad R, Liou RF, Blumenthal T: Functional analysis of a *C. elegans trans*-splice acceptor. Nucl Acids Res 21: 913-919 (1993).

47. Conrad R, Lea K, Blumenthal T: SL 1 trans-splicing specified by AU-rich synthetic RNA inserted at the 5′ end of *Caenorhabditis elegans* pre-mRNA. RNA 1: 164–170 (1995).

48. Crispino JD, Blencowe BJ, Sharp PA: Complemetation by SR proteins of pre-mRNA splicing reactions depleted of U1snRNP. Science 265: 1866–1869 (1994).

49. Crispino JD, Sharp PA: A U6 snRNA:pre-mRNA interaction can be rate-limiting for U1-independent splicing. Genes Devel 9: 2314–2323 (1995).

50. Csank C, Taylor FM, Martindale DW: Nuclear pre-mRNA introns: analysis and comparison of intron sequences from *Tetrahymena thermophia* and other eukaryotes. Nucl Acid Res 18: 5133–5141 (1990).

51. Czarnecka E, Nagao RT, Key JL, Gurley JB: Characterization of *Gmhsp26-A*, a stress gene encoding a divergent heat shock protein of soybean: heavy metal-induced inhibition of intron processing. Mol Cell Biol 8: 1113–1122 (1988).

52. Dangeard P: Recherches sur la structure des noyaux chez quelques Angiospermes. Botaniste 28: 291 (1937).

53. Dean C, Favreau M, Bond-Nutter D, Bedbrook J, Dunsmuir P: Sequences downstream of translation start regulate quantitative expression of two *Petunia rbcS* genes. Plant Cell 1: 201–208 (1989).

54. Delorme V, Giranton J-L, Hatzfeld Y, Friry A, Heizmann P, Ariza MJ, Dumas C, Gaude T, Cock JM: Characterization of the *S*-locus genes, *SLG* and *SRK*, of the *Brassica S3 halotype*: identification of a membrane-localized protein encoded by the *S* locus receptor kinase gene. Plant J 7: 429–440 (1995).

55. Dietrich MA, Prenger JP, Guilfoyle TJ: Analysis of the genes encoding the largest subunit of RNA polymerase II in *Arabidopsis* and soybean. Plant Mol Biol 15: 207–223 (1990).

56. Dreyfuss G, Matunis MJ, Pinol-Roma S, Burd CG: hnRNP proteins and the biogenesis of mRNA. Annu Rev Biochem 62: 289–231 (1993).

57. Eftimiu-Heim P: Micronucléoles et caryocinèse chez les Cucurbitacées. Botaniste 28: 5 (1937).

58. Eperon LP, Graham IR, Griffiths AD, Eperon IC: Effects of RNA secondary structure on alternative splicing of pre-mRNA: is folding limited to a region behind the transcribing RNA polymerase. Cell 54: 393–401 (1988).

59. Fakan S: Perichromatin fibrils are in situ forms of nascent transcripts. Trends Cell Biol 4: 86–90 (1994).

60. Ferreira JA, Carmo-Fonseca M, Lamond AI: Differential interactions of splicing snRNPs with coiled bodies and interchromatin granules during mitosis and assembly of daughter cell nuclei. J Cell Biol 126: 11–23 (1994).

61. Field DJ, Friesen, JD: Functionally redundant interactions between U2 and U6 spliceosomal snRNAs. Genes Devel 10: 489–501 (1996).

62. Filipowicz W, Kiss T: Structure and function of nucleolar snRNPs. Mol Biol Rep 18: 149–156 (1993).

63. Filipowicz W, Gniadkowski M, Klahre U, Liu H-X: Pre-mRNA splicing in plants. In: Lamond A (ed) Pre-mRNA Processing, pp. 65–78. R.G. Landes Publishers, Georgetown, TX (1995).

64. Frey M, Tavantzis SM, Saedler H: The maize *En-1/Spm* element transposes in potato. Mol Gen Genet 217: 172–177 (1989).

65. Frey MR, Matera AG: Coiled bodies contain U7 small nuclear RNA and associate with specific DNA sequences in interphase human cells. Proc Natl Acad Sci USA 92: 5915–5919 (1995).

66. Fu H, Park WD: Sink-and vascular-associated sucrose synthase functions are encoded by diferent gene classes in potato. Plant Cell 7: 1369–1385 (1995).

67. Fu H, Kim SY, Park WD: High-level tuber expression and sucrose inducibility of a potato *Sus4* sucrose synthase gene require 5' and 3' flanking sequences and the leader intron. Plant Cell 7: 1387–1394 (1995).

68. Fu H, Kim SY, Park WD: A potato *Sus3* sucrose synthase gene contains a context-dependent 3' element and a leader intron with both positive and negative tissue-specific effects. Plant Cell 7: 1395–1403 (1995).

69. Fu X-D: Specific commitment of different pre-mRNAs to splicing by single SR proteins. Nature 365: 82–85 (1993).

70. Fu X-D: The superfamily of arginine/serine-rich splicing factors. RNA 1: 663–680. (1995).

71. Fu X-D, Mayeda A, Maniatis T, Krainer AR: General splicing factors SF2 and SC35 have equivalent activities *in vitro* and both affect alternative 5' and 3' splice site selection. Proc Natl Acad Sci USA 89: 11224-11228 (1992).

72. Fuller-Pace FV: RNA helicases: modulators of RNA structure. Trends Cell Biol 4: 271–274 (1994).

73. Fütterer J, Potrykus I, Valles Brau MP, Dasgupta I, Hull R, Hohn T: Splicing in a plant pararetrovirus. Virology 198: 663–676 (1994).

74. Gall JG: Spliceosomes and snurposomes. Science 252: 1499–1500 (1991).

75. Gallie DR, Young TE: The regulation of gene expression in transformed maize aleurone and endosperm protoplasts. Analysis of promoter activity, intron enhancement, and mRNA untranslated regions on expresion. Plant Physiol 106: 929–939 (1994).

76. Gierl A: The *En/Spm* transposable elements of maize. Curr Top Microbiol Immunol 204: 145–159 (1995).

77. Giranton J-L, Ariza MJ, Dumas C, Cock JM, Gaude T: The *S* locus receptor kinase gene encodes a soluble glycoprotein corresponding to the SRK extracellular domain in *Brassica oleracea*. Plant J 8: 827–834 (1995).

78. Glyn MCP, Leitch AR: The distribution of a spliceosome protein in cereal (Triticaceae) interphase nuclei from cells with different metabolic activities and through the cell-cycle. Plant J 8: 531–540 (1995).

79. Gniadkowski M, Hemmings-Mieszczak M, Klahre U, Liu H-X, Filipowicz W: Characterisation of intronic uridine-rich sequence elements acting as possible targets for nuclear proteins during pre-mRNA splicing in *Nicotiana plumbaginifolia*. Nucl Acids Res 24: 619–627 (1996).

80. Goguel V, Wang Y, Rosbash M: Short artificial hairpins sequester splicing signals and inhibit yeast pre-mRNA splicing. Mol Cell Biol 13: 6841–6848 (1993).

81. Goodall GJ, Filipowicz W: The AU-rich sequences present in the introns of plant nuclear pre-mRNAs are required for splicing. Cell 58: 473–483 (1989).

82. Goodall GJ, Filipowicz W: The minimum functional length of pre-mRNA introns in monocots and dicots. Plant Mol Biol 14: 727–733 (1990).

83. Goodall GJ, Filipowicz W: Different effects of intron nucleotide composition and seconday structure on pre-mRNA splicing in monocot and dicot plants. EMBO J 10: 2635–2644 (1991).

84. Goodall GJ, Kiss T, Filipowicz W: Nuclear RNA splicing and small nuclear RNAs and their genes in higher plants. Oxf Surv Plant Cell Mol Biol 7: 255–296 (1991).

85. Görlach J, Raesecke H-R, Abel G, Wehrli R, Amrhein N, Schmid J: Organ-specific differences in the ratio of alternatively spliced chorismate synthase (*LeCS2*) transcripts in tomato. Plant J 8: 451–456 (1995).

86. Gozani O, Patton JG, Reed R: A novel set of of spliceosome associated proteins and the essential splicing factor PSF bind stably to pre-mRNA prior to catalytic step II of the splicing reaction EMBO J 13: 3356–3367 (1994).

87. Gozani O, Feld R, Reed R: Evidence that sequence-independent binding of highly conserved U2 snRNP proteins upstream of the branch site is required for assembly of spliceosomal complex A. Genes Devel 10: 233–243 (1996).

88. Grotewold E, Athma P, Peterson P: Alternatively spliced products of the maize *P* gene encode proteins with homology to the DNA-binding domain of *myb*-like transcription factors. Proc Natl Acad Sci USA 88: 4587–4591 (1991).

89. Gui J-F, Lane WS, Fu X-D: A serine kinase regulates intracellular localization of splicing factors in the cell-cycle. Nature 369: 678–682 (1994).

90. Hall SL, Padgett RA: Requirement of U12 snRNA for in vivo splicing of a minor class of eukaryotic nuclear pre-mRNA introns. Science 271: 1716–1718 (1996).

91. Hanley BA, Schuler MA: Developmental expression of plant snRNAs Nucl Acids Res 19: 6319–6325 (1991).

92. Hartmuth K, Barta A: In vitro processing of a plant prre-mRNA in HeLa cell nuclear extract. Nucl Acids Res 14: 7513–7528 (1986).

93. Hayashi M, Tsugeki R, Kondo M, Mori H, Nishimura M: Pumpkin hydroxypyruvate reductases with and without a putative C-terminal signal for targeting to microbodies may be produced by alternative splicing. Plant Mol Biol 30: 183–189 (1996).

94. Herschlag D: RNA chaperones and the RNA folding problem. J Biol Chem 270: 20871–20874 (1995).

95. Hershberger RJ, Benito M-I, Hardeman KJ, Warren C, Chandler VL, Walbot V: Characterization of the major transcripts encoded by the regulatory *MuDR* transposable element of maize. Genetics 140: 1087–1098 (1995).

96. Hirose T, Sugita M, Sugiura M: cDNA structure, expression and nucleic acid-binding properties of three RNA-binding proteins in tobacco: occurence of tissue-specific alternative splicing. Nucl Acids Res 21: 3981–3987 (1993).

97. Hodges PE, Beggs JD: RNA splicing-U2 fulfils a commitment. Curr Biol 4: 264–267 (1994).

98. Hoffman BE, Grabowski PJ: U1 snRNP targets an essential splicing factor, U2AF65, to the 3' splice site by a network of interactions spanning the exon. Genes Devel 6: 2554–2568 (1992).

99. Hopf N, Plesofsky-Vig N, Brambl R: The heat shock response of pollen and other tissues of maize. Plant Mol Biol 19: 623–630 (1992).

100. Huang MT, Gorman CM: Intervening sequences increase efficiency of RNA 3′ processing and accumulation of cytoplasmic RNA. Nucl Acids Res 18: 937–947 (1990).

101. Hunt AG, Mogen BD, Chu NM, Chua N-H: The SV40 small t intron is accurately and efficiently spliced in tobacco cells. Plant Mol Biol 16: 375–379 (1991).

102. Izaurralde E, Mattaj IW: RNA export. Cell: 81 153–159 (1995).

103. Izaurralde E, Lewis J, McGuigan C, Jankowska M, Darzynkiewicz E, Mattaj IW: A nuclear cap binding protein complex involved in pre-mRNA splicing. Cell 78: 657–668 (1994).

104. Jackson DA, Hassan AB, Errington RJ, Cook PR: Visualisation of focal sites of transcription within human nuclei. EMBO J 12: 1059–1065 (1993).

105. Jiménez-García LF, Segura-Valdez M de L, Ochs RL, Rothblum LI, Hannan R, Spector DL: Nucleogenesis: U3 snRNA-containing prenucleolar bodies move to sites of active pre-rRNA transcription after mitosis. Mol Biol Cell 5: 955–966 (1994).

106. Kandels-Lewis S, Seraphin B: Role of U6 snRNA in 5′ splice site selection. Science 262: 2035–2039 (1993).

107. Keith B, Chua N-H: Monocot and dicot pre-mRNAs are processed with different efficiencies in transgenic tobacco. EMBO J 5: 2419–2425 (1986).

108. Keller W: 3′ end cleavage and polyadenylation of nuclear messenger RNA precursors. In: Lamond A (ed). Pre-mRNA Processing, pp. 113–128. R.G. Landes Publishers, Georgetown, TX (1995).

109. Kim SH, Smith J, Claude A, Lin R-J: The purified yeast pre-mRNA splicing factor PRP2 is an RNA dependent NTPase. EMBO J 11: 2319–2326 (1992).

110. Kiss T, Marshallsay C, Filipowicz W: Alteration of the RNA polymerase specificity of U3 snRNA gene during evolution and in vitro. Cell 65: 517–526 (1991).

111. Kiss-László Z, Blanc S, Hohn T: Splicing of a cauliflower mosaic virus 35S RNA is essential for viral infectivity. EMBO J 14: 3552–3562 (1995).

112. Kohtz JD, Jamison SF, Wil, CL, Zuo P, Lührmann R, Garcia-Blanco MA, Manley, JL: Protein-protein interactions and 5′ splice site recognition in mammalian mRNA precursors. Nature 368: 119–124 (1994).

113. Konarska MM, Sharp P: Electrophoretic separation of complexes involved in the splicing of precursors to mRNA. Cell 46: 845–855 (1986).

114. Kopriva S, Cossu R, Bauwe H: Alternative splicing results in two different transcripts for H-protein of the glycine cleavage system in the C$_4$ species Flaveria trinervia. Plant J 8: 435–441 (1995).

115. Korning PG, Hebsgaard SM, Rouzé P, Brunak S: Cleaning the GenBank Arabidopsis thaliana data set. Nucl Acids Res 24: 316–320 (1996).

116. Krainer AR, Mayeda A, Kozak D, Binns G: Functional expression of cloned human splicing factor SF2: homolgy to RNA-binding proteins U1 70K, and Drosophila splicing regulators. Cell 66: 383–394 (1991).

117. Krämer A: The structure and function of proteins involved in mammalian pre-mRNA splicing. Annu Rev Biochem 65: 367–409 (1996).

118. Kulesza H, Simpson GG, Waugh R, Beggs JD, Brown JWS: Detection of a plant spliceosomal protein analogous to the yeast splicing component, PRP8. FEBS Lett 318: 4–6 (1993).

119. Kumar V, Trick M: Expression of the S-locus receptor kinase multigene family in Brassica oleracea. Plant J 6: 807–813 (1994).

120. Lafontaine JG: A light and electron microscope study of small, spherical nuclear bodies in meristematic cells of Allium cepa, Vicia faba and Raphanus sativus. J Cell Biol 26: 1–17 (1965).

121. Lafontaine JG, Luck BT, Gugg S: Nucleolus-associated bodies in meristematic cells of two plant species (Cicer arietinum and Leucaena glauca) with different ploidy levels. Can J Bot 69: 1329–1336 (1991).

122. Lafontaine JG, Chamberland H: Relationship of nucleolus associated bodies with the nucleolar organizer tracks in plant interphase nuclei (Pisum sativum). Chromosoma 103: 545–553 (1995).

123. Lamond AI, Carmo-Fonseca M: Localisation of splicing snRNPs in mammalian cells. Mol Biol Rep 18: 127–133 (1993).

124. Lamond AI, Carmo-Fonseca M: The coiled body. Trends in Cell Biol. 3: 198–204 (1993).

125. Lamond AI, Konarska MM, Sharp PA: A mutational analysis of spliceosome assembly: evidence for splice site collaboration during spliceosome formation. Genes Devel 1: 532–543 (1987).

126. Larkin R, Guilfoyle T: The second largest subunit of RNA polymerase II from Arabidopsis thaliana. Nucl Acids Res 21: 1038 (1993).

127. Lavigueur A, La Branche H, Kornblihtt AR, Chabot B: A splicing enhancer in the human fibronectin alternate ED1 exon interacts with SR proteins and stimulates U2 snRNP binding. Genes Devel 7: 2405–2417 (1993).

128. Lazar G, Schaal T, Maniatis T, Goodman HM: Identification of a plant serine-arginine-rich protein similar to the mammalian splicing factor SF2/ASF. Proc Natl Acad Sci USA 92: 7672–7676 (1995).

129. Lazarowitz SG: Geminiviruses: genome structure and gene function. Crit Rev Plant Sci 11: 327–349 (1992).

130. Lazarowitz SG, Pinder AJ, Damsteegt VD, Rogers SG: Maize streak virus genes essential for systemic spread and symptom development. EMBO J 8: 1023–1032 (1989).

131. Leader DJ, Sanders JF, Waugh R, Shaw P, Brown JWS: Molecular characterisation of plant U14 small nucleolar RNA genes: closely linked genes are transcribed as polycistronic U14 transcripts. Nucl Acids Res 22: 5196–5203 (1994).

132. Lesser CF, Guthrie C: Mutations in U6 snRNA that alter splice site specificity: implications for the active site Science 262: 1982–1988 (1993).

133. Li H, Bingham PM: Arginine/serine-rich domains of the su(wa) and tra RNA processing regulators target proteins to a subnuclear compartment implicated in splicing. Cell 67: 335–342 (1991).

134. Lin C-H, Patton JG: Regulation of alternative 3′ splice site selction by constitutive splicing factors. RNA 1: 234–245 (1995).

135. Liu H-X, Goodall GJ, Kole R, Filipowicz W: Effects of secondary structure on pre-mRNA splicing: Hairpins sequestering the 5′ but not the 3′ splice site inhibit intron processing in Nicotiana plumbaginifolia. EMBO J 14: 377–388 (1995).

136. Liu H-X, Filipowicz W: Mapping of branch point nucleotides in mutant pre-mRNAs expressed in plant cells. Plant J 9: 369–380 (1996).

137. Lopato S, Mayeda A, Krainer A, Barta A: Pre-mRNA splicing in plants: characterization of SR splicing factors. Proc Natl Acad Sci USA 93: 3074–3079 (1996)

138. Lou H, McCullough AJ, Schuler MA: Expression of maize *Adh1* intron mutants in tobacco nuclei. Plant J 3: 393–403 (1993).

139. Lou H, McCullough AJ, Schuler MA: 3′ splice site selection in dicot plant nuclei is position independent. Mol Cell Biol 13: 4485–4493 (1993).

140. Luck BT, Lafontaine JG: An ultracytochemical study of nuclear bodies in meristematic plant cells (*Cicer arietinum*). Can J Bot 60: 611–619 (1982).

141. Luehrsen KR, Walbot V: Intron creation and polyadenylation in maize are directed by AU-rich RNA. Genes Devel 8: 1117–1130 (1994).

142. Luehrsen KR, Walbot V: Addition of A- and U-rich sequence increases the splicing efficiency of a deleted form of a maize intron. Plant Mol Biol 24: 449–463 (1994).

143. Luehrsen KR, Taha S, Walbot V: Nuclear pre-mRNA processing in higher pants. Prog Nucl Acid Res Mol Biol 47: 149–193 (1994).

144. Lugert T, Werr W: A novel DNA-binding domain in the *Shrunken* initiator-binding protein (IBP1). Plant Mol Biol 25: 493–506 (1994).

145. Maas C, Laufs J, Grant S, Korfhage C, Werr W: The combination of a novel stimulatory element in the first exon of the maize *shrunken-1* gene with the following intron enhances reporter gene expression up to 1000-fold. Plant Mol Biol 16: 199–207 (1991).

146. MacMillan AM, Query CC, Allerson CR, Chen S, Verdine GL, Sharp PA: Dynamic association of proteins with the pre-mRNA branch region. Genes Devel 8: 3008–3020 (1994).

147. Madhani HD, Guthrie C: Randomization-selection analysis of snRNAs in vivo: evidence for a tertiary interaction in the spliceosome. Genes Devel 8: 1071–1086 (1994).

148. Madhani HD, Guthrie C: Dynamic RNA-RNA interactions in the spliceosome. Annu Rev Genet 28: 1–26 (1994).

149. Magrelli A, Langenkemper K, Dehio C, Schell J, Spena A: Splicing of a *rolA* transcript of *Agrobacterium rhizogenes* in *Arabidopsis*. Science 266: 1986–1988 (1994).

150. Malatesta M, Zancanaro C, Martin TE, Chan EKL, Amalria FJ, Lührmann R, Vogel P, Fakan S: Is the coiled body involved in nucleolar functions? Exp Cell Res 211: 415–419 (1994).

151. Mascarenhas D, Mettler IJ, Pierce DA, Lowe HW: Intron-mediated enhancement of heterologous gene expression in maize. Plant Mol Biol 15: 913–920 (1990).

152. Masson P, Rutherford G, Banks JA, Fedoroff N: Essential large transcripts of maize Spm transposable element are generated by alternative splicing. Cell 58: 755–765 (1989).

153. Matera AG, Tycowski KT, Steitz JA, Ward DC. Organization of small nucleolar ribonucleoproteins (snoRNPs) by fluorescence in situ hybridization and immunocytochemistry. Mol Biol Cell 5: 1289–1299 (1994).

154. Matera AG, Ward DC: Nucleoplasmic organization of small nuclear ribonucleoproteins in cultured human cells. J Cell Biol 121: 715–727 (1993).

155. Mattaj IW: Splicing in space. Nature 372: 727–728 (1994).

156. Maquat LE: When cells stop making sense: effects of nonsense codons on RNA metabolism in vertebrate cells. RNA 1: 453–465 (1995).

157. Maxwell ES, Fournier MJ: The small nucleolar RNAs. Annu Rev Biochem 64: 897–934 (1995).

158. Mayeda A, Krainer AR: Regulation of alternative pre-mRNA splicing by hnRNP A1 and splicing factor SF2. Cell 68: 365–375 (1992).

159. McCullough AJ, Lou H, Schuler MA: In vivo analysis of plant pre-mRNA splicing using an autonomously replicating vector. Nucl Acids Res 19: 3001–3009 (1991).

160. McCullough AJ, Lou H, Schuler MA: Factors affecting authentic 5′ splice site selection in plant nuclei. Mol Cell Biol 13: 1323–1331 (1993).

161. Medina MA, Moreno Díaz de la Espina S, Martin M, Fernandez-Gómez ME: Interchromatin granules in plant nuclei. Biol Cell 67: 331–339 (1989).

162. Meier UT, Blobel G: NAP57, a mammalian nucleolar protein with a putative homolog in yeast and bacteria. J Cell Biol 127: 1505–1514 (1994).

163. Mermoud JE, Cohen PT, Lamond AI: Regulation of mammalian spliceosome assembly by a protein phosphorylation mechanism. EMBO J 13: 5679–5688 (1994).

164. Michaud S, Reed R: A functional association between the 5′ and 3′ splice site is established in the earliest prespliceosome complex (E) in mammals. Genes Devel 7: 1008–1020 (1993).

165. Min H, Chan RG, Black DL: The generally expressed hnRNP F is involved in a neural specific pre-mRNA splicing event. Genes Devel 9: 2659–2671 (1995).

166. Montag K, Salamini F, Thompson RD: *ZEMa*, a member of a novel group of MADS box genes, is alternatively spliced in maize endosperm. Nucl Acids Res 23: 2168–2177 (1995).

167. Moreno Díaz de la Espina S, Sanchez Pinan A, Risueño MC: Localisation of acid phosphatase activity, phosphate ions and inorganic cations in plant nuclear coiled bodies. Cell Biol Int Rep 6: 601–607 (1982).

168. Moreno Díaz de la Espina S, Sanchez Pina A, Risueõ MC, Medina FJ, Fernández-Gómez ME: The role of plant coiled bodies in nuclear RNA metabolism. Electron Microsc 2: 240–241 (1980).

169. Moreno Díaz de la Espina S, Mínguez A, Vázquez-Nin GH, Echeverría OM: Fine structural organization of a non-reticulate plant cell nucleus. Chromosoma 101: 311–321 (1992).

170. Moore MJ, Sharp PA: Evidence of two active sites in the spliceosome provided by stereochemistry of pre-mRNA. Nature 365: 364–368 (1993).

171. Moore MJ, Query CC, Sharp PA: Splicing of precursors to messenger RNAs by the spliceosome. In: Gestland R, Atkins J. (eds) The RNA World, pp. 303–358. Cold Spring Harbor Lab Press (1993).

172. Mount SM, Burks C, Hertz G, Stormo GD, White O, Fields C: Splicing signals in *Drosophila*: intron size, information content and consensus sequences. Nucl Acids Res 20: 4255–4262 (1992).

173. Müller KJ, Romano N, Gerstner O, Garcia-Maroto F, Pozzi C, Salamini, Rohde W: The barley *Hooded* mutation is caused by a duplication in a homebox gene intron. Nature 374: 727–730 (1995).

174. Mullineaux PM, Guerineau F, Accotto G-P: Processing of complementary sense RNAs of *Digitaria* steak virus in its host and in transgenic tobacco. Nucl Acids Res 18: 7259–7265 (1990).

175. Nash J, Walbot V: *Bronze-2* gene expression and intron splicing patterns in cells and tissues of *Zea mays* L. Plant Physiol 100: 464–471 (1992).

176. Nash J, Luehrsen KR, Walbot V: Bronze-2 gene of maize: reconstruction of a wild-type allele and analysis of transcription and splicing. Plant Cell 2: 1039–1049 (1990).

177. Neel H, Weil D, Giansante C, Dautry F: In vivo cooperation between introns during pre-mRNA processing. Genes Devel 7: 2194–2205 (1993).

38

178. Nelson KK, Green MR: Mammalian U2 snRNP has a sequence-specific RNA-binding activity. Genes Devel 3: 1562–1571 (1989).

179. Nelson KK, Green MR: Mechanism for activation of cryptic splice site activation during pre-mRNA splicing. Proc Natl Acad Sci USA 87: 6253–6257 (1990).

180. Nesic D, Maquat LE: Upstream introns influence the efficiency of final intron removal and RNA 3' end formation. Genes Devel 8: 363–375 (1994).

181. Newman A: Activity in the spliceosome. Cur Biol 4: 462–464 (1994).

182. Newman A, Norman C: U5 interacts with exon sequences at 5' and 3' splice sites. Cell 68: 743–754 (1992).

183. Nilsen TW: RNA-RNA interaction in the spliceosome: unravelling the ties that bind. Cell 78: 1–4 (1994).

184. Niwa M, Berget SM: Polyadenylation precedes splicing in vitro. Gene Expr 1: 5–15 (1991).

185. Niwa M, Berget SM: Mutation of the AAUAAA polyadenylation signal depresses in vitro splicing of proximal but not distal introns. Genes Devel 5: 2086–2095 (1991).

186. Niwa M, Rose SD, Berget SM: In vitro polyadenylation is stimulated by the presence of an upstream intron. Genes Devel 4: 1552–1559 (1990).

187. Niwa M, MacDonald CC, Berget SM: Are vertebrate exons scanned during splice site selection? Nature 360: 277–280 (1992).

188. Norris SR, Meyer SE, Callis J: The intron of *Arabidopsis thaliana* polyubiquitin genes is conserved in location and is a quantitative determinant of chimeric gene expression. Plant Mol Biol 21: 895–906 (1993).

189. Nussaume L, Harrison K, Klimyuk V, Martinessen R, Sundresan V, Jones JDG: Analysis of splice donor and acceptor site function in a transposable gene trap derived from the maize element *Activator*. Mol Gen Genet 249: 91–101 (1995).

190. Ochs RL, Stein TW Jr, Tan EM: Coiled bodies in the nucleolus of breast cancer cells. J Cell Sci 107:385–399 (1994).

191. Ohno M, Sakamoto H, Shimura Y: Preferential excision of the 5' proximal intron from mRNA precursors with two introns as mediated by the cap structure. Proc Natl Acad Sci USA 84: 5187–5191 (1987).

192. Okagaki RJ, Sullivan TD, Schiefelbein JW, Nelson OE: Alternative 3' splice acceptor sites modulate enzymic activity in derivative alleles of the maize *bronze1-mutable 13* allele. Plant Cell 4: 1453–1462 (1992).

193. O'Keefe RT, Mayeda A, Sadowski CL, Krainer AR, Spector DL: Disruption of pre-mRNA splicing splicing in vivo results in reorganization of splicing factors. J Cell Biol 124: 249–260 (1994).

194. Orozco BM, McClung CR, Werneke JM, and Ogren WL: Molecular basis of the ribulose-1,5-bisphosphate carboxylase/oxygenase activase mutation in *Arabidopsis thaliana* is a guanine to adenine transition at the 5'-splice junction of intron 3. Plant Physiol 102: 227–232 (1993).

195. Oubridge C, Ito N, Evans PR, Teo CH, Nagai K: Crystal structure at 1.92 Å resolution of the RNA-binding domain of the U1A spliceosomal protein complexed with an RNA hairpin. Nature 372: 432–438 (1994).

196. Oritz DF, Strommer JN. The *Mu1* maize transposable element induces tissue-specific aberrant splicing and polyadenylation in two *Adh* mutants. Mol Cell Biol 10: 2090–2095 (1990).

197. Osteryoung KW, Sundberg H, Vierling E: Poly(A) tail length of a heat shock protein RNA is increased by severe heat stress, but intron splicing is unaffected. Mol Gen Genet 239: 323–333 (1993).

198. Palfi Z, Bach M, Solymosy F, Lührmann R: Purification of the major UsnRNPs from broad bean extracts and characterization of their protein consituents. Nucl Acids Res 17: 1445–1458 (1989).

199. Parker R, Guthrie C: A point mutation in the conserved hexanucleotide at a yeast 5' splice junction uncouples recognition, cleavage and ligation. Cell 41: 107–118 (1985).

200. Parker R, Siliciano PG: Evidence for an essential non-Watson-Crick interaction between the first and last nucleotides of a nuclear pre-mRNA intron. Nature 361: 660–662 (1993).

201. Parker RA, Siliciano PG, Guthrie C: Recognition of the TACTAAC box during mRNA splicing in yeast involves base-pairing to the U2-like snRNA. Cell 49: 229–239 (1987).

202. Pautot V, Brzezinski R, Tepfer M: Expression of a mouse metallothionein gene in transgenic plant tissue. Gene 77: 133–140 (1989).

203. Peletier M: Recherches cytologiques sur l'*Aesculus hippocastanum* L. Botaniste 27: 279–321 (1935).

204. Pereira A, Saedler H: Transpositional behavior of the maize *En/Spm* element in transgenic tobacco. EMBO J 8: 1315–1321 (1989).

205. Peterhans A, Datta SK, Datta K, Goodall GJ, Potrykus I, Paszkowski J: Recognition efficiency of Dicotyledoneae-specific promoter and RNA processing signals in rice. Mol Gen Genet 222: 361–368 (1990).

206. Plumpton M, McGarvey M, Beggs JD: A dominant negative mutation in the conserved RNA helicase motif 'SAT' causes splicng factor PRP2 to stall in spliceosomes. EMBO J 13: 879–887 (1994).

207. Potashkin J, Naik K, Wentz-Hunter K: U2AF homolog required for splicing in vivo. Science 262: 573–575 (1993).

208. Purugganan MD: Transposable elements as introns: evolutionary connections. Trends Ecol Evol 8: 239–243 (1993).

209. Puvion-Dutilleul F, Besse S, Chan EK, Tan EM, Puvion E: p80 coilin: a component of coiled bodies and interchromatin granule-associated zones. J Cell Sci 108: 1143–1153 (1995).

210. Query CC, Moore MJ, Sharp PA: Branch nucleophile selection in pre-mRNA splicing: Evidence for the bulged duplex model. Genes Devel. 8: 587–597 (1994).

211. Raska I, Ochs RL, Andrade LEC, Chan EKL, Burlingame R, Peebles C, Groul D, Tan EM: Association between the nucleolus and the coiled body. J Struct Biol 104: 120–127 (1990).

212. Raska I, Andrade LEC, Ochs RL, Chan EKL, Chang C-M, Roos G, Tan EM: Immunological and ultrastructural studies of the nuclear coiled body with auoimmune antibodies. Exp Cell Res. 195: 27–37 (1991).

213. Reddy ASN, Czernik AJ, Gynheung A, Pooviah BW: Cloning of the cDNA for U1 small nuclear ribonucleoprotein particle 70K protein from *Arabidopsis thaliana*. Biochim Biophys Acta 1171: 88–92 (1992).

214. Reed R, Maniatis T: The role of the mammalian branchpoint sequence in pre-mRNA splicing. Genes Devel 2: 1268–1276 (1988).

215. Risueño MC, Medina FJ: Ultrastructural, cytochemical and autoradiographic characterization of coiled bodies in the plant cell nucleus. Biol Cell 44: 229–238 (1982).

216. Robberson BL, Cote GJ, Berget SM: Exon definition may facilitate splice site selection in RNAs with mutiple exons. Mol Cell Biol 10: 84–94 (1990).

217. Roesler KR, Shorrosh BS, Ohlrogge JB: Structure and expression of an *Arabidopsis* acetyl-coenzyme A carboxylase gene. Plant Physiol 105: 611–617 (1994).

218. Rosbash M, Singer RH: RNA travel: tracks from DNA to cytoplasm. Cell 75: 399–401 (1993).

219. Roscigno RF, Garcia-Blanco M: SR proteins escort the U4/U6.U5 tri-snRNP to the spliceosome. RNA 1: 692–706 (1995).

220. Roth MB: Spheres, coiled bodies and nuclear bodies. Curr Opin Cell Biol 7: 325–328 (1995).

221. Roth MB, Zahler AM, Stolk JA: A conserved family of nuclear phosphoproteins localized to sites of polymerase II transcription. J Cell Biol 115: 587–596 (1991).

222. Ruis BL, Kivens WJ, Siliciano PG: The interaction between the first and last intron nucleotides in the second step of pre-mRNA splicing is independent of other conserved intron nucleotides. Nucl Acids Res 22: 5190–5195 (1994).

223. Rundle SJ, Zielinski RE: Alterations in barley ribulose-1,5bisphosphate carboxylase/oxygenase activase gene expression during development and in response to illumination. J Biol Chem 266: 14802–14807 (1991).

224. Ruskin B, Green MR: An RNA processing activity that debranches RNA lariats. Science 229: 135–140 (1985).

225. Rymond BC, Rosbash M: Yeast pre-mRNA splicing. In: Jones EW, Pringle JR, Broach JR (eds) The Molecular and Cellular Biology of the Yeast Saccharomyces, pp. 143–194. Cold Spring Harbor Laboratory Press, Cold Spring Harbour, NY (1992).

226. van Santen VL, Spritz RA: Splicing of plant pre-mRNAs in animal systems and vice versa. Gene 56: 253–265 (1987).

227. Sato S, Willson C, Dickinson HG: Origin of nucleolus-like bodies found in the nucleoplasm and cytoplasm of *Vicia faba* meristematic cells. Biol Cell 64: 321–329 (1988).

228. Sawa H, Shimura Y: Requirement of protein factors and ATP for the dissasembly of the spliceosome after mRNA splicing reaction. Nucl Acids Res 19: 6819–6821 (1991).

229. Schalk HJ, Matzeit V, Schiller B, Schell J, Gronenborn B: Wheat dwarf virus, a geminivirus of graminaceous plants needs splicing for replication. EMBO J 8: 359–364 (1989).

230. Schwer B, Guthrie C: PRP16 is an RNA-dependent ATPase that interacts transiently with the spliceosome. Nature 349: 494–499 (1991).

231. Schwer B, Guthrie C: A conformational rearrangement in the spliceosome is dependent on PRP16 and ATP hydrolysis. EMBO J. 11: 5033–5039 (1992).

232. Seraphin B: Sm and Sm-like proteins belong to a large family: identification of proteins of U6 as well as the U1, U2 U4 and U5snRNPs. EMBO J 14: 2089–2098 (1995).

233. Seraphin B, Kretzner L, Rosbash M: A U1snRNA:pre-mRNA basepairing interaction is required early in yeast spliceosome assembly but does not uniquely define the 5′ cleavage site. EMBO J 7: 2533–2538 (1988).

234. Seraphin B, Rosbash M: Identification of functional U1snRNA pre-mRNA complexes committed to spliceosome assembly and splicing. Cell 59: 349–58 (1989).

235. Sharp PA: Split genes and RNA splicing. Cell 77: 805–815 (1994).

236. Shaw PJ, Jordan EG: The nucleolus. Annu Rev Cell Devel Biol 11: 93–121 (1995).

237. Sheen J: Molecular mechanisms underlying the differential expression of maize pyruvate, orthophosphate dikinase genes. Plant Cell 3: 225–245 (1991).

238. Simpson CG, Brown JWS: Efficient splicing of an AU-rich antisense intron sequence. Plant Mol Biol 21: 205–211 (1993).

239. Simpson CG, Leader DJ, Brown JWS: In: Croy RRD (ed) Plant Molecular Biology Labfax, pp. 183–251. BIOS Scientific Publishers, Oxford (1993).

240. Simpson CG, Simpson GG, Clark G, Leader DJ, Vaux P, Waugh R, Brown JWS: Splicing of plant pre-mTNAs. Proc Royal Soc Edinburgh 99b: 31–50 (1992).

241. Simpson CG, Clark G, Davidson D, Smith P, Brown JWS: Mutation of putative branchpoint consensus sequences in plant introns reduces splicing efficiency. Plant J 9: 381–389 (1996).

242. Simpson GG, Vaux P, Clark G, Waugh R, Beggs JD, Brown JWS: Evolutionary conservation of the spliceosomal protein, U2B″. Nucl Acids Res 19: 5213–5217 (1991).

243. Simpson GG, Clark G, Rothnie H, Boelens W, Van Venrooij W, Brown JWS: Molecular characterisation of the spliceosomal proteins U1A and U2B″ from higher plants. EMBO J 14: 4540–4550 (1995).

244. Singh R, Valcarcel J, Green MR: Distinct binding specificities and functions of higher eukaryotic polypyrimidine tract-binding proteins. Science 268: 1173–1176 (1995).

245. Sinibaldi RM, Mettler IJ: Intron splicing and intron-mediated enhanced gene expression in monocots. Progr Nucl Acid Res Mol Biol 42: 229–257 (1992).

246. Smith CWJ, Patton JG, Nadal-Ginard B: Alternative splicing in the control of gene expression. Annu Rev Genet 23: 527–577 (1989).

247. Smith CWJ, Porro EB, Patton JG, Nadal-Ginard B: Scanning from an independently specified branchpoint defines the 3′ splice site of mammalian introns. Nature 342: 243–247 (1989).

248. Smith CWJ, Chu TT, Nadal-Ginard B: Scanning and competition between AGs are involved in 3′ splice site selection in mammalian introns. Mol Cell Biol 13: 4939–4952 (1993).

249. Smith KP, Carter KC, Johnson CV, Lawrence JB: U2 and U1 snRNA gene loci associate with coiled bodies. J Cell Biochem 59: 473–485 (1995).

250. Sontheimer EJ, Steitz JA: The U5 and U6 small nuclear RNAs as active site components of the spliceosome. Science 262: 1989–1996 (1993).

251. Solnick D, Lee SI: Amount of RNA secondary structure required to induce an alternative splice. Mol Cell Biol 7: 3194–3198 (1987).

252. Solymosy F, Pollák T: Uridylate-rich small nuclear RNAs (UsnRNAs), their genes, pseudogenes, and UsnRNPs in plants: structure and function. A comparative approach. Crit Rev Plant Sci 12: 275–369 (1993).

253. Spector DL: Macromolecular domains within the cell nucleus. Annu Rev Cell Biol 9: 265–315 (1993).

254. Staknis D, Reed R: SR proteins promote the first specific recognition of Pre-mRNA and are present together with the U1 small nuclear ribonucleoprotein particle in a general splicing enhancer complex. Mol Cell Biol 14: 7670–7682 (1994).

255. Stein JC, Howlett B, Boyes DC, Nasrallah ME, Nasrallah JB: Molecular cloning of a putative receptor protein kinase gene encoded at the self-incompatibility locus of *Brassica oleracea*. Proc Natl Acad Sci USA 88: 8816–8820 (1991).

256. Sun JS, Manley JL: A novel U2-U6 snRNA structure is necessary for mammalian mRNA splicing. Genes Devel 9: 843–854 (1995).

257. Sun Q, Mayeda A, Hampson RK, Krainer AR, Rottman FM: General splicing factor SF2/ASF promotes alternative splicing by binding to an exonic splicing enhancer. Genes Devel 7: 2598–2608 (1993).

258. Swanson MS: Functions of nuclear pre-mRNA/mRNA binding proteins. In: Lamond A (ed) Pre-mRNA Processing, pp. 17–29. R.G. Landes Publishers, Georgetown, TX (1995).

259. Takahashi T, Naito S, Komeda Y: Isolation and analysis of the expression of two genes for the 81 Kd heat-shock proteins from *Arabidopsis*. Plant Physiol 99: 383 (1992).

260. Tamaoki M, Tsugawa H, Minami E, Kayano T, Yamamoto N, KanoMurakami Y, Matsuoka M: Alternative RNA products from a rice homeobox gene. Plant J 7: 927–938 (1995).

261. Tanaka A, Mita S, Ohta S, Kyozuka J, Schimamoto J, Nakamura K: Enhancement of foreign gene expression by a dicot intron in rice but not in tobacco is correlated with an increased level of mRNA and an efficient splicing of the intron. Nucl Acids Res 18: 6767–6770 (1990).

262. Tantikanjana T, Nasrallah ME, Stein JC, Chen C-H, Nasrallah JB: An alternative transcript of the *S* locus glycoprotein gene in a class II pollen-recessive self-incompatibility haplotype of *Brassica oleracea* encodes a membrane-anchored protein. Plant Cell 5: 657–666 (1993).

263. Tarn W-Y, Steitz JA: SR proteins can compensate for the loss of U1 snRNP functions in vitro. Genes Devel 8: 2704–2717 (1994).

264. Tarn W-Y, Steitz JA: Modulation of 5' splice site choice in pre-messenger RNA by two distinct steps. Proc Natl Acad Sci USA 92: 2504–2508 (1995).

265. Tarn W-Y, Steitz JA: A novel spliceosome containing U11, U12 and U5 snRNPs excises a minor class (AT-AC) intron in vitro. Cell 84: 801–811(1996).

266. Thiessen H, Etzerodt M, Reuter R, Schneider C, Lottspeich F, Argos P, Lührmann R, Philipson L: Cloning of the human cDNA for the U1 RNA-associated 70K protein EMBO J 5: 3209–3217 (1986).

267. Tian M, Maniatis T: A splicing enhancer complex controls alternative splicing of doublesex pre-mRNA. Cell 74: 105–114 (1993).

268. Turner BM, Franchi L: Identification of protein antigens associated with the nuclear matrix and with clusters of interchromating granules in both interphase and mitotic cells. J Cell Sci 87: 269–282 (1987).

269. Tsai DE, Harper DS, Keene JD: U1-snRNP-A protein selects a ten nucleotide consensus sequence from a degenerate RNA pool presented in various structural contexts. Nucl Acids Res 19: 4931–4936 (1991).

270. Tycowski KT, Shu M-D, Steitz JA: A mammalian gene with introns instead of exons generating stable RNA products. Nature 379: 464–466 (1996).

271. Umen JG, Guthrie C: The second catalytic step of pre-mRNA splicing. RNA 1: 869–885 (1995).

272. Valcarel J, Singh R, Green MR: Mechanisms of regulated pre-mRNA splicing. In: Lamond A (ed) Pre-mRNA Processing, pp. 97–112. R.G. Landes Publishers, Georgetown, TX (1995).

273. Varagona MJ, Purugganan M, Wessler SR: Alternative splicing induced by insertion of retrotransposons into the maize waxy gene. Plant Cell 4: 811–820 (1992).

274. Vasil V, Clancy M, Ferl RJ, Vasil IK: Increased gene expresion by the first intron of maize *shrunken-1* locus in grass species. Plant Physiol 91: 1575–1579 (1989).

275. Waigmann E, Barta A: Processing of chimeric introns in dicot plants: evidence for a close cooperation between 5' and 3' splice sites. Nucl Acids Res 20: 75–81 (1992).

276. Wang Z-Y, Zheng F-Q, Shen G-Z, Gao J-P, Snustad P, LI M-G, Zhang J-L, Hong M-M: The amylose content in rice endosperm is related to the post-transcriptional regulation of the *waxy* gene: Plant J 7: 613–622 (1995).

277. Wansink DG, Schul W, van der Kraan I, van Steensel B, van Driel R, de Jong, L: Fluorescent labelling of nascent RNA reveals transcription by RNA polymerase II scattered throughout the nucleus. J Cell Biol 122: 283–293 (1993).

278. Wassarmann DA, Steitz JA: Interactions of small nuclear RNAs with precursor messenger RNA during in vitro splicing. Science 257: 1918–1925 (1992).

279. Weil CF, Wessler SR: The effects of plant transposable element insertion on transcription initiation and RNA processing. Annu Rev Plant Physiol Plant Mol Biol 41: 527–552 (1990).

280. Weiner AM: mRNA splicing and autocatalytic introns: distant cousins or the products of chemical determinism? Cell 72: 161–164 (1993).

281. Werneke JM, Chatfield JM, Ogren WL: Alternative mRNA slicing generates the two ribulosebisphosphate carboxylase/oxygenase activase polypeptides in spinach and *Arabidopsis*. Plant Cell 1: 815–825 (1989).

282. Wells SE, Neville M, Haynes M, Wang J, Igel H, Ares M: *CUS1*, a suppressor of cold sensitive U2 snRNA mutations, is a novel yeast splicing factor homologous to human SAP 145. Genes Devel 10: 220–232 (1996).

283. Wessler S: The maize transposable Ds1 element is alternatively spliced from exon sequences. Mol Cell Biol 11: 6192–6196 (1991).

284. White O, Soderlund C, Shanmugan P, Fields C: Information contents and dinucleotide compositions of plant intron sequences vary with evolutionary origin. Plant Mol Biol 19: 1057–1064 (1992).

285. Wiebauer K, Herrero J-J, Filipowicz W: Nuclear pre-mRNA processing in plants: distinct modes of 3'-splice site selection in plants and animals. Mol Cell Biol 8: 2042–2051 (1988).

286. Williams LM, Jordan EG, Barlow PW: The ultrastructure of nuclear bodies in interphase plant cell nuclei. Protoplasma 118: 99–103 (1983).

287. Williams LM, Charest PM, Fitzgerald GJ, Lafontaine J-G: A comparison of nuclease-gold and protease-gold complex labeling over the nucleolus and nuclear bodies of *Pisum sativum* root tip cells. Biol Cell 54: 65–72 (1985).

288. Winter J, Wright R, Duck N, Gasser C, Fraley R, Shah D: The inhibition of petunia hsp 70 mRNA processing during CdCl$_2$ stress. Mol Gen Genet 211: 315 (1988).

289. Wise JA: Guides to the heart of the spliceosome. Science 262: 1978–1979 (1993).

290. Wu JA, Manley J: Mammalian pre-mRNA branch site selection by U2snRNP involves base-pairing. Genes Devel 3: 1553–1561 (1989).

291. Wu JV, Maniatis T: Specific interactions between proteins implicated in splice site selection and regulated alternative splicing. Cell 75: 1061–1070 (1993).

292. Wuarin J, Schibler U: Physical isolation of nascent RNA chains transcribed by RNA polymerase II: evidence for co-transcriptional splicing. Mol Cell Biol 14: 4855–4871 (1994).

293. Wyatt JR, Sontheimer EJ, Steitz JA: Site-specific crosslinking of mammalian U5 snRNP to the 5' splice site before the first step of pre-mRNA splicing. Genes Devel 6: 2542–2553 (1992).

294. Xing Y, Johnson CV, Moen PT, McNeill JA, Lawrence JB: Non random gene organization: structural rearrangements of specific pre-mRNA transcription and splicing with SC-35 domains. J Cell Biol 131: 1635–1647 (1995).

295. Xue J, Rask L: The unusual 5' splicing border GC is used in myrosinase genes of the Brassicaceae. Plant Mol Biol 29: 167–171 (1995).

296. Yoshimatsu T, Nagawa F: Control of gene expression by artificial introns in *Saccharomyces cerevisiae*. Science 244: 1346–1348 (1989).

297. Yost HJ, Lindquist S: RNA splicing is interupted by heat shock protein synthesis. Cell 45: 185–193 (1986).

298. Zachar Z, Kramer J, Mims IP, Bingham PM: Evidence for channeled diffusion of pre-mRNAs during nuclear RNA transport in metazoans. J Cell Biol 121: 729–742 (1993).

299. Zahler AM, Stolk JA, Lane WS Roth MB: SR proteins: a conserved family of pre-mRNA splicing factors. Genes Devel 6: 837–847 (1992).

300. Zahler AM, Neugebauer KM, Lane WS, Roth MB: Distinct functions of SR proteins in alternative splicing. Science 260: 219–222 (1993).

301. Zamore PD, Green MR: Biochemical characterization of U2 snRNP auxillary factor: an essential pre-mRNA splicing factor with a novel intranuclear distribution. EMBO J 10: 207–214 (1991).

302. Zamore PD, Patton JG, Green MR: Cloning and domain structure of the mammalian splicing factor U2AF. Nature 355: 609–614 (1992).

303. Zhang M, Zamore PD, Carmo-Fonseca M, Lamond AI, Green MR: Cloning and intracellular localization of the U2 small nuclear ribonucleoprotein auxillary factor small subunit. Proc Natl Acad Sci USA 89: 8769–8773 (1992).

304. Zhang G, Taneja K, Singer RH, Green MR: Localization of pre-mRNA splicing in mammalian nuclei. Nature 372: 809–812 (1994).

305. Zhuang Y, Weiner AM: A compensatory base change in human U2snRNA can suppress a branch site mutation. Genes Devel 3: 1545–1552 (1989).

306. Zhuang Y, Leung H, Weiner AM: The 5' splice site of simian virus 40 large T antigen can be improved by increasing the base complementarity to U1 snRNA. Mol Cell Biol 7: 3018–3020 (1987).

307. Zhuang T, Goldstein AM, Weiner AM: UACUAAC is the preferred branchsite for mammalian mRNA splicing. Proc Natl Acad Sci USA 86: 2752–2756 (1989).

Plant Molecular Biology **32**: 43–61, 1996.
© 1996 *Kluwer Academic Publishers.*

Plant mRNA 3′-end formation

Helen M. Rothnie
Friedrich Miescher-Institut, P.O. Box 2543, CH-4002 Basel, Switzerland

Key words: 3′-end formation, mRNA processing, polyadenylation, plant gene expression

Contents

Abstract	43	FUEs	50
Introduction	43	Cleavage site	51
Cis-acting components of plant poly(A)		Downstream elements	51
signals	44	A role for base composition and	
Plant nuclear genes	44	secondary structure?	51
Pea *rbcS*-E9	44	Monocots vs. dicots	53
Maize 27 kDa zein	45	A model for a plant 3′-end processing	
Wheat histone H3	45	complex	53
Agrobacterium T-DNA genes	46	Trans-acting factors	55
Plant pararetroviruses	47	Poly(A) polymerase	55
Plant pararetroviral poly(A) site		Other factors	56
regulation	47	Transcription termination	57
Modular architecture of plant poly(A)		Future perspectives	57
signals	49	Acknowledgements	57
NUEs	49	References	57

Abstract

Our understanding of how the 3′ ends of mRNAs are formed in plants is rudimentary compared to what we know about this process in other eukaryotes. The salient features of plant pre-mRNAs that signal cleavage and polyadenylation remain obscure, and the biochemical mechanism is as yet wholly uncharacterised. Nevertheless, despite the lack of universally conserved *cis*-acting motifs, a common underlying architecture is emerging from functional analyses of plant poly(A) signals, allowing meaningful comparison with components of poly(A) signals in other eukaryotes. A plant poly(A) signal consists of one or more near-upstream elements (NUE), each directing processing at a poly(A) site a short distance downstream of it, and an extensive far-upstream element (FUE) that enhances processing efficiency at all sites. By analogy with other systems, a model for a plant 3′-end processing complex can be proposed. Plant poly(A) polymerases have been isolated and partially characterised. These, together with hints that some processing factors are conserved in different organisms, opens promising avenues toward initial characterisation of the *trans*-acting factors involved in 3′-end formation of mRNAs in higher plants.

Introduction

Post-transcriptional processing of pre-mRNA is a fundamental step in eukaryotic gene expression. The primary transcript is capped at the 5′-end, introns are removed by splicing, and the mature 3′-end is formed by an endonucleolytic cleavage followed by addition of a poly(A) tail to the 3′-end of the upstream cleavage product. Poly(A) tails play an important role in many aspects of RNA function and metabolism, particularly in mRNA translation and turnover [4, 40, 42, 57, 104, 105]. In addition, 3'-end formation is a necessary prerequisite for transcription termination [19, 32, 33, 35, 84, 96]. mRNA 3'-end formation has been shown to be a post-transcriptional processing event in metazoan and yeast systems, and it is highly likely that the 3'-ends of plant mRNAs are also formed by RNA processing rather than by transcription termination (see later).

Most of our knowledge of the process of 3'-end formation has come from studies in vertebrate systems. Cleavage and polyadenylation reactions can be faithfully reproduced *in vitro* in cell-free systems, and biochemical characterisation and cloning of the factors involved is now enabling the precise molecular details of these reactions to be examined. In vertebrates, the site of cleavage and polyadenylation is flanked by the highly conserved AAUAAA motif and a U- or GU-rich downstream element. These signals are specifically recognised and contacted by RNA-binding proteins within large multi-subunit factors which assemble on the pre-mRNA to form the processing complex (reviewed in [77, 124, 125]).

Far less is known about mRNA 3'-end processing in plants. The discovery of the importance of the AAUAAA motif for polyadenylation in animal systems naturally prompted a search for similar motifs in plant genes. It soon became apparent that AAUAAA is not universally conserved as a poly(A) signal; analysis of plant 3'-UTRs found an exact match of this sequence at an appropriate position in less than 40% of cases [34, 52, 59]. In addition to sequence analyses, experimental data showed that animal poly(A) signals were not properly recognised in plant cells [52], again suggesting functional differences in the *cis*-acting sequences controlling 3'-end formation. Another distinguishing feature is that animal genes normally have a single poly(A) site, whereas in plants the position of cleavage can be quite heterogeneous within a single transcription unit, leading to the production of mRNA populations with a variety of end points (e.g. [11, 12, 21, 26, 93, 116]). The most extreme example of this 3'-end heterogeneity described to date comes from a nuclear gene in *Nicotiana plumbaginifolia* encoding a chloroplast RNA-binding protein, where 14 distinct 3'-end processing sites were identified [64]. All of these features suggest that the processes of mRNA cleavage and polyadenylation in plants might differ mechanistically from those characterised in vertebrate systems. This

chapter reviews the current state of knowledge of plant polyadenylation signals and addresses possible models of the processing mechanism, drawing parallels with other systems where possible.

Cis-acting components of plant poly(A) signals

Our current understanding of what constitutes a plant poly(A) signal is based on sequence comparisons [34, 59, 82] and data from a rather small number of functional studies (reviewed in [51, 132]). The following sections provide a detailed summary of the mutational analyses of plant poly(A) signals carried out to date. These data are mainly derived from nuclease mapping analysis of the 3' ends of transcripts expressed either stably *in planta* or transiently in plant protoplasts. Efficiency of poly(A) site use is calculated by measuring the proportion of transcripts processed at the test poly(A) site versus those which read through to be processed further downstream, either at cryptic sites in the wild-type 3'-flanking sequence or at a second poly(A) site introduced downstream of the test site to 'trap' readthrough transcripts. Efficiency of the test site can thus be expressed as its ability to compete with the downstream site.

Plant nuclear genes

To date, surprisingly few poly(A) signals from nuclear plant genes have been functionally analysed in detail. In fact, there is only one example from dicot plants, that of the pea *rbcS*-E9 gene encoding the small subunit of ribulose-1,5-bisphosphate carboxylase, and two from monocots: the poly(A) signals of the maize 27 kDa zein gene, and the wheat histone H3 gene.

Pea rbcS-*E9*
The 3' end of the pea *rbcS*-E9 gene (fused to the chloramphenicol acetyl transferase (CAT) reporter gene) has been extensively studied as a transgene in tobacco. The majority of transcripts are processed at one of three poly(A) sites, with a fourth site further downstream being activated upon deletion of upstream sequences [50]. Extensive deletion and linker scanning analysis revealed the modular nature of sequences controlling use of these sites [53, 85, 86], and led to the definition of distinct functional elements. Each processing site is under the control of a short sequence just upstream of it (Fig. 1). This has been termed the near

upstream element (NUE). More extensive regions further upstream (far upstream element, FUE) affect the overall processing efficiency at all sites. The NUE of site 1 has been studied in detail and appears to extend over nine nucleotides (AAAUGGAAA [70]). Mutational analysis of this sequence showed that: (1) no single point mutant was able to inactivate the signal, (2) creating a canonical AAUAAA within NUE 1 increased efficiency of use of site 1, and (3) the sequence context surrounding the NUE is important [70]. The FUE extends over at least 60 nt (Fig. 1) and although deletion of the entire region drastically affects processing at sites 1–3 [85], no specific sequence motifs responsible for FUE function can be defined by linker scanning analysis [86]. The only discernable features of the FUE are its decided U-richness and several UG-containing motifs similar to those seen in other FUEs (see below). The role of specific sequences downstream of the poly(A) sites in conferring specificity and efficiency of processing remains unclear (see below).

Maize 27 kDa zein

S1 nuclease analysis of total RNA isolated from maize endosperm using a probe covering the region 3′ to the 27 kDa zein gene identified two closely spaced poly(A) sites [133]. Sequences influencing 3′-end processing were analysed in maize protoplasts. There are two copies of an AAUGAA motif just upstream of the poly(A) sites (Fig. 1). The downstream copy directs processing at the wild-type sites. If this copy of the motif is mutated, the upstream AAUGAA can take over to direct processing at a cryptic site present upstream of the wild-type sites. Both of these motifs require the presence of sequences further upstream to effect efficient processing [133]. This situation is analogous to that in the *rbcS*-E9 case, with the two AAUGAA motifs representing NUEs and the upstream sequences the FUE. Mutations to the G residue of this NUE demonstrated that AAUAAA was less efficiently used and that G is the optimal nucleotide at this position in the context of this poly(A) signal [131]. Mutations within the FUE which reduced processing efficiency implicated several regions as being functionally important (Fig. 1) and these are conserved in other zein genes [133]. Deletion of sequences downstream of the processing sites had little effect. Also, mutation of the nucleotides surrounding the poly(A) site itself did not greatly disturb processing although the transcripts ended at slightly more positions than in the wild type. Spacing of the various components of the poly(A) signal is critical; an increase of just 17 nt between the FUE and the two AAUGAA motifs was sufficient to reduce processing at the wild-type site and shift it to the upstream site. Increasing the spacing still further resulted in loss of processing at either site. Similarly, increasing the distance between the two copies of AAUGAA reduced processing at the wild-type site and resulted in processing at a new site at an appropriate distance downstream of the upstream AAUGAA [131].

Wheat histone H3

One class of metazoan mRNAs which are non-polyadenylated and whose 3′ ends are formed by a distinct specialised mechanism are those encoding histones. However, in plants [16] and some lower eukaryotes (yeast, tetrahymena; see [9]), histone mRNAs do have poly(A) tails. Instead of the characteristic inverted repeat and purine tract that signal 3′-end processing of animal histone mRNAs (reviewed in [9]), a highly conserved motif (G/AAUG(G)AAAUG) is found 17–27 nt upstream of the site of poly(A) addition in plant histone mRNAs [16, 91]. The *cis*-acting sequences controlling 3′-end formation of the wheat histone H3 transcript have been analysed in transformed sunflower cells. The poly(A) site maps to the same position in wheat and sunflower [118], suggesting that this monocot poly(A) signal is correctly recognised in dicot cells. Mutation of 2 nucleotides within the conserved motif (AAU<u>G</u>GA<u>A</u>AUG→AAU<u>U</u>GA<u>C</u>AUG) virtually abolished processing [91], consistent with the notion that this conserved sequence is the NUE of plant histone poly(A) signals. A less well defined region further upstream was found to greatly enhance processing efficiency [91]. Like other FUEs, this region contains UG-rich stretches (Fig. 1), and similar sequences are conserved upstream of many histone poly(A) sites [91]. The region downstream of the poly(A) site is also conserved in plant histone genes and contains numerous repeats of the sequence GATT. Although deletion analyses have implicated these sequences as being important for 3′-end formation [89], interpretation of these results was complicated by the concomitant drop in the level of detectable transcripts, and it remains possible that these sequences are required for H3 promoter function [89]. In fact, an internal deletion of this region that removes several GATT motifs but which did not affect overall transcript abundance had only a very slight negative effect on processing at the H3 site [91].

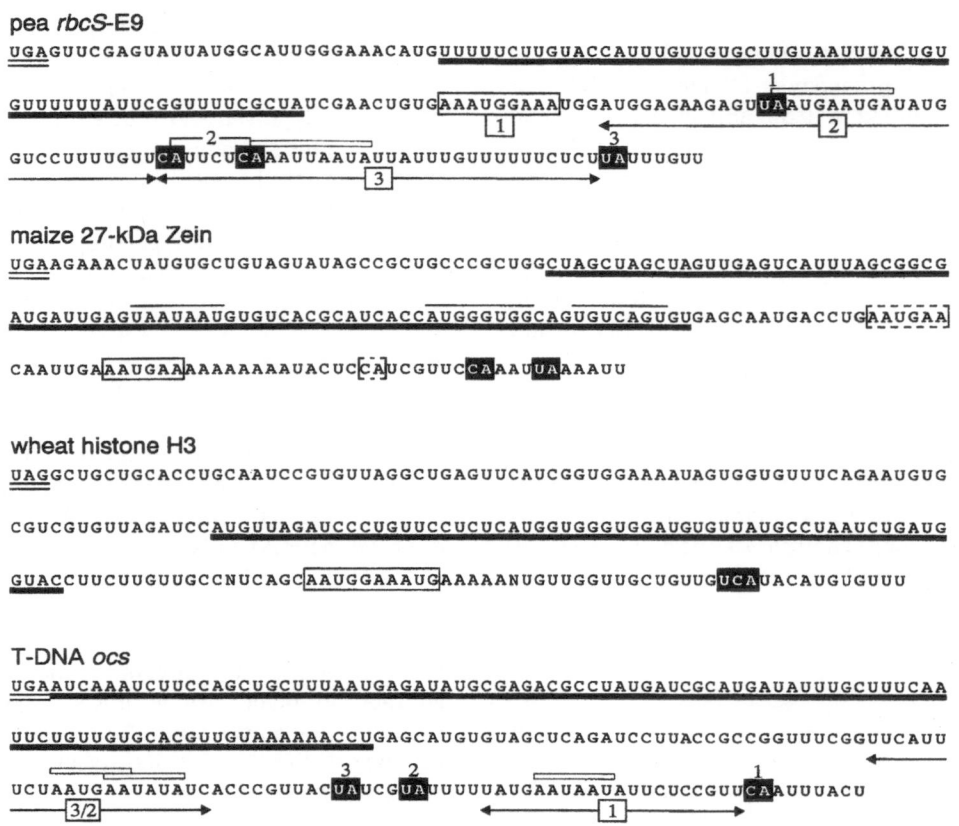

Figure 1. Poly(A) signals of plant nuclear genes and T-DNA genes. The 3'-UTRs of 4 genes whose poly(A) signals have been functionally analysed is shown. Sequence from the translational stop codon (double underlined) to just beyond the major poly(A) site(s) (black boxes) is given. The extent of FUEs as defined by gross deletion analysis is underlined. NUEs defined by point mutation are boxed, those defined by linker scanning analysis are shown by double-headed arrows, with bars above the sequence indicating likely candidate motifs for the functional NUE in each case. The cryptic poly(A) site in the zein gene and its NUE are enclosed in dashed boxes. Motifs in the zein FUE shown by mutational analysis to affect FUE function are indicated by thin lines above the sequence. The multiple poly(A) sites and their associated NUEs in the pea *rbcS*-E9 and the *ocs* sequences are numbered. Sequence data: pea *rbcS*-E9 [86]; maize 27 kDa zein [24, 133]; wheat histone H3 [91, 117, 118]; octopine synthase [76].

Agrobacterium *T-DNA genes*

Poly(A) signals originating from transcription units of the Ti plasmid of *Agrobacterium tumefaciens* are often used in plant genetic engineering. The 3' ends of several T-DNA transcripts have been mapped: nopaline synthase (*nos*) [8, 28], 'transcript 7' [31], and octopine synthase (*ocs*) [25], but only the latter has been functionally analysed [31, 55, 76]. Transcripts from the wild-type *ocs* gene are processed at one of two sites around 190 nt downstream from the end of the *ocs* ORF (Fig. 1). The more downstream of these sites is the major site in the wild-type *ocs* gene and is referred to as site 1. In chimaeric genes where the CAT reporter gene is fused to the *ocs* 3'-end region, the upstream site is used to a similar extent as site 1 and resolves

into two sites (2 and 3) [76]. Linker scanning mutation analysis identified short regions containing NUEs for each of these sites and suggested that the NUE for site 1 could be AAUAAU. The NUE controlling sites 2 and 3 contained the sequence AAUGAAUAUA. As in the case of plant nuclear genes, a region far upstream of the processing sites (FUE) is required for efficient 3'-end formation, with deletion of sequences upstream of −83 (relative to site 1) decreasing use of sites 1–3 and revealing the presence of cryptic downstream sites [76]. Although this latter study concluded from a deletion analysis that sequences downstream of the processing sites are involved in poly(A) site selection, careful examination of this and other data [31] rather suggest that specific downstream sequences are not required as part of the poly(A) signal. Both studies include mutations in which the region surrounding site 1 and everything downstream of it is deleted. In both cases the majority of transcripts are now processed at site 2/3, which still has its NUE and FUE, but now has a completely different downstream flanking sequence. Thus, this site at least does not depend on a specific downstream component. The conclusion that downstream sequences play a role in 3′-site selection in this gene came from an analysis in which the 3′ ends of transcripts from some deletion mutants mapped not to the *ocs* sequences but within Tn9 sequences which remain downstream of the CAT reporter gene in these constructs. A similar phenomenon was seen using the same reporter gene context fused to the *rbcS*-E9 3′ region [85]. These transcripts were non-polyadenylated. In both studies, removal of a short segment of the Tn9 sequence restored 3′-end processing to the *ocs* or *rbcS*-E9 region of the construct (even in the downstream deletion mutants) thus cautioning that while these unusual 3′-end processing events are intriguing (and might indicate an effect on transcription termination), they have been seen associated only with certain Tn9 sequences in particular contexts and there is no direct evidence that specific downstream sequences form part of the poly(A) signal itself in either the *ocs* or the *rbcS*-E9 genes.

Plant pararetroviruses

The poly(A) signals of three plant pararetroviruses have been functionally analysed. Cauliflower mosaic virus (CaMV) and figwort mosaic virus (FMV) are caulimoviruses; rice tungro bacilliform virus (RTBV) is a badnavirus. Caulimoviruses infect exclusively dicot plants, while badnaviruses have been found in both monocots and dicots. Both groups are classified as plant pararetroviruses, i.e. non-integrating DNA viruses with a double-stranded circular genome, replicating via reverse transcription (reviewed in [99]). The process of reverse transcription requires the production of a terminally redundant transcript covering the whole genome. This in turn requires that the poly(A) site, which lies just downstream of the transcription start site, is ignored on the first pass and used efficiently when it is encountered at the 3′ end of the transcript. In this respect, the plant pararetroviruses share a regulatory problem with animal retro- and pararetro-viruses (see below).

In all three viruses studied, the *cis*-acting sequences required to signal 3′-end formation are entirely contained within the terminal redundancy (Fig. 2) ([107–109], H.R., manuscript in preparation), i.e. no specific sequences either upstream of the transcription start site, or downstream of the processing site are required to signal accurate and efficient processing. In CaMV and RTBV, an AAUAAA motif situated 14 or 20 nt upstream of the poly(A) site, respectively, is an essential component of the poly(A) signal ([85, 108, 109] H.R., manuscript in preparation). In FMV, although a perfect AAUAAA motif is present, it does not seem to be recognised by the processing machinery. Instead, the NUE is most probably a UAUAAA motif situated 15 nt upstream of the poly(A) site [107]. The FUEs of these viral poly(A) sites are fairly extensive (Fig. 2). In CaMV, a tandem repeat of the motif UUUGUA 13 nt upstream of the AAUAAA is particularly important [100, 108]. In FMV, deletion of a region containing similar motifs drastically reduced processing [107]. A perfect UUUGUA motif is present in the FUE of the RTBV poly(A) signal, and related sequences are conserved in other plant pararetroviruses (not shown). However, this element alone is not sufficient to define the FUE and further upstream sequences are clearly required for maximal processing efficiency [43, 85, 100, 107, 108]. The spatial arrangement of poly(A) site components was also shown to be important. In the CaMV poly(A) signal, altering the distance between the AAUAAA and the UUUGUA repeat reduced the efficiency of processing, and, unlike in animal poly(A) signals, this U-rich element does not function in a downstream position [100].

Plant pararetroviral poly(A) site regulation

As described above, a feature common to all retroid viruses is the production of a terminally redundant RNA as the template for reverse transcription. In retroviruses, genomic RNA is transcribed from an integrated provirus flanked by identical long terminal repeats (LTRs) of the form U3-R-U5. The U3-R and R-U5 boundaries are defined by the start of transcription and the poly(A) site, respectively. The presence of identical R-U5 boundaries in both LTRs requires that recognition of the poly(A) signal is regulated so that transcripts are not prematurely processed. Similarly, in pararetroviruses, transcription proceeds around the circular DNA genome, production of the full length terminally redundant transcript requiring a bypass of the poly(A) signal the first time it is encountered. Retroviruses and pararetroviruses have evolved a number of strategies to achieve poly(A) site regulation (reviewed in [99]). In many

48

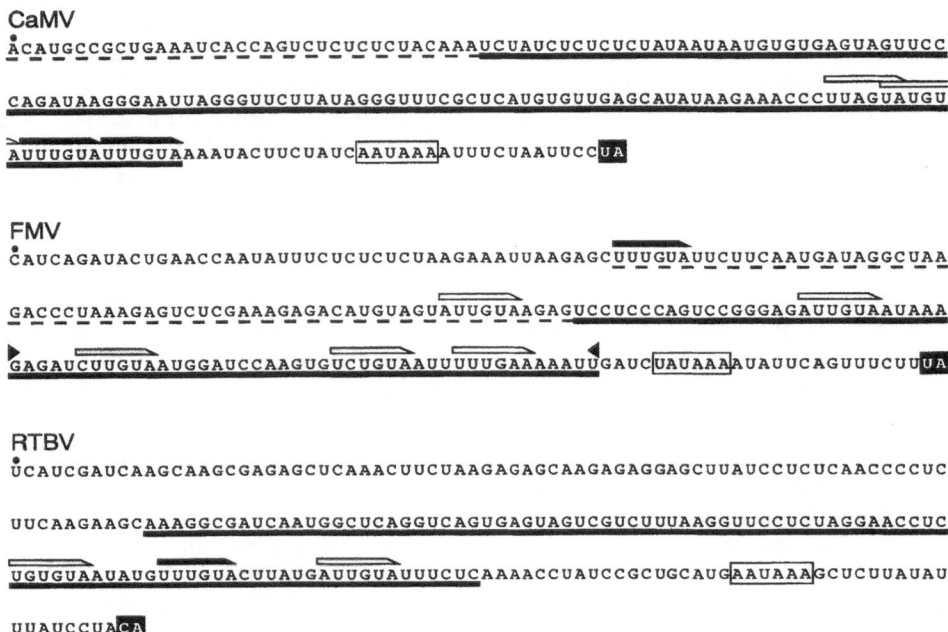

Figure 2. Plant pararetroviral poly(A) signals. The sequences of the terminal redundancies in cauliflower mosaic virus (CaMV), figwort mosaic virus (FMV), and rice tungro bacilliform virus (RTBV) pregenomic RNAs are shown. This region is delimited by the transcription start site (black dots) and the poly(A) site (black boxes). NUEs are boxed, FUEs underlined, with a dashed underline indicating sequences affecting processing efficiency only slightly. Perfect and one base mismatches of UUUGUA are indicted with black and white arrows, respectively. A internal deletion in the FMV sequence causing a dramatic loss in processing is delimited with black triangles. Sequence data: CaMV [108], FMV [107], RTBV [47].

cases the problem is solved by ensuring that elements essential to the recognition of the poly(A) signal lie upstream of the transcription start site. Consequently, a functional poly(A) signal will only be formed at the 3'-end of the full length transcript. The plant pararetroviral poly(A) signals which have been analysed to date have all their *cis*-acting elements within the terminal redundancy, thus do not appear to regulate poly(A) site use with a requirement for upstream sequences. When placed 3' to a reporter gene, all these poly(A) sites function very efficiently, ruling out default inefficiency in the site as a mechanism of bypassing the 5' poly(A) site. In the one study where this question was directly addressed, it was found that efficiency of recognition of the CaMV poly(A) site increased with the distance from the promoter [109]. Promoter-proximal occlusion of the poly(A) site has also been reported in the case of human immunodeficiency virus (HIV-1) [128], spleen necrosis virus (SNV) [56] and ground squirrel hepatitis virus (GSHV) [18]. The mechanism by which proximity to the promoter might inhibit a poly(A) site has not been elucidated. Recent studies on the HIV-1 poly(A) site, in the context of the whole provirus,

have revealed that promoter proximity *per se* is in fact unlikely to be the explanation for poly(A) site bypass. The determining factor in inhibition of the promoter-proximal poly(A) site in HIV-1 is the presence of the major splice donor, which is situated just beyond the end of the LTR, i.e. downstream of the poly(A) site [2]. Interaction between the 3'-end processing machinery and splicing factors (possibly U1snRNP) has been postulated to be responsible for poly(A) site occlusion in this case [2]. Given the recent finding that the pregenomic RNAs of the plant pararetroviruses can also be spliced [39, 63], it will be interesting to examine the effect, if any, of splicing signals on poly(A) site regulation in these viruses. This regulatory mechanism in HIV-1 had previously gone undetected due to analysis of the isolated viral poly(A) signal in heterologous contexts. The promoter-proximal poly(A) site in CaMV was also analysed in the absence of downstream viral sequences [109], and it is now known that a 5'-splice site is situated in the 35S RNA leader, 291 nt downstream of the poly(A) site [63]. A distinction from the HIV-1 case is that the circular nature of the CaMV genome means that the 5'-splice site will be downstream of the poly(A) signal at both the 5′- and 3′-ends of the primary transcript. Thus, it is difficult to envisage how this splice site is involved in differential poly(A) site recognition; further studies specifically addressing this issue will be required. In RTBV, the RNA processing signals are reversed, with the 5′-splice site 120 nt upstream of the poly(A) site. Upstream splicing signals can inhibit downstream poly(A) sites [127] and there are several examples of poly(A) sites being inefficiently used if they are located within an intron [37, 69, 92]. The effect of the surrounding intron on processing at the RTBV poly(A) site is currently under investigation.

In CaMV, the efficiency of poly(A) site bypass is not 100%, and the 'short-stop' RNA arising from processing at the promoter-proximal site is readily detected in transfected protoplasts and to a lesser extent in infected plants [109]. A short-stop RNA is also detected in plants infected with FMV [107]. In CaMV-infected plants, the proportion of short-stop RNA varies in resistant and susceptible host plant species [110]. A higher ratio of short-stop:35S RNA was a feature of host plants showing only mild or undetectable symptoms. The significance of this observation and the function, if any, of the short-stop RNA in the virus life cycle remain unclear.

The very high amounts of RTBV short-stop RNA produced in transfected rice protoplasts suggest that the RTBV poly(A) signal is not inhibited by a promoter-proximal position (HR, manuscript in preparation). However, a short-stop RNA cannot be detected in RTBV-infected rice plants, suggesting either that the short-stop RNA is unstable in infected plants, or that the poly(A) site bypass is highly efficient. This could depend on the presence of viral or host factors that are lacking in the transient expression system, or perhaps on the relative juxtaposition of RNA processing signals in the viral context.

Modular architecture of plant poly(A) signals

Components of animal, plant, and yeast poly(A) signals are represented in Fig. 3. The modular architecture of a plant poly(A) signal can be described by three features: the NUE, the FUE and the cleavage site itself, although the extent and nature of these sequences can vary. Plant poly(A) signals appear to be much more diffuse, redundant, and complex than their animal counterparts. In this respect they more closely resemble those of the yeast *Saccharomyces cerevisiae*,

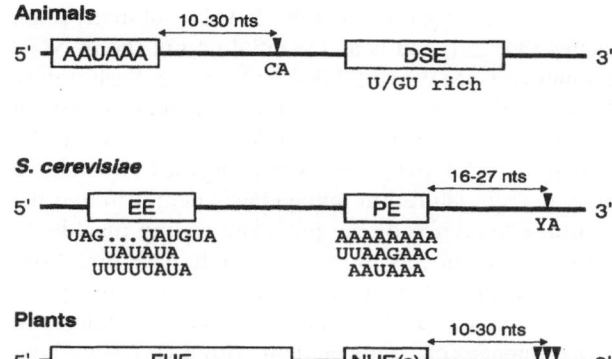

Figure 3. Modular architecture of poly(A) signals. Schematic representation of the components of poly(A) signals in animals, yeast (*S. cerevisiae*), and plants. DSE, downstream element; EE, efficiency element; PE, positioning element; NUE, near-upstream element; FUE, far-upstream element. Representative sequence motifs characteristic of each class of element are indicated. The cleavage/polyadenylation site is indicated by an arrowhead. Multiple arrow-heads in the plant signal represent the possibility of multiple cleavage sites downstream of multiple NUEs. The model for the yeast poly(A) signal is based on that proposed in [45]. In yeast, a PE can be followed by one or several poly(A) sites; only the major poly(A) site is indicated. The drawing is not to scale.

where it has also proved impossible to define a single universal signal (reviewed in [124]). The most recent model describing *S. cerevisiae* poly(A) sites proposes a three component signal [45], with efficiency- and positioning-elements (EE, PE) looking very much like the equivalents of plant FUEs and NUEs (Fig. 3). Note that this is distinct from the model proposed for the fission yeast *Schizosaccharomyces pombe* [49]. Although there are no absolute consensus motifs in the elements comprising either *S. cerevisiae* or plant poly(A) signals, common features are beginning to emerge. The following sections describe the characteristics of plant poly(A) signal components that can be gleaned from the analyses summarised above, bearing in mind that much of the data comes from the poly(A) signals of compact genomes (viruses, T-DNA) and thus generalisations should be made with care.

NUEs

Plant poly(A) signal NUEs characterised to date range from AAUAAA-like motifs to rather unrelated sequences (Figs. 1 and 2). Although often present in a variant form, AAUAAA is nevertheless found as a consensus in alignments of the 3'-UTRs of many plant genes [34, 59], and is an essential part of the poly(A) signals of CaMV and RTBV (see above). Preliminary analysis of a poly(A) signal from another class of plant DNA virus also implicates an AAUAAA motif as the functional NUE [81]. However, using the CaMV signal as a model, saturation mutagenesis demonstrated the extreme tolerance of the plant processing machinery to variations in this motif. All single base mutations were recognised with upwards of 60% of the wild-type efficiency (Fig. 4), and only drastic changes in the sequence (e.g. replacement with a G+C-rich hexanucleotide) could reduce the level of processing to that of the deletion mutant [100]. Very similar results were obtained in a recent study of AAUAAA saturation mutagenesis in *S. cerevisiae* [45] (Fig. 4). In addition, AAUAAA is not always optimal in a given poly(A) signal [131] and can be completely nonfunctional even if present [107]. Other studies indicate that NUEs can exceed 6 nt, can be only distantly related to AAUAAA, and suggest the critical importance of the surrounding sequence context (see above). Thus, an NUE is still hard to spot by sequence inspection alone, although AAUAAA-like motifs appear to represent at least one class of such elements. The NUE is probably the target for a nuclear 3'-end processing factor (discussed below). How can this be reconciled with the apparent heterogeneity of NUE sequences? Although this question will remain unresolved until we know more of the biochemistry of plant 3'-end processing, possible answers include: extreme variation and/or redundancy in the signals recognised by factors, consensus sequences which we have yet to decipher, or distinct classes of elements that are recognised by a battery of different protein factors.

FUEs

All plant poly(A) signals characterised so far have a requirement for an FUE, the function of which is to enhance overall processing efficiency. This is presumably achieved via interaction with protein factors belonging to the processing complex (see below). The FUEs of different plant poly(A) signals are interchangeable (e.g. the CaMV FUE can substitute for the FUEs of the zein, FMV, and *rbcS*-E9 poly(A) signals and vice versa [86, 107, 131]), demonstrating their functional conservation despite differences in primary structure, and supporting the notion that a basic 3'-end processing machinery is universally conserved between dicots and monocots. These results also show

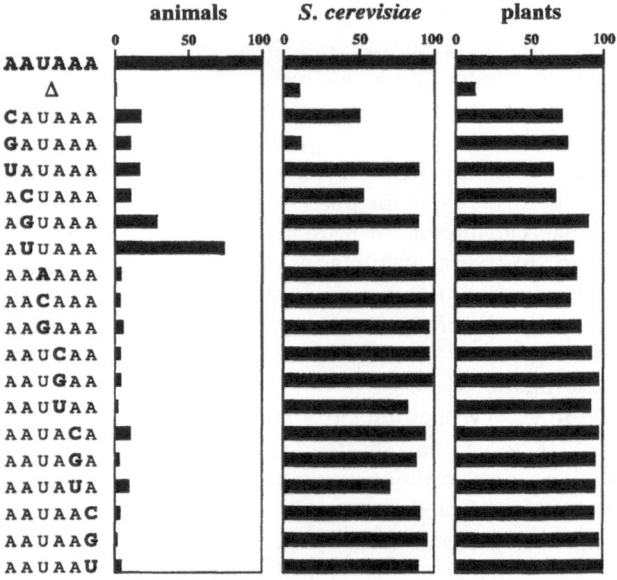

Figure 4. A comparison of the effects of mutations in the AAUAAA motif in animal, yeast (*S. cerevisiae*), and plant poly(A) signals. The figure summarises the effect on processing efficiency of mutations in the AAUAAA motif in poly(A) signals from animal (SV40-late [36, 113]), yeast (a derivative of the *cyc1* gene [45]), and plant (CaMV [100]) poly(A) signals. Δ: in the animal and plant panels, this represents deletion of AAUAAA; in *S.cerevisiae*, the AAUAAA was replaced with GTCACG. Mutated nucleotides are indicated in large bold type. The black bars indicate the efficiency of processing of mutant derivatives relative to that obtained with AAUAAA (set to 100% in each system).

that the FUE itself does not determine the 3'-end profile of a given transcription unit. Heterologous FUEs can also induce processing of a cryptic poly(A) site if introduced upstream of its NUE [100, 107, 108]. Such 'induction' assays have proved useful in showing activity of isolated FUE fragments, and in demonstrating that the effect of multiple FUE sequences is additive. For example, the CaMV UUUGUA motif was shown to be able to induce recognition of a cryptic site in the *nos* poly(A) signal in an additive and orientation-dependent manner [100]. Elements of the FUEs of CaMV and FMV have also been shown to augment each other [107].

What are the active sequences within the FUEs of plant poly(A) signals? No conserved sequence element common to all FUEs can be discerned, although they all contain U- or UG-rich sequences (Figs. 1 and 2). The element UUUGUA is important for proper functioning of the CaMV site [100, 108] and related motifs can be found in a number of other FUEs, often repeated several times. This sequence is particularly intriguing since it bears striking homology to several elements in the upstream regions of animal viral poly(A) signals, thus making it a good candidate for recognition by a conserved processing factor. A number of different classes of animal viruses are now known to require upstream elements (USEs) in addition to AAUAAA as part of their poly(A) signals [29, 30, 95, 101, 112, 115, 122], and the recent finding of a USE in a cellular gene poly(A) signal raises the possibility that this may be a more general phenomenon [87]. Some of these animal USEs and upstream elements from the poly(A) signal of the yeast Ty element [48] contain motifs similar to the plant UUUGUA element (Table 1). A loose consensus of UUUGUA can in fact be suggested as the common denominator in all these signals, but it should be stressed that most of the sequences represented in Table 1 are of viral origin.

UUUGUA motifs represent only one class of FUE element, however. In some FUEs it is hard to find sequences related to this motif, and FUEs which lack any such sequence are clearly functional. As with NUEs, there may be redundancy in *cis*-acting signals recognised by factors, as yet undiscovered consensus sequences, or multiple functional elements that are the targets for distinct protein factors.

Cleavage site

Examination of the poly(A) sites in Figs. 1 and 2 reveals a consensus of YA at the cleavage site. This has also been found in compilations of many other sequences [34, 59] and is similar to the situation in animals and yeast (reviewed in [124]). It has been suggested that the cleavage site itself be defined as a *cis*-acting component of the poly(A) signal in plants, based on the ability of particular cleavage sites to be used in conjunction with different NUEs [86]. However, in most cases where the original cleavage site is removed or mutated, cleavage still occurs at an appropriate position downstream of a functional FUE/NUE, even in the absence of a suitable YA dinucleotide [43, 76, 81, 86, 133], although with less precision (i.e. a cluster of end points are seen). Also, alterations in NUE position results in different cleavage site choice [131]. Taken together, these observations argue that although certain nucleotides in certain contexts are preferred as substrates for the cleavage reaction, the cleavage site is predominantly determined by the position of the NUE, and it is simply a matter of semantics whether this site is included in the description of the poly(A) signal.

Downstream elements

Much evidence suggests that specific downstream sequences are not a general feature of plant poly(A) signals. It is clear that specific sequences downstream of the cleavage site are not required as part of the poly(A) signals of the plant pararetroviruses and the maize zein gene. In the case of *ocs*, *rbcS*-E9 and histone H3, a direct contribution of a downstream element is questionable, for reasons outlined above. Although other studies have shown that deletion of sequences 3′ of the poly(A) site affects gene expression levels (*ocs* [55]; potato wound-inducible proteinase inhibitor II gene [1]), a direct effect on 3′-end processing at specific poly(A) sites was not demonstrated. In *S. cerevisiae*, the role of elements downstream of poly(A) sites also remains unclear, although cases where downstream elements are not required have been described (reviewed in [124]; again note the distinction from *S. pombe*, where 3′-end processing signals flank the processing site [49]). The DSE of animal poly(A) signals contributes to specification of the poly(A) site through binding of a specific factor (see below). In plants, the *cis*-acting determinants seem to be present exclusively upstream of the cleavage site, making it unlikely that *trans*-acting factors specifically recognise downstream sequences. However, if the 3′-end is indeed formed by processing rather than transcription termination, the substrate for the cleavage reaction will contain RNA sequences downstream of the poly(A) site, and this region may still have a role to play in providing an adequate scaffold for the processing complex which assembles on the upstream sequences. Thus, factors such as sequence composition and/or potential secondary structure in the downstream region might influence processing efficiency.

A role for base composition and secondary structure?

Although consensus motifs in the NUEs and FUEs of plant poly(A) signals defy precise definition, they are clearly present and are probably the major determinants directing assembly of the processing complex (see below). However, other features may have a role to play in selecting poly(A) sites in plant pre-mRNAs. One such feature is sequence composition. The 3′-UTRs (Fig. 1) and the viral sequences (Fig. 2) discussed here are all U-rich. Some are also enriched in A residues. UA-richness is a characteristic feature of plant introns (see Simpson and Filipowicz, this volume). UA-rich sequences within introns might be

Table 1. A compilation of UUUGUA-like motifs in the upstream elements of poly(A) signals from plant, animal and yeast sources.

Plants	*rbcS*-E9	u u u c U U G U A c c a	[86]
		c c a U U U G U u g u g	
		g u g c U U G U A a u u	
	ocs	u c u g U U G U g u g c	[76]
		c a c g U U G U A a a a	
	CaMV	c c c U U a G U A u g u	[108]
		u a g U a U G U A u u u	
		g u a U U U G U A u u u	
		g u a U U U G U A a a a	
	FMV	a g c U U U G U A u u c	[107]
		a g u a U U U G U A a g a	
		g a g a U U U G U A a u a	
		g a u c U U U G U A a u g	
		g u g U c U G U A a u u	
		a u u U U U G a A a a a	
	RTBV	c u c U g U G U A a u a	[47]
		a u g U U U G U A c u u	
		a u g a U U U G U A u u u	
Animals	GSHV	u u a U U U G U A u u a	[101, 102]
	SV40 late	u u a U U U G U g a a a	[112]
		a a a U U U G U g a u g	
		u u a U U U G U A a c c	
Yeast	Ty	u a u U c U G U A u a c	[48]
		u a g U a U G U A g a a	
Consensus		– – – U U U G U A – – –	

the targets for specific factors, and help to define intron-exon boundaries. It has been suggested that relative UA-richness is a general signal for RNA processing in maize, defining both the extent of introns and, in the absence of a 5′ splice site, signalling polyadenylation [73]. Despite a bias towards UA-rich sequences in plant poly(A) signals, this feature alone is unlikely to be sufficient, given the sometimes crippling effects of relatively minor mutations, and that nucleotides other than A or U have been implicated by point mutation as being functionally important in some *cis*-acting elements (see above). However, an expanded description of a plant poly(A) signal could perhaps be a suitable FUE/NUE combination in a generally U- or UA-rich context. Another feature which has been discussed as playing a role in yeast poly(A) signals (see [124]), and which also might be important in plants, is RNA secondary structure. To date, this issue has not been directly addressed, but the long leader regions of the plant pararetrovirus pregenomic RNAs (which include the poly(A) signals) are predicted to form extensive stem-loop structures [38]. Such a structure has been confirmed by chemical and enzymatic probing in the case of CaMV and, interestingly, sequences involved in RNA processing (both 3′-end formation and splicing) exist in alternative secondary structure conformations (M. Hemmings-Mieszczak and T. Hohn, manuscript in preparation). Interestingly, the potential for base pairing and formation of secondary structure exists between the AAUAAA and the UUUGUA elements of the CaMV poly(A) signal. However, given the multitude of functions of the leader sequence in the viral life cycle, it remains to be seen if RNA secondary structure plays any role in 3′-end processing.

Monocots vs. dicots

Several lines of evidence suggest that a basic 3′-end processing machinery is conserved between monocots and dicots. First, poly(A) sites utilised in the maize zein and wheat histone H3 genes are the same whether they are expressed in monocot or dicot cells [118, 131]. Second, poly(A) signal mutations in CaMV and RTBV behave similarly in monocot and dicot transient expression systems (unpublished data), suggesting that the critical *cis*-acting signals recognised by the two cell types are the same. Third, the overall features of poly(A) site components are conserved, as is the common FUE/NUE architecture. Although these observations strongly suggest a high degree of conservation, a report describing inefficient processing of a monocot poly(A) site in tobacco [61], and the observation that the FUE sequences in the maize zein gene do not much resemble those of FUEs active in dicots, indicate that specific distinctions may have evolved in the 3′-end processing systems of the two classes. Further work will be required to discover what these differences might be.

A model for a plant 3′-end processing complex

Models depicting the assembly of 3′-end processing complexes on mammalian, yeast, and plant pre-mRNAs are shown in Fig. 5. There is a great deal of evidence substantiating the animal model, and a fair amount is known about the biochemistry of 3′-end processing in *S. cerevisiae*. The model shown for plants is purely hypothetical, but is based on the idea that a common 3′-end processing mechanism is conserved in all eukaryotes. To put a model for plants in context, a brief summary of what is known of the biochemistry of 3′-end formation in other eukaryotes is required (see [62, 77, 124] for recent comprehensive review of this area and references to the primary literature).

In mammalian cells, AAUAAA and the DSE are contacted by components of the multi-subunit factors 'cleavage and polyadenylation specificity factor' (CPSF) and 'cleavage stimulation factor' (CstF), respectively (Fig. 5). Specificity of poly(A) polymerase (PAP) for AAUAAA-containing substrates is conferred by its interaction with CPSF. The cleavage complex comprises CPSF, CstF, and two cleavage factors (CF I and CF II). The presence of PAP is also required for cleavage of some substrates. Subsequent polyadenylation is dependent on

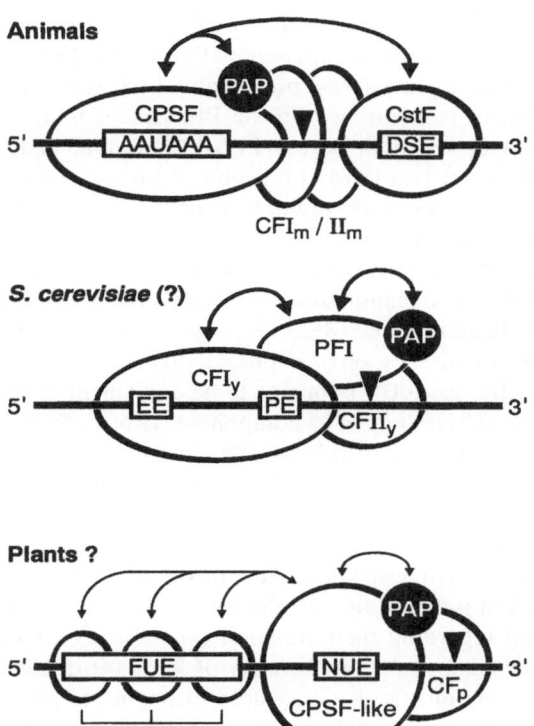

Figure 5. 3′-end processing complexes. Models depicting the assembly of 3′-end processing factors on pre-mRNAs in animals, yeast (*S. cerevisiae*), and plants are shown. The latter is highly speculative (see text). The pre-mRNA is shown as a black line, with an arrow-head indicating the location of the cleavage site. Abbreviations for *cis*-acting elements as in Fig. 3. Protein factors are represented as ellipses: CPSF, cleavage and polyadenylation specificity factor; CstF, cleavage stimulation factor; CF, cleavage factor; PAP, poly(A) polymerase; PF I, polyadenylation factor I. Note that the cleavage factors in mammals and yeast are not equivalent and are distinguished as in ref [62], i.e. as CFI_m, $CFII_m$, and CFI_y, $CFII_y$, respectively. Accordingly, the putative cleavage factor in plants is called CF_p. Established and putative protein-protein interactions are indicated with thick and thin double-headed arrows, respectively. An additional component involved in polyadenylation of mammalian mRNAs, poly(A) binding protein II (PAB II; [90, 123, 126]) is not shown.

an AAUAAA-containing substrate, CPSF, PAP, and poly(A)-binding protein II (PAB II). PAB II is required for conferring processivity to the PAP enzyme and in controlling the length of the poly(A) tail added. In *S. cerevisiae*, biochemical fractionation has identified four chromatographically separable components required for 3′-end processing (Fig. 5). Two cleavage factors, CF I and CF II, are sufficient for cleavage (note that yeast CFs I and II are not related to the mammalian factors with the same names). Polyadenylation requires

CF I, a polyadenylation factor (PF I), and yeast PAP. An RNA-binding component within CF I is thought to contact the pre-mRNA, presumably by specific recognition of elements within the PE, EE, or both. PF I interacts with both CF I and PAP. Other than PAP, the N-terminal two-thirds of which is highly conserved between yeast and vertebrates, processing factors in yeast and mammals are not obviously related. Some limited homology between components of yeast CF I and PF I, and mammalian CstF and CPSF, respectively, have been reported [88, 119], suggesting some evolutionary conservation of 3'-processing factors. Mechanistically, yeast CF I appears to be equivalent to mammalian CPSF; one of its components (RNA 15; [83]) is likely to be responsible for RNA binding, and CF I is required for both steps of the reaction, although part of the function of CPSF (interaction with PAP) is taken over by PF I in yeast.

Given what we know about the various components of plant poly(A) signals, the inferences which can be made regarding their function (see above), and our knowledge of the biochemisty of 3'-end formation in other organisms, it is possible to begin to speculate about the nature of the processing complex in plants.

It seems very likely that the NUE of a plant poly(A) signal is the functional equivalent of the animal AAUAAA motif and the PE in yeast, and is a target for recognition by a 3'-end processing factor. In animal cells, a key initial step is the recognition of the pre-mRNA substrate by CPSF (Fig. 5), which binds specifically to RNAs containing AAUAAA and involves all six nucleotide positions in the interaction (reviewed in [77, 124, 125]). This explains the extreme conservation of the animal poly(A) signal, and predicts that if a CPSF-like factor exists in plants its sequence specificity must be a lot more relaxed than that of its animal equivalent. In this respect, a plant CPSF-like factor might have features in common with yeast CF I, which also has to accomodate a high level of heterogeneity in its target sequence. If the NUE is indeed the target for recognition for a CPSF-like factor, specificity might be conferred by the cooperative interaction of the NUE, FUE, and their cognate binding factors. Alternatively, different classes of NUEs could be recognised by distinct factors and several forms of processing complex might exist.

A subtle difference in the functions of the animal AAUAAA and the plant NUE might be that the latter plays a greater role in determining the exact position of cleavage. From the studies cited above, a spatial relationship between NUE and cleavage site can be inferred. In animal poly(A) signals, cleavage site location is determined by both the AAUAAA and the DSE [17, 75], i.e. sequences which flank the cleavage site on both sides. Presumably, cleavage factors are aligned appropriately through interaction with the CPSF-CstF complex which has assembled on the pre-mRNA (Fig. 5). In plant poly(A) signals, positioning of the cleavage factor would have to be guided solely by upstream determinants.

What kind of factors might contact the FUE? It is tempting to think that the similarity in the sequences typical of plant FUEs and the DSEs of mammalian poly(A) signals (i.e. U- and/or GU-rich elements) might indicate some similarity in their binding factors. Although there is no evidence that this is the case, FUE-binding factors are depicted as CstF-like in Fig. 5 to signify equivalence of function with CstF in promoting stability of the processing complex. The possibility of multiple binding sites and factors is indicated (Fig. 5) to represent the additive properties of plant FUEs (see above), the implication being that the stability of the processing complex increases with the number of binding factors bound. This feature suggests a difference from the CstF-DSE interaction, where the strength of binding of a single CstF complex will depend on the sequence of the DSE. The additive effects of upstream elements in some animal and yeast retroid element poly(A) signals with sequence elements similar to those of plant FUEs [48, 102] (and see Table 1) suggest that proteins recognising these sequences might be universally conserved. Such proteins would not be expected to be required for processing of less highly regulated poly(A) sites, explaining why they may have been overlooked in biochemical characterisation of mammalian and yeast 3'-end processing extracts so far. Novel processing components do not need to be invoked to explain the effect of all animal virus USEs however; in HIV-1, a CPSF-USE interaction stimulates processing by helping to overcome the poor context of the AAUAAA [41]. Interaction of CPSF with elements upstream of AAUAAA has also been implicated in cytoplasmic polyadenylation; the upstream elements in this case being U-rich cytoplasmic polyadenylation elements (CPEs, reviewed in [124]).

A protein which has been shown to influence 3'-end processing via interaction with upstream sequences is the A protein of U1snRNP (U1A). In mammals, U1A autoregulates its own expression by direct inhibition of PAP, the interaction being mediated by the presence of two cognate U1A binding sites in the U1A mRNA, just upstream of its poly(A) signal [10, 44]. The recent cloning of U1A from potato and *Arabidopsis* revealed that this autoregulatory mechanism is not conserved in plants [114], thus precluding use of the U1A-PAP interaction as a tool to investigate plant PAP. U1A has also been reported to stimulate 3'-end processing via interaction of U1snRNP with the UUUGUA-like USE of the SV40 late poly(A) signal [74]. How U1A exerts its stimulatory effect in this case has not been elucidated, and other studies have found no interaction between either plant or human U1A protein and the UUUGUA elements of the CaMV or SV40 late poly(A) signals [114].

Other factors which might be expected to be found in the plant 3'-end processing complex include a cleavage factor (Fig. 5) and of course PAP itself (see below). The identity and properties of the endonuclease activity which carries out the cleavage reaction itself remains unknown in both the mammalian and the yeast systems, but it is likely that cleavage is determined by protein-protein interactions once specific contacts with the RNA have been made by other components, rather than by direct recognition of the RNA by the cleavage factor. Thus, we might expect a separable cleavage factor also in plants.

Trans-acting factors

The major stumbling block in the elucidation of the biochemistry of 3'-end formation in higher plants is the lack of an assay system which can accurately reproduce the processes of RNA cleavage and polyadenylation *in vitro*. However, as discussed above, it seems reasonable to assume that assembly of a processing complex mediated by recognition of *cis*-acting elements on the pre-mRNA somehow resembles this process in other eukaryotes. The following sections describe what is known of plant processing factors to date, and suggests possible strategies for their future isolation and characterisation.

Poly(A) polymerase

One component of the 3'-end formation process which can safely be assumed to be present in higher plants is PAP itself. PAP enzymes have now been cloned from several sources (reviewed in [124]). Although there is not enough direct nucleotide sequence homology to use either the mammalian or yeast PAP clones as molecular probes to isolate a plant homologue (unpublished data), PAP is the one component of the processing machinery

which can be purified without the need for a functional 3'-processing extract. In fact, PAP activities in plant tissues were first reported over twenty years ago [13, 79, 103, 111], and have been described in a diverse range of plants such as cotton [46], maize [78], wheat [13, 66, 68, 111], pea [7], mung bean [80], cow pea [120, 121], and tobacco [22]. Plant PAPs have been purified to apparent electrophoretic homogeneity from mung bean hypocotyls [106], germinating wheat embryos [65], and cow pea seedlings [120]. These enzymes differ in their apparent molecular weights and subunit structures. The cow pea and wheat enzymes are similar in size (63 and 64 kDa, respectively), and appear to be monomeric. Mung bean PAP has a molecular mass of 120 kDa as determined by gel permeation chromatography, but appears as a single band of 30 kDa on SDS-PAGE thus suggesting that this particular enzyme is a tetramer [106]. Despite the differences in size and subunit structure, these PAP activities have similar biochemical properties, both to each other and to mammalian and yeast PAP. All are activated by Mn^{2+}, act on exogenous RNA primers (with a free 3'-OH), and have a pH optimum of ca. 8.0. PAP activities partially purified from tobacco cell suspension cultures [22] and maize seedlings [78, 79] share these features and have estimated molecular weights of 65–70kD.

Recently, a PAP activity isolated from the leaves of young pea plants has been reported that differs in many aspects from the previously characterised enzymes [23]. The pea enzyme is apparently comprised of several chromatographically separable components (PAP-I, -II, -III), each of which being inactive alone. A major difference from other plant PAPs is that this pea enzyme requires Mg^{2+} as the divalent cation (in fact it is inactive in the presence of Mn^{2+}), and its activity could not be measured on exogenously added substrate RNAs. Instead, co-purifying RNAs in the form of hnRNP particles acted as substrates. PAP activities with similar properties (i.e. separable components, a requirement for Mg^{2+}, and using endogenous substrates) were recovered from extracts prepared from leaves of young maize and Indian mustard plants [23]. PAP-I and -III isolated from different plants were interchangeable, suggesting that this form of PAP is functionally conserved. One component of the pea activity (PAP-I) cross-reacts with a monoclonal antibody raised against PAP from *S. cerevisiae*. This latter enzyme is a single polypeptide and requires no other components for activity [71, 72]. The relationship between yeast PAP and this unusual pea PAP remains to be determined. In addition, PAP with more 'classical' properties (i.e. requiring exogenous primer and Mn^{2+}) can also be isolated from young pea and maize seedlings [7, 78, 79], leading to a need for further investigation to resolve the identities and roles of these different activities.

PAP is known to be post-translationally modified by phosphorylation in mammals [97] and in *Xenopus* oocytes [3]. In the latter case, phosphorylation is regulated during oocyte maturation. Regulated phosphorylation of mammalian PAP has not been reported, but it seems likely that the multiple phosphorylation targets in the C-terminal portion of the protein might be involved in modulating PAP activity, perhaps (like in *Xenopus*) in early development (see below). Although there is currently no data on the phosphorylation state of plant PAP enzymes, other post-translational modifications have been observed; there is some evidence that wheat PAP is a glycoprotein [60]. From some sources, more that one PAP activity can be resolved [22, 120]; these activities have the same catalytic properties and differ only in their behaviour in some chromatographic separation steps. This may also be indicative of alternatively modified forms of the enzyme.

An interesting feature of PAP is its importance in early development in many species, in both the plant and animal kingdoms. In early plant embryos and in maturing oocytes of both vertebrates and invertebrates, translational control of stored mRNAs is an important regulatory mechanism. This is achieved by controlling the length of the poly(A) tail in the cytoplasm, and thereby the translatability of the RNA (reviewed in [20, 98, 129, 130]). PAP is obviously of central importance to this process. In *Xenopus* oocytes, there are strong indications that the PAP activity in the cytoplasm is the same as the nuclear enzyme [3]. In germinating plant embryos, an increase in PAP activity is paralleled by an increase in the amount of poly(A)$^+$ RNA [66, 67, 80]. This process depends on active translation [66] and there is good evidence that PAP is itself translated from a stored mRNA in early embryos of wheat [65, 68] and mung bean [80]. This parallels similar regulation of PAP activity in vertebrates; a maternal mRNA encoding PAP is present in *Xenopus* oocytes [3].

Like oocyte maturation in animals, development of plant embryos is under hormonal control, and PAP lies on one such signal transduction pathway. PAP activity is stimulated by giberellic acid (GA_3) and this effect is antagonised by abscissic acid or auxin [5, 6, 67]. Thus, PAP plays a pivotal role in coordinating events during early development in both plants and animals, underlining the importance of this enzyme in both the production of mRNA at all stages of development, and in regulating translation during embryogenesis.

Clearly, adequate information and tools are available to allow the cloning of a plant PAP in the near future. Amino acid sequencing of purified PAP proteins will be the next logical step towards the further characterisation of this enzyme in plants.

Other factors

Although no 3'-end processing factors other than PAP have been functionally characterised in higher plants, there are several indications that at least some of the known mammalian and yeast factors may be conserved in plants. Screening the databanks, especially expressed sequence tag (EST) libraries, reveals several plant cDNA sequences with striking homologies to components of both mammalian and yeast 3'-end processing factors. For example, the recent cloning of the 160 kDa subunit of bovine CPSF revealed several homologous sequences in the database, including an EST from rice [58]. In the continuing absence of a functional *in vitro* 3'-end processing extract, such homologies may provide the only immediate entry point into the plant polyadenylation machinery. Of course, such an approach can only complement biochemical analysis, but identification of even some of the factors by alternative means would be a step in the right direction. As described above, it should now be relatively easy to clone plant PAP. Although in mammalian and yeast sytems, PAP is not the factor which makes the primary contact with the pre-mRNA, it contacts other factors in the complex and can be used to isolate these components using techniques exploiting protein-protein interactions. The success of this approach was recently demonstrated with the cloning of FIP1, a factor interacting with yeast PAP [94]. FIP1 was cloned using the two-hybrid system, i.e. solely on the basis of its interaction with PAP, but it has in fact turned out to be a component of PF I, a factor previously characterised by biochemical fractionation to be part of the yeast 3'-end processing machinery (see above).

Even if a nuclear extract is not competent to effect processing of an exogenously added RNA, it may still contain components which specifically recognise polyadenylation signals thus allowing partial characterisation of such factors. Some preliminary data showing proteins in nuclear extracts interacting with the FUE of the maize zein gene poly(A) signal have been recently reported [132].

Transcription termination

An underlying assumption in this article is that the 3'-end of plant mRNAs is formed by processing of the primary transcript, rather than by transcription termination *per se*. In animals, transcription termination occurs at a variable distance (from a few hundred nucleotides to over several kb) downstream of the poly(A) site, and is dependent on functional 3'-end processing (reviewed in [96]). In yeast, transcription termination and 3'-end processing appear to be very tightly coupled and it took the development of an *in vitro* processing extract to demonstrate that the 3'-ends of yeast pre-mRNAs are in fact produced by post-transcriptional cleavage [14, 15]. Also in plants, there is evidence that polyadenylation and transcription termination are linked [54], but nuclear run-on analysis detects primary transcripts to about 300 nt beyond the poly(A) site [27], confirming that 3'-end processing of plant mRNAs is post-transcriptional, and suggesting that transcription termination occurs downstream at a distance intermediate between the distances found in yeast and metazoa.

Future perspectives

Progress in understanding plant RNA processing mechanisms has lagged behind that in other eukaryotic systems, mainly due to the lack of *in vitro* assays, for either 3'-end processing or splicing (see Simpson and Filipowicz, this volume). However, the accumulated evidence suggests that these mechanisms are fundamentally the same across all eukaryotes. What is of interest now is to discover the specific adaptations which have evolved in different organisms, although much ground work still remains to be done in plants. The major directions for further study of plant mRNA 3'-end formation are: intensive characterisation of the *cis*-acting sequence requirements of poly(A) signals of plant origin, investigation of the relationship between polyadenylation and transcription termination, and characterisation of the *trans*-acting factors involved. Understanding poly(A) signal recognition and mRNA 3'-end formation in plants will not only increase our understanding of plant gene expression and allow evolutionary comparisons of the mechanisms involved, but may also benefit plant genetic engineering by allowing rational design of transgene transcriptional control signals.

Acknowledgements

Critical reading of the manuscript by Thomas Hohn, Nick Proudfoot and Elmar Wahle is gratefully acknowledged. Thanks also to Arthur Hunt and Ann Depicker for providing pre-prints. Special thanks go to Mike Rothnie for preparation of the figures.

References

1. An G, Mitra A, Choi HK, Costa MA, An K, Thornburg RW, Ryan CA: Functional analysis of the 3' control region of the potato wound-inducible proteinase inhibitor II gene. Plant Cell 1: 115– 122 (1989).
2. Ashe MP, Griffin P, James W, Proudfoot NJ: Poly(A) site selection in the HIV-1 provirus: inhibition of promoter-proximal polyadenylation by the downstream major splice donor site. Genes Devel 9: 3008–3025 (1995).
3. Ballantyne S, Bilger A, Astrom J, Virtanen A, Wickens M: Poly(A) polymerases in the nucleus and cytoplasm of frog oocytes: dynamic changes during oocyte maturation and early development. RNA 1: 64–78 (1995).
4. Beelman CA, Parker R: Degradation of mRNA in eukaryotes. Cell 81: 179–183 (1995).
5. Berry M, Sacher RC: Hormonal regulation of poly(A) polymerase activity by gibberellic acid in embryo-less half-seeds of wheat (*Triticum aestivum*). FEBS Lett 132: 109–113 (1981).
6. Berry M, Sacher RC: Expression of conserved message of poly(A) polymerase through hormonal control in wheat aleurone layers. FEBS Lett 141: 164–168 (1982).
7. Berry M, Sacher RC: Regulation of poly(A) polymerase activity and poly(A)$^+$RNA levels by auxin in pea epicotyls. FEBS Lett 154: 139–144 (1983).
8. Bevan M, Barnes WM, Chilton M: Structure and transcription of the nopaline synthase gene region of T-DNA. Nucl Acids Res 11: 369–385 (1982).
9. Birnstiel ML, Busslinger M, Strub K: Transcription termination and 3' processing: the end is in site! Cell 41: 349–359 (1985).
10. Boelens WC, Jansen EJR, van Venrooij J, Stripecke R, Mattaj IW, Gunderson SI: The human U1 snRNP-specific U1A protein inhibits polyadenylation of its own pre-mRNA. Cell 72: 881–892 (1993).
11. Boutry M, Chua N: A nuclear gene encoding the beta subunit of the mitochondrial ATP synthase in *Nicotiana plumbaginifolia*. EMBO J 4: 2159–2165 (1985).
12. Broglie KE, Gaynor JJ, Broglie RM: Ethylene-regulated gene expression: molecular cloning of the genes encoding an endochitinase from *Phaseolus vulgaris*. Proc Natl Acad Sci USA 83: 6820–6824 (1986).
13. Burkard G, Keller EB: Poly(A) polymerase and poly(G) polymerase in wheat chloroplasts. Proc Natl Acad Sci USA 71: 389– 393 (1974).
14. Butler JS, Platt T: RNA processing generates the mature 3' end of yeast CYC1 messenger RNA *in vitro*. Science 242: 1270–1274 (1988).
15. Butler JS, Sadhale PP, Platt T: RNA processing *in vitro* produces mature 3' ends of a variety of *Saccharomyces cerevisiae* mRNAs. Mol Cell Biol 10: 2599–2605 (1990).

58

16. Chaubet N, Chaboute M, Clément B, Ehling M, Philipps G, Gigot C: The histone H3 and H4 mRNAs are polyadenylated in maize. Nucl Acids Res 16: 1295–1304 (1988).

17. Chen F, MacDonald CC, Wilusz J: Cleavage site determinants in the mammalian polyadenylation signal. Nucl Acids Res 23: 2614–2620 (1995).

18. Cherrington J, Russnak R, Ganem D: Upstream sequences and cap proximity in the regulation of polyadenylation in ground squirrel hepatitis virus. J Virol 66: 7589–7596 (1992).

19. Connelly S, Manley JL: A functional mRNA polyadenylation signal is required for transcription termination by RNA polymerase II. Genes Devel 2: 440–452 (1988).

20. Curtis D, Lehmann R, Zamore PD: Translational regulation in development. Cell 81: 171–178 (1995).

21. Czarnecka E, Gurley WB, Nagao RT, Mosquera LA, Key JL: DNA sequence and transcript mapping of a soybean gene encoding a small heat shock protein. Proc Natl Acad Sci USA 82: 3726–3730 (1985).

22. D'Alessandro M, Srivastava BIS: Poly(A) polymerase and poly(ADP-ribose) polymerase activities in normal and crown gall tumor tissue cultures of tobacco. FEBS Lett 188: 239–242 (1985).

23. Das Gupta J, Li Q, Thomson AB, Hunt AG: Characterization of a novel plant poly(A) polymerase. Plant Sci 110: 215–226 (1995).

24. Das OP, Ward K, Ray S, Messing J: Seuence variation between alleles reveals two types of copy correction at the 27-kDa zein locus of maize. Genomics 11: 849–856 (1991).

25. De Greve H, Dhaese P, Seurinck J, Lemmers M, Van Montagu M, Schell J: J Mol Appl Genet 1: 499–512 (1982).

26. Dean C, Tamaki S, Dunsmuir P, Favreau M, Katayama C, Dooner H, Bedbrook J: mRNA transcripts of several plant genes are polyadenylated at multiple sites in vivo. Nucl Acids Res 14: 2229–2240 (1986).

27. Depicker A, Ingelbrecht I, van Houdt H, De Loose M, Van Montagu M: Posttranscriptional reporter transgene silencing in transgenic tobacco. Proceedings of University of Nottingham Easter School: 'Mechanisms and Application of Gene Silencing' 27–31 March 1995, Nottingham, England (1995).

28. Depicker A, Stachel S, Dhaese P, Zambryski P, Goodman HM: Nopaline synthase: transcript mapping and DNA sequence. J Mol Appl Genet 1: 561–573 (1982).

29. DeZazzo JD, Imperiale MJ: Sequences upstream of AAUAAA influence poly(A) site selection in a complex transcription unit. Mol Cell Biol 9: 4951–4961 (1989).

30. DeZazzo JD, Kilpatrick JE, Imperiale MJ: Involvement of long terminal repeat U3 sequences overlapping the transcription control region in human immunodeficiency virus type 1 mRNA 3' end formation. Mol Cell Biol 11: 1624–1630 (1991).

31. Dhaese P, De Greve H, Gielen J, Seurinck J, Van Montagu M, Schell J: Identification of sequences involved in the polyadenylation of higher plant nuclear transcripts using Agrobacterium T-DNA genes as models. EMBO J 2: 419–426 (1983).

32. Edwalds-Gilbert G, Prescott J, Falck-Pedersen E: 3' RNA processing efficiency plays a primary role in generating termination-competent RNA polymerase II elongation complexes. Mol Cell Biol 13: 3472–3480 (1993).

33. Eggermont J, Proudfoot NJ: Poly(A) signals and transcriptional pause sites combine to prevent interference between RNA polymerase II promoters. EMBO J 12: 2539–2548 (1993).

34. Elliston K Messing J: The molecular architecture of plant genes: a phylogenetic perspective. In Kahl G (ed) Architecture of Eucaryotic Genes, pp. 21–56. VCH Verlag, Weinheim, Germany (1988).

35. Enriquez-Harris P, Levitt N, Briggs D, Proudfoot NJ: A pause site for RNA polymerase II is associated with termination of transcription. EMBO J 10: 1833–1842 (1991).

36. Fitzgerald M, Shenk T: The sequence 5'-AAUAAA-3' forms part of the recognition site for polyadenylation of late SV40 mRNAs. Cell 24: 251–260 (1981).

37. Furth PA, Choe W-T, Rex JH, Byrne JC, Baker CC: Sequences homologous to 5' splice sites are required for the inhibitory activity of papillomavirus late 3' untranslated regions. Mol Cell Biol 14: 5278–5289 (1994).

38. Fütterer J, Gordon K, Bonneville JM, Sanfaçon H, Pisan B, Penswick J, Hohn T: The leading sequence of caulimovirus large RNA can be folded into a large stem-loop structure. Nucleic Acids Res 16: 8377–8390 (1988).

39. Fütterer J, Potrykus I, Valles-Brau MP, Dasgupta I, Hull R, Hohn T: Splicing in a plant pararetrovirus. Virology 198: 663–670 (1994).

40. Gallie DR: The cap and poly(A) tail function synergistically to regulate mRNA translational efficiency. Genes Devel 5: 2108–2116 (1991).

41. Gilmartin GM, Fleming ES, Oetjen J, Graveley BR: CPSF recognition of an HIV-1 mRNA 3'-processing enhancer: multiple sequence contacts involved in poly(A) sie definition. Genes Devel 9: 72–83 (1995).

42. Green PJ: Control of mRNA stability in higher plants. Plant Physiol 102: 1065–1070 (1993).

43. Guerineau F, Brooks L, Mullineaux P: Effect of deletions in the cauliflower mosaic virus polyadenylation sequence on the choice of the polyadenylation sites in tobacco protoplasts. Mol Gen Genet 226: 141–144 (1991).

44. Gunderson SI, Beyer K, Martin G, Keller W, Boelens WC, Mattaj IW: The human U1A snRNP protein regulates polyadenylation via a direct interaction with poly(A) polymerase. Cell 76: 531–541 (1994).

45. Guo Z, Sherman F: 3'-end forming-signals of yeast mRNA. Mol Cell Biol 15: 5983–5990 (1995).

46. Harris B, Dure L: Polyadenylylation of stored mRNA in cotton seed germination. Prog Nucl Acid Res Mol Biol 19: 113–118 (1976).

47. Hay JM, Jones MC, Blakebrough ML, Dasgupta I, Davies JW, Hull R: An analysis of the sequence of an infectious clone of rice tungro bacilliform virus, a plant pararetrovirus. Nucl Acids Res 19: 2615–2621 (1991).

48. Hou W, Russnak R, Platt T: Poly(A) site selection in the yeast Ty retroelement requires an upstream region and sequence-specific titratable factor(s) in vitro. EMBO J 13: 446–452 (1994).

49. Humphrey T, Birse CE, Proudfoot NJ: RNA 3' end signals of the S. pombe ura4 gene comprise a site determining and efficiency element. EMBO J 13: 2441–2451 (1994).

50. Hunt AG: Identification and characterisation of cryptic polyadenylation sites in the 3' region of a pea ribulose-1,5-bisphosphate carboxylase small subunit gene. DNA 7: 329–336 (1988).

51. Hunt AG: Messenger RNA 3' end formation in plants. Annu Rev Plant Physiol Plant Mol Biol 45: 47–60 (1994).

52. Hunt AG, Chu NM, Odell JT, Nagy F, Chua N: Plant cells do not properly recognize animal gene polyadenylation signals. Plant Mol Biol 8: 23–35 (1987).

53. Hunt AG, MacDonald MH: Deletion analysis of the polyadenylation site of a pea ribulose-1,5-bisphosphate

carboxylase small-subunit gene. Plant Mol Biol 13: 125–138 (1989).

54. Ingelbrecht I, Breyne P, Vancompernolle K, Jacobs A, Van Montagu M, Depicker A: Transcriptional interference in transgenic plants. Gene 109: 239–242 (1991).

55. Ingelbrecht ILW, Herman LMF, Dekeyser RA, Van Montagu MC, Depicker AG: Different 3′ end regions strongly influence the level of gene expression in plant cells. Plant Cell 1: 671–680 (1989).

56. Iwasaki K, Temin HM: The efficiency of RNA 3′-end formation is determined by the distance between the cap site and the poly(A) site in spleen necrosis virus. Genes Devel 4: 2299–2307 (1990).

57. Jackson RJ, Standart N: Do the poly(A) tail and 3′ untranslated region control mRNA translation. Cell 62: 15–24 (1990).

58. Jenny A, Keller W: Cloning of cDNAs encoding the 160 kDa subunit of the bovine cleavage and polyadenylation specificity factor. Nucl Acids Res 23: 2629–2635 (1995).

59. Joshi CP: Putative polyadenylation signals in nuclear genes of higher plants: a compilation and analysis. Nucl Acids Res 15: 9627–9640 (1987).

60. Kapoor R, Verma N, Saluja D, Lakhani S, Sacher RC: Purification and characterization of a poly(A) polymerase from germinated wheat embryos: enzyme glycosylation. Plant Sci 89: 167– 176 (1993).

61. Keith B, Chua N: Monocot and dicot pre-mRNAs are processed with different efficiencies in transgenic tobacco. EMBO J 5: 2419–2425 (1986).

62. Keller W: No end yet to messenger RNA 3′ processing. Cell 81: 829–832 (1995).

63. Kiss-László Z, Blanc S, Hohn T: Splicing of cauliflower mosaic virus 35S RNA is essential for viral infectivity. EMBO J 14: 3552–3562 (1995).

64. Klahre U, Hemmings-Mieszczak M, Filipowicz W: Extreme heterogeneity of polyadenylation sites in mRNAs encoding chloroplast RNA-binding proteins in Nicotiana plumbaginifolia. Plant Mol Biol 28: 569–574 (1995).

65. Lakhani S, Kapoor R, Verma N, Sacher RC: Evidence for the de novo synthesis of poly(A) polymerase in germinated wheat embryos. Phytochemistry 28: 1031–1035 (1989).

66. Lakhani S, Sacher RC: Regulation of poly(A) polymerase activity and poly(A)+ RNA in germinated wheat embryos under conditions of pathogenesis. Biochim Biophys Acta 825: 303–315 (1985).

67. Lakhani S, Sacher RC: Hormonal regulation of poly(A)+ RNA and poly(A) polymerase activity in wheat embryos. Plant Sci 42: 191–200 (1985).

68. Lakhani S, Thiru AN, Sacher RC: Synthesis of poly(A) polymerase from conserved messenger RNA in germinating excised embryos of wheat. Phytochemistry 22: 1561–1566 (1983).

69. Levitt N, Briggs D, Gil A, Proudfoot NJ: Definition of an efficient synthetic poly(A) site. Genes Devel 3: 1019–1025 (1989).

70. Li Q, Hunt AG: A near-upstream element in a plant polyadenylation signal consists of more than six nucleotides. Plant Mol Biol 28: 927–934 (1995).

71. Lingner J, Kellermann J, Keller W: Cloning and expression of the essential gene for poly(A) polymerase from S. cerevisiae. Nature 354: 496–498 (1991).

72. Lingner J, Radtke I, Wahle E, Keller W: Purification and characterization of poly(A) polymerase from Saccharomyces cerevisiae. J Biol Chem 266: 8741–8746 (1991).

73. Luehrsen KR, Walbot V: Intron creation and polyadenylation in maize are directed by AU-rich RNA. Genes Devel 8: 1117–1130 (1994).

74. Lutz CS, Alwine JC: Direct interaction of the U1snRNP-A protein with the upstream efficiency element of the SV40 late polyadenylation signal. Genes Devel 8: 567–586 (1994).

75. MacDonald CC, Wilusz J, Shenk T: The 64-kilodalton subunit of the CstF polyadenylation factor binds to pre-mRNAs downstream of the cleavage site and influences cleavage site location. Mol Cell Biol 14: 6647–6654 (1994).

76. MacDonald MH, Mogen BD, Hunt AG: Characterization of the polyadenylation signal from the T-DNA-encoded octopine synthase gene. Nucl Acids Res 19: 5575–5581 (1991).

77. Manley JL: A complex assembly catalyzes polyadenylation of mRNA precursors. Curr Opin Genet Devel 5: 222–228 (1995).

78. Mans RJ, Huff NJ: Utilization of ribonucleic acid and deoxyoligomer primers for polyadenylic acid synthesis by adenosine triphosphate: polynucleotidylexotransferase from maize. J Biol Chem 250: 3672–3678 (1975).

79. Mans RJ, Walter TJ: Transfer RNA-primed oligoadenylate synthesis in maize seedlings. II. Primer, substrate and metal specificities and size of product. Biochim Biophys Acta 247: 113- -121 (1971).

80. Mathur M, Saluja D, Sacher RC: De novo synthesis of poly(A) polymerase in mung bean hypocotyls, involving stored mRNA. Phytochemistry 28: 1037–1042 (1989).

81. Merits A, Zelenina DA, Mizenina OA, Chernov BK, Morozov SY: Poly(A) addition site mapping and polyadenylation signal analysis in a plant circovirus replication-related gene. Virology 211: 345–349 (1995).

82. Messing J, Geraghty D, Heidecker G, Hu N, Kridl J, Rubenstein I: Plant gene structure. In Kosuge T, Meredith P, Hollaender A (eds) Genetic Engineering of Plants: An Agricultural Perspective, pp. 211–227. Plenum Press, New York (1983).

83. Minvielle-Sebastia L, Preker PJ, Keller W: RNA14 and RNA15 proteins as components of a yeast pre-mRNA 3′-end processing factor. Science 266: 1702–1705 (1994).

84. Miralles VJ: Termination of transcription in an in vitro system is dependent on a polyadenylation sequence. Nucl Acids Res 19: 3593–3599 (1991).

85. Mogen BD, MacDonald MH, Graybosch R, Hunt AG: Upstream sequences other than AAUAAA are required for efficient messenger RNA 3′-end formation in plants. Plant Cell 2: 1261–1272 (1990).

86. Mogen BD, MacDonald MH, Leggewie G, Hunt AG: Several distinct types of sequence elements are required for efficient mRNA 3′ end formation in a pea rbcS gene. Mol Cell Biol 12: 5406–5414 (1992).

87. Moreira A, Wollerton M, Monks J, Proudfoot NJ: Upstream sequence elements enhance poly(A) site efficiency of the C2 complement gene and are phylogenetically conserved. EMBO J 14: 3809–3819 (1995).

88. Murthy KGK, Manley JL: The 160-kD subunit of human cleavage-polyadenylation specificity factor coordinates premRNA 3′-end formation. Genes Devel 9: 2672–2683 (1995).

89. Nakayama T, Ohtsubo N, Mikami K, Kawata T, Tabata T, Kanazawa H, Iwabuchi M: Cis-acting sequences that modulate transcription of wheat histone H3 gene and 3′ processing of H3 premature mRNA. Plant Cell Physiol 30: 825–832 (1989).

90. Nemeth A, Krause S, Blank D, Jenny A, Jenö P, Lustig A, Wahle E: Isolation of genomic and cDNA clones encoding

60

bovine poly(A) binding protein II. Nucl Acids Res 23: 4034–4041 (1995).

91. Ohtsubo N, Iwabuchi M: The conserved 3′-flanking sequence, AATGGAAATG, of the wheat histone H3 gene is necessary for the accurate 3′-end formation of mRNA. Nucl Acids Res 22: 1052–1058 (1994).

92. Peterson ML: Regulated immunoglobulin (Ig) RNA processing does not require specific cis-acting sequences: non Ig RNA can be alternatively processed in B cells and plasma cells. Mol Cell Biol 14: 7891–7898 (1994).

93. Pichersky E, Brock TG, Nguyen D, Hoffman NE, Piechulla B, Tanksley SD, Green BR: A new member of the CAB gene family: structure, expression and chromosomal location of Cab-8, the tomato gene encoding the Type III chlorophyll a/b-binding polypeptide of photosystem I. Plant Mol Biol 12: 257–270 (1989).

94. Preker PJ, Lingner J, Minvielle-Sebastia L, Keller W: The FIP1 gene encodes a component of a yeast pre-mRNA polyadenylation factor that directly interacts with poly(A) polymerase. Cell 81: 379–389 (1995).

95. Prescott JC, Falck-Pedersen E: Sequence elements upstream of the 3′ cleavage site confer substrate strength to the adenovirus L1 and L3 polyadenylation sites. Mol Cell Biol 14: 4682–4693 (1994).

96. Proudfoot N: How RNA polymerase II terminates transcription in higher eukaryotes. Trends Biochem Sci 14: 105–110 (1989).

97. Raabe T, Murthy KGK, Manley JL: Poly(A) polymerase contains multiple functional domains. Mol Cell Biol 14: 2946–2957 (1994).

98. Richter JD: Translational control in development: a perspective. Devel Genet 14: 407–411 (1993).

99. Rothnie HM, Chapdelaine Y, Hohn T: Pararetroviruses and retroviruses: a comparative review of viral structure and gene expression strategies. Adv Virus Res 44: 1–67 (1994).

100. Rothnie HM, Reid J, Hohn T: The contribution of AAUAAA and the upstream element UUUGUA to the efficiency of mRNA 3′-end formation in plants. EMBO J 13: 2200–2210 (1994).

101. Russnak R, Ganem D: Sequences 5′ to the polyadenylation signal mediate differential poly(A) site use in hepatitis B viruses. Genes Devel 4: 764–776 (1990).

102. Russnak RH: Regulation of polyadenylation in hepatitis B viruses: stimulation by the upstream activating signal PS1 is orientation-dependent, distance-independent, and additive. Nucleic Acids Res 19: 6449–6456 (1991).

103. Sacher RC: Biosynthesis of ribonucleic acid and polyadenylic acid in tobacco leaf homogenates. Biochim Biophys Acta 169: 58–66 (1968).

104. Sachs A, Wahle E: Poly(A) tail metabolism and function in eukaryotes. J Biol Chem 268: 22955–22958 (1993).

105. Sachs AB: Messenger RNA degradation in eukaryotes. Cell 74: 413–421 (1993).

106. Saluja D, Mathur M, Sacher RC: Purification of poly(A) polymerase from mung bean hypocotyls: subunit structure, molecular properties and characterization of the reaction product. Plant Sci 60: 27–38 (1989).

107. Sanfaçon H: Analysis of figwort mosaic virus (plant pararetrovirus) polyadenylation signal. Virology 198: 39–49 (1994).

108. Sanfaçon H, Brodmann P, Hohn T: A dissection of the cauliflower mosaic virus polyadenylation signal. Genes Devel 5: 141–149 (1991).

109. Sanfaçon H, Hohn T: Proximity to the promoter inhibits recognition of cauliflower mosaic virus polyadenylation signal. Nature 346: 81–84 (1990).

110. Sanfaçon H, Wieczorek A: Analysis of cauliflower mosaic virus RNAs in Brassica species showing a range of susceptibility to infection. Virology 190: 30–39 (1992).

111. Sasaki K, Tazawa T: Polyriboadenylate synthesizing activity in chromatin of wheat seedlings. Biochem Biophys Res Commun 52: 1441–1449 (1973).

112. Schek N, Cooke C, Alwine JC: Definition of the upstream efficiency element of the simian virus 40 late polyadenylation signal by using in vitro analyses. Mol Cell Biol 12: 5386–5393 (1992).

113. Sheets MD, Ogg SC, Wickens MP: Point mutations in AAUAAA and the poly (A) addition site: effects on the accuracy and efficiency of cleavage and polyadenylation in vitro. Nucl Acids Res 18: 5799–5805 (1990).

114. Simpson GG, Clark GP, Rothnie HM, Boelens W, van Venrooij W, Brown JWS: Molecular characterisation of the spliceosomal proteins U1A and U2B″ from higher plants. EMBO J 14: 4540–4550 (1995).

115. Sittler A, Gallinaro H, Jacob M: Upstream and downstream cis-acting elements for cleavage at the L4 polyadenylation site of adenovirus-2. Nucl Acids Res 22: 222–231 (1994).

116. Sullivan TD, Christensen AH, Quail PH: Isolation and characterization of a maize chlorophyll a/b binding protein gene that produces high levels of mRNA in the dark. Mol Gen Genet 215: 431–440 (1989).

117. Tabata T, Fukasawa M, Iwabuchi M: Nucleotide sequence and genomic organisation of a wheat histone H3 gene. Mol Gen Genet 196: 397–400 (1984).

118. Tabata T, Terayama C, Mikami K, Uchiyama H, Iwabuchi M: An accurate transcription of wheat histone genes in sunflower cells. Plant Cell Physiol 28: 73–82 (1987).

119. Takagaki Y, Manley JL: A polyadenylation factor subunit is the human homologue of the Drosophila suppressor of forked protein. Nature 372: 471–474 (1994).

120. Tarui Y, Minamikawa T: Purification and properties of poly(A) polymerase from Vigna unguiculata. Plant Cell Physiol 29: 835–842 (1988).

121. Tarui Y, Minamikawa T: Poly(A) polymerase from Vigna unguiculata seedlings. Eur J Biochem 186: 591–596 (1989).

122. Valsamakis A, Zeichner S, Carswell S, Alwine JC: The human immunodeficiency virus type 1 polyadenylylation signal: a 3′ long terminal repeat element upstream of the AAUAAA necessary for efficient polyadenylylation. Proc Natl Acad Sci USA 88: 2108–2112 (1991).

123. Wahle E: Poly(A) tail length control is caused by termination of processive synthesis. J Biol Chem 270: 2800–2808 (1995).

124. Wahle E: 3′-end cleavage and polyadenylation of mRNA precursors. Biochim Biophys Acta 1261: 183–194 (1995).

125. Wahle E, Keller W: The biochemistry of 3′-end cleavage and polyadenylation of messenger RNA precursors. Annu Rev Biochem 61: 419–440 (1992).

126. Wahle E, Lustig A, Jenö P, Maurer P: Mammalian poly(A)-binding protein II. J Biol Chem 268: 2937–2945 (1993).

127. Wasserman KM, Steitz JA: Association with terminal exons in pre-mRNAs: a new role for the U1 snRNP? Genes Devel 7: 647–659 (1993).

128. Weichs an der Glon C, Monks J, Proudfoot NJ: Occlusion of the HIV poly(A) site. Genes Devel 5: 244–253 (1991).

129. Wickens M: Forward, backward, how much, when: mechanisms of poly(A) addition and removal and their role in early development. Semin Devel Biol 3: 399–412 (1992).

130. Wormington M: Poly(A) and translation: development control. Curr Opin Cell Biol 5: 950–954 (1993).

131. Wu L, Ueda T, Messing J: Sequence and spatial requirements for the tissue- and species-independent 3′-end processing mechanism of plant mRNA. Mol Cell Biol 14: 6829–6838 (1994).

132. Wu L, Ueda T, Messing J: The formation of mRNA 3′- ends in plants. Plant J 8: 323–329 (1995).

133. Wu L, Ueda U, Messing J: 3′-end processing of the maize 27 kDa zein mRNA. Plant J 4: 535–544 (1993).

Plant Molecular Biology **32**: 63–78, 1996.
© 1996 *Kluwer Academic Publishers.*

Control of mRNA stability in higher plants

Michael L. Abler & Pamela J. Green*
*MSU-DOE Plant Research Laboratory and Department of Biochemistry, Michigan State University, East Lansing, MI 48824–1312, USA (*author for correspondence)*

Key words: mRNA degradation, mRNA turnover, mRNA half-life measurements, instability sequences, RNA-binding proteins, translation, mRNA decay mechanisms

Contents

Abstract	63	*Cis-* and *trans*-acting determinants of	
Introduction	64	mRNA stability	69
Measuring mRNA stability	64	*Cis*-acting elements that control	
Nuclear run-on transcription	64	inherent mRNA stability	69
Kinetics of mRNA decay	65	Putative *cis*-acting determinants of	
Transcriptional inhibitors	65	regulated mRNA stability	70
RNA electroporation	65	*Trans*-acting determinants of mRNA	
In vitro measurements	66	stability	71
Approach to steady state	66	Plant mRNA decay mechanisms	72
Pulse-chase	66	Translation and mRNA stability	74
Regulated promoters	66	Conclusion and future prospects	75
Effects of endogenous and exogenous		Acknowledgements	75
stimuli on mRNA stability	67	References	75

Abstract

The degradation rates of different mRNAs in higher plants can vary over a broad range and are regulated by a variety of endogenous and exogenous stimuli. During the past several years, efforts to better understand the control of mRNA stability in plants have increased considerably and this has led to improved methodologies and important mechanistic insights. In this review, we highlight some of the most interesting examples of plant transcripts that are controlled at the level of mRNA decay and discuss what has been learned from their study. Experiments that implicate or demonstrate the involvement of particular *cis-* and *trans*-acting factors in mRNA decay pathways are a major focus, as are those experiments that have led to mechanistic models. Emphasis is also placed on studies that address the relationship between translation and mRNA stability. Our current knowledge indicates that some of the determinants and pathways for mRNA decay may differ in plants compared to other eukaryotes, whereas others appear to be similar. This knowledge, coupled with the availability of biochemical, molecular and genetic approaches to elucidate plant mRNA decay mechanisms, should continue to lead to findings of novel and general significance.

Introduction

The control of mRNA stability is one of the most prominent forms of post-transcriptional gene regulation in eukaryotic cells. Although, historically, mRNA degradation has received less attention than mRNA synthesis, clearly both processes act together to establish the steady-state levels of different mRNAs and determine how fast those levels can change. Because mRNA degradation is the downstream process, its control can enhance, diminish or override regulation (or the lack thereof) exerted at the transcriptional level. Available data indicate that the average mRNA in plants and vertebrates survives and often continues to be translated for several hours before it is degraded [6, 74, 80]. However, mRNAs that degrade in a matter of minutes [33, 52, 57, 80] or remain intact for days or even weeks [6] are known to exist in higher eukaryotes. A range of mRNA stability has also been reported for bacteria and lower eukaryotes such as yeast, but in these organisms, the average transcripts degrade more rapidly than in higher eukaryotes. This feature makes it easier for microorganisms to quickly adapt to changing environmental conditions.

As sessile organisms, plants might benefit considerably from control at the level of mRNA stability, particularly when rapid responses to exogenous or endogenous stimuli are required. For example, if the optimal response to an environmental stimulus is to rapidly shut down the synthesis of a given protein, then having the protein encoded by a very unstable mRNA will be a big advantage. Conversely, for proteins that are needed at relatively constant levels, the buffering capacity of stable mRNAs would be the most beneficial. By differentially modulating the stability of individual transcripts in response to various signals, a further level of regulation can be achieved.

Interest in the control of inherent and regulated mRNA stability in higher plants has grown appreciably in recent years for several reasons. First, research continues to reveal genes that are apparently regulated at the level of mRNA stability [24, 76]. Second, the methodologies and model systems that are available for measuring mRNA stability have improved, and it is apparent that plants may provide opportunities to study unique mechanisms or to apply unique approaches. Finally, the study of mRNA stability in plants can have applied as well as basic significance because in some cases limitations at the level of mRNA stability can hinder the expression of foreign genes introduced into plants for crop improvement [17]. In

this chapter, our discussion focuses on the stability of nuclear-encoded transcripts in higher plants. Our goal is not to present a comprehensive review, but rather to concentrate on the recent molecular analyses that provide the greatest mechanistic insights and a few less developed examples with high future potential. In addition to the progress made in plants, marked advances have resulted from studies of mRNA stability in yeast and animal systems, as highlighted in several recent reviews [14, 63, 66, 77]. We do not include investigations probing the role of mRNA stability in gene silencing because this topic is dealt with elsewhere in this volume (chapter by Baulcombe). Readers are also encouraged to see other chapters in this volume for discussions on mRNA stability in *Chlamydomonas* and in plant organelles (chapters by Rochaix and Sugita and Sugiura, respectively).

Measuring mRNA stability

A number of methods have been used to measure and compare the stability of different mRNAs. Some approaches directly measure mRNA half-lives, whereas others rely on the measurement of other parameters to evaluate mRNA stability. Each method has its strengths and limitations, factors which must be considered when selecting a method to evaluate mRNA stability and when comparing data obtained with different methods. Therefore, we begin with descriptions of several methods used to investigate mRNA stability and briefly discuss their strengths and possible limitations.

Nuclear run-on transcription

The most common method of finding evidence for regulation at the level of mRNA stability is to carry out nuclear run-on transcription experiments. During this procedure, isolated nuclei are allowed to continue transcription (i.e. run-on) in the presence of a labeled nucleotide. The amount of label incorporated into RNAs transcribed from the gene of interest is quantitated by hybridization and considered to represent the transcriptional activity at the time the nuclei were harvested. Generally, these experiments are used to compare multiple genes under the same condition or the same gene under different conditions. Large differences in transcript accumulation that cannot be accounted for by similar differences in transcriptional activity in run-on assays are generally attributed to differences in mRNA stability. However, one cannot be sure that discrepancies between run-ons and mRNA accumulation are due to control at the level of mRNA stability without further experiments, particularly in the case of small discrepancies. This is because run-on assays are not always highly quantitative and they are based on assumptions that may be difficult to confirm (e.g. that transcriptional activity *in vivo* is faithfully recapitulated by the transcriptional activity of isolated nuclei; see [83] for further discussion). Nevertheless, transcripts with large differences in mRNA accumulation that correspond to small differences in transcription warrant further investigation by one or more of the methods below because they are the most likely to be controlled at the level of mRNA stability.

Kinetics of mRNA decay

Degradation of mRNA is assumed to be a stochastic process, much like radioactive decay. Consequently, in the absence of synthesis, the change in mRNA concentration at any given time (dC/dt) is a first order process, dependent on the amount of mRNA present at that time and the decay rate constant k_d) for the transcript of interest.

$$dC/dt = -k_d C. \tag{1}$$

However, mRNA decay is usually discussed in terms of the half-life of a transcript, or the time required for half of the existing mRNA molecules to be degraded. The half-life can be derived from the above equation and represented as:

$$t_{1/2} = \ln 2/k_d. \tag{2}$$

Thus the half-life is inversely proportional to the decay constant. While this analysis of mRNA decay appears to be generally true, there are known exceptions. For example, the mammalian c-*myc* and c-*fos* mRNAs are subject to removal of their poly(A) tails prior to decay of the body of the transcript [33]. Those mRNAs then decay with biphasic kinetics, composed of a lag phase wherein deadenylation occurs, followed by first-order decay of the body of the mRNA. The extent to which complex decay mechanisms alter mRNA half-life measurements is not known, and it may be negligible in most cases, yet it is important to realize that mRNA decay kinetics may not always be straightforward.

Experimentally, half-lives are most often measured by stopping transcription of the mRNA in question (see below) and monitoring the disappearance of the transcript over time by RNA hybridization. A semilogarithmic plot of mRNA concentration as a function of time then yields a straight line with the decay constant as the slope, as defined by:

$$\ln(C/C_0) = -k_d t \tag{3}$$

where C_0 is the initial mRNA concentration, C is the mRNA concentration at time t, and k_d is the decay constant. Half-lives can be read off the graph or calculated from Eq. 2.

Transcriptional inhibitors

The decay rates of endogenous transcripts are often measured by treating plant cells with transcriptional inhibitors such as actinomycin D, cordycepin, or α-amanitin and monitoring the disappearance of the mRNA over time via RNA blot hybridization. Such experiments have proven extremely useful in providing direct measurements of mRNA half-lives, and have provided a large proportion of the direct half-life data in plants [24, 43, 45, 57, 60, 69]. Actinomycin D inhibits transcription by intercalating into the DNA, whereas cordycepin acts as a chain-terminating adenosine analogue. Thus, cordycepin affects polyadenylation as well as transcription. α-Amanitin is an inhibitor of eukaryotic RNA polymerases II and III, blocking transcription by binding to the polymerase. It has been reported that the use of transcriptional inhibitors may increase the observed half-life for some transcripts, presumably by inhibiting the synthesis of a labile factor involved in mRNA degradation [62], and a few such cases have been identified in plants [16, 21, 98]. Often, changes in mRNA half-life in response to a given stimulus can still be observed in these cases, although the magnitude of the effect may be dampened relative to that in the absence of the inhibitor [21, 98]. This technique is rarely used for very stable transcripts, due to the toxicity of extended exposure to the inhibitors.

RNA electroporation

Electroporation of *in vitro* synthesized transcripts into plant protoplasts has also been used as a means to measure mRNA half-lives. Major advantages of this technique are that it avoids the use of transcriptional inhibitors, and the state of the 5′ cap and poly(A) tail can be easily controlled. The half-life of the electroporated transcript can be determined by isolating RNA from the protoplasts at intervals following electroporation and detecting the remaining electroporated transcript on RNA blots [23, 25, 26]. Another approach has been to monitor the functional half-life (in contrast to the chemical half-life) of the RNA by monitoring protein activity levels over time [23]. A potential limitation of electroporation is that it is currently unknown how many transcripts have similar relative stabilities when they are introduced into the cytoplasm via electroporation rather than via normal export pathways from the nucleus. Alternatives that avoid this potential problem (albeit with their own limitations) would be to electroporate DNA into protoplasts and monitor decay of the encoded transcripts following inhibitor treatment or after transcription naturally ceases [56].

In vitro *measurements*

The establishment of *in vitro* mRNA decay systems is an additional approach that has shown promise for examining the control of mRNA stability in mammalian cells [66] and, more recently, in plants. An oat polysome-based *in vitro* mRNA decay system was found to faithfully mimic the *in vivo* order of decay and the half-lives of oat phytochrome A, β-tubulin, and actin [9]. In addition, *in vitro* systems containing polysomes or an S150 extract from soybean seedlings or mature petunia leaves have been shown to produce degradation products of the soybean *SRS4* transcripts that appear to be identical with putative decay intermediates known to be produced *in vivo* [78, 79]. The soybean and petunia systems have begun to provide insights into *SRS4* mRNA decay mechanisms [78, 79], but they have not yet been checked with other transcripts. With further testing, it seems likely that plant *in vitro* systems will prove to be important tools for comparing mRNA decay rates, particularly if they can be shown to recognize *cis*-acting sequences that affect mRNA stability *in vivo*.

Approach to steady state

Another way to estimate mRNA half-life is by measuring the rate at which a given transcript accumulates to a steady-state level (reviewed in [34]). In an approach to steady-state experiment, transcripts are continuously labeled *in vivo* and the radioactivity incorporated into the mRNA of interest is monitored over time by molecular hybridization. When the synthesis and decay rates of the mRNA are equal, the transcript is at its steady-state level, and the total amount of radioactiv-

ity in the pool of that transcript will remain constant thereafter. The decay constant can be derived from the equation [32]:

$$\ln(1 - C_t/C_\infty) = -k_\mathrm{d}t \qquad (4)$$

where C_t represents mRNA specific activity at time t and C_∞ is the amount of radioactivity approached asymptotically. The half-life of the mRNA can then be calculated using Eq. 2. In order for the experiment to be valid, the supply of radioactive precursor must be constant and, if necessary, adjustments must be made for changes in specific activity that occur due to cell division. Also, it may be difficult to detect low-abundance and short-lived mRNAs due to low specific activity of their transcript pools. Recently, an approach to steady-state calculation was used to confirm mRNA half-life measurements made upon inhibitor treatment [45].

Pulse-chase

Transcriptional pulse-chase methods, where a radiolabeled nucleotide is supplied to cells to produce a 'pulse' of labeled mRNA and subsequently 'chased' with excess cold nucleotide, have not been widely used. It can be extremely difficult to detect rare or unstable transcripts using the transcriptional pulse-chase method because the large reservoir of unlabeled nucleotides in the cells limits the amount of incorporated label. The reservoir of unlabeled nucleotides also makes it difficult to achieve a rapid chase. Incorporation of radioactive nucleotides into the mRNA during the chase results in an apparent half-life that is longer than the true half-life, and the effect is greatest on transcripts with short half-lives [74].

Regulated promoters

A particularly advantageous method for measuring mRNA half-lives involves the transformation of plant cells with one or more genes under the control of a regulated promoter, a technique used extensively in other systems [13, 29, 73]. Specifically, the use of a repressible promoter that stops transcription of the transgene(s) in response to a stimulus will allow the decay of the corresponding mRNA to be followed over time, in the absence of transgene mRNA synthesis. Regulating transcription in this manner only alters the transcription of the transgene(s) and therefore should have little or no effect on other cellular processes. As long as the promoter is strong enough to produce a sufficient pool of transcript at the start of the experiment, the decay rate of the transgene mRNA can be determined after the promoter is shut off. Under these conditions, the repressible promoter overcomes the primary drawbacks of the transcriptional inhibitor, electroporation, and pulse-chase methods. One such promoter developed for use in plants is called Top10 [91]. Top10 contains seven tetracycline operator sequences joined to a TATA box. The promoter is dependent on the TetVP16 fusion protein for transcriptional activity [29]. TetVP16 consists of the operator-binding portion of the bacterial tetracycline repressor fused to the acidic activation domain of the herpes simplex virus transcription factor, VP16. In plant [91] and mammalian [29] cells TetVP16 is rapidly inactivated in the presence of tetracycline, resulting in a cessation of transcription from the Top10 promoter. Tetracycline can be applied to excised plant tissues by vacuum infiltration [91], to intact plants via root uptake in liquid growth systems [91], or to cultured plant cells by addition to the growth medium [27]. Recent studies with stably transformed tobacco cells have successfully used Top10 constructs to evaluate mRNA sequences that control the rate of *SAUR-AC1* transcript decay [27]. Although the use and improvement of regulated promoter systems may represent the trend for the future, most of our current knowledge of mRNA half-lives derives from the other approaches discussed above.

Effects of endogenous and exogenous stimuli on mRNA stability

A large number of reports provide evidence for post-transcriptional regulation of mRNA abundance in plants, based on differences between nuclear run-on and mRNA accumulation data (discussed above) [11, 30, 61, 76, 89, 94]. However, the step in gene expression that is the target of the regulation has thus far been determined in relatively few instances. In cases where discrepancies between transcription rates and mRNA accumulation levels have been investigated further, differences in mRNA stability have often been found.

Plant hormones regulate gene expression at many levels [24], and there are indications of hormonal regulation at the level of mRNA stability as well. In dark-adapted *Lemna gibba* plants, cytokinin treatment increases fivefold the level of transcripts encoding the major chlorophyll *a/b*-binding protein of light-harvesting complex II (LHCP), as measured by RNA

blot hybridization [19]. Similarly, in dark-adapted plants subjected to a pulse of red light two hours prior to harvest, a threefold increase in LHCP mRNA was observed in response to cytokinin treatment. Nuclear run-on transcription experiments carried out with nuclei isolated from dark-adapted or red-light-pulsed plants showed at most a 50% increase in transcription in response to cytokinin [19]. The difference in mRNA accumulation levels may be due to a stabilization of LHCP transcript in the presence of cytokinin.

Another probable example of hormonal regulation at the level of mRNA stability derives from studies of ethylene regulation of gene expression in ripening tomato fruit. Most genes in this system are controlled by ethylene at the transcriptional level. One exception is the E17 gene (subsequently called *eri*, for ethylene-responsive proteinase inhibitor [51]), which shows no difference in transcription rates when fruits are treated with exogenous ethylene, yet the mRNA increases sixfold [48], indicating that the transcript may be stabilized in the presence of ethylene. Half-life measurements of the tomato E17 and *Lemna* LHCP mRNAs with and without hormone treatment would be the most definitive way to confirm that the differences in transcript accumulation are due to effects on mRNA stability. It has also been suggested, based on the accumulation of wheat *Em* transcript in the presence of abscisic acid (ABA) and α-amanitin, that ABA regulates *Em* mRNA abundance at the level of transcript stability [93], though direct evidence is again lacking.

Heat shock is another stimulus known to affect the expression of many genes [86]. For example, α-amylase transcripts in barley aleurone layers are unstable during heat shock [2, 7]. Transcripts for both the high pI and low pI isozymes decayed with a maximum half-life of 30 min when the incubation temperature was raised from 25 °C to 40 °C [7]. No transcription inhibitors were used in the time course analysis, hence the estimate of the maximum half-life. It has been proposed that the short half-lives of the transcripts during heat shock are due to active destabilization of otherwise stable mRNAs [39, 58, 59]. Heat shock also decreases the accumulation of other secretory protein transcripts in barley aleurone layers [7] and some wound-inducible mRNAs encoding extracellular proteins in carrot root disks [8]. These data support the hypothesis that heat shock-induced disruption of the endoplasmic reticulum (ER) leads to the rapid decay of mRNA that is normally translated on ER-bound polysomes [2]. Unfortunately, none of the critical comparisons were carried out under the same conditions with respect to inhibitors, i.e. experiments in the absence of heat shock involved cordycepin treatment, whereas those in the presence of heat shock were performed without inhibitors [8, 38]. Therefore, it is unclear whether the transcripts are actively destabilized by the heat shock or stabilized in the presence of transcriptional inhibitors. Interestingly, heat shock has been found to stabilize transcripts electroporated into plant protoplasts, as discussed by D.R. Gallie in this volume.

Sucrose starvation effects on α-amylase transcripts in cultured rice cells provide an excellent example of an exogenous stimulus affecting mRNA stability. Yu and coworkers compared the decay rates of rice α-amylase mRNA (using actinomycin D as a transcriptional inhibitor) in cells incubated in sucrose-containing or sucrose-free medium. The results of these experiments demonstrated that the pool of α-amylase mRNA is more stable in the absence of sucrose than in its presence [70]. Recent work with probes specific for eight rice α-amylase transcripts revealed that the levels of all of the mRNAs were quite low in cells provided with sucrose. In sucrose-starved cells two transcripts, $\alpha amy3$ and $\alpha amy8$, account for 90% of the α-amylase mRNA pool while the other transcripts, represented by $\alpha amy7$, are weakly expressed. Although the half-lives of the three transcripts are different, all are approximately fourfold more stable in the absence of sucrose (S.-M. Yu, personal communication).

Light effects on gene expression have been studied extensively, and it is not surprising that light likely affects the stability of some light-responsive mRNAs. The ferredoxin-encoding genes *Fed-1* from pea [15] and *FedA* from *Arabidopsis* [89] are of particular interest because they show increased RNA accumulation in light-grown versus dark-adapted leaves that cannot be accounted for by changes in transcription. *Arabidopsis FedA* transcript is twenty times as abundant in the light as in the dark, whereas the transcriptional activity, as measured by nuclear run-ons, was only twofold higher in the light than in the dark. The large difference in mRNA accumulation coupled with the small difference in transcriptional activity suggests that the *FedA* transcript is either stabilized in the light or destabilized in the dark.

Similarly for *Fed-1*, nuclear run-on transcription cannot explain the increase in transcript abundance in the light [15]. Moreover, the sequences responsible for light regulation of *Fed-1* have been delineated by examining the light-responsiveness of various chimeric reporter genes in mature leaves of transgenic tobacco

[15]. These experiments showed that increased accumulation of pea *Fed-1* mRNA in the light is dependent on an internal light regulatory element that includes portions of the 5′-untranslated region (UTR) and the coding region of the transcript. Neither the 5′-UTR nor the coding region alone is sufficient for induction of the light response; however, a 230 nucleotide *Fed-1* fragment consisting of the 5′-UTR and the first third of the coding region fused to reporter sequences shows about fourfold higher transcript accumulation in the light than in the dark, similar to wild-type *Fed-1*. Blocking translation of chimeric gene transcripts by the insertion of nonsense codons abolishes the light response conferred by *Fed-1* sequences, demonstrating the requirement of an open reading frame for increased mRNA accumulation in the light [16]. Taken together, these data strongly indicate that the pea *Fed-1* mRNA is stabilized in the light.

Control at the level of mRNA stability may contribute to the regulation of the light-responsive genes encoding the small subunit of ribulose bisphosphate carboxylase/oxygenase (*rbcS*), but the nature of the effect differs depending on the gene family members being examined, the conditions under which the plants are examined, the developmental stage of the plant, and the plant species. When petunia plants are placed in darkness, the ensuing decrease in *rbcS* transcripts parallels the decrease in transcription, indicating a lack of post-transcriptional control under these conditions [81]. However, re-exposure of petunia plants to light after 48 h of darkness induces the accumulation of the *rbcS A* subfamily of mRNAs to a higher level than would be expected from the increase in transcription, suggesting a stabilization of the mRNAs in the light [81]. Potato *rbcS* transcripts decay more rapidly in the dark than in the light in the presence of cordycepin. This indicates that the transcripts are actively destabilized in the dark [21] or stabilized in the light. Interestingly, different potato *rbcS* transcripts disappear at different rates in the dark, an effect that was masked by cordycepin, which stabilized the transcripts in the dark [21]. Differential decay of individual *rbcS* transcripts has also been reported in *Lemna gibba* [64, 75]. The half-lives of two of the transcripts, *SSU1* (ca. 8 h) and *SSU5B* (ca. 5 h), were determined using cordycepin to inhibit transcription [64]. The higher stability of the *SSU1* mRNA at least partially explains its greater accumulation in the cytoplasm than *SSU5B*, despite the fourfold higher transcriptional activity of *SSU5B*. In mature soybean, discrepancies between mRNA accumulation levels and transcription rates suggest that *rbcS* transcripts are less stable in the dark than in the light [81], similar to the results in potato. In contrast, discrepancies in transcription rates and mRNA accumulation data from soybean seedlings suggest *rbcS* mRNAs are more stable in the dark than in the light [71]. The degradation rates of the soybean *rbcS* transcripts in light and dark conditions at the different developmental stages have not yet been determined; however, the simplest explanation of the soybean data is that differential control of mRNA stability occurs during development.

A particularly interesting case of regulated mRNA stability is that of a proline-rich protein, PvPRP1, mRNA. In cells of common bean (*Phaseolus vulgaris*) the *PvPRP1* mRNA is destabilized in response to treatment with fungal elicitors [98]. After actinomycin D treatment, the *PvPRP1* transcript has a half-life of 60 h in the absence of fungal elicitors and 18 h in the presence of fungal elicitors. However the half-life of *PvPRP1* in elicitor-treated cells is much greater in the presence of actinomycin D (18 h) than in the absence of actinomycin D (maximum of 45 min) [98]. Therefore, although the *PvPRP1* mRNA is clearly destabilized in response to elicitors, the magnitude of the effect cannot be calibrated exactly, due to the stabilizing effect of actinomycin D on the transcript.

Cis- and *trans*-acting determinants of mRNA stability

Most of the work characterizing the *cis*- and *trans*-acting factors responsible for mRNA stability has been done with unstable transcripts, that is, those with a half-life of about an hour or less. One reason for this is that genes with unstable transcripts often encode proteins whose expression must be tightly controlled, such as those genes involved in growth and differentiation. Unstable transcripts include phytochrome [35, 69] and certain auxin-induced transcripts in plants [28, 45, 52], and protooncogene and cytokine transcripts in mammalian cells [33, 66]. The rapid decay of unstable transcripts is considered to be an active process for two main reasons. First, data indicate that most transcripts are relatively stable in eukaryotic cells [6,74]. For example, in one study aimed at isolating genes for unstable transcripts (GUTs), only a small percentage of tobacco cDNAs were found to correspond to unstable transcripts [80], supporting the hypothesis that stability rather than instability is likely to be the default. Second, nearly all of the sequences that have been

shown to affect mRNA stability act to decrease mRNA half-life [67, 76] or serve to block the function of an instability sequence [4]. However, in mammalian cells, elements that stabilize the α2-globin mRNA in erythroid cells have recently been identified in the 3′-UTR of the transcript [92]. In transgenic tobacco, it has been found that the fusion of a reporter gene (driven by the CaMV 35S promoter) to 3′ ends from different plant genes results in large differences in mRNA accumulation [42]. Other experiments carried out using rice protoplasts demonstrate that the 5′-UTR of the wheat *Em* transcript contains an element that enhances gene expression assayed at the protein level [50]. While these differences have not been shown to be the result of differences in mRNA stability, it is quite possible that *cis*-acting elements within the 5′ and 3′ ends of these plant transcripts serve to stabilize the chimeric transcripts.

Cis-*acting elements that control inherent mRNA stability*

The two *cis*-acting mRNA stability determinants that are common to virtually all plant mRNAs are the m^7Gppp cap at the 5′ end and the poly(A) tail at the 3′ end. Gallie and coworkers found that the 5′ cap stabilizes transcripts two- to fourfold, and the poly(A) tail stabilizes transcripts two- to three-fold by electroporating capped or uncapped mRNAs and mRNAs with or without a poly(A) tail into tobacco protoplasts [23, 25, 26]. The stabilizing effects of the cap and tail are not synergistic, but they are additive [23]. Although it is unclear how the mechanisms responsible for removal of the cap and poly(A) tail are regulated in plants, these *cis*-acting components are considered to play a general role in determining a transcript's overall stability.

To begin to understand the kinds of sequences that control the stability of specific plant mRNAs, the focus has been on unstable mRNAs such as those encoded by the *Small Auxin-Up RNA* (*SAUR*) genes of soybean and *Arabidopsis* [28, 52]. *SAUR* transcripts are among the most unstable plant mRNAs known, with half-lives of 10 to 50 min, depending on how the half-lives are measured [20]. The most detailed studies of *SAUR* mRNA stability have been carried out on the *SAUR-AC1* gene of *Arabidopsis* [27]. As a first step toward delineating the sequences responsible for instability of the *SAUR-AC1* mRNA, the auxin-regulated promoter region of the gene was removed and the sequences downstream of the promoter, including the coding region and 3′-UTR, were placed under the control of the CaMV

35S promoter. The resulting 35S-*SAUR* chimeric gene was expressed at a very low level in both the absence and presence of auxin, consistent with the presence of instability sequences in the *SAUR-AC1* mRNA that function in an auxin-independent manner. Subsequent experiments showed that both the coding region and the 3'-UTR of *SAUR-AC1* contribute to low mRNA levels. Effects on mRNA stability were assayed using chimeric genes under the control of the tetracycline-repressible Top10 promoter. Messenger RNA half-life analysis following tetracycline treatment demonstrated that the *SAUR-AC1* coding region does not contain elements that decrease mRNA stability. In contrast, the 3'-UTR was found to act as a potent mRNA instability determinant.

The most highly conserved sequence in the 3'-UTRs of *SAUR* transcripts is the DST (downstream) element [28, 52, 95]. As illustrated in Fig. 1, this element consists of three conserved regions (subdomains) separated by two variable regions. The ATAGAT and the T-GTA in the second and third subdomain are invariant in all reported *SAUR* DST sequences. It had previously been shown that a synthetic element consisting of two copies of the soybean *SAUR 15A* DST is sufficient to destabilize reporter transcripts in stably transformed tobacco cells when placed downstream of the coding sequences [57]. These data may provide clues as to the regions of the *SAUR* 3'-UTR that contribute to its ability to function as an instability element. It is interesting that two copies of the DST element were required to target reporter transcripts for rapid decay in stably transformed tobacco cells, yet the endogenous *SAUR* genes contain only one copy of DST [28, 52, 95]. The difference may be due to the DST element's relative position in the 3'-UTR, or some redundancy of portions of the DST element may be required for instability. The latter hypothesis is supported by the occurrence of multiple subdomains of the DST element in the 3'-UTR of *SAUR-AC1* (Fig. 1).

Another *cis*-acting element that functions in plants is related to mammalian instability determinants with multiple AUUUA pentamers. The 3'-UTRs of many unstable mammalian lymphokine, cytokine, and protooncogene mRNAs contain several AUUUA repeats, and the AUUUA repeats have been determined to function as important instability determinants in some transcripts [33]. To test whether the AUUUA motif is recognized as an instability determinant in plants, a synthetic element containing multiple overlapping AUUUA pentamers (Fig. 1) was inserted into the 3'-UTR of the β-glucuronidase and β-globin report-

er transcripts [60]. Half-life measurements in stably transformed tobacco cell lines showed that the AUUUA repeat made reporter transcripts degrade rapidly compared to transcripts with no insert or with a control insert containing interspersed G and C nucleotides. Sequences with the same A+U content but lacking the AUUUA motif had little effect on the stability of reporter transcripts, indicating that it is not simply regions rich in adenine and uridine residues that are being recognized as instability determinants. It is also likely that the AUUUA element functions in intact plants, as reporter transcripts containing the instability determinant accumulate to much lower levels in transgenic plants than do transcripts containing a control insert interspersed with G and C residues [60]. As in mammalian cells, the mere presence of an AUUUA sequence is not sufficient to destabilize mRNAs in plants [33, 90]. In plants, the minimal sequence requirements for AUUUA-mediated instability are not known, but in some mammalian examples the element must be flanked by U residues on both sides [46, 99]. Curiously, AUUUA sequences do not appear to function as mRNA instability determinants in yeast [54].

Putative cis-*acting determinants of regulated mRNA stability*

In addition to the aforementioned *cis*-acting sequences that control inherent mRNA stability, progress has recently been made toward identifying possible *cis*-acting determinants that differentially regulate mRNA stability. One of the earliest and most significant examples comes from analysis of the light-regulated pea *Fed-1* mRNA mentioned earlier. This transcript contains an internal light-response element that appears to stabilize *Fed-1* mRNA in the light [15, 16]. Light induction of *Fed-1* mRNA accumulation in the mature leaves of transgenic plants requires a 38-nucleotide region of the 5'-UTR and sequences that extend into the coding region of the transcript [15]. Specifically, in the presence of the 5'-UTR, translation of 47 codons of *Fed-1* is sufficient to confer light-responsiveness on a *Fed-1-GUS* fusion transcript, whereas translation of only 5 codons is not [16]. In the absence of the 5'-UTR, the coding region is not light-responsive [15], indicating an interaction between a *cis*-acting element in the 5'-UTR and the requirement for translation.

Although the *Fed-1* light-regulatory region is the most highly resolved, several reports provide hints as

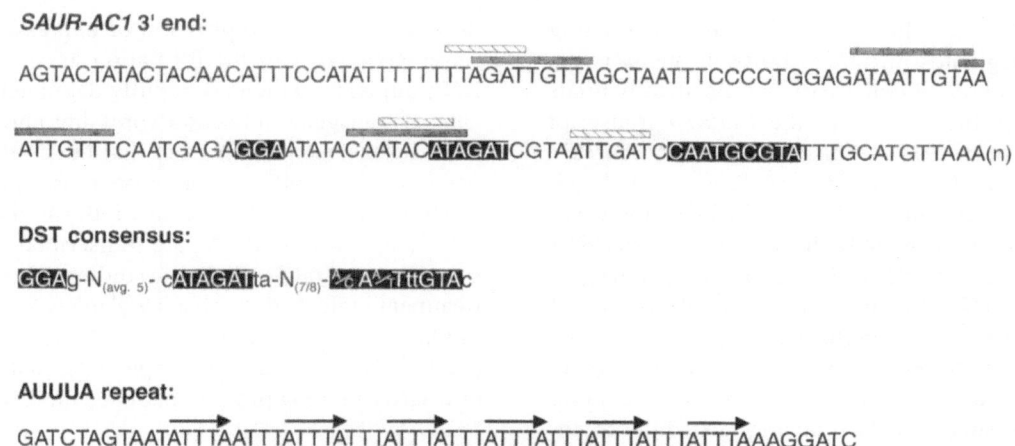

Figure 1. Known functional *cis*-acting determinants of mRNA instability in plants. The sequences of the 3′ end of the *SAUR-AC1* [28] transcript, DST element [57] and AUUUA repeat [60] are shown. White letters on a black background in the *SAUR-AC1* 3′ end sequence [28] show the endogenous DST element. Hashed rectangles depict sequences similar to the conserved ATAGAT subdomain of DST, while the black rectangles depict sequences similar to the conserved GTA subdomain of DST. Arrows in the AUUUA repeat indicate the 11 copies of the pentamer motif in the sequence.

to the identity of other *cis*-acting sequences that may regulate mRNA stability in response to particular stimuli. For example, a 27-nucleotide sequence in the 3′-UTR of the *PvPRP1* mRNA has been implicated in elicitor-mediated regulation of *PvPRP1* mRNA stability because an elicitor-regulated RNA-binding protein interacts with this sequence *in vitro* [97] (see below). The 27 nucleotide sequence is U-rich and contains a copy of the AUUUA sequence. Other candidates for mRNA stability determinants are regulatory sequences located downstream of protein-coding regions. These include 3′ sequences that mediate sucrose inducibility, root cortex-specific expression, or wound inducibility of the potato *Sus4* [22], oilseed rape *AX92* [18], and potato inhibitor II [84] genes, respectively. It should be noted, however, that some 3′ end elements may function at the level of transcription, especially those that are eventually localized to the region downstream of the poly(A) addition site [12, 47].

Trans-*acting determinants of mRNA stability*

Similar to the situation for *cis*-acting elements, both general and specific *trans*-acting factors are considered to be important for the control of mRNA stability. The cytoplasmic poly(A) binding proteins (PABPs), which bind along the poly(A) tail of mRNAs, stand out among the general *trans*-acting factors because of the known roles of the yeast PABP and the multiplicity

of PABPs in plants. PABP has recently been shown to inhibit mRNA decapping in yeast, thereby preventing transcript degradation by 5′ to 3′ exonuclease activity [10]. Yeast PABP alters poly(A) tail metabolism in several ways, including increasing the rate of deadenylation [10]. The increased deadenylation rate may occur due to an interaction with the yeast poly(A) nuclease (PAN), which is dependent on PABP for its activity [49]. The deadenylation rate has been shown in mammalian cells and yeast to be an important factor in determining the half-life of some mRNAs [54, 72].

In plants, the situation is likely to be more complex than in yeast because characterization of PABP polypeptides and genes indicates that plants contain several differentially regulated PABPs [3, 37, 96]. Polypeptides of about 70 kDa (the expected size for PABPs) have been identified from tobacco, *Arabidopsis*, and pea extracts that bind to poly(A) [96]. *Arabidopsis* contains at least four genes (*PAB1*, *2*, *3*, and *5*) that encode PABPs on the basis of amino acid sequence homology [3, 37], and the binding of the *PAB5* gene product to poly(A) has been demonstrated directly [3]. *PAB3* and *PAB5* are floral-specific and *PAB1* is preferentially expressed in roots and to a lesser extent in young flowers [3]. The differential expression of multiple PABPs may provide plants with an additional method of regulating mRNA stability by allowing different interactions to occur between PABP and other proteins, such as poly(A) nucleases [49, 68].

A promising candidate for a specific *trans*-acting factor that regulates mRNA stability in plants is the 50 kDa protein (called PRP-BP) that specifically binds a sequence in the 3'-UTR of the *PvPRP1* transcript of common bean [97]. As discussed earlier, *PvPRP1* mRNA is destabilized when bean cells are treated with fungal elicitor. Binding activity of PRP-BP increases markedly in elicitor-treated cells prior to rapid *PvPRP1* mRNA degradation, suggesting that binding of the protein to the 3'-UTR functions in the destabilization of *PvPRP1* mRNA [97]. Perhaps the most interesting aspect of the binding activity is that it is apparently regulated by the redox state of the protein. If cellular extracts are treated with the reducing agent dithiothreitol (DTT) and then assayed for binding to the *PvPRP* 3'-UTR, extracts from elicited or unelicited cells exhibit comparably high levels of PRP-BP activity [97]. Conversely, treatment of cellular extracts with oxidizing agents inhibits PRP-BP activity *in vitro*. Taken together, the data suggest that exposure of bean cells to fungal elicitors causes a reduction of PRP-BP, whereupon the protein binds to the *PvPRP1* 3'-UTR, thereby destabilizing the transcript [97].

PRP-BP and other yet to be identified RNA-binding proteins that trigger decay of specific plant mRNAs may themselves function as endoribonucleases or they may facilitate recognition of the transcript by ribonucleases (RNases). In any event, RNA-degrading activities must figure prominently in mRNA decay pathways. Many plant RNases have been studied and characterized, but those RNases that play a role in mRNA decay have not been differentiated from those with other roles in RNA metabolism, albeit several secretory RNases likely fit into the latter category [1, 31]. Recently, a genetic approach to this problem was undertaken, in which mutant *Arabidopsis* plants with altered RNase profiles were isolated using a substrate-based gel assay (M.L. Abler and P.J. Green, unpublished data). The isolated mutants affect several different RNases by increasing, decreasing, or eliminating particular activities. By examining the metabolism of reporter transcripts and endogenous RNAs in these mutants it should be possible to learn whether the affected RNases participate in the mRNA decay process.

Although *trans*-acting determinants of mRNA stability are usually thought to be proteins, RNAs can also function in this capacity in at least some antisense mechanisms. More often than not, sense mRNA accumulates to low levels in plants containing antisense transcripts. The simplest hypothesis to explain these results is that repression of expression by antisense RNA involves the formation of sense:antisense RNA duplexes, which are rapidly degraded [5]. In the case of transgenic tobacco expressing antisense *rbcS* RNA, it has been determined that the antisense RNA decreases the stability of the sense transcript [43]. Nuclear run-on transcription assays indicate that the *rbcS* sense and antisense RNAs are transcribed at about the same rate, and mRNA half-life studies after cordycepin treatment indicate that sense *rbcS* mRNA is less stable in plants expressing antisense RNA than in wild-type plants [43]. It is tempting to speculate that a dsRNase or dsRNA-binding protein mediates this effect.

Plant mRNA decay mechanisms

One of the most effective strategies for elucidating mRNA decay mechanisms is to analyze the structure and formation of decay intermediates. Unfortunately, for most plant and other eukaryotic mRNAs, no degradation intermediates can be detected on RNA gel blots, implying that once degradation is initiated, it is extremely rapid. However, two plant mRNAs, oat *PHYA* and soybean *SRS4*, appear to be exceptions to this rule, and the analysis of these transcripts has provided most of our insight into the mechanistic nature of mRNA decay pathways in higher plants.

In both soybean and transgenic petunia, a series of discrete fragments of the *SRS4* transcript are evident on RNA gel blots [82]. Several observations argue that these fragments are bonafide degradation intermediates. First, the same *SRS4* RNA fragments accumulate when the gene is transcribed under the control of either the CaMV 35S or the *SRS4* promoter [82]. Second, the fragments are polysome associated [82]. Third, experiments adding tracer RNAs to homogenized samples indicate that the fragments do not arise during RNA purification steps *in vitro* [82]. Finally, the addition of *in vitro* synthesized *SRS4* RNA to a cell-free mRNA decay system containing polysomes [78] or an S150 extract [79] generates the same degradation products observed *in vivo*. S1 nuclease and primer extension mapping of the fragments indicates that most major proximal products with an intact 5' end can be matched with a distal product with an intact 3' end [79], indicating that each pair of products probably arose from endonuclease cleavage of the full-length *SRS4* mRNA, possibly directed by local secondary structure [79]. Further analysis showed that this endonuclease cleavage is independent of decapping or deadenylation of

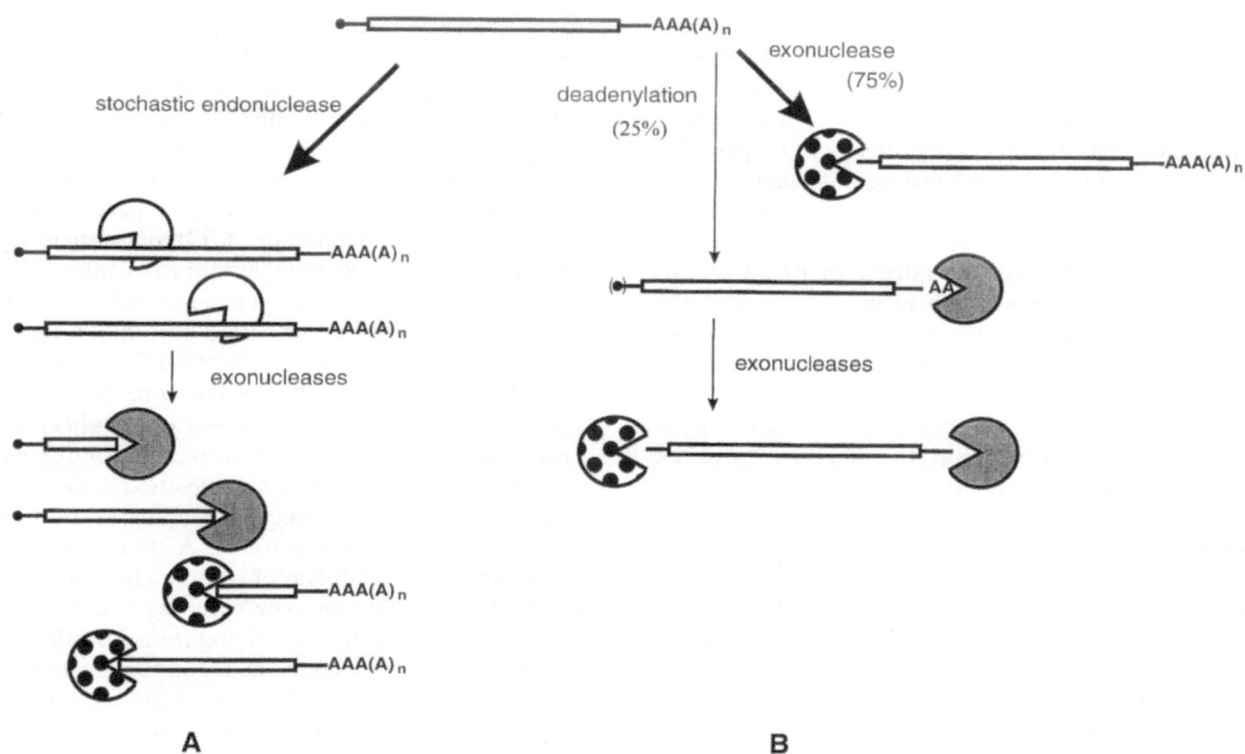

Figure 2. Proposed mechanisms of *SRS4* and *PHYA* mRNA decay in plants. The 5′ cap is represented by a black circle, solid lines indicate the 5′ leader and 3′-untranslated regions, coding regions are represented with rectangles, and AAA(A)$_n$ represents the poly(A) tail of the mRNA. White enzymes are stochastic endonucleases, gray and spotted enzymes are exonucleases. A depicts stochastic endonucleolytic decay without prior deadenylation as proposed for *SRS4* mRNA [79]. B depicts exonucleolytic decay pathways proposed for *PHYA* mRNA [36]. Parentheses around the 5′ cap indicate that the presence or absence of the cap has not been determined. Heavy arrows in A and B represent the major degradation pathways for the soybean *SRS4* [79] and oat *PHYA* [36] transcripts, respectively.

the transcript. These and other data led to the model in Fig. 2A for the decay of *SRS4* RNA [79]. In this model, decay is initiated by a stochastic endonuclease and then the proximal and distal products are subjected to 3′ → 5′ or 5′ → 3′ exonuclease digestion, respectively.

Similar to the situation for the *SRS4* mRNA, evidence indicates that fragments of the oat *PHYA* mRNA that can be observed on RNA gel blots are *in vivo* generated degradation intermediates. The *PHYA* mRNA fragments are observed with different isolation procedures when other endogenous or exogenously added RNAs remain intact [36]. *PHYA* mRNA fragments are also present in RNA isolated from a polysome-based *in vitro* system, and are associated with polysomes *in vivo* as well. However, unlike the *SRS4* RNA fragments which accumulate as discrete products, the *PHYA* fragments form a continuous distribution that ranges in size from 4.2 kb to about 200 nucleotides

[36]. The analysis of these fragments with different probes has led to the proposal that they arise from two pathways that primarily involve exoribonuclease activities, as shown in Fig. 2B. In this model, about 75% of the molecules are degraded before removal of the poly(A) tail by a 5′ → 3′ exonuclease, whereas 25% are deadenylated prior to degradation by the combined action of 5′ → 3′ and 3′ → 5′ exoribonucleases [36]. Although an alternative model involving a stochastic endonuclease can be invoked to explain *PHYA* mRNA decay, this seems less likely due to the continuous distribution of RNA fragments [36].

The decay of *SRS4* mRNA and most of the *PHYA* mRNA differs significantly from the best understood mechanism for eukaryotic mRNA decay, the deadenylation-dependent decapping pathway elucidated in yeast [13, 14]. In this yeast pathway, the poly(A) tail is removed, perhaps by a poly(A) nuclease [49], and this event triggers the removal of the

5' cap by a decapping enzyme. The decapped mRNA is then degraded by XRN1, a $5' \rightarrow 3'$ exoribonuclease [13, 41, 53]. Only the deadenylated fraction (25%) of the *PHYA* mRNA would appear to be a good candidate for decay by this type of pathway. A deadenlyation-independent decapping pathway also exists in yeast [55]; the existence of such a pathway in plants could explain the decay of the majority of the *PHYA* mRNA, but the dependence of *PHYA* mRNA decay on decapping remains to be investigated. The mechanism responsible for the decay of *SRS4* mRNA appears to be novel. In the future it will be important to determine whether the *SRS4* or the *PHYA* mRNA decay pathways are representative of the decay pathways of most other plant mRNAs that do not give rise to visible degradation intermediates.

Translation and mRNA stability

Accumulating evidence indicates that mRNA decay mechanisms are often coupled to translation in plants [76] and other organisms [66]. With respect to general determinants, the cap and poly(A) tail, which contribute to the stability of mRNAs, are even more important to assure efficient translation (see the chapter by Gallie in this volume). Therefore, mechanisms that remove these elements would be expected to decrease gene expression at the levels of both translation and mRNA stability. Another observation that links translation and mRNA stability is the stabilizing effect that the protein synthesis inhibitor cycloheximide has on a number of unstable mRNAs [76]. This could occur because translation of a labile *trans*-acting factor, or translation of the mRNA itself, is required for rapid degradation. The decay of mRNA is likely to take place, at least in part, on polyribosomes because the decay intermediates of the *SRS4* and *PHYA* mRNAs have been found to be polysome associated [36, 82] and most *in vitro* decay systems are polysome based [9, 66, 78]. Unfortunately, it has been difficult to develop *in vitro* decay systems that are also translation competent, so most direct data about the influence of translation on mRNA stability have come from other approaches. For example, in the case of the pea *Fed-1* gene, several mutations that block translational initiation or elongation have been examined for their effects on light-regulated transcript accumulation [16]. Such mutations abolish the light response of *Fed1*, consistent with the model of translation being a requirement for the stabilization of *Fed1* in the light (discussed above). Restoration of the reading

frame, even when different amino acids are substituted, restores the light response [16].

Compelling evidence for a strong connection between translation and mRNA stability in plants comes from studies of the effect of premature nonsense codons that occur naturally or have been engineered into plant genes. Key examples have been the major Kunitz trypsin inhibitor (*KTi3*) transcript in soybeans, and the phytohemagglutinin *PHA* (also called *Pdlec1*) transcript in common bean. A KTi3-null line of soybean contains far less transcript than a line of soybean producing normal levels of KTi3, even though the KTi3 gene is transcribed at the same rate in both lines, as measured by nuclear run-on analysis [44]. The only differences between the normal and null genes are a transversion and a two-base deletion in the latter. The deletion causes a frameshift leading to four stop codons in the new reading frame. A similar situation was described for a cultivar of common bean in which the phytohemagglutinin gene contains a single base pair deletion that shifts the reading frame, resulting in a premature nonsense codon at codon 53. The level of this mutant mRNA is greatly reduced compared to that of wild type [88]. The *PHA* promoter was shown to be fully functional by exchanging it with the homologous region of a normal gene and assaying lectin accumulation in seeds of transgenic tobacco [87]. Repair of the frameshift mutation led to normal levels of mRNA and protein in transgenic tobacco [87]. Taken together, the above data strongly suggested that premature translation termination was causing destabilization of the *KTi3* and *PHA* transcripts.

More recent data have directly established that premature nonsense codons decrease *PHA* mRNA accumulation at the level of stability [85a]. Half-life measurements in stably transformed tobacco cell lines showed that transcripts containing the *PHA* frameshift mutation discussed above (leading to a nonsense codon at codon 53) or simply a nonsense mutation in codon 53 were markedly less stable than wild-type transcripts. Transgenic tobacco and *Arabidopsis* plants also showed lower accumulation of nonsense-containing transcripts relative to wild-type, as would be expected for less stable mRNAs [85a]. These experiments establish the existence of a nonsense-mediated decay pathway in plants. At present it is unknown how many transcripts are substrates for the same or related pathways, but it is likely that many plant transcripts may be subject to accelerated decay when nonsense mutations arise within them. Several plant transcripts in addition to KTi3 have been reported to decrease in abundance due to nonsense or frameshift mutations [85, 88]. In addition, nonsense-mediated decay is known to occur in other eukaryotic organisms, and mutations blocking the pathway in yeast and *Caenorhabditis elegans* increase the accumulation of multiple nonsense-containing transcripts [63, 65]. The purpose of this mechanism may be to help organisms avoid the production of potentially deleterious, truncated proteins.

To further characterize nonsense-mediated decay in plants, *PHA* transcripts containing nonsense codons at different positions within the coding region were examined. Transcripts with nonsense codons at 20%, 40%, and 60% of the way through the coding region were found to be as unstable as the original frameshifted transcript, while a *PHA* mRNA with a nonsense codon 80% of the way through the coding region was at least as stable as wild-type *PHA* mRNA [85a]. Apparently the relevant degradation machinery recognizes the 60% stop codon as abnormal, but makes no distinction between the 80% stop codon and the normal stop codon. Possible explanations for the above data are that the degradation machinery may recognize a *cis* element between 60% and 80% of the transcript that is normally masked by ribosomes, or there may be a requirement for translation of the region between 60% and 80% of the coding region in order to maintain transcript stability. Interestingly, the insertion of segments of rare codons early in the *PHA*-coding region is insufficient to destabilize the mRNA (A. van Hoof and P.J. Green, unpublished), suggesting that rare codons cannot substitute for premature nonsense codons to induce rapid decay. Although it has been suggested that rare codons cause mRNA instability of transcripts in plants [56] and other eukaryotes [40], support to date derives only from circumstantial evidence.

Conclusions and future prospects

The past five years has been perhaps the most fruitful period yet in terms of research in the area of plant mRNA stability. The first *cis*-acting determinants of rapid decay have been identified and found to include seemingly novel elements (from SAUR transcripts) as well as those that function in other eukaryotes (AUUUA repeats and premature nonsense codons). It is now clear that stimuli that have unique effects on plants, such as light and elicitor treatment, can act at the level of mRNA stability, and important candidates for *cis*- and *trans*-acting determinants that mediate these effects have also been identified. mRNA decay

pathways have been proposed and our understanding of the coupling between translation and mRNA decay has increased dramatically. These advances are due in part to the development of improved methodology, but also to increased enthusiasm now that it has become apparent that the control of mRNA stability likely contributes to the expression of many plant genes. The future promises to be even more exciting as these directions are continued and additional tools are employed to elucidate mRNA decay mechanisms. For example, genetic screens for mRNA decay mutants are feasible in plants, particularly those plants with small genomes like *Arabidopsis* where loss of function mutations can be recovered at high frequency. As discussed earlier, one genetic screen has already identified several mutations that affect various RNase activities, some of which could function in mRNA decay. Selection schemes can also be designed to isolate mutations that stabilize a selectable marker transcript containing a specific instability sequence. Depending on which instability sequence is chosen for these experiments, the results could be unique to plants, or they could be of broad significance. Together with *in vitro* studies, these efforts should greatly enhance our understanding of pathways of mRNA decay and how those pathways are regulated.

Acknowledgements

We are grateful to our many colleagues who submitted preprints and reprints, to Drs Michael Sullivan and Matthew Tanzer for comments on the manuscript, to Ambro van Hoof for help with Fig. 2 and comments on the manuscript, and to Ms Karen Bird for editorial assistance. Work in the authors' laboratory was supported by grants from the DOE (DE-FG02–91ER20021) and USDA (9301155) to P.J.G. M.L.A was supported, in part, by an NIH postdoctoral fellowship (5F32GM16886).

References

1. Bariola PA, Green PJ: Plant Ribonucleases. In: Riordan JF, D'Alessio G (eds) Ribonucleases: Structure and Function, Academic Press, Orlando, FL (1996), In press.
2. Belanger FC, Brodl MR, Ho TD: Heat shock causes destabilization of specific mRNAs and destruction of endoplasmic reticulum in barley aleurone cells. Proc Natl Acad Sci USA 83: 1354–1358 (1986).
3. Belostotsky DA, Meagher RB: Differential organ-specific expression of three poly(A)-binding-protein genes from *Ara-*

76

bidopsis thaliana. Proc Natl Acad Sci USA 90: 6686–6690 (1993).

4. Binder R, Horowitz JA, Basilion JP, Koeller DM, Klausner RD, Harford JB: Evidence that the pathway of transferrin receptor mRNA degradation involves an endonucleolytic cleavage within the 3′ UTR and does not involve poly(A) tail shortening. EMBO J 13: 1969–1980 (1994).

5. Bourque JE: Antisense strategies for genetic manipulations in plants. Plant Sci 105: 125–149 (1995).

6. Brock ML, Shapiro DJ: Estrogen stabilizes vitellogenin mRNA against cytoplasmic degradation. Cell 34: 207–214 (1983).

7. Brodl MR, Ho TD: Heat shock causes selective destabilization of secretory protein mRNAs in barley aleurone cells. Plant Physiol 96: 1048–1052 (1991).

8. Brodl MR, Ho TD: Heat shock in mechanically wounded carrot root disks causes destabilization of stable secretory protein mRNA and dissociation of endoplasmic reticulum lamellae. Physiol Plant 86: 253–262 (1992).

9. Byrne DH, Seeley KA, Colbert JT: Half-lives of oat mRNAs *in vivo* and in a polysome-based *in-vitro system.* Planta 189: 249–256 (1993).

10. Caponigro G, Parker R: Multiple functions for poly(A)-binding protein in mRNA decapping and deadenylation in yeast. Genes Devel 9: 2421–2432 (1995).

11. De Rocher EJ, Bohnert HJ: Development and environmental stress employ different mechanisms in the expression of a plant gene family. Plant Cell 5: 1611–1625 (1993).

12. Dean C, Favreau M, Bond-Nutter D, Bedbrook J, Dunsmuir P: Sequences downstream of translation start regulate quantitative expression of two petunia *rbcS* genes. Plant Cell 1: 201–208 (1989).

13. Decker CJ, Parker R: A turnover pathway for both stable and unstable mRNAs in yeast: evidence for a requirement for deadenylation. Genes Devel 7: 1632–1643 (1993).

14. Decker CJ, Parker R: Mechanisms of mRNA degradation in eukaryotes. Trends Biochem Sci 19: 336–340 (1994).

15. Dickey LF, Gallo-Meagher M, Thompson WF: Light regulatory sequences are located within the 5′ portion of the *Fed-1* message sequence. EMBO J 11: 2311–2317 (1992).

16. Dickey LF, Nguyen T-T, Allen GC, Thompson WF: Light modulation of ferredoxin mRNA abundance requires an open reading frame. Plant Cell 6: 1171–1176 (1994).

17. Diehn SH, De Rocher EJ, Green PJ: Problems that can limit the expression of foreign genes in plants: lessons to be learned from *B.t.* toxin genes. In: Setlow JK (ed) Genetic Engineering: Principles and Methods, Plenum Press, New York (1996), In press.

18. Dietrich RA, Radke SE, Harada JJ: Downstream DNA sequences are required to activate a gene expressed in the root cortex of embryos and seedlings. Plant Cell 4: 1371–1382 (1992).

19. Flores S, Tobin EM: Cytokinin modulation of LHCP mRNA levels: the involvement of post-transcriptional regulation. Plant Mol Biol 11: 409–415 (1988).

20. Franco AR, Gee MA, Guilfoyle TJ: Induction and superinduction of auxin-responsive mRNAs with auxin and protein synthesis inhibitors. J Biol Chem 265: 15845–15849 (1990).

21. Fritz CC, Herget T, Wolter FP, Schell J, Schreier PH: Reduced steady-state levels of *rbcS* mRNA in plants kept in the dark are due to differential degradation. Proc Natl Acad Sci USA 88: 4458–4462 (1991).

22. Fu H, Kim SY, Park WD: High-level tuber expression and sucrose inducibility of a potato *Sus4* sucrose synthase gene

23. Gallie DR: The cap and poly(A) tail function synergistically to regulate mRNA translational efficiency. Genes Devel 5: 2108–2116 (1991).

24. Gallie DR: Posttranscriptional regulation of gene expression in plants. Annu Rev Plant Physiol Plant Mol Biol 44: 77–105 (1993).

25. Gallie DR, Feder JN, Schimke RT, Walbot V: Posttranscriptional regulation in higher eukaryotes: the role of the reporter gene in controlling expression. Mol Gen Genet 228: 258–264 (1991).

26. Gallie DR, Lucas WJ, Walbot V: Visualizing mRNA expression in plant protoplasts: factors influencing efficient mRNA uptake and translation. Plant Cell 1: 301–311 (1989).

27. Gil P, Green PJ. Multiple regions of the *Arabidopsis SAUR-AC1* gene control transcript abundance: the 3′ untranslated region functions as an mRNA instability determinant. EMBO J 15: 1678–1686 (1996).

28. Gil P, Liu Y, Orbovic V, Verkamp E, Poff KL, Green PJ: Characterization of the auxin-inducible *SAUR-AC1* gene for use as a molecular genetic tool in *Arabidopsis.* Plant Physiol 104: 777–784 (1994).

29. Gossen M, Bujard H: Tight control of gene expression in mammalian cells by tetracycline-responsive promoters. Proc Natl Acad Sci USA 89: 5547–5551 (1992).

30. Green PJ: Control of mRNA stability in higher plants. Plant Physiol 102: 1065–1070 (1993).

31. Green PJ: The ribonucleases of higher plants. Annu Rev Plant Physiol Plant Mol Biol 45: 421–445 (1994).

32. Greenberg JR: High stability of messenger RNA in growing cultured cells. Nature 240: 102–104 (1972).

33. Greenberg ME, Belasco JG: Control of the decay of labile protooncogene and cytokine mRNAs. In: Belasco JG, Brawerman G (eds) Control of Messenger RNA Stability, pp. 199–218. Academic Press, San Diego, CA (1993).

34. Hargrove JL, Hulsey MG, Beale EG: The kinetics of mammalian gene expression. BioEssays 13: 667–674 (1991).

35. Higgs DC, Barnes LJ, Colbert JT: Abundance and half-life of the distinct oat phytochrome A3 and A4 mRNAs. Plant Mol Biol 29: 367–377 (1995).

36. Higgs DC, Colbert JT: Oat phytochrome A mRNA degradation appears to occur via two distinct pathways. Plant Cell 6: 1007–1019 (1994).

37. Hilson P, Carroll KL, Masson PH: Molecular characterization of PAB2, a member of the multigene family coding for poly(A)-binding proteins in *Arabidopsis thaliana.* Plant Physiol 103: 525–533 (1993).

38. Ho TD: On the mechanism of hormone controlled enzyme formation in barley aleurone layers. Ph.D. thesis, Michigan State University, East Lansing, MI (1976).

39. Ho TD, Varner JE: Response of barley aleurone layers to abscisic acid. Plant Physiol 57: 175–178 (1976).

40. Hoekema A, Kastelein RA, Vasser M, DeBoer HA: Codon replacement in the *PGK1* gene of *Saccharomyces cerevisiae*: experimental approach to study the role of biased codon usage in gene expression. Mol Cell Biol 7: 2914–2924 (1987).

41. Hsu CL, Stevens A: Yeast cells lacking 5′ → 3′ exoribonuclease 1 contain mRNA species that are poly(A) deficient and partially lack the 5′ cap structure. Mol Cell Biol 13: 4826–4835 (1993).

42. Ingelbrecht IW, Herman LMF, Dekeyser RA, Van Montagu MC, Depicker AG: Different 3′ end regions strongly influence

the level of gene expression in plant cells. Plant Cell 1: 671–680 (1989).

43. Jiang C-Z, Kliebenstein D, Ke N, Rodermel S: Destabilization of *rbcS* sense transcripts by antisense RNA. Plant Mol Biol 25: 569–576 (1994).

44. Jofuku KD, Schipper RD, Goldberg RB: A frameshift mutation prevents Kunitz trypsin inhibitor mRNA accumulation in soybean embryos. Plant Cell 1: 427–435 (1989).

45. Koshiba T, Ballas N, Wong LM, Theologis A: Transcriptional regulation of *PS-IAA4/5* and *PS-IAA6* early gene expression by indoleacetic acid and protein synthesis inhibitors in pea (*Pisum sativum*). J Mol Biol 253: 396–413 (1995).

46. Lagnado CA, Brown CY, Goodall GJ: AUUUA is not sufficient to promote poly(A) shortening and degradation of an mRNA: the functional sequence within AU-rich elements may be UUAUUUA(U/A)(U/A). Mol Cell Biol 14: 7984–7995 (1994).

47. Larkin JC, Oppenheimer DG, Pollock S, Marks MD: Arabidopsis *GLABROUS1* gene requires downstream sequences for function. Plant Cell 5: 1739–1748 (1993).

48. Lincoln JE, Fischer RL: Diverse mechanisms for the regulation of ethylene-inducible gene expression. Mol Gen Genet 212: 71–75 (1988).

49. Lowell JE, Rudner DZ, Sachs AB: 3′-UTR-dependent deadenylation by the yeast poly(A) nuclease. Genes Devel 6: 2088–2099 (1992).

50. Marcotte WRJ, Russell SH, Quatrano RS: Abscisic acid-responsive sequences from the Em gene of wheat. Plant Cell 1: 969–976 (1989).

51. Margossian LJ, Federman AD, Giovannoni JJ, Fischer RL: Ethylene-regulated expression of a tomato fruit ripening gene encoding a proteinase inhibitor I with a glutamic residue at the reactive site. Proc Natl Acad Sci USA 85: 8012–8016 (1988).

52. McClure BA, Hagen G, Brown CS, Gee MA, Guilfoyle TJ: Transcription, organization, and sequence of an auxin-regulated gene cluster in soybean. Plant Cell 1: 229–239 (1989).

53. Muhlrad D, Decker CJ, Parker R: Deadenylation of the unstable mRNA encoded by the yeast *MFA2* gene leads to decapping followed by 5′ → 3′ digestion of the transcript. Genes Devel 8: 855–866 (1994).

54. Muhlrad D, Parker R: Mutations affecting stability and deadenylation of the yeast *MFA2* transcript. Genes Devel 6: 2100–2111 (1992).

55. Muhlrad D, Parker R: Premature translational termination triggers mRNA decapping. Nature 370: 578–581 (1994).

56. Murray EE, Rocheleau T, Eberle M, Stock C, Sekar V, Adang M: Analysis of unstable RNA transcripts of insecticidal crystal protein genes of *Bacillus thuringiensis* in transgenic plants and electroporated protoplasts. Plant Mol Biol 16: 1035–1050 (1991).

57. Newman TC, Ohme-Takagi M, Taylor CB, Green PJ: DST sequences, highly conserved among plant *SAUR* genes, target reporter transcripts for rapid decay in tobacco. Plant Cell 5: 701–714 (1993).

58. Nolan RC, Ho TD: Hormonal regulation of α-amylase expression in barley aleurone layers. Plant Physiol 88: 588–593 (1988).

59. Nolan RC, Ho TD: Hormonal regulation of gene expression in barley aleurone layers. Planta 174: 551–560 (1988).

60. Ohme-Takagi M, Taylor CB, Newman TC, Green PJ: The effect of sequences with high AU content on mRNA stability in tobacco. Proc Natl Acad Sci USA 90: 11811–11815 (1993).

61. Or E, Boyer SK, Larkins BA: *opaque2* modifiers act post-transcriptionally and in a polar manner on-zein gene expression in maize endosperm. Plant Cell 5: 1599–1609 (1993).

62. Peltz SW, Brewer G, Bernstein P, Hart PA, Ross J: Regulation of mRNA turnover in eukaryotic cells. Crit Rev Euk Gene Exp 1: 99–126 (1991).

63. Peltz SW, He F, Welch E, Jacobson A: Nonsense-mediated mRNA decay in yeast. Prog Nucl Acid Res Mol Biol 47: 271–298 (1994).

64. Peters JL, Silverthorne J: Organ-specific stability of two *Lemna rbcS* mRNAs is determined primarily in the nuclear compartment. Plant Cell 7: 131–140 (1995).

65. Pulak R, Anderson P: mRNA surveillance by the *Caenorhabditis elegans smg* genes. Genes Devel 7: 1885–1897 (1993).

66. Ross J: mRNA stability in mammalian cells. Microbiol Rev 59: 423–450 (1995).

67. Sachs AB: Messenger RNA degradation in eukaryotes. Cell 74: 413–421 (1993).

68. Sachs AB, Deardorff JA: Translation initiation requires the PAB-dependent poly(A) ribonuclease in yeast. Cell 70: 961–973 (1992).

69. Seeley KA, Byrne DH, Colbert JT: Red light-independent instability of oat phytochrome mRNA *in vivo*. Plant Cell 4: 29–38 (1992).

70. Sheu J-J, Jan S-P, Lee H-T, Yu S-M: Control of transcription and mRNA turnover as mechanisms of metabolic repression of α-amylase gene expression. Plant J 5: 655–664 (1994).

71. Shirley BW, Meagher RB: A potential role for RNA turnover in the light regulation of plant gene expression: ribulose–1,5-bisphosphate carboxylase small subunit in soybean. Nucl Acids Res 18: 3377–3385 (1990).

72. Shyu A-B, Belasco JG, Greenberg ME: Two distinct destabilizing elements in the c-*fos* message trigger deadenylation as a first step in rapid mRNA decay. Genes Devel 5: 221–231 (1991).

73. Shyu A-B, Greenberg ME, Belasco JG: The c-*fos* transcript is targeted for rapid decay by two distinct mRNA degradation pathways. Genes Dev 3: 60–72 (1989).

74. Siflow CD, Key JL: Stability of polysome-associated poly-adenylated RNA from soybean suspension culture cells. Biochemistry 18: 1013–1018 (1979).

75. Silverthorne J, Tobin EM: Post-transcriptional regulation of organ-specific expression of individual *rbcS* mRNAs in *Lemna gibba*. Plant Cell 2: 1181–1190 (1990).

76. Sullivan ML, Green PJ: Post-transcriptional regulation of nuclear-encoded genes in higher plants: the roles of mRNA stability and translation. Plant Mol Biol 23: 1091–1104 (1993).

77. Surdej P, Riedl A, Jacobs-Lorena M: Regulation of mRNA stability in development. Annu Rev Genet 28: 263–282 (1994).

78. Tanzer MM, Meagher RB: Faithful degradation of soybean *rbcS* mRNA *in vitro*. Mol Cell Biol 14: 2640–2650 (1994).

79. Tanzer MM, Meagher RB: Degradation of the soybean ribulose–1,5-bisphosphate carboxylase small-subunit mRNA, SRS4, initiates with endonucleolytic cleavage. Mol Cell Biol 15: 6641–6652 (1995).

80. Taylor CB, Green PJ: Identification and characterization of genes with unstable transcripts (*GUTs*) in tobacco. Plant Mol Biol 28: 27–38 (1995).

81. Thompson DM, Meagher RB: Transcriptional and post-transcriptional processes regulate expression of RNA encoding the small subunit of ribulose–1,5-bisphosphate carboxylase differently in petunia and in soybean. Nucl Acids Res 18: 3621–3629 (1990).

78

82. Thompson DM, Tanzer MM, Meagher RB: Degradation products of the mRNA encoding the small subunit of ribulose–1,5-bisphosphate carboxylase in soybean and transgenic petunia. Plant Cell 4: 47–58 (1992).

83. Thompson WF, White MJ: Physiological and molecular studies of light-regulated nuclear genes in higher plants. Annu Rev Plant Physiol Plant Mol Biol 42: 423–466 (1991).

84. Thornburg RW, An G, Cleveland TE, Johnson R, Ryan CA: Wound-inducible expression of a potato inhibitor II-chloramphenicol acetyltransferase gene fusion in transgenic tobacco plants. Proc Natl Acad Sci USA 84: 744–748 (1987).

85. Vancanneyt G, Rosahl S, Willmitzer L: Translatability of a plant-mRNA strongly influences its accumulation in transgenic plants. Nucl Acids Res 18: 2917–2921 (1990).

85a. Van Hoof A, Green PJ: Premature nonsense codons decrease the stability of phytophemagglutinin mRNA in a position-dependent manner. Plant J, in press.

86. Vierling E: The roles of heat shock proteins in plants. Annu Rev Plant Physiol Plant Mol Biol 42: 579–620 (1991).

87. Voelker TA, Moreno J, Chrispeels MJ: Expression analysis of a pseudogene in transgenic tobacco: a frameshift mutation prevents mRNA accumulation. Plant Cell 2: 255–261 (1990).

88. Voelker TA, Staswick P, Chrispeels MJ: Molecular analysis of two phytohemagglutinin genes and their expression in *Phaseolus vulgaris* cv. Pinto, a lectin-deficient cultivar of the bean. EMBO J 5: 3075–3082 (1986).

89. Vorst O, Van Dam F, Weisbeek P, Smeekens S: Light-regulated expression of the *Arabidopsis thaliana ferredoxin A* gene involves both transcriptional and post-transcriptional processes. Plant J 3: 793–803 (1993).

90. Walker EL, Weeden NF, Taylor CB, Green PJ, Coruzzi GM. Molecular evolution of duplicate copies of genes encoding cytosolic glutamine synthetase in *Pisum sativum*. Plant Mol Biol 29: 1111–1125 (1996).

91. Weinmann P, Gossen M, Hillen W, Bujard H, Gatz C: A chimeric transactivator allows tetracycline-responsive gene expression in whole plants. Plant J 5: 559–569 (1994).

92. Weiss IM, Liebhaber SA: Erythroid cell-specific mRNA stability elements in the α2-globin 3' nontranslated region. Mol Cell Biol 15: 2457–2465 (1995).

93. Williamson JD, Quatrano RS: ABA-regulation of two classes of embryo-specific sequences in mature wheat embryos. Plant Physiol 86: 208–215 (1988).

94. Wolfraim LA, Langis R, Tyson H, Dhindsa RS: cDNA sequence, expression, and transcript stability of a cold acclimation-specific gene, *cas18*, of alfalfa (*Medicago falcata*) cells. Plant Physiol 101: 1275–1282 (1993).

95. Yamamoto KT, Mori H, Imaseki H: cDNA cloning of indole–3-acetic acid-regulated genes: Aux22 and SAUR from mung bean (*Vigna radiata*) hypocotyl tissue. Plant Cell Physiol 33: 93–97 (1992).

96. Yang J, Hunt AG: Immunological characterization of plant polyadenylate-binding proteins. Plant Sci 99: 161–170 (1994).

97. Zhang S, Mehdy MC: Binding of a 50-kD protein to a U-rich sequence in an mRNA encoding a proline-rich protein that is destabilized by fungal elicitor. Plant Cell 6: 135–145 (1994).

98. Zhang S, Sheng J, Liu Y, Mehdy MC: Fungal elicitor-induced bean proline-rich protein mRNA down-regulation is due to destabilization that is transcription and translation dependent. Plant Cell 5: 1089–1099 (1993).

99. Zubiaga AM, Belasco JG, Greenberg ME: The nonamer UUAUUUAUU is the key AU-rich sequence motif that mediates mRNA degradation. Mol Cell Biol 15: 2219–2230 (1995).

Plant Molecular Biology **32:** 79–88, 1996.
© 1996 *Kluwer Academic Publishers.*

RNA as a target and an initiator of post-transcriptional gene silencing in trangenic plants

David C. Baulcombe
The Sainsbury Laboratory, John Innes Centre, Norwich NR4 7UH, UK

Key words: RNA degradation, antisense RNA, RNA-dependent, RNA polymerase, DNA methylation, cosuppression, virus resistance

Contents

Abstract	79	Transgene methylation and PTGS	83
Introduction	79	The influence of endogenous gene	
Examples of post-transcriptional gene		transcription on PTGS	84
silencing (PTGS)	80	The influence of viruses on PTGS	84
Mechanisms of PTGS	81	The influence of transgene	
A role for antisense RNA	81	transcription on PTGS	84
How antisense RNA might mediate		One or more mechanisms to initiate	
PTGS	82	PTGS?	84
Initiation of PTGS	82	Somatic instability of PTGS	87
Direct transcription of antisense RNA	83	References	87

Abstract

Post-transcriptional gene silencing in transgenic plants is the manifestation of a mechanism that suppresses RNA accumulation in a sequence-specific manner. The target RNA species may be the products of transgenes, endogenous plant genes or viral RNAs. For an RNA to be a target it is necessary only that it has sequence homology to the sense RNA product of the transgene. There are three current hypotheses to account for the mechanism of post transcriptional gene silencing. These models all require production of an antisense RNA of the RNA targets to account for the specificity of the mechanism. There could be either direct transcription of the antisense RNA from the transgene, antisense RNA produced in response to over expression of the transgene or antisense RNA produced in response to the production of an aberrant sense RNA product of the transgene. To determine which of these models is correct it will be necessary to find out whether transgene methylation, which is frequently associated with the potential of transgenes to confer post-transcriptional gene silencing, is a cause or a consequence of the process.

Introduction

It is now well established that transgenes in plants may suppress expression of homologous endogenous genes or transgenes (reviewed recently in [5, 15, 27, 28, 35, 36, 46]). This homology-dependent gene silencing represents one of the most puzzling and potentially important phenomena in transgenic plants. It is puzzling because the current paradigms of gene regulation do not explain the mechanisms that could cause gene silencing. It may be important because these unknown mechanisms could represent genetic controls involved in plant growth, development and response to environmental factors. Gene silencing is also important because, potentially, it is a powerful mechanism of inactivating genes that reduce yield or quality of products from plants. Conversely, gene silencing is a complication for the exploitation of transgenic plants because uncontrolled it could suppress the expression of host genes and transgenes that are necessary for efficient crop production.

Gene silencing may operate at the level of transcription if the transgene shares homology with the promoter of the silenced gene. Homology in transcribed regions leads to post-transcriptional gene silencing (PTGS), which is the subject of this review. In the examples of PTGS discussed here the silencer transgene and the silenced genes are transcribed in the same orientation although, as discussed elsewhere [5], it is possible that PTGS and antisense suppression are related processes.

There are two phases of the process of PTGS. One phase is the suppression of RNA accumulation. RNA is involved at this phase as a target molecule. There is also an initiation phase which determines whether PTGS will be active. This initiation phase may also involve RNA. Before assessing the current information about these two phases it is first necessary to define how a gene silencing phenomenon is placed into the post-transcriptional rather than transcriptional category.

Examples of post-transcriptional gene silencing (PTGS)

The known and likely examples of PTGS in plants involve many different genes, plants and constructs (Table I). It is therefore likely that PTGS is a general phenomenon that could be produced in all plant species with most genes. The most reliable indicator of a post-transcriptional rather than a transcriptional mechanism is from transcription run-off analysis with isolated nuclei. In the absence of run-off data it may be inferred that the gene silencing is post-transcriptional, rather than transcriptional, if the silencer transgene and the silenced genes share homology in the transcribed region. However, if there is homology in both the promoter and the transcribed region there could be either transcriptional silencing or PTGS. For example, with the maize *A1* cDNA expressed in transgenic petunia, there was transcriptional silencing of the *A1* transgenes [39]. The silencer and silenced transgenes in these lines were identical in the promoter region as well as in the transcribed region and it is likely that this promoter homology led to the transcriptional gene silencing. The converse of this example is provided by tobacco plants carrying transgenes with 35S promoter of CaMV (35S)

coupled to the *uidA* reporter gene of *Escherichia coli* (GUS). These transgenes had PTGS activity despite the sequence identity in the 35S promoter region of the silencer and silenced loci [14].

It is also possible to use viruses to determine whether a gene silencing mechanism is transcriptional or post-transcriptional. This approach to the analysis of gene silencing is based on the finding that the mechanism of PTGS has the potential to suppress RNA viruses as well as RNA from nuclear genes. The antiviral potential of PTGS was established by transgenic expression of viral cDNAs in the sense orientation relative to the virus genome. In many instances the lines carrying these viral cDNA transgenes were specifically resistant to the virus from which the transgene was derived [33, 40, 49]. The resistance in these lines was associated with low level accumulation of the transgene RNA and the potential of the transgene to silence homologous loci at the post transcriptional level. Production of virus encoded proteins was not required for the resistance of silencing phenotypes [32, 40, 50]. Transgenic resistance associated with PTGS is referred to as homology-dependent resistance to reflect the specificity of the resistance mechanism for viruses with extreme sequence similarity to the sense RNA product of the transgene [40].

The association of PTGS and virus resistance was established initially in lines transformed with fragments of viral cDNA. Subsequently, using constructs based on the genome of potato virus X (PVX), it was shown that PTGS of non viral genes had the potential to suppress virus accumulation provided that there was sequence homology of the virus and the transgene. For example, tobacco plants displaying PTGS of GUS or neomycin phosphotransferase (NPT) were specifically resistant to PVX.GUS or PVX.NPT respectively and tomato plants with PTGS of polygalacturonase (PG) were specifically resistant to PVX.PG [14]. These data have implications for the analysis of the mechanism of PTGS that are discussed in more detail below. In addition these data illustrate how virus constructs may be useful for the diagnosis of PTGS in situations where it is difficult to carry out a nuclear run-off assay or where independent confirmation of the run-off data is required.

Table 1. Examples of post-transcriptional gene silencing in transgenic plants

Target genes	Plant	Construct[2]	Transcription assay[3]	References
Non-viral				
Chalcone synthase	petunia	nos	yes	[41, 51, 52]
Dihydroflavonol-4-reductase	petunia	nos	no	[52]
Homeotic gene-fbp1	petunia	nos	no	[1]
Polygalacturonase	tomato	nos	no	[47, 48]
Phytoene synthase	tomato	CaMV	no	[16]
Nopaline synthase	*Nicotiana*	CaMV	no	[18]
Neomycin phosphotransferase	*Nicotiana*	CHS	yes	[26]
β-glucuronidase (GUS)	*Nicotiana*	nos	yes	[13, 14, 25]
β-1, 3-glucanase	*Nicotiana*	nos	yes	[11]
Chitinase	*Nicotiana*	chitinase	no	[23]
S-adenosyl-L-methionine synthetase	*Nicotiana*	SAM	no	[6]
Nitrate reductase	*Nicotiana*	CaMV	no	[9]
Nitrite reductase	*Nicotiana*	NiR	no	[54]
Acetohydroxyacid synthase	*Nicotiana*	AHA	no	[7]
rolB	*Arabidopsis*	nos	yes	[12]
Viral[1]				
Potato leafroll virus	potato	nos	no	[29]
Tobacco etch virus	*Nicotiana*	CaMV, TML	yes	[17, 32, 33]
Potato virus X	*Nicotiana*	nos	yes	[40]
Tomato spotted wilt	tomato	nos	no	[44]
Potato virus Y	*Nicotiana*	nos	yes	[49]

[1] It is likely that there are many other examples of PTGS in transgenic plants expressing viral cDNAs [4]. The diagnostic features of these examples are that there is homology-dependent resistance and an irregular relationship between virus resistance and the accumulation of the viral transgene RNA.
[2] All constructs have the 35S promoter although there is PTGS of chalcone synthase in petunia transformed with promoterless constructs. The abbreviations refer to the transcriptional terminator/poly (A) addition site as follows: nos, nopaline synthase of *Agrobacterium tumefaciens*; CaMV, cauliflower mosaic virus; CHS, chalcone synthase; SAM, *S*-adenosyl-L-methionine synthase; NiR, nitrite reductase; AHA, acetohydroxyacid synthase; LMT, large tumour morphology gene of *Agrobacterium tumefaciens*.
[3] Indicates whether a transcription run-off assay has been carried out.

Mechanisms of PTGS

A role for antisense RNA

A transgene locus conferring PTGS can suppress host and viral RNAs in a sequence specific manner [33, 40] depending on sequence homology of the transgene and its RNA target. In principle, this specificity could result from a direct interaction of the transgene and the target RNA or because the transgene influences accumulation of a factor that interacts with the target RNA. That factor could be either protein or RNA.

It is unlikely that the transgene DNA has the potential to interact with the target RNAs directly because, at least with the viral RNAs, they are not in the nucleus [33, 40]. It is similarly unlikely that the transgene influ-

ences a protein that mediates the silencing mechanism because PTGS is conferred by transgenes encoding untranslatable RNAs [32, 40, 50]. It is also fundamentally unlikely that every sequence with the potential to confer PTGS encodes a protein that can bind to its RNA template. This leaves the RNA product of the transgene as the most plausible mediator of PTGS.

If this RNA mediator has the same polarity as the target RNA there would be a potential for sequence-specific interaction in regions of RNA secondary structure. The RNA mediator could displace the base-paired RNA. However there is no evidence that the target sequences of PTGS are those with RNA secondary structure. An antisense RNA mediator with polarity opposite to that of the target RNA could easily explain the specificity of PTGS of nuclear and viral genes. Others have also come to the conclusion that an antisense RNA is a necessary cofactor in PTGS [20, 33, 40].

There have been some attempts to detect antisense RNA associated with PTGS. In some instances there was evidence for antisense RNA but not, so far, correlated with PTGS [51].

There are gaps in the reported searches for antisense RNA which do not account for the possibilities that the molecules may be small, heterodisperse or covalently associated with proteins. Given the almost inescapable conclusion that an antisense RNA is needed to confer sequence specificity on PTGS, a major research priority should be a search for these molecules. The subsequent sections of this review assess and interpret the known information about PTGS in terms of this proposed involvement of antisense RNA.

How antisense RNA might mediate PTGS

Antisense RNAs and oligonucleotides are known potent inhibitors of gene expression at several levels [42, 55]. These molecules anneal to the complementary sense RNA or DNA. Depending on the target sequence the antisense RNAs can inhibit translation or stability of corresponding sense RNAs. In principle, PTGS could be due to inhibition at either of these levels. A block at the level of translation could cause reduced accumulation of the target RNA because translation has an indirect effect on RNA stability [19]. A direct destabilization of sense RNA could be the consequence of duplex formation with the antisense RNA because double-stranded RNA may be a substrate for a double-stranded RNAase [42].

The effect of PTGS on accumulation of viral, as opposed to cellular, RNAs could be the consequence of antisense inhibition at either of these levels. However it might be expected that viral RNAs would be suppressed to a greater extent than cellular RNA because the duplex of viral and antisense RNA inhibits several processes specific for accumulation of viral RNAs. For example, an inhibition of translation could affect RNA stability indirectly, as discussed above. It could also have a direct effect on virus accumulation if the target gene encodes a protein required for accumulation of virus RNA. In addition, a region of duplex sense/antisense RNA could prevent a replicase enzyme having access to the viral genome and thereby interfere directly with the process of virus replication. Data showing that viral RNAs may be more sensitive to PTGS than cellular RNA was generated in the analysis of tobacco plants transformed with the RNA-dependent

RNA polymerase (RdRp) gene of PVX. These lines displayed PTGS of the transgene and were extremely resistant to PVX [40]. In one of these lines the transgene had the potential to reduce accumulation of cellular RNA by a factor of 8. In striking contrast, the suppression of viral RNA accumulation was at least ten thousand fold [34].

There is currently very little direct information about the target mechanisms of PTGS although there is evidence consistent with targeted degradation of RNA. For example, in tomato exhibiting PTGS of PG [48] and tobacco displaying PTGS of a tobacco etch virus (TEV) transgene [17] there were discrete-sized sense RNAs that were homologous to, and smaller than, the normal transgene RNA. These RNAs may be degradation products produced during the PTGS mechanism. From the discrete size of these RNAs it was suggested that the degradation mechanism preferentially cleaves defined sequence motifs or structures within the target region determined by the silencer transgene [17, 48]. However these putative motifs or structures could not be identified from the sequence at or around the cleavage sites [17].

The potential of cytoplasmic viruses of the potex- and poty-viral groups to be a target of PTGS implies that the mechanism is also cytoplasmic. Consistent with a cytoplasmic site of action is the finding that PTGS of β-glucanase in tobacco has no effect on the accumulation of the unspliced precursor of the β-glucanase mRNA [10]. However this analysis of pre mRNA is also consistent with suppression of maturation or transport of the nuclear pre-mRNA and it remains possible that the mechanism of PTGS is active in several compartments of the cell.

Initiation of PTGS

It is generally observed that PTGS is a feature of only a small proportion of lines transformed with any one construct although a modified 35S GUS construct did induce variable PTGS in all lines tested [13]. Typically the silencing was evident in 5–20% of lines. There are currently three models to account for the reason why the level of PTGS varies between lines (Fig. 1). All three models predict the formation of an antisense RNA to mediate the PTGS, as described above. The first proposes that a silencing transgene is integrated in the plant genome adjacent to endogenous promoters [20]. This endogenous promoter would direct transcription of the transgene directly into antisense RNA. The second model (Fig. 1) invokes a sensing mechanism that can detect high levels of transgene RNA. According to this model, once the mechanism senses accumulation of the transgene RNA above a certain threshold level, the transgene RNA would be copied into antisense RNA by an endogenous RdRp [33]. In the third model (Fig. 1) the distinctive feature of the silencing transgenes is that they produce an aberrant RNA which is a template for the host-encoded RdRp. The production of the aberrant RNA would therefore lead to accumulation of antisense RNA [14]. The following sections assess the evidence for and against these hypothetical mechanisms of PTGS.

Direct transcription of antisense RNA

It would be predicted, if PTGS is mediated by antisense RNA produced directly from promoters adjacent to the transgene, that transgenes designed to produce antisense RNA would confer PTGS of nuclear genes and virus resistance. Consistent with this prediction it has been found that nuclear genes are suppressed by antisense transgenes [42]. However virus resistance is conferred only rarely by transgenes designed to produce the antisense strand of positive strand RNA genomes [8, 22, 29]. It is more often found that antisense constructs do not confer resistance and therefore that antisense RNA produced by direct transcription from a transgene does not have the potential to confer the type of PTGS associated with virus resistance.

Further evidence against the direct transcription of antisense RNA is for transcription run-off analysis in petunia lines displaying PTGS of chalcone synthase (CHS) [51]. There was very little transcription of CHS antisense RNA and no relationship of these low levels of antisense CHS with the degree of PTGS. Other data that are not easily reconciled with the direct transcription of antisense RNA include the observations that PTGS may be transgene dosage-dependent [11, 17, 40, 54] or affected by expression of host genes with homology to the transgene [6, 9, 48]. Together, these various data indicate that direct transcription of antisense RNA is the least plausible of the three models.

Transgene methylation and PTGS

There are now several reports that transgene methylation is associated with PTGS [14, 21, 26, 49]. For example, in transgenic tobacco there was a correlation between the level of PTGS and the degree of methylation of an NPT transgene: the methylation was high in the progeny of a transformed plant that displayed the gene silencing phenotype [26]. Other examples in which there was an association of PTGS and transgene methylation were also in tobacco. These include lines displaying homology-dependent resistance and PTGS of a PVY transgene [49] or PTGS of a GUS transgene [14]. The methylation of the GUS transgene was in the 3' part of the gene which, as RNA, was the target of the gene silencing mechanism. This spatial correlation of transgene methylation and the target of PTGS reinforces the notion that these are associated phenomena [14].

As yet there is no clear indication whether transgene methylation is a cause or an effect of PTGS. However, PTGS of a GUS transgene in tobacco is relieved by the application of 5-azacytidine [43]. This agent is an inhibitor of DNA methylation, suggesting that the loss of PTGS was due to suppression of transgene methylation.

A role of transgene methylation as a causal factor in PTGS is consistent with the observations from tobacco and petunia that PTGS is more frequent in lines with multiple homologous transgenes than in lines with a single-copy insert [11, 17, 24, 25, 40, 54]. It is thought that multiple copies of homologous DNA at a single locus or at unlinked positions in the genome are able to interact through ectopic pairing [36]. The analyses of both plant and fungal systems indicates that ectopic pairing then leads to methylation of the interacting DNA [2, 3, 38, 39, 45]. An observation with a fungal system links this association of repeated DNA, transgene methylation and the aberrant RNA model of PTGS. It was shown in *Ascobolus immersus* that duplication of the 3' part of the *met2* gene led to methylation of the corresponding region of the intact endogenous gene. This methylation of the transcribed region of *met2* disrupted transcription so that the accumulated RNA was truncated in the 3' region [3].

English *et al.* [14] pointed out similarities in this fungal system and in transgenic tobacco with PTGS of GUS. There were multiple copies of the GUS transgenes in these lines and hypermethylation at the 3' end of the transgenes [14, 24, 25]. The same 3' region was the target, as RNA, of the mechanism of PTGS suggesting that the methylation of the transgene somehow influenced a sequence specific RNA degradation. The coincidence of the DNA methylation and target of PTGS could be explained if there was aberrant RNA produced from the methylated DNA, as in *Ascobolus immersus*, and if the aberrant RNA was the template for antisense RNA production [14].

The influence of endogenous gene transcription on PTGS

In some examples the initiation of PTGS is affected by transcription of host genes with homology to the silencer transgene. For example, constructs with the 35S promoter to transcribe PG and phytoene synthase (PE) sequences were not silenced until the endogenous PG and PE genes were activated during fruit ripening [47, 48]. After that stage both the endogenous genes and the homologous transgenes were subject to PTGS. Similarly in tobacco with nitrate reductase [9] or S-adenosyl methonine synthetase (SAM synthetase)[6] transgenes, the level of PTGS was increased under conditions in which host gene expression was enhanced. Clearly this effect of host gene expression is more easily reconciled with the threshold model than either of the alternative hypotheses.

The influence of viruses on PTGS

Homology-dependent resistance varies in strength between lines. In the most extreme examples there is no detectable accumulation of the target virus anywhere in the inoculated plant [31, 40, 44, 53]. In other examples the resistance is weaker and there may be virus accumulation at least on the inoculated leaf. One of these intermediate types of homology-dependent resistance involves TEV [17, 33]. Initially the plant was susceptible to the TEV and the viral cDNA transgene was expressed at a high level. The initially infected leaves showed the normal TEV symptoms. However, in the upper leaves of the plant the symptoms were attenuated or absent and there was little or no TEV accumulation. Accumulation of the transgene RNA was suppressed in this asymptomatic tissue and there was resistance to secondary infection provided that the inoculated virus was similar to the sense RNA of the transgene. This induced resistance was referred to as 'recovery' [33], although that term is not strictly accurate because the resistant tissue never showed signs of disease. It was suggested that the resistant state in these infected plants was related to PTGS mediated by an RNA threshold. The combined accumulation of homologous RNAs produced by the transgene and the virus would have contributed to this threshold.

Viral RNA accumulation may also influence endogenous plant gene (i.e. non-transgene) expression. When *Nicotiana benthamiana* was inoculated with a TMV genome carrying part of the *N. benthamiana* phytoene desaturase (PDS) gene there was evidence for suppression of the endogenous PDS expression [30]. On the inoculated leaf and the first leaves to be systemically infected there were mild mosaic symptoms typical of wild-type TMV. In contrast, in the upper parts of the plant there was photobleaching consistent with suppression of the host PDS sequence and consequent loss of carotenoid protection against photo-oxidation. This suppression of PDS was likely due to virus-induced PTGS of the homologous host genes and, like the recovery phenomenon, could be more easily explained by the threshold model than the direct antisense and aberrant RNA models.

The influence of transgene transcription on PTGS

The threshold hypothesis of PTGS leads to the prediction that high-level transcription of a transgene would be both necessary and sufficient for PTGS. Consistent with this prediction is the fact that most reports of PTGS involve the 35S promoter of CaMV which is transcribed at a high level throughout the plant (Table 1). Analyses of transcription in the examples of PTGS associated with homology-dependent resistance to TEV and PVY are also consistent with the threshold hypothesis. Transcription of the viral transgenes was higher in the resistant lines displaying PTGS than in susceptible lines in which the transgene RNA accumulated at high levels [17, 33]. The expression of PTGS in tobacco lines with single copy GUS transgenes is also consistent with the threshold model. In these lines the PTGS developed during later stages of plant growth [13]. There were high levels of GUS and GUS mRNA in the juvenile plants and it was suggested that these high levels of transgene expression exceeded the threshold necessary for activation of PTGS.

One or more mechanisms to initiate PTGS?

It is difficult to reconcile these various observations with any one of the three models of the mechanism for the initiation of PTGS. The direct production of antisense RNA does not fit the run-off analysis of antisense RNA production, methylation data or the influence of host gene transcription, virus infection or transgene transcription on PTGS. The threshold model does not obviously tie in with the transgene methylation data. It is also difficult to reconcile the threshold model with the finding that there is PTGS of chalcone synthase in petunia due to a transgene construct that is transcribed at a very low level, if at all [51]. Nor does the threshold model fit easily with several examples of PTGS with

35S constructs in which the silencer transgenes are transcribed at a similar or lower levels than loci that do not confer PTGS [12].

The aberrant RNA model can accommodate the methylation data and can explain why there is not necessarily a correlation of transgene transcription and PTGS. However this model is not consistent with the influence of endogenous genes on PTGS. Nor does it explain PTGS of GUS in tobacco lines with single copies of the GUS transgene [13]. There is no potential for ectopic pairing of homologous DNA in these lines and it is not obvious why these single-copy transgenes would produce aberrant RNA.

The failure of any one model to accommodate all of the experimental data must mean there are several mechanisms or that the models need further refinement. Two refinements to the aberrant RNA model have been recently described which may remove some of the inconsistencies referred to above and which may also unite aspects of the aberrant RNA model with the threshold model [5].

The first of these refinements addresses the initiation of PTGS when the genome carries multiple homologous sequences. According to the original model, the production of aberrant RNA follows from ectopic pairing of homologous DNA and the consequent methylation of transcribed regions [14]. The proposed refinement to the model is that the ectopic pairing would be a non-reciprocal interaction of homologous sequences. Only one of the interacting DNAs need be transcribed and methylated consequent to the ectopic pairing. This locus, referred to as the receptor locus, would be the template for abberant RNA production. The second silencer locus need not be transcribed or methylated by the interaction. The only requirement of the silencer locus is that it has the potential to pair ectopically with the receptor locus. There is a precedent for non-reciprocal ectopic pairing of DNA from the analysis of transcriptional gene silencing of the nopaline synthase promoter [37].

This refinement to the model was proposed to account for the potential of a non-transcribed chalcone synthase transgene to silence the endogenous gene. The modification could also accommodate the influence of endogenous genes on PTGS of PG [48] and PE [47] in tomato and SAM synthetase [6] and nitrate reductase in tobacco [9]. According to the modified model the transgenes in these examples would carry out the silencer role and the endogenous homologues the receptor function. The model thus allows for PTGS independently of transcription of the transgene.

The second refinement to the aberrant RNA model is similar to the first but proposes that, in certain situations, the role of the silencer locus could be carried out by RNA. This variation is proposed to account for the influence of viruses on PTGS. It would require that there is the potential for the viral RNA or the transgene RNA to interact with the homologous sequence in the nucleus. This RNA:DNA interaction would lead to DNA methylation and would initiate production of the aberrant RNA as proposed originally.

A precedent for DNA methylation due to an RNA:DNA interaction is provided by the analysis of plants carrying transgenes based on viroid cDNA [56]. Whenever those plants were infected and produced high levels of the viroid RNA there was methylation of the viroid cDNA transgenes. The transgenes were not methylated in the uninfected plants.

For RNA-directed methylation of DNA there must be the potential for the RNA to interact directly with the homologous DNA. Viroid RNAs accumulate in the nucleus and this type of interaction could occur throughout the plant. In contrast, viral RNAs accumulate in the cytoplasm of infected cells. Consequently, the potential for an interaction of viral RNA with homologous DNA would be restricted to dividing cells during the phase of nuclear membrane breakdown. This proposed requirement for an interaction of viral RNA with homologous DNA means that viruses would not activate PTGS in the inoculated tissue in which there would be little or no cell division. The PTGS should be manifest only in cells that undergo division and emerge from the meristem after the inoculated virus has spread through the plant. The development of the recovery phenotype in young leaves of plants displaying the induced form of homology-dependent resistance to TEV is consistent with this prediction [33]. Similarly, the virus-induced silencing of the host PDS gene is restricted to tissue that develops after inoculation and is also consistent with the proposed interaction of viral RNA and DNA in dividing cells [30].

An RNA:DNA interaction could also account for PTGS in the absence of virus infection when there is only a single transgene present in the plant genome and therefore no opportunity for ectopic pairing of homologous DNA [13]. In this situation the RNA:DNA interaction would involve the transgene and its RNA product instead of viral RNA. The suggestion of this type of interaction brings together the threshold and aberrant RNA hypotheses of PTGS because the likelihood of the RNA:DNA interaction would increase with the accumulation of the silencer RNA. It is likely

86

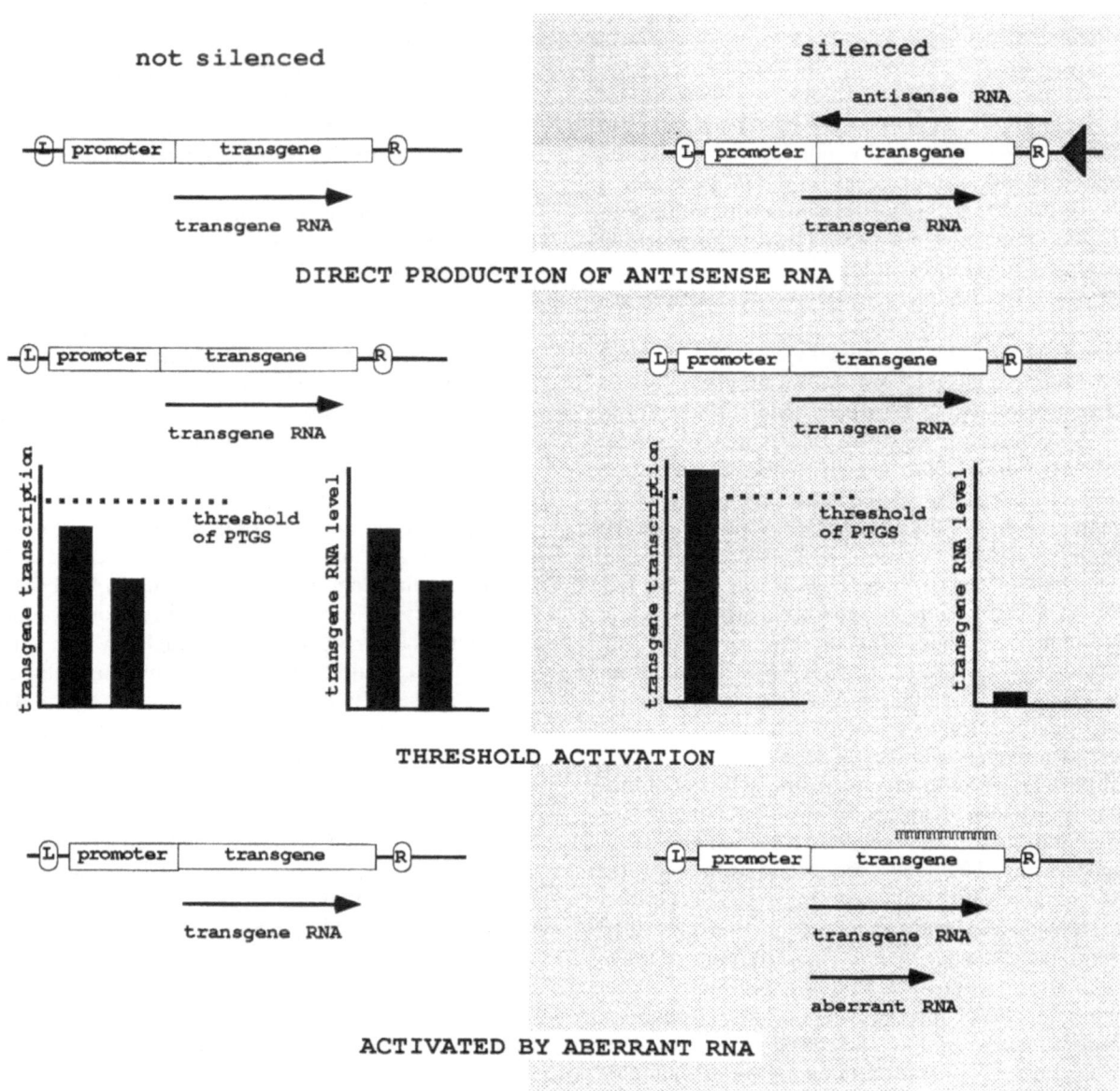

Figure 1. Hypothetical mechanisms of PTGS and homology-dependent resistance. These models are based on a transgene construct in which a promoter coupled to a transgene between the left (L) and right (R) borders of the T-DNA of *Agrobacterium tumefaciens* Ti plasmid is transformed into a plant genome. The direct production of antisense RNA model proposes that there is no PTGS when the T-DNA is inserted in the plant genome unless the insert is adjacent to a plant promoter (triangle). This promoter would produce the antisense RNA that could feature in the mechanism of cosuppression and gene silencing. The threshold model proposes that PTGS is activated by high level transcription of the transgene. If transcription is below this threshold the final RNA accumulation may be either high, medium or low corresponding to the transcription rate. However if the transcription is above the threshold the final accumulation of transgene RNA (and other RNAs with homology to the transgene) would be very low. The transcription rate may be affected by features of the genome or chromatin at or close to the transgene insert. In the aberrant RNA model it is proposed that the transgene in the silenced state produces an aberrant RNA which could activate the cosuppression mechanism. A factor that could influence production of the aberrant RNA is methylation (m) in the transgene. If the transgene is not silenced there may be either no methylation or methylation that does not affect production of aberrant RNA.

that accumulation of silencer RNA in dividing cells, as for the virus-induced PTGS, and for the same reasons, would lead to stronger silencing than RNA accumulation in differentiated cells.

Somatic instability of PTGS

One of the more striking features of PTGS is somatic instability. Flowers of petunia, in which there is PTGS of CHS, are not usually uniformly white as would be expected if the anthocyanin pigmentation pathway was blocked [41, 52]. There is often a variegated phenotype in which white and pigmented tissues are mottled or display spectacular patterns. The PTGS associated with homology-dependent resistance to PVY is somatically unstable and was lost when the transgenic plants are cultured *in vitro* and allowed to regenerate back into mature plants [50]. This instability, and the consequent unpredictability of a transgene phenotype, is a serious constraint to the practical application of transgenic plants.

The only clue to the basis of the instability of PTGS is from the transgene methylation data. Instability of transgene methylation could lead either to loss or to suppression of PTGS. There would be the potential for loss of PTGS if there is a failure to maintain the methylation of the transgene during cell division and if, as predicted by the aberrant RNA model, methylation is required for PTGS. There is also the potential for an increase of transgene methylation to suppress PTGS. The increased methylation could lead to transcriptional suppression of the transgene [39, 45]. This transcriptional suppression would lead to inhibition of PTGS if, as in the aberrant RNA and threshold models, sense RNA production is necessary for the silencing phenotype.

References

1. Angenent GC, Franken J, Busscher M, Weiss D, van Tunen AJ: Co-suppression of the petunia homeotic gene *fbp2* affects the identity of the generative meristem. Plant J 5: 33–44 (1994).
2. Assaad FF, Tucker KL, Signer ER: Epigenetic repeat-induced gene silencing (RIGS) in *Arabidopsis*. Plant Mol Biol 22: 1067–1085 (1993).
3. Barry C, Faugeron G, Rossignol J-L: Methylation induced premeiotically in *Ascobulus*: coextension with DNA repeat lengths and effect on transcript elongation. Proc Natl Acad Sci USA 90: 4557–4561 (1993).
4. Baulcombe DC: Replicase mediated resistance: a novel type of virus resistance in transgenic plants? Trends Microbiol 2: 60–63 (1994).
5. Baulcombe DC, English JJ: Ectopic pairing of homologous DNA and post-transcriptional gene silencing in transgenic plants. Curr Opin Biotechnol 173–180 (1996).
6. Boerjan W, Bauw G, Van Montagu M, Inzé D: Distinct phenotypes generated by overexpression and supression of *S*-adenosyl-L-methionine synthetase reveal developmental patterns of gene silencing in tobacco. Plant Cell 6: 1401–1414 (1994).
7. Brandle JE, Mchugh SG, James L, Labbe H, Miki BL: Instability of transgene expression in-field grown tobacco carrying the *csrl-l* gene for sulfonylurea herbicide resistance. Bio/technology 13: 994–998 (1995).
8. Day AG, Bejarano ER, Buck KW, Burrell M, Lichtenstein CP: Expression of an antisense viral gene in transgenic tobacco confers resistance to the DNA virus tomato golden mosaic-virus. Proc Natl Acad Sci USA 88: 6721–6725 (1991).
9. de Borne FD, Vincentz M, Chupeau Y, Vaucheret H: Co-suppression of nitrate reductase host genes and transgenes in transgenic tobacco plants. Mol Gen Genet 243: 613–621 (1994).
10. de Carvalho Niebel F, Frendo P, Van Montagu M, Cornelissen M: Post-transcriptional cosuppression of β-glucanase genes does not affect accumulation of transgene nuclear mRNA. Plant Cell 7: 347–358 (1995).
11. de Carvalho F, Gheysen G, Kushnir S, Van Montagu M, Inzé D, Castresana C: Suppression of β-1, 3-glucanase transgene expression in homozygous plants. EMBO J 11: 2595–2602 (1992).
12. Dehio C, Schell J: Identification of plant genetic loci involved in posttranscriptional mechanism for meiotically reversible transgene silencing. Proc Natl Acad Sci USA 91: 5538–5542 (1994).
13. Elmayan T, Vaucheret H: Single copies of a 35S-driven transgene can undergo post-transcriptional silencing at each generation or can be transcriptionally inactivated *in trans* by a 35S silencer. Plant J 9: 787–797 (1996).
14. English JJ, Mueller E, Baulcombe DC: Suppression of virus accumulation in transgenic plants exhibiting silencing of nuclear genes. Plant Cell 8: 179–188 (1996).
15. Flavell RB, O'Dell M, Metzlaff M, Bonhomme S, and Cluster PD: Developmental regulation of co-suppression in *Petunia hybrida*. In: Meyer P (ed) *Current Topics in Microbiology and Immunology Vol. 197: Gene Silencing in Higher Plants and Related Phenomena in Other Eukaryotes*, Springer-Verlag, Berlin (1995).
16. Fray RG, Grierson D: Identification and genetic analysis of normal and mutant phytoene synthase genes of tomato by sequencing, complementation and co-suppression. Plant Mol Biol 22: 589–602 (1993).
17. Goodwin J, Chapman K, Parks TD, Wernsman EA, Dougherty WG: Genetic and biochemical dissection of transgenic RNA-mediated resistance. Plant Cell 8: 95–105 (1996).
18. Goring DR, Thomson L, Rothstein SJ: Transformation of a partial nopaline synthase gene in tobacco suppresses the expression of a resident wild-type gene. Proc Natl Acad Sci USA 88: 1770–1774 (1991).
19. Green PJ: Control of mRNA stability in higher plants. Plant Physiol 102: 1065–1070 (1993).
20. Grierson D, Fray RG, Hamilton AJ, Smith CJS, Watson CF: Does co-suppression of sense genes in transgenic plants involve antisense RNA? Trendo Biotechnol 9: 122–123 (1991).
21. Grierson D, Tucker GA, Keen J, Ray J, Bird CR, Schuch W: Sequencing and identification of a cDNA clone for tomato polygalacturonase. Nucl Acids Res 14: 8595–8603 (1986).
22. Hammond J, Kamo KK: Effective resistance to potyvirus infection conferred by expression of antisense RNA in transgenic plants. Mol Plant-Microbe Interact 8: 674–682 (1995).

88

23. Hart CM, Fischer B, Neuhaus JM, Meins F Jr: Regulated inactivation of homologous gene-expression in transgenic nicotiana-sylvestris plants containing a defense-related tobacco chitinase gene. Mol Gen Genet 235: 179–188 (1992).

24. Hobbs SLA, Kpodar P, DeLong CMO: The effect of T-DNA copy number, position and methylation on reporter gene expression in tobacco transformants. Plant Mol Biol 15: 851–864 (1990).

25. Hobbs SLA, Warkentin TD, DeLong CMO: Transgene copy number can be positively or negatively associated with transgene expression. Plant Mol Biol 21: 17–26 (1993).

26. Ingelbrecht I, Van Houdt H, Van Montagu M, Depicker A: Posttranscriptional silencing of reporter transgenes in tobacco correlates with DNA methylation. Proc Natl Acad Sci USA 91: 10502–10506 (1994).

27. Jorgensen R, Que QD, English J, Cluster P, Napoli C: Sense-suppression of flower color genes as a sensitive reporter of epigenetic states of gene-expression in plant development. Plant Physiol 108: 14 (1995).

28. Jorgensen RA: Cosuppression, flower color patterns and metastable gene expression states. Science 268: 686–691 (1995).

29. Kawchuk LM, Martin RR, Mcpherson J: Sense and antisense RNA-mediated resistance to potato leafroll virus in Russet Burbank potato plants. Mol Plant-Microbe Interact 4: 247–253 (1991).

30. Kumagai MH, Donson J, Della-Cioppa G, Harvey D, Hanley K, Grill LK: Cytoplasmic inhibition of carotenoid biosynthesis with virus-derived RNA. Proc Natl Acad Sci USA 92: 1679–1683 (1995).

31. Lindbo JA, Dougherty WG: Pathogen-derived resistance to a potyvirus: immune and resistant phenotypes in transgenic tobacco expressing altered forms of a potyvirus coat protein nucleotide sequence. Mol Plant-Microbe Interact 5: 144–153 (1992).

32. Lindbo JA, Dougherty WG: Untranslatable transcripts of the tobacco etch virus coat protein gene sequence can interfere with tobacco etch virus replication in transgenic plants and protoplasts. Virology 189: 725–733 (1992).

33. Lindbo JA, Silva-Rosales L, Proebsting WM, Dougherty WG: Induction of a highly specific antiviral state in transgenic plants: implications for regulation of gene expression and virus resistance. Plant Cell 5: 1749–1759 (1993).

34. Longstaff M, Brigneti G, Boccard F, Chapman S, Baulcombe DC: Extreme resistance to potato virus X infection in plants expressing a modified component of the putative viral replicase. EMBO J 12: 379–386 (1993).

35. Matzke MA, Matzke AJM: How and why do plants inactivate homologous (trans)genes? Plant Physiol 107: 6679–685 (1995).

36. Matzke, MA, Matzke, AJM, Sheid OM: Inactivation of repeated genes: DNA-DNA interaction? In: Paszkowski J (ed) *Homologous Recombination and Gene Silencing in Plants*. Kluwer Academic Publishers, Dordrecht (1995).

37. Matzke MA, Neuhuber F, Matzke AJM: A variety of epistatic interactions can occur between partially homologous transgene loci brought together by sexual crossing. Mol Gen Genet 236: 379–386 (1993).

38. Meyer P, Heidmann I: Epigenetic variants of a transgenic petunia line show hypermethylation in transgene DNA: an indication for specific recognition of foreign DNA in transgenic plants. Mol Gen Genet 243: 390–399 (1994).

39. Meyer P, Heidmann I, Niedenhof I: Differences in DNA-methylation are associated with a paramutation phenomenon in transgenic petunia. Plant J 4: 89–100 (1993).

40. Mueller E, Gilbert JE, Davenport G, Brigneti G, Baulcombe DC: Homology-dependent resistance: transgenic virus resistance in plants related to homology-dependent gene silencing. Plant J 7: 1001–1013 (1995).

41. Napoli C, Lemieux C, Jorgensen RA: Introduction of a chimeric chalcone synthase gene into *Petunia* results in reversible co-suppression of homologous genes *in trans*. Plant Cell 2: 279–289 (1990).

42. Nellen W, Lichtenstein C: What makes an mRNA anti-sensitive? Trends Biochem Sci 18: 419–423 (1993).

43. Palmgren G, Mattson O, Okkels FT: Treatment of *Agrobacterium* or leaf disks with 5-azacytidine increases transgene expression in tobacco. Plant Mol Biol 21: 429–435 (1993).

44. Pang SZ, Slightom JL, Gonsalves D: Different mechanisms protect transgenic tobacco against tomato spotted wilt and impatiens necrotic spot tospoviruses. Bio/technology 11: 819–824 (1993).

45. Park Y-D, Papp I, Moscone EA, Iglesias VA, Vaucheret H, Matzke AJM, Matzke MA: Gene silencing mediated by promoter homology occurs at the level of transcription and results in meiotically heritable alterations in methylation and gene activity. Plant J 9: 183-194 (1996).

46. Phillips RL, Matzke MA, Oono K: Treasure your exceptions. Plant Cell 7: 1522–1527 (1995).

47. Seymour GB, Fray RG, Hill P, Tucker GA: Down-regulation of two non-homologous endogenous tomato genes with a single chimaeric sense gene construct. Plant Mol Biol 23: 1–9 (1993).

48. Smith CJS, Watson CF, Bird CR, Ray J, Schuch W, Grierson D: Expression of a truncated tomato polygalacturonase gene inhibits expression of the endogenous gene in transgenic plants. Mol Gen Genet 224: 447–481 (1990).

49. Smith HA, Powers H, Swaney S, Brown C, Dougherty WG: Transgenic potato virus Y resistance in potato: evidence for an RNA-mediated cellular response. Phytopathology 85: 864–870 (1995).

50. Smith HA, Swaney SL, Parks TD, Wernsman EA, Dougherty WG: Transgenic plant virus resistance mediated by untranslatable sense RNAs: expression, regulation, and fate of nonessential RNAs. Plant Cell 6: 1441–1453 (1994).

51. van Blokland R, van der Geest N, Mol JNM, Kooter JM: Transgene-mediated suppression of chalcone synthase expression in *Petunia hybrida* results from an increase in RNA turnover. Plant J 6: 861–877 (1994).

52. van der Krol AR, Mur LA, Beld M, Mol JNM, Stuitje AR: Flavonoid genes in petunia: addition of a limited number of gene copies may lead to a suppression of gene expression. Plant Cell 2: 291-299 (1990).

53. van der Vlugt RAA, Ruiter RK, Goldbach RW: evidence for sense RNA-mediated protection to PVY[N] in tobacco plants transformed with the viral coat protein cistron. Plant Mol Biol 20: 631–639 (1992).

54. Vaucheret H, Palauqui JC, Elmayan T, Moffatt B: Molecular and genetic analysis of nitrite reductase co-suppression in transgenic tobacco plants. Mol Gen Genet 248: 311–317 (1995).

55. Wagner EGH, Simons RW: Antisense RNA control in bacteria, phage and plasmids. Annu Rev Microbiol (1994).

56. Wassenegger M, Heimes S, Riedel L, Sänger HL: Methylation of plant genome-integrated viroid cDNAs is induced upon replication of the corresponding viroid RNA: a novel mechanism of a RNA-directed sequence-specific *de novo* methylation of genes. Cell 76: 567–576 (1994).

Plant Molecular Biology **32**: 89–106, 1996.
© 1996 *Kluwer Academic Publishers.*

RNA structure and the regulation of gene expression

Petra Klaff*, Detlev Riesner & Gerhard Steger
Institut für Physikalische Biologie, Heinrich-Heine-Universität Düsseldorf, D-40225 Düsseldorf, Germany
(*author for correspondence)

Key words: RNA processing, mRNA stability, translational efficiency, structural motifs, metastable structures

Contents

Abstract   89
Introduction   90
Structural motifs   90
  Secondary structures   90
    Double-stranded segments   90
    Hairpins and loops   93
    Stem-loop structures within the 5'
      mRNA untranslated region   94
    Stem-loop structures within the 3'
      mRNA untranslated region   95
  Tertiary structures   95
    Bifurcations   95
    Triple helices   95
    Pseudoknots   96
    tRNA-like structures and
      pseudoknots at the 3' terminus of
      plant viral RNA   96
    Structural basis of frameshifting   97

Ribozymes   97
  RNase P   98
  'Hammerhead' and 'hairpin'
    ribozymes   98
  Ribosomal leader RNAs   98
Structural transitions and metastable
  structures   99
  Prediction of RNA structure formation
    and rearrangement   99
  Examples of the *in vivo* importance
    of structural transitions and
    metastable structures   99
  Tetrahymena group I intron   99
  Bacterial mRNA   100
  Regulation of plasmid copy number   100
  Viroid replication and processing   100
Acknowledgements   101
References   101

Abstract

RNA secondary and tertiary structure is involved in post-transcriptional regulation of gene expression either by exposing specific sequences or through the formation of specific structural motifs. An overview of RNA secondary and tertiary structures known from biophysical studies is followed by a review of examples of the elements of RNA processing, mRNA stability and translation of the messenger. These structural elements comprise sense-antisense double-stranded RNA, hairpin and stem-loop structures, and more complex structures such as bifurcations, pseudoknots and triple-helical elements. Metastable structures formed during RNA folding pathway are also discussed. The examples presented are mostly chosen from plant systems, plant viruses, and viroids. Examples from bacteria or fungi are discussed only when unique regulatory properties of RNA structures have been elucidated in these systems.

Introduction

RNA sequence motifs involved in the post-transcriptional regulation of gene expression, such as the Shine-Dalgarno sequence or the polyadenylation signal, were characterized several years ago, but only more recently has it become clear that either the sequence motifs have to be exposed for interaction with a specific neighboring structure, or that the secondary and tertiary structures are more critical for regulatory function than the sequence *per se*. However, RNA structure acts as a regulation signal not by itself but by interacting with protein factors or, in some cases, by base pairing with another RNA and then interacting with a protein. In this review, we will concentrate not so much on whole molecules but on distinct structural elements, and we will discuss those examples where the functional relevance of these elements has been shown.

Any consideration of the regulation of gene expression in plants by RNA structure must cover not only plant mRNA but also plant virus RNA, for two reasons: (1) most plant viruses are RNA viruses and, therefore, viral RNA structure has always been of interest, and (2) in some cases, viral RNA was easier to handle experimentally, and attracted earlier the attention of many researchers.

We will first describe the structural elements of RNA as they are known from general studies, giving preferential attention to those types of RNA structural elements which are known to be involved in post-transcriptional regulation. We will proceed from the simple and well-known structures like sense-antisense double strands, hairpin and stem-loop structures to more complex structures like bifurcations (junctions), pseudoknots and triple-helical motifs. Finally, we will consider structures that are not in thermodynamic equilibrium but are controlled by the kinetics of RNA synthesis and structure folding.

Structural motifs

An RNA molecule can form highly specific structures which are stabilized by a variety of interactions. The major interactions contributing to such structures are base stacking and hydrogen bonding between the bases. These interactions result in the common Watson-Crick base pairs as well as unusual base pairs like wobble and Hoogsteen base pairs, and their reverse forms. In addition, the backbone might interact with bases or with other sugar and phosphate groups via hydrogen bonding and stacking of sugar or phosphate groups [for a review see 18].

Any secondary structure can be presented in a planar graph: the nucleic acid backbone is drawn as a circle, nucleotides are positioned at equal distances on the circumference, and base pairs are marked by straight lines (chords) connecting their nucleotides. These connecting lines should not cross one another. If they do so, the structure contains a tertiary structural element like a pseudoknot (see below).

Secondary structures

For a summary, elements of RNA secondary structure are presented in Fig. 1. The simplest secondary structure is the perfectly paired double strand formed by two complementary chains. The secondary structure of any single RNA chain has to consist of at least a base-paired region (helix) bridged by a loop of single-stranded nucleotides, the so-called hairpin. Hairpin loops are known to serve as nucleation sites for RNA folding. The double-stranded region can be interrupted by symmetric and asymmetric internal loops as well as one-sided loops, called bulge loops. The smallest symmetric internal loop consisting of two opposing nucleotides is a mismatch. A hairpin loop bridges a single helix whereas internal and bulge loops connect two helices. Loops connecting more than two helices are called bifurcations or junctions. A structure without bifurcations is often called a stem-loop structure.

Double-stranded segments

The stability of a double-stranded segment is mainly determined by the stacking interactions of neighboring base pairs. Within the cell, double-helical RNA elements occur either as long double-stranded molecules or as the helical parts of more complex secondary structures. The discussion here will focus on sense-antisense RNA interactions and their regulatory role in gene expression [for reviews, see 63, 88, 92].

Naturally occurring sense-antisense dsRNAs vary from <50 to >1000 bp in length, those constructed artificially (see below) from about 20 to several hundred. The lower limit is determined by the thermodynamic stability and varies with the G:C content. The structure of the sense-antisense dsRNA is the A-form, right-handed double helix. At the level of atomic coordinates, the details of the double helices will also depend on the sequence, but at present, this degree of

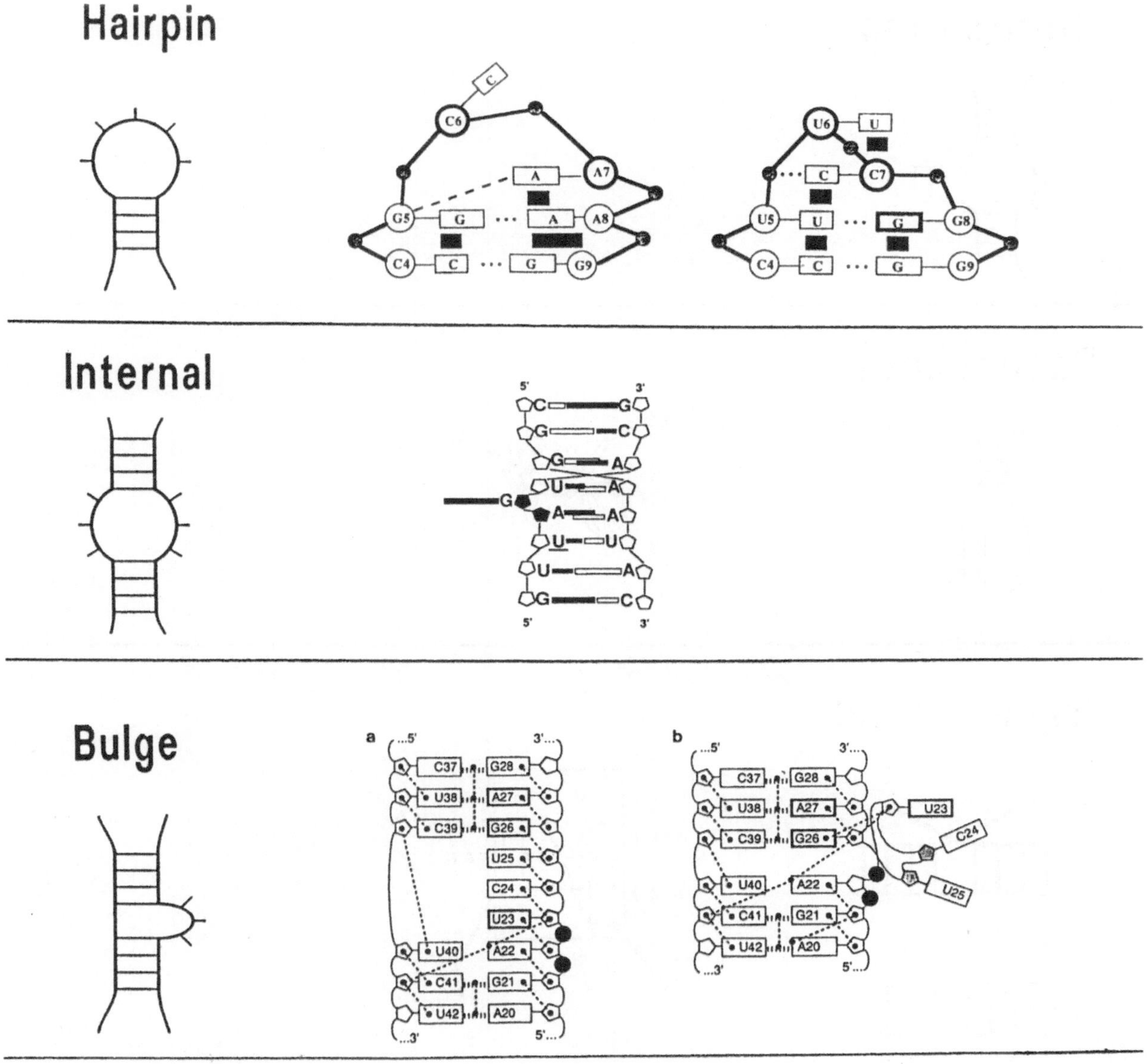

Hairpin

Internal

Bulge

Figure 1. Secondary structure elements of RNA. Left column: schematic representations; right column: examples selected from the literature, for which the three-dimensional structure has been determined by NMR studies. As a hairpin element, a so-called tetraloop is presented [51, 133]; the E-loop of eukaryotic 5S RNA is shown as an internal loop [18, 134]; as a bulge loop, the TAT region of HIV RNA is depicted, which is uncomplexed (a) or bound to argininamide (b) [103]. A pseudoknot is schematically shown in two- and three-dimensional models, subsequently confirmed by NMR [96, 104]. As an example of triple-stranded RNA, a domain of the *Tetrahymena* group I intron is presented as determined by NMR [19] or modeling based on comparative sequence analysis [81]. Reproduced from the literature with permission.

detail does not need to be considered in the mechanistic models. Furthermore, the structure might depend on the number of base pairs, but only in stems with less than 8–10 bp. Sense-antisense RNA complexes are not restricted to the formation of homogeneous double strands but might contain internal loops.

Regulation of gene expression by the formation of double stranded regions between a sense and an antisense RNA was first discovered in prokaryotes, where it is involved in the control of transcription, translation and mRNA decay [for a review see 136]. The mechanisms of many examples are well understood.

92

Bifurcation

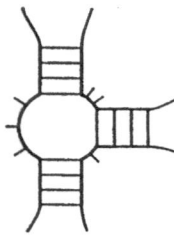

Pseudoknot

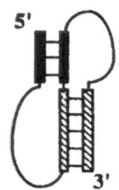

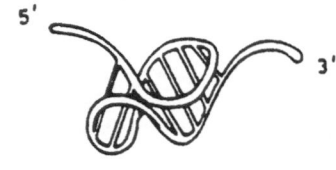

Triple strand

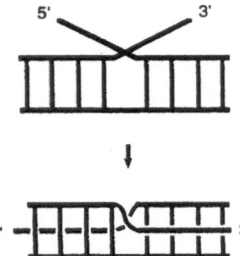

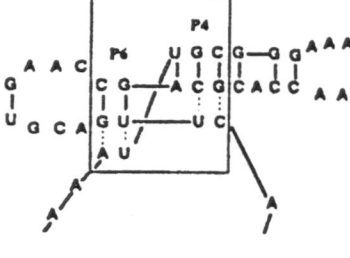

Figure 1. Continued.

In eukaryotes, however, and especially in plants, natural antisense RNA and the principles of its biological activity is still poorly understood. To date, only a few plant genes known to give rise to sense and antisense transcripts have been identified. Maize has an RNA complementary to the α-tubulin mRNA [31]. The antisense RNA is expressed in tissues in which the α-tubulin mRNA is present at a very low level, and the antisense transcript is the same size as the sense transcript. But, the two transcripts may not necessar-

ily originate from the same gene, because α-tubulin is encoded by a multigene family with high homology between the members. A similar finding has been made in barley. Here several RNA types complementary to α-amylase mRNA have been detected. The complementarity, however, is not perfect, indicating that the antisense transcript must be encoded by another gene or pseudogene [112]. Although, in both examples, the expression of sense and antisense RNA follows a certain developmental pattern, there is no evidence for the underlying regulatory mechanism. A more precise model of the regulatory function of natural antisense RNA has been presented for a developmentally expressed gene in *Dictyostelium discoideum*, the prespore-specific gene EB4-PSV, which encodes a protein associated with the membrane of prespore vesicles [54]. The constitutively transcribed mRNA only accumulates when cells form aggregates as a step in the formation of the prespore/prestalk pattern. In disaggregated cells, the mRNA is unstable, and antisense RNA derived from the same locus can be detected. Expression of the antisense RNA is developmentally controlled at the transcriptional level. The model implies regulation of mRNA accumulation by the extent of double strand formation with the endogenous antisense RNA. These RNA double strands are unstable, subject to rapid degradation by a, so far putative, double-strand-specific RNase.

In contrast to the restricted number of reports on natural regulation of gene expression by antisense RNA in plants, this principle is widely used to manipulate plant gene expression by biotechnological means. The expression of antisense RNA complementary to endogenous mRNAs in transgenic organisms usually inhibits gene expression [for a review see 44]. The mRNA does not accumulate to its normal level in the presence of its antisense molecule; with the mRNA encoding the small subunit of ribulose bisphosphate carboxylase (*rbcS*) a fivefold acceleration of mRNA degradation could be observed in *in vivo* degradation experiments [56]. Moreover, processes can be affected which occur later in the expression pathway then translation [21, 131]. Plants expressing antisense RNA have been constructed for various purposes, for example to elucidate protein functions within developmental pathways like flower development [98] or fruit ripening [131]. Furthermore, this approach has been shown to be useful in mediating protection of plants against pathogens like tobacco mosaic virus [101], geminiviruses [6] or viroids [77].

Despite the small number of examples of endogenous antisense-RNA-regulated genes in plants, it is most likely that dsRNA, which could be formed by sense-antisense RNAs, plays a regulatory role in cellular processes. The presence of RNases specific for dsRNA indicates at least the transient presence of the appropriate substrate. dsRNA-specific nucleases have been characterized in tobacco anthers [78]. These RNases are differentially expressed during pollen development from the microspore stage to maturity. Furthermore, the failure to detect RNA double strands

either in plants expressing endogenous antisense RNA or in plants genetically engineered to do so requires the existence of such enzymes.

In addition, dsRNA is probably involved in plant signal transduction. Several reports have described plant analogues of mammalian dsRNA-dependent protein kinases [22, 53, 66, 67, 75]. These enzymes are stimulated by plant pathogens like bacteria, tobacco mosaic virus or viroids, and have been assigned a role in the signal transduction of pathogen defense. Furthermore, the tobacco kinase can phosphorylate exogenous histones in a dsRNA-dependent manner, suggesting an additional role for dsRNA and the inducible protein kinase in healthy plant cells. The interaction of viroids with a kinase has been suggested recently as the primary pathogenic event [28].

Hairpins and loops
The role of hairpins and loops as structural signals in the regulation of plant gene expression has mainly been analyzed in plant organelles, especially chloroplasts. Since chloroplasts resemble prokaryotes in certain respects, several principles of bacterial gene regulation involving structural elements of RNA can also be observed in chloroplasts from green algae and higher plants.

Stem-loop structures have to be considered at two levels of molecular 'accuracy'. The first level is the well-known secondary structure of base pairing scheme, including as structural elements double-helical stems, hairpin loops, internal loops and bulge loops (Fig. 1). Without further information, one has to assume that the bases in the loops have a high degree of freedom and are accessible for interactions with proteins and other compounds. Computer programs are available to predict these secondary structures with good accuracy [34, 80, 118, 123, 141, 145, 146].

It needs to be emphasized here that a secondary structure is not always the structure of lowest free energy but might also be a suboptimal or metastable structure. Computer programs yield not only the structure of lowest free energy but also suboptimal [34, 123, 141, 145] or metastable structures (see below). The latter are not merely of theoretical interest, but might have an advantage in fast folding after transcription. Figure 2 illustrates a sequence whose structure of lowest free energy is one stem-loop (left panel); however, if the first hairpin (Fig. 2, right panel) formed during synthesis is sufficiently stable and the activation barrier

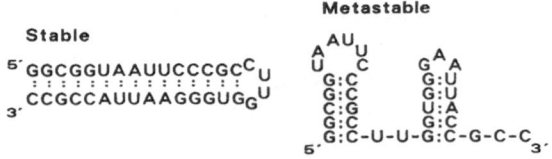

Figure 2. Stable and metastable structures formed by the same sequence.

sufficiently high, when RNA synthesis is complete, a metastable structure with two stem-loops remains.

A new level of RNA structure complexity has become evident from investigations applying experimental methods with atomic resolution. These showed that the simple assumption of flexible loops does not hold in all cases; in several examples looped segments are highly ordered structures. Examples redrawn from the original literature are shown in Fig. 1. In addition to loops, structural elements like pseudoknots and triple strands (see below) are typical components of RNA tertiary structure and, by definition, were not contained in secondary structure prediction. In a few cases, such as loops of four bases (the so-called tetraloops) and pseudoknots, it is possible to predict these higher-ordered structures, but in most cases it is not, and they all have to be borne in mind as potential structures improving or preventing a specific interaction.

Stem-loop structures within the 5′ mRNA untranslated region

The 5′-untranslated region of the chloroplast *psbA* mRNA, encoding the D1 protein of photosystem II, mediates translational regulation in the green alga *Chlamydomonas reinhardtii* as well as in higher plants [24, 25, 121, 122]. In both types of organism the 5′-untranslated region of this mRNA can form a stem-loop structure as shown by chemical mapping in spinach [60] and by mutational analysis in *C. reinhardtii* [79]. Furthermore, in *C. reinhardtii* the structural requirements for translation were analyzed in transgenic chloroplasts. According to these studies, the Shine-Dalgarno consensus sequence is necessary along with the immediately adjacent elements, which have been identified as binding message-specific proteins.

Mutations affecting the secondary structure reduce translational efficiency. The mechanism underlying this phenomenon is not yet known. A detailed study of the structure and sequence requirements of the 5′-untranslated region for translational initiation has also been performed in *Euglena gracilis*. Mutations affect-

ing the sequence and structure of the leader of *rbcL* mRNA (encoding the large subunit of ribulose bisphosphate carboxylase) and of *atpH* mRNA (encoding the CF0III subunit of the ATP synthase complex) were analyzed for their ability to form translation initiation complexes [8, 61]. In contrast to *rbcL* mRNA, the *atpH* mRNA has a Shine-Dalgarno sequence positioned, relative to the AUG codon, in a location equivalent to that of a bacterial Shine-Dalgarno sequence. In *atpH* mRNA, the sequence surrounding the AUG can generate a number of conformations, which show a higher degree of base pairing than observed in *rbcL* mRNA. The Shine-Dalgarno sequence has to be transiently single-stranded to interact with the 30S subunit being stabilized by the codon-anticodon interactions after the binding of fMet-tRNA. The subunit binding shifts the equilibrium between the various RNA conformations inducing additional molecular change into the single-stranded conformation and, therefore, entry into the translation process. With *rbcL* mRNA there is no primary sequence information specifying the translational start site. The AUG codon is located within an unstructured or weakly structured region of the message, allowing easy access for the 30S subunit. Strong secondary-structure elements introduced synthetically close to the AUG codon significantly reduce initiation complex formation. Here, the basic information required to direct the 30S subunit to the translation start site is an AUG codon in a highly unstructured region of the RNA.

With plant nuclear mRNAs, most studies on the role of 5′ leaders have been performed on the translational activity of plant viral RNAs [for a review see 37]. Such leaders can enhance the translation of reporter genes [26, 38]. Translational enhancement was found in leaders with no or little secondary structure [57], but in cell-free systems using the chloramphenicol transferase mRNA, recognition of an AUG codon in a suboptimal sequence context was higher when the adjacent downstream sequence could form a hairpin structure than when it was unstructured [62]. The choice of start codons in the mRNA of barley yellow dwarf virus (BYDV), which encodes overlapping genes, is also regulated by secondary structure [29]. The ribosome pausing at the second AUG enhances initiation at the first AUG. The first AUG, however, is often bypassed for two reasons: the suboptimal context and the formation of secondary structures involving the AUG.

Stem-loop structures within the 3' mRNA untranslated region

Mono- and polycistronic protein-coding transcription units found in the genome of higher-plant chloroplasts are generally flanked by inverted repeat sequences at their 3' ends which can form stem-loop structures. The position of the inverted repeats corresponds with the 3' ends of the mature mRNAs. It has been suggested that these elements function as transcription terminators [52, 59, 147], and in some cases it has been shown that the 3' end structure is partially effective in transcription termination. An example is the mRNA-encoding subunit IV of the cytochrome b6f complex, *petD*, in spinach, where the inverted repeat is followed by a uridine-rich track similar to those of *Escherichia coli* rho-independent terminators [for a review see 95]. Transcription termination is achieved very efficiently at plastid tRNA genes [125]. The organization of the tobacco [120] and the spinach plastid genome reveals that several transcription units are followed by tRNA genes.

The function of the 3' inverted repeats is two-fold. Effective processing has to be performed because transcription termination occurs downstream of the mature end of chloroplast mRNAs. The inverted repeat sequences serve as processing signals *in vitro*. Alteration of the 3' sequence of *psbA* mRNA without destroying the hairpin element does not impair the processing event, whereas product accumulation is changed [1]. The half-life of the 3' end of the *psbA* mRNA *in vitro* can be both lengthened and shortened by mutagenesis, indicating the second function of these structural elements, i.e., the inverted repeat structures also serve as general stabilizing elements protecting the 5'-located mRNA against exonucleolytic degradation. This has been shown *in vivo* in *C. reinhardtii*. The *C. reinhardtii atpB* mRNA encoding the β-subunit of the ATP synthase complex contains 3' sequences which form a complex stem-loop structure. Deletions of part or the whole element in transformed *C. reinhardtii* cells decreases *atpB* mRNA accumulation, whereas transcription rates are unaffected [126].

Tertiary structures

Bifurcations

In more complex RNA secondary structures, several helices may originate from one loop. This arrangement is called a bifurcation, a junction or a multibranched loop. Figure 1 shows a two-dimensional representation of a bifurcation. One should not regard the bifurcation loop as a flexible part of the molecule. In tRNAs where the three-dimensional structure is known, two stems form a continuous stack and the position of the other stems is fixed. In that sense, bifurcations are not elements of secondary structure, but might be part of a specific tertiary structure. Other known elements of a tertiary structure are triple strands and pseudoknots (Fig. 1). The loops which are formed by pseudoknots or triple strands can be as small as a few nucleotides, or large, leading to long range interactions. It is obvious that these interactions define the relative orientations of helices that are distant in sequence but need close proximity for proper function.

Structures with a bifurcation(s) are found, e.g. in ribozymes (discussed below) and within the splicing complex of pre-mRNAs. Here we will discuss an example of *trans*-splicing described in chloroplasts of *C. reinhardtii*; *trans*-splicing is also known to occur in organelles of other plants. In chloroplasts, most introns belong to class I or class II introns on the basis of their primary and secondary structure [83, 116]. A third class of introns, designated class III, has been described in *E. gracilis*, which lacks the characteristic primary and secondary structure [20]. In *C. reinhardtii*, the gene encoding the 80 kDa chlorophyll a apoprotein of photosystem I (*psaA*) consists of three exons, two of which are in opposite orientation at different locations of the chloroplast genome. Thus, their transcription must be discontinuous, and *trans*-splicing is required to form the mature mRNA [for a review see 111]. In addition to several nuclear genes, one chloroplast gene has been identified that is essential for the *psaA trans*-splicing. This gene (*tscA*) encodes a 450 nt RNA [43]. According to the model of *psaA trans*-splicing, the *tscA*-RNA completes the catalytic core structure of the group II intron by allowing the formation of the characteristic wheel-like secondary structure with six protruding domains [43; see also contribution by Rochaix, this volume].

Triple helices

RNA base triples are formed by unpaired nucleotides which are hydrogen bonded to an already base-paired nucleotide. One or two hydrogen bonds to the Watson-Crick base pair may be formed by the third base, which can be located in the major or the minor groove. Furthermore, stacking interactions contribute to the stability of base triples.

In biological systems, base triples have been proposed for several RNAs [82, 138, 139]. One of the functions they are attributed is stabilization of the three-dimensional shape of the RNA molecules by creating long-range interactions between a base pair and a distant single-stranded nucleotide. One well characterized example is the tertiary structure of tRNA, where several base triples can be observed. In tRNA^{Phe}, three nucleotides in the junction loop are complexed with Watson-Crick pairs of the D-stem, where they bind within the major groove. Two of the triples, A9:A23-U12 and G46:G22-C13 are formed by two hydrogen bonds between the purine and the Watson-Crick pair (Fig. 3). In contrast, the third base of the triple G45:G10-C25 only forms one hydrogen bond [18]. Different base triples are found in other tRNAs like tRNA^{Asp} [138]. Instead of the G45:G10-U25 base triple of tRNA^{Phe}, an A46:G22-Ψ13 is formed. Here the adenosine is located in the major groove and forms one hydrogen bond to the G-Ψ base pair. In tRNA^{Asp} an unusual base triple is found: A21 is bound to the tertiary reverse-Hoogsteen pair A14:U8 by the formation of base-base and base-sugar hydrogen bonds. These base triples located at the beginning of the D-loop at the hinge region of two helices are involved in stabilizing the characteristic L shape of the RNA. Another molecule known to form base triples is the self-splicing intron of *Tetrahymena*, where the base triples are important for catalytic activity [82]. From genetic evidence, strong interactions between adjacent base triples, G-C:U and C-G:C, formed between the stem P4 and junction J5/7, result from stacking interactions between the parallel, stacked base triples. Such an element is likely to consist of a classical double-stranded helix and a third strand with approximate A-form geometry interacting in its major groove. Triple helices involving DNA are believed to participate in homologous recombination [for a review see 14].

As with the antisense RNA concept, the principle of triple helix formation can be used as a regulatory tool for biotechnological purposes. This approach has not yet been applied in plants, but it has been shown *in vitro* as well as *in vivo* in animal systems that gene expression can be repressed by triple-strand formation [for a review see 71]. In most of these studies, transcription is affected. Initiation and elongation can be blocked by binding a single-stranded oligodesoxyribonucleotide to the DNA for T7 transcription *in vitro* or polymerase II transcription in HeLa cell culture [72, 100, 107, 144]. In HIV, triple strand formation of an RNA-DNA hybrid with DNA or RNA oligonucleotides inhibits RNase H cleavage and initiation of plus-strand DNA synthesis *in vitro*. In cell culture experiments, the addition of the triple-strand-forming oligonucleotide efficiently inhibits retrovirus replication indicating that the triple strand can also be formed *in vivo* [135]. The triple strand approach may also be useful in plant systems.

Pseudoknots

One tertiary structure possibility arises from the base pairing of the nucleotides of a loop with a complementary sequence outside this loop. When the number of tertiary base pairs is lower than about six, a 'pseudoknot' is formed; with longer base-paired regions, for example above a full helix turn of 11 nucleotides, a 'real knot' would be formed; this has never been observed in an RNA structure. Depending on the type of loop(s) and single-stranded regions involved in the formation of the tertiary interaction, several different pseudoknots may be classified [for a review see 140]. In the H(airpin)-type pseudoknot, nucleotides from a hairpin loop form base pairs with the single-stranded region 3' adjacent to the hairpin helix; both helices are coaxially stacked. The possibility for such H-pseudoknot formation depends on the length of its two helices, the length of the three unpaired nucleotide stretches connecting both helices and the type of unpaired nucleotides [96, 104, 143]. Pseudoknot formation depends on the presence of divalent cations, and the gain in thermodynamic stability by pseudoknot formation is only a little above the stability of one of the helices. The presence of pseudoknots has been confirmed in a wide variety of RNAs [for review see 23]; we would like to concentrate on those described in plant viral RNAs.

tRNA-like structures and pseudoknots at the 3' terminus of plant viral RNA. A number of plant viral RNAs contain a tRNA-like element at their 3' end, which in most cases is a substrate for aminoacyl-tRNA synthetase, for RNase P, and, after removal of the 3' adenosine, for adenylation by CTP/ATP:tRNA nucleotidyl transferase [for reviews see 42, 48]. The terminal amino-acid-accepting helix of tRNA is mimicked in these tRNA-like elements by an H-pseudoknot [110]. In addition to the tRNA-like element, several viral RNAs contain further pseudoknot(s) upstream [7, 33]. The tRNA-like structure and the additional pseudoknots are critical for viral RNA replication into (−) strands by the viral RNA-dependent RNA poly-

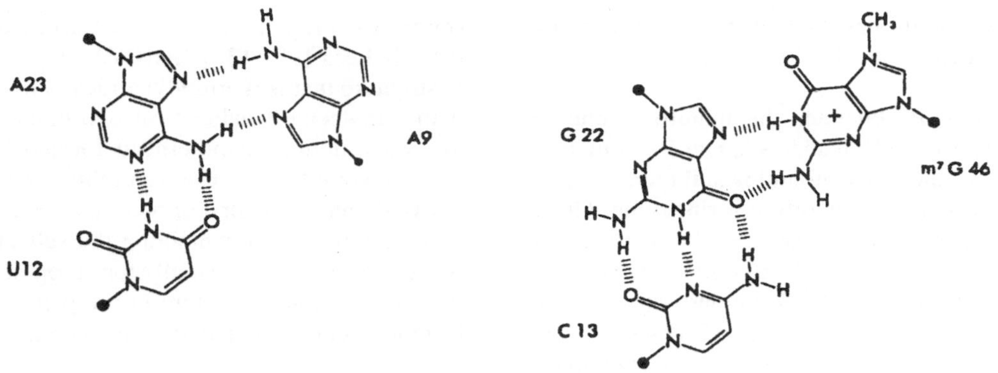

Figure 3. Basetriples of tRNA[Phe] of yeast.

merase [65, 129] and may promote efficient translation [39, 40, 68].

Structural basis of frameshifting. A number of eukaryotic positive-strand viruses contain overlapping reading frames [for a review see 9; see also Maia *et al.*, this volume]. To translate those proteins, a purposeful shift in the reading frame is programmed into the RNA. At a certain signal, encoded by both the sequence and a tertiary structure, ribosomes are induced to move into a reading frame −1 in 5′ direction. At this new position, they continue to translate the protein of the new reading frame. In plants, three luteoviruses [BYDV, beet western yellow virus (BWYV) and potato leaf roll virus (PLRV)] and one dianthovirus (red clover necrotic mosaic virus (RCNMV) have been described that show this kind of frameshifting [11, 41, 58, 102]. In BWYV, frameshifting has to occur to switch from an orf encoding a 39 kDa protein to one encoding a 60 kDa protein. In BWYV, PLRV and RCNMV, frameshifting is necessary to synthesize the putative RNA-dependent RNA polymerase.

The frameshifting signal consists of two elements: (1) the 'slippery sequence', which is a heptanucleotide, and (2) a structural element, stem-loop or pseudoknot located downstream of the slippery sequence. The slippery sequence usually contains two homopolymeric triplets of the general sequence XXXYYYZ, which would allow two tRNAs bound to the ribosome to base pair with two of the three positions of the anticodon after frameshifting [55]. An exception to this rule is the slippery sequence of RCNMV which is GAAUUUU. The structural element located downstream is necessary for efficient ribosomal frameshifting. It has been suggested that ribosomes may be slowed or stalled at

these structures, allowing frameshifting to occur [55]. In plants, both stem-loop structures and pseudoknots have been described. In BWYV, a simple pseudoknot is predicted [41]. In PLRV, a stem loop structure was first described downstream of the slippery sequence [102]; however, it is also possible that a pseudoknot is formed [64]. On the basis of sequence analysis, in BYDV also a stem-loop structure or a pseudoknot can be formed [11]. In RCNMV, a stem loop is predicted. For the last two viruses, there is no experimental evidence for the kind of structural signal [58].

In plant viruses, the efficiency of the frameshifting described so far is low. *In vitro* efficiencies of 1.4% (BWYV), and 1 or 2% in PLRV, depending on the isolate, have been reported. *In vivo*, using transient expression assays in potato protoplasts, the same frequency of 1% was observed for PLRV, when the frameshifting signal was inserted into a reporter gene. It was proposed that the frameshifting efficiency is reminiscent of the stability of the secondary structure located downstream of the slippery sequence, ensuring a functionally relevant high ratio between the fusion protein and the translation products derived by frameshifting [30, 41].

Ribozymes
In the past 15 years, RNAs have been detected that can catalyze reactions similar to true enzymes. Most of these highly specific reactions are cleavage of RNA and/or ligation of RNAs in *cis* or *trans* [for reviews see 116, 128]. These catalytic RNAs or ribozymes may be divided into several classes depending on their reactions, the mechanism and/or the ribozyme structure. Some ribozymes, especially 'hammerhead' and 'hairpin' ribozymes, may have functions similar to antis-

ense RNA but should have the advantage of catalytic turnover [for reviews see 99, 105, 115, 127].

RNase P. RNase P generates the mature 5′ end of tRNAs by removing 5′ leader sequences from pre-tRNAs. *In vitro*, the RNA subunit is sufficient to catalyze this reaction and is therefore a ribozyme. Other natural substrates, however, are also known [2], for example, the above-described tRNA-like elements at the end of viral RNAs. The secondary structure of RNase P is well established [for a review see 12] and there are models (including a loop-loop pseudoknot) for its three-dimensional structure and interaction with the substrate [49, 94, 137]. RNAse P activity was established mainly in bacteria but is also known from plants and plant organelles [73, 74].

'Hammerhead' and 'hairpin' ribozymes. Several plant pathogenic RNAs can undergo a site-specific self-cleavage, at least *in vitro*. One viroid belongs to this group of RNAs, the avocado sunblotch viroid (ASBVd), as do the virusoids from lucerne transient streak virus (vLTSV), velvet tobacco mottle virus (vVTMoV), *Solanum nodiflorum* mottle virus (vSN-MV), and subterranean clover mottle virus (vSCMoV), the satellite RNAs of BYDV (sBYDV) and of tobacco ringspot virus (sTRSV), and the small satellite RNAs of arabis mosaic virus (sArMV) and chicory yellow mottle virus (sCYMV). Viroids are independently replicating, noncoding, circular RNAs varying in size from 246 to 463 nt; they exist as highly base-paired, rod-like secondary structures and infect a wide variety of plant species [for a review see 27]. Plant virus satellite RNAs vary from about 250 to about 1500 nt, depend on their 'helper' virus for replication and encapsidation, and may alter the disease of their helper virus [for a review see 114]. During replication, these RNAs produce multimeric forms which can undergo autocatalytic cleavage via two different secondary/tertiary structural elements: the (+) strand of sTRSV and sArMV, the (+) and (−) strand of vLTSV and the (+) strands of the other virusoids self-cleave using a 'hammerhead' structure; (+) and (−) strands of ASBVd self-cleave via a 'double hammerhead', and the (−) strand of sTRSV and of sArMV self-cleave and self-ligate using a structure called a 'hairpin' or 'paperclip' ribozyme. The secondary structure of hammerhead ribozymes consists of a bifurcation of three helices; the self-cleavage site is located in the bifurcation loop. Three-dimensional models for the hammerhead structure show a high degree of stacking in the central loop region and a special arrangement of the three helices [97, 132]. ASBVd may form a similar structure that is thermodynamically highly unstable with a molecule smaller than two units; thus, cleavage via a double hammerhead formed by two units seems favorable [35]. The paperclip ribozyme consists of two domains, the first of two small helices separated by an internal loop containing the self-cleavage site, while the second is a small stem-loop structure. The three-dimensional structure of the paperclip seems to be more complex than that of the hammerhead [3, 13].

Ribosomal leader RNAs

Long-range interactions involved in RNA structure formation have been proposed for ribosomal RNAs. In ribosomal RNAs, leader sequences, which are not present in the mature molecule, have been shown to be of major importance for the formation of functional ribosomes in yeast and in prokaryotes like *E. coli*. In yeast, the external transcribed spacer (ETS) as well as the internal transcribed spacer (ITS) of the ribosomal RNA precursor transcript affect the accumulation of ribosomal subunits. Partial deletions of the ETS upstream of the 17S rRNA gene and of ITS1, located between the 17S and the 5.8S rRNA genes, prevent the accumulation of functional 40S ribosomal subunits without affecting the formation of 60S subunits [90]. This finding suggests that both spacers contain information required for the production of 17S rRNA but not for 26S rRNA. This information may be the formation of a (long-range) secondary structure and its involvement in the removal of the ETS from the primary transcript. Thereby it may function as a *cis*-regulating element and/or the binding site for ribosomal proteins. In the case of ITS2, located between the 5.8S and the 26S rRNA gene, the presence of the spacer is required at some stage of 60S subunit formation [91]. Structural alterations within the ITS2 region, which did not alter processing significantly, reduced the growth rate of cells that exclusively express the mutant rRNA genes [93]. Accordingly, in *E. coli*, point mutations in the leader region of pre-rRNA (i.e. 5′ ETS) caused growth defects, although rRNA processing was not disturbed and normal amounts of 30S subunits were produced [130]. For plants, similar observations have not yet been made. However, on the basis of the evolutionary conservation of ribosomal genes, it is most likely that similar roles of RNA leader sequences in the formation of highly ordered structures or macromolecule complexes will be identified.

Structural transitions and metastable structures

All the RNA structures discussed so far can undergo structural transitions. Under particular conditions they are stabilized or destabilized, and in most cases, the different structures are present in thermal equilibrium. Consequently, the structural transition alters the concentration of a particular structural state, but all structures involved in an equilibrium are available for binding to regulatory factors and the binding will shift the equilibrium.

Since the early experiments in 1966 of Fresco and coworkers [36], it has been known, however, that RNA can also exist in metastable conformations, which, by definition, are not the state of lowest free energy and are not in equilibrium with the state of lowest free energy. These structures are trapped in a conformation that would require a high activation energy to undergo the transition into the most stable structure of lowest free energy. Such a trapping event can be induced *in vitro* by particular protocols changing temperature and ionic conditions. *In vivo* metastable RNA structures are generated as a consequence of sequential folding of the RNA strand during transcription; hairpins are formed in the not yet completed strand, and might have such a stability under physiological conditions that they cannot be transformed into more stable but global structures after completion of the transcription process.

There are several systems in which the relevance of RNA folding for its biological function has been elucidated. The data, however, are derived from either *Tetrahymena*, bacteria, phages or viroids. Because they represent a mechanism of gene regulation by the kinetics of structure formation, they will be discussed briefly after a short introduction to computational methods for the prediction of RNA structure formation and rearrangement.

Prediction of RNA structure formation and rearrangement

The computational methods can predict the optimal secondary structure of an RNA or even the structure distribution of an RNA; the algorithms, however, are based on thermodynamics which do not consider activation barriers during rearrangement or refolding of structures and give no information about the pathway of folding into optimal structures. When such information is necessary, a kinetic approach has to be used. Kinetic approaches to RNA structure prediction were introduced by Martinez [76], Mironov *et al.* [84] and

Mironov and Kister [85; see also 86, 87]. These authors proposed a Monte Carlo method for constructing secondary structure(s) based on rate constants for iterative addition of complete helical regions to an already existing structure. This approach, however, is close to a gradient-descent method due to the high activation barriers involved in manipulating complete helical segments. Consequently, it tends to halt in local minima of the energy landscape of RNA secondary structure, i.e., it cannot describe the kinetically controlled pathway of refolding from metastable into thermodynamically optimal structures. This restriction was circumvented by taking into account the closing and opening of single base pairs instead of helices and using a special Monte Carlo method termed 'simulated annealing' [119]. This algorithm can accurately simulate kinetic folding processes and predict folding times. Additionally, polymerase elongation rates for the simulation of sequential folding during RNA synthesis may be taken into account.

A quite different mathematical approach, a genetic algorithm, was used by van Batenburg *et al.* [4] and Gultyaev *et al.* [46] to obtain insight into the same problem. Simulations using this method were also able to predict relevant metastable structures of certain RNAs. The analysis of free energies for intermediate foldings allows estimation of the ranges of kinetic refolding barriers. The folding kinetics of the proposed metastable structures can be calculated afterwards using a model of a multistep refolding process with elementary steps of double-helical stem formation or disruption [47].

Examples of the in vivo importance of structural transitions and metastable structures Tetrahymena group I intron

The question whether the RNA structure of long transcripts is kinetically or thermodynamically favored was approached by an analysis of the *Tetrahymena* group I intron [for a review see 15] using self-splicing as an indicator for proper folding of the pre-rRNA. *In vitro* transcription of the intron alone results in fully active molecules suggesting that the active structure is both kinetically and thermodynamically favored [16]. Natural *Tetrahymena* precursor RNAs can adopt two alternative secondary structures within the exon sequences, in one of which the 5' splice site is inhibited by the formation of a stable conserved hairpin, resulting in a concomitant loss of splicing [142]. Transcription of the pre-RNA with the T7 system showed that the pre-RNAs containing the natural ribosomal exon sequences do not splice at the maximum rate but regain full activity after passing through a denaturation-renaturation cycle [32]. Thus, the structure of a population of pre-RNA molecules after transcription does not correspond to the active conformation. The most consistent model is that RNA folding is controlled by sequence-dependent interactions within a minimal structural domain, which might be supported by protein-RNA interactions *in vivo*.

Bacterial mRNA. In bacteria, two examples indicate the importance of the kinetics of RNA folding for translational regulation. The expression of a transposase encoded by IS10, a mobile genetic element in *E. coli*, is translationally controlled. Transcription of the IS10 transposase mRNA starting from a promoter outside IS10 results in little protein expression. The element responsible is called 'protection from outside transcription' (*pot*) and is located on the mRNA. Low expression of the IS10 transposase results from regulation at the translational level and is based on mRNA folding which sequesters the ribosome-binding site. Mutations that increase ribosome binding to the mRNA slow the kinetics of *pot* structure formation but have only a small effect on the final inhibitory structure of the *pot* RNA. The findings suggest the formation of kinetic intermediates which allow translation while the final stable structure does not. Therefore, the velocity of RNA structure formation may be able to regulate translation [70]. A similar model has been proposed for the translational regulation of the expression of phage MS2 maturation protein. The mRNA forms a cloverleaf-like structure consisting of three stem-loops enclosed by a long-distance interaction (LDI). The 3' moiety of the LDI contains the Shine-Dalgarno region, whereas the complementary element is located 80 nucleotides upstream. The authors suggest that translation only starts on the mRNA when it is not folded into its thermodynamically optimal structure, i.e., when the base pairing between the LDI and its complementary region has not yet taken place. Base pairing between these regions might be delayed by the intervening sequence. The model is supported by mutational studies varying the length of the intervening sequence: its reduction in length reduces expression, while an increase enhances protein A synthesis. Stabilization of the LDI does not diminish translation [45].

Regulation of plasmid copy number. Metastable structures of RNA molecules and the kinetics of their refolding have also been shown to be important in regulating the replication of the *E. coli* ColE1 plasmids. RNAII, an RNA species encoded by ColE1 plasmids, serves as a primer for plasmid DNA replication by forming an RNA-DNA hybrid which is cleaved by RNase H yielding a 3' hydroxyl terminus. Primer formation is negatively regulated by interaction of RNAII with its antisense RNAI of 108 nucleotides [for a review see 136]. The inhibitory effect of ColE1 RNAI on primer formation is based mainly on the kinetics of the duplex formation with RNAII. The biological activity of RNAI and RNAII depends on the folding of the molecules. If RNAII is transcribed under the control of the T7 late promoter by inducible T7 polymerase, the RNA is overexpressed. However, an increase in RNAII does not result in a higher copy number of the ColE1 plasmid, indicating that RNAII transcribed by T7 polymerase does not serve as a primer for plasmid replication. The folding of RNAII is crucial to the formation of persistent hybrids between DNA and the priming RNA. It is proposed that the transcriptional velocity is critical for facilitating the proper folding of RNAII and the subsequent interaction with the plasmid DNA. This model is supported by the finding that overproduction of RNAII by *E. coli* polymerase indeed results in an increase in ColE1 plasmid copy number [17].

Theoretical work also emphasizes the role of kinetic rather than equilibrium features of RNAII in the efficiency of replication inhibition. Genetic algorithm studies of the folding pathway of the wild-type RNAII in comparison to copy number mutants revealed the transient formation of a metastable structure with a considerable half-life and which is increased in the mutants. This delays the formation of the stable mutant RNAII structure. Thus, inhibition of plasmid replication by formation of the sense-antisense duplex is sensitive to the kinetics of RNA folding [47].

Viroid replication and processing. Viroids are a good example where the phenomenon of metastable structure was investigated in the context of viroid transcription and processing *in vitro* by physicochemical methods, in plant cell extracts by biochemical methods and inside the plant cell by site-directed mutagenesis. During viroid replication, the circular [by definition (+) strand] viroid is transcribed into an oligomeric (−) strand RNA, which acts as a template for the synthesis of an oligomeric (+) strand RNA [for a review see 10]. Both transcription steps are catalyzed by a host enzyme, the DNA-dependent RNA polymerase II [89, 117]. The (+) strand oligomeric RNA is cleaved enzymatically to unit-length molecules which are then

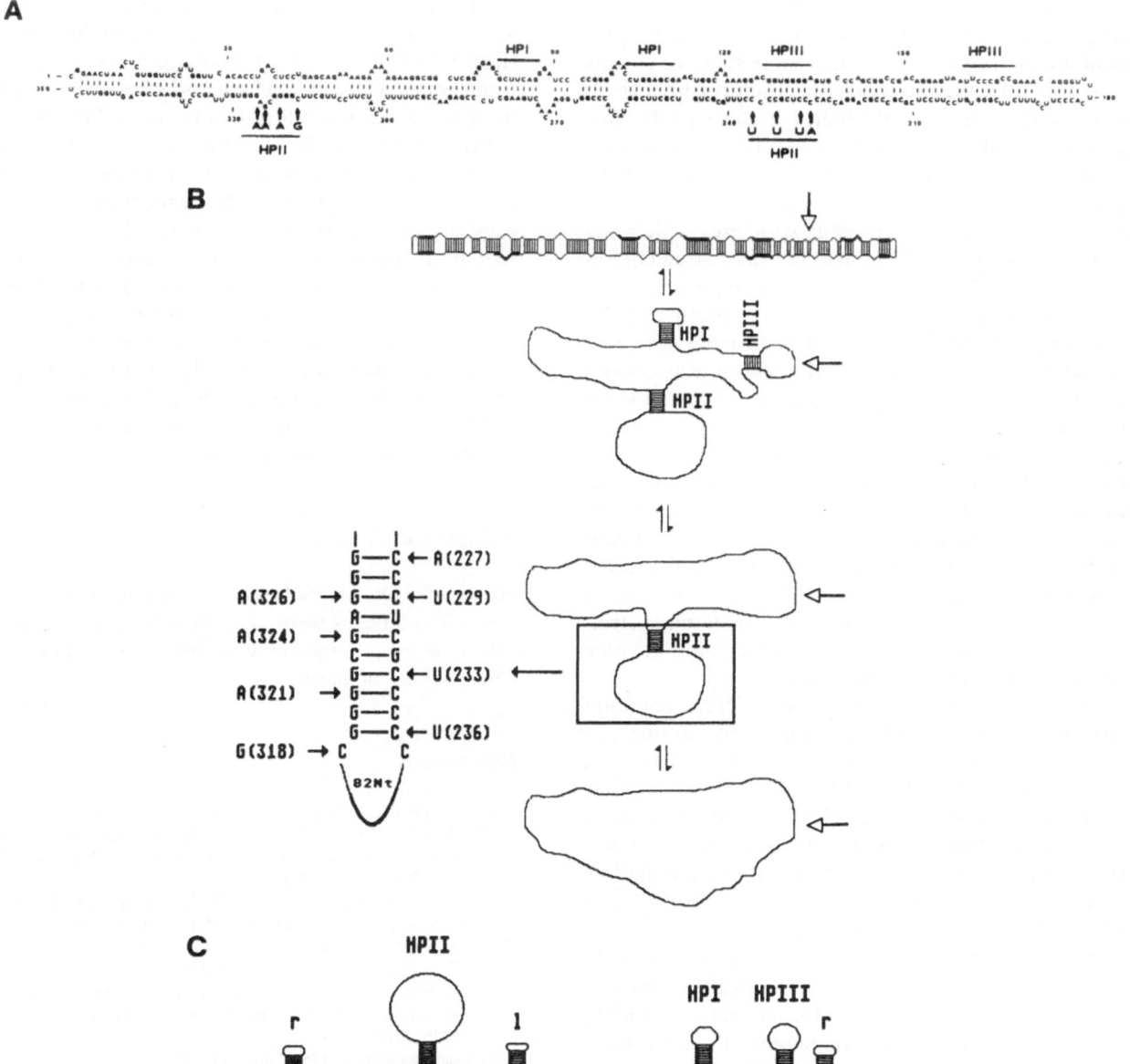

Figure 4. Secondary structures of the potato spindle tuber viroid (PSTVd). A. Native, rod-like structure of circular PSTVd. B. Equilibrium conformations during thermal denaturation of circular PSTVd. C. Multihairpin structure of the linear PSTVd after transcription. Hairpins I, II and III are designated HPI, HPII and HPIII. Filled arrows directed at the lower strand of the rod-like native structure (A) and at the enlarged hairpin II (B) indicate nucleotide exchanges. Open arrows label the position of the 5′ end at nt 147 and the 3′ end at nt 146 of the linear PSTVd transcript. Hairpin structures forming at the left and the right terminus of the native structure are designated by l and r, respectively [69].

ligated to the mature viroid circles. This brief description is restricted to the mechanism found with viroids of the so-called potato spindle tuber viroid (PSTVd) class.

Under native conditions, viroids form a rod-like structure (Fig. 4A). As the first step of thermal tran-

ition, all base pairs are disrupted and very stable hairpins are newly formed (HPI, HPII, HPIII; Fig. 4B). The transition may be viewed as a switch from an extended to a branched structure with a marked loss of base pairing. At higher temperatures, the stable hairpins dissociate independently from each other in the order of their individual thermal stabilities [for a review see 108]. HPI and HPII are formed from regions with highly homologous sequences among a large series of viroids. Consequently, the phylogenetic argument indicates a functional relevance of the hairpins, although the thermodynamically favored structure contains none of the stable hairpins under physiological conditions. Thus, any functional role for the hairpins has to be attributed to intermediate metastable structures which may be formed during replication of the RNA strand before the synthesis of the whole strand is completed.

Temperature-gradient gel electrophoresis (TGGE) is a simple method to analyze metastable structures experimentally [109, 113, 124]. Viroid transcription has been simulated by *in vitro* transcription of longer-than-unit length viroid RNAs from cloned viroid cDNA using T7 polymerase. The structure of the newly synthesized RNA was analyzed by TGGE without further treatment, and it could be shown in several studies [5, 50, 124] that the structure formed deviates drastically from the most stable structure, but can be rearranged to it by specific incubation protocols. Figure 4C gives a very simple representation of such a metastable structure containing only HPI, II, III and the terminal hairpins of the rod-like conformation.

The functional relevance of a HPII-containing structure of viroid RNA was studied by site-directed mutagenesis [69]. The mutations (Fig. 4B) were designed to destabilize HPII (except the mutation at position 318 that stabilizes HPII), but to introduce as little perturbation as possible into the native structure (Fig. 4A). Infectivity tests showed that the mutations in the central region of HPII reverted to the wild-type sequence, whereas the mutations in the peripheral regions of HPII remained genetically stable. Thus, the integrity of the center of HPII is critical for infectivity. It was suggested that HPII might act as a binding site for host cell transcription factors. To differentiate between sequence and structure effects, the mutations which reverted to the wild-type sequence were analyzed with respect to the time course of reversion and the sequence variation during reversion [106]. It could be concluded that HPII is a functional element in the (−) strand replication intermediate, and that it is essential for template activity during (+) strand synthesis. G:U pairs or reversions to other base pairs are tolerated for short-term replication of PSTVd but the exact wild-type sequence proved to be superior with regard to fitness and replicability of PSTVd.

The relevance of detailed structural features of the (+) strand replication intermediate became evident when the processing reaction was analyzed either in an *in vitro* reaction using the heterologous enzyme RNase T1 [124] or in a homologous nuclear extract from potato cells [5]. There must be two cleavage reactions to form the unit-length linear intermediate, which is ligated to the circle of exact length. In both types of system, the most stable structure was not the active substrate structure; the 'processing structure' exposes both cleavage sites for specific or partially specific recognition by the cleavage enzyme, and is then transformed into a structure in which both cleavage sites are in the closest spatial relationship favoring the subsequent ligation to circles. For processing in the homologous nuclear extract, only one of four possible substrate RNA structures was the active conformation, and this structure was metastable as isolated but stabilized by binding of cellular factors.

Acknowledgements

We thank Dr M. Schmitz for stimulating discussions. The authors were supported by grants from the Deutsche Forschungsgemeinschaft and the Fonds der Chemischen Industrie.

References

1. Adams CC, Stern DB: Control of mRNA stability in chloroplasts by 3′ inverted repeats: effects of stem and loop mutations on degradation of *psbA* mRNA *in vitro*. Nucl Acids Res 18: 6003–6010 (1990).
2. Alifano P, Rivellini F, Piscitelli C, Arraiano CM, Bruni CB, Carlomagno MS: Ribonuclease E provides substrates for ribonuclease P-dependent processing of a polycistronic mRNA. Genes Devel 8: 3021–3031 (1994).
3. Anderson P, Monforte J, Tritz R, Nesbitt S, Hearst J, Hampel A: Mutagenesis of the hairpin ribozyme. Nucl Acids Res 22: 1096–1100 (1994).
4. van Batenburg FH, Gultyaev AP, Pleij CW: An APL-programmed genetic algorithm for the prediction of RNA secondary structure. J Theor Biol 174: 269–280 (1995).
5. Baumstark T, Riesner D: Only one of four possible secondary structures of the central conserved region of potato spindle tuber viroid is a substrate for processing in a potato nuclear extract. Nucl Acids Res 23: 4246–4254 (1995).
6. Bejarano ER, Lichtenstein CP: Expression of TGMV antisense RNA in transgenic tobacco inhibits replication of BCTV but not ACMV geminiviruses. Plant Mol Biol 24: 241–248 (1994).
7. van Belkum A, Abrahams JP, Pleij CW, Bosch L: Five pseudoknots are present in the 204 nucleotides long 3′ noncoding region of the tobacco mosaic virus RNA. Nucl Acids Res: 7673–7686 (1985).

8. Betts L, Spremulli L: Analysis of the role of the Shine-Dalgarno sequence and mRNA secondary structure on the efficiency of translational initiation in *Euglena gracilis* chloroplast *atpH* mRNA. J Biol Chem 269: 26456–26463 (1994).

9. Bierley I: Ribosomal frame shifting on viral RNAs. J Gen Virol 76: 1885–1892 (1995).

10. Branch A, Robertson HD: A replication cycle for viroids and other small infectious RNAs. Science 223: 450–455 (1984).

11. Brault V, Miller WA: Translational frameshifting mediated by a viral sequence in plant cells. Proc Natl Acad Sci USA 89: 2262–2266 (1992).

12. Brown JW, Haas ES, Gilbert DG, Pace NR: The Ribonuclease P database. Nucl Acids Res 22: 3660–3662 (1994).

13. Butcher SE, Burke JM: A photo-cross-linkable tertiary structure motif found in functionally distinct RNA molecules is essential for catalytic function of the hairpin ribozyme. Biochemistry 33: 992–999 (1994).

14. Camerini-Otero RD, Hsieh P: Parallel DNA triplexes, homologous recombination and other homology dependent DNA interactions. Cell 73: 217–223 (1993).

15. Cech TR: Self-splicing of group I introns. Annu Rev Biochem 59: 543–568 (1990).

16. Cech TR, Zaug AJ, Grabowski PJ: The intervening sequence of the ribosomal RNA precursor is converted to a circular RNA in isolated nuclei of *Tetrahymena*. Cell 23: 486–496 (1981).

17. Chao MY, Kan MC, Lin-Chao S: RNAII transcribed by IPTG-induced T7 RNA polymerase is non-functional as a replication primer for ColE1-type plasmids in *Escherichia coli*. Nucl Acids Res 23: 1691–1695 (1995).

18. Chastein M, Tinoco I Jr: Structural elements in RNA. Prog Nucl Acids Res Mol Biol 41: 131–173 (1991).

19. Chastain M, Tinoco I Jr: A base-triple structural domain in RNA. Biochemistry 31: 12733–12741 (1992).

20. Christopher DA, Hallick RB: *Euglena gracilis* chloroplast ribosomal protein operon: a new chloroplast gene for ribosomal protein L5 and description of a novel organelle intron category designated group III. Nucl Acids Res 17: 7591–7608 (1989).

21. Cornelissen M, Vandewiele M: Both RNA level and translation efficiency are reduced by anti-sense RNA in transgenic tobacco. Nucl Acids Res 17: 833–843 (1989).

22. Crum CJ, Hu J, Hiddinga HJ, Roth DA: Tobacco mosaic virus infection stimulates the phosphorylation of a plant protein associated with double-stranded RNA-dependent protein kinase activity. J Biol Chem 263: 13440–13443 (1988).

23. Dam E ten, Pleij K, Draper D: Structural and functional aspects of RNA pseudoknots. Biochemistry 31: 11665–11676 (1992).

24. Danon A, Mayfield SP: ADP-dependent phosphorylation regulates RNA-binding *in vitro*: implications in light-modulated translation. EMBO J 13: 2227–2235 (1994).

25. Danon A, Mayfield SPY: Light-regulated translational activators: identification of gene specific mRNA-binding proteins. EMBO J 10: 3993–4001 (1991).

26. Day MJD, Ashurst JL, Mathias SF, Watts JW, Wilson TMA, Dixon RA: Plant viral leaders influence expression of a reporter gene in tobacco. Plant Mol Biol 23: 97–109 (1993).

27. Diener TO (ed): The Viroids. Plenum Press, New York (1987).

28. Diener TO, Hammond RW, Black T, Katze MG: Mechanism of viroid pathogenesis: Differential activation of the interferon-induced, double-stranded RNA activated, Mr

29. Dinesh-Kumar SP, Miller WA: Control of start codon choice on a plant viral RNA encoding overlapping genes. Plant Cell 5: 679–692 (1993).

30. Dinman JD, Icho T, Wickner RB: A −1 ribosomal frameshift in a doublestranded RNA virus of yeast forms a *gag-pol* fusion protein. Proc Natl Acad Sci USA 88: 174–178 (1991).

31. Dolfini S, Consonni G, Mereghetti M, Tonelli C: Antiparallel expression of the sense and antisense transcripts of maize α-tubulin genes. Mol Gen Genet 241: 161–169 (1993).

32. Emerick VL, Woodson SA: Self-splicing of the *Tetrahymena* pre-mRNA is decreased by misfolding during transcription. Biochemistry 32: 14062–14067 (1993).

33. Felden B, Florentz C, Giége R, Westhof E: Solution structure of the 3′ end of brome mosaic virus genomic RNAs. Conformational mimicry with canonical tRNAs. J Mol Biol 235: 508–531 (1994).

34. Fontana W, Konings DAM, Stadler PF, Schuster P: Statistics of RNA secondary structures. Biopolymers 33: 1389–1404 (1993).

35. Forster AC, Davies C, Sheldon CC, Jeffries AC, Symons RH: Self-cleaving viroid and newt RNAs may only be active as dimers. Nature 334: 265–267 (1988).

36. Fresco JR, Adams A, Ascione R, Henley D, Lindahl T: Tertiary structure in transfer ribonucleis acids. Cold Spring Harbor Symp Quant Biol 31: 527–537 (1966).

37. Gallie DR: Posttranscriptional regulation of gene expression in plants. Annu Rev Plant Physiol Plant Mol Biol 44: 77–105 (1993).

38. Gallie DR, Sleat DE, Watts JW, Turner PC, Wilson TMA: A comparison of eukaryotic viral 5′ leader sequences as enhancers of mRNA expression *in vivo*. Nucl Acids Res 15: 8693–8711 (1987).

39. Gallie DR, Walbot V: RNA pseudoknot domain of tobacco mosaic virus can functionally substitute for a poly(A) tail in plant and animal cells. Genes Devel 4: 1149–1157 (1990).

40. Gallie DR, Feder JN, Schimke RT, Walbot V: Functional analysis of the tobacco mosaic virus tRNA-like structure in cytoplasmic gene regulation. Nucl Acids Res 19: 5031–5036 (1991).

41. Garcia A, Cuin JV, Pleij CWA: Differential response to frame shift signals in eukaryotic and prokaryotic translational systems. Nucl Acids Res 21: 401–406 (1993).

42. Giége R, Florentz C, Dreher TW: The TYMV tRNA-like structure. Biochimie 75: 569–582 (1993).

43. Goldschmidt-Clermont M, Girard-Bascou J, Choquet Y, Rochaix R-D: A small chloroplast RNA may be required for trans-splicing in *Chlamydomonas reinhardtii*. Cell 65: 135–143 (1991).

44. Green PJ, Pines O, Inouye M: The role of antisense RNA in gene regulation. Annu Rev Biochem 55: 569–597 (1986).

45. Groeneveld H, Thimon K, van Duin J: Translational control of maturation protein synthesis in phage MS2: A role for the kinetics of RNA folding? RNA 1: 79–88 (1995).

46. Gultyaev AP, van Batenburg FH, Pleij CW: The computer simulation of RNA folding pathways using a genetic algorithm. J Mol Biol 250: 37–51 (1995).

47. Gultayaev AP, van Batenberg FHD, Pleij CWA: The influence of a metastable structure in plasmid primer RNA on antisense RNA binding kinetics. Nucl Acids Res 23: 3718–3725 (1995).

48. Haenni A-L, Joshi S, Chapeville F: tRNA-like structures in the genomes of viruses. Prog Nucl Acid Res Mol Biol 27: 85–104 (1982).

68 000 protein kinase by viroid strains of varying pathogenicity. Biochimie 75: 533–538 (1993).

104

49. Harris ME, Nolan JM, Malhotra A, Brown JW, Harvey SC, Pace NR: Use of photoaffinity crosslinking and molecular modeling to analyze the global architecture of ribonuclease P RNA. EMBO J 13: 3953–3963 (1994).

50. Hecker R, Wang Z, Steger G, Riesner R: Analysis of RNA structures by temperature gradient gelelectrophoresis: viroid replication and processing. Gene 72: 59–74 (1988).

51. Heus H, Pardi A: Structural features that give rise to the unusual stability of RNA hairpins containing GNRA loops. Science 253: 191–194 (1991).

52. Heinemeyer W, Alt J, Herrmann RG: Nucleotide sequence of the clustered genes for apocytochrom b6 and subunit 4 of the cytochrome b6f complex in the spinach plastid genome. Curr Gen 8: 543–549 (1984).

53. Hiddinga JH, Crum cJ, Hu J, Roth DA: Viroid-induced phosphorylation of a host protein related to dsRNA-dependent protein kinase. Science 241: 451–453 (1988).

54. Hildebrandt M, Nellen W: Differential antisense transcription form the EB4 gene locus: implications on the antisense-mediated regulation of mRNA stability. Cell 69: 197–204 (1992).

55. Jacks T, Madhani HD, Masiarz FR, Varmus HE: Signals for ribosomal frameshifting in Rous sarcoma virus gag-pol region. Cell 55: 447–458 (1988).

56. Jiang C-Z, Kliebenstein D, Ne N, Rodermel S: Destabilization of rbcS sense transcripts by antisense RNA. Plant Mol Biol 25: 569–576 (1994).

57. Joblin SA, Gehrke L: Enhanced translation of chimeric messenger RNAs containing a plant viral untranslated leader sequence. Nature 325: 622–625 (1987).

58. Kim KH, Lommel SA: Identification and analysis of the site of −1 ribosomal frameshifting in red clover necrotic mosaic virus. Virology 200: 574–582 (1994).

59. Kirsch W, Seyer P, Herrmann RG: Nucleotide sequence of the clustered genes for two P700 chlorophyll a apoproteins of the photosystem I reaction center and the ribosomal protein S14 of the spinach plastid chromosome. Curr Genet 10: 843–855 (1986).

60. Klaff P, Guissem W: A 43 kD light-regulated chloroplast RNA-binding protein interacts with the psbA 5′ untranslated region. Photosyn Res 46: 235–248 (1995).

61. Koo JS, Spremulli LL: Effect of the secondary structure in the Euglena gracilis chloroplast ribulose-bisphosphate carboxylase/oxygenase messenger RNA on translational initiation. J Biol Chem 269: 7501–7508 (1994).

62. Kozak M: Context effects and inefficient initiation at non-AUG codons in eukaryotic cell-free translation systems. Mol Cell Biol 9: 507–5080 (1989).

63. van der Krol AR, Mol JNM, Stuitje AR: Antisense genes in plants: an overview. Gene 72: 45–50 (1988).

64. Kujawa AB, Drugeon G, Hulanicka D, Haenni A-L: Structural requirements for efficient translational frameshifting in the synthesis of the putative viral RNA-dependent RNA polymerase. Nucl Acids Res 21: 2165–2171 (1993).

65. Lahser FC, Marsh LE, Hall TC: Contributions of the brome mosaic virus RNA-3 3′-nontranslated region to replication and translation. J Virol 67: 3295–3303 (1993).

66. Langland JO, Jin S, Bertram BL, Roth DA: Identification of a plant-encoded analog of PKR, the mammalian double-stranded RNA-dependent protein kinase. Plant Physiol 108: 259–1267 (1995).

67. Lawton MA, Yamamoto RT, Hanks SK, Lamb CJ: Molecular cloning of plant transcripts encoding protein kinase homologs. Proc Natl Acad Sci USA 86: 3140–3144 (1989).

68. Leathers V, Tanguay R, Kobayashi M, Gallie DR: A phylogenetically conserved sequence within viral 3′ untranslated RNA pseudoknots regulates translation. Mol Cell Biol 13: 5331–5347 (1993).

69. Loss P, Schmitz M, Steger G, Riesner D: Formation of a thermodynamically metastable structure containing hairpin II is critical for infectivity of potato spindle tuber viroid RNA. EMBO J 10: 719–727 (1991).

70. Ma CK, Kolesnikow T, Rayner JC, Simons EL, Yim H, Simons RW: Control of translation by mRNA secondary structure: the importance of the kinetics of structure formation. Mol Microbiol 14: 1033–1047 (1994).

71. Maher LL, III: DNA triple-helix formation: an approach to artificial gene repressors? BioEssays 14: 807–815 (1992).

72. Maher LL, III: Inhibition of T7 RNA polymerase initiation by triple-helical DNA complexes: a model for artificial gene repression. Biochemistry 31: 7587–7594 (1992).

73. Marchfelder A, Brennicke A: Characterization and partial purification of tRNA processing activities from potato mitochondria. Plant Physiol 105: 1247–1254 (1994).

74. Marchfelder A, Brennicke A: Plant mitochondrial RNase P and E. coli RNase P have different substrate specificities. Biochem Mol Biol Int 29: 621–633 (1993).

75. Martin GB, Brommonschenkel SH, Chunwongse J, Frary A, Ganal MW, Spivey R, Wu T, Earle ED, Tanksley SD: Map-based cloning of a protein kinase gene conferring disease resistance in tomato. Science 262: 1432–1436 (1993).

76. Martinez HM: An RNA folding rule. Nucl Acids Res 12: 323–334 (1984).

77. Matousek J, Schröder ARW, Trnena L, Reimers M, Baumstark T, Dedic P, Vlasak J, Becker I, Kreuzaler F, Fladung M, Riesner D: Inhibition of viroid infection by antisense RNA expression in transgenic plants. Biol Chem Hoppe-Seyler 375: 765–777 (1994).

78. Matousek J, Trnena L, Oberhauser R, Lichtenstein CP, Nellen W: dsRNA degrading nucleases are differentially expressed in tobacco anthers. Biol Chem Hoppe-Seyler 375: 261–269 (1994).

79. Mayfield SP, Cohen A, Danon A, Yohn CB: Translation of the psbA mRNA of Chlamydomonas reinhardtii requires a structured RNA element contained within the 5′ untranslated region. J Cell Biol 127: 1537–1545 (1994).

80. McCaskill: The equilibrium partition function and base pair binding probabilities for RNA secondary structure. Biopolymers 29: 1105–1119 (1990).

81. Michel F, Westhof E: Modelling of the three-dimensional architecture of group I catalytic introns based on comparative sequence analysis. J Mol Biol 216: 585–610 (1990).

82. Michel F, Ellington AD, Couture S, Szostak JW: Phylogenetic and genetic evidence for base triples in the catalytic domain of group introns. Nature 347: 578–580 (1990).

83. Michel F, Janquier A, Dujon B: Comparison of fungal mitochondrial introns reveals extensive homologies in RNA secondary structure. Biochimie 64: 867–881 (1982).

84. Mironov AA, Dyakonova LP, Kister AE: A kinetic approach to the prediction of RNA secondary structures. J Biomol Struct Dyn 2: 953–962 (1985).

85. Mironov A, Kister A: RNA secondary structure formation during transcription. J Biomol Struct Dyn 4: 1–9 (1986).

86. Mironov AA, Lebedev VF: A kinetic model of RNA folding. Biosystems 30: 49–56 (1993).

87. Mironov AA, Alexandrov NN, Bogodarova NYu, Grigorjev A, Lebedev VF, Lunovskaya LV, Truchan ME, Pevzner PA:

DNASUN: a package of computer programs for the biotechnology laboratory: Comput Appl Biosci 11: 331–335 (1995).

88. Mol JNM, van der Krol AR, van Tunen AJ, van Blokland R, de Lange P, Stuitje AR: Regulation of plant gene expression by antisense RNA. FEBS Lett 268: 427–430 (1990).

89. Mühlbach HP, Sänger HL: Viroid replication is inhibited by α-Amanitin. Nature 278: 185–188 (1979).

90. Musters W, Boon K, van der Sande CAFM, van Heerikhuizen H, Planta RJ: Functional analysis of transcribed spacers of yeast ribosomal RNAs. EMBO J 9: 3989–3996 (1990).

91. Musters W, Planta RJ, van Heerikhuizen H, Raué H: Functional analysis of the transcribed spacers of Saccharomyces cerevisiae ribosomal DNA: it takes a precursor to form a ribosome. In: Hill WE, Dahlberg AE, Garrett RA, Moore RA, Schlessinger PB, Warner JR (eds) The Ribosome: Structure, Function and Evolution, pp. 435–442. American Society for Microbiology, Washington, DC (1990).

92. Nellen W, Lichtenstein C: What makes an mRNA anti-sensitive? Trends Biochem. Sci 18: 419–423 (1993).

93. Nues RW, Rientjes JMJ, Morré SA, Mollee E, Planta RJ, Venema J, Raué HA: Evolutionary conserved structural elements are critical for processing of internal transcribed spacer 2 from Saccharomyces cerevisiae precursor ribosomal RNA. J Mol Biol 250: 24–36 (1995).

94. Oh BK, Pace NR: Interaction of the 3'-end of tRNA with ribonuclease P RNA. Nucl Acids Res 22: 4087–4094 (1994).

95. Platt T: Transcription termination and the regulation of gene expression. Annu Rev Biochem 55: 339–372 (1986).

96. Pleij CWA, Reitveld K, Bosch L: A new principle of RNA folding based on pseudoknotting. Nucl Acids Res 13: 1717–1731 (1985).

97. Pley HW, Flaherty KM, McKay DB: Three-dimensional structure of a hammerhead ribozyme. Nature 372: 68–74 (1994).

98. Pnueli L, Hareven D, Rounsley SD, Yanofsky MF, Lifschitz E: Isolation of the tomato AGAMOUS gene TAG1 and analysis of its homeotic role in transgenic plants. Plant Cell 6: 163–167 (1994).

99. Poeschla E, Wong-Staal F: Antiviral and anticancer ribozymes. Curr Opin Oncol 6: 601–606 (1994).

100. Postel EH, Flint SJ, Kessler DJ, Hogan ME: Evidence that a triplex-forming oligodeoxynucleotide binds to c-myc promotor in HeLa cells, thereby reducing c-myc mRNA levels. Proc Natl Acad Sci USA 88: 8227–8231 (1991).

101. Powell PA, Stark DM, Sanders PR, Beachy RN: Protection against tobacco mosaic virus in transgenic plants that express tobacco mosaic virus antisense RNA. Proc Natl Acad Sci USA 86: 6949–6952 (1989).

102. Pürfer D, Tacke E, Schmitz J, Kull B, Kaufmann A, Rhode W: Ribosomal frameshifting in plants: a novel signal directs the −1 frameshift in the synthesis of the putative viral replicase of potato leafroll luteovirus. EMBO J 11: 1111–1117 (1992).

103. Puglisi JD, Tan R, Calnan BJ, Frankel AD, Willamson JR: Conformation of the TAR RNA-arginine complex by NMR spectroscopy. Science 257: 76–80 (1992).

104. Puglisi JD, Wyatt JR, Tinoco I: Conformation of an RNA pseudoknot. J Mol Biol 214: 437–453 (1990).

105. Pyle AM: Ribozymes: a distinct class of metalloenzymes. Science 261: 709–714 (1993).

106. Qu F, Heinrich C, Loss P, Steger G, Tien P, Riesner D: Multiple pathways of reversion in viroids for conservation of structural elements. EMBO J 12: 2129–2139 (1993).

107. Rando RF, DePaolis L, Durland RH, Jayaraman K, Kessler DJ, Hogan ME: Inhibition of T7 and T3 RNA polymerase directed transcription elongation in vitro. Nucl Acids Res 22: 678–685 (1994).

108. Riesner D: Structure formation of viroids. In: Diener TO (ed) The Viroids, pp. 63–98. Plenum Press, New York (1987).

109. Riesner D, Henco K, Steger G: Temperature-gradient electrophoresis: a method for analysis of conformational transitions and mutations in nucleic acids and proteins. In: Chrambach A, Dunn MJ, Radola BJ (eds) Advances in Electrophoresis, vol 4, pp. 169–250, VCH, Weinheim (1991).

110. Rietveld K, van Poelgeest R, Pleij CWA, Boom JH van, Bosch L: The tRNA-like structure at the 3' terminus of turnip yellow mosaic virus RNA. Differences and similarities with canonical tRNA. Nucl Acids Res 10: 1929–1946 (1982).

111. Rochaix J-D: Post-transcriptional steps in the expression of chloroplast genes. Annu Rev Cell Biol 8: 1–28 (1992).

112. Rogers JC: RNA complementary to α-amylase mRNA in barley. Plant Mol Biol 11: 125–138 (1988).

113. Rosenbaum V, Riesner D: Temperature-gradient gel electrophoresis: thermodynamic analysis of nucleic acids and proteins in purified form and in cellular extracts. Biophys Chem 26: 235–246 (1987).

114. Roossinck MJ, Sleat D, Palukaitis P: Satellite RNAs of plant viruses: structures and biological effects. Microbiol Rev 56: 265–279 (1992).

115. Rossi JJ: Practical ribozymes. Making ribozymes work in cells. Curr Biol 4: 469–471 (1994).

116. Saldanha R, Mohr G, Belfort M, Lambowitz AM: Group I and group II introns. FASEB J 7: 15–24 (1993).

117. Schindler I-M, Mühlbach HP: Involvement of nuclear DNA-dependent RNA polymerases in potato spindle tuber viroid replication: a reevaluation. Plant Sci 84: 221–229 (1992).

118. Schmitz M, Steger G: Base-pair probability profiles of RNA secondary structures. Comp Appl Biosci 8: 389–399 (1992).

119. Schmitz M, Steger G: Description of RNA folding by 'simulated annealing'. J Mol Biol 255: 254–266 (1996).

120. Shinozaki K, Ohme M, Tanaka M, Wakasuki T, Hashida N, Matsubayasha T, Zaita N, Chungwongse J, Obokata J, Yamaguchi-Shinozaki K, Otho C, Torazawa K, Meng BY, Sugita M, Deno H, Kamogashira T, Yamada K, Kusuda J, Takaiwa F, Kata A, Todoh N, Shimada H, Sugiura M: The complete nucleotide sequence of the tobacco chloroplast genome; its gene organisation and expression. EMBO J 5: 2043–2049 (1986).

121. Staub JM, Maliga P: Accumulation of D1 polypeptide in tobacco plastids via the untranslated region of the psbA mRNA. EMBO J 12: 601–606 (1993).

122. Staub JM, Maliga P: Translation of psbA mRNA is regulated by light via the 5' untranslated region in tobacco plastids. Plant J 6: 547–553 (1994).

123. Steger G, Hofmann H, Förtsch J, Gross HJ, Randles JW, Sänger HL, Riesner D: Conformational transitions in viroids and virusoids: comparison of results from energy minimization algorithm and from experimental data. J Biomol Struct Dyn 2: 543–571 (1984).

124. Steger G, Baumstark T, Mörchen M, Tabler M, Tsagris M, Sänger HL, Riesner D: Structural requirements for viroid processing by RNase T1. J Mol Biol 227: 719–737 (1992).

125. Stern DB, Gruissem W: Control of plastid gene expression: 3' inverted repeats act as mRNA processing signals and stabilizing elements, but do not terminate transcription. Cell 51: 1145–1157 (1987).

126. Stern DB, Radwanski ER, Kindle K: A 3' stem-loop structure of the Chlamydomonas chloroplast atpB gene regulates mRNA accumulation in vivo. Plant Cell 3: 285–297 (1991).

106

127. Sullivan SM: Development of ribozymes for gene therapy. J Invest Dermatol 103: 85–89 (1994).
128. Symons RH: Ribozymes. Crit Rev Plant Sci 10: 189–234 (1991).
129. Takamatsu N, Watanabe Y, Meshi T, Okada Y: Mutational analysis of the pseudoknot region in the 3′ noncoding region of tobacco mosaic virus RNA. J Virol 64: 3686–3693 (1990).
130. Theissen G, Thelen L, Wagner R: Some base substitutions in the leader of Escherichia coli ribosomal RNA operon affect the structure and function of ribosomes–evidence for a transient scaffold function of the rRNA leader. J Mol Biol 233: 203–218 (1993).
131. Theologis A, Oeller PW, Wong LM, Rottmann WH, Gantz DM: Use of a tomato mutant constructed with reverse genetics to study fruit ripening, a complex developmental process. Devel Genet 14: 282–295 (1993).
132. Tuschl T, Gohlke C, Jovin TM, Westhof E, Eckstein F: A three-dimensional model for the hammerhead ribozyme based on fluorescence measurements. Science 266: 785–789 (1994).
133. Varani G, Cheong C, Tinoco Jr I: Structure of an unusually stable RNA hairpin. Biochemistry 30: 3280–3289 (1991).
134. Varani G, Wimberly B, Tinoco Jr I: Conformation and dynamics of an RNA internal loop. Biochemistry 28: 7760–7772 (1989).
135. Volkmann S, Jendis J, Faruendorf A, Mölling K: Inhibition of HIV-1 reverse transcription by triple-helix forming oligonucleotides with viral RNA. Nucl Acids Res 23: 1204–1212 (1995).
136. Wagner EGH, Simons RW: Antisense RNA control in bacteria, phages and plasmids. Annu Rev Microbiol 48: 713–742 (1994).
137. Westhof E, Altman S: Three-dimensional working model of M1 RNA, the catalytic RNA subunit of ribonuclease P from Escherichia coli. Proc Natl Acad Sci USA 91: 5133–5137 (1994).
138. Westhof E, Dumas P, Moras D: Cristallographic refinement of yeast aspartic acid transfer RNA. J Mol Biol 184: 119–145 (1985).
139. Westhof E, Romby P, Romaniuk PJ, Ebel J-P, Ehresmann C, Ehresmann B: Computer modeling from solution data of spinach chloroplast and of Xenopus laevis somatic and oocyte 5S rRNAs. J Mol Biol 207: 417–431 (1989).
140. Westhof E, Michel F: Prediction and experimental investigation of RNA secondary and tertiary folding. In: Nagai K, Mattaj IW (eds) RNA-Protein interactions, pp. 25–51. IRL Press, Oxford (1994).
141. Williams AL, Tinoco I Jr: A dynamic programming algorithm for finding alternative RNA secondary structures. Nucl Acids Res 14: 299–315 (1986).
142. Woodson SA, Cech TR: Alternative structures in the 5′ exon affect both forward and reverse self-splicing of the Tetrahymena intervening sequence RNA. Biochemistry 30: 2042–2050 (1991).
143. Wyatt JR, Puglisi JD, Tinoco I: RNA pseudoknots: stability and loop size requirements. J Mol Biol 214: 455–470 (1990).
144. Young SL, Krawczyk SH, Matteucci MD, Toole JJ: Triple helix formation inhibits transcription elongation in vitro. Proc Natl Acad Sci USA 88: 10023–10026 (1991).
145. Zuker M: On finding all suboptimal foldings of an RNA molecule. Science 244: 48–52 (1989).
146. Zuker M, Stiegler P: Optimal computer folding of large RNA sequences using thermodynamics and auxiliary information. Nucl Acids Res 9: 133–148 (1981).
147. Zurawski G, Perrot B, Bottomley W, Whitfeld PR: The structure of the gene for the large subunit of the ribulose-1,5-bisphosphate carboxylase from spinach chloroplast DNA. Nucl Acids Res 9: 3251–3270 (1981).

Plant Molecular Biology **32**: 107–144, 1996.
© 1996 *Kluwer Academic Publishers.*

The plant translational apparatus

Karen S. Browning
Department of Chemistry and Biochemistry, University of Texas at Austin, Austin, TX 78712, USA

Key words: Protein synthesis, initiation factors, elongation factors, ribosomes, plants, translation

Contents

Abstract	107	Initiation of plant protein synthesis	121
Introduction	107	Elongation	121
Initiation	108	eEF1	124
eIF1	108	eEF1: the subunits	124
eIF2	109	eEF1A	124
eIF2 cofactors	111	eEF1Bα, eEF1Bβ and eEF1Bγ	126
eIF3	111	eEF2	127
eIF4A	112	Termination	129
eIF4B	114	Plant ribosomes	130
eIF1A (eIF4C)	116	Protein synthesis and the cytoskeleton	133
eIF5A (eIF4D)	116	Epilogue	134
eIF4F and eIFiso4F	117	Acknowledgements	134
eIF5	120	References	134
eIF6	121		

Abstract

Protein synthesis in both eukaryotic and prokaryotic cells is a complex process requiring a large number of macromolecules: initiation factors, elongation factors, termination factors, ribosomes, mRNA, amino-acylsynthetases and tRNAs. This review focuses on our current knowledge of protein synthesis in higher plants.

Abbreviations: eIF, eukaryotic initiation factor; eEF, eukaryotic elongation factor; EST, expressed sequence tag; eRF, eukaryotic release factor; GUS, β-glucoronidase; HCR, heme-controlled repressor; PKR, double-stranded RNA-activated protein kinase; SDS-PAGE, sodium dodecyl sulfate polyacrylamide gel electrophoresis.

Introduction

Protein synthesis in both eukaryotic and prokaryotic cells is a complex process requiring a large number of macromolecules: initiation factors, elongation factors, termination factors, ribosomes, mRNA, amino-acylsynthetases and tRNAs. The regulation of the coordinate expression of all of the components required for protein synthesis and the regulation of their activities must also be very complex. Eukaryotic protein synthesis has been studied extensively in wheat germ, mammals and yeast. Numerous reviews have been published detailing what is known about the factors required and the intermediate steps in the translation process in mammalian cells [118–120, 152–154, 191–194, 202, 218, 226a, 251–253, 294, 306, 330, 342] and yeast cells [7, 125, 226]. This review focuses on our current knowledge of protein synthesis in higher

108

plants and, in particular, wheat germ. It is not intended to be a comprehensive review of the translational apparatus in all eukaryotic systems. Specific examples from yeast or mammalian cells are given for comparison purposes and where there is a lack of information from higher plants. The mechanism of protein synthesis and the macromolecules needed for protein synthesis are very similar among all eukaryotes. However, there are some aspects that make each system unique.

In the past 10–15 years a tremendous amount of information about plant protein synthesis has been gathered. However, there are still some significant gaps in our knowledge, particularly about how plants regulate protein synthesis during development and stress. With the variety of molecular biology tools now available, we should be able to make great strides in the next 10–15 years in increasing our understanding of this extremely complicated cellular process in plants.

A new nomenclature for the initiation and elongation factors is used in this review (see Tables 1 and 5). This nomenclature was adopted by a IUBMB committee in April 1995 and will be published in Biochemie (J. Hershey, personal communication). The major changes for eukaryotic factors are: (1) the omission of the '−' in the names of the factors; (2) re-instatement of the eIF6 designation for the anti-reassociation factor; (3) re-naming of the subunits of eIF4F, eIF4G (large subunit) and eIF4E (cap-binding protein); and (4) a new system for naming the subunits of elongation factor 1.

Initiation

A summary of what is known about the physical and functional properties of the factors required for the initiation of protein synthesis in higher plants is given in Tables 1 and 2, and an outline for the intermediate steps in the initiation process is given in Fig. 1.

Each of the initiation factors will be discussed in terms of its physical and functional properties, regulation, amino acid sequence, and gene structure and/or expression.

eIF1

eIF1 has not been purified from plants; however, the gene for this factor, GOS2, has been identified in rice [79]. eIF1 isolated from rabbit reticulocytes is a single polypeptide (M_r ca. 15 000, [21, 274, 329]). The function of eIF1 is unclear as it only slightly

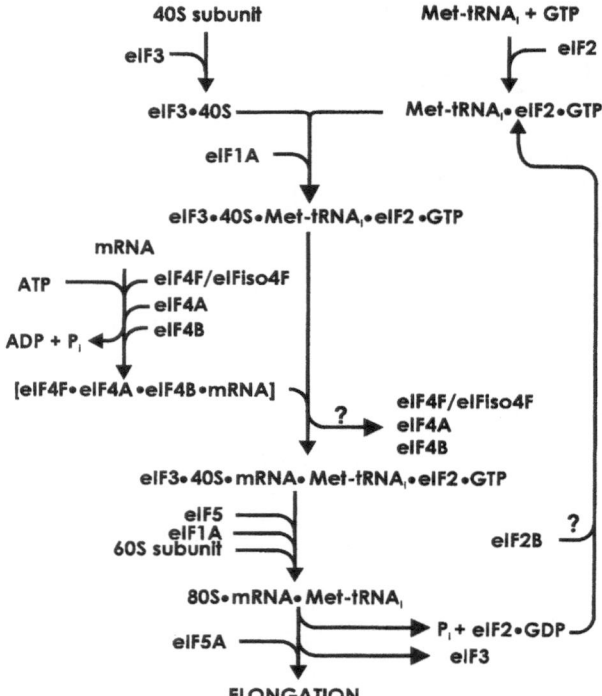

Figure 1. Intermediate steps in the initiation of eukaryotic protein synthesis.

stimulates polypeptide synthesis at a number of steps [274]. Recently, the cDNA for mammalian eIF1 was sequenced [145] and found to be the counterpart of the yeast gene *SUI1* [348] and the rice gene *GOS2* [79]. The yeast *SUI1* gene product, in conjunction with eIF2•Met-tRNA$_i$, is essential for the selection of the correct AUG start site by the ribosome [348]. The genes for rice eIF1 (GOS2) and yeast eIF1 (SUI1) are single-copy genes [79, 348]. The promoter for the rice eIF1 gene (GOS2) is very strong and the mRNA is expressed in all tissues tested. The GOS2 promoter contains a TGACG motif that specifically binds a *cis*-acting nuclear protein from rice, ASF-1, that has similar DNA-binding properties to the tobacco transcription factor TGA-1a [79]. The rice GOS2 promoter is also active in other monocot species based on its ability to express a fusion protein [79].

More work is needed to understand the interaction of eIF1 with eIF2•Met-tRNA$_i$•GTP and how these proteins facilitate the selection of the correct AUG by the 40S ribosomal subunit.

Table 1. Initiation factors from higher plants.

Factor	$M_r{}^a$	Function	Plant source [ref.]	DNA sequence information
eIF1	12 732	Unknown in plants	not found	rice (GOS 2 gene) X51910 [79]
eIF2	α 42 000;	Forms ternary complex with GTP	wheat germ [165]	wheat (all subunits, A. Metz
	β 38 000	and Met-tRNA$_i$		and K. Browning, unpublished)
	γ 50 000	Binds Met-tRNA$_i$ to 40S subunit		
eIF3	10 subunits^c	Binds mRNA to 40S subunits	wheat germ [165]	None known
eIF4A	46 000	ATP-dependent unwinding of mRNA	wheat germ [165]	*Arabidopsis* X65052, X65053 [197]
		Binds mRNA to 40S units		maize U17979 [138]
				rice D12627 [215]
				tobacco^d
				wheat Z21510 [195]
eIF4B	59 000	ATP-dependent unwinding of mRNA	wheat germ [34]	wheat (S. Malmström and
		Binds mRNA to 40S subunit		K. Browning, unpublished)
eIF1A/eIF4C	17 600	Unclear	wheat germ [309]	wheat L08060 [81]
eIF5A/eIF4D^e	16 700	Unclear	rice [188]	alfalfa X59441 [234]
				tobacco X63541, X63542, X63543 [57]
eIF4F			wheat germ [40]	eIF4G (p220)
eIF4G^f	220 000	ATP-dependent unwinding of mRNA;		wheat (L. Allen and
				K. Browning, unpublished)
eIF4E^g	26 000	Binds mRNA to 40S subunit		eIF4E (p26)
				Arabidopsis (K. Ruud and
				K. Browning, unpublished)
				wheat Z12616 [198]
				rice U34597
				tobacco (J. Combe and
				D. Twell, unpublished)
eIFiso4F			wheat, maize,	eIFiso4G (p86)
eIFiso4G^f	86 000	ATP-dependent unwinding of mRNA	cauliflower [40]	wheat M95746, M95747 [4]
eIFiso4E^g	28 000	Binds mRNA to 40S subunit		maize (L. Allen, Y.F. Chen
				and L. Morejohn, unpublished)
				Arabidopsis (R. Ahmed and
				K. Browning, unpublished)
				eIFiso4E (p28)
				wheat M95818 [4]
				rice U34598
eIF5	48 918	Joining of 60S subunit	wheat, partially purified [165, 282]	kidney bean 47221 [93a]
eIF6	25 000	Prevents association of 60S 40S subunits	wheat germ [260]	None known

^{a} Based on SDS-PAGE gel mobility. The M_r for eIF1, eIF5 and eIF5A are calculated from deduced amino acid sequence.

^{b} The reference indicates the most current method of purification. See text for additional references.

^{c} Subunits for wheat germ eIF3: 116 000; 107 000; 87 000; 83 000; 56 000; 45 000; 41 000; 36 000; 34 000; 28 000. The 41 000 and 28 000 are present in less than stoichiometric amounts relative to the other subunits [122].

^{d} tobacco eIF4A genes: 4A4, X79006; 4A8, X79005; 4A10, X79008; 4A12, X79007 [32]; tobacco eIF4A cDNAs: 4A2, X61205; 4A3, X61206; 4A6, X79139; 4A7, X79137; 4A8, X79004; 4A9, X79135; 4A10, X79009; 4A11, X79136; 4A13, X79140; 4A14, X79141; 4A15, X79138 [31, 224, 225].

^{e} Contains the unique amino acid hypusine.

^{f} In the new nomenclature the large subunit of eIF4F, previously designated eIF-4Fγ, is now designated eIF4G.

^{g} This subunit is a m^{7}G cap-binding protein. In the new nomenclature the small subunit of eIF4F, previously designated eIF–4Fα, eIF–4E or CBP, is now designated eIF4E.

eIF2

Plant eIF2 has been isolated from wheat germ [22, 66, 165, 279, 285, 296] and consists of three non-identical subunits with molecular weights of ca. 38 000 (p38), 42 000 (p42) and 50 000 (p50) as measured by SDS-PAGE (see Fig. 2). Rabbit reticulocyte eIF2 also consists of three non-identical subunits with molecular weights of ca. 36 000 (α), 38 000 (β) and 55 000 (γ) [192]. Initially, it was thought that the p38, p42 and

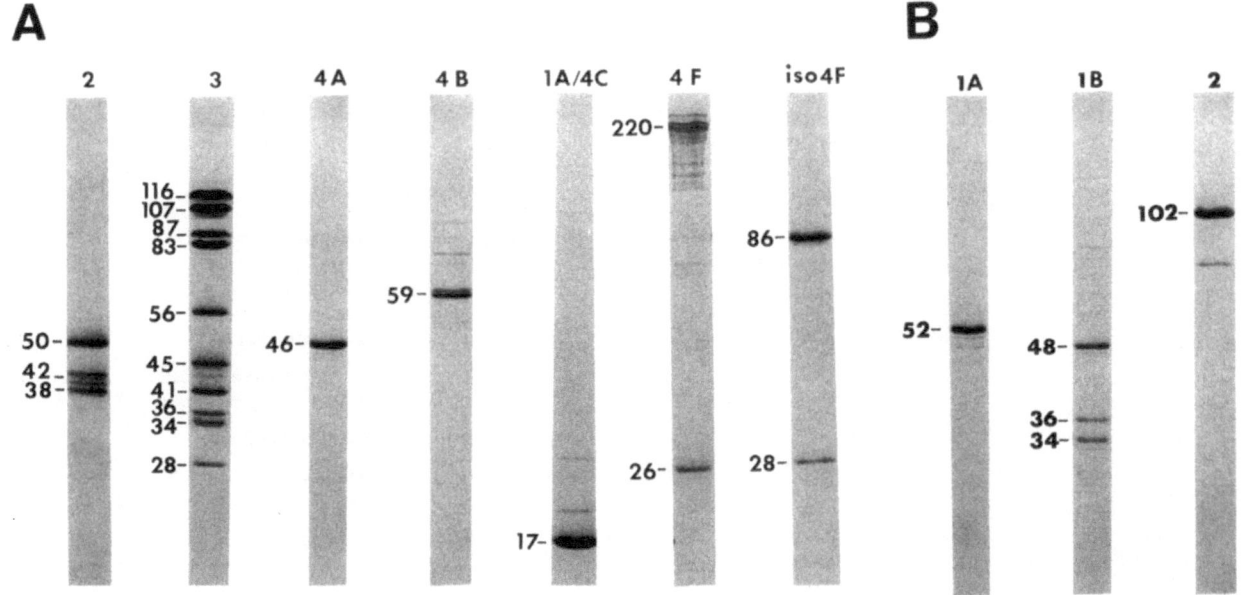

Figure 2. SDS-polyacrylamide gel analysis of purified initiation and elongation factors from wheat germ. The purified wheat germ initiation and elongation factors were electrophoresed by SDS PAGE as described [165]. The molecular weights of the factors and subunits are indicated. The p42 subunit of eIF2 (functional equivalent to mammalian eIF2α) always appears as a doublet or triplet; it is not known if this is due to degradation or isoelectric state. eIF4B also appears as a doublet; it is not known if this is due to degradation or isoelectric state.

p50 subunits of wheat germ eIF2 were the functional equivalents of the α, β and γ subunits, respectively, of mammalian eIF2. However, recent evidence obtained by molecular cloning and cDNA sequencing indicates that the p42 subunit of wheat germ eIF2 is the functional equivalent of the α-subunit of mammalian eIF2, the p38 subunit is the functional equivalent of the β-subunit and the p50 subunit is the functional equivalent of the γ-subunit (A. Metz and K. Browning, manuscript in preparation). The subunit assignment of wheat germ eIF2 based on SDS-PAGE mobility led to confusion with regard to the possible regulatory effect of phosphorylation of the p42 subunit on the activity of plant eIF2 (see below).

eIF2 binds GDP or GTP, and in the presence of GTP binds Met-tRNA$_i$. eIF2 binds GDP more tightly that GTP. However, mammalian and wheat germ eIF2 differ significantly in their relative binding affinities for GTP and GDP. Rabbit reticulocyte eIF2 has a K_d 100-fold higher for GTP than for GDP [328]. Wheat germ eIF2 has only a 10- to 20-fold higher K_d for GTP than for GDP [22, 66, 166, 285] and in the presence of Met-tRNA$_i$ the K_d for GTP is only 2 to 4-fold higher [20]. This difference in the relative binding affinities for GDP and GTP led to speculation on the need

for eIF2B, the recycling factor for eIF2•GDP [285], since an eIF2B equivalent has not been isolated yet from higher plants (see below). eIF2 also binds mRNA [94, 142, 237, 258, 295, 321]. The binding of mRNA to wheat germ eIF2 inhibits the binding of GDP or GTP and the subsequent binding of Met-tRNA$_i$ [20]. Conversely, the binding of GDP, GTP, or GTP and Met-tRNA$_i$ inhibit the binding of mRNA [20]. The competition between mRNA and the guanine nucleotides may be due to overlapping binding sites or may be due to a conformational change in one or more of the subunits of eIF2 [20]. Unlike mammalian eIF2, wheat germ eIF2 does not appear to bind ATP (K. Browning, unpublished data).

Regulation of the activity of eIF2 in mammalian cells has been studied extensively. Two kinases, heme controlled repressor (HCR) and double-stranded RNA-activated protein kinase (PKR), have been identified that specifically phosphorylate the α-subunit of mammalian eIF2 and regulate its activity (for reviews of mammalian eIF2α and kinases see [61, 67, 245, 266, 339]). A yeast eIF2α kinase (GCN2) is induced during amino acid starvation by the transcription factor GCN4. The GCN2 kinase phosphorylates the α-subunit of yeast eIF2 resulting in a slow-down of yeast protein synthesis (for reviews of yeast GCN4 and yeast eIF2 see [125, 226]). Because mammalian and yeast eIF2 bind GDP so tightly, a guanine exchange factor designated GEF or eIF2B is required for the removal of GDP from eIF2. eIF2B is a complex that consists of five non-identical subunits and several yeast genes that are part of the exquisitely complicated GCN4 regulatory pathway have been shown to be subunits of yeast eIF2B (for a review of eIF2B see [243]). eIF2B binds to eIF2•GDP, catalyzing the release of GDP from eIF2, allowing eIF2 to bind GTP and Met-tRNA$_i$ again. The phosphorylation of the α-subunit of mammalian and yeast eIF2 causes eIF2 to bind eIF2B more tightly, effectively sequestering the eIF2B and allowing the pool of inactive eIF2•GDP to build up. The lack of functional eIF2•GTP•Met-tRNA$_i$ prevents the initiation process and protein synthesis is inhibited.

At this time it is not clear whether or not the activity of eIF2 in higher plants is also regulated in a similar manner. Several mammalian eIF2α-kinases [22, 189, 279, 285] and a wheat germ casein kinase [41] phosphorylate the p42 subunit of wheat eIF2, the functional equivalent to mammalian eIF2α. However, no effect of the phosphorylation on protein synthesis activity could be conclusively demonstrated *in vitro*. Recently a plant kinase which is immunologically and enzymatically similar to mammalian PKR (mPKR), has been purified from barley leaves [158]. This plant kinase, designated pPKR, phosphorylates the p42 subunit of wheat germ eIF2 and inhibits protein synthesis *in vitro* [157]. This is the first evidence that higher plants may also regulate protein synthesis by phosphorylation of the α-subunit (p42) of plant eIF2.

A recycling factor comparable to mammalian or yeast eIF2B has not been identified in higher plants as mentioned above. Because the wheat germ eIF2 does not exhibit as great a difference between the binding affinities for GDP and GTP as mammalian eIF2, it has been suggested that a recycling factor may not be necessary [285]. However, analysis of the EST[1] database with the DNA sequences for two subunits of

yeast eIF2B (*GCN3* and *GCD2*), suggest that there are *Arabidopsis* and rice ESTs that encode the functional equivalents of these proteins. It is premature to abandon the search for the plant equivalent to eIF2B at this time.

The cDNAs for the subunits of wheat eIF2, p42 (equivalent to mammalian eIF2α; incomplete cDNA), p38 (equivalent to mammalian eIF2β; full-length cDNA) and p50 (equivalent to mammalian eIF2γ, incomplete cDNA) have been obtained and sequenced (A. Metz and K. Browning, unpublished data). They are very similar to their mammalian and yeast counterparts, reflecting a high degree of structure/function conservation. The deduced amino acid sequence from an EST for *Arabidopsis* eIF2α contains the consensus phosphorylation site of mammalian and yeast eIF2α. The gene structure for the subunits of eIF2 and the regulation of their expression have yet to be elucidated in higher plants.

eIF2 cofactors

Two factors have been isolated from wheat germ that stimulate the formation of eIF2•GDP and eIF2•GTP•Met-tRNA$_i$ complexes [166, 223]. One of these factors, Co-eIF2α, is a small protein (M_r ca. 15 000) and the other one, Co-eIF2β, is a larger protein (M_r ca. 83 000); both factors are single polypeptides. Interestingly, both of these factors are heat stable and their activities are not synergistic, suggesting that they act independently [166, 223]. The relationship of these factors to other initiation factors (e.g. eIF1 or the subunits of eIF2B) is not known at this time.

eIF3

Wheat germ eIF3 has been isolated [54, 60, 165, 278, 296] and shown to be a large stable complex that migrates in sucrose gradients as a 15S particle [160]. Wheat germ eIF3 contains 10 non-identical subunits (ranging in M_r from 28 000 to 116 000) in close to 1:1 ratios (see Fig. 2 and Table 1 [162]). Mammalian eIF3 is also a large, stable complex composed of 8 non-identical subunits [193]. The structure of wheat germ eIF3 visualized in electron micrographs suggests a cone-like structure with either three or four appendages (see Fig. 9, [229]).

Attempts to remove one or more of the subunits of wheat germ eIF3 and reassociate the subunits into

---

[1] An EST (expressed sequence tag) is a type of cDNA sequence deposited in the GenBank dbEST database (URL:http//www.ncb.nlm.nih.gov/dbEST/index.html). These sequences are from a single-pass DNA sequencing reaction, are only 200–300 nucleotides in length and frequently contain sequencing errors [2]. The EST sequence is compared to sequences in GenBank to determine if it can be matched with a known sequence(s), if so a notation is made in the GenBank entry. There are several EST sequence projects, in *Arabidopsis*, rice, maize, *C. elegans* and man to name a few. The purpose is to create partial sequence catalogs of all the expressed genes in an organism. The sequences are usually

public domain and if a sequence of interest to a researcher is found, that cDNA clone may be obtained for further analysis.

an active complex have been unsuccessful [122]. Partial dissociation of the complex occurs in 2 M urea with subunits p116, p83 and p36 being the most easily removed [122]. Two forms of wheat germ eIF3 have been isolated from wheat germ that differ in their ability to support *in vitro* polypeptide synthesis. The less active form contains lower amounts of two of the subunits, p116 and p36, suggesting that these subunits are essential for activity [162]. Polyclonal antibodies to wheat germ eIF3 are directed primarily to four of the 10 subunits (p116, p87, p56 and p36). These antibodies inhibit the ability of eIF3 to support *in vitro* protein synthesis and prevent the binding of mRNA to 40S ribosomal subunits [160]. A monoclonal antibody directed against p36 also inhibits protein synthesis and binding of mRNA to 40S ribosomal subunits, providing further evidence that p36 is essential for the activity of eIF3 [160]. The p107 subunit of wheat germ eIF3 is phosphorylated by a plant casein kinase, but no effect on activity is observed *in vitro* [41].

Wheat germ eIF3 binds to the 40S ribosomal subunit in the absence of other factors. It also enhances the binding of ternary complex, eIF2•GTP•Met-tRNA$_i$, to the 40S ribosomal subunit [60] and is required for the subsequent binding of mRNA to the 40S ribosomal subunit [160]. Wheat germ eIF3 binds mRNA in the absence of 40S ribosomal subunits. The binding of eIF3 to mRNA, measured by retention on nitrocellulose, is inhibited by any single-stranded nucleic acid (RNA or DNA), but the binding is not affected by NTPs, m^{7}GTP, m^{7}GpppG or GDP [19]. Biophysical analysis of the binding of wheat germ eIF3 to mRNA showed that the interaction is independent of the m^{7}G cap structure, does not require ATP hydrolysis and that the interaction is primarily ionic in nature [50]. At low concentrations of NTPs ($\leqslant 10$ μM), the binding of mRNA to eIF3 is enhanced possibly through a conformational change, but at higher concentrations the NTP acts more like competitive inhibitor for the mRNA-binding site of eIF3 [50]. The interaction of mRNA analogues with either wheat germ eIF4F or eIFiso4F (an isoenzyme form of plant eIF4F, see below) enhances the binding of mRNA to eIF3 [48]. These results suggest a model where the interaction of wheat germ eIF3 with NTP induces a conformation in the eIF3 to enhance subsequent binding of eIF3•NTP to the eIF4F•mRNA complex that is then competent to interact with the 40S ribosomal subunit (see Fig. 9, [50]).

Only recently have any eIF3 subunits been cloned and sequenced. Two yeast genes, PRT1 and GCD10,

have been identified as subunits of eIF3 [103, 211]. Interestingly, yeast GCD10 is a translational repressor of GCN4 whose expression is also controlled by the phosphorylation of eIF2α by the GCN2 kinase [103, 125]. The PRT1 and GCD10 gene products encode 90 000 and 55 000 kDa proteins, respectively. Whether these correspond to the p87 and p56 subunits of wheat eIF3 is not known at this time. The GCD10 protein is also a mRNA binding protein and contains a RNA binding motif (RRM) [103]. ESTs for mustard and *Arabidopsis* PRT1-like proteins are present in Gen-Bank; ESTs for a GCD10-like protein have not been identified.

eIF4A

eIF4A has been isolated from wheat germ and like its mammalian counterpart, is a single polypeptide of 50 000 Da [165, 280]. Both plant and mammalian eIF4A have been isolated in a complex with eIF4F depending upon the isolation procedure [88, 335]. Mammalian eIF4A has been shown to cycle in and out of the eIF4F complex during the initiation process [347]. Wheat germ eIF4A binds ATP [281] and catalyzes the hydrolysis of ATP [164]. ATP hydrolysis catalyzed by eIF4A alone is not stimulated by the presence of mRNA [164], nor does wheat germ eIF4A bind mRNA or mRNA analogues [13].

The role of eIF4A is to work in concert with eIF4F and eIF4B to catalyze the ATP-dependent movement and/or unwinding of secondary structures in the 5′-untranslated region of mRNAs prior to binding of the 40S ribosomal subunit [13]. The ATP hydrolysis catalyzed by wheat germ eIF4A is greatly stimulated by the presence of mRNA and eIF4F or eIFiso4F [164]. ATP enhances the binding affinity of wheat germ eIF4A for wheat germ eIFiso4F and appears to enhance protein-protein interactions rather than protein-mRNA interactions [13]. Wheat germ eIF4A catalyzes the ATP-dependent unwinding of double-stranded RNA in the presence of eIF4F or eIFiso4F ([34] and D. Goss, personal communication). A direct interaction of the large subunit of eIFiso4F with wheat germ eIF4A can be demonstrated in the yeast two-hybrid system (A. Metz and K. Browning, manuscript in preparation).

Wheat germ eIF4A substitutes poorly for rabbit reticulocyte eIF4A in ATP-hydrolysis assays containing other reticulocyte components. However, rabbit reticulocyte eIF4A is fully active when substituted for wheat eIF4A in ATP-hydrolysis assays containing wheat germ components [1]. Furthermore, rabbit

Table 2. Isoforms and abundance of plant initiation and elongation factors.

Factor	Number of isoforms[a]	% of total protein[b]
eIF2[c]		0.5
p38 (β)	ca. 4	
p42 (α)	2	
p50 (γ)	ca. 1–2	
eIF3[d]		1.3
p116	1	
p107	1	
p87	ca. 1–2	
p83	ca. 2–3	
p56	ca. 1–2	
p45	ca. 1–2	
p41	ca. 2–3	
p36	ca. 3–4	
p34	ca. 1–2	
p28	ca. 2–3	
eIF4A[c,e]	2	0.4
eIF4B[c]	ca. 8–10	unknown
eIF4F[c]		0.08
p220 (eIF4G)	ND[f]	
p26 (eIF4E)	ca. 2	
eIFiso4F[c]		0.25
p86 (eIFiso4G)	ND[f]	
p28 (eIFiso4E)	ca. 2	
eEF1A[c]	ND[f]	4.9
eEF2[c]	ca. 2	1.5

[a] Determined by two-dimensional gel electrophoresis.
[b] As reported by Browning *et al.* [36].
[c] D. Gallie and K. Browning, manuscript in preparation.
[d] As reported by Heufler *et al.* [122].
[e] As reported by Webster *et al.* [40].
[f] ND, not determined. Too basic to accurately determine the number of isoforms.

reticulocyte eIF4A is functional in a wheat germ polypeptide synthesis system deficient in eIF4A, whereas, wheat germ eIF4A is non-functional in a rabbit reticulocyte polypeptide synthesis system deficient in eIF4A [1]. It appears some portion of the components of the mammalian protein synthesis machinery have diverged in evolution to such an extent that plant eIF4A is no longer able to interact with them; whereas, the mammalian eIF4A is still able to interact with the components of the plant protein synthesis machinery.

eIF4A has been cloned and sequenced from a number of plant sources, wheat germ [195], *Arabidopsis* [197], maize [138], rice [215] and tobacco [31, 32,

224, 225]. The amino acid sequences are highly conserved among higher plants (see Fig. 3 and Table 3) and their counterparts from other eukaryotes. For comparison purposes, wheat eIF4A is 61% and 70% similar to yeast and mouse eIF4A, respectively [195]. Wheat eIF4A over-expressed in *E. coli* and purified has the same activity in polypeptide synthesis *in vitro* as native wheat germ eIF4A (K. Ruud and K. Browning, unpublished data). The eIF4A family contains several conserved amino acid motifs (AXXXXGKT, DEAD, SAT, HGIGRXXR) involved in ATP binding, ATP hydrolysis and double-stranded RNA unwinding [232, 233]. Other RNA unwinding proteins (helicases) contain these amino acid motifs and eIF4A is the prototype of the helicase family [173, 334]. Note that all the plant eIF4As contain an additional SAT motif just beyond the DEAD box with the exception of tobacco eIF4A3. The SAT motif is essential for the double-stranded RNA unwinding activity of eIF4A [233]. The function of this second SAT motif in plant eIF4A is unknown. The spacings between the amino acid motifs are maintained, suggesting a strong conservation of a structure/function that is necessary for its enzymatic activities and interaction with other components of the translational machinery.

The tobacco eIF4A gene family has been extensively studied and found to have two distinct gene families, NeIF4A2 and NeIF4A3, containing at least 10 differentially expressed genes [31, 32, 224, 225]. The large number of tobacco eIF4A genes is in contrast to yeast and mouse which have only two active eIF4A genes [174, 213]. NeIF4A3 diverges the most from other plant eIF4A sequences and may represent another class of eIF4A-like proteins. It has not yet been determined if NeIF4A3 is a functional protein synthesis initiation factor. Variable expression of the eIF4A genes is observed in several tissue types [225]. There appears to be at least one eIF4A-like gene product that localizes to the chloroplast and associates with chloroplast ribosomes [225]. There is also an eIF4A gene that is specifically expressed in pollen [31]. Several forms of tobacco eIF4A with different mobilities by SDS PAGE have been detected immunologically, corroborating the number of actively transcribed genes. Either the amino acid sequence variations affect SDS PAGE mobility, or some of the eIF4A polypeptides are post-translationally modified [225]. However, the only post-translational modification reported for any eukaryotic eIF4A, is the hypoxia induced phosphorylation of a threonine residue of maize eIF4A [335]. The increase in the phosphorylation of maize eIF4A during

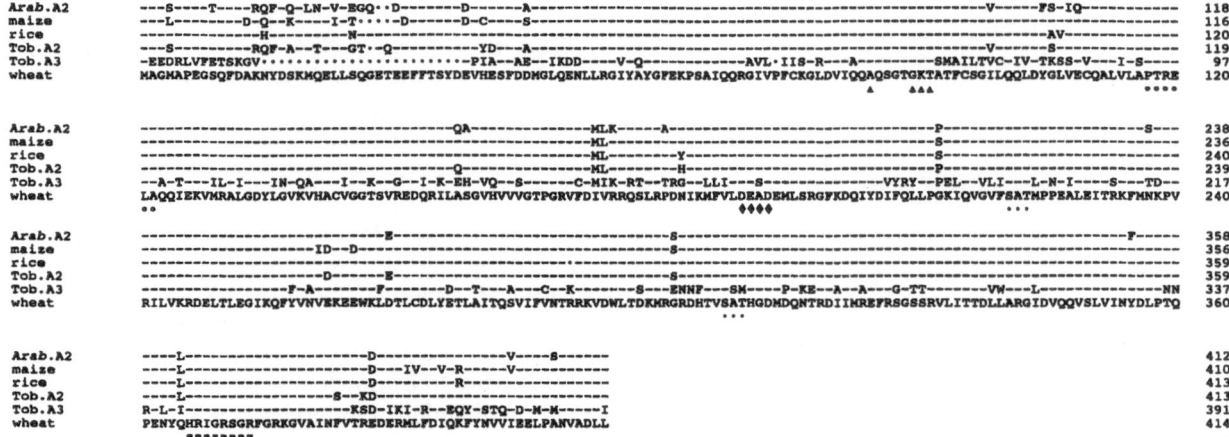

Figure 3. Alignment of eIF4A amino acid sequences from plants. The amino acid sequences were compared using MACAW v2.0.3 [275]. Only 1 of the 2 known *Arabidopsis* eIF4A sequences is used for the comparison. Only 2 of the 11 known tobacco eIF4A sequences are used: Tob. A2 which is >90% similar to all other tobacco eIF4As and Tob. A3 which is the most divergent (see Table 3). The conserved motifs are as indicated (see text). The dashes indicate the identical amino acid is present as compared to the wheat eIF4A. The dots indicate a gap inserted by the alignment program. The GenBank accession numbers for the eIF4A sequences used are: *Arabidopsis* A2 cDNA, X65053 [197]; maize cDNA, U17979 [138]; rice cDNA, D12627 [215]; tobacco A2 gene, X61205 [224]; tobacco A3 gene, X61206 [224]; wheat cDNA, Z21510 [195].

Table 3. Comparison of amino acid sequences of plant eIF4A[a].

	Arab. A2	Maize	Rice	Tob.A2	Tob.A3	Wheat
Arab. A2	100	90	91	93	61	91
maize		100	94	93	60	93
rice			100	94	60	97
Tob.A2				100	61	94
Tob. A3					100	60
wheat						100

[a] GenBank accession numbers are as follows: *Arab. A2*, X65053 [197]; maize, U17979 [138]; rice, D12627 [215]; Tob.A2, X61205 [224]; Tob. A3, X61206 [224]; wheat, Z21510 [195]. Similarity was determined using the ALIGN function of Microgenie v7.01 (Beckman).

hypoxia coincides with a drop in intracellular pH and a shut down of protein synthesis; however, only ca. 50% of the eIF4A is ever phosphorylated [335].

The genes for four of the tobacco eIF4A polypeptides have been cloned and all contain three introns in the coding sequences. The intron positions and intron/exon boundaries are conserved within the tobacco genes [32]. A fourth intron is located in the 5'-untranslated region, reminiscent of the intron located in the 5'-untranslated region of eEF1A genes (see eEF1A) from *Arabidopsis* [72], tomato [286] and soybean [3]. The purpose of such a large gene family for eIF4A in tobacco is unknown, but may be a result of gene duplication events. These duplicated genes perhaps have evolved for specialized eIF4A functions

(i.e. the pollen-specific eIF4A) or are simply redundant genes to insure against lethal mutations. Recently the sequence for maize eIF4A cDNA was reported and mapping indicated only two loci [138]. Whether this represents a monocot versus dicot difference in plant eIF4A gene structure remains to be determined.

eIF4B

An initiation factor that has approximately the same molecular mass as mammalian eIF4B (ca. 80 000 kDa) was isolated from wheat germ [163] and incorrectly given the designation eIF4B in early reports [1, 38, 42, 128, 164, 165]. This factor has since been correctly identified as an isozyme form of eIF4F and termed eIFiso4F (see below, [34]). Another initiation factor was subsequently isolated from wheat germ that has similar enzymatic properties to mammalian eIF4B [39, 283]. This initiation factor was named eIF4B by Seal *et al.* [283] and eIF4G by Browning *et al.* [39]. eIF4G was correctly renamed eIF4B when eIFiso4F was identified [34]. References in the literature to eIF4G [35, 37, 39, 128] actually refer to eIF4B and references in the literature to eIF4B prior to 1989 [1, 38, 42, 128, 164, 165] are actually describing eIFiso4F.

Wheat germ eIF4B is a single polypeptide with a molecular weight of 59 000 (see Fig. 2, [39]). The molecular weight for mammalian eIF4B is 80 000 as measured by SDS-PAGE and is a homodimer [108]. The predicted molecular weights from the amino acid sequences of mammalian and yeast eIF4B are 65 000 [199] and 49 000, respectively [6, 69]. Wheat germ eIF4B has a number of isoelectric states (see Table 2), suggesting that is present in multiple phosphorylation forms like mammalian eIF4B [119].

The role of eIF4B is to act in conjunction with eIF4F and eIF4A to enhance the ATP-dependent unwinding of secondary structures in the 5'-untranslated region of mRNAs prior to binding of the mRNA to the 40S ribosomal subunit. It has been shown that wheat germ eIF4B: (1) stimulates the RNA-dependent ATP hydrolysis activity of wheat germ eIF4A and eIF4F about 2-fold, but is not required for it [34]; (2) stimulates the ATP-dependent unwinding of double-stranded RNA about 2-fold, but is not required for it [34]; and (3) cross-links to mRNA only in the presence of ATP, eIF4A and eIF4F [34]. In contrast, mammalian eIF4B and eIF4A, but not eIF4F, are required for ATP-dependent unwinding of double-stranded RNA and yeast eIF4A is able to substitute for the mammalian eIF4A [137]. However, yeast eIF4A and yeast eIF4B are not able to catalyze this reaction and apparently require an additional factor(s) [8]. Given the evidence from the wheat system, this missing factor is probably eIF4F. All three mammalian factors, eIF4B, eIF4F, eIF4A, are necessary for ATP hydrolysis activity, whereas in wheat germ, eIF4A and eIF4F are required and wheat germ eIF4B only stimulates ATP hydrolysis activity [34]. These data would suggest that there are fundamental differences between plants, mammals and yeast with respect to the requirement for eIF4B in the ATP-dependent unwinding of mRNA.

Biophysical data show that the binding of oligonucleotides to wheat germ eIF4B is not affected by the presence of a m^7G cap structure or an initiation codon [284]. The binding affinity of wheat germ eIF4B for oligonucleotides is 10-fold higher than that of wheat germ eIFiso4F under the same ionic conditions [284]. ATP also appears to bind at the same site as oligonucleotides, since the affinity of wheat germ eIF4B for oligonucleotides decreases in the presence of ATP [284].

Recently it has been reported that mammalian and yeast eIF4B have RNA annealing and strand displacement activities [8]. It is not known at this time if wheat germ eIF4B has such an activity. Wheat germ eIF4B has also been shown to form a complex(s) with poly(A) by gel shift analysis and the addition of wheat germ eIF4B, eIF4A and eIF4F reversed poly(A) inhibition of protein synthesis in wheat germ extracts [97]. Furthermore, eIF4B appears to stimulate the formation of poly(A)-binding protein (PAB)•poly(A) complexes (D. Gallie, personal communication). The interaction of eIF4B with poly(A) or PAB have not yet been demonstrated in mammalian or yeast systems. Mammalian eIF4B also catalyzes the recycling of the cap-binding protein [247], an activity that has not been demonstrated for wheat germ eIF4B.

A partial cDNA sequence for wheat germ eIF4B has been obtained (S. Malmström and K. Browning, manuscript in preparation). This cDNA is missing ca. 50–70 amino acids from the N-terminus that should contain the two consensus RNA binding motifs (RRM) reported for the mammalian [212] and yeast [6, 69] eIF4B polypeptides. Comparison of the amino acid sequences for wheat eIF4B to yeast and mammalian eIF4B shows a very low degree of similarity, <20%. There are only two regions between wheat germ eIF4B and mammalian eIF4B and only one between wheat germ eIF4B and yeast eIF4B that can be aligned with MACAW (see Fig. 4). This lack of conservation is surprising given the high degree of similarity among other initiation factors (i.e. eIF4A, eIF1A and eIF5A); however, the subunits of eIF4F also have a low degree of similarity (see below). It is perhaps this low degree of similarity that may explain some of the differences in enzymatic properties between plant, mammalian and yeast eIF4B.

The gene structure of eIF4B in plants or mammals is unknown at this time. Slight differences in cDNA sequences suggest that multiple forms may exist, but this has not been rigorously confirmed (S. Malmström and K. Browning, unpublished observation). There is only one gene in yeast for eIF4B and disruption of this gene gives a slow growth, temperature-sensitive phenotype [6, 69]. This suggests that in yeast either

116

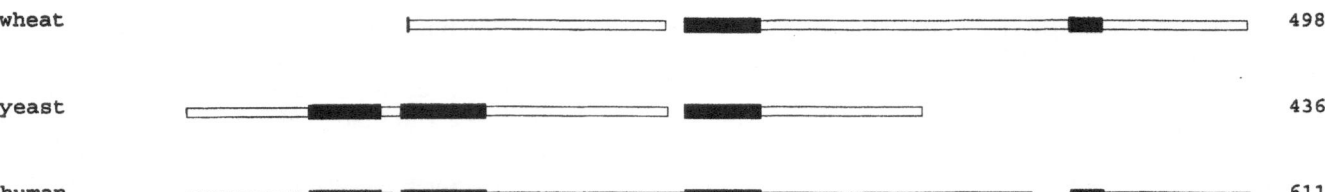

wheat   498

yeast   436

human   611

Figure 4. Schematic of alignment of the amino acid sequences of wheat eIF4B, yeast eIF4B and human eIF4B. The amino acid of the wheat eIF4B is incomplete (see text). About 50–75 amino acids are missing from the N-terminus, which should contain the consensus RNA binding motifs (S. Malmström and K. Browning, manuscript in preparation). The alignment is with MACAW v2.0.3 [275]. GenBank accession numbers used for the alignment are: yeast eIF4B gene, X71996 [6, 69] and human eIF4B cDNA, X55733 [199].

eIF4B is not an essential gene, or other compensating genes are present that have not been identified.

eIF1A (eIF4C)

eIF1A has been isolated from wheat germ and like its mammalian counterpart is a small (M_r 14 000), heat stable, single polypeptide [165, 277, 309]. eIF1A has pleiotropic effects on the initiation of protein synthesis stimulating (usually only about 2-fold) a number of intermediate steps: AUG-directed Met-puromycin synthesis, AUG-directed binding of Met-tRNA$_i$ to 40S ribosomal subunits and 80S ribosomes, dissociation of 80S ribosomes into 40S and 60S ribosomal subunits, as well as overall polypeptide synthesis [21, 277, 308, 309]. eIF1A does not appear to be a limiting factor for initiation in wheat germ extracts [309]. In *in vitro* systems from wheat germ and rabbit reticulocytes, rabbit reticulocyte eIF1A and wheat germ eIF1A are able to substitute for each other, suggesting a highly conserved structure [309]. More recently, mammalian eIF1A has been shown to be an RNA-binding protein [338]. Whether wheat germ eIF1A has a similar activity has not been determined.

A combination of protein sequencing and cDNA sequencing was used to obtain the complete sequence for wheat germ eIF1A [81]. This sequence is 68% and 56% similar to mammalian and yeast eIF1A, respectively. The highest amount of similarity (77%) is in the N-terminal 109 amino acids (see Fig. 5). The high degree of similarity was expected given the ability of mammalian and wheat germ eIF1A to substitute for each other in polymerization assays. Wheat germ eIF1A over-expressed in *Escherichia coli* and purified has the same activity in polypeptide synthesis *in vitro* as native eIF1A (S. Lax and K. Browning, unpublished data). eIF1A is very hydrophilic in both the N- and C-terminal regions suggesting a dipolar structure

[81]. Interestingly, there are no obvious motifs in the sequence that would suggest that eIF1A is a RNA-binding protein.

There are no other complete sequences for plant eIF1A known at this time, nor is the gene structure known.

eIF5A (eIF4D)

eIF5A has been isolated from rice and, like mammalian eIF5A, is a low-molecular-weight single polypeptide (M_r 18 000) [188]. eIF5A is the only protein that contains a modified lysine residue (hypusine) that is found in all eukaryotic eIF5As (see reviews [230, 231]). The post-translational modification of eIF5A occurs by the transfer of the butylamine from spermine to the ε-amino group of Lys–50 and subsequent hydroxylation to form hypusine [230, 231].

The function of eIF5A is unclear. The only *in vitro* activity that has been demonstrated is the stimulation of AUG-directed Met-puromycin formation, suggesting that it may participate in the formation of the first peptide bond [274]. No stimulation of polypeptide synthesis *in vitro* has ever been observed.

Studies in yeast showed that protein synthesis continues long after the intracellular concentration of eIF5A had been depleted, suggesting that eIF5A may not be absolutely required for protein synthesis [144]. However, a link between the hypusine formation and the ability of cells to proliferate was established [111, 231]. Furthermore, mammalian eIF5A may be a cofactor in HIV-1 viral replication [23]. eIF5A binds the HIV-1 Rev transactivator protein and is required for the Rev-mediated translocation of HIV-1 mRNAs from the nucleus to the cytoplasm [259]. It has been shown also that HIV-1 viral proteins require eIF5A for their translation [46] and eIF5A is expressed at a higher level in HIV-1-infected cells consistent with a need for eIF5A

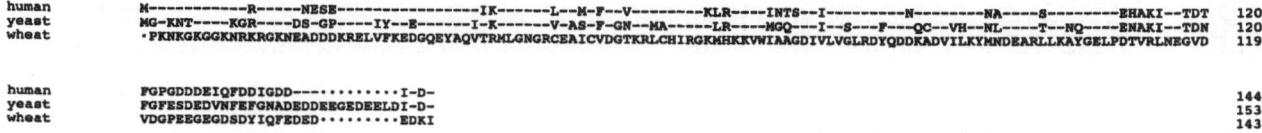

```
human    M----------R-----NESE----------------IK------L--M-F--V--------KLR----INTS--I---------N--------NA----S--------EHAKI--TDT  120
yeast    MG-KNT----KGR----DS-GP----IY--E------I-K------V-AS-F-GN--MA----LR----MGQ---I--S---F----QC--VH--NL----T--NQ----EHAKI--TDN  120
wheat    ·PKNKGKGGKNRKRGKNEADDDKRELVFKEDGQEYAQVTRMLGNGRCEAICVDGTKRLCHIRGKMHKKVWIAAGDIVLVGLRDYQDDKADVILKYMNDEARLLKAYGELPDTVRLNEGVD  119


human    FGPGDDDEIQFDDIGDD----···········I-D-                                                                                       144
yeast    FGFESDEDVNFEFGNADEDDEEGEDEELDI-D-                                                                                         153
wheat    VDGPEEGEGDSDYIQFEDED·········EDKI                                                                                         143
```

Figure 5. Alignment of eIF1A amino acid sequences of wheat eIF1A, yeast eIF1A and human eIF1A. The amino acid sequences were compared using MACAW v2.0.3 [275]. The dashes indicate the identical amino acid is present compared to the wheat eIF1A. The dots indicate a gap inserted by the alignment program. The GenBank accession numbers for the eIF1A sequences used are: wheat cDNA, L08060 [81]; yeast gene, U11585 [337]; human cDNA, L18960 [81].

for viral replication [23]. A significant portion of eIF5A accumulates in the nucleus [259]. More recently a class of cellular mRNAs (hypusine-dependent messenger nucleic acids, *hymns*) was identified in mammalian cells. These *hymns* disappear and reappear in polysomes in parallel to the inhibition and re-instatement of hypusine formation, respectively, and encode enzymes required for cell proliferation [112]. These results suggest that the hypusinated form of eIF5A binds specifically to certain mRNAs, whose protein product is required for the transition into S phase (the start of DNA replication, necessary for cell proliferation to begin), and facilitates their translation. It does perhaps explain the lack of stimulation of *in vitro* protein synthesis by eIF5A, since one of its roles appears to be that of a 'guide' protein for specific mRNAs to the ribosomes and therefore might not affect overall protein synthesis. However, the exact mechanism of this 'guide' protein is far from understood. Given the conservation of the hypusine modification of eIF5A in all eukaryotic cells, a similar type of function most likely exists for plant eIF5A.

A cDNA clone for eIF5A from alfalfa [234] and two cDNAs and a gene for eIF5A from tobacco [57] have been sequenced. The amino acid sequences for the plant eIF5A are about 95% similar, compared to 59% and 51% for yeast and human eIF5A, respectively (see Fig. 6). There appear to be at least two genes (NeIF5A1 and NeIF5A2) in tobacco and although both genes are expressed in all tissue types, they have different patterns of expression [57]. NeIF5A2 appears to be expressed more or less equally in all tissue types; however, NeIF5A1 shows preferential expression in tissues with high photosynthetic activity (leaves and sepal) and very low expression in non-photosynthetic tissue (roots and fruit) [57].

In light of the more recent studies on the role of eIF5A, one could speculate that perhaps one form of plant eIF5A may facilitate the translation of mRNAs encoding the enzymes that carry out photosynthesis. It

is perhaps worth determining if hypusine formation in plants is coordinated with photosynthetic activity.

eIF4F and eIFiso4F

The greatest differences between the translation process in plants and other eukaryotes appear to reside in the structure and function of eIF4F. eIF4F isolated from mammalian cells is a complex containing three subunits: M_r 220 000, function unknown; M_r 45 000, eIF4A; and M_r 24 000, a m^7G cap-binding protein [193]. Mammalian eIF4F has also been isolated in a form that does not contain eIF4A [88]. eIF4F isolated from wheat germ contains two subunits: M_r 220 000, function unknown and M_r 26 000, a m^7G cap-binding protein (see Fig. 2, [163, 165]). In addition, an isozyme form of eIF4F was isolated from wheat germ. This factor, designated eIFiso4F [40, 163, 165] or CSF (cap-site factor, [283]), is a complex containing two subunits: M_r 86 000, function unknown, and M_r 28 000, a m^7G cap-binding protein (see Fig. 2).

As discussed in the section on eIF4B, there was confusion as to the identity of eIFiso4F and it was initially designated eIF4B. All papers prior to 1989 that refer to eIF4B actually refer to eIFiso4F [1, 38, 42, 128, 164, 165]. The protein or gene appears to be present in all higher plants tested: wheat [40], maize [40], cauliflower [40], *Arabidopsis* (R. Ahmed and K. Browning, unpublished data) and barley (F. Mueller-Uri, personal communication). An isozyme form of eIF4F has not been found in any other eukaryotes. The designations of p220 and p26 for the subunits of eIF4F, and p86 and p28 for the subunits of eIFiso4F will be used in this review, rather than the new nomenclature indicated in Table 1.

Antibodies raised to wheat germ eIF4F do not react with wheat germ eIFiso4F, and conversely, antibodies raised to wheat germ eIFiso4F do not react with wheat germ eIF4F, indicating that the subunits are not degradation products of each other [38]. eIFiso4F has the same functional properties as eIF4F: (1) it substitutes

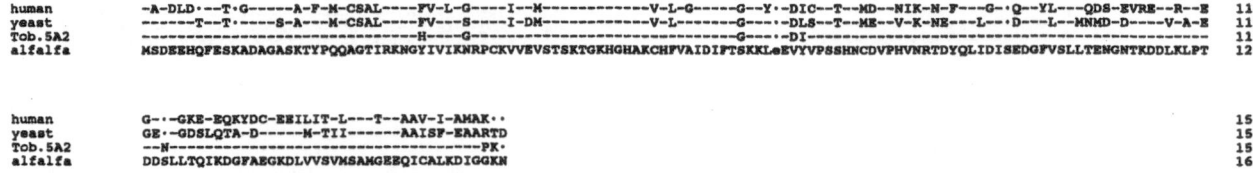

```
human      -A-DLD·--T·G-----A-F-M-CSAL----FV-L-G----I--M-------------V-L-G------G--Y·-DIC--T--MD--NIK-N-F---G-·Q--YL---QDS-EVRE--R--E  116
yeast      -----T--T·-----S-A--M-CSAL----FV---S----I-DM-------------V-L-------G----·-DLS--T--ME--V-K-NE---L--·D---L--MNMD-D-----V-A-E  117
Tob.5A2    --------------------------------H----G-------------------------G---·-DI------------------------------------------------------  119
alfalfa    MSDEEHQFESKADAGASKTYPQQAGTIRKNGYIVIKNRPCKVVEVSTSKTGKHGHAKCHFVAIDIFTSKKLeEVYVPSSHNCDVPHVNRTDYQLIDISEDGFVSLLTENGNTKDDLKLPT  120
```

```
human      G-·-GKE-EQKYDC-EEILIT-L---T--AAV-I-AMAK··                          154
yeast      GE·-GDSLQTA-D-----M-TII------AAISF-EAARTD                          157
Tob.5A2    --N----------------------------------PK·                           159
alfalfa    DDSLLTQIKDGFAEGKDLVVSVMSAMGEEQICALKDIGGKN                          161
```

Figure 6. Alignment of eIF5A amino acid sequences of tobacco, alfalfa, yeast and human. The amino acid sequences were compared using MACAW v2.0.3 [275]. The dashes indicate the identical amino acid is used compared to the alfalfa eIF5A. The dots indicate a gap inserted by the alignment program. The GenBank accession numbers for the eIF5A sequences used are: tobacco gene, X63541 [57]; alfalfa cDNA, X59441 [234]; yeast gene, J05455 [272]; human cDNA, M23419 [293].

for eIF4F in an *in vitro* translation system deficient in eIF4F; (2) it substitutes for eIF4F in supporting the binding of mRNA to 40S ribosomal subunits; (3) it exhibits RNA-dependent ATP hydrolysis activity; and (4) it exhibits ATP-dependent RNA unwinding activity in the presence of eIF4A [1, 38, 163, 164]. The amount of eIFiso4F in wheat germ extracts is 3- to 5-fold higher than the amount of eIF4F (see Table 2, [36]). The ratio of the subunits of eIFiso4F is 1:1 measured by quantitation of Coomassie blue staining on SDS-PAGE gels [165]. Originally, the ratio of the large subunit to small subunit of wheat germ eIF4F is reported to be 1:4 [165]; however, more recent analysis suggests that the ratio is probably 1:1 (L. Allen and K. Browning, unpublished observation). The large subunit of eIF4F is very susceptible to degradation, and obtaining preparations without degradation products has proven very difficult even with the addition of numerous protease inhibitors. The degradation products of the large subunit are able to bind p26, but are not accurately measured by scanning of stained SDS-PAGE gels (L. Allen and K. Browning, unpublished data).

Biophysical data showed that the affinity of wheat germ eIF4F is lower than eIFiso4F for hypermethylated cap groups [47], and while wheat germ eIF4F is more sensitive to the presence of hair-pin structures, eIFiso4F prefers linear structures [49]. Interestingly, the p28 subunit of eIFiso4F is found to resemble more closely the mammalian cap-binding protein, eIF4E, in its interactions with cap analogues more than the p26 subunit of wheat germ eIF4F [47]. These results suggest that the cap-binding proteins of eIF4F and eIFiso4F may use a discriminatory mechanism for selection of capped mRNAs for translation. Both wheat germ eIF4F and eIFiso4F form complex(s) with poly(A) by gel shift analysis and stimulate the formation of PAB•poly(A) complexes. eIF4B stimulates these interactions in a synergistic manner (D. Gallie, personal communication). A combination of eIF4A,

eIF4B and eIF4F or eIFiso4F reverse poly(A) inhibition of protein synthesis in wheat germ extracts [97].

Recently, the p86 subunit of eIFiso4F has been shown to interact in an ATP-independent manner with maize microtubules in binding assays *in vitro* and to promote the bundling of microtubules [26]. Furthermore, p86 was shown by immunofluorescence to colocalize with microtubules in fixed root-tip cells from maize [26]. The role of this interaction with microtubules in protein synthesis is unknown at this time; however, one may speculate upon the implications (see Protein synthesis and the cytoskeleton).

The cDNAs for the subunits for wheat eIF4F and eIFiso4F have been cloned and sequenced ([4, 198] and L. Allen and K. Browning, unpublished data). Based on the deduced amino acid sequences, the predicted molecular weights for both of the wheat cap-binding proteins is about 24 000 Da; however, the p28 and p26 nomenclature is retained to prevent confusion. The sequences for rice cap-binding proteins, p28 and p26, have also been determined (E. Aliyeva, A. Metz and K. Browning, unpublished data). The similarity between p28 and p26 from either wheat or rice is about 50% (see Table 4). The similarity of rice p26 and wheat p26 or rice p28 and wheat p28 is about 80% (see Table 4). The plant cap-binding proteins are also similar to cap-binding proteins from other eukaryotes (see Table 4). The most striking feature of all cap-binding proteins is the conservation of the position and number of tryptophan residues thought to be involved in the binding of the m7G cap group. Another interesting feature of the plant cap-binding proteins is the potential for phosphorylation. The phosphorylation state of mammalian cap-binding protein, eIF4E, has been shown to be very important for activity [251, 254]. The site of this phosphorylation was originally misidentified as Ser-53 [78, 167]. Recently the site of phosphorylation has been shown to be Ser-209 [141, 182]. Ser-209 of mammalian cap-binding protein is located at the C-

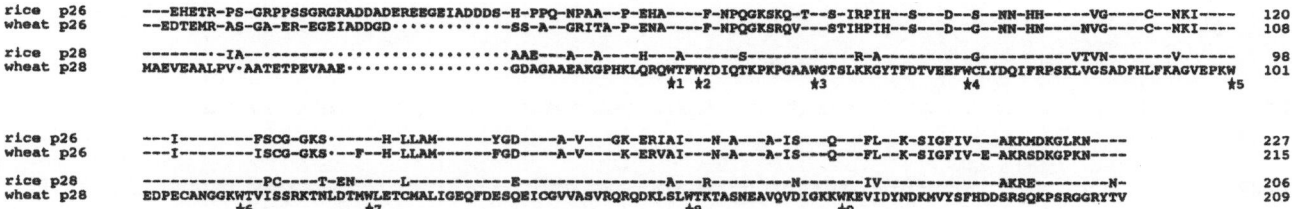

```
rice  p26  ---EHETR-PS-GRPPSSGRGRADDADEREEGEIADDDS-H-PPQ-NPAA--P-EHA----F-NPQGKSKQ-T--S-IRPIH--S---D--S--NN-HH-----VG----C--NKI----   120
wheat p26  --EDTEMR-AS-GA-ER-EGEIADDGD············SS-A--GRITA-P-ENA----F-NPQGKSRQV--STIHPIH--S---D--G--NN-HN----NVG----C--NKI----   108

rice  p28  ---------IA---·------------············AAE---A--A----H---A------S------------R-A----------G-----------VTVN------V------    98
wheat p28  MAEVEAALPV·AATETPEVAAE················GDAGAAEAKGPHKLQRQWTFWYDIQTKPKPGAAWGTSLKKGYTFDTVEEFWCLYDQIFRPSKLVGSADFHLFKAGVEPKW   101
                                                    ★1 ★2        ★3              ★4                                    ★5

rice  p26  ---I---------FSCG-GKS·------H-LLAM------YGD----A-V----GK-ERIAI---N-A---A-IS----Q---FL--K-SIGFIV--AKKMDKGLKN----   227
wheat p26  ---I---------ISCG-GKS·--F--H-LLAM------FGD----A-V----K-ERVAI---N-A---A-IS----Q---FL--K-SIGFIV-E-AKRSDKGPKN----   215

rice  p28  -------------PC----T-EN-----L-----------E------------A---R--------N------IV--------AKRE--------N--          206
wheat p28  EDPECANGGKWTVISSRKTNLDTMWLETCMALIGEQFDESQEICGVVASVRQRQDKLSLWTKTASNEAVQVDIGKKWKEVIDYNDKMVYSFHDDSRSQKPSRGGRYTV   209
                ★6          ★7              ★8        ★9
```

Figure 7. Alignment of cap-binding proteins from plants. The amino acid sequences were compared using MACAW v2.0.3 [275]. The conserved tryptophan residues are indicated (★). The dashes indicate the identical amino acid is used compared to the wheat p28. The dots indicate a gap inserted by the alignment program. The GenBank accession numbers for the cap-binding proteins used are: wheat p26 cDNA, Z12616 [198]; rice p26 cDNA, U34597; wheat p28 cDNA, M95818 [4]; rice p28 cDNA, U34598.

Table 4. Comparison of amino acid sequences of eukaryotic cap-binding proteins (% similarity).[a]

	Wheat p26	Wheat p28	Rice p26	Rice p28	Yeast	*Drosophila*	*Xenopus*	Human
Wheat p26	100	49	78	48	31	32	40	39
Wheat p28		100	47	82	31	29	35	35
Rice p26			100	47	33	33	39	39
Rice p28				100	32	29	36	36
Yeast					100	34	31	31
Drosophila						100	41	44
Xenopus							100	89
Human								100

[a] GenBank accession numbers are as follows: wheat p26, Z12616 [198]; wheat p28, M95818 [4]; rice p26, U34597; rice p28, U34598; yeast, M15436 [5]; *Drosophila*, U16139 [117]; *Xenopus*, D31837 [331]; human, M15353 [265]. Similarity was determined using the ALIGN function of Microgenie v7.01 (Beckman).

terminus, a region of very poor conservation among cap-binding proteins; however, there are serine and threonine residues in this region of both wheat p28 and p26 (see Fig. 7). Multiple isoforms of p26 and p28 have been detected on two-dimensional blots ([335], and K. Browning and D. Gallie, unpublished observations). These isoforms suggest that phosphorylation may occur, although the sites of phosphorylation and the effects of phosphorylation on protein synthesis are unknown for the plant cap-binding proteins.

The large subunits of eIF4F and eIFiso4F are less similar to each other than the cap-binding proteins (see Fig. 8A). Cloning and sequencing of two full-length cDNAs for the large subunit of wheat eIFiso4F showed that the molecular weight of this polypeptide is ca. 86 000 kDa and is referred to as p86 [4]. Cloning of the large subunit of wheat eIF4F yielded cDNAs that are about 2/3 full-length (L. Allen and K. Browning, unpublished data). A schematic presentation of a comparison of the large subunits of wheat eIF4F and eIFiso4F with large subunits from yeast and humans is shown in Fig. 8B. There are five regions of similarity, but the overall similarity is quite low (ca. 25–30%).

Recently, a region in the N-terminal half of mammalian p220 was shown to be involved in the binding of the cap-binding protein [178]. A similar sequence is also present very near the N-terminus of wheat p86 and has been shown experimentally to be required for interaction with p28 and protein synthesis activity *in vitro* (A. Metz and K. Browning, manuscript in preparation).

The individual subunits of eIFiso4F have been overexpressed in *E. coli* and purified [320]. The individual subunits alone do not have activity in polypeptide synthesis *in vitro*, but when mixed in equal molar amounts, reconstitute a fully active complex [320]. The eIFiso4F complex formed from expressed subunits is indistinguishable from the native complex in several *in vitro* assays ([320] and D. Goss, personal communication). The only difference that can be observed is an alteration in the CD spectra compared to the native eIFiso4F, suggesting a change in the protein conformation for the expressed complex (D. Goss, personal communication). CD measurements of the native eIFiso4F at different pHs show that there are significant changes in the α-helix and β-sheet content of the protein based on the pH (D. Goss, personal communication). The bind-

120

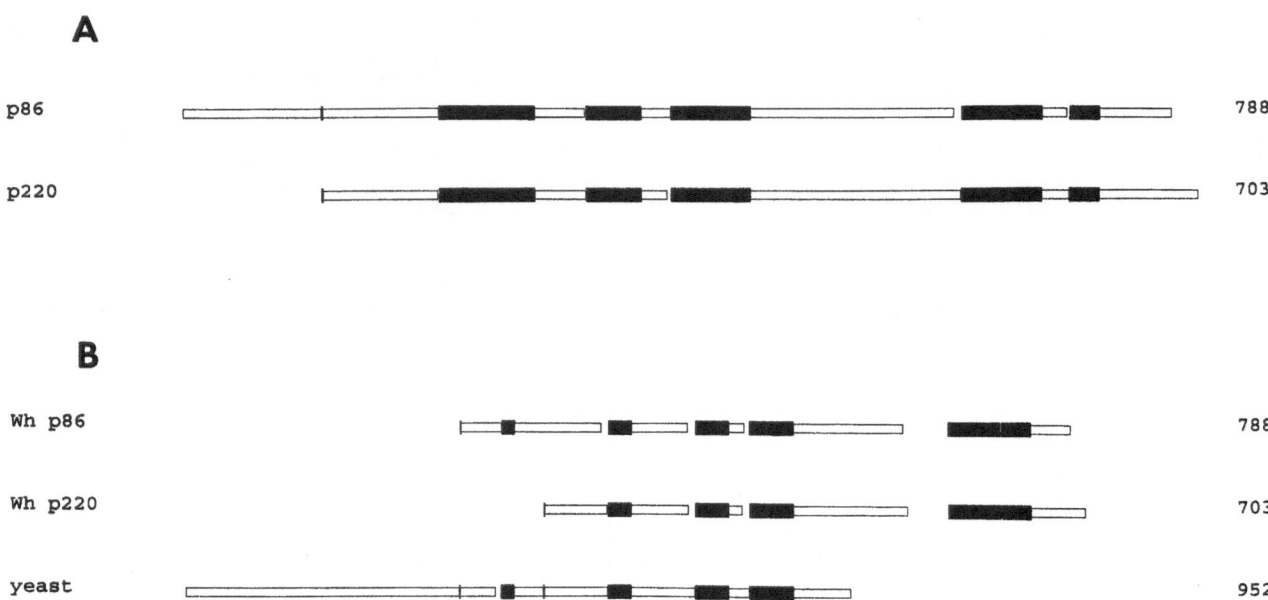

Figure 8. Schematic of the alignment of the large subunits of eIFiso4F and eIF4F. A. Scheme of the alignment of the p86 subunit of wheat eIFiso4F and partial p220 subunit of wheat eIF4F by MACAW v2.0.3 [275]. The wheat p220 amino acid sequence is incomplete. About 300 amino acids are missing from the N-terminus (L. Allen and K. Browning, unpublished data). B. Scheme of the alignment of the wheat p86, partial wheat p220 with large subunits of yeast and human eIF4F. The GenBank accession numbers are: wheat p86 cDNA, M95746 [4]; yeast gene, L16923 [107]; human cDNA, D12686 [343].

ing of cap analogues and capped oligonucleotides also caused significant changes in the protein conformation, with the capped oligonucleotides having a greater effect (D. Goss, personal communication). These results suggest that eIFiso4F is a very dynamic protein undergoing continuous changes in conformation as it binds mRNAs and interacts with other proteins.

The gene structure of eIF4F or eIFiso4F is unknown at this time. However, the presence of cDNAs for p86 with slightly different sequences and restriction patterns, in both wheat [4] and maize (Y. Chen and L. Morejohn, personal communication), suggests that there will be multiple genes for p86. Multiple ESTs with slightly different sequences were also observed for the rice p28 and p26, again suggesting the presence of more than one gene (E. Aliyeva, A. Metz and K. Browning, unpublished observation). The gene organization and expression of these proteins is currently being studied in our laboratory.

eIF5

Very little is known about plant eIF5. eIF5 purified from mammalian and yeast cells has a molecular weight of 58 000; however, it elutes from molecular sieving columns at a position corresponding to a molecular weight of globular protein of ca. 160 000, suggesting it may be a dimer [56, 64, 74]. Mammalian and yeast eIF5 catalyze the hydrolysis of the bound GTP in the eIF2•GTP•Met-tRNA$_i$ ternary complex bound to the 40S ribosomal subunit. Hydrolysis of GTP is necessary for the release of eIF2•GDP, thereby allowing the joining of the 60S ribosomal subunit [55]. Recently, it has been shown that eIF5 forms a stable complex with eIF2 [59]. Yeast and mammalian eIF5 have been cloned and sequenced and are 60% similar [56, 74]. eIF5 contains a GTPase superfamily motif, consistent with its role in GTP hydrolysis [74].

eIF5 has only been partially purified from wheat germ [165, 282]. A cDNA for kidney bean eIF5 was recently reported [193a].

eIF6

eIF6 from wheat germ is a small (M_r 23 000), acidic polypeptide. Very little work has been done on this factor since it was first isolated from wheat germ [260] and mammals [248, 316]. eIF6 exhibits ribosome anti-reassociation activity, by binding to the 60S ribosomal subunit and preventing it from associating with the 40S ribosomal subunit [260]. Whether eIF6 is one of the 'acidic ribosomal proteins' is unknown at this time. eIF6 has not been cloned or sequenced from any eukaryote. When or how it is released from the 60S subunit is also not known. It will no doubt show up in the yeast system sooner or later as a translation mutant.

Initiation of plant protein synthesis

A proposed model of the binding of mRNA to the 40S ribosomal subunits in higher plants is shown in Fig. 9. This model incorporates what is known from functional and biophysical *in vitro* analyses of the interactions and activities of wheat germ initiation factors eIF4F, eIFiso4F, eIF4A, eIF4B, eIF3 and PAB (poly(A)-binding protein) as described in the above sections.

Step A. eIF4F and eIFiso4F discriminate between mono- and hypermethylated cap groups on the 5' end of mRNA [47]. eIFiso4F prefers hypermethylated cap groups and more linear RNA structures [49]. The cap-binding proteins (p26 and p28) bind to the m^7G of mRNA in an ATP-independent manner, as shown by cross-linking *in vitro* [38]. The cap-binding protein undergoes a conformational change upon the binding of the cap group (D. Goss, personal communication).

Step B. The binding of eIF4A to eIFiso4F is enhanced by the presence of ATP [13] and the hydrolysis of ATP by eIF4A is stimulated by the presence of mRNA and eIFiso4F [164]. eIF4A interacts with the m^7G of mRNA only in the presence of eIFiso4F and ATP, as shown by cross-linking *in vitro* [38].

Step C. eIF4B stimulates the ATP hydrolysis and RNA unwinding activities of eIF4A and eIFiso4F [34, 39]. eIF4B interacts with the m^7G of mRNA only in the presence of eIFiso4F, eIF4A and ATP, as shown by cross-linking *in vitro* [34,39]. eIF4B and eIFiso4F each form complexes with poly(A) *in vitro* and stimulate the binding of PAB to poly(A) *in vitro* (D. Gallie, personal communication).

Step D. The cap-binding protein probably disengages once the mRNA•eIFiso4F complex is stabilized by the interactions with eIF4A, eIF4B and PAB. The factor complex proceeds to unwind mRNA in preparation for binding of the 40S ribosomal subunit.

Step E. The 40S ribosomal subunit with eIF2•GTP•Met-tRNA$_i$ and eIF3 already bound interacts with the mRNA•factor complex. The presence of RNA enhances the interaction of eIF3 with eIFiso4F [48]. eIF3 also interacts with NTP to induce a conformational change that probably enhances the interaction of eIF3 with the mRNA•factor complex [50]. The wheat germ eIF3 molecule is depicted as one of two possible forms visualized by electron microscopy [229].

Step F. The mRNA•factor•40S ribosomal subunit complex is then correctly aligned to await the binding of the 60S subunit and commence elongation. When the initiation factors are released and in what order is not known at this time. The interactions of other initiation factors (eIF1, eIF1A and eIF5A) are also not depicted as their functions in plant initiation have not been fully elucidated. However, it is highly likely that they are part of the 40S complex that binds the mRNA•factor complex.

Elongation

Elongation factors from a wide variety of eukaryotic cells are very similar to each other in structure and function, and are similar to their prokaryotic counterparts. The process of elongation has been extensively studied in eukaryotes and many reviews on the process of elongation and elongation factors have been written [119, 192, 202, 218, 244, 255, 291]. A summary of the physical and functional properties of the elongation factors is given in Table 5, and an outline of the steps in elongation is given in Fig. 10.

eEF1

The 'traditional' role of eEF1 is to bring the correct aminoacyl-tRNA to the A site on the ribosome in preparation for the next peptide bond to be formed during elongation. The mechanism used to 'read ahead' on the mRNA and have the next aminoacyl-tRNA ready for that codon is not known at this time.

Plant eEF1, like that of other eukaryotes, is composed of four non-identical subunits (p52, p48, p36,

122

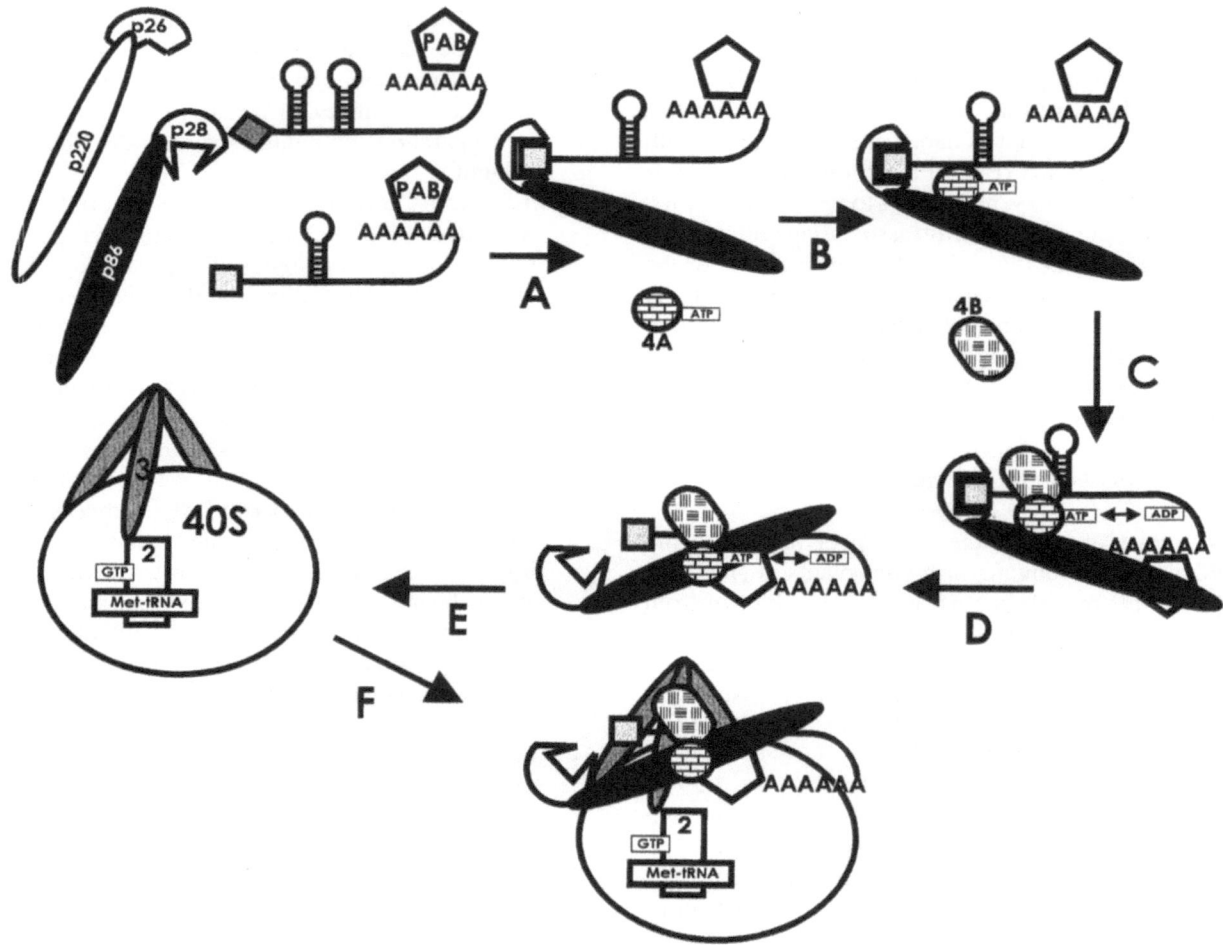

Figure 9. Model of intermediate steps in the binding of mRNA to the 40S ribosomal subunit in higher plants. (See section Initiation of plant protein synthesis for an explanation of the steps.)

p34, see Fig. 2 and Table 5). eEF1 has been isolated from wheat [27, 85, 106, 161, 282], rice [84] and cauliflower [84], as well as numerous other eukaryotic sources [192]. There are conflicting reports on the stoichiometry of the eEF1 complex. Studies on eEF1 complex formation using purified subunits from rice indicate a 1:1:1:1 stoichiometry [87]. Studies on *Artemia* eEF1 complex formation also using purified subunits, indicate an $A_2:B\alpha_1:B\beta_1:B\gamma_1$ stoichiometry [135]. Janssen *et al.* [135] present an intriguing model for the structure of the eEF1 complex in *Artemia*. This model suggests that the eEF1Bγ subunit is the anchor in the endoplasmic reticulum. eEF1Bγ has been previously shown to specifically bind to membranes and to tubulin [133]. The eEF1Bα and eEF1Bβ subunits bind to the eEF1Bγ-subunit creating two binding sites

for eEF1A, such that two eEF1A molecules are able to interact with one ribosome. It was shown that one of the eEF1A molecules more easily dissociates in the presence of aminoacyl-tRNA and GTP, while the other remains tightly bound [135]. This type of arrangement would facilitate the exchange of GTP and thus increase the rate of translation. In support of this model is the recent finding that two molecules of GTP on two molecules of prokaryotic EF1A are hydrolyzed for each elongation cycle in *E. coli* [276]. While this model is appealing from the standpoint of protein synthesis and the role of eEF1 in it, it does not address the apparent and growing evidence for association of eEF1A with the cytoskeleton (see Protein synthesis and the cytoskeleton).

Table 5. Elongation factors from higher plants.

Factor	M_r[a]	Function	Plant source [ref.][b]	DNA sequence information
eEF1A (EF-1α)	52 000	Bind aminoacyl-tRNA and GTP	wheat germ [165]; rice [84] cauliflower [84]	*Arabidopsis* X16430, X16431, X16432 [11, 71, 72, 172] barley L11740 [303], Z50789, Z23130 [82] carrot X60302, D12709 [146] maize D45408, D45407 [22a] soybean X56856 [3] tobacco D63396 [156], U04632 [352] tomato X14449 [242], X53043 [286] wheat M90077 [196]
eEF1B	γ 48 000 β 36 000[c] α 34 000[d]	Recycle eEF1A•GDP	wheat germ [165] rice [84] cauliflower [84]	γ-subunit None known β-subunit *Arabidopsis* X74733, X74734 rice D23674 [185], L36094 [155] α-subunit rice D12821 [184] wheat D13147 [219]
eEF2[e]	102 000	Translocation	wheat germ [165]	None known

[a] Based on SDS-PAGE gel mobility.
[b] The reference indicates the most current method of purification. See text for additional references.
[c] The old *Artemia* and mammalian designation was EF-1δ. The old plant designation was EF-1β.
[d] The old *Artemia* and mammalian designation was EF-1β. The old plant designation was EF-1β'.
[e] Contains unique diphthamide modification.

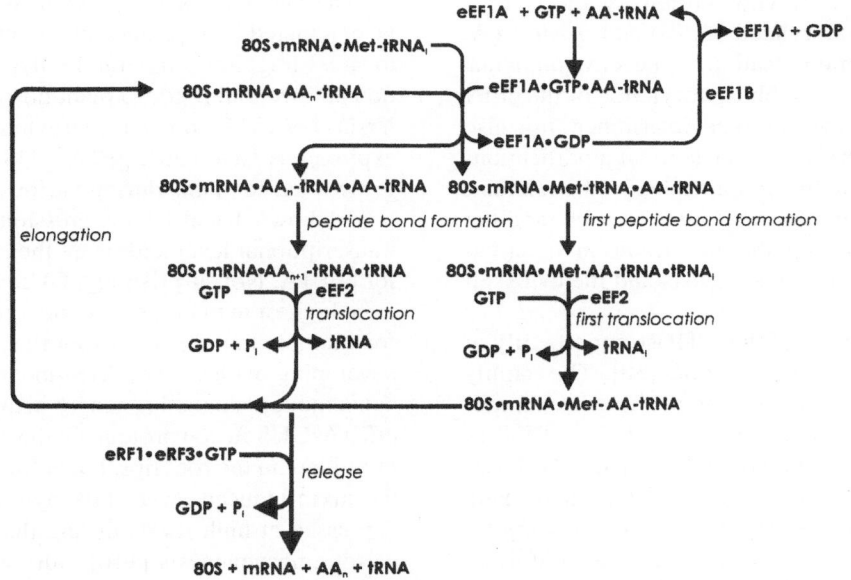

Figure 10. Intermediate steps in elongation of eukaryotic protein synthesis.

eEF1: the subunits

eEF1A. Elongation factor 1A is probably the most extensively studied factor involved in protein synthesis and yet it is still not completely understood. eEF1A is turning up in the most interesting places and may have more to do in the cell than bring GTP and aminoacyl-tRNA to the elongating ribosome. eEF1A is also one of the more abundant proteins in eukaryotic cells, including wheat germ, representing up to 5% of the soluble protein [36, 170, 290, 307]. eEF1A has been isolated from wheat germ [84, 161, 282] and is similar in molecular weight (52 000) and enzymatic properties to eEF1A from other eukaryotes. eEF1A from yeast, brine shrimp and mammals contain post-translation modifications (eight methyl and two glycerylphosphorylethanolamine modifications); however, there does not appear to be a direct correlation between these types of modifications and activity in protein synthesis [53, 80]. Phosphorylation *in vitro* and *in vivo* of mammalian eEF1A, eEF1Bα and eEF1Bβ by protein kinase C in response to phorbol esters, stimulates poly(U)-directed polyphenylalanine synthesis [326, 327]. Casein kinase II phosphorylates the α- and β-subunits of the mammalian eEF1B complex only when GDP is bound to eEF1A, suggesting a conformational change in the complex is necessary to expose the phosphorylation sites on eEF1Bα and eEF1Bβ [227]. These observations suggest that the phosphorylation of the subunits of eukaryotic eEF1 may play a role in the regulation of eukaryotic protein synthesis. Carrot eEF1A is phosphorylated *in vivo* and *in vitro* (W. Boss, personal communication). The environmental conditions or effectors of phosphorylation of the eEF1 complex in plants have yet to be determined. It is also not known if the other post-translational modifications of eEF1A (methyl or glycerylphosphorylethanolamine modifications) occur in plants; however, such modifications could have roles in the 'non-traditional' activities of eEF1A (see Protein synthesis and the cytoskeleton).

eEF1A is a member of the GTPase superfamily of proteins reviewed by Bourne *et al.* [30]. This family of proteins has three classes, one of these is exemplified by the bacterial elongation factor EF1A (EF-Tu), the prokaryotic counterpart of eEF1A. These GTPases share a similar cycle, the release of GDP (with the help of additional factors, EF1B (EF-Ts) for prokaryotes and eEF1B for eukaryotes) to an 'inactive' conformation. Upon preferential binding of GTP over GDP in the guanine nucleotide binding site the protein assumes an 'active' conformation. The GTPase superfamily shares many structural features that contribute to their role as 'molecular switches' in signal transduction of cells [30]. Perhaps this role as a 'molecular switch' is why eEF1A is turning up in unexpected places, doing unexpected things.

eEF1A, in addition to its more well-known role in the GTP-dependent binding of aminoacyl-tRNAs, is reported to be associated with the cytoskeleton of both plants and mammals, both with microfilaments and microtubules (reviewed by Condeelis [68]). The association of plant protein synthesis machinery with the cytoskeleton will be discussed further in a later section of this review (see Protein synthesis and the cytoskeleton). Recently, a phosphatidylinositol 4-kinase activator (PIK-A49) from carrot cells that binds actin and enhances actin bundling was shown to have >90% similarity to eEF1A from plants [346]. The actin binding protein, ABP-50, from *Dictyostelium* was also determined to be eEF1A [344]. Both ABP-50 and PIK-A49 exhibit eEF1A activity in supporting polyphenylalanine synthesis that is functionally indistinguishable from rabbit eEF1A [344, 346]. More recently, phosphorylation of PIK-A49 by a calcium-dependent protein kinase (CDPK) was shown to be necessary for PIK-A49 activity as an activator of phosphotidylinosital–4 kinase; however, the effect of the phosphorylation on eEF1A activity or protein synthesis was not determined in carrot cells [345].

The role of plant eEF1A expression is beginning to be elucidated in developmental control, in responding to stress (e.g. low temperature, hypoxia, fungal infection and wounding) or to plant hormones (e.g. auxin). Ursin *et al.* [315] conducted an extensive analysis of the expression of a tomato eEF1A-GUS construct in transgenic tobacco plants during various stages of growth. It was shown that eEF1A expression is regulated at the transcriptional level and/or by the stability of mRNA for eEF1A. Not surprisingly, GUS activity was found to be highest in plant tissues undergoing rapid growth, for example in meristems, root tips, young leaves and developing ovules [315]. Treatment of seedlings with auxin induces dramatic changes in the expression of the eEF1A-GUS fusion protein. GUS staining is no longer prominent in the root tips, but is found at the region of the auxin-induced curve of the hypocotyl [315]. Genes expressed at high levels during the globular stage of carrot embryogenesis [146], during cold response of barley [82] and a broadbean nodule-specific transcript [239] all were found to encode eEF1A. The amount of stable eEF1A mRNA in soybean seedings increases in response to light [3]. The levels of bean eEF1A expression drops during the necrotrophic phase of fungal infection [181]. Potato tubers show a biphasic increase in protein synthesis in response to wounding that is coordinated with an increase in the transcription of eEF1A mRNA and accumulation of eEF1A protein [206]. Hypoxia in potato tubers and maize roots initially induces a shut-down of protein synthesis, most likely due to the drop in intracellular pH, followed by a resumption of protein synthesis 12–16 h later of proteins necessary for alcohol fermentation [12, 324, 336]. The failure of eEF1A to dissociate from polysomes in hypoxic potato tubers appears to be the cause of the initial shut-down of protein synthesis and the resumption of protein synthesis correlates with increased expression of eEF1A [325]. All these observations taken together suggest a rapid response of the expression of eEF1A to changes in the protein synthesis needs of the plant, either due to environmental conditions or to internal signals for development.

A comparison of amino acid sequences for eEF1A from several plant species is shown in Fig. 11. Quite obviously the sequences among plant species are highly conserved, >95%. For comparison purposes, the similarity of the wheat eEF1A amino acid sequence to yeast and human eEF1A is 73% and 77%, respectively [196]. Scattered differences in the amino acid sequences are observed and most of the differences are present in more than one species suggesting that the changes are not necessarily random. Some differences appear to be monocot versus dicot, some are between gene and cDNA (e.g. carrot and barley) and some may be artifactual due to sequencing errors. Even though eEF1A is a multi-gene family in plants, it may be useful for establishing phylogenetic relationships as was recently done for the single eEF1A gene of the moth subfamily, *Heliothine* [65].

There is one amino acid change in particular in one of the barley eEF1A sequences that bears comment. Analysis of eEF1A mutants from the fungus, *Podospora anserina*, that have increased translational fidelity and eEF1A mutants from *Saccharomyces cerevisiae* that have decreased translational fidelity, allow identification of a region of eEF1A that is potentially involved in translational fidelity [288]. There is a single-base mutation in this region (see Fig. 12) in the deduced amino acid sequence of an eEF1A cDNA isolated from a cold-tolerant strain of barley [82]. This suggests that one of the roles of multiple genes for eEF1A may be to provide plants with the ability to adjust the translational fidelity under a variety of adverse conditions.

The genes for plant eEF1A have been extensively studied in *Arabidopsis* [11, 70–72, 172]. eEF1A is encoded by four actively transcribed genes in *Arabidopsis* and the gene organization does not appear to change among different varieties of *Arabidopsis* [11]. Three of the four genes (A1, A2 and A3) are encoded within a 10 kb fragment and possibly arise from a gene duplication event. The spacial arrangement of the fourth gene, A4, to the other genes is not known at this time. Three of the genes, A1, A2 and A4 contain a telo box element, for binding to nuclear scaffolds, that is frequently associated with highly expressed genes in eukaryotes [11]; however, deletion of the telo box does affect the expression of GUS fused to the A1 promoter in transient assays [71] or transgenic plants [70] suggesting a higher-order need for this element, perhaps in the nucleus. Further analysis of the A1 gene showed a *cis*-acting domain (TEF1 box), ca. 100 nucleotides upstream of the transcription start site. The TEF1 box specifically binds *trans*-acting nuclear factors from *Arabidopsis* [71]. Similar sequences were found in the A2, A3 and A4 promoters. These signals appear to be conserved among angiosperms since the *trans*-acting nuclear factors are present in several plant species, both mono- and dicot [72]. Two conserved introns are also found in all the *Arabidopsis* genes, one within the coding sequence and one within the 5′-untranslated region. The presence of the intron in the 5′-untranslated region is necessary for maximal promoter activity [70, 71]. The genes from tomato and soybean for eEF1A appear to have a similar genomic structure to that of *Arabidopsis* and also belong to a multigene family [3, 72, 286]. Expression of GUS fused to the *Arabidopsis* eEF1A A1 promoter showed that elements to the 5′ and 3′ side of the transcription start site were involved in the level of and site of expression of eEF1A during growth [70].

Lastly, another interesting place that eEF1A has turned up, is as an indicator of the lysine content of maize, barley and sorghum endosperm [110]. eEF1A is one of several mRNAs overexpressed in *opaque 2* mutants of maize, which have a significantly higher lysine content over wild-type maize. The *opaque 2* gene encodes a transcription factor that controls the expression of zein storage proteins. The exact mechanism of how the expression of eEF1A and other non-zein proteins is increased in these mutants is not known at this time [110].

126

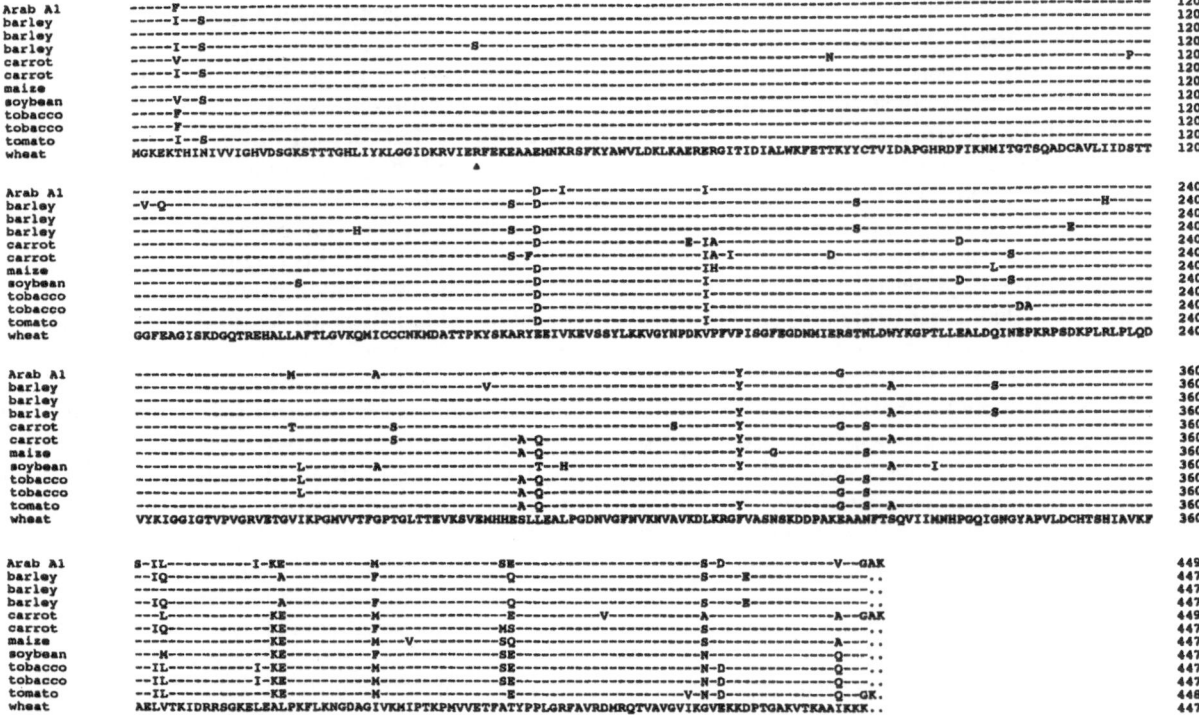

Figure 11. Alignment of the amino acid sequences for plant eEF1A. The amino acid sequences were compared using MACAW v2.0.3 [275]. The dashes indicate the identical amino acid is used compared to the wheat eEF1A. The dots indicate a gap inserted by the alignment program. Only one sequence per plant species is shown, unless there were differences in the reported amino acid sequences (e.g. barley, also see Table 5 for a listing of all known plant eEF1A sequences). The GenBank accession numbers for the eEF1A sequences shown are: *Arabidopsis* A1 gene, X16430 [11]; barley cDNA L11740 [303]; barley gene, Z50789; barley cDNA (cold-tolerant), Z23130 [82]; carrot gene X60302, carrot cDNA D12709 [146]; maize gene, D45408 [22a]; soybean gene, X56856 [3]; tobacco cDNA, D63396 [156]; tomato cDNA, X14449 [242]; wheat cDNA, M90077 [196].

eEF1Bα, eEF1Bβ and eEF1Bγ. Compared with the literature for eEF1A there is a paucity of knowledge, not only in plants, but other eukaryotes as well, concerning the eEF1B subunits α, β and γ which are required for recycling eEF1A•GDP.

It was originally thought that eEF1B contains only two subunits [27]. The 'β'-subunit appears as doublet by SDS PAGE, and the second polypeptide was thought to be a degradation product or modified form [51, 133]. However, early studies with eEF1B from wheat indicated that even though the two eEF1'β' subunits have similar peptide maps, they are distinct subunits that differ in their ability to interact with eEF1Bγ [84, 85]. Recently, cloning and sequencing has shown that the doublet consists of two similar, but non-identical subunits with indistinguishable guanine nucleotide exchange activities [319]. The 'new' subunit was given the designation of 'EF-1δ' in *Artemia* [9, 319]. The plant EF-1'β' doublet subunits were termed 'EF-1'β'

P. anserina	
Normal	RTIEKFEK
Increased Fidelity Mutation	RTIEEFEK
S. cerevisiae	
Normal	RTIEKFEK
Decreased Fidelity Mutation	RTIKKFEK
Barley	
Normal	RVIERFEK
Cold-Tolerant Barley Mutation	RVIESFEK

Figure 12. eEF1A amino acid motif involved in fidelity of translation. Comparison of a mutation in the eEF1A of a cold-resistant strain of barley (Z23130 [82]), to yeast and *Podospora* mutants involved in the fidelity of translation [288].

and 'EF-1β′' [84]. Molecular cloning and sequencing of the cDNAs encoding rice 'EF-1β' showed it to be the functional equivalent of *Artemia* 'EF-1δ' and the plant 'EF-1β′' to be the functional equivalent of *Artemia*

'EF-1β' [184, 185]. The new nomenclature for these proteins is indicated in Table 5 (J. Hershey, personal communication). The subunit previously designated 'EF-1β' in *Artemia* and animals and 'EF-1β″' in rice is now given the designation eEF1Bα. The subunit previously designated 'EF-1δ' in *Artemia* and animals and 'EF-1β' in rice is now given the designation eEF1Bβ. This new nomenclature is used for this review.

Amino acid sequences of plant cDNAs for the α- and β-subunits are limited to rice [155, 184, 185], wheat [219] and *Arabidopsis* [105]. A comparison of the plant amino acid sequences for eEF1Bα and eEF1Bβ is shown in Fig. 13 and summarized in Table 6. At this point in time very little is known of the gene structure and expression of eEF1Bα or eEF1Bβ in any eukaryote. An *Arabidopsis* gene sequence deposited for eEF1Bβ indicates the presence of at least three introns [105], but nothing is known about its expression. There are minor differences in the deduced amino acid sequences for the *Arabidopsis* eEF1Bα gene and cDNA, suggesting that more than one gene may be present. The sequences of two rice eEF1Bβ cDNAs from different tissues (anther and endosperm) are quite different (see Fig. 13), suggesting that more than one gene may be present and expressed differentially. The rice endosperm eEF1Bβ [155] was originally reported to be similar to rice eEF1Bα; however, the alignment by MACAW (see Fig. 13) suggests it is more similar to the β-subunit of rice eEF1.

The amino acid sequences of rice and wheat eEF1Bα and rice and *Arabidopsis* eEF1Bβ are compared with the respective subunits from other species in Fig. 14. The C-terminal portion of the molecules is highly conserved among different eukaryotes [9]. The C-terminal portion of the *Artemia* eEF1Bα has been shown to carry out the guanine-nucleotide exchange reaction on eEF1A [319]. This indicates that the guanine exchange domain is located in the C-terminus and is highly conserved during evolution. The N-termini of the eEF1Bα subunits are reasonably well conserved among all eukaryotes; however, the N-termini of the eEF1Bβ subunit are poorly conserved between plants and animals (see Fig. 14). This suggests that the changes in the amino acid sequence of the N-termini of eEF1Bβ occurred after the divergence of plants and animals. Further evidence for this hypothesis is the absence in rice or *Arabidopsis* eEF1Bβ sequences of a leucine zipper motif found in human, *Artemia* and *Xenopus* eEF1Bβ [9]. This indicates that the function of the leucine zipper must have arisen after plants and animals diverged during evolution. The apparent

sequence divergence of eEF1Bβ would make it an excellent candidate for phylogenetic analysis.

The phosphorylation of *Artemia* eEF1Bα Ser-89 by casein kinase II inhibits its ability to catalyze guanine-nucleotide exchange activity on eEF1A [136]. In addition, phosphorylation of Ser-89 inhibits the formation of an *Artemia* eEF1AB complex [318]. Conspicuously missing in the rice and wheat eEF1Bα sequence is a counterpart to Ser-89 of *Artemia* eEF1Bα. Wheat eEF1Bα is not phosphorylated by casein kinase II *in vitro*; however, there is a serine residue(s) in wheat eEF1Bβ that is phosphorylated [86, 184].

eEF1Bγ has not yet been sequenced from any plant sources, although potential EST sequences for *Arabidopsis* are in the EST database. The precise function of eEF1Bγ is not known. During meiosis in *Xenopus*, the γ-subunit is phosphorylated by the M-phase promoting factor, which is a regulator of the G2- to M-phase transition [134]. eEF1Bγ is reported to bind membranes and tubulin [133], stimulate the guanine-nucleotide exchange activity of eEF1Bα and eEF1Bβ [133] and suggested to be the 'anchor' for the eEF1 complex in the endoplasmic reticulum [135]. eEF1Bγ is encoded by a multigene family in yeast whose amino acid sequences are only ca. 65% similar [150], in contrast to 100% identity of the amino acid sequences of the two yeast genes for eEF1A [208]. One of the eIF1Bγ genes (*tef3*) was isolated as Ca^{2+}-dependent membrane binding protein [143] and as a extragenic suppressor of a cold-sensitive yeast mutant deficient in 40S ribosomal subunit assembly [256]. It would appear eEF1Bγ, like eEF1A, is also going to have some surprises and be a very interesting molecule to study.

Elongation factor 2

eEF2 catalyzes the GTP-dependent translocation of the peptidyl-tRNA from the A site to the P site on the ribosome, resulting in the movement of the mRNA three nucleotides to the 3′ side and release of deacylated tRNA from the P site (for a review see [218]). eEF2 is also a member of the GTPase superfamily [30] and in contrast to eEF1A, eEF2 does not need a recycling factor for the exchange of GDP for GTP. The 'pretranslocation' ribosome favors the binding of eEF2 and the 'post-translocation' ribosome favors the binding of eEF1A. These two binding sites apparently overlap slightly at the ribosomal interface where both the 18S rRNA and 28S rRNA are exposed. Both rRNAs are believed to have a direct role in the functioning of the ribosome during translation (for a review see [217]).

128

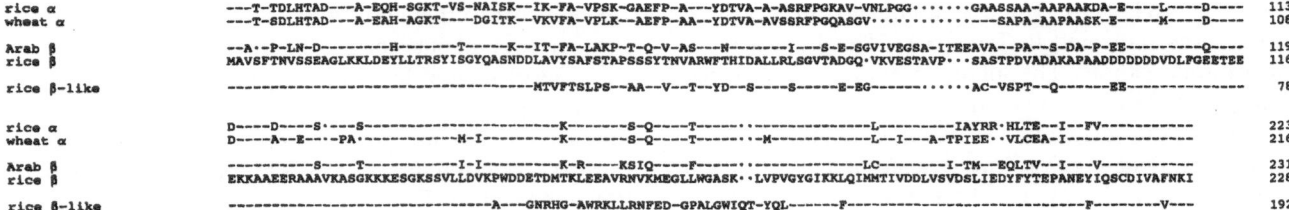

Figure 13. Alignment of the amino acid sequences for plant eEF1Bα and eEF1Bβ. The amino acid sequences were compared using MACAW v2.0.3 [275]. The dashes indicate the identical amino acid is used compared to the wheat eEF1Bβ. The dots indicate a gap inserted by the alignment program. The GenBank accession numbers for the sequences used are: rice α cDNA, D12821 [184]; wheat α cDNA, D13147 [219]; *Arabidopsis* β cDNA, X74733; rice β cDNA, D23674 [185]; rice β-like cDNA, L36094 [155].

Table 6. Comparison of amino acid sequences of plant eEF1Bα and eEF1Bβ (% similarity).[a]

	Rice α	Wheat α	Rice β	*Arabidopsis* β	Rice β-like
Rice α	100	76	58	62	51
Wheat α		100	56	58	48
Rice β			100	69	70
Arabidopsis β				100	59
Rice β-like					100

[a] GenBank accession numbers are as follows: rice α, D12821 [184]; wheat α, D13147 [219]; rice β, D23674 [185]; *Arabidopsis* β, X74733; rice β-like, L36094 [155]. Similarity was determined using the ALIGN function of Microgenie v7.01 (Beckman).

eEF2 is post-translationally modified at His-715 to yield the unique residue, diphthamide, found only in eEF2 [187, 257, 323]. This unique diphthamide residue is the target for the specific and covalent attachment of ADP-ribose to eEF2 by several bacterial toxins (e.g. diphtheria toxin, [25, 221]). The net effect of ADP-ribosylation by bacterial toxins is the inhibition of translocation activity of eEF2 that ultimately leads to cell death [24, 228]. This diphthamide modification is present in all eukaryotes including eEF2 from wheat germ, which can be ADP-ribosylated by diphtheria toxin [161]. It has been shown in yeast through mutagenesis of His-715 and gene deletion, that His-715 and diphthamide modification is not necessary for eEF2 activity or viability in yeast [149, 241]. This is in contrast to mammalian eEF2 where modification of the corresponding His residue to aspartic acid, lysine or arginine results in an inactive protein [220]. Interestingly, there are at least six yeast genes necessary to carry out the diphthamide modification [25] and yeast mutants that are resistant to diphtheria toxin have defects in this pathway [62, 63, 186, 236]. Natural mutations of the His-715 do not appear to occur, even though viability apparently is not affected. Why such a complex modification pathway would be

conserved when viability is not at stake is not clear. However, an answer may be beginning to emerge. Cellular ADP-ribosyltransferases have recently been identified, and although their mechanism of action is similar to the bacterial toxins, they ADP-ribosylate only a small fraction of the eEF2 at any given time (reviewed in [131]). A model has been proposed by Iglewski [131] to explain both the conservation of the diphthamide modification and the presence of cellular ADP-ribosyltransferases. Translation is known to pause during translocation of secreted polypeptides across the endoplasmic reticulum; however, the mechanism of pausing is unknown. This model suggests that the ADP-ribosylation of eEF2 is the means by which translation is temporarily halted for translocation to occur and then translation resumes when the ADP-ribose is removed. Unfortunately, an enzyme capable of removing ADP-ribose from eEF2 has not been identified as yet. It is an intriguing model that will bear more scrutiny.

eEF2 from mammals is phosphorylated by a Ca^{2+}/calmodulin-dependent kinase (eEF2 kinase) that causes inhibition of eEF2 translocation activity [209, 261–264]. Wheat germ eEF2 is phosphorylated by eEF2 kinase from rabbit reticulocytes that results in

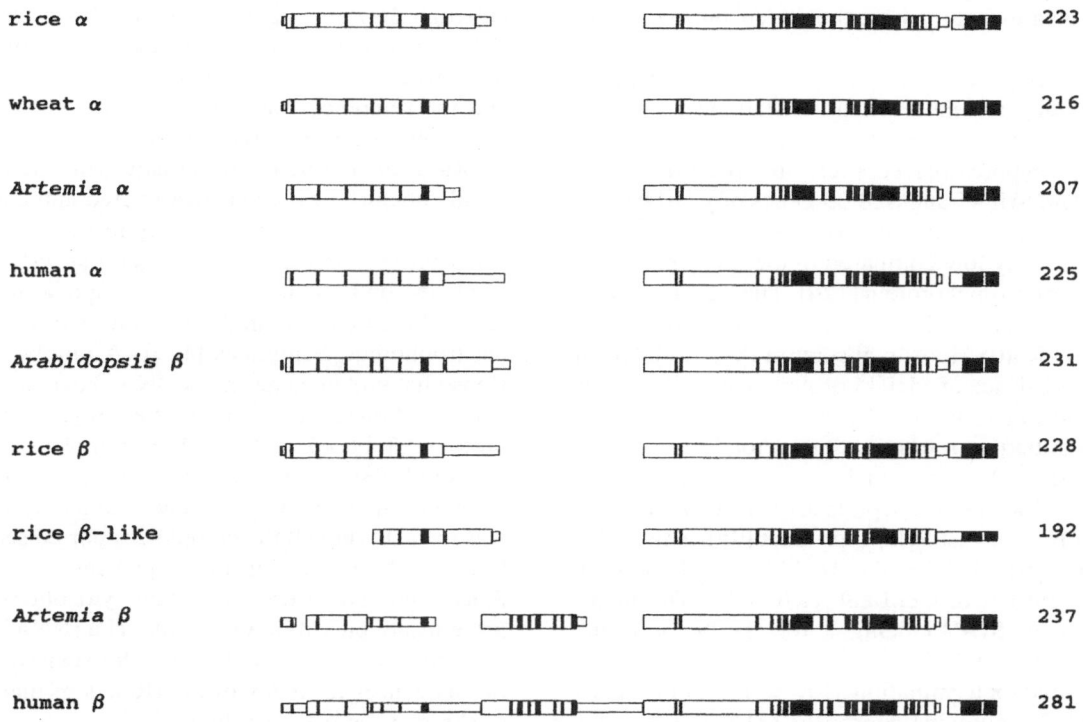

Figure 14. Schematic of the alignment of plant eEF1Bα and eEF1Bβ with eEF1Bα and eEF1Bβ from other eukaryotes. The amino acid sequences were compared using MACAW v2.0.3 [275]. The GenBank accession numbers used are: rice α cDNA, D12821 [184]; wheat α cDNA, D13147 [219]; *Artemia* α cDNA, X51871 [180]; human α cDNA, X60489 [267]; *Arabidopsis* β cDNA, X74733; rice β cDNA, D23674 [185]; rice β-like cDNA, L36094 [155]; *Artemia* β cDNA, S47630 (PIR accession number) [9]; human β cDNA, Z21507 [26].

inhibition of eEF2 activity [292]. However, an eEF2 kinase activity could not be identified in plant tissues at various stages of growth [292]. This implies that such a kinase does not exist in plants; however, plants have been notoriously difficult about yielding their kinases and it may be premature to give up the search for a plant eEF2 kinase.

eEF2 has been isolated from wheat germ as a single polypeptide of M_r ca. 102 000 (see Fig. 2) and displays activities similar to its mammalian and yeast counterparts [161, 280, 314]. However, a cDNA or gene for plant eEF2 has not been reported. There are *Arabidopsis* and rice sequences in the EST database that have similarity to mammalian eEF2. The expression and regulation of the eEF2 genes and eEF2 activity in plants is of particular interest given the inter-relationship of ribosomes, eEF2 and plant ribosome inactivating proteins (RIPSs, see Plant ribosomes). Because plant ribosomes are more resistant to RIPs, it is possible that there are functional differences between eEF2 from plants and mammals and how plant eEF2 activity is regulated. We know plant eEF2 can be ADP-ribosylated

by diphtheria toxin. Whether there are comparable 'toxins' in pathogenic organisms of plants that affect eEF2 activity will be of great interest.

Termination

The process of releasing the polypeptide chain from the ribosome is begun when the ribosome encounters one of three termination codons in the A site. A release factor binds to the ribosome, the peptidyl-tRNA is hydrolyzed and the nascent polypeptide is released along with the uncharged tRNA. The ribosome dissociates into its subunits, ready to begin another round of initiation. The hydrolysis of GTP is necessary for termination, although the exact point of hydrolysis in the process is unknown. Prokaryotic release factors, RF1 and RF2, are codon specific and require the presence of RF3 that stimulates release activity in a GTP-dependent manner. Mammalian release factor activity appears to function as dimer, requires GTP for activity and is not codon specific (see reviews [52, 192,

317]). Release factors have not been purified from higher plants.

Yeast has proven particularly useful in identifying potential genes involved in translation termination (for reviews see [302, 310]). One class of mutations, omnipotent suppressors, suppress all three termination codons. The SUP45 and SUP35 genes of yeast belong to this class of mutants. The expression of SUP45 was found to be very low compared to the level of expression of L3 ribosomal protein [124]. The SUP45 protein is also shown genetically and immunologically to associate with the 40S ribosomal subunit [89, 301]. Given the low abundance of SUP45 protein and the low ratio to 40S ribosomal subunits the factor was thought to be either an initiation or termination factor.

Recently, a termination factor was isolated from rabbit reticulocytes and peptide sequences indicated it is similar to the yeast SUP45 product [95]. This factor is now designated eRF1. An *Arabidopsis* SUP45-like cDNA is present in GenBank (X69375). The major problem with SUP45 being a release factor is the absence of any GTP-binding sites, since GTP is known to be required for termination. However, the other yeast gene, SUP35, contains consensus GTP-binding sites in the C-terminal domain that are similar to eEF1A [341], and there is also genetic and immunological evidence for an association with the 40S ribosomal subunit [351]. The SUP35 gene product also is expressed at a comparable level to SUP45. These results suggest SUP35 is also a termination factor. Recently, biochemical evidence was obtained that a SUP35-like protein from *Xenopus*, now designated eRF3, enhances the release activity of eRF1 through the formation of a eRF1•eRF3•GTP complex that interacts with the stop codon of mRNA [351]. A SUP35-like plant EST is not present in the database.

The availability of the sequences for eRF1 and eRF3 will allow the elucidation of the mechanism of eukaryotic termination of polypeptide chain elongation. Whether additional eukaryotic RFs exist with a preference for a particular termination codon(s) (e.g. like the prokaryotic RF1 and RF2) remains to be determined.

Plant ribosomes

The function of ribosomes from all eukaryotes is the assembly of amino acids into protein molecules using mRNA as a blueprint. Plant ribosomes have an architecture similar to their mammalian counterpart although they are slightly smaller [43–45, 204, 205]. For more extensive reviews of eukaryotic ribosomes, their structure, function and biogenesis (see [218, 342] and the references therein). An excellent review on chloroplast ribosomes was recently published [114].

Ribosomal proteins from plant cytosolic large and small subunits have been fractionated and catalogued by two-dimensional gel electrophoresis [271, 287]. The ribosomal proteins from several species of higher plants, both monocot and dicot, appear to be conserved based on immunological and two-dimensional electrophoretic techniques [109]. A number of plant ribosomal protein genes and cDNAs have been identified based on similarity to other eukaryotic ribosomal proteins (Tables 7 and 8). Only complete sequences (as could best be determined from the GenBank submissions) are compiled in Tables 7 and 8. Every effort was made to find all the complete reported sequences; however, I apologize if any sequences were omitted. A large number of ESTs for plant cytosolic ribosomal proteins are present in GenBank. The rRNAs of plant ribosomes will not be included in this review, although the sequences for many of the rRNAs of many plants are known and being studied.

Most of the plant ribosomal protein genes appear to have two or more copies of the gene, to have high levels of expression in rapidly growing tissue and to have their expression developmentally or environmentally regulated [29, 101, 102, 139, 148, 159, 168, 183, 298, 299, 304, 305, 322].

The maize S14 ribosomal protein gene family has 3–6 members and the expression of the S14 mRNA is developmentally regulated [159]. The largest amount of expression is found in the endosperm prior to the beginning of seed storage protein synthesis, suggesting a need for an increased number of ribosomes [159]. The maize S13 ribosomal protein gene is also found to be highly expressed in rapidly growing tissues and expression appears to parallel DNA synthesis [139]. The expression of potato S19, L27 and L7 all increase during tuberization [304, 305]. S15a is highly expressed in mitotically active tissues (young leaves, meristem and flower buds) of *Brassica napus* [29]. Tobacco L2 mRNA expression was also shown to be correlated with actively growing tissues [183]. Expression of L27 ribosomal protein mRNA increases at least 10-fold after decapitation of pea seedling axillary buds and belongs to a multigene family in a specific class of genes expressed during the transition from dormancy to growth [298, 299]. Expression of tobacco L25 and L34 ribosomal protein mRNAs are highly expressed in

Table 7. Compilation of small ribosomal subunit protein sequences from higher plants.

Ribosomal protein	GenBank accession number [ref.]
S4	
potato	X76651 [33]
S4e	
cotton	X79300 [313]
S6	
tobacco	X68050 [113]
S8	
rice	D38010 [210]
S10	
Arabidopsis	X80694
S11	
Arabidopsis	J05216 [100]; L28828 [171]; L07877 [176]
maize	X55967 [168]
soybean	M31024 [100]; L28831 [171]
S13	
maize	X62455 [139]
pea	Z25509
S14	
maize	P19950[a], P19951[a] [159]
S15	
Arabidopsis	Z23161 [269]
rice	D10962 [147]
S15a	
Arabidopsis	L27461 [28]
Brassica napus	X59983, X59984 [29]
S16	
cotton	X75954 [312]
Lupinus polyphyllus	X51766 [333]
rice	L36313
S18	
Arabidopsis	Z28962, Z28701, Z28702, Z23165 [322]
S19	
Arabidopsis	X78035
potato	NOT IN GENBANK [304]
S20	
rice	D12632 [214]
S21	
rice	D12633 [214]
S25	
tomato	X76714 [340]
S28	
Arabidopsis	L09755 [130]
maize	X82124
S31	
rice	D38011

[a] SWISS-PROT accession number.

Table 8. Compilation of large ribosomal subunit protein sequences from higher plants.[1]

Ribosomal protein	GenBank accession number [ref.]
L2	
Arabidopsis	X86765
tobacco	X62500 [183]
tomato	X64562
L3	
Arabidopsis	M32654,M32655 [148]
rice	D12630 [216]
L5	
alfalfa	X78284
L7	
potato	NOT IN GENBANK [304]
L7a	
rice	D12631 [216]
L16	
Arabidopsis	X81798, X81799, X81800
L17	
barley	X62724, X62725 [179]
tobacco	L18915 [102]
L18	
Arabidopsis	U15741
L19	
tobacco	Z31720 [203]
L25	
tobacco	L18908 [101]
L27	
pea	X70702 [299]; U10043, U10044, U10045, U10046 [298]
potato	Z30162 [305]
L32	
maize	X75646
L34	
pea	U10047
tobacco	L27089, L27107 [101]
L37	
tomato	X79074 [340]
L37a	
Brassica rapa	L21897, Z24739
L38	
tomato	X69979 [93]
L41	
cotton	X75423 [311]
P0	
rice	D21130 [123]
P2	
maize	U29383

rapidly growing tissue and are induced by wounding and hormone treatment [101].

A mutation in the PFL locus of *Arabidopsis* gives rise to a stunted growth phenotype that suggests a generalized effect of the gene on growth, particularly the meristems. The PFL gene has been identified as a gene for ribosomal protein S18 and is the first mutation reported for a S18 ribosomal protein in eukaryotes [322]. The only other reported mutations in eukaryotic ribosomal proteins are in *Drosophila*, these mutations also give rise to a stunted or minute phenotype [10, 151]. The promoter for the PFL locus is active in cells with high mitotic activity (e.g. meristems) or wounded cells, but not cells with high metabolic rates [322]. Two other copies of the S18 gene are also found in *Arabidopsis*. The presence of multiple copies of ribosomal protein genes in plants and their high expression in mitotically active tissues appear to assure that there is an ample supply of ribosomal proteins for assembly into ribosomes during rapid growth. This is in contrast to mammals which usually only have one active gene for ribosomal proteins. It is this compensating presence of the two other gene copies for S18 that allows the mutation at the PFL locus to give rise to a recessive, reduced growth phenotype rather be lethal [322].

The genes for S11 from *Arabidopsis* and soybean show some upstream elements that are similar to promoter elements found in *Arabidopsis* genes for S15 and eEF1A [70, 171, 269] suggesting that the same elements may control expression of both ribosomal proteins and eEF1A. Coordinate expression of plant ribosomal genes and the presence of multiple genes for most plant ribosomal proteins appears to be ubiquitous. There are some ribosomal protein genes that are differentially expressed during endosperm development [159], growth upon dormancy [299] and wounding [322]. At least some of the ribosomal proteins are translated from stored mRNAs during early gemination of maize. Some of these mRNAs are stored in a mature form (S4 and S6), whereas other mRNAs (L3 and L6) may be in an unspliced form [18]. A pre-existing pool of acidic ribosomal proteins may also exist in dry axes of maize embryos [18]. It appears that plants have evolved several mechanisms to regulate the components of ribosomes during growth and development and under environmental stress.

There have been studies showing the effects of auxins on the increased expression of ribosomal protein mRNAs [99] and rRNAs [98, 99, 190] suggesting that this plant hormone induces coordinate expression of plant ribosomal proteins and RNA [75] during peri-ods of rapid growth. Auxins have also been shown to have an effect on phosphorylation of ribosomal proteins in embryonic maize tissue [240]. The phosphorylation state of four ribosomal proteins (37, 31, 16 and 14.5 kDa) of maize axes changes during germination [238]. One of these proteins (31 kDa) appears to correspond to the small ribosomal protein S6 and its phosphorylation can be induced by heat stress or hypoxia [238]. The remaining three proteins (37, 16 and 14.5 kDa) are acidic large subunit proteins that are antigenically related to the yeast acidic ribosomal protein homologues [238]. The presence of several acidic ribosomal proteins has been documented for plant ribosomes [104, 123]. The acidic ribosomal proteins are characterized by an acidic pI, a repeated amino acid motif and the presence of two pools of protein, one on the ribosomes (phosphorylated form) and one free in the cytosol (unphosphorylated) [14]. One member of the class of acidic ribosomes is p40 and has been isolated and characterized in *Arabidopsis* [104]. This protein appears to be involved in the regulation of ribosome activity during elongation and possibly association of polysomes with the cytoskeleton [104].

Plants, as do other eukaryotes, may use phosphorylation/dephosphorylation of ribosomal proteins to alternately increase or decrease the rate of protein synthesis during periods of stress. The effects of heat shock on the assembly of tomato ribosomes has been analyzed [271]. Heat stress in wheat is reported to increase the phosphorylation of the large ribosomal protein(s) with a concomitant increase in the rate of polyphenylalanine synthesis when tested *in vitro* [90, 91]. However, heat-shock of tomato protoplasts was shown to cause the dephosphorylation of the small ribosomal protein S6 and which coincided with a decrease in protein synthetic activity [270]. These observations reflect the complex process for the regulation of protein synthesis by alternately increasing and decreasing the activity of ribosomal subunits. Hypoxia in maize seedling roots causes an increase in the amount of free ribosomes and ribosomal subunits and a decrease in the level of large polysomes [12]. A decrease in the phosphorylation of a 31 kDa protein (presumably S6 based on molecular weight) was also observed in hypoxic maize seedling roots [12]. Several genes with similarity to ribosomal protein S6 kinase have been reported from *Arabidopsis*. The *atpk1* and *atpk2* genes have similarity to S6 kinase; however, they do not phosphorylate S6 *in vitro*, but rather two small acidic ribosomal proteins [350]. Two more *Arabidopsis* genes, *atpk19* and *atpk6*, encoding a S6 kinase-like gene product are induced by cold and salt stress [201]; however, the substrates for these gene products were not determined. A gibberellin-induced kinase in oat aleurone cells also has similarity to the *Arabidopsis atpk1* gene [129]. The presence of cold-, salt- and hormone-inducible kinases with similarity to S6 kinases suggests that environmental stress may induce a signal transduction response to turn on the genes that alter the activity of the ribosomes in order to acclimate the plant to the change in environment.

The ribosome inactivating proteins (RIPs) are cyto-toxic proteins that are unique to plants (see Barbieri *et al.* [15] and references therein). These proteins are expressed differentially and appear to be a protective mechanism where plants defend themselves against bacteria, viruses, fungi and herbivores. There are several classes of RIPs [15, 207]. RIPs are toxic to many types of cells and are the basis for a number of potential therapeutic methods for targeting and destroying cancer cells. The mechanism of inactivation is a unique *N*-glycosidase activity that cleaves the glycosidic bond of A_{4324} in a conserved loop and stem of 28S rRNA in the 60S ribosomal subunit. This cleavage interferes with the binding of eEF2 and causes protein synthesis to stop. How plants protect their own ribosomes from inactivation is not clear at this time.

More recently, another class of RIPs has been identified, a jasmonate-induced protein (JIP). This protein (JIP60) is induced by methyl jasmonate in response to environmental stress or wounding [58, 249, 250]. JIP60 cleaves intact mammalian and plant ribosomes into subunits by an unknown mechanism. Only ribosomes from plant cells treated with methyl jasmonate for at least 48 hr are susceptible to the inhibition by JIP60. The mechanism for targeting ribosomes in these cells is unknown. JIP60 irreversibly inhibits protein synthesis by degradation of polysomes and promotes cell death. By regulating the amount of methyl jasmonate present in the damaged cells and surrounding tissue, a plant is perhaps able to control the spread of pathogens or allow wounded tissue to die, apoptosis for plants as it were.

Protein synthesis and the cytoskeleton

There is a long-standing 'guilt by association' of the components of protein synthesis (initiation factors, elongation factors, mRNA and polysomes) with the cytoskeletal framework [16, 17, 92, 121, 126, 133, 169, 177, 222, 235, 289]. However, the precise molecular mechanisms of these interactions are unknown. Plant polysomes are reported to interact with components of the cytoskeleton and membranes [76, 77, 132, 300, 349] and recently an acidic ribosomal protein (p40) from *Arabidopsis* has been shown to associate with polysomes and cytoskeleton [104]. There have been a number of recent reports on the association of plant initiation factors and elongation factors with microfilaments and/or microtubules.

eEF1A seems to be the most commonly reported factor associated with the cytoskeleton and it has a variety of activities (binding, bundling and severing) with the components of the cytoskeleton (reviewed by Condeelis [68]). Plant eEF1A has been described as: (1) a phosphatidyl–4-kinase activator that binds actin and enhances actin bundling [346]; (2) a calcium/calmodulin-sensitive microtubule bundling factor [83]; (3) having a role in the organization of perinuclear microtubules during the transition from M to G_1 phase of the cell cycle [115, 156]; and (4) a microtubule-severing protein at low molar ratios to polymerized tubulin and a microtubule stiffening and bundling protein at higher molar ratios [175].

Recently, the large subunit (p86) of wheat eIFiso4F was reported to bind and bundle microtubules and promote growth by end-to-end annealing *in vitro* [26, 127]. The biological significance of these observations was demonstrated by the co-localization of p86 with microtubule bundles in maize root cells [26]. Amino-acid sequence analysis of p86 indicates its N- and C-termini have similarity to the microtubule-binding domains of *kat*, a kinesin-like protein from *Arabidopsis* [200]. However, it does not appear that p86 is a kinesin motor because the binding of p86 to microtubules *in vitro* is not ATP-dependent [26].

The role of mRNA localization (reviewed by St. Johnston [297]) and translational control during development (reviewed by Curtis *et al.* [73]), appear to be intimately associated with the cytoskeleton as a framework for movement/localization/sequestration of mRNAs. There are two possibilities, largely speculative, on the possible function of eIFiso4F association with microtubules. eIFiso4F could be functioning to bind mRNAs through the cap-binding protein subunit (p28) and be a component of a 'mRNA transport particle'. Such particles may be transported by motor proteins to the site of translation. Conversely, eIFiso4F could also be localized at sites of translation, waiting to receive mRNAs from a 'mRNA transport particle'. These 'transport particles' are of unknown biochemical composition and have only been observed

134

by light microscopy [297]. Interestingly, *vasa*, a protein known to participate in mRNA localization in *Drosophila* [140, 246, 332] is a member of the eIF4A-like DEAD-box family of RNA helicases [96, 116].

The next few years should bring some exciting new discoveries about the regulation and localization of protein synthesis in plant cells and the relationship of the cytoskeleton to this process.

Epilogue

Plant protein synthesis is similar to protein synthesis in other eukaryotes in many respects, but has several notable differences: (1) plants have an isoenzyme form of eIF4F, which has a distinctly different large subunit (86 kDa vs. 220 kDa) and a distinctly different cap-binding protein; (2) plant eIF4A is phosphorylated in response to hypoxic conditions and the phosphorylation appears to be coordinate with the inhibition of protein synthesis; and (3) plants have a distinctly different factor requirement for RNA-dependent ATP hydrolysis and ATP-dependent double-stranded RNA unwinding requiring only eIF4F and eIF4A or eIFiso4F and eIF4A.

The study of the regulation of protein synthesis in higher plants is still in its infancy. We do not know if the same mechanisms of regulation well defined in other eukaryotes (phosphorylation of eIF2α, the subunits of eIF4F, eEF1A, eEF2) even occur in plant cells. However, the plant protein synthesis factors whose counterparts are phosphorylated in other eukaryotic systems can be phosphorylated by the kinases from other eukaryotes, indicating that the phosphorylation sites are present. However, the plant counterparts to these kinases have been elusive. The recent report of a plant version of the eIF2α kinase, pPKR, suggests that plants do use phosphorylation as a control mechanism. The EST projects will certainly facilitate finding the cDNAs for these kinases, which will in turn enable researchers to elucidate the mechanisms of protein synthesis regulation in plants.

The apparent interaction of translational components with the cytoskeleton and increased expression of protein synthesis components in actively growing tissue or in response to external stimuli are important clues to our understanding of protein synthesis in plants. We can expect that signal transduction pathways in plants and the pathways of protein synthesis regulation will cross in many unexpected ways. The journey of the next 5 years in plant protein synthesis be every bit as exciting as the past 5 years have been.

Acknowledgements

I would like to dedicate this review to Dr Joanne M. Ravel upon the occasion of her retirement. As a mentor and friend she has taught me more than can be enumerated. I would also like to thank Drs Julia Bailey-Serres, Wendy Boss, Eric Davies, Dan Gallie, Dixie Goss, Gisela Kramer, Cris Kuhlemeier, Lou Morejohn and Joanne Ravel for reading and commenting on all or portions of this manuscript. Lastly, I want to apologize for any omissions of work that may have occurred. Given the vastness of material and the fact that plant protein synthesis has never been reviewed, the gathering of references was an arduous task despite the miracles of MEDLINE and GenBank on Internet. K.S.B. is supported by a grant from the National Science Foundation (MCB 9406601).

References

1. Abramson RD, Browning KS, Dever TE, Lawson TG, Thach RE, Ravel JM, Merrick WC: Initiation factors that bind mRNA: A comparison of mammalian factors with wheat germ factors. J Biol Chem 263: 5462–5467 (1988).
2. Adams MD, Kelley JM, Gocayne JD, Dubnick M, Polymeropoulos MH, Xiao H, Merril CR, Wu A, Olde B, Moreno RF, Kerlavage AR, McCombie WR, Venter JC: Complementary DNA sequencing: expressed sequence tags and human genome project. Science 252: 1651–1656 (1991).
3. Aguilar F, Montandon P-E, Stutz E: Two genes encoding the soybean translation elongation factor eEF-1α are transcribed in seedling leaves. Plant Mol Biol 17: 351–360 (1991).
4. Allen ML, Metz AM, Timmer RT, Rhoads RE, Browning KS: Isolation and sequence of the cDNAs encoding the subunits of the isozyme form of wheat protein synthesis initiation factor 4F. J Biol Chem 267: 23232–23236 (1992).
5. Altmann M, Handschin C, Trachsel H: mRNA cap-binding protein: cloning of the gene encoding protein synthesis initiation factor eIF–4E from *Saccharomyces cerevisiae*. Mol Cell Biol 7: 998–1003 (1987).
6. Altmann M, Müller PP, Wittmer B, Ruchti F, Lanker S, Trachsel H: A *Saccharomyces cerevisiae* homologue of mammalian translation initiation factor 4B contributes to RNA helicase activity. EMBO J 12: 3997–4003 (1993).
7. Altmann M, Trachsel H: The yeast *Saccharomyces cerevisiae* system: a powerful tool to study the mechanism of protein synthesis initiation in eukaryotes. Biochimie 76: 853–861 (1994).
8. Altmann M, Wittmer B, Méthot N, Sonenberg N, Trachsel H: The *Saccharomyces cerevisiae* translation initiation factor Tif3 and its mammalian homologue, eIF–4B, have RNA annealing activity. EMBO J 14: 3820–3827 (1995).

9. Amons R, Guerrucci M-A, Karssies RH, Morales J, Cormier P, Möller W, Bellé R: The leucine-zipper in elongation factor EF-1δ, a guanine-nucleotide exchange protein, is conserved in *Artemia* and *Xenopus*. Biochim Biophys Acta 1218: 346–350 (1994).

10. Andersson S, Saeboe-Larssen S, Lambertsson A, Merriam J, Jacobs-Lorena M: A *Drosophila* third chromosome Minute locus encodes a ribosomal protein. Genetics 137: 513–520 (1994).

11. Axelos M, Bardet C, Liboz T, Le Van Thai A, Curie C, Lescure B: The gene family encoding the *Arabidopsis thaliana* translation elongation factor EF-1α: Molecular cloning, characterization and expression. Mol Gen Genet 219: 106–112 (1989).

12. Bailey-Serres J, Freeling M: Hypoxic stress-induced changes in ribosomes of maize seedling roots. Plant Physiol 94: 1237–1243 (1990).

13. Balasta ML, Carberry SE, Friedland DE, Perez RA, Goss DJ: Characterization of the ATP-dependent binding of wheat germ protein synthesis initiation factors eIF-(iso)4F and eIF–4A to mRNA. J Biol Chem 268: 18599–18603 (1993).

14. Ballestra JPG, Remacha M, Naranda T, Santos C, Bermejo B, Jimenez-Diaz A, Ortiz-Reyes B: The acidic ribosomal proteins and the control of protein synthesis in yeasts. In: Brown AJP, Tuite MF, McCarthy JEG (eds) Protein Synthesis and Targeting in Yeast, pp. 67–80. Springer-Verlag, Berlin (1993).

15. Barbieri L, Battelli MG, Stirpe F: Ribosome-inactivating proteins from plants. Biochim Biophys Acta 1154: 237–282 (1993).

16. Bassell GJ: High resolution distribution of mRNA within the cytoskeleton. J Cell Biochem 52: 127–133 (1993).

17. Bektas M, Nurten R, Gürel Z, Sayers Z, Bermek E: Interactions of eukaryotic elongation factor 2 with actin: a possible link between protein synthetic machinery and cytoskeleton. FEBS Lett 356: 89–93 (1994).

18. Beltrán-Peña E, Ortíz-López A, Sánchez de Jiménez E: Synthesis of ribosomal proteins from stored mRNAs early in seed germination. Plant Mol Biol 28: 327–336 (1995).

19. Benkowski LA: Elements That Regulate the Translational Efficiency of mRNA. Ph.D. Dissertation, University of Texas at Austin (1993).

20. Benkowski LA, Ravel JM, Browning KS: mRNA binding properties of wheat germ protein synthesis initiation factor 2. Biochem Biophys Res Commun 214: 1033–1039 (1995).

21. Benne R, Hershey JWB: The mechanism of action of protein synthesis initiation factors from rabbit reticulocytes. J Biol Chem 253: 3078–3087 (1978).

22. Benne R, Kasperaitis M, Voorma HO, Ceglarz E, Legocki AB: Initiation factor eIF–2 from wheat germ: purification, functional comparison to eIF–2 from rabbit reticulocytes and phosphorylation of its subunits. Eur J Biochem 104: 109–117 (1980).

22a. Berberich T, Sugawara K, Harada M, Kusano T: Molecular cloning, characterization and expression of an elongation factor 1α gene in maize. Plant Mol. Biol. 29: 611–615 (1995).

23. Bevec D, Klier H, Holter W, Tschachler E, Valent P, Lottspeich F, Baumruker T, Hauber J: Induced gene expression of the hypusine-containing protein eukaryotic initiation factor 5A in activated human T lymphocytes. Proc Natl Acad Sci USA 91: 10829–10833 (1994).

24. Bodley JW: Does diphtheria toxin have nuclease activity? Science 250: 832 (1990).

25. Bodley JW, Veldman SA: ADP-ribosylating toxins and G proteins. In: Moss J, Vaughan M (eds) Insights into Signal Transduction, pp. 21–30. American Society for Microbiology, Washington, D.C. (1990).

26. Bokros CL, Hugdahl JD, Kim HH, Hanesworth VR, Van Heerden A, Browning KS, Morejohn LC: Function of the p86 subunit of eukaryotic initiation factor (iso)4F as a microtubule-associated protein in plant cells. Proc Natl Acad Sci USA 92: 7120–7124 (1995).

27. Bollini R, Soffientini AN, Bertani A, Lanzani GA: Some molecular properties of the elongation factor EF1 from wheat embryos. Biochemistry 13: 5421–5425 (1974).

28. Bonham-Smith PC, Moloney MM: Nucleotide and protein sequences of a cytoplasmic ribosomal protein S15a gene from *Arabidopsis thaliana*. Plant Physiol 106: 401–402 (1994).

29. Bonham-Smith PC, Oancia TL, Moloney MM: Cytoplasmic ribosomal protein S15a from *Brassica napus*: molecular cloning and developmental expression in mitotically active tissues. Plant Mol Biol 18: 909–919 (1992).

30. Bourne HR, Sanders DA, McCormick F: The GTPase superfamily: conserved structure and molecular mechanism. Nature 349: 117–126 (1991).

31. Brander KA, Kuhlemeier C: A pollen-specific DEAD-box protein related to translation initiation factor eIF–4A from tobacco. Plant Mol Biol 27: 637–649 (1995).

32. Brander KA, Mandel T, Owttrim GW, Kuhlemeier C: Highly conserved genes coding for eukaryotic translation initiation factor eIF–4A of tobacco have specific alterations in functional motifs. Biochim Biophys Acta 1261: 442–444 (1995).

33. Braun HP, Emmermann M, Mentzel H, Schmitz UK: Primary structure and expression of a gene encoding the cytosolic ribosomal protein S4 from potato. Biochim Biophys Acta 1218: 435–438 (1994).

34. Browning KS, Fletcher L, Lax SR, Ravel JM: Evidence that the 59-kDa protein synthesis initiation factor from wheat germ is functionally similar to the 80-kDa initiation factor 4B from mammalian cells. J Biol Chem 264: 8491–8494 (1989).

35. Browning KS, Fletcher L, Ravel JM: Evidence that the requirements for ATP and wheat germ initiation factors 4A and 4F are affected by a region of satellite tobacco necrosis virus RNA that is 3′ to the ribosomal binding site. J Biol Chem 263: 8380–8383 (1988).

36. Browning KS, Humphreys J, Hobbs W, Smith GB, Ravel JM: Determination of the amounts of the protein synthesis initiation and elongation factors in wheat germ. J Biol Chem 265: 17967–17973 (1990).

37. Browning KS, Lax SR, Humphreys J, Ravel JM, Jobling SA, Gehrke L: Evidence that the 5′-untranslated leader of mRNA affects the requirement for wheat germ initiation factors 4A, 4F and 4G. J Biol Chem 263: 9630–9634 (1988).

38. Browning KS, Lax SR, Ravel JM: Identification of two messenger RNA cap binding proteins in wheat germ. Evidence that the 28-kDa subunit of eIF–4B and the 26-kDa subunit of eIF–4F are antigenically distinct polypeptides. J Biol Chem 262: 11228–11232 (1987).

39. Browning KS, Maia DM, Lax SR, Ravel JM: Identification of a new protein synthesis initiation factor from wheat germ. J Biol Chem 262: 538–541 (1987).

40. Browning KS, Webster C, Roberts JKM, Ravel JM: Identification of an isozyme form of protein synthesis initiation factor 4F in plants. J Biol Chem 267: 10096–10100 (1992).

41. Browning KS, Yan TFJ, Lauer SJ, Aquino LA, Tao M, Ravel JM: Phosphorylation of wheat germ initiation factors and ribosomal proteins. Plant Physiol 77: 370–373 (1985).

136

42. Butler JS, Clark JM Jr: Eucaryotic initiation factor 4B of wheat germ binds to the translation initiation region of a messenger ribonucleic acid. Biochemistry 23: 809–815 (1984).

43. Cammarano P, Felsani A, Gentile M, Gualerzi C, Romeo A, Wolf G: Formation of active hybrid 80-S particles from subunits of pea seedlings and mammalian liver ribosomes. Biochim Biophys Acta 281: 625–642 (1972).

44. Cammarano P, Pons S, Romeo A, Galdieri M, Gualerzi C: Characterization of unfolded and compact ribosomal subunits from plants and their relationship to those of lower and higher animals: evidence for physicochemical heterogeneity among eucaryotic ribosomes. Biochim Biophys Acta 281: 571–596 (1972).

45. Cammarano P, Romeo A, Gentile M, Felsani A, Gualerzi C: Size heterogeneity of the large ribosomal subunits and conservation of the small subunits in eucaryote evolution. Biochim Biophys Acta 281: 597–624 (1972).

46. Campbell LH, Borg KT, Haines JK, Moon RT, Schoenberg DR, Arrigo SJ: Human immunodeficiency virus type 1 Rev is required in vivo for binding of poly(A)-binding protein to Rev-dependent RNAs. J Virol 68: 5433–5438 (1994).

47. Carberry SE, Darzynkiewicz E, Goss DJ: A comparison of the binding of methylated cap analogues to wheat germ protein synthesis initiation factors 4F and (iso)4F. Biochemistry 30: 1624–1627 (1991).

48. Carberry SE, Goss DJ: Interaction of wheat germ protein synthesis initiation factors eIF–3, eIF–(iso)4F, and eIF–4F with mRNA analogues. Biochemistry 30: 6977–6982 (1991).

49. Carberry SE, Goss DJ: Wheat germ initiation factors 4F and (iso)4F interact differently with oligoribonucleotide analogues of rabbit α-globin mRNA. Biochemistry 30: 4542–4545 (1991).

50. Carberry SE, Goss DJ: Characterization of the interaction of wheat germ protein synthesis initiation factor eIF–3 with mRNA oligonucleotide and cap analogues. Biochemistry 31: 296–299 (1992).

51. Carvalho JF, Carvalho M, Merrick WC: Purification of various forms of elongation factor 1 from rabbit reticulocytes. Arch Biochem Biophys 234: 591–602 (1984).

52. Caskey CT: Peptide chain termination. Trends Biochem Sci 5: 234–237 (1980).

53. Cavallius J, Zoll W, Chakraburtty K, Merrick WC: Characterization of yeast EF-1α: non-conservation of post-translational modifications. Biochim Biophys Acta 1163: 75–80 (1993).

54. Ceglarz E, Goumans H, Thomas A, Benne R: Purification and characterization of protein synthesis initiation factor eIF–3 from wheat germ. Biochim Biophys Acta 610: 181–188 (1980).

55. Chakrabarti A, Maitra U: Function of eukaryotic initiation factor 5 in the formation of an 80 S ribosomal polypeptide chain initiation complex. J Biol Chem 266: 14039–14045 (1991).

56. Chakravarti D, Maitra U: Eukaryotic translation initiation factor 5 from Saccharomyces cerevisiae. Cloning, characterization, and expression of the gene encoding the 45,346-Da protein. J Biol Chem 268: 10524–10533 (1993).

57. Chamot D, Kuhlemeier C: Differential expression of genes encoding the hypusine-containing translation initiation factor, eIF–5A, in tobacco. Nucl Acids Res 20: 665–669 (1992).

58. Chaudhry B, Muller-Uri F, Cameron-Mills V, Gough S, Simpson D, Skriver K, Mundy J: The barley 60 kDa jasmonate-induced protein (JIP60) is a novel ribosome-inactivating protein. Plant Journal 6: 815–824 (1994).

59. Chaudhuri J, Das K, Maitra U: Purification and characterization of bacterially expressed mammalian translation initiation factor 5 (eIF–5): demonstration that eIF–5 forms a specific complex with eIF–2. Biochemistry 33: 4794–4799 (1994).

60. Checkley JW, Cooley LL, Ravel JM: Characterization of initiation factor eIF–3 from wheat germ. J Biol Chem 256: 1582–1586 (1981).

61. Chen J-J, London IM: Regulation of protein synthesis by heme-regulated eIF–2α kinase. Trends Biochem Sci 20: 105–108 (1995).

62. Chen JY, Bodley JW: Biosynthesis of diphthamide in Saccharomyces cerevisiae. Partial purification and characterization of a specific S-adenosylmethionine:elongation factor 2 methyltransferase. J Biol Chem 263: 11692–11696 (1988).

63. Chen JY, Bodley JW, Livingston DM: Diphtheria toxin-resistant mutants of Saccharomyces cerevisiae. Mol Cell Biol 5: 3357–3360 (1985).

64. Chevesich J, Chaudhuri J, Maitra U: Characterization of mammalian translation initiation factor 5 (eIF–5). Demonstration that eIF–5 is a phosphoprotein and is present in cells as a single molecular form of apparent M_r 58,000. J Biol Chem 268: 20659–20667 (1993).

65. Cho S, Mitchell A, Regier JC, Mitter C, Poole RW, Friedlander TP, Zhao S: A highly conserved nuclear gene for low-level phylogenetics: elongation factor-1α recovers morphology-based tree for heliothine moths. Mol Biol Evol 12: 650–656 (1995).

66. Clarke RD, Ranu RS: Characterization of wheat germ initiation factor eIF–2. Mol Cell Biochem 74: 129–135 (1987).

67. Clemens MJ: Regulation of eukaryotic protein synthesis by protein kinases that phosphorylate initiation factor eIF–2. Plant Mol Biol Rep 19: 201–210 (1994).

68. Condeelis J: Elongation factor 1α, translation and the cytoskeleton. Trends Biochem Sci 20: 169–170 (1995).

69. Coppolecchia R, Buser P, Stotz A, Linder P: A new yeast translation initiation factor suppresses a mutation in the eIF–4A RNA helicase. EMBO J 12: 4005–4011 (1993).

70. Curie C, Axelos M, Bardet C, Atanassova R, Chaubet N, Lescure B: Modular organization and developmental activity of an Arabidopsis thaliana EF-1α gene promoter. Mol Gen Genet 238: 428–436 (1993).

71. Curie C, Liboz T, Bardet C, Gander E, Médale C, Axelos M, Lescure B: Cis and trans-acting elements involved in the activation of Arabidopsis thaliana A1 gene encoding the translation elongation factor EF-1α. Nucl Acids Res 19: 1305–1310 (1991).

72. Curie C, Liboz T, Montané M-H, Rouan D, Axelos M, Lescure B: The activation process of Arabidopsis thaliana A1 gene encoding the translation elongation factor EF-1α is conserved among angiosperms. Plant Mol Biol 18: 1083–1089 (1992).

73. Curtis D, Lehmann R, Zamore PD: Translational regulation in development. Cell 81: 171–178 (1995).

74. Das K, Chevesich J, Maitra U: Molecular cloning and expression of cDNA for mammalian translation initiation factor 5. Proc Natl Acad Sci USA 90: 3058–3062 (1993).

75. Datta N, LaFayette PR, Kroner PA, Nagao RT, Key JL: Isolation and characterization of three families of auxin down-regulated cDNA clones. Plant Mol Biol 21: 859–869 (1993).

76. Davies E, Comer EC, Lionberger JM, Stankovic B, Abe S: Cytoskeleton-bound polysomes in plants. III. Polysome-cytoskeleton-membrane interactions in corn endosperm. Cell Biol Int 17: 331–340 (1993).

77. Davies E, Fillingham BD, Oto Y, Abe S: Evidence for the existence of cytoskeleton-bound polysomes in plants. Cell Biol Int 15: 973–981 (1991).

78. De Benedetti A, Rhoads RE: Overexpression of eukaryotic protein synthesis initiation factor 4E in HeLa cells results in aberrant growth and morphology. Proc Natl Acad Sci USA 87: 8212–8216 (1990).

79. De Pater BS, Van der Mark F, Rueb S, Katagiri F, Chua NH, Schilperoort RA, Hensgens LA: The promoter of the rice gene GOS2 is active in various different monocot tissues and binds rice nuclear factor ASF–1. Plant J 2: 837–844 (1992).

80. Dever TE, Costello CE, Owens CL, Rosenberry TL, Merrick WC: Location of seven post-translational modifications in rabbit elongation factor 1α including dimethyllysine, trimethyllysine, and glycerylphosphorylethanolamine. J Biol Chem 264: 20518–20525 (1989).

81. Dever TE, Wei C-L, Benkowski LA, Browning K, Merrick WC, Hershey JWB: Determination of the amino acid sequence of rabbit, human, and wheat germ protein synthesis factor eIF–4C by cloning and chemical sequencing. J Biol Chem 264: 3212–3218 (1989).

82. Dunn MA, Morris A, Jack PL, Hughes MA: A low-temperature-responsive translation elongation factor 1α from barley (Hordeum vulgare L.). Plant Mol Biol 23: 221–225 (1993).

83. Durso NA, Cyr RJ: A calmodulin-sensitive interaction between microtubles and a higher plant homolog of elongation factor–1α. Plant Cell 6: 893–905 (1994).

84. Ejiri S: Purification and characterization of polypeptide chain elongation factor 1 from plants. Meth Enzymol 118: 140–153 (1986).

85. Ejiri S, Ebata N, Kawamura R, Katsumata R: Occurance of four subunits in high molecular weight forms of polypeptide chaing elongation factor 1 from wheat embryo. J Biochem (Tokyo) 94: 319–322 (1983).

86. Ejiri S, Honda H: Effect of cyclic AMP and cyclic GMP on the autophosphorylation of elongation factor 1 from wheat embryos. Biochem Biophys Res Commun 128: 53–60 (1985).

87. Ejiri S, Kawamura R, Katsumata T: Interactions among four subunits of elongation factor 1 from rice embryo. Biochim Biophys Acta 1217: 266–272 (1994).

88. Etchison D, Milburn S: Separation of protein synthesis initiation factor eIF–4A from a p220-associated cap binding complex activity. Mol Cell Biol 76: 15–25 (1987).

89. Eustice DC, Wakem LP, Wilhelm JM, Sherman F: Altered 40 S ribosomal subunits in omnipotent suppressors of yeast. J Mol Biol 188: 207–214 (1986).

90. Fehling E, Weidner M: Temperature characteristics and adaptive potential of wheat ribosomes. Plant Physiol 80: 181–186 (1986).

91. Fehling E, Weidner M: Adaptive potential of wheat ribosomes toward heat depends on the large ribosomal subunit and ribosomal protein phosphorylation. Plant Physiol 87: 562–565 (1988).

92. Fey EG, Ornelles DA, Penman S: Association of RNA with the cytoskeleton and the nuclear matrix. J Cell Sci 5: 99–119 (1986).

93. Fleming AJ, Mandel T, Roth I, Kuhlemeier C: The patterns of gene expression in the tomato shoot apical meristem. Plant Cell 5: 297–309 (1993).

93a. Floyd B, Bartlett SG: Nucleotide sequence of a cDNA encoding eucaryotic initiation factor 5 in bean. Plant Physiol 109: 1126 (1995).

94. Flynn A, Shatsky IN, Proud CG, Kaminski A: The RNA-binding properties of protein synthesis initiation factor eIF–2. Biochim Biophys Acta 1219: 293–301 (1994).

95. Frolova L, Le Goff X, Rasmussen HH, Cheperegin S, Drugeon G, Kress M, Arman I, Haenni A-L, Celis JE, Philippe M, Justesen J, Kisselev L: A highly conserved eukaryotic protein family possessing properties of polypeptide chain release factor. Nature 372: 701–703 (1994).

96. Fujiwara Y, Komiya T, Kawabata H, Sato M, Fujimoto H, Furusawa M, Noce T: Isolation of a DEAD-family protein gene that encodes a murine homolog of Drosophila vasa and its specific expression in germ cell lineage. Proc Natl Acad Sci USA 91: 12258–12262 (1994).

97. Gallie DR, Tanguay R: Poly(A) binds to initiation factors and increases cap-dependent translation in vitro. J Biol Chem 269: 17166–17173 (1994).

98. Gantt JS, Key JL: Auxin-induced changes in the level of translatable ribosomal protein messenger ribonucleic acids in soybean hypocotyl. Biochemistry 22: 4131–4139 (1983).

99. Gantt JS, Key JL: Coordinate expression of ribosomal protein mRNAs following auxin treatment of soybean hypocotyls. J Biol Chem 260: 6175–6181 (1985).

100. Gantt JS, Thompson MD: Plant cytosolic ribosomal protein S11 and chloroplast ribosomal protein CS17. Their primary structures and evolutionary relationships. J Biol Chem 265: 2763–2767 (1990).

101. Gao J, Kim SR, Chung YY, Lee JM, An G: Developmental and environmental regulation of two ribosomal protein genes in tobacco. Plant Mol Biol 25: 761–770 (1994).

102. Gao J, Kim SR, Lee JM, An G: Nucleotide and protein sequences of 60S ribosomal protein L17 from tobacco (Nicotiana tabacum L.). Plant Physiol 103: 1027–1028 (1993).

103. Garcia-Barrio MT, Naranda T, Vazquez de Aldana CR, Cuesta R, Hinnebusch AG, Hershey JWB, Tamame M: GCD10, a translational repressor of GCN4, is the RNA-binding subunit of eukaryotic translation initiation factor–3. Genes Devel 9: 1781–1796 (1995).

104. Garcia-Hernandez M, Davies E, Staswick PE: Arabidopsis p40 homologue. A novel acidic protein associated with the 40S subunit of ribosomes. J Biol Chem 269: 20744–20749 (1994).

105. Gidekel M, Jiminez B, Herrera-Estrella L: Isolation and characterization of an elongation factor–1β gene from Arabidopsis thaliana. GenBank X74733–X74734 (1993).

106. Golinska B, Legocki AB: Purification and some properties of elongation factor 1 from wheat germ. Biochim Biophys Acta 324: 156–170 (1973).

107. Goyer C, Altmann M, Lee HS, Blanc A, Deshmukh M, Woolford JL Jr, Trachsel H, Sonenberg N: TIF4631 and TIF4632: two yeast genes encoding the high-molecular-weight subunits of the cap-binding protein complex (eukaryotic initiation factor 4F) contain an RNA recognition motif-like sequence and carry out an essential function. Mol Cell Biol 13: 4860–4874 (1993).

108. Grifo JA, Tahara SM, Morgan MA, Shatkin AJ, Merrick WC: New initiation factor activity required for globin mRNA translation. J Biol Chem 258: 5804–5810 (1983).

109. Gualerzi C, Janda HG, Passow H, Stoffler G: Studies on the protein moiety of plant ribosomes. J Biol Chem 249: 3347–3355 (1974).

110. Habben JE, Moro GL, Hunter BG, Hamaker BR, Larkins BA: Elongation factor 1α is highly correlated with the lysine content of maize endosperm. Proc Natl Acad Sci USA 92: 8640–8644 (1995).

138

111. Hanauske-Abel HM, Park MH, Hanauske AR, Popowicz AM, Lalande M, Folk JE: Inhibition of the G1-S transition of the cell cycle by inhibitors of deoxyhypusine hydroxylation. Biochim Biophys Acta 1221: 115–124 (1994).

112. Hanauske-Abel HM, Slowinska B, Zagulska S, Wilson RC, Staiano-Coico L, Hanauske A-R, McCaffrey T, Szabo P: Detection of a sub-set of polysomal mRNAs associated with modulation of hypusine formation at the G1-S boundary: proposal of a role for eIF–5A in onset of DNA replication. FEBS Lett 366: 92–98 (1995).

113. Hansen G, Estruch JJ, Spena A: Tobacco cDNA encoding the ribosomal protein S6. Nucl Acids Res 20: 5230 (1992).

114. Harris EH, Boynton JE, Gillham NW: Chloroplast ribosomes and protein synthesis. Microbiol Rev 58: 700–754 (1994).

115. Hasezawa S, Nagata T: Microtubule organizing centers in plant cells: localization of a 49 kDa protein that is immunologically cross-reactive to a 51 kDa protein from sea urchin centrosomes in synchronized tobacco BY–2 cells. Protoplasma 176: 64–74 (1993).

116. Hay B, Jan LY, Jan YN: A protein component of *Drosophila* polar granules is encoded by *vasa* and has extensive sequence similarity to ATP-dependent helicases. Cell 55: 577–587 (1988).

117. Hernández G, Sierra JM: Translation initiation factor eIF–4E from *Drosophila*: cDNA sequence and expression of the gene. Biochim Biophys Acta 1261: 427–431 (1995).

118. Hershey JWB: Overview: phosphorylation and translation control. Enzyme 44: 17–27 (1990).

119. Hershey JWB: Translational control in mammalian cells. Annu Rev Biochem 60: 717–755 (1991).

120. Hershey JWB: Expression of initiation factor genes in mammalian cells. Biochimie 76: 847–852 (1994).

121. Hesketh JE, Pryme IF: Interaction between mRNA, ribosomes and the cytoskeleton. Biochem J 277: 1–10 (1991).

122. Heufler C, Browning KS, Ravel JM: Properties of the subunits of wheat germ initiation factor 3. Biochim Biophys Acta 951: 182–190 (1988).

123. Hihara Y, Umeda M, Hara C, Toriyama K, Uchimiya H: Nucleotide sequence of a rice acidic ribosomal phosphoprotein P0 cDNA. Plant Physiol 105: 753–754 (1994).

124. Himmelfarb HJ, Maicas E, Friesen JD: Isolation of the SUP45 omnipotent suppressor gene of *Saccharomyces cerevisiae* and characterization of its gene product. Mol Cell Biol 5: 816–822 (1985).

125. Hinnebusch AG: Translational control of *GCN4*: An *in vivo* barometer of initiation-factor activity. Trends Biochem Sci 19: 409–414 (1994).

126. Howe JG, Hershey JWB: Translation initiation factor and ribosome association with the cytoskeletal framework fraction of HeLa cells. Cell 37: 85–93 (1984).

127. Hugdahl JD, Bokros CL, Morejohn LC: End-to-end annealing of plant microtubules by the p86 subunit of eukaryotic initiation factor-(iso)4F. Plant Cell 7: 2129–2138 (1995).

128. Humphreys J, Browning KS, Ravel JM: Identification of a kinase in wheat germ that phosphorylates the large subunit of initiati on factor 4F. Plant Physiol 88: 483–486 (1988).

129. Huttly AK, Phillips AL: Gibberellin-regulated expression in oat aleurone cells of two kinases that show homology to MAP kinase and a ribosomal protein kinase. Plant Mol Biol 27: 1043–1052 (1995).

130. Hwang I, Goodman HM: Cloning of an *Arabidopsis* ribosomal protein S28 cDNA. Plant Physiol 102: 1357–1358 (1993).

131. Iglewski WJ: Cellular ADP-ribosylation of elongation factor 2. Mol Cell Biochem 138: 131–133 (1994).

132. Ito Y, Abe S, Davies E: Co-localization of cytoskeleton proteins and polysomes with a membrane fractions from peas. J Exp Bot 45: 253–259 (1994).

133. Janssen GMC, Moller W: Elongation factor $1\beta\gamma$ from *Artemia*: purification and properties of its subunits. Eur J Biochem 171: 119–129 (1988).

134. Janssen GMC, Morales J, Schipper A, Labbé J-C, Mulner-Lorillon O, Bellé R, Möller W: A major substrate of maturation promoting factor identified as elongation factor 1 $\beta\gamma\delta$ in *Xenopus laevis*. J Biol Chem 266: 14885–14888 (1991).

135. Janssen GMC, Van Damme HTF, Kriek J, Amons R, Möller W: The subunit structure of elongation factor 1 from *Artemia*. Why two α-chains in this complex. J Biol Chem 269: 31410–31417 (1994).

136. Janssen GMC, Maessen GDF, Amons R, Moller W: Phosphorylation of elongation factor 1β by an endogenous kinase affects its catalytic nucleotide exchange activity. J Biol Chem 263: 11063–11066 (1988).

137. Jaramillo M, Browning K, Dever TE, Blum S, Trachsel H, Merrick WC, Ravel JM, Sonenberg N: Translation initiation factors that function as RNA helicases from mammals, plants and yeast. Biochim Biophys Acta 1050: 134–139 (1990).

138. Jayachandran S, Bailey-Serres J: Nucleotide sequence of a cDNA for the maize protein synthesis initiation factor 4A. Plant Physiol 108: 1317–1318 (1995).

139. Joanin P, Gigot C, Philipps G: cDNA nucleotide sequence and expression of a maize cytoplasmic ribosomal protein S13 gene. Plant Mol Biol 21: 701–704 (1993).

140. Jongens TA, Hay B, Jan LY, Jan YN: The germ cell-less gene product: a posteriorly localized component necessary for germ cell development in *Drosophila*. Cell 70: 569–584 (1992).

141. Joshi B, Cai A-L, Keiper BD, Minich WB, Mendez R, Beach CM, Stepinski J, Stolarski R, Darzynkiewicz E, Rhoads RE: Phosphorylation of eukaryotic protein synthesis initiation factor 4E at Ser-209. J Biol Chem 270: 14597–14603 (1995).

142. Kaempfer R, Van Emmelo J, Fiers W: Specific binding of eukaryotic initiation factor 2 to satellite tobacco necrosis virus RNA at a 5′-terminal sequence comprising the ribosome binding site. Proc Natl Acad Sci USA 78: 1542–1546 (1981).

143. Kambouris NG, Burke DJ, Creutz CE: Cloning and genetic characterization of a calcium- and phospholipid-binding protein from *Saccharomyces cerevisiae* that is homologous to translation elongation factor–1γ. Yeast 9: 151–163 (1993).

144. Kang HA, Hershey JWB: Effect of initiation factor eIF–5A depletion on protein synthesis and proliferation of *Saccharomyces cerevisiae*. J Biol Chem 269: 3934–3940 (1994).

145. Kasperaitis MAM, Voorma HO, Thomas AAM: The amino acid sequence of eukaryotic translation initiation factor 1 and its similarity to yeast initiation factor SUI1. FEBS Lett 365: 47–50 (1995).

146. Kawahara R, Sunabori S, Fukuda H, Komamine A: A gene expressed preferentially in the globular stage of somatic embryogenesis encodes elongation-factor 1α in carrot. Eur J Biochem 209: 157–162 (1992).

147. Kidou S, Umeda M, Kato A, Uchimiya H: Plant cDNA homologue to rat insulinoma gene encoding ribosomal protein S15. Nucl Acids Res 21: 2013 (1993).

148. Kim Y, Zhang H, Scholl RL: Two evolutionarily divergent genes encode a cytoplasmic ribosomal protein of *Arabidopsis thaliana*. Gene 93: 177–182 (1990).

149. Kimata Y, Harashima S, Kohno K: Expression of non-ADP-ribosylatable, diphtheria toxin-resistant elongation factor 2 in *Saccharomyces cerevisiae*. Biochem Biophys Res Commun 191: 1145–1151 (1993).

150. Kinzy TG, Ripmaster TL, Woolford JL,Jr.: Multiple genes encode the translation elongation factor EF-1γ in *Saccharomyces cerevisiae*. Nucl Acids Res 22: 2703–2707 (1994).

151. Kongsuwan K, Yu Q, Vincent A, Frisardi MC, Rosbash M, Lengyel JA, Merriam J: A *Drosophila* Minute gene encodes a ribosomal protein. Nature 317: 555–558 (1985).

152. Kozak M: The scanning model for translation: an update. J Cell Biol 108: 229–241 (1989).

153. Kozak M: A consideration of alternative models for the initiation of translation in eukaryotes. Crit Rev Biochem Mol Biol 27: 385–402 (1992).

154. Kozak M: Regulation of translation in eukaryotic systems. Annu Rev Cell Biol 8: 197–225 (1992).

155. Krishnan HB: Nucleotide and primary sequences of a rice endosperm cDNA are extensively homologous to elongation factor 1β'. Biochem Biophys Res Commun 209: 1026–1031 (1995).

156. Kumagai F, Hasezawa S, Takahashi Y, Nagata T: The involvement of protein synthesis elongation factor 1α in the organization of microtubules on the perinuclear region during the cell cycle transition from M phase to G1 phase in tobacco BY–2 cells. Bot Acta, in press (1995).

157. Langland JO, Langland LA, Browning KS, Roth DA: Phosphorylation of plant eukaryotic initiation factor–2 by the plant encoded double-stranded RNA-dependent protein kinase, pPKR, and inhibition of protein synthesis in vitro. J Biol Chem 271: 4539–4544 (1996).

158. Langland JO, Jin S, Jacobs BL, Roth DA: Identification of a plant-encoded analog of PKR, the mammalian double-stranded RNA-dependent protein kinase. Plant Physiol 108: 1259–1267 (1995).

159. Larkin JC, Hunsperger JP, Culley D, Rubenstein I, Silflow CD: The organization and expression of a maize ribosomal protein gene family. Genes Devel 3: 500–509 (1989).

160. Lauer SJ, Browning KS, Ravel JM: Characterization of initiation factor 3 from wheat germ. 2. Effects of polyclonal and monoclonal antibodies on activity. Biochemistry 24: 2928–2931 (1985).

161. Lauer SJ, Burks E, Irvin JD, Ravel JM: Purification and characterization of three elongation factors, EF-1α, EF-1βγ and EF-2, from wheat germ. J Biol Chem 259: 1644–1648 (1984).

162. Lauer SJ, Burks EA, Ravel JM: Characterization of initiation factor 3 from wheat germ. 1. Effects of proteolysis on activity and subunit composition. Biochemistry 24: 2924–2928 (1985).

163. Lax S, Fritz W, Browning K, Ravel J: Isolation and characterization of factors from wheat germ that exhibit eukaryotic initiation factor 4B activity and overcome 7-methylguanosine 5'-triphosphate inhibition of polypeptide synthesis. Proc Natl Acad Sci USA 82: 330–333 (1985).

164. Lax SR, Browning KS, Maia DM, Ravel JM: ATPase activities of wheat germ initiation factors 4A 4B and 4F. J Biol Chem 261: 15632–15636 (1986).

165. Lax SR, Lauer SJ, Browning KS, Ravel JM: Purification and properties of protein synthesis initiation and elongation factors from wheat germ. Meth Enzymol 118: 109–128 (1986).

166. Lax SR, Osterhout JJ, Ravel JM: Factors from wheat germ that enhance the activity of eukaryotic initiation factor eIF–2: isolation and characterization of Co-eIF-β. J Biol Chem 257: 8233–8237 (1982).

167. Lazaris-Karatzas A, Montine KS, Sonenberg N: Malignant transformation by a eukaryotic initiation factor subunit that binds to mRNA 5' cap. Nature 345: 544–547 (1990).

168. Lebrun M, Freyssinet G: Nucleotide sequence and characterization of a maize cytoplasmic ribosomal protein S11 cDNA. Plant Mol Biol 17: 265–268 (1991).

169. Lenk R, Ransom L, Kaufmann Y, Penman S: A cytoskeletal structure with associated polyribosomes obtained from HeLa cells. Cell 10: 67–78 (1977).

170. Lenstra JA, Bloemendal H: The major proteins from HeLa cells. Eur J Biochem 130: 419 (1983).

171. Lenvik TR, Key JL, Gantt JS: Ribosomal protein S11 genes from *Arabidopsis* and soybean. Plant Physiol 105: 1027–1028 (1994).

172. Liboz T, Bardet C, Le Van Thai A, Axelos M, Lescure B: The four members of the gene family encoding the *Arabidopsis thaliana* translation elongation factor *EF-1α* are actively transcribed. Plant Mol Biol 14: 107–110 (1990).

173. Linder P, Lasko PF, Ashburner M, Leroy P, Nielsen PJ, Nishi K, Schnier J, Slonimski PP: Birth of the D-E-A-D box. Nature 337: 121–122 (1989).

174. Linder P, Slonimski PP: An essential yeast protein, encoded by duplicated genes *TIF1* and *TIF2* and homologous to the mammalian translation initiation factor eIF–4A, can suppress a mitochondrial missense mutation. Proc Natl Acad Sci USA 86: 2286–2290 (1989).

175. Littlepage LE, Garman ME, Mendenhall JM, Morejohn LC: Plant microtubule (MT) severing, stiffening and bundling by wheat germ protein synthesis elongation factor 1α (EF-1α). Mol Biol Cell 6: 258a (1995).

176. Lu G, Wu K, Ferl RJ: A cDNA for *Arabidopsis* cytosol ribosomal protein S11. Plant Physiol 102: 695–696 (1993).

177. Luby-Phelps K: Effect of cytoarchitecture on the transport and localization of protein synthetic machinery. J Cell Biochem 52: 140–147 (1993).

178. Mader S, Lee H, Pause A, Sonenberg N: The translation initiation factor eIF–4E binds to a common motif shared by the translation factor eIF–4γ and the translational repressors 4E-binding proteins. Mol Cell Biol 15: 4990–4997 (1995).

179. Madsen LH, Kreiberg JD, Gausing K: A small gene family in barley encodes ribosomal proteins homologous to yeast YL17 and L22 from archaebacteria, eubacteria, and chloroplasts. Curr Genet 19: 417–422 (1991).

180. Maessen GDF, Amons R, Maassen JA, Moller W: Primary structure of elongation factor 1β from *Artemia*. FEBS Lett 208: 77–83 (1986).

181. Mahe A, Grisvard J, Dron M: Fungal-and Plant-specific gene markers to follow the bean anthracnose infection process and normalize a bean chitinase mRNA induction. Mol Plant-Microbe Inter 5: 242–248 (1992).

182. Makkinje A, Xiong H, Li M, Damuni Z: Phosphorylation of eukaryotic protein synthesis initiation factor 4E by insulin-stimulated protamine kinase. J Biol Chem 270: 14824–14828 (1995).

183. Marty I, Meyer Y: cDNA nucleotide sequence and expression of a tobacco cytoplasmic ribosomal protein L2 gene. Nucl Acids Res 20: 1517–1522 (1992).

184. Matsumoto S, Oizumi N, Taira H, Ejiri S: Cloning and sequencing of the cDNA encoding rice elongation factor 1β'. FEBS Lett 311: 46–48 (1992).

140

185. Matsumoto S, Terui Y, Xi S, Taira H, Ejiri S: Cloning and characterization of the cDNA encoding rice elongation factor 1β. FEBS Lett 338: 103–106 (1994).

186. Mattheakis LC, Shen WH, Collier RJ: DPH5, a methyltransferase gene required for diphthamide biosynthesis in *Saccharomyces cerevisiae*. Mol Cell Biol 12: 4026–4037 (1992).

187. Maxwell ES, Robinson EA, Henriksen O: Proceedings: elongation factor 2: amino acid sequence at the site of ADP-ribosylation. J Biochem (Tokyo) 77: 9p–9b (1975).

188. Mehta AM, Saftner RA, Mehta RA, Davies PJ: Identification of posttranslationally modified 18-kilodalton protein from rice as eukaryotic translation initiation factor 5A. Plant Physiol 106: 1413–1419 (1994).

189. Mehta HB, Dholakia JN, Roth WW, Parekh BS, Montelaro RC, Woodley CL, Wahba AJ: Structural studies on the eukaryotic chain initiation factor 2 from rabbit reticulocytes and the brine shrimp *Artemia* embryos: phosphorylation by the heme-controlled repressor and casein kinase II. J Biol Chem 261: 6705–6711 (1986).

190. Melanson DL, Ingle J: Regulation of ribosomal RNA accumulation by auxin in artichoke tissue. Plant Physiol 62: 761–765 (1978).

191. Merrick WC: Overview: mechanism of translation initiation in eukaryotes. Enzyme 44: 7–16 (1990).

192. Merrick WC: Mechanism and regulation of eukaryotic protein synthesis. Microbiol Rev 56: 291–315 (1992).

193. Merrick WC: Eukaryotic protein synthesis: An *in vitro* analysis. Biochimie 76: 822–830 (1994).

194. Merrick WC, Dever TE, Kinzy TG, Conroy SC, Cavallius J, Owens CL: Characterization of protein synthesis factors from rabbit reticulocytes. Biochim Biophys Acta 1050: 235–240 (1990).

195. Metz AM, Browning KS: Sequence of a cDNA encoding wheat eukaryotic protein synthesis initiation factor 4A. Gene 131: 299–300 (1993).

196. Metz AM, Timmer RT, Allen ML, Browning KS: Sequence of a cDNA encoding the α-subunit of wheat translation elongation factor 1. Gene 120: 315–316 (1992).

197. Metz AM, Timmer RT, Browning KS: Sequences for two cDNAs encoding *Arabidopsis thaliana* eukaryotic protein synthesis initiation factor 4A. Gene 120: 313–314 (1992).

198. Metz AM, Timmer RT, Browning KS: Isolation and sequence of a cDNA encoding the cap binding protein of wheat eukaryotic protein synthesis initiation factor 4F. Nucl Acids Res 20: 4096 (1992).

199. Milburn SC, Hershey JWB, Davies MV, Kelleher K, Kaufman RJ: Cloning and expression of eukaryotic initiation factor 4B cDNA: sequence determination identifies a common RNA recognition motif. EMBO J 9: 2783–2790 (1990).

200. Mitsui H, Nakatani K, Yamaguchi-Shinozaki K, Shinozaki K, Nishikawa K, Takahashi H: Sequencing and characterization of the kinesin-related genes *katB* and *katC* of *Arabidopsis thaliana*. Plant Mol Biol 25: 865–876 (1994).

201. Mizoguchi T, Hayashida N, Yamaguchi-Shinozaki K, Kamada H, Shinozaki K: Two genes that encode ribosomal-protein S6 kinase homologs are induced by cold or salinity stress in *Arabidopsis thaliana*. FEBS Lett 358: 199–204 (1995).

202. Moldave K: Eukaryotic protein synthesis. Annu Rev Biochem 54: 1109–1149 (1985).

203. Monke G, Sonnewald U: Elevated mRNA levels of the ribosomal protein L19 and a calmodulin-like protein in assimilate-accumulating transgenic tobacco plants. Plant Physiol 107: 1451–1452 (1995).

204. Montesano L, Glitz DG: Wheat germ cytoplasmic ribosomes. J Biol Chem 263: 4939–4944 (1988).

205. Montesano L, Glitz DG: Wheat germ cytoplasmic ribosomes. J Biol Chem 263: 4932–4938 (1988).

206. Morelli JK, Shewmaker CK, Vayda ME: Biphasic stimulation of translational activity correlates with induction of translation elongation factor 1 subunit α upon wounding in potato tubers. Plant Physiol 106: 897–903 (1994).

207. Mundy J, Leah R, Boston R, Endo Y, Stirpe F: Genes encoding ribosomal inactvating proteins. Plant Mol Biol Rep (CPGN Supplement) 12: s60–s62 (1994).

208. Nagashima K, Kasai M, Nagata S, Kaziro Y: Structure of the two genes coding for polypeptide chain elongation factor 1α (EF-1α) from *Saccharomyces cerevisiae*. Gene 45: 265–273 (1986).

209. Nairn AC, Palfrey HC: Identification of the Major Mr 100,000 substrate for calmodulin-dependent protein kinase III in mammalian cells as elongation factor–2. J Biol Chem 262: 17299–17303 (1987).

210. Nakamura I, Kameya N, Aoki T, Tada T, Norita E, Kanzaki H, Uchimiya H: Nucleotide sequence of a rice cDNA encoding a homolog of the eukaryotic ribosomal protein S8. Plant Physiol 107: 1463–1464 (1995).

211. Naranda T, MacMillan SE, Hershey JWB: Purified yeast translational initiation factor eIF-3 is an RNA-binding protein complex that contains the PRT1 protein. J Biol Chem 269: 32286–32292 (1994).

212. Naranda T, Strong WB, Menaya J, Fabbri BJ, Hershey JWB: Two structural domains of initiation factor eIF–4B are involved in binding to RNA. J Biol Chem 269: 14465–14472 (1994).

213. Nielsen PJ, Trachsel H: The mouse protein synthesis initiation factor 4A gene family includes two related functional genes which are differentially expressed. EMBO J 7: 2097–2105 (1988).

214. Nishi R, Hashimoto H, Uchimiya H, Kato A: The primary structure of two proteins from the small ribosomal subunit of rice. Biochim Biophys Acta 1216: 113–114 (1993).

215. Nishi R, Kidou S, Uchimiya H, Kato A: Isolation and characterization of a rice cDNA which encodes the eukaryotic initiation factor 4A. Biochim Biophys Acta 1174: 293–294 (1993).

216. Nishi R, Kidou S, Uchimiya H, Kato A: The primary structure of two proteins from the large ribosomal subunit of rice. Biochim Biophys Acta 1216: 110–112 (1993).

217. Noller HF: Ribosomal RNA and translation. Annu Rev Biochem 60: 191–227 (1991).

218. Nygård O, Nilsson L: Translational dynamics: interactions between the translational factors, tRNA and ribosomes during eukaryotic protein synthesis. Eur J Biochem 191: 1–17 (1990).

219. Oizumi N, Matsumoto S, Taira H, Ejiri S: Nucleotide sequences of the cDNA encoding wheat elongation factor 1β'. Nucl Acids Res 20: 5225 (1992).

220. Omura F, Kohno K, Uchida T: The histidine residue of codon 715 is essential for function of elongation factor 2. Eur J Biochem 180: 1–8 (1989).

221. Oppenheimer NJ, Bodley JW: Diphtheria toxin. Site and configuration of ADP-ribosylation of diphthamide in elongation factor 2. J Biol Chem 256: 8579–8581 (1981).

222. Ornelles DA, Fey EG, Penman S: Cytochalasin releases mRNA from the cytoskeletal framework and inhibits protein synthesis. Mol Cell Biol 6: 1650–1662 (1986).

223. Osterhout JJ, Lax SR, Ravel JM: Factors from wheat germ that enhance the activity of eukaryotic initiation factor eIF–2: isolation and characterization of Co-eIF–2α. J Biol Chem 258: 8285–8289 (1983).

224. Owttrim GW, Hofmann S, Kuhlemeier C: Divergent genes for translation initiation factor eIF–4A are coordinately expressed in tobacco. Nucl Acids Res 19: 5491–5496 (1991).

225. Owttrim GW, Mandel T, Trachsel H, Thomas AA, Kuhlemeier C: Characterization of the tobacco eIF–4A gene family. Plant Mol Biol 26: 1747–1757 (1994).

226. Pain VM: Translational control during amino acid starvation. Biochimie 76: 718–728 (1994).

226a. Pain VM: Initiation of protein synthesis in eukarytotic cells. Eur J Biochem 236: 747–771 (1996).

227. Palen E, Venema RC, Chang Y-WE, Traugh JA: GDP as a regulator of phosphorylation of elongation factor 1 by casein kinase II. Biochemistry 33: 8515–8520 (1994).

228. Pappenheimer AM Jr: Diphtheria toxin. Annu Rev Biochem 46: 69–94 (1977).

229. Paredes AM: Physical characterization of initiation factor eIF–3 by image enhancement and electron microscopy. Master's Thesis, University of Texas at Austin (1988).

230. Park MH, Wolff EC, Folk JE: Is hypusine essential for eukaryotic cell proliferation? Trends Biochem Sci 18: 475–479 (1993).

231. Park MH, Wolff EC, Folk JE: Hypusine: its post-translational formation in eukaryotic initiation factor 5A and its potential role in cellular regulation. Biofactors 4: 95–104 (1993).

232. Pause A, Méthot N, Svitkin Y, Merrick WC, Sonenberg N: Dominant negative mutants of mammalian translation initiation factor eIF–4A define a critical role for eIF–4F in cap-dependent and cap-independent initiation of translation. EMBO J 13: 1205–1215 (1994).

233. Pause A, Sonenberg N: Mutational analysis of a DEAD box RNA helicase: the mammalian translation initiation factor eIF–4A. EMBO J 11: 2643–2654 (1992).

234. Pay A, Heberle-Bors E, Hirt H: Isolation and sequence determination of the plant homologue of the eukaryotic initiation factor 4D cDNA from alfalfa, *Medicago sativa*. Plant Mol Biol 17: 927–929 (1991).

235. Penman S: Rethinking cell structure. Proc Natl Acad Sci USA 92: 5251–5257 (1995).

236. Perentesis JP, Genbauffe FS, Veldman SA, Galeotti CL, Livingston DM, Bodley JW, Murphy JR: Expression of diphtheria toxin fragment A and hormone-toxin fusion proteins in toxin-resistant yeast mutants. Proc Natl Acad Sci USA 85: 8386–8390 (1988).

237. Perez-Bercoff R, Kaempfer R: Genomic RNA of Mengovirus. V. Recognition of common features by ribosomes and eucaryotic initiation factor 2. J Virol 41: 30–41 (1982).

238. Perez-Mendez A, Aguilar R, Briones E, Sanchez-de-Jimenez E: Characterization of ribosomal protein phosphorylation in maize axes during germination. Plant Sci 94: 71–79 (1993).

239. Perlick AM, Puhler A: A survey of transcripts expressed specifically in root nodules of broadbean (*Vicia faba* L.). Plant Mol Biol 22: 957–970 (1993).

240. Pérez L, Aguilar R, Méndez AP, Sánchez de Jiménez E: Phosphorylation of ribosomal proteins induced by auxins in maize embryonic tissues. Plant Physiol 94: 1270–1275 (1990).

241. Phan LD, Perentesis JP, Bodley JW: *Saccharomyces cerevisiae* elongation factor 2. Mutagenesis of the histidine precursor of diphthamide yields a functional protein that is resistant to diphtheria toxin. J Biol Chem 268: 8665–8668 (1993).

242. Pokalsky AR, Hiatt WR, Ridge N, Rasmussen R, Houck CM, Shewmaker CK: Structure and expression of elongation factor 1α in tomato. Nucl Acids Res 17: 4661–4673 (1989).

243. Price N, Proud C: The guanine nucleotide-exchange factor, eIF–2B. Biochimie 76: 748–760 (1994).

244. Proud CG: Peptide-chain elongation in eukaryotes. Mol Biol Rep 19: 161–170 (1994).

245. Proud CG: PKR: A new name and new roles. Trends Biochem Sci 20: 241–246 (1995).

246. Raff JW, Whitfield WGF, Glover DM: Two distinct mechanisms localise cyclin B transcripts in synctial *Drosophila* embryos. Development 110: 1249–1261 (1990).

247. Ray BK, Lawson TG, Abramson RD, Merrick WC, Thach RE: Recycling of messenger RNA cap binding proteins mediated by eukaryotic initiation factor 4B. J Biol Chem 261: 11466–11470 (1986).

248. Raychaudhuri P, Stringer EA, Valenzuela DM, Maitra U: Ribosomal subunit anitassociation activity in rabbit reticulocyte. J Biol Chem 259: 11930–11935 (1984).

249. Reinbothe S, Mollenhauer B, Reinbothe C: JIPs and RIPs: the regulation of plant gene expression by jasmonates in response to environmental cues and pathogens. Plant Cell 6: 1197–1209 (1994).

250. Reinbothe S, Reinbothe C, Lehmann J, Becker W, Apel K, Parthier B: JIP60, a methyl jasmonate-induced ribosome-inactivating protein involved in plant stress reactions. Proc Natl Acad Sci USA 91: 7012–7016 (1994).

251. Rhoads RE: Protein synthesis, cell growth and oncogenesis. Curr Opin Cell Biol 3: 1019–1024 (1991).

252. Rhoads RE: Regulation of eukaryotic protein synthesis by initiation factors. J Biol Chem 268: 3017–3020 (1993).

253. Rhoads RE, Joshi B, Minich WB: Participation of initiation factors in the recruitment of mRNA to ribosomes. Biochimie 76: 831–838 (1994).

254. Rhoads RE, Joshi-Barve S, Rinker-Schaeffer C: Mechanism of action and regulation of protein synthesis initiation factor 4E: effects on mRNA discrimination. cellular growth rate, and oncogenesis. Prog Nucl Acid Res 46: 183–219 (1993).

255. Riis B, Rattan SIS, Clark BFC, Merrick WC: Eukaryotic protein elongation factors. Trends Biochem Sci 15: 420 (1990).

256. Ripmaster TL, Vaughn GP, Woolford JL Jr: DRS1 to DRS7, novel genes required for ribosome assembly and function in *Saccharomyces cerevisiae*. Mol Cell Biol 13: 7901–7912 (1993).

257. Robinson EA, Henriksen O, Maxwell ES: Elongation factor 2. Amino acid sequence at the site of adenosine diphosphate ribosylation. J Biol Chem 249: 5088–5093 (1974).

258. Rosenfeld MG, Barrieux A: A. Characterization of GTP-dependent Met-tRNA binding protein. J Biol Chem 252: 3843–3847 (1977).

259. Ruhl M, Himmelspach M, Bahr GM, Hammerschmid F, Jaksche H, Wolff B, Aschauer H, Farrington GK, Probst H, Bevec D, Hauber J: Eukaryotic initiation factor 5A is a cellular target of the human immunodeficiency virus type 1 Rev activation domain mediating *trans*-activation. J Cell Biol 123: 1309–1320 (1993).

260. Russell DW, Spremulli LL: Purification and characterization of a ribosome dissociation factor (eukaryotic initiation factor 6) from wheat germ. J Biol Chem 254: 8796–8800 (1979).

261. Ryazanov AG: Ca²⁺/calmodulin-dependent phosphorylation of elongation factor 2. FEBS Lett 214: 331–334 (1987).

262. Ryazanov AG, Davydova EK: Mechanism of elongation factor 2 (EF-2) inactivation upon phosphorylation: Phos-

142

phorylated EF-2 is unable to catalyze translocation. FEBS Lett 251: 187–190 (1989).

263. Ryazanov AG, Shestakova EA, Natapov PG: Phosphorylation of elongation factor 2 by EF-2 kinase affects rate of translation. Nature 334: 170–173 (1988).

264. Ryazanov AG, Spirin AS: Phosphorylation of elongation factor 2: a key mechanism regulating gene expression in vertebrates. New Biol 2: 843–850 (1990).

265. Rychlik W, Domier LL, Gardner PR, Hellmann GM, Rhoads RE: Amino acid sequence of the mRNA cap-binding protein from human tissues. Proc Natl Acad Sci USA 84: 945–949 (1987).

266. Samuel CE: The eIF–2α protein kinases, regulators of translation in eukaryotes from yeasts to humans. J Biol Chem 268: 7603–7606 (1993).

267. Sanders J, Maassen JA, Amons R, Möller W: Nucleotide sequence of human elongation factor–1β cDNA. Nucl Acids Res 19: 4551 (1991).

268. Sanders J, Raggiaschi R, Morales J, Moller W: The human leucine zipper-containing guanine-nucleotide exchange protein elongation factor–1δ. Biochim Biophys Acta 1174: 87–90 (1993).

269. Sangwan V, Lenvik TR, Gantt JS: The *Arabidopsis thaliana* ribosomal protein S15 (rig) gene. Biochim Biophys Acta 1216: 221–226 (1993).

270. Scharf KD, Nover L: Heat-shock induced alterations of ribosomal protein phosphorylation in plant cell cultures. Cell 30: 427–437 (1982).

271. Scharf KD, Nover L: Control of ribosome biosynthesis in plant cell cultures under heat shock conditions. II. Ribosomal proteins. Biochim Biophys Acta 909: 44–57 (1987).

272. Schnier J, Schwelberger HG, Smit-McBride Z, Kang HA, Hershey JWB: Translation initiation factor 5A and its hypusine modification are essential for cell viability in the yeast *Saccharomyces cerevisiae*. Mol Cell Biol 11: 3105–3114 (1991).

273.

274. Schreier MH, Erni B, Staehelin T: Initiation of mammalian protein synthesis. I. Purification and characterization of seven initiation factors. J Mol Biol 116: 727–753 (1977).

275. Schuler GD, Altschul SF, Lipman DJ: A workbench for multiple alignment construction and analysis. Proteins 9: 180–190 (1991).

276. Scoble J, Bilgin N, Ehrenberg M: Two GTPs are hydrolysed on two molecules of EF-Tu for each elongation cycle during code translation. Biochimie 76: 59–62 (1994).

277. Seal SN, Schmidt A, Marcus A: A heat-stable protein synthesis initiation factor from wheat germ. J Biol Chem 257: 8634–8637 (1982).

278. Seal SN, Schmidt A, Marcus A: Fractionation and partial characterization of the protein synthesis system of wheat germ. J Biol Chem 258: 866–871 (1983).

279. Seal SN, Schmidt A, Marcus A: Wheat Germ eIF–2 and Co-eIF–2: Resolution and functional characterization in in vitro protein synthesis. J Biol Chem 258: 10573–10576 (1983).

280. Seal SN, Schmidt A, Marcus A: Fractionation and partial characterization of the protein synthesis system of wheat germ. I. Resolution of two elongation factors and five initiation factors. J Biol Chem 258: 859–865 (1983).

281. Seal SN, Schmidt A, Marcus A: Eukaryotic initiation factor 4A is the component that interacts with ATP in protein chain initiation. Proc Natl Acad Sci USA 80: 6562–6565 (1983).

282. Seal SN, Schmidt A, Marcus A: The wheat germ protein synthesis system. Meth Enzymol 118: 128–140 (1986).

283. Seal SN, Schmidt A, Marcus A, Edery I, Sonenberg N: A wheat germ cap-site factor functional in protein chain initiation. Arch Biochem Biophys 246: 710–715 (1986).

284. Sha M, Balasta ML, Goss DJ: An interaction of wheat germ initiation factor 4B with oligoribonucleotides. J Biol Chem 269: 14872–14877 (1994).

285. Shaikhin SM, Smailov SK, Lee AV, Kozhanov EV, Iskakov BK: Interaction of wheat germ translation initiation factor 2 with GDP and GTP. Biochimie 74: 447–454 (1992).

286. Shewmaker CK, Ridge NP, Pokalsky AR, Rose RE, Hiatt WR: Nucleotide sequence of an EF-1α genomic clone from tomato. Nucl Acids Res 18: 4276 (1990).

287. Sikorski MM, Przbyl D, Legocki AB, Nierhaus KH: Group fractionation of wheat germ ribosomal proteins. Plant Sci Lett 30: 303–320 (1983).

288. Silar P: Is translational accuracy an out-dated topic. Trends Genet 10: 71–72 (1994).

289. Singer RH: Spatial organization of mRNA within cells. J Cell Biochem 52: 125–126 (1993).

290. Slobin LI: The role of eucaryotic elongation factor Tu in protein synthesis. Eur J Biochem 110: 555–563 (1980).

291. Slobin LI: Polypeptide chain elongation. In: Trachsel H (ed) Translation in Eukaryotes, pp. 149–175. CRC, Ann Arbor, MI (1991).

292. Smailov SK, Lee AV, Iskakov BK: Study of phosphorylation of translation elongation factor 2 (EF-2) from wheat germ. FEBS Lett 321: 219–223 (1993).

293. Smit-McBride Z, Dever TE, Hershey JWB, Merrick WC: Sequence determination and cDNA cloning of eukaryotic initiation factor 4D, the hypusine-containing protein. J Biol Chem 264: 1578–1583 (1989).

294. Sonenberg N: Regulation of translation and cell growth by eIF–4E. Biochimie 76: 839–846 (1994).

295. Sonenberg N, Shatkin A: Nonspecific effect of m7GMP on protein-RNA interactions. J Biol Chem 253: 6630–6632 (1978).

296. Spremulli LL, Walthall BJ, Lax SR, Ravel JM: Partial purification of the factors required for the initiation of protein synthesis in wheat germ. J Biol Chem 254: 143–148 (1979).

297. St Johnston D: The intracellular localization of messenger RNAs. Cell 81: 161–170 (1995).

298. Stafstrom JP, Devitt ML: Nucleotide sequence of four ribosomal protein L27 cDNAs from growing axillary buds of pea. Plant Physiol 107: 1031–1032 (1995).

299. Stafstrom JP, Sussex IM: Expression of a ribosomal protein gene in auxillary buds of pea seedlings. Plant Physiol 100: 1494–1502 (1992).

300. Stankovic B, Abe S, Davies E: Co-localization of polysomes, cytoskeleton, and membranes with protein bodies from corn endosperm. Protoplasma 177: 66–72 (1993).

301. Stansfield I, Grant GM, Akhmaloka, Tuite MF: Ribosomal association of the yeast SAL4 (SUP45) gene product: implications for its role in translation fidelity and termination. Mol Microbiol 6: 3469–3478 (1992).

302. Stansfield I, Tuite MF: Polypeptide chain termination in *Saccharomyces cerevisiae*. Curr Genet 25: 385–395 (1994).

303. Sutton F, Kenefick DG: Nucleotide sequence of a cDNA encoding an elongation factor (EF-1α) from barley primary leaf. Plant Physiol 104: 807 (1994).

304. Taylor MA, Arif SA, Kumar A, Davies HV, Scobie LA, Pearce SR, Flavell AJ: Expression and sequence analysis of cDNAs induced during the early stages of tuberization in different organs of the potato plant (*Solanum tuberosum* L.). Plant Mol Biol 20: 641–651 (1992).

305. Taylor MA, Davies HV: Nucleotide sequence of a cDNA clone for a 60S ribosomal protein L27 gene from potato (*Solanum tuberosum* L.). Plant Physiol 105: 1025–1026 (1994).

306. Thach RE: Cap recap: the involvement of eIF–4F in regulating gene expression. Cell 68: 177–180 (1992).

307. Thiele D, Cottrelle P, Iborra R, Buhler J-M, Sentenac A, Fromageot P: Elongation factor 1α from *Saccharomyces cerevisiae*: rapid large-scale purification and molecular characterization. J Biol Chem 260: 3084–3089 (1985).

308. Thomas AAM, Benne R, Voorma HO: Initiation of eukaryotic protein synthesis. FEBS Lett 128: 177–184 (1981).

309. Timmer RT, Lax SR, Hughes DL, Merrick WC, Ravel JM, Browning KS: Characterization of wheat germ protein synthesis initiation factor eIF–4C and comparison of eIF–4C from wheat germ and rabbit reticulocytes. J Biol Chem 268: 24863–24867 (1993).

310. Tuite MF, Stansfield I: Translation: knowing when to stop. Nature 372: 614–615 (1994).

311. Turley RB, Ferguson DL, Meredith WR Jr: Isolation and characterization of a cDNA encoding ribosomal protein L41 from cotton (*Gossypium hirsutum* L.). Plant Physiol 105: 1449–1450 (1994).

312. Turley RB, Ferguson DL, Meredith WR Jr: Isolation and characterization of a cDNA encoding ribosomal protein S16 from cotton (*Gossypium hirsutum* L.). Plant Physiol 106: 1219–1220 (1994).

313. Turley RB, Ferguson DL, Meredith WR Jr: A cDNA encoding ribosomal protein S4e from cotton (*Gossypium hirsutum* L.). Plant Physiol 108: 431–432 (1995).

314. Twardowski T, Legocki AB: Purification and some properties of elongation factor 2 from wheat germ. Biochim Biophys Acta 324: 171–183 (1973).

315. Ursin VM, Irvine JM, Hiatt WR, Shewmaker CK: Developmental analysis of elongation factor–1α expression in transgenic tobacco. Plant Cell 3: 583–591 (1991).

316. Valenzuela DM, Chaudhuri A, Maitra U: Eukaryotic ribosomal subunit anti-association activity of calf liver is contained in a single polypeptide chain protein of M_r = 25,500 (eukaryotic initiation factor 6). J Biol Chem 257: 7712–7719 (1982).

317. Valle RPC, Haenni AL: Peptide chain termination. In: Trachsel H (ed) Translation in Eukaryotes, pp. 177–191. CRC Press, Ann Arbor, MI (1991).

318. Van Damme H, Amons R, Janssen G, Möller W: Mapping the functional domains of the eukaryotic elongation factor 1βγ. Eur J Biochem 197: 505–511 (1991).

319. Van Damme HTF, Amons R, Karssies R, Timmers CJ, Janssen GMC, Möller W: Elongation factor 1β of *Artemia*: localization of functional sites and homology to elongation factor 1δ. Biochim Biophys Acta 1050: 241–247 (1990).

320. Van Heerden A, Browning KS: Expression in *Escherichia coli* of the two subunits of the isozyme form of wheat germ protein synthesis initiation factor 4F. Purification of the subunits and formation of an enzymatically active complex. J Biol Chem 269: 17454–17457 (1994).

321. Van Heugten HAA, Kasperaitis MAM, Thomas AAM, Voorma HO: Evidence that eukaryotic initiation factor (eIF) 2 is a cap-binding protein that stimulates cap recognition by eIF–4B and eIF–4F. J Biol Chem 266: 7279–7284 (1991).

322. Van Lijsebettens M, Vanderhaeghen R, De Block M, Bauw G, Villarroel R, Van Montagu M: An S18 ribosomal protein gene copy at the *Arabidopsis* PFL locus affects plant development

323. by its specific expression in meristems. EMBO J 13: 3378–3388 (1994).

323. Van Ness BG, Howard JB, Bodley JW: ADP-ribosylation of elongation factor 2 by diphtheria toxin. NMR spectra and proposed structures of ribosyl-diphthamide and its hydrolysis products. J Biol Chem 255: 10710–10716 (1980).

324. Vayda ME, Schaeffer HJ: Hypoxic stress inhibits the appearance of wound-response proteins in potato tubers. Plant Physiol 88: 805–809 (1988).

325. Vayda ME, Shewmaker CK, Morelli JK: Translational arrest in hypoxic potato tubers is correlated with the aberrant association of elongation factor EF-1α with polysomes. Plant Mol Biol 28: 751–757 (1995).

326. Venema RC, Peters HI, Traugh JA: Phosphorylation of elongation factor 1 (EF-1) and valyl-tRNA synthetase by protein kinase C and stimulation of EF-1 activity. J Biol Chem 266: 12574–12580 (1991).

327. Venema RC, Peters HI, Traugh JA: Phosphorylation of valyl-tRNA synthetase and elongation factor 1 in response to phorbol esters is associated with stimulation of both activities. J Biol Chem 266: 11993–11998 (1991).

328. Voorma HO: Initiation: Met-tRNA binding. In: Trachsel H (ed) Translation In Eukaryotes, pp. 97–108. CRC Press, Ann Arbor, MI (1991).

329. Voorma HO, Thomas A, Goumans H, Amesz H, van der Mast C: Isolation and purification of initiation factors of protein synthesis from rabbit reticulocyte lysate. Meth Enzymol 60: 124–135 (1979).

330. Voorma HO, Thomas AAM, Van Heugten HAA: Initiation of protein synthesis in eukaryotes. Plant Mol Biol Rep 19: 139–145 (1994).

331. Wakiyama M, Saigoh M, Shiokawa K, Miura K: mRNA encoding the translation initiation factor eIF–4E is expressed early in *Xenopus* embryogenesis. FEBS Lett 360: 191–193 (1995).

332. Wang C, Dickson LK, Lehmann R: Genetics of nanos localization in *Drosophila*. Dev Dyn 199: 103–115 (1994).

333. Warskulat U, Perrey R, Wink M: Molecular cloning of a cDNA from *Lupinus polyphyllus* cell cultures encoding a ribosomal protein (rps16). Plant Mol Biol 16: 739–740 (1991).

334. Wasserman DA, Steitz JA: Alive with DEAD proteins. Nature 349: 463–464 (1991).

335. Webster C, Gaut RL, Browning KS, Ravel JM, Roberts JKM: Hypoxia enhances phosphorylation of eukaryotic initiation factor 4A in maize root tips. J Biol Chem 266: 23341–23346 (1991).

336. Webster C, Kim C-Y, Roberts JKM: Elongation and termination reactions of protein synthesis on maize root tip polyribosomes studied in a homologous cell-free system. Plant Physiol 96: 418–425 (1991).

337. Wei C, Kainuma M, Hershey JWB: Characterization of yeast translation initiation factor 1A and cloning of its essential gene. J Biol Chem 270: 22788–22794 (1995).

338. Wei C-L, MacMillan SE, Hershey JWB: Protein synthesis initiation factor eIF–1A is a moderately abundant RNA-binding protein. J Biol Chem 270: 5764–5771 (1995).

339. Wek RC: eIF–2 kinases: regulators of general and gene-specific translation initiation. Trends Biochem Sci 19: 491–496 (1994).

340. Werner R, Guitton MC, Muehlbach HP: A tomato cDNA encodes a protein homologous to the eukaryotic ribosomal protein S25. Plant Physiol Biochem 33: 373–377 (1995).

144

341. Wilson PG, Culbertson MR: SUF12 Suppressor protein of yeast a fusion protein related to the EF-1 family of elongation factors. J Mol Biol 199: 559–573 (1988).

342. Wool IG: Eukaryotic ribosomes, structure, function, biogenesis, and evolution. In: Trachsel H (ed) Translation in Eukaryotes, pp. 3–33. CRC Press, Ann Arbor, MI (1991).

343. Yan R, Rychlik W, Etchison D, Rhoads RE: Amino acid sequence of the human protein synthesis initiation factor eIF–4γ. J Biol Chem 267: 23226–23231 (1992).

344. Yang F, Demma M, Warren V, Dharmawardhane S, Condeelis J: Identification of an actin-binding protein from *Dictyostelium* as elongation factor 1α. Nature 347: 494–496 (1990).

345. Yang W, Boss WF: Regulation of phosphatidylinositol 4-kinase by the activator PIK-A49. J Biol Chem 269: 3852–3857 (1994).

346. Yang W, Burkhart W, Cavallius J, Merrick WC, Boss WF: Purification and characterization of a phosphatidylinositol 4-kinase activator in carrot cells. J Biol Chem 268: 392–398 (1993).

347. Yoder-Hill J, Pause A, Sonenberg N, Merrick WC: The p46 subunit of eukaryotic initiation factor (eIF)–4F exchanges with eIF–4A. J Biol Chem 268: 5566–5573 (1993).

348. Yoon H, Donahue TF: The *sui1* suppressor locus in *Saccharomyces cerevisiae* encodes a translation factor that functions during tRNA$_i^{Met}$ recognition of the start codon. Mol Cell Biol 12: 248–260 (1992).

349. You W, Abe S, Davies E: Cosedimentation of pea root polysomes with the cytoskeleton. Cell Biol Int Rep 16: 663–673 (1992).

350. Zhang SH, Broome MA, Lawton MA, Hunter T, Lamb CJ: atpk1, a novel ribosomal protein kinase gene from *Arabidopsis*. II. Functional and biochemical analysis of the encoded protein. J Biol Chem 269: 17593–17599 (1994).

351. Zhouravleva G, Frolova L, Le Goff X, Le Guellec R, Inge-Vechtomov S, Kisselev L, Philippe M: Termination of translation in eukaryotes is governed by two interacting polypeptide chain release factors, eRF1 and eRF3. EMBO J 14: 4065–4072 (1995).

352. Zhu JK, Damsz B, Kononowicz AK, Bressan RA, Hasegawa PM: A higher plant extracellular vitronectin-like adhesion protein is related to the translational elongation factor–1α. Plant Cell 6: 393–404 (1994).

Plant Molecular Biology **32:** 145–158, 1996.
© 1996 *Kluwer Academic Publishers.*

Translational control of cellular and viral mRNAs

Daniel R. Gallie
Department of Biochemistry, University of California, Riverside, CA 92521–0129, USA

Key words: 5' leader, 3'-untranslated region, plant virus, poly(A) tail, cap, initiation factors

Contents

Abstract   145
Introduction   145
Several factors are required to facilitate
    translation   146
The cap and poly(A) tail are functionally
    co-dependent regulators of translation   146
Functional alternatives to the poly(A) tail   147
Histone mRNAs   149
Functional alternatives to the cap
    structure are still dependent on the
    poly(A) tail for function   150
mRNAs without either a cap or poly(A)
    tail still require an interaction between
    the termini for efficient translation   150

Structural features of the 5'-untranslated
    leader that can influence translation   151
The 5'-untranslated leader can contain
    translational enhancers   152
The role of small upstream open reading
    frames in regulating translation   152
How environmental stress impacts
    translation   153
Conclusion   154
Acknowledgements   155
References   155

Abstract

We are becoming increasingly aware of the role that translational control plays in regulating gene expression in plants. There are now many examples in which specific mechanisms have evolved at the translational level that directly impact the amount of protein produced from an mRNA. All regions of an mRNA, i.e., the 5' leader, the coding region, and the 3'-untranslated region, have the potential to influence translation. The 5'-terminal cap structure and the poly(A) tail at the 3' terminus serve as additional elements controlling translation. Many viral mRNAs have evolved alternatives to the cap and poly(A) tail that are functionally equivalent. Nevertheless, for both cellular and viral mRNAs, a co-dependent interaction between the terminal controlling elements appears to be the universal basis for efficient translation.

Introduction

Although gene expression was long thought to be regulated primarily at the level of transcription, we now know that regulation can and does take place at every step during gene expression. This includes splicing, 3'-end processing, and polyadenylation of a pre-mRNA in the nucleus; nucleocytoplasmic transport; and trans-

lation and mRNA turnover in the cytoplasm. The structural characteristics of plant cellular mRNAs are quite conserved. Plant mRNAs are monocistronic and contain 5'- and 3'-untranslated regions (UTRs) in addition to the coding region. All plant cellular mRNAs that have been examined contain a cap, an inverted and methylated GTP at the 5' terminus [$m^7G(5')ppp(5')N$], and terminate in a poly(A) tail. The only known exceptions are histone mRNAs in *Volvox* and *Chlamydomonas* [22, 71] which terminate in a stem loop structure that is conserved throughout metazoans [40, 63]. There are a number of viral mRNAs which also deviate from most cellular mRNAs in that they naturally lack a cap or poly(A) tail. All regions of an mRNA, i.e., the cap, the 5'-UTR, the coding region, the 3'-UTR, and the poly(A) tail have the potential to influence translational efficiency and message stability.

Several factors are required to facilitate translation

The process of translation is composed of three phases: (1) initiation, in which the 80S ribosome assembles at the start codon of an mRNA and initiates the formation of the first peptide bond of the nascent protein; (2) elongation, which involves the translocation of the ribosome down the open reading frame resulting in the production of the protein; and (3) termination, which results in the release of both the completed protein and the 80S ribosome. Initiation is considered to be the rate-limiting step [42] and consequently, it is this phase that is most often subject to regulation.

The function of the initiation factors (eIFs) is to bring the mRNA, ribosomal subunits, and initiator Met-tRNA together to form a competent translation complex at the initiation codon (Fig. 1). Only those factors that associate with the 5'-terminal cap are discussed here (for a thorough discussion of initiation factors, see the review by Browning in this issue). Although 40S scanning is inhibited by secondary structure present in the 5' leader, three initiation factors can promote the binding and scanning of the 40S subunit. eIF-4F (or its isoform called eIF-iso4F) binds to the cap structure (m^7GpppN, where N is any nucleotide) at the 5' terminus. Following eIF-4F binding, eIF-4A and eIF-4B bind to the mRNA through association with eIF-4F and the cap. Together, eIF-4A and eIF-4B function as an ATP-dependent RNA helicase that unwinds any secondary structure present in the 5' leader, a necessary prerequisite to 40S subunit scanning as secondary structure inhibits this process. After the removal of the secondary structure, the 40S subunit (to which eIF-2, GTP, and the initiator Met-tRNA are bound) binds at or close to the 5' terminus of the mRNA and scans the untranslated leader. Once the 40S subunit has located the initiation codon, the 60S subunit joins to create the translationally competent 80S ribosome.

The cap and poly(A) tail are functionally co-dependent regulators of translation

The observation that the cap and poly(A) tail are bifunctional, synergistic regulators of translational efficiency has lead to the development of the co-dependent model of translation [26] (Fig. 2). Their role as regulators of translation is quantitatively greater than their effect on mRNA stability [26]. The co-dependency between the cap and the poly(A) tail means that, in the absence of a 5' cap (or functionally analogous structure, see below), the poly(A) tail fails to enhance translation. The cap also facilitates translation and, in some species, it will do so to a small extent even for mRNAs lacking a poly(A) tail (or functionally analogous structure, see below). However, its function is increased by over an order of magnitude when a poly(A) tail or functional equivalent is present. The synergy between the cap and poly(A) tail suggests that these elements, in conjunction with their associated proteins, are in communication during translation initiation [26]. The mechanism underlying the interaction between the termini appears to involve both the binding of the cap-associated initiation factors with the poly(A) tail and the mutual stabilization of RNA-binding between these initiation factors and the poly(A)-binding (PAB) protein. In a survey of purified translation initiation factors from wheat germ, eIF-4F, eIF-iso4F, and eIF-4B were found to bind specifically to poly(A) in an *in vitro* gel retardation assay [33]. No other initiation factor examined formed a complex with poly(A). Binding to poly(A) was specific as no binding to other RNAs including an A-rich (50% A residues) RNA was observed [33]. eIF-4F and eIF-4B mutually enhance their binding to poly(A) when both are present suggesting a cooperative stabilization in RNA binding (H. Le and D. Gallie, unpublished observations). Addition of exogenous poly(A) to an *in vitro* translation lysate prepared from wheat germ preferentially repressed the translation of uncapped mRNAs relative to that of capped mRNAs [33]. Supplementation of the lysate with eIF-4F and eIF-4B reversed this exogenous, poly(A)-mediated translational repression. These data are consistent with the hypothesis that the cap-associated factors eIF-4F and eIF-4B bind to poly(A). In addition to direct binding to poly(A), the cap-associated factors stabilize PAB protein binding to poly(A) RNA. Individually, eIF-4B and eIF-4F (and eIF-iso4F) promote PAB protein binding and the combined effect of these eIFs on PAB protein is synergistic (H. Le and D. Gallie, unpublished observations).

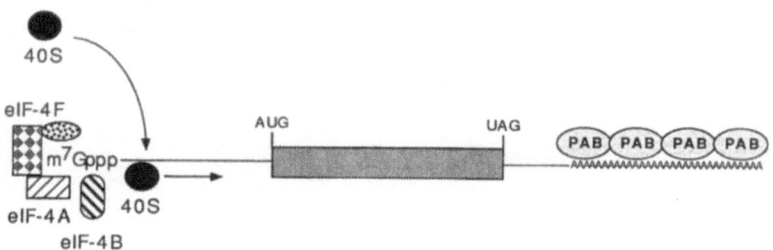

Figure 1. A schematic diagram of a typical mRNA and the proteins that bind to the terminal regulatory elements. The initiation factors, eIF-4F, eIF-4A, and eIF-4B are shown associated with the 5′-terminal cap structure and the poly(A)-binding (PAB) protein is shown bound to the poly(A) tail at the 3′ terminus. After initiation factor binding, the 40S ribosomal subunit at or close to the 5′ terminus scans down the 5′ leader in search for the AUG initiation codon.

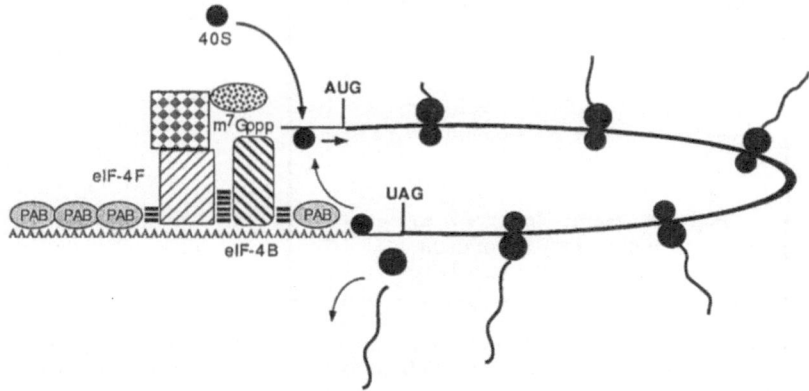

Figure 2. The co-dependent model of translation. eIF-4F and eIF-4B are shown bound to both the 5′-terminal cap structure and the poly(A) tail. Protein-protein contacts between eIF-4F/eIF-4B, eIF-4F/PAB, and eIF-4B/PAB that stabilize protein/mRNA binding are shown as the multiple thick lines between the proteins. This stable complex maintains the close physical proximity of the termini of the mRNA which allows the efficient recycling of ribosomes. The 60S subunit is shown dissociating from the mRNA upon translation termination whereas the 40S subunit recycles back to the 5′ terminus to participate in another round of initiation.

Taken together, these observations suggest that the 5′-terminal cap, the cap-associated eIFs, PAB protein, and the poly(A) tail form a large complex that is stabilized by protein-RNA and protein-protein interactions (Fig. 2). Therefore, one consequence of the interaction between the termini may be to stabilize the association of the translational machinery with an mRNA as means of committing that machinery to the mRNA for as long as its physical integrity is maintained. This ensures that the mRNA will be used in multiple rounds of translation.

Functional alternatives to the poly(A) tail

As mentioned above, all known plant cellular mRNAs are polyadenylated. Although plant viral mRNAs are not cellular mRNAs, they are translated by the translational machinery of the cell and therefore must com-

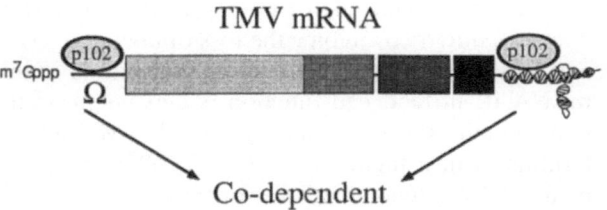

Figure 3. A schematic diagram of the tobacco mosaic viral genome. The genomic RNA serves as a capped mRNA that naturally lacks a poly(A) tail. The p102 protein that binds specifically to the 5′ leader (Ω) and the 3′-UTR is also shown. Efficient translation from this mRNA is dependent on the functional interaction between the termini.

pete with cellular mRNAs for this machinery. Although many plant viral mRNAs are polyadenylated, there are several that terminate in an alternative structure and represent the only known non-polyadenylated mRNAs in plants. Tobacco mosaic virus (TMV) genomic RNA

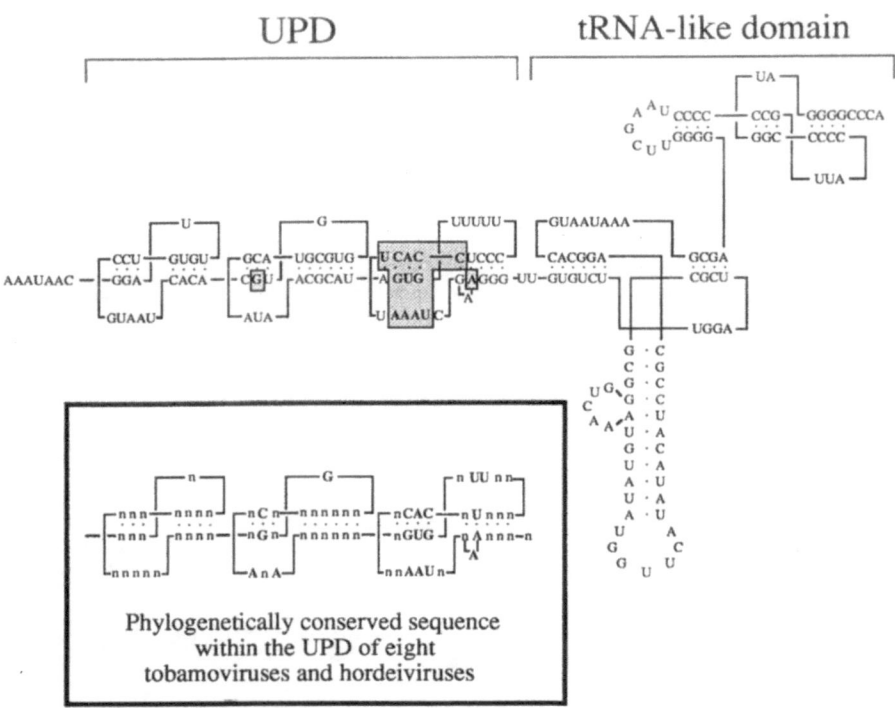

Figure 4. The sequence and higher order structure of the tobacco mosaic virus 3'-UTR. The upstream pseudoknot domain (UPD) and the tRNA-like domain are indicated. The primary sequence within the UPD that is required for regulating translation is shown in the shaded boxes. The sequence that is absolutely conserved in eight viruses is shown in the boxed UPD.

serves as a capped, non-polyadenylated mRNA that is efficiently translated (Fig. 3). If the co-dependent model of translation holds true for mRNAs in general, two related questions arise with respect to the mechanism by which TMV mRNA is translated. How does this mRNA overcome the loss of the poly(A) tail and how does the cap facilitate the translation of this mRNA if, in fact, cap function is dependent on the poly(A) tail? Instead of a poly(A) tail, TMV mRNA terminates in a highly structured 3'-UTR composed of five RNA pseudoknots that encompass a 177 base region (Fig. 4). The 3'-terminal 105 base domain contains two of the five pseudoknots, mimics the three-dimensional shape of tRNAs, and is required in viral RNA replication. Immediately upstream of this region are the remaining three pseudoknots in a domain called the upstream pseudoknot domain (UPD). Many members of the tobamovirus family, of which TMV is the type strain, have been sequenced and all contain an UPD composed of three RNA pseudoknots. Virtually all of the phylogenetically conserved primary sequence is present in the pseudoknot within the UPD that is immediately upstream of the tRNA-like structure (Fig. 4). The UPD is responsible for enhancing translational efficiency in a way that is functionally similar to a poly(A) tail [35, 56]. Both the conserved primary sequence within the 3'-distal pseudoknot of the UPD and its higher-ordered structure are necessary for the function of the UPD in regulating translation [56]. A 102 kDa protein (p102) that binds to the TMV 5' leader (described below) also binds to the 3'-distal pseudoknot of the UPD [96a]. p102 is a highly conserved protein within the plant kingdom including monocots, dicots, and even gymnosperms. Not only is its molecular weight and antigenicity conserved but it also has retained specific RNA-binding activity in all of these species [96a]. p102 expression is developmentally regulated, being expressed in elevated levels in the seed and in the meristematic regions. p102 is a heat-shock protein as p102 expression is induced in heat-stressed wheat coleoptiles.

As observed for a poly(A) tail, the TMV 3'-UTR is co-dependent on the cap at the 5' terminus in order to function as a regulator of translation [56]. Not only does the TMV 3'-UTR require a cap for its own function, but it stimulates the function of the cap by an

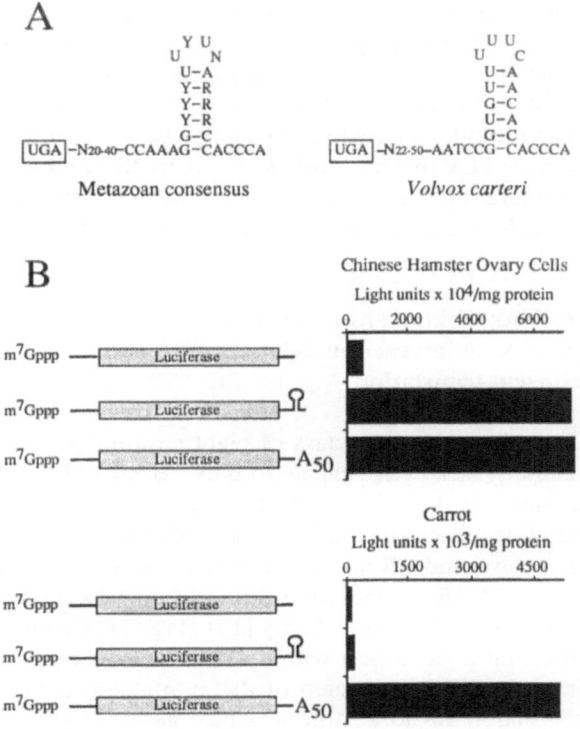

A

Metazoan consensus   Volvox carteri

B   Chinese Hamster Ovary Cells
                       Light units x 10⁴/mg protein

Figure 5. The 3′-terminal stem-loop structure of the cell-cycle-regulated histone mRNAs regulates translation in animal cells but not in higher plant species. A. The consensus sequence of the stem-loop structure found in metazoan histone mRNAs and the stem-loop structure found in histone mRNA from the lower plant *Volvox carteri*. B. The impact of the metazoan histone stem-loop structure on expression from luciferase reporter mRNA in Chinese hamster ovary cells and carrot protoplasts. *luc* mRNA with an equivalent length 3′-UTR served as the negative control and a poly(A)⁺ *luc* mRNA served as the positive control. The mRNAs were synthesized *in vitro* and delivered to the cells using electroporation. Following incubation of the cells, luciferase activity was measured to determine the extent of translation from each *luc* mRNA construct.

order of magnitude, data suggesting that cap function can be enhanced by regulatory elements other than the poly(A) tail. Identical results were obtained with the 3′-UTR of brome mosaic virus [29], another naturally poly(A)⁻ mRNA that also terminates in a 3′-UTR composed of RNA pseudoknots and a tRNA-like structure. These viral mRNAs, therefore, have evolved functional alternatives to a poly(A) tail that nevertheless require interaction with the 5′ cap in order to facilitate translation.

Histone mRNAs

The metazoan cell-cycle regulated histone mRNAs represent the only known class of cellular mRNAs that naturally lack a poly(A) tail. Instead, these histone mRNAs terminate in a stem-loop structure that is highly conserved from *C. elegans* to humans [40, 63] (Fig. 5). Expression of the cell-cycle regulated histones is tightly coupled to nuclear DNA synthesis during the S phase of the mitotic cell cycle [88] with subsequent destabilization of the mRNA during the G_2 phase [39]. Studies focusing on the 3′-terminal stem-loop structure of the cell-cycle regulated histone mRNAs have demonstrated its involvement in multiple steps of histone gene expression, including 3′-end processing [5, 64, 68, 69, 70, 100], nucleocytoplasmic transport [20, 104], and cytoplasmic mRNA stability [9, 58, 80]. A 45 kDa protein associated with polysomes has been identified that specifically recognizes the histone stem-loop structure [81, 82, 105]. mRNAs that naturally lack a poly(A) tail, such as the cell-cycle regulated histone mRNAs, present an apparent paradox in that, according to the co-dependent model of translation, the absence of the poly(A) tail should render the cap largely non-functional, and consequently, the mRNA should be translationally incompetent. However, the histone terminal stem loop is functionally similar to a poly(A) tail in that it is sufficient and necessary to regulate translational efficiency in CHO cells (Fig. 5) and, like a poly(A) tail, it is co-dependent on the cap in its function as a regulator of translation [29a]. This means that the histone stem-loop regulates translation but only if a cap is present at the 5′ terminus. It is important to note that although these cell-cycle-regulated histone mRNAs have evolved an alternative to the poly(A) tail they have nevertheless maintained a functional co-dependency between the terminal regulatory elements to promote efficient translation, in this case, between the cap and the histone stem-loop.

Plant histone mRNAs are polyadenylated [11, 12, 50, 107] and therefore do not contain the stem-loop structure present in animal histone mRNAs. Either the histone stem-loop evolved after the divergence of plant and animals or it was lost from plants during their subsequent evolution. The observation that histone mRNAs of *Chlamydomonas* [22] and the multicellular algae, *Volvox carteri* [71, 72], are not polyadenylated but contain the phylogenetically-conserved stem-loop present in metazoans (Fig. 5) supports the later hypothesis. Moreover, the conservation of the histone stem-loop suggests that the regulation of histone

150

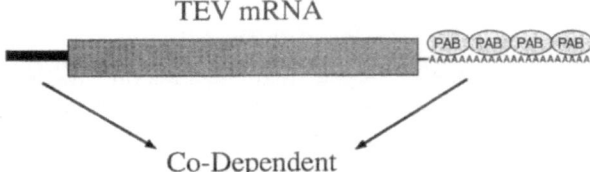

TEV mRNA

Co-Dependent

Figure 6. A schematic diagram of the tobacco etch viral genome. The genomic RNA serves as a poly(A)$^+$ mRNA that naturally lacks a cap. The 5' leader, which confers cap-independent translation on reporter mRNAs, is shown as a thick line. The PAB protein is shown bound to the poly(A) tail. Efficient translation from this mRNA is dependent on the functional interaction between the 5' leader and the poly(A) tail.

expression may also be conserved between metazoans and these algal species.

If the histone stem-loop regulatory mechanism had once been present in the evolution of plants, the specific nuclear and polysomal proteins that recognize this structure in animals [81, 105] may still be present in higher plants, particularly if they were required to facilitate other cellular processes. However, expression from reporter mRNAs terminating in the histone stem-loop was not improved relative to poly(A)$^-$ mRNA in carrot protoplasts (Fig. 5), suggesting that this regulatory element is not recognized in higher plants [29a]. These observations suggest that the histone stem-loop regulatory mechanism evolved before plants and animals diverged but was subsequently lost in the evolution of higher plants.

Functional alternatives to the cap structure are still dependent on the poly(A) tail for function

In addition to viral mRNAs that are naturally capped but lack a poly(A) tail, there are other viral mRNAs that are naturally polyadenylated but lack a cap. Tobacco etch virus (TEV) is an example of the latter type of viral mRNA (Fig. 6). Considering that the poly(A) tail is co-dependent with the cap to enhance translation, the absence of a cap in TEV mRNA might be expected to render the poly(A) tail ineffective, resulting in an inefficiently translated mRNA. However, the TEV 5' leader confers cap-independent translation on reporter mRNAs [10, 34], suggesting that the TEV 5' leader is the functional equivalent to a cap. A prediction of the co-dependent hypothesis is that the TEV leader and the poly(A) tail might synergistically enhance the translation of the coding region present between these elements. This prediction was, in fact, borne out: an

interaction between the TEV 5' leader and a poly(A) tail, resulting in a synergistic increase in translational efficiency, was observed for reporter mRNAs in both plant and animal cells [34] (Fig. 6). Consequently, as seen for those alternatives to a poly(A) tail, an mRNA such as TEV that has evolved an alternative to the cap structure still requires a functional interaction between the ends of the message for its efficient translation.

mRNAs without either a cap or poly(A) tail still require an interaction between the termini for efficient translation

There is yet another class of plant viral mRNAs that naturally lack both a cap and a poly(A) tail and yet are competitive for the translational machinery in a cell. Satellite tobacco necrosis virus (STNV) is a 1239 base monocistronic mRNA that terminates in a 619 nucleotide 3'-UTR composed of multiple stem-loop structures and RNA pseudoknots [15] (Fig. 7). Deletions affecting a stem-loop within a 110–150 base region immediately downstream of the stop codon reduced translation 10- to 20-fold in wheat germ lysate [98]. This region was co-dependent on the STNV 29 nucleotide 5' leader in order to direct cap-independent translation. Neither the STNV 5' leader nor the 3'-UTR functioned well alone. The absence of either was overcome by the addition of a cap to the mRNA [98]. The 5' leader from tobacco mosaic virus, which is a translational enhancer (see below), could partially substitute for the STNV 5' leader in activating the STNV 3'-UTR during translation [15], an observation suggesting some degree of functional similarity between the STNV and TMV 5' leaders. Deletions within the STNV mRNA 3'-UTR increased the mRNA's requirement for eIF-4F, suggesting the 3'-UTR influences an early step in translation initiation. No viral protein was required for the regulation.

The 5677 nucleotide genomic RNA of barley yellow dwarf virus (BYDV) serves as a naturally uncapped and poly(A)$^-$ mRNA for the first two open reading frames of this polycistronic mRNA (Fig. 7). The second cistron is translated by a ribosomal frameshift mechanism [6] whereas the distal cistrons are translated from subgenomic mRNAs. A *cis*-acting element present within a 500 nucleotide region within what functions as the 3'-UTR for the 5' proximal cistrons is required to confer cap-independent translation in an *in vitro* translation extract derived from wheat germ but not from rabbit reticulocytes [102]. Although

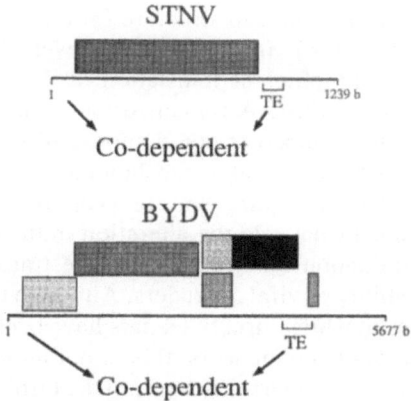

STNV

TE 1239 b

Co-dependent

BYDV

TE 5677 b

Co-dependent

Figure 7. A schematic diagram of the satellite tobacco necrosis and barley yellow dwarf viral genomes. Both viral RNAs serve as mRNAs that naturally lack a cap and a poly(A) tail. The open reading frames that code for expressed viral proteins are shown as shaded boxes. The region that functions as the translational enhancer (TE) in the 3'-UTR is indicated. Efficient translation from this mRNA is dependent on the functional interaction between the 5' leader and the TE.

this BYDV 3' element is ca. 4500 nucleotides from the 5' terminus, nevertheless, it is co-dependent on the 141 nucleotide 5' leader to enhance translation. No viral protein or other sequence within the viral genome is required. Both STNV and BYDV mRNAs have evolved a translational strategy that completely obviates the need for a cap and poly(A) tail, and yet like the standard elements present in cellular mRNAs, expression from these unusual viral mRNAs still requires an interaction between the 5' and 3' regulatory elements.

Structural features of the 5'-untranslated leader that can influence translation

The 5' leader can influence translation passively; for example, both the length and degree of secondary structure present within the leader can affect the rate of initiation. In other cases, the 5' leader actively regulates translation, often by serving as a binding site for proteins that mediate the regulation.

In maize, translation increased as a function of leader length [36]. Although expression was virtually unaffected by increasing the length of the leader from 14 to 29 bases, a 3 to 4-fold increase in expression was observed when the length of the 5' leader increased from 29 to 74 bases [36]. 5' leaders longer than 74 bases or shorter than 14 bases were not examined in this study. However, extremely short 5' leaders can

impair the fidelity of translation, at least in an *in vitro* translation lysate derived from wheat germ: positioning the AUG start codon within a few bases of the 5' terminus can cause the scanning 40S ribosomal subunits to skip over this first AUG codon thereby reducing the frequency of initiation at this AUG [53]. In a survey of 79 higher plant nuclear genes, 5' leaders ranged from 9 to 193 bases in length with the majority (53%) being 40–80 bases long [45].

Plant 5' leaders are often AU rich: 32% of the genes surveyed were >70% AU, 39% were 61–70% AU, 21% were 51–60% AU [45]. The degree of AU richness influences the potential for stable secondary structure within the leader which in turn impacts the translational efficiency of an mRNA [51, 84]. Both the position and the stability of the secondary structure dictates the extent to which 40S ribosomal subunits are impeded in their scanning of the 5' leader [52]. Secondary structure close to the 5' terminus can prevent 40S subunit binding altogether. This is a consequence of the failure of eIF-4F, eIF-4A, and eIF-4B to bind to the cap structure, a necessary prerequisite for 40S subunit binding to the mRNA [89]. If secondary structure is present sufficiently distal to the 5' terminus to allow initiation factor and, consequently, 40S subunit binding, it nevertheless can serve as an impediment to 40S subunit scanning.

The presence of secondary structure has a much greater impact on scanning 40S subunits when it is present within the 5' leader than it does for translocating 80S ribosomes when present within the coding region. A hairpin with a free energy of -61 kcal/mol was sufficient to block 40S subunit scanning when the structure was positioned within the 5' leader but it did not prevent ribosome translocation when positioned within the coding region [52]. Even an RNA duplex with a free energy of -345 kcal/mol present within the coding region of a lysozyme mRNA construct was insufficient to block translation [60]. The ability of the translating ribosome to melt out secondary structure is probably a consequence of the hydrolysis of GTP at each translocation step which serves to drive translation forward. Although the presence of secondary structure may not bring translocating ribosomes to a permanent halt, it can slow their progression. Ribosomal pausing has been suggested to occur at sites which include rare codons or stable secondary structure and has been observed to occur in both nuclear and chloroplast mRNAs [48, 49, 94, 106].

The 5'-untranslated leader can contain translational enhancers

In addition to the length and structure of the leader, the 5'-untranslated region can serve as the site for regulatory sequences that actively control the rate of translation initiation. A number of examples are now known. The majority of the leaders investigated have come from viral mRNAs. The first leaders examined for their ability to enhance translation were the 68 base leader (called Ω) from TMV genomic RNA and the 36 base leader from alfalfa mosaic virus (AMV) RNA 4 [31, 44]. The Ω leader has been used in a number of studies [16, 38, 66, 67, 93] and has been shown to enhance the translation of genes delivered either as mRNA [28, 30, 31, 32, 36], plasmid DNA for transient assays [17], or integrated into the genome in transgenic plants where it enhanced translation 4 to 6-fold [17]. Ω enhances the translational efficiency of foreign mRNAS in the absence of any viral gene product and does so to a greater extent in dicots than in monocots [30]. The sequence of Ω is highly structured: three copies of an eight base direct repeat and one copy of a 25 base poly(CAA) region comprise 72% of the leader. Mutational analysis of Ω identified the poly(CAA) region as the primary element responsible for the enhancement *in vivo* [36]. The same 102 kDa protein that binds specifically to the UPD within the TMV 3'-UTR also binds to the 25 base poly(CAA) region present within Ω and may mediate the enhancement associated with this sequence.

As discussed above, some plant viral mRNAs naturally lack a cap and yet are efficiently translated. In addition to the 5' leader from tobacco etch virus (TEV), those from several viruses, including potato virus X [86, 109], potato virus Y [59], potato virus S [99]; turnip mosaic potyvirus [2]; potato leafroll luteovirus [96]; pea seed borne mosaic virus [74]; cowpea mosaic virus [97, 101] have been shown to confer efficient translation to reporter genes in a cap-independent manner. Except for the co-dependent interaction of the TEV 5' leader and the poly(A) tail [34], the mechanism by which these viral leaders promote efficient translation has not been elucidated.

The 5' leaders of at least some viral mRNAs may enhance translation through altering the requirement for initiation factors. mRNAs containing the AMV RNA 4 leader had a reduced requirement for eIF-4A, eIF-4B, and eIF-3 in wheat germ lysate [8]. AMV RNA 4 has a cap at the 5' terminus which is the binding site for eIF-4F. Removal of the cap increased the amount of eIF-4F needed for efficient translation, demonstrating the importance of the cap structure for the translation of this mRNA [24]. In contrast, the level of eIF-4F required for the efficient translation of the naturally uncapped STNV mRNA remained the same whether the mRNA was capped or not, demonstrating that this viral mRNA has evolved a translational strategy that has eliminated the requirement for a cap. It remains to be determined what role the alteration in the requirement for initiation factors plays in the translational enhancer ability of viral 5' leaders. Although there are many cases in which viral 5' leaders have been shown to be translational enhancers, this is not an universal function of viral leaders. Those from the turnip yellow mosaic viral genomic or coat mRNAs do not enhance translation, at least in the species examined which were non-hosts for the virus [32, 43].

Although the translational enhancers present in viral leaders appear to exhibit no tissue-specific activity, there is at least once example of a cellular mRNA in which the untranslated regions do regulate expression post-transcriptionally in a tissue-specific manner. α-Amylase is expressed and secreted from the aleurone layer to the endosperm of cereal grains during germination and represents one of the major hydrolytic enzymes produced in this tissue. Together, the 5' leader and 3'-UTR of the barley α-amylase mRNA enhanced the translation of reporter mRNA when it was delivered and translated in aleurone protoplasts isolated from maize and barley [37]. The α-amylase untranslated regions did not increase reporter mRNA translation in protoplasts isolated from maize or carrot suspension cells [37]. Neither did they function to increase expression in animal cells or in *in vitro* translation lysates. Therefore, the preferential translation of mRNAs containing the α-amylase untranslated regions occurs specifically in that tissue which is responsible for the high level of α-amylase expression.

The role of small upstream open reading frames in regulating translation

A consequence of 40S subunit scanning from the 5' terminus is that the ribosomal subunit will encounter the 5'-most proximal AUG codon first. As initiation predominately occurs at the first AUG encountered and the frequency with which eukaryotic ribosomes re-initiate is low, the presence of small upstream open reading frames (uORFs) generally reduces expression from distal cistrons. The generic effect of small uORFs was quantitated in carrot: a single small uORF positioned upstream of the *uidA* gene reduced expression 2- to 6-fold, whereas two uORFs reduced expression to 30-fold [87]. There are several plant viral RNAs, such as those from the caulimovirus family, that serves as polycistronic mRNAs (see the review by Hohn and Fütterer in this issue). These viral mRNAs appear to have evolved specific mechanisms to increase expression from distal cistrons that require viral protein and *cis*-acting elements [25, 91] and consequently are exceptions to the rule.

The yeast amino acid biosynthetic regulatory gene, *GCN4*, contains four small uORFs within its mRNA and is the paradigm for the regulatory role of small uORFs during translation. These uORFs exert fine control over the expression of GCN4 according to the availability of amino acids and demonstrate the regulatory potential of small uORFs [41]. Some plant cellular mRNAs are now known to contain one or more small upstream open reading frames that influence expression from the main ORF. These mRNAs tend to be highly regulated genes in which maximal levels of expression are not required. The 235 nucleotide 5′ leader of the maize *Lc* mRNA contains a 38 codon, small uORF that represses translation of the main coding region by 25- to 30-fold. Both the presence of rare codons and the peptide itself contribute to the translational repression [14]. The 280 nucleotide 5′ leader of *Opaque-2* mRNA from maize contains three small uORFs that repress translation 5-fold. The first two uORFs mediate the repression and are required in *cis* in order to function [62]. The 5′ leader from the plasma membrane proton-ATPase (*pma1*) gene of *Nicotiana plumbaginifolia* also contains a short uORF, 10 codons in length, that impacts the expression from the H$^+$-ATPase coding region [65]. Mutation of this uORF resulted in a 2.7-fold increase in translation of the main open reading frame. Similar observations have been made for a second member of this gene family, *pma3*, in which a 6 codon uORF is present in the 5′ leader (M. Lukaszewicz and M. Boutry, personal communication).

How environmental stress impacts translation

One way in which plants respond to abiotic stress, including heat-shock, hypoxia, and water stress, is a reduction in global translation concomitant with the selective translation of a class of stress-induced or pre-existing mRNAs. The translational response involves the alteration of the pre-existing translation-

al machinery as well as requiring specific sequence information present within some mRNAs.

Exposure to heat shock results in profound changes at almost every level of gene expression including transcription, splicing, nucleocytoplasmic transport, translation, and protein turnover. Although many of the molecular events involved in heat shock gene induction are remarkably conserved in eukaryotes, the control of translation upon thermal stress varies considerably among species. In yeast, the heat-shock response appears to involve transcription mechanisms only whereas in *Xenopus* oocytes, the heat-shock response is mediated entirely at the translational level [3, 4]. In mammalian cells, the translation of non-heat-shock mRNAs is not fully repressed during thermal stress [19]. In *Drosophila*, a 10 min heat-shock is sufficient to release normal cellular mRNAs from polysomes which are replaced by heat-shock-specific mRNAs [95]. Non-*hsp* mRNAs are not destroyed, but rather they are maintained in the cell where they are recruited for translation during recovery from the thermal stress [95]. As *hsp* mRNAs are actively recruited onto polysomes and translated throughout the period of heat shock [61], *hsp* mRNAs have evolved mechanisms to escape the global repression of translation.

The heat-shock response in plant species lies between the extremes observed in *Xenopus* oocytes and yeast. Gene expression is controlled at both the transcriptional and translational level. The translational efficiency of non-*hsp* mRNAs decreases as a function of the length and severity of the thermal stress [27]. The translation of mRNAs that are capped and polyadenylated is proportionately repressed to a greater extent than those mRNAs lacking one or both of these structural elements, suggesting that it is the interaction between the cap and the poly(A) tail that is targeted by the thermal stress. Non-*hsp* mRNAs disassemble from polysomes upon thermal stress resulting in the global repression of protein synthesis [47]. Only the *hsp* mRNAs escape translational repression. Non-*hsp* mRNAs are not degraded following heat shock [47, 76] and their stability increases as a function of the severity of the heat shock [27]. They have been observed to localize in heat-shock granules (HSGs) that contain HSPs including two major HSPs, HSP70 and HSP17 [75, 76] and associate with the cytoskeleton surrounding the nucleus in perinuclear complexes in plants [76], vertebrates [13], and in *Drosophila* [1, 55]. Why are *hsp* mRNAs translationally competent during heat shock? The 5′ leader of the maize *hsp70* gene was sufficient and necessary for reporter mRNAs containing this sequence to escape the global repression of translation during heat shock [85]. Interestingly however, a survey of the 5' leaders from 140 eukaryotic *hsp* mRNAs from animals, plants, and fungi found that they were not significantly different from non-*hsp* mRNAs in leader length, A+T content, the sequence context around the initiation codon, or the presence of upstream, nonfunctional AUGs [46].

The reprogramming of translation upon thermal stress correlates with modifications in the translational machinery. Changes in phosphorylation of several initiation factors in response to heat-shock have been observed in animal cells [19]. Two of the best studied examples are (1) the dephosphorylation of eIF-4F in HeLa [18] and Ehrlich cells [54] resulting in impaired cap-binding activity and reduced eIF-4F association with the cap [54, 83, 108] and (2) the phosphorylation of eIF-2α which prevents recycling of this factor and thereby inhibits initiator Met-tRNA binding to the scanning 40S ribosomal subunit [73]. No significant changes in the phosphorylation status of eIF-2 was observed in heat-shocked wheat coleoptiles (D. Gallie and K. Browning, unpublished observations), data suggesting that the role that eIF-2 plays in the observed changes in translational activity in animal cells may differ significantly in plants.

Animals and plants may also differ in the regulation of eIF-4A protein activity. Although this factor does not appear to be phosphorylated in animals, phosphorylation does occur in hypoxically treated maize roots [103] and correlates with the global repression of translation [90]. Phosphorylation of eIF-4A was also observed in heat-shocked wheat coleoptiles to the same extent observed following hypoxia (D. Gallie and K. Browning, unpublished observations), observations suggesting that the phosphorylation of eIF-4A may be part of the programmatic response to stress in general.

eIF-4B is present in 8–10 isoforms in animals [19] and plants (D. Gallie and K. Browning, unpublished observations). The acidic cluster (4–6 isoforms) represents phosphorylated isoforms whereas the 4 isoforms within the basic cluster result from an as yet undetermined modification [21]. Dephosphorylation of eIF-4B correlates with a reduction in its activity. In both plants and animals, eIF-4B undergoes dephosphorylation upon thermal stress [19; D. Gallie and K. Browning, unpublished observations]. Therefore, the modifications of the translational machinery in plants as a consequence of heat-shock appear to differ in several ways from that observed in animal cells.

Hypoxia, like heat stress, results in the disassembly of pre-existing polysomes followed by the polysomal recruitment of mRNAs encoding anaerobic proteins [23, 90]. Hypoxia caused an increase in expression from the alcohol dehydrogenase-1 gene at both the transcriptional and the translational levels concomitant with a decrease in the size of polysomes containing non-hypoxically-induced mRNAs such as that of the mitochondrial adenine nucleotide translocator gene [23].

In the dessication-tolerant moss, the reprogramming of protein synthesis following water stress is mediated partly at the level of mRNA recruitment into polysomes [77, 78, 79, 92]. Although actively translating mRNA can be trapped in polysomes following rapid dessication in this species, significant changes in polysomal-associated mRNA occur after rehydration even though the population of cellular mRNAs does not substantially change.

Environmental stimuli can also alter the transcriptional initiation site or the selection of the polyadenylation site. Of the two initiation sites used in the transcription of the tomato phenylalanine-lyase gene, an increase at the 3'-distal site was observed following stresses such as wounding and exposure to light [57]. The three polyadenylation sites of the nuclear-encoded, chloroplast RNA-binding protein in *Mesembryanthemum crystallinum* are differentially utilized upon salt stress resulting in a shift in the relative amounts of each form of mRNA [7]. Changes in the transcription initiation or polyadenylation sites affects the length and primary sequence of the 5' leader and 3'-untranslated region. Such changes have the potential to affect gene expression but whether they do in the case of these genes remains to be determined experimentally.

Conclusions

Even though there are now many genes known to be subject to translational control, we have only begun to appreciate the full extent of this type of gene regulation in plants. Mechanistic patterns are beginning to emerge, such as the co-dependency between the termini as the basis for efficient translation regardless if the mRNA contains a cap, poly(A) tail, or alternative regulatory elements. A second example is the impact of upstream open reading frames to modulate translation. Nevertheless, many additional examples of translational control are needed if we are to gain a complete understanding of the ways in which mechanisms present in unrelated genes are similar and those ways in which mechanisms present in related genes differ. Moreover, our understanding of the details of the translational regulatory mechanisms is still in its infancy. The identification and characterization of those specific RNA-binding proteins that mediate the regulation from an RNA element will be critical to elucidate their role in translational control. What is apparent already is that the control of translation constitutes an important component of gene expression in plants.

Acknowledgements

I thank those colleagues who provided unpublished results for this review. The support of the USDA (93–37100–8939, 93–37301–9124, and 95–37100–1618) is gratefully acknowledged.

References

1. Arrigo A-P: Cellular localization of HSP23 during *Drosophila* development and following subsequent heat shock. Devel Biol 122: 39–48 (1987).

2. Basso J, Dallaire P, Charest PJ, Devantier Y, Laliberte J-F: Evidence for an internal ribosome entry site within the 5′ non-translated region of turnip mosaic potyvirus RNA. J Gen Virol 75: 3157–3165 (1994).

3. Bienz M: Developmental control of the heat shock response in *Xenopus*. Proc Natl Acad Sci 81: 3138–3142 (1984).

4. Bienz M, Gurdon JB: The heat shock response in *Xenopus* oocytes is controlled at the translational level. Cell 29: 811–819 (1982).

5. Bond UM, YarioTA, Steitz JA: Multiple processing-defective mutations in a mammalian histone pre-mRNA are suppressed by compensatory changes in U7 RNA both in vivo and in vitro. Genes Devel 5: 1709–1722 (1991).

6. Brault V, Miller WA: Translational frameshifting mediated by a viral sequence in plant cells. Proc Natl Acad Sci USA 89: 2262–2266 (1992).

7. Breiteneder H, Michalowski CB, Bohnert HJ: Environmental stress-mediated differential 3′ end formation of chloroplast RNA-binding protein transcripts. Plant Mol Biol 26: 833–849 (1994).

8. Browning KS, Lax SR, Humphreys J, Ravel JM, Jobling SA, Gehrke L: Evidence that the 5′-untranslated leader of mRNA affects the requirement for wheat germ initiation factors 4A, 4F, and 4G. J Biol Chem 263: 9630–9634 (1988).

9. Capasso O, Bleecker GC, Heintz N: Sequences controlling histone H4 mRNA abundance. EMBO J 6: 1825–1831 (1987).

10. Carrington JC, Freed DD: Cap-independent enhancement of translation by a plant potyvirus 5′ nontranslated region. J Virol 64: 1590–1597 (1990).

11. Chaboute M-E, Chaubet N, Clement B, Gigot C, Philipps G: Polyadenylation of histone H3 and H4 mRNAs in dicotyledonous plants. Gene 71: 217–223 (1988).

12. Chaubet N, Chaboute M-E, Clement B, Ehling M, Philipps G, Gigot C: The histone H3 and H4 mRNAs are polyadenylated in maize. Nucl Acids Res 16: 1295–1304 (1988).

13. Collier NC, Schlesinger MJ: The dynamic state of heat shock proteins in chicken embryo fibroblasts. J Cell Biol 103: 1495–1507 (1986).

14. Damiani RD, Wessler S: An upstream open reading frame represses expression of *Lc*, a member of the *R/B* family of maize transcriptional activators. Proc Natl Acad Sci USA 90: 8244–8248 (1993) .

15. Danthinne X, Seurinck J, Meulewaeter F, Van Montagu M, Cornelissen M: The 3′ untranslated region of satellite tobacco necrosis virus RNA stimulates translation in vitro. Mol Cell Biol 13: 3340–3349 (1993).

16. De Haan P, Gielen JJL, Prins M, Wijkamp IG, van Schepen A, Peters D, van Grinsven MQJM, Goldbach R: Characterization of RNA-mediated resistance to tomato spotted wilt virus in transgenic tobacco plants. Bio/Technology 10: 1133–1137 (1992).

17. Dowson Day MJ, Ashurst JL, Mathias SF, Watts JW, Wilson TMA, Dixon RA: Plant viral leaders influence expression of a reporter gene in tobacco. Plant Mol Biol 23: 97–109 (1993).

18. Duncan R, Milburn SC, Hershey JWB: Regulated phosphorylation and low abundance of HeLa cell initiation factor eIF-4F suggest a role in translational control. J Biol Chem 262: 380–388 (1987).

19. Duncan R, Hershey JWB: Protein synthesis and protein phosphorylation during heat stress, recovery, and adaptation. J Cell Biol 109: 1467–1481 (1989).

20. Eckner R, Ellmeier W, Birnstiel ML: Mature mRNA 3′ end formation stimulates RNA export from the nucleus. EMBO J 10: 3513–3522 (1991).

21. Fabbri BJ, Hershey JWB: Comparison of in vivo and in vitro phosphorylation patterns of eIF-4B. In: Translational Control, p. 66. Cold Spring Harbor Laboratory, Cold Spring Harbor, NY (1996).

22. Fabry S, Muller K, Lindauer A, Park PB, Cornelius T, Schmitt R: The organization, structure, and controlling elements of *Chlamydomonas* histone genes reveal features linking plant and animal genes. Curr Genet 28: 333–345 (1995).

23. Fennoy SL, Bailey-Serres J: Post-transcriptional regulation of gene expression in oxygen-deprived roots of maize. Plant J 7: 287–295 (1995).

24. Fletcher L, Corbin SD, Browning KS, Ravel JM: The absence of a m⁷G cap on β-globin mRNA and alfalfa mosaic virus RNA 4 increases the amounts of initiation factor 4F required for translation. J Biol Chem 265: 19582–19587 (1990).

25. Fütterer J, Hohn T: Translation of a polycistronic mRNA in the presence of the cauliflower mosaic virus transactivator protein. EMBO J 10: 3887–3896 (1991).

26. Gallie DR: The cap and poly(A) tail function synergistically to regulate mRNA translational efficiency. Genes Devel 5: 2108–2116 (1991).

27. Gallie DR, Caldwell C, Pitto L: Heat shock disrupts cap and poly(A) tail function during translation and increases mRNA stability of introduced reporter mRNA. Plant Physiol 108: 1703–1713 (1995).

28. Gallie DR, Feder JN, Schimke RT, Walbot V: Post-transcriptional regulation in higher eukaryotes: the role of the reporter gene in controlling expression. Mol Gen Genet 228: 258–264 (1991).

29. Gallie DR, Kobayashi M: The role of the 3′-untranslated region of non-polyadenylated plant viral mRNAs in regulating translational efficiency. Gene 142: 159–165 (1994).

156

29a. Gallie DR, Lewis NJ, Marzluff WF: The histone 3'-terminal stem-loop is necessary for translation in Chinese hamster ovary cells. Nucl Acids Res, in press (1996).

30. Gallie DR, Lucas WJ, Walbot V: Visualizing mRNA expression in plant protoplasts: factors influencing efficient mRNA uptake and translation. Plant Cell 1: 301–311 (1989).

31. Gallie DR, Sleat DE, Watts JW, Turner PC, Wilson TMA: The 5'-leader sequence of tobacco mosaic virus RNA enhances the expression of foreign gene transcripts in vitro and it in vivo. Nucl Acids Res 15: 3257–3273 (1987).

32. Gallie DR, Sleat DE, Watts JW, Turner PC, Wilson TMA: A comparison of eukaryotic viral 5'-leader sequences as enhancers of mRNA expression in vivo. Nucl Acids Res 15: 8693–8711 (1987).

33. Gallie DR, Tanguay R: Poly(A) binds to initiation factors and increases cap-dependent translation in vitro. J Biol Chem 269: 17166–17173 (1994).

34. Gallie DR, Tanguay R, Leathers V: The tobacco etch viral 5' leader and poly(A) tail are functionally synergistic regulators of translation. Gene 165: 233 (1995).

35. Gallie DR, Walbot V: RNA pseudoknot domain of tobacco mosaic virus can functionally substitute for a poly(A) tail in plant and animal cells. Genes Devel 4: 1149–1157 (1990).

36. Gallie DR, Walbot V: Identification of the motifs within the tobacco mosaic virus 5'-leader responsible for enhancing translation. Nucl Acids Res 20: 4631–4638 (1992).

37. Gallie DR, Young TE: The regulation of gene expression in transformed maize aleurone and endosperm protoplasts. Plant Physiol 106: 929–939 (1994).

38. Gielen JJL, de Haan P, Kool AJ, Peters D, van Grinsven MQJM, Goldbach RW: Engineered resistance to tomato spotted wilt virus, a negative-strand RNA virus. Bio/technology 9: 1363–1367 (1991).

39. Harris ME, Bohni R, Schneiderman MH, Ramamurthy L, Schumperli D, Marzluff WF: Regulation of histone mRNA in the unperturbed cell cycle: evidence suggesting control at two posttranscriptional steps. Mol Cell Biol 11: 2416–2424 (1991).

40. Hentschel CC, Birnstiel ML: The organization and expression of histone gene families. Cell 25:301–313 (1981).

41. Hinnebusch AG: Translational control of GCN4: an in vivo barometer of initiation-factor activity. Trends Biochem Sci 19: 409–414 (1994).

42. Jagus R, Anderson WF, Safer B: The regulation of initiation of mammalian protein synthesis. Prog Nucl Acids Res Mol Biol 25: 127–185 (1981).

43. Jobling SA, Cuthbert CM, Rogers SG, Fraley RT, Gehrke L: In vitro transcription and translational efficiency of chimeric SP6 messenger RNAs devoid of 5' vector nucleotides. Nucl Acids Res 16: 4483–4498 (1988).

44. Jobling SA, Gehrke L: Enhanced translation of chimaeric messenger RNAs containing a plant viral untranslated leader sequence. Nature 325: 622–625 (1987).

45. Joshi CP: An inspection of the domain between putative TATA box and translation start site in 79 plant genes. Nucl Acids Res 16: 6643–6653 (1987).

46. Joshi CP, Nguyen HT: 5' untranslated leader sequences of eukaryotic mRNAs encoding heat shock induced proteins. Nucl Acids Res 23: 541–549 (1995).

47. Key JL, Lin CY, Chen YM: Heat shock proteins of higher plants. Proc Natl Acad Sci USA 78: 3526–3530 (1981).

48. Kim J-K, Hollingsworth MJ: Localization of in vivo ribosome pause sites. Anal Biochem 206: 183–188 (1992).

49. Kim J, Klein PG, Mullet JE: Ribosomes pause at specific sites during synthesis of membrane-bound chloroplast reaction center protein D1. J Biol Chem 266: 14931–14938 (1991).

50. Koning AJ, Tanimoto EY, Kiehne K, Rost T, Comai L: Cell-specific expression of plant histone H2A genes. Plant Cell 3: 657–665 (1991).

51. Kozak M: Influences of mRNA secondary structure on initiation by eukaryotic ribosomes. Proc Natl Acad Sci USA 83: 2850–2854 (1986).

52. Kozak M: Circumstances and mechanisms of inhibition of translation by secondary structure in eucaryotic mRNAs. Mol Cell Biol 9: 5134–5142 (1989).

53. Kozak M: A short leader sequence impairs the fidelity of initiation by eukaryotic ribosomes. Gene Expression 1: 111–115 (1991).

54. Lamphear BJ, Panniers R: Heat shock impairs the interaction of cap-binding protein complex with 5' mRNA cap. J Biol Chem 266: 2789–2794 (1991).

55. Leicht BG, Biessmann H, Palter KB, Bonner JJ: Small heat shock proteins of Drosophila associate with the cytoskeleton. Proc Natl Acad Sci USA 83: 90–94 (1986).

56. Leathers V, Tanguay R, Kobayashi M, Gallie DR: A phylogenetically conserved sequence within viral 3' untranslated RNA pseudoknots regulates translation. Mol Cell Biol 13: 5331–5347 (1993).

57. Lee SW, Heinz R, Robb J, Nazar RN: Differential utilization of alternate initiation sites in a plant defense gene responding to environmental stimuli. Eur J Biochem 226: 109–114 (1994).

58. Levine BJ, Chodchoy N, Marzluff WF, Skoultchi AI: Coupling of replication type histone mRNA levels to DNA synthesis requires the stem-loop sequence at the 3' end of the mRNA. Proc Natl Acad Sci USA 84: 6189–6193 (1987).

59. Levis C, Astier-Manifacier S: The 5' untranslated region of PVY RNA even located in an internal position enables initiation of translation. Virus Genes 7: 367–379 (1993).

60. Lingelbach K, Dobberstein B: An extended RNA/RNA duplex structure within the coding region of mRNA does not block translational elongation. Nucl Acids Res 16: 3405–3414 (1988).

61. Lindquist S: The heat shock response. Annu Rev Biochem 55: 1151–1191 (1986).

62. Lohmer S, Maddaloni M, Motto M, Salamini F, Thompson RD: Translation of the mRNA of the maize transcriptional activator Opaque-2 is inhibited by upstream open reading frames present in the leader sequence. Plant Cell 5: 65–73 (1993).

63. Marzluff WF: Histone 3' ends: essential and regulatory functions. Gene Exp 2: 93–97 (1992).

64. Melin L, Soldati D, Mital R, Streit A, Schumperli D: Biochemical demonstration of complex formation of histone pre-mRNA with U7 small nuclear ribonucleoprotein and hairpin binding factors. EMBO J 11: 691–697 (1992).

65. Michelet B, Lukaszewicz M, Dupriez V, Boutry M: A plant plasma membrane protein-ATPase gene is regulated by development and environment and shows signs of a translational regulation. Plant Cell 6: 1375–1389 (1994).

66. Millar AJ, Short SR, Hiratsuka K, Chua N-C, Kay SA: Firefly luciferase as a reporter of regulated gene expression in higher plants. Plant Mol Biol Rep 10: 324–337 (1992).

67. Millar AJ, Carre IA, Strayer CA, Chua N-C, Kay SA: Circadian clock mutants in Arabidopsis identified by luciferase imaging. Sci 267: 1161–1163 (1995).

68. Mowry KL, Oh R, Steitz JA: Each of the conserved sequence elements flanking the cleavage site of mammalian histone pre-mRNAs has a distinct role in the 3'-end processing reaction. Mol Cell Biol 9: 3105–3108 (1989).

69. Mowry KL, Steitz JA: Both conserved signals on mammalian histone pre-mRNAs associate with small nuclear ribonucleoproteins during 3'-end formation in vitro. Mol Cell Biol 7: 1663–1672 (1987).

70. Mowry KL, Steitz JA: Identification of the human U7 snRNP as one of several factors involved in the 3'-end maturation of histone premessenger RNA's. Science 238: 1682–1687 (1987).

71. Muller K, Lindauer A, Bruderlein M, Schmitt R: Organization and transcription of *Volvox* histone-encoding genes: similarities between algal and animal genes. Gene 93: 167–175 (1990).

72. Muller K, Schmitt R: Histone genes of *Volvox carteri*: DNA sequence and organization of two H3-H4 gene loci. Nucl Acids Res16:4121–4136 (1988).

73. Murtha-Riel P, Davies MV, Scherer BJ, Choi S-Y, Hershey JWB, Kaufman RJ: Expression of a phosphorylation-resistant eukaryotic initiation factor 2 α-subunit mitigates heat shock inhibition of protein synthesis. J Biol Chem 268: 12946–12951 (1993).

74. Nicolaisen M, Johansen E, Poulsen GP, Borkhardt B: The 5' untranslated region from pea seedborne mosaic potyvirus RNA as a translational enhancer in pea and tobacco protoplasts. FEBS Lett303: 169–172 (1992).

75. Nover L, Scharf K-D, Neumann D: Formation of cytoplasmic heat shock granules in tomato cell cultures and leaves. Mol Cell Biol 3: 1648–1655 (1983).

76. Nover L, Scharf K-D, Neumann D: Cytoplasmic heat shock granules are formed from precursor particles and are associated with a specific set of mRNAs. Mol Cell Biol 9: 1298–1308 (1989).

77. Oliver MJ, Bewley JD: Plant desiccation and protein synthesis. V. Stability of poly(A)$^-$ and poly(A)$^+$ RNA during desiccation and their synthesis upon rehydration in the desiccation-tolerant moss *Tortula ruralis* and the intolerant moss *Cratoneuron filicinum*. Plant Physiol 74: 917–922 (1984).

78. Oliver MJ, Bewley JD: Plant desiccation and protein synthesis. VI. Changes in protein synthesis elicited by desiccation of the moss *Tortula ruralis* are effected at the translational level. Plant Physiol 74: 923–927 (1984).

79. Oliver MJ: Influence of protoplasmic water loss on the control of protein synthesis in the desiccation-tolerant moss *Tortula ruralis*. Plant Physiol 97: 1501–1511 (1991).

80. Pandey NB, Marzluff WF: The stem-loop Structure at the 3' end of histone mRNA is necessary and sufficient for regulation of histone mRNA stability. Mol Cell Biol 7: 4557–4559 (1987).

81. Pandey NB, Sun J-H, Marzluff WF: Different complexes are formed on the 3' end of histone mRNA with nuclear and polysomal proteins. Nucl Acids Res 19: 5653–5659 (1991).

82. Pandey NB, Williams AS, Sun J-H, Brown VD, Bond U, Marzluff WF: Point mutations in the stem-loop at the 3' end of mouse histone mRNA reduce expression by reducing the efficiency of 3' end formation. Mol Cell Biol 14: 1709–1720 (1994).

83. Panniers R, Stewart EB, Merrick WC, Henshaw EC: Mechanism of inhibition of polypeptide chain initiation in heat-shocked Ehrlich cells involves reduction of eukaryotic initiation factor 4F activity. J Biol Chem 260: 9648–9653 (1985).

84. Pelletier J, Sonenberg N: Insertion mutagenesis to increase secondary structure within the 5' noncoding region of a eukaryotic mRNA reduces translational efficiency. Cell 40: 515–526 (1985).

85. Pitto L, Gallie DR, Walbot V: The role of the leader sequence during thermal repression of translation in maize, tobacco, and carrot protoplasts. Plant Physiol 100: 1827–1833 (1992).

86. Pooggin MM, Skryabin KG: The 5'-untranslated leader of potato virus X RNA enhances the expression of a heterologous gene in vivo. Mol Gen Genet 234: 329–331 (1992).

87. Putterill JJ, Gardner RC: Initiation of translation of the β-glucuronidase reporter gene at internal AUG codons in plant cells. Plant Sci 62: 199–205 (1989).

88. Robbins E, Borun TW: The cytoplasmic synthesis of histones in HeLa cells and its temporal relationship to DNA replication. Proc Natl Acad Sci USA 58: 1977–1983 (1967).

89. Rozen F, Edery I, Meerovitch K, Dever TE, Merrick WC, Sonenberg N: Bidirectional RNA helicase activity of eucaryotic translation initiation factors 4A and 4F. Mol Cell Biol 10: 1134–1144 (1990).

90. Sachs MM, Freeling M, Okimoto R: The anaerobic proteins of maize. Cell 20: 761–767 (1980).

91. Scholthof HB, Wu FC, Gowda S, Shepard RJ: Regulation of cauliflower gene expression and the involvement of *cis*-acting elements on both viral transcripts. Virology 190: 403–412 (1992).

92. Scott HB, Oliver MJ: Accumulation and polysomal recruitment of transcripts in response to dessication and rehydration of the moss *Tortula ruralis*. J Exp Bot 45: 577–583 (1994).

93. Shaul O, Galili G: Increased lysine synthesis in tobacco plants that express high levels of bacterial dihydrodipicolinate synthase in their chloroplasts. Plant J 2: 203–209 (1992).

94. Stollar NE, Kim J-K, Hollingsworth MJ: Ribosomes pause during the expression of the large ATP synthase gene cluster in spinach chloroplasts. Plant Physiol 105: 1167–1177 (1994).

95. Storti RV, Scott MP, Rich A, Pardue ML: Translational control of protein synthesis in response to heat shock in *D. melanogaster* cells. Cell 22: 825–834 (1980).

96. Tacke E, Prufer D, Salamini F, Rohde W: Characterization of a potato leafroll luteovirus subgenomic RNA: differential expression by internal translation initiation and UAG suppression. J Gen Virol 71: 2265–2272 (1990).

96a. Tanguay RL, Gallie DR: Isolation and characterization of the 102-kilodalton RNA-binding protein that binds to the 5' and 3' translational enhancers of tobacco mosaic virus RNA. J Biol Chem, in press (1996).

97. Thomas AAM, Ter Haar E, Wellink J, Voorma HO: Cowpea mosaic virus middle component RNA contains a sequence that allows internal binding of ribosomes and that requires eukaryotic initiation factor 4F for optimal translation. J Virol 65: 2953–2959 (1991).

98. Timmer RT, Benkowski LA, Schodin D, Lax SR, Metz AM, Ravel JM, Browning KS: The 5' and 3' untranslated regions of satellite tobacco necrosis virus RNA affect translational efficiency and dependence on a 5' cap structure. J Biol Chem 268: 9504–9510 (1993).

99. Turner R, Bate N, Twell D, Foster GD: Analysis of a translational enhancer upstream from the coat protein open reading frame of potato virus S. Arch Virol 134: 321–333 (1994).

100. Vasserot AP, Schaufele FJ, Birnstiel ML: Conserved terminal hairpin sequences of histone mRNA precursors are not involved in duplex formation with the U7 RNA but act as a target site for a distinct processing factor. Proc Natl Acad Sci USA 86: 4345–4349 (1989).

158

101. Verver J, LeGall O, van Kammen A, Wellink J: The sequence between nucleotides 161 and 512 of cowpea mosaic virus M RNA is able to support internal initiation of translation *in vitro*. J Gen Virol 72: 2339–2345 (1991).

102. Wang S, Miller WA: A sequence located 4.5 to 5 kilobases from the 5′ end of the barley yellow dwarf virus (PAV) genome strongly stimulates translation of uncapped mRNA. J Biol Chem 270: 13446–13452 (1995).

103. Webster C, Gaut RL, Browning KS, Ravel JM, Roberts JKM: Hypoxia enhances phosphorylation of eukaryotic initiation factor 4A in maize root tips. J Biol Chem 266: 23341–23346 (1991).

104. Williams AS, Ingledue III TC, Kay BK, Marzluff WF: Changes in the stem-loop at the 3′ terminus of histone mRNA affects its nucleocytoplasmic transport and cytoplasmic regulation. Nucl Acids Res 22: 4660–4666 (1994).

105. Williams AS, Marzluff WF: The sequence of the stem and flanking sequences at the 3′ end of histone mRNA are critical determinants for the binding of the stem-loop binding protein. Nucl Acids Res 23: 654–662 (1995).

106. Wolin S, Walter P: Ribosome pausing and stacking during translation of a eukaryotic mRNA. EMBO J 7: 3559–3569 (1988).

107. Wu S-C, Gyorgyey J, Dudits D: Polyadenylated H3 histone transcripts and H3 histone variants in alfalfa. Nucl Acids Res 17: 3057–3063 (1989).

108. Zapata JM, Maroto FG, Sierra JM: Inactivation of mRNA cap-binding Protein complex in *Drosophila melanogaster* embryos under heat shock. J Biol Chem 266: 16007–16014 (1991).

109. Zelenina DA, Kulaeva OI, Smirnyagina EV, Solovyev AG, Miroshnichenko NA, Fedorkin ON, Rodionova NP, Morozov SY, Atabekov JG: Translation enhancing properties of the 5′-leader of potato virus X genomic RNA. FEBS Lett 296: 267–270 (1992).

Plant Molecular Biology **32**: 159–189, 1996.
© 1996 *Kluwer Academic Publishers.*

Translation in plants – rules and exceptions

Johannes Fütterer[1],* & Thomas Hohn[2]
[1]*Institute of Plant Sciences, ETHZ, Universitätstrasse 2, CH–8092 Zürich, Switzerland (*author for correspondence);* [2]*Friedrich Miescher Institute, PO Box 2543, CH–4002, Basel, Switzerland*

Key words: Plant virus, leader, caulimovirus, luteovirus, frameshift, readthrough, internal initiation

Contents

Abstract	159	Short ORFs in the 5' leader sequence	165
Introduction	159	RNAs with more than one coding ORF	168
The rules	160	Viral transactivation of polycistronic	
Cap-dependent ribosome binding and		translation	170
translation initiation	160	Leader sequences and	
Cap-dependence	160	cap-independent, internal ribosome	
Scanning	161	entry	171
Translation-initiation site selection	162	Effect of the 3'-untranslated region	173
Translation elongation	162	Cotranslational effects	174
Translation termination	163	Translation under special conditions	178
The exceptions	163	Conclusions	178
Variations of initiation site selection	163	Acknowledgements	179
Non-AUG start codons	163	References	179
RNAs with multiple translation			
initiation sites	164		

Abstract

Translation processes in plants are very similar to those in other eukaryotic organisms and can in general be explained with the scanning model. Particularly among plant viruses, unconventional mRNAs are frequent, which use modulated translation processes for their expression: leaky scanning, translational stop codon readthrough or frameshifting, and transactivation by virus-encoded proteins are used to translate polycistronic mRNAs; leader and trailer sequences confer (cap-independent) efficient ribosome binding, usually in an end-dependent mechanism, but true internal ribosome entry may occur as well; in a ribosome shunt, sequences within an RNA can be bypassed by scanning ribosomes. Translation in plant cells is regulated under conditions of stress and during development, but the underlying molecular mechanisms have not yet been determined. Only a small number of plant mRNAs, whose structure suggests that they might require some unusual translation mechanisms, have been described.

Introduction

Plant cells harbor three different translation machineries: the eukaryotic ribosomes in the cytoplasm and systems for translation in mitochondria and plastids. The transcriptional and post-transcriptional regulation of chloroplast gene expression has been reviewed recently [233, 282]. This review deals exclusively with translation events and mRNAs in the cytoplasm and concentrates on *cis*-active sequences that influence efficiency and fidelity of translation. Although the translation of only a few nuclear plant genes has been analyzed so far, it is very safe to assume that translation of plant mRNAs in general proceeds according to the rules established for other eukaryotic systems. The 'exceptions' described mostly apply to viral RNAs encoded by cytoplasmic RNA viruses or by plant pararetroviruses in the nucleus.

The rules

Most of the different translation processes have been elaborated in mammalian cell systems, in extracts derived from such cells, and more recently in yeast cells. However, an extract derived from wheat germ has been used in many experiments, in parallel with the most widely used extract of rabbit reticulocytes. In addition, in recent years a variety of experiments have been performed with transiently or stably transformed plants or plant cells. It was concluded from these experiments that most translational features are similar in plant and animal cells ([127, 320] for review). Some plant translation factors have been characterized and the genes cloned (see review by Browning, this issue). Some of the data suggesting general similarities between translation in mammalian and in plant cells are discussed below.

Cap-dependent ribosome binding and translation initiation

The vast majority of eukaryotic mRNAs are capped at the 5' end, polyadenylated at the 3' end, and monocistronic, i.e. only one open reading frame (ORF) is translated into a protein per mRNA molecule. Translation of such mRNAs is well explained by the scanning model, which states that 40S ribosomal subunits – associated with eukaryotic initiation factors (eIF) 1A, 3 and the 'ternary complex' of eIF2, GTP and initiator methionine tRNA – bind near the capped 5' end of the mRNA, which in turn is associated with a set of factors of the eIF4 group. The function of the members of the eIF4 group seems to be to prepare the RNA for ribosome binding by removing hindering secondary structures and promoting the linear migration of the bound ribosomes along the RNA in search of a translation start site (scanning). At such a start site, usually an AUG codon, the 60S ribosomal subunit joins on and the resulting 80S subunit begins the synthesis of

the encoded polypeptide [162, 202, 208-210, 238, 239, 278, 279, 331].

In the original scanning model, the 40S ribosomal subunit together with a number of initiation factors was defined as the scanning entity [202]. The precise order of the association of factors and ribosomal subunits with each other and with the mRNA is still disputed, and it is possible that more than one pathway can lead to translation initiation, depending on the prevalent conditions [76, 324]. Particularly, ribosome-independent RNA binding of eIF2 has been observed [26, 188]. Alternative hypotheses with factors of the eIF4 group [314] or with eIF2 [332] as the scanning factor have been proposed. According to these latter models, the 40S subunit binds to the mRNA at a later step (possibly only directly at the AUG start codon). We feel that it is more difficult to explain some of the features of eukaryotic translation with these models than with the 40S ribosome scanning model, particularly some aspects of reinitiation of translation at downstream ORFs (see below). In the following, therefore, we discuss translation mainly in terms of the original scanning model.

Cap-dependence

The direct influence of the cap in plant translation systems has been demonstrated in many *in vitro* translation experiments with wheat germ extracts and by transfection of protoplasts with *in vitro* synthesized RNAs containing or lacking a cap structure [16, 48, 53, 102, 120, 123, 240, 327, 334, 337]. The competitivity of AMV (for virus acronyms, see Table 1) RNAs for the translation machinery was directly related to cap accessibility [134]. The cap dependence of *in vitro* systems in general depends on the extract used, on the precise reaction conditions [202, 224], on the concentrations of the different initiation factors [102, 324] and on the specific RNAs ([218, 217, 334]; see also the review by Gallie, this issue). Stimulatory effects of the cap on expression in the range of 2- to 100-fold were observed. The wheat germ extract is generally more cap-dependent than the rabbit reticulocyte lysate [224]. *In vivo*, the effect of the cap structure was partially to stabilize the RNA but the main influence was clearly on translation efficiency [123, 124].

The importance of the RNA 5' end for translation was also indicated by the strong inhibitory effect of an inserted stem structure (-125 kJ/mol) at the immediate 5' end of an RNA encoded by a transfected nuclear gene [112, 114].

Table 1. Virus acronyms.

AMV	Alfalfa mosaic	*Alfalfa*
BMV	Brome mosaic	*Bromo*
BWYV	Beet western yellows	*Luteo*
BWYV-ST9	BWYV associated RNA	*Satellite*
BYDV	Barley yellow dwarf	*Luteo*
BYDV-RMV	Barley yellow dwarf	*Luteo*
BYDV-RPV	Barley yellow dwarf	*Luteo*
BYV	Beet yellows	*Clostero*
CaMV	Cauliflower mosaic	*Caulimo*
CarMV	Carnation mottle	*Carmo*
CCFV	Cardamine chlorotic fleck	*Carmo*
CERV	Carnation etched ring	*Caulimo*
CfMV	Cocksfoot mottle	*Sobemo*
CNV	Cucumber necrosis	*Tombus*
CPMV	Cowpea mosaic	*Como*
CRSV	Carnation ringspot	*Diantho*
CTV	Citrus tristeza	*Clostero*
CyRSV	Cymbidium ringspot	*Tombus*
EMV	Eggplant mosaic	*Tymo*
FMV	Figwort mosaic	*Caulimo*
KYMV	Kennedya yellow mosaic	*Tymo*
MCMV	Maize chlorotic mottle	*Sobemo*
MNSV	Melon necrotic spot	*Carmo*
OYMV	Osirio yellow mosaic	*Tymo*
PCV RNA1	Peanut clump	*Sobemo*
PEMV	Pea enation mosaic	*Enamo*
PLRV	Potato leafroll	*Luteo*
PClSV	Peanut chlorotic streak	*Caulimo*
PPV	Plum pox	*Poty*
PVM	Potato M	*Carla*
RCNMV	Red clover necrotic mosaic	*Diantho*
RDV	Rice dwarf	*Phytoreo*
RTBV	Rice tungro bacilliform	*Badna*
RYMV	Rice yellow mottle	*Sobemo*
SBMV	Southern bean mosaic	*Sobemo*
SBWMV	Soil-borne wheat mosaic	*Sobemo*
SDV	Soybean dwarf	*Luteo*
SoClM	Soybean chlorotic mottle	*Caulimo*
STMV	Satellite tobacco mosaic	*Satellite*
STNV	Satellite tobacco necrosis	*Satellite*
TBSV	Tomato bushy stunt	*Tombus*
TCV	Turnip crinkle	*Carmo*
TEV	Tobacco etch	*Poty*
TMV	Tobacco mosaic	*Tobamo*
TNV	Tobacco necrosis	*Necro*
TRV	Tobacco rattle	*Tobra*
TuMV	Turnip mosaic	*Poty*
TYMV	Turnip yellow mosaic	*Tymo*

Scanning

The ribosome migrates along the RNA from the first point of ribosome association to the translation initiation site. Evidence for such a scanning process is obtained from the of insertion of strong secondary structures or additional initiation sites into the scanned region. The scanning ribosome or its associated factors are apparently unable to melt stem structures of a given stability [14, 199]. The required minimal stability has not been determined for plant cells, but hairpins with a free energy below -180 kJ/mol [16, 112, 114, 204] were sufficient to significantly inhibit translation of downstream ORFs in a variety of plant protoplast and *in vitro* systems. Similarly, the introduction of alternative initiation sites reduces downstream translation because the ribosome 'wastes' its initiation capacity at these sites (see below). Both findings show that the ribosome encounters the inhibiting features on the way to the normal initiation site, i.e. it most likely migrates along the RNA. The region upstream of the main initiation site is termed the 'leader sequence' or '5′ untranslated region', although the latter designation could be misleading since in some cases such a region might actually contain short ORFs which may be translated (see below).

Only a small number of leader sequences of nuclear plant genes have been systematically compared [186]. The analysis is complicated by the frequent lack of precise RNA data for plant genes. However, it can be concluded so far that the leader sequences of most plant genes (like those of other eukaryotic genes) are shorter than 100 nucleotides, lack strong secondary structures and short ORFs and are thus suited to the ribosome-scanning mechanism. A few exceptions will be discussed below. The leader sequences of many RNA viruses seem to have primary or secondary structures that allow a particularly efficient translation, often even in the absence of a cap structure. These features are discussed in detail by Gallie (this volume). In general, longer leaders (e.g. 80 nucleotides) result in higher translation rates than shorter leaders (e.g. 10 nucleotides). The exact difference depends on the assay system and salt conditions [201]; in transfected plant protoplasts an approximately twofold effect was reported [126], while the effects in the reticulocyte lysate can be greater [206]. However, mRNAs with leader sequences less than 10 nucleotides long can still be efficiently translated in plant cells (G. Chen *et al.*, unpublished observation) and are actually found in some coat protein mRNAs of RNA viruses (e.g. BMV).

Translation-initiation site selection

Upon initiation, the ternary complex dissociates from the ribosome as eIF2-GDP. Since, according to the scanning model, the ternary complex is delivered to the initiation codon together with the ribosome, ribosomes can initiate translation only once unless a new ternary complex can be recruited. Due to the scanning process, this one-and-only initiation site is the one located closest to the 5' end of the RNA. Translation usually begins at an AUG codon, and the efficiency of AUG codon recognition is modulated by the sequence context of this codon [198]. Sequence compilation of plant translation initiation regions has suggested that the most frequent (and therefore regarded as optimal) AUG context for plant genes is similar to that of mammalian genes (AACAATGGC [55, 186, 227]). The most crucial positions in both cases are a purine at position − 3 and a guanine at position +4 (where the A of the AUG is +1). For mammalian cells, an influence of positions +5 and +6 has been recently documented [32, 144]; other positions seem to be less important. Start codons that deviate from the optimal context at one or more of the crucial positions may be recognized less efficiently and allow the passage of ribosomes to further downstream start codons (leaky scanning). The influence of sequence context on initiation in plant cells has been directly verified by mutagenesis studies *in vitro*, in plant protoplasts, and in transgenic plant cells [6, 77, 87, 146, 203, 220, 253, 328]. Differences in initiation efficiency between codons in optimal or suboptimal contexts strongly depend on the conditions of the assay system [87, 203]. While some workers concluded that the context of the initiation codon in plants is of minor importance [227], others have shown that reduction of the Mg^{2+} concentration from 3 mM to below 2 mM produces significant codon context discrimination, similar to that in mammalian systems [86, 203]. However, it is unclear whether these optimal *in vitro* conditions also apply to translation *in vivo*. The (suboptimal) codon context of soybean lipoxygenase ORFs (AAAG.ATG.TTT) was found to give 10 times higher β-glucuronidase (GUS) expression rates than the (also unfavorable) context in the standard GUS expression plasmid pBI 211 [192]. In transgenic plants, translation initiation patterns from two consecutive, in-frame AUG codons were found to be drastically different in different parts of the plant (and at different developmental stages) [161].

The efficiency of start-codon recognition can be influenced by features that lead to a prolonged pause of the scanning ribosome at the position of the codon. A secondary structure element located 14 nt downstream of the start codon was found to increase initiation efficiency at weak AUG and at non-AUG codons [205]. A similar pause results from a downstream initiation event and can cause more efficient recognition of a properly spaced upstream AUG [87]. In contrast, initiation at a downstream codon may be negatively influenced by overlapping translation from an upstream start site, possibly because of interference by the translating 80S ribosomes [211].

Translation elongation

Decoding of the RNA occurs at the aminoacyl tRNA site (A site) of the ribosome in 3-base steps by an amino-acylated tRNA with appropriate anticodon. Peptide elongation occurs by transfer of the nascent peptide from the tRNA of the previous decoding step, which is located at the peptidyl tRNA site (P site), to the aminoacyl tRNA at the A site. The mRNA(codon)-tRNA-peptide complex is then translocated to the P site and the A site becomes free for interaction with the next tRNA. A third site (exit or E site) on the ribosome interacts with the unloaded tRNA before release. The binding of a tRNA to its codon differs in the A and P sites [272] and interactions between the tRNAs at the different ribosomal sites influence the kinetics of decoding [222, 312]. Translation elongation is promoted by eukaryotic elongation factors eEF1 and eEF2 and requires the hydrolysis of GTP [281]. Unlike the scanning process, ribosome translocation during peptide synthesis is relatively insensitive to inhibition by RNA secondary structure. Even basepairing between long sense and anti-sense RNA stretches can be disrupted, provided it begins more than about 20 nt downstream of the AUG codon and, therefore, does not interfere with the formation of the 80S ribosome [185, 221, 300]. Structural elements may nevertheless cause transient pausing of translating ribosomes (see below). The speed of translation elongation can vary between less than 2 to about 10 amino acids per second per translating ribosome in eukaryotic cells, depending on the cellular conditions and the mRNA ([289] for review). While initiation is generally the rate-limiting step in translation, elongation can become limiting with some particular ORFs [62, 94] or under certain physiological conditions [289]; for example, elongation rates are reduced at the slightly acidic pH that is induced by oxygen deprivation [354]. For plants, regulation at the level of elongation has been proposed for two oat seed proteins [37] and in response to heat-shock [10], oxygen deprivation [99, 348, 354], wounding [248], and light [28, 305, 310].

The nascent peptide is transported through a ribosome tunnel to the surface [71, 364], where it can interact with signal recognition particles or other cytoplasmic factors [359]. In bacteria, slowing down of elongation can be caused by the nascent peptide [145, 226], and the penultimate amino acid in particular may influence the termination step [245, 249]. These effects are probably caused by interactions between the peptide and translation factors, rRNAs or the tunnel through which it is transported to the surface of the ribosome [154], and it is likely that similar mechanisms exist in eukaryotes. Ribosomal pause sites not ascribed to secondary structure elements have been detected for a number of genes [193, 194, 263, 362]. It has been proposed that codons that are recognized by rare tRNAs ('rare codons' [47]) interfere with efficient translation [371] and could be a regulatory feature of gene expression [66, 291]. The speed of decoding could also influence the folding of the nascent peptide chain [213], and possibly the targeting [263]. In microorganisms, rare codons tend to be avoided in highly expressed genes and adjustment of the codon bias can result in increased expression of a gene [78, 315]. In yeast [166] and *Escherichia coli* [59, 135], a number of rare codons near the initiation site is required to substantially reduce expression of the respective ORF.

It is likely that all these mechanisms also apply to translation in plant cells. A compilative analysis of protein-coding sequences revealed that the codon usage in plants differs from other organisms, and differences can also be discerned between plant families [49, 252]. In plants, particularly in monocots, highly expressed genes show no special preference for frequent codons [49, 252]. Nevertheless, bacterial genes have been modified to comply with the plant codon usage and were indeed expressed to a much higher level [107, 212, 267]. During this mutagenesis, however, signals leading to potential aberrant RNA processing were also altered, and it is still not clear which of the features of the new RNA sequence are responsible for the elevated expression.

Translation termination

For the stop codon (UAA, UAG, UGA) usually no complementary tRNAs are available and release factors associate with the ribosome; this causes release of the polypeptide chain and termination of translation [45, 69, 338, 344, 345]. In all eukaryotes studied so far, the three stop codons are recognized by only one release factor, eRF1 [106, 372] or SUP45 in yeast [317]. Recognition or termination activity is stimulated by a second protein factor, eRF3 [372] or SUP 35 in yeast ([317, 318] for review). The immediate sequence context of stop codons can modulate recognition efficiency. Hierarchies of termination efficiency have been established for *E. coli* [271], mammals [236], and yeast [33], and sequence comparisons suggest the existence of similar context effects in other eukaryotes, including plants [9, 42, 43, 55, 344]. The fate of the ribosome after translation termination remains unclear. However, at least some ribosomal subunits must resume scanning after translation termination since in some cases ribosomes are able to reinitiate translation at other downstream start sites. It has been proposed that the reinitiation capacity is modulated by sequences surrounding the stop codons [142, 242], but it is not known whether this occurs at the level of ribosome-bound factors or at the level of ribosome-RNA association.

The exceptions

Translation of simple, monocistronic RNAs is easily explained by the scanning model. The translation of exceptional RNAs with unconventional initiation sites or encoding more than one protein or peptide, discussed in the following sections, is not fundamentally different but requires a few modifications or additions to the original model. Most of these RNAs are derived from plant viruses [168, 243, 285].

Variations of initiation site selection

Non-AUG start codons
Translation in mammalian and insect cells can commence at codons which differ from AUG in one position. In artificial constructs, most of the possible AUG derivatives are active to some extent [203, 237, 262]. In natural mRNAs, CUG is mainly found as an alternative start codon [36, 54, 151, 273, 342, 343], but indica-

tions of the activity of AUC and ACG have also been presented [19, 72]. In *Saccharomyces cerevisiae*, non-AUG codons are recognized very inefficiently [65]. Non-AUG recognition can be enforced by mutations in eIF2β [89] and eIF2γ [90]. For plants, the activity of such codons was analyzed with artificial constructs in wheat germ extracts [203] and in transfected protoplasts [138]. In both systems, initiation at non-AUG codons was more context dependent than that at AUG codons, and in the wheat germ extract recognition was much higher at high Mg^{2+} concentrations [203]. For protoplasts, a hierarchy of activity was found of CUG (30% efficiency of an AUG codon), GUG, ACG (15%), UUG, AUA, AUC, AUU (2 to 5%), and AAG, AGG (<0.1%) [138].

It has been inferred that non-AUG initiation occurs in a mutated version of a CaMV RNA at the start of the *pol* gene [297]. This example is important in that it shows that mutation of the start codon of an ORF does not necessarily abolish ORF expression completely.

An AUU codon opens the first ORF of rice tungro bacilliform virus (RTBV). Translation efficiency at this codon is low in RNAs where the position is reached in a normal scanning mechanism but reaches about 10% of AUG efficiency at its natural position downstream of the long RTBV leader sequence [116]. This leader sequence contains a number of features that inhibit scanning and most likely a specific mechanism is required for ribosomes to reach the region downstream of the leader (see below). This mechanism may also lead to enhanced recognition of the AUU codon.

No other plant genes with translation initiation exclusively at non-AUG codons have been described. In mammals, most ORFs with a non-AUG initiation site also have a normal AUG initiation site further downstream. Additional non-AUG initiation is found mainly for ORFs encoding regulatory proteins that are located downstream of a long, structured leader sequence. The protein variants with N-terminal extensions may have regulatory effects different to the 'normal' one [2, 46] and initiation at the additional sites can be regulated by the cellular conditions ([151, 152] for review). Similar translation events may occur in plants, but these rare and probably important events have not yet been detected because few plant genes and their protein products have been analyzed to the required extent. In soil-borne wheat mosaic virus (SBWMV), an N-terminal extension variant of a 29 kDa protein with unknown function may be produced by initiation at an as yet unidentified, non-AUG codon upstream of the first AUG codon on RNA 2 [303].

80S ribosome formation at an AUU codon in the leader sequence of TMV RNA (Ω) has been reported [101, 173, 340]. The codon is in frame with the TMV 126 kDa protein but it is not known whether it serves as an initiation site for a variant of this protein. An ORF fused to the Ω leader was found to be translated also from an upstream AUU codon with high efficiency [323]. The 5' leader sequence of AMV RNA 3 even binds two 80S ribosomes, presumably also at AUU codons [268]; one 80S ribosome also binds to an as yet unassigned position in the leaders of BMV RNA3 [5] and TYMV RNA [101]. The relevance of these binding events for translation of the respective RNAs remains unclear. Disome or trisome formation is not involved in the translation stimulating effect of these leader sequences [126, 309].

Non-AUG initiation can occur at inconspicuous positions of an mRNA: in the leader sequence of the CaMV 35S RNA, efficient initiation at a CUG codon that leads to translation of a short ORF has been detected (discussed by Gordon *et al.* [138]). Again, it is not known whether this translation event has any functional importance, but the example shows that caution is required when the possible translation events on one RNA are only deduced from sequence data.

RNAs with multiple translation initiation sites
If the translation start site for an ORF is not the most 5'-proximal on a given RNA, three different possibilities exist for the route by which a translation-competent ribosomal subunit may reach this codon (Fig. 1).

1. Most simply, but least compatible with the scanning model, ribosomes avoid upstream initiation sites by jumping or by directly binding to a downstream site.

2. In line with the scanning model, the upstream codon may be avoided by a fraction of ribosomes due to effects of the sequence context (leaky scanning). A special case for leaky scanning are start codons that are so close to the RNA 5' end that the decoding site of the 43S subunit is already downstream when the ribosome is bound to the 5' end [206, 298].

3. In cases where the upstream ORF terminates before the downstream initiation site, ribosomes may continue scanning and reinitiate translation; reinitiation requires the recruitment of initiation factors by the scanning ribosomes.

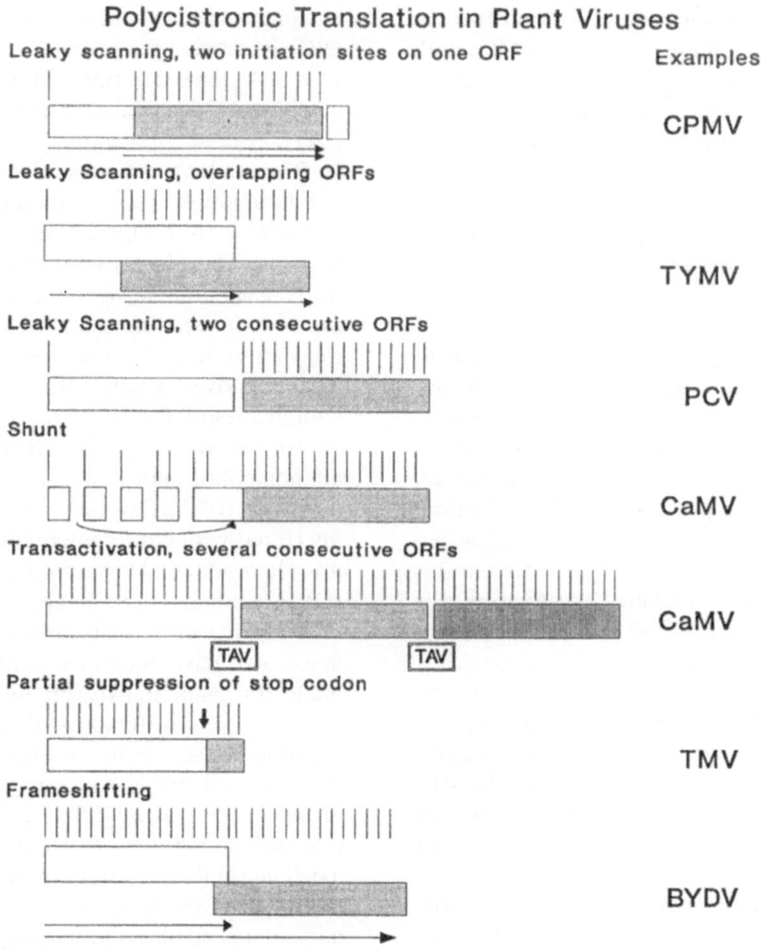

Polycistronic Translation in Plant Viruses

Figure 1. Strategies of polycistronic in plant viruses. The arrangement of the ORFs is shown schematically. The series of line marks above the ORFs represent the AUG codons. The horizontal arrows represent the movements of translating ribosomes, the bend arrow the movement of the shunting ribosome or complex. The vertical arrow points to the site of the partially suppressed stop codon. Examples for viruses using the polycistronic translation strategies are given at the right.

Examples of all these mechanisms are known for specific mammalian RNAs [207] and at least some have been described for RNAs translated in plants.

RNAs with multiple initiation sites can be grouped in three classes (Fig. 1):

1. RNAs that contain one or more short ORFs (sORFs) in their leader region; the sORF may terminate before or overlap the main ORF;
2. RNAs that contain multiple, overlapping long ORFs;
3. RNAs that contain multiple, non-overlapping long ORFs.

The distinction between a 'coding, long' ORF and a sORF is somewhat arbitrary. Usually a sORF is shorter

than 50 codons and the putative translation product is not known to have any function besides possibly regulating the translation efficiency in *cis*.

Short ORFs in the 5' leader sequence

Systematic comparison of the 5' leader sequences of 79 plant genes showed that 6 of these contain upstream AUG(s) [186]. Our own, unsystematic inspection of plant gene sequences also showed that about 10% contain one or more AUG codons in the 5' leader sequence. In many cases, it remains to be seen which of these AUG codons are present in functional mRNAs; some may be removed by splicing. Unspliced RNAs are not necessarily degraded in plant cells [136] and may be represented in cDNA preparations. In other cases, multiple transcription start sites may lead to the production of mRNAs with different 5′ leader sequences. The effect of upstream sORFs on translation has been studied for a few nuclear-encoded mRNAs: The mRNA for the maize *Lc* protein contains a 38 codon long sORF (with a total of three AUG codons) in its 256 nucleotide long leader region, which ends 62 nucleotides upstream of the Lc initiation codon [73]. As expected from the scanning model, the presence of this sORF causes a reduction in *Lc* translation. In an assay that used the transcription-transactivating activity of the *Lc* protein for quantification of *Lc* translation, a 30-fold repression was observed in biolistically transfected maize aleurone cells. The repressing effect depended greatly on the intactness of the sORF sequence; all point mutations alleviated the effect at least partially. The mutations also included some which led to conservative amino acid exchanges or were even silent [73]. Examples in which the coding sequence of an sORF is important for its repressing effect are documented for yeast [242, 358] and mammalian cells ([50, 80, 133] for review) but the stringent sequence requirements observed for the *Lc* leader sORF are unique. It is noteworthy, that the mutated versions of the sORF apparently had almost no inhibitory effect on downstream translation although they still contained AUG codons and should therefore be translated.

The mRNA leader of the maize *opaque-2* gene contains three, partially overlapping sORFs which inhibit translation of the downstream *opaque-2* ORF about fivefold in tobacco protoplasts [225]. Individual sORFs showed a similar inhibitory effect. When the leader sequence was modified such that either the first or second sORF was elongated to overlap the downstream, long ORF, inhibition was increased. This is taken as evidence that at least the first two sORF initiation sites are actually used as efficient translation start sites and that the normally observed initiation at the opaque-2 ORF AUG is a reinitiation event [225]. As stated above, it is possible that even a poorly translated overlapping ORF may interfere with downstream initiation [211] and it cannot be excluded that the opaque-2 ORF is normally translated by leaky scanning, which is inactivated in the overlapping ORF configuration.

The leader sequences of several plasma membrane proton ATPases from *Nicotiana plumbaginifolia*, tomato and *Arabidopsis thaliana* also have sORFs. The sORF in the *pma1* RNA is 10 codons long, contains two AUG codons in suboptimal sequence context and ends about 60 nucleotides upstream of the coding ORF. Removal of any single AUG codon had no effect on translation but removal of both increased translation about twofold in wheat germ extracts and in protoplasts [241].

The leader sequences of several *Arabidopsis thaliana* homeobox genes contain sORFs. For the homeobox gene 1 (*ATH1*), the leader is longer than 600 nucleotides. The 5′ halves of the probably two different mRNA versions have not yet been characterized but the 3′ half contains a number of sORFs and inhibits translation in an *in vitro* system about fivefold [276]. It is unknown whether the leader is involved in the complex regulation of *ATH1* activity.

An sORF in the leader of the soybean aminoalcohol-phosphotransferase 1 (*AAPT1*) inhibited translation of the downstream ORF in yeast and thereby interfered with complementation of a yeast mutant by this plant gene [84]. Again, it is unknown whether the sORF regulates *AAPT1* expression in plants.

Negative effects of sORFs on downstream translation also have been observed in artificial constructs made for plant transformation [31, 275, 284]. The effects of several features of an upstream sORF were tested in constructs in which a reporter ORF was preceded by an artificially designed sORF. The length of the sORF was modulated by sequence multiplication and the number of potential start codons was varied. The inhibitory effect increased with the length of the sORF. Even the shortest possible sORF, consisting only of an AUG codon, reduced downstream translation by about a factor of two. Intermediate length sORFs (around 30 codons) reduced fivefold and sORFs longer than 100 codons completely abolished downstream translation [113]. In plant protoplasts, sORFs with additional internal AUG codons were not more inhibitory than those without, probably because the first start codon was in an optimal sequence context and thus all approaching ribosomes initiated there. If the sORF overlapped the downstream ORF, translation was completely precluded. A limited number of sORF sequence variants in the coding sequence or around the stop codon were also tested, but no further features influencing the inhibitory potential of the sORF could be discerned [113]. From all these results it is clear, that sORFs can be inhibitory because they are translated and thus reduce translation initiation at the downstream start codon. However, initiation at an sORF does not completely preclude translation reinitiation at a further downstream ORF by the same ribosome [105, 113, 275]. This process should require the recruitment of a new set of initiation factors. In the artificial constructs described by Fütterer and Hohn [113], the reinitiation efficiency was dependent on the length of the sORF and reinitiation was observed despite an intercistronic distance of only 16 nucleotides.

The effect of sORFs on translation of a further downstream ORF has also been studied in mammalian and yeast systems. A similar length dependence of the inhibitory effect of the sORF was observed in mammalian cells [228]. Besides, on the sORF's coding sequence (see above), the efficiency of downstream translation was also dependent on sequences downstream of the sORF's termination codon [142] and on the distance between sORF and downstream ORF. Distances greater than 50 [200] or 200 [1] nucleotides were required and in the yeast system, downstream translation was strongly influenced by the physiological state of the cells ([1, 165, 260, 358] for reviews). Tissue-specific effects of sORFs on downstream translation have also been observed in mammalian cells [375]. It is likely that similar effects also exist in plants. So far the only safe fact seems to be that an sORF-containing leader always causes lower expression than an sORF-free leader. In particular, for long leader sequences with several sORFs, the multitude of possible activities of ribosomes, such as translation, leaky scanning, stalling, translation-termination-dependent dissociation from the RNA, non-productive scanning (due to the lack of initiation factors), sequence- or distance-dependent reinitiation etc., makes a precise prediction of the degree of inhibition impossible.

Up to 14 AUG codons are found in the leader sequences of the pregenomic RNAs of the plant pararetroviruses, and the leaders of the CaMV 35S RNA [17, 109] and the RTBV 35S RNA [58] are indeed inhibitory to downstream translation. For CaMV, inhibition was up to 100-fold in some protoplast systems derived from non-host plants [17, 109]. In other protoplast systems, including some from virus host plants, the inhibition was much less severe (around two- to six-fold; [109]). These leader sequences can also form extensive secondary structures [108, 147, 158]. It is possible that most of the translation inhibitory effect of the leader was caused by secondary structure rather than by the sORFs, as was shown for the c-sis proto-oncogene RNA with its long, sORF-containing leader [170]. It is, however, likely in CaMV that some of the sORFs inhibit downstream translation since they are inhibitory when located in the supposedly unstructured leaders of truncated mRNAs. The first three CaMV sORFs are very short (2–5 codons) and their

start codons are in suboptimal sequence context. Consequently, they are only slightly inhibitory for translation of a reporter gene fused to sORFs located further downstream [108, 114]. The longer sORFs in the 3′ half of the leader, at least one of which has a start codon in optimal sequence context, are more inhibitory [111]. Translation of a reporter ORF positioned in the center of the leader follows the scanning model. Either enough ribosomes avoid initiation at the upstream sORFs because of the unfavorable sequence context or they regain initiation capacity after sORF translation [112, 114]. The first longer ORF downstream of the CaMV 35S RNA leader is thought to be translated by modified scanning which was termed ribosome shunt [114]. A similar process might be active on the RTBV 35S RNA [116]. In the shunt, initially scanning ribosomes are transferred directly from a donor to an acceptor site without linear scanning of the intervening region. The ribosome migration on the 35S RNA leader has been studied by insertion of strong stem-loop structures and additional ORFs at various sites. (It is noteworthy that in all cases where no ribosome shunt was involved, these additional elements produced exactly the effect that would have been predicted by the scanning model. Therefore, these experiments are also another confirmation of the general features of the scanning model in plant cells.) The shunt could also be detected in *trans*, albeit at low efficiency [114] and was independent of viral proteins. Since shunt efficiency showed a certain cell-type dependence, involvement of cellular factors is likely [109]. Although mutagenesis analyses have defined regions in the leader sequence of the CaMV and RTBV 35S RNAs that are important for the shunt process [58, 111, 114, 116], the structural features that allow a shunt are only poorly defined so far. It is assumed that long-range RNA interactions are involved, probably with the assistance of associated proteins. Computer programs predicted a stable stem-loop structure for almost the complete CaMV and RTBV leader sequences [108, 158], and for CaMV such a structure was also predicted by an analysis of folding parameters of the growing RNA molecule [147]. Direct analysis of the *in vitro* structure of the leader confirms the presence of the predicted structure or a similar one (M. Hemmings-Mieszczak, G. Steger, and T. Hohn, unpublished observations). In an alternative folding of the CaMV 35S RNA leader sequence, the presence of pseudo-knots in the acceptor as well as in the donor regions was noted (K. Gordon, pers. comm.). Which of these structures, if any, is important for the ribosome shunt has still to be determined. The location of the ribosome acceptor region was defined for CaMV and for RTBV with some precision. In both cases, it lies immediately downstream of a CT-rich region and for RTBV it involves the AUU start codon of the first ORF [114, 116]. The efficiency of recognition of this AUU start codon is increased above that of the same codon reached by scanning, possibly because a shunted ribosome has more time for recognition before scanning is resumed [116]. Recently, translation of CaMV 35S RNA leader constructs in wheat germ extracts produced results that suggest that a ribosome shunt can also occur in this heterologous *in vitro* system (W. Schmid-Puchta, D. Dominguez and T. Hohn, unpublished observations).

Ribosome shunt-like mechanisms have been described for translation of Sendai virus RNA [72], the adenovirus tripartite leader (A. Yueh and R.J. Schneider, pers. comm.) and for a papova-virus (G. Hobom, pers. comm.).

RNAs with more than one coding ORF

RNAs that encode more than one protein in consecutive or overlapping ORFs have not been described for plants; however, a number of plant viruses use such RNAs (Fig. 1) and apply one or several unusual mechanism(s) for their translation (Fig. 2).

In the most easily explainable cases, translation initiation at two start sites is regulated by the rules of leaky scanning. This requires that the first AUG is in an unfavorable sequence context. The second initiation site can be in the same or in a different reading phase, giving rise to an N-terminal deletion variant of the first protein, or a completely different protein, respectively. The degree of leakiness of an AUG codon can be high; a potential start codon in the leader of PPV was found to be recognized very inefficiently [280] while for two ORFs of the luteoviruses PLRV and BYDV-PAV ratios between 100:1 and 1:7 for translation of two overlapping ORFs have been described [86, 87, 322]. This mainly depends on the sequence context [87] and on the conditions of the translation system (cation concentration [89]). A protoplast system was more discriminating (100-fold difference) between an optimal and a suboptimal start codon than the reticulocyte lysate [87]. In those cases where leaky scanning has been observed or postulated, the first start codon is usually in an unfavorable sequence context (Table 2).

As described above, the ratio of expression from two overlapping ORFs depends not only on the direct

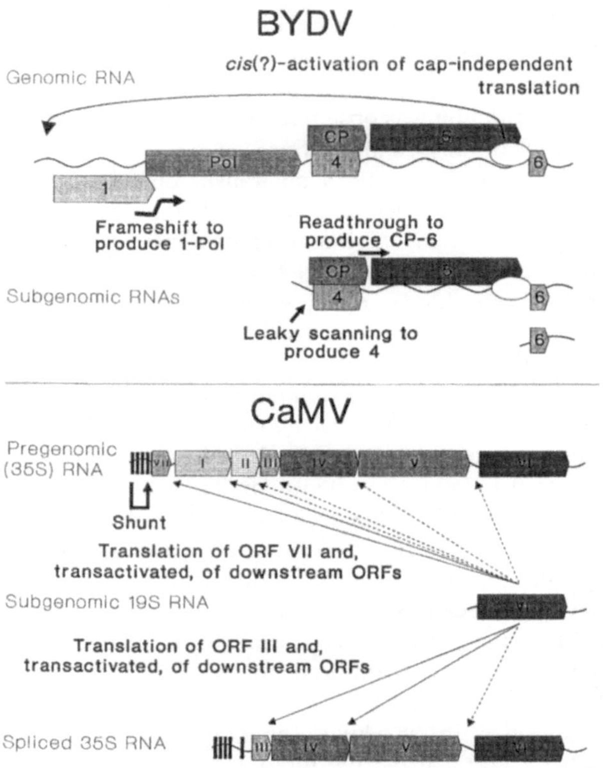

Figure 2. Expression strategies of luteo- and caulimoviruses exemplified by BYDV and CaMV, respectively. The genomic RNA and the subgenomic mRNAs are shown including their ORFs. *BYDV*: ORF 1, CP and ORF 6 are translated as first ORFs from the three RNAs. Pol is translated as ORF1-Pol fusion protein upon a frameshift linking the two ORFs from the full length RNA. ORF 4, which overlaps with CP is translated upon leaky scanning and ORF 5 is translated as CP-ORF 5 fusion protein by readthrough of an amber stop codon from the longer subgenomic RNA. A site close to the 3' end of the two larger RNAs promotes translation in *cis*. *CaMV*: Three mRNAs were identified, the pregenomic 35S RNA, the subgenomic 19S RNA and a spliced version of the 35S RNA. The first ORF (ORF VII) of the 35S RNA is translated after a shunt mechanism. Whether translation of the first ORF (ORF III) from the spliced RNA also requires shunting or another mechanism (leaky scanning, reinitiation) is not known. Translation of the second and probably consecutive ORFs from unspliced and spliced 35S RNA occurs upon transactivation involving the translation product of ORF VI (transactivator) derived from the 19S RNA. The transactivator is the only larger CaMV protein obtained by classical translation.

sequence context of the start codons but may also be influenced by structural features of the RNA. In mammalian cells, mutual influences of overlapping ORFs during the elongation phase of translation have been described, probably caused by the different speed of decoding of the two overlapping reading phases [94]. A similar effect may account for observations made with start codon mutations of the overlapping TYMV ORFs

Table 2. Leaky start codons of plant viral ORFs. The immediate sequence context of start codons known or supposed to be leaky is presented. Nucleotides that conform to the general eukaryotic start codon consensus sequence (A/G)NNAUGG are highlighted in bold. The distance to the next downstream AUG codon and the ORFs in whose translation the start codons are involved are listed. References to viruses not further mentioned in this review can be found in the compilation by Miller *et al.* [243].

	ORFs	Leaky AUG codon	Distance to second AUG (nts)	Ref.
Overlapping ORFs				
Luteovirus				
PLRV-S,A,C	ORF0/ORF1	CAU.**AUG**.AU	128	243
BWYV		UUG.**AUG**.CA	137	
RPV	"	CGC.**AUG**.UU	146	
BYDV-PAV, MAV, RMV	ORF3/ORF4	UGA.**AUG**.AA	14–41	
BYDV-RPV	"	UUA.**AUG**.AG	20	
SDV	"	**AGU.AUG.GU**	11	
BWYV	"	UUA.**AUG**.AA	26	
PLRV	"	UUA.**AUG**.AG	20	
Enamovirus				
PEMV RNA1	ORF0/ORF1	UUU.**AUG**.CA	77	
PEMV RNA2	25K/27K	UAU.**AUG**.AC	11	
Carmovirus				
MCMV	p31.6/p50	UUC.**AUG**.CC	14	
Tymovirus				
TYMV	replicase	CAA.**AUG**.AG	2	247
EMV	"	UCA.**AUG**.CC	2	
KYMV	"	CUG.**AUG**.UC	2	
OYMV	"	UUC.**AUG**.UC	2	
Tombusvirus				
CNV, TBSV, CyRSV	21K/20K	UUC.**AUG**.GA	27	283
Phytoreovirus				
RDV (S12 RNA)	p34/p10.5	**A**UA.**AUG**.UU	267	
Satellite virus				
STMV	p6.8/p17.5	UUU.**AUG**.CU	104	244
Two initiation sites in one ORF				
Potyvirus				
PPV	Polyprotein	UUU.**AUG**.CA	106	280
Comovirus				
CPMV M-RNA	Polyprotein	**A**CA.**AUG**.UU	349	350
Phytoreovirus				
RDV (S12 RNA)	10.5K ORF	UUA.**AUG**.CU	19	244
Badnavirus				
RTBV (spliced RNA)	ORF IV	UCA.**AUG**.GC	71	115
Independent ORFs				
Furovirus				
PCV RNA2	coat protein/p39	CUU.**AUG**.UC	617	164
Badnavirus				
RTBV	ORFs I and II	**AA**U.**A**UU.**GA**	562	158
	ORFs II and III	UAC.**AUG**.AG	325	

for a 69 kDa and a 206 kDa protein, where it was found that elimination of the first AUG had no positive effect on translation from the second one, while elimination of the second AUG increased translation from the first one [355]. This suggests that translation (initiation or elongation) of the downstream ORF is also rate limiting for the overlapping upstream ORF. Initiation efficiency at the upstream BYDV coat protein (CP) ORF was positively influenced by efficient initiation of the 17 kDa ORF starting 40 nucleotides downstream. Probably the pause resulting from initiation at the downstream AUG results in ribosome stacking and provides more time for recognition of the upstream AUG [87].

Leaky scanning is at least partially responsible for translation of the 95 kDa protein of CPMV M RNA, which initiates about 350 nucleotides downstream of the 105 kDa protein; no AUG codon in any of the three reading phases is found in this intervening region [350]. Translation of RNAs of the rice dwarf phytoreovirus [321], the satellite tobacco mosaic virus [244] and luteo-, tymo- and tombus viruses also probably involves leaky scanning ([243] for review).

In at least two cases, leaky scanning is the mechanism of translation of subsequent, non-overlapping ORFs. An ORF on PCV RNA 2 is located downstream of the 620 nucleotides long coat protein ORF but is accessible to scanning ribosomes because this long upstream region is devoid of AUG codons apart from the one opening the CP ORF. Leaky scanning was deduced from the negative effect on downstream translation of the insertion of stem structures or of additional AUG codons into the upstream region [164]. A similar, even more extreme case is found in RTBV, where the 900 nucleotides encoding ORFs I and II upstream of ORF III contain only one AUG codon that opens ORF II and has an unfavorable sequence context. A similarly peculiar bias is also observed in the genomes of the other badnaviruses. Improvement of the efficiency of RTBV ORF I translation by mutating its AUU initiation codon to AUG drastically reduced ORF II and III expression, as expected for a leaky scanning mechanism (unpublished observations).

If the upstream ORF is opened by an efficiently recognized start codon and/or if no particular bias against the presence of additional AUG codons in any of the three reading phases exists, a following ORF will normally not be translated because leaky scanning is impossible and reinitiation of translation is inefficient. In *in vitro* systems, translation efficiencies for downstream ORFs of 1–20% have been found [137], but in plants or protoplasts downstream ORFs are expressed with considerably lower efficiency. This has generally been experienced with constructs designed to express a marker ORF downstream of another ORF in transgenic plants [8, 174, 374]. An up to 1500-fold reduction of expression efficiency of the downstream ORF was observed [8]; however, even low expression efficiencies can be sufficient to cause phenotypes, such as antibiotic resistance [8, 174] or GUS staining [374]. In one exceptional case, the introduction of a complete bacterial arylsulfatase ORF upstream of a GUS ORF seemed not to influence GUS translation in transiently or stably transformed plant cells [63]. However, in the respective dicistronic constructs, GUS translation begins at an in-frame AUG accidentally present considerably upstream of the AUG that is used in the monocistronic reference construct, and leads to different GUS proteins with possibly different enzyme stabilities or activities. In transient expression experiments in a variety of plant protoplast systems, expression of a downstream ORF was at or below the detection limit [34, 112, 139, 294]. In *Orychophragmus violaceus* protoplasts, detectable expression of a CAT ORF downstream of a GUS ORF was only observed when the intercistronic distance was increased to more than 300 nucleotides [112].

Viral transactivation of polycistronic translation

The pregenomic RNA of the caulimoviruses and probably also a number of internally spliced derivatives of these RNAs serve as polycistronic mRNAs for a number of viral proteins [34, 88, 110, 139, 196, 294]. The respective ORFs closely follow each other without long intercistronic distances, are often opened by efficiently recognized start codons, and usually also contain internal AUG codons. In plant protoplasts and in transgenic plants, most of the downstream ORFs are indeed not or only poorly expressed unless the virus encoded transactivator (TAV) is present (Fig. 2). Transactivation activity has been demonstrated for the ORFs VI of CaMV [34, 83, 112] and FMV [139, 294, 295] and it likely resides also in the corresponding ORFs of CERV, SoCMV, and PClSV [83]. The TAV protein specifically enhances the translation of a downstream ORF. Expression was obtained from totally artificial, polycistronic constructs as long as the ORF organization allowed a reinitiation mechanism, i.e. when long ORF overlaps were avoided [112, 113]. In the artificial polycistronic RNAs, transactivation was particularly efficient when the first ORF was around 30 codons long; shorter and longer ORFs were less effective [113]. Transactivation was observed for several ORFs following such a short ORF. The polar effects of the insertion of stem-loop structures into polycistronic mRNAs and the specificity for non-overlapping ORFs suggest that transactivation causes enhanced reinitiation of translation. The dependence of transactivation efficiency on first ORF length suggests, that the transactivator directly or indirectly acts on the translating (or terminating) ribosome [112, 113]. The optimal length of the first ORF of 30 codons may be significant since a 30 amino acid nascent peptide is just long enough to emerge from the translating ribosome [231, 359]. It appears possible that at this stage of translation a structural change of the ribosome occurs which leads to loss of a residual reinitiation capacity and also to a loss of transactivation responsiveness. The absence of a requirement for *cis*-active sequences does not necessarily implicate the lack of such sequences in the CaMV genome. The presence of the CaMV ORF VII supported transactivation [34] although an artificial ORF of similar length did not [113]. The expression constructs also contained the CaMV polyadenylation signal, which contributes the terminal 200 nucleotides of the transcripts. Studies with FMV suggested that *cis*-active sequences are required for transactivation in this case. These sequences are located at the end of the FMV 35S RNA leader sequence [140] and in the ORF VI coding region, which was effective as a 3'-untranslated region in the respective plasmid constructs [295]. For the former, however, no distinction was made between sequences that allow ribosome access to the region downstream of the leader (a FMV ribosome shunt) and sequences that are specifically involved in transactivation, nor was the precise configuration of the sORFs with respect to the reporter ORF discussed. In CaMV, the TAV also stimulates translation directly downstream of the leader [114]. This increased expression was dependent on the shunt process and was apparently not caused by the action of the TAV on ribosomes that migrated through the leader sequence. In the case of the stimulating sequences in the 3'-'untranslated' region, it remains to be seen whether the effects are due to stimulation of translation or to RNA stabilization by TAV-induced translation of ORFs in this region. Strong reductions of mRNA levels have been reported for constructs that contained inefficiently translated 3' ORFs [8, 174] and are in general often observed for RNAs with premature stop codons and, therefore, long untranslated 3' regions in plants [347] and other eukaryotes ([232, 259] for review).

The transactivating function could be localized to the central third of the TAV protein [83]. To achieve normal levels of transactivation with a peptide comprising only this portion (mini-TAV), 100-fold more DNA encoding the TAV polypeptide had to be transfected. It is not known whether this lower efficiency is due to a defect in transactivation itself or to problems with protein folding and stability. The mini-TAV was only active in *Nicotiana plumbaginifolia* protoplasts, while the full-length TAV was active in a number of dicot plant protoplasts and also in maize protoplasts [34, 139]. TAV was found to be associated with polysomes and, in an overlay binding assay, also with an 18 kDa ribosomal or ribosome-associated protein from plants and also from yeast (A. Himmelbach, Y. Chapdelaine and T. Hohn, unpublished observations). An RNA-binding activity of TAV residing outside the minimal transactivating region may enhance activity by increasing TAV concentration near the RNA [83]. Crossing of transgenic *Arabodopsis* lines containing either the transactivator or a dicistronic GUS reporter construct revealed that transactivation can also be obtained in transgenic plants [374]. No tissue or development specificity was observed. *Arabidopsis* and tobacco plants expressing the transactivator are not easy to obtain and show abnormal phenotypes [18, 325, 374]. At present it is unclear whether these phenotypes are caused by transactivation or some other activity of the protein. Recently, TAV-dependent transactivation of polycistronic translation was also described for *Saccharomyces cerevisiae* [299]. This latter finding opens up many possibilities for the study of the mechanism and for the characterization of the cellular partners involved.

Leader sequences and cap-independent, internal ribosome entry

Downstream ORFs on a polycistronic mRNA could in principle also be translated in a scanning-independent mechanism by ribosomes that enter the RNA at an internal position (internal ribosome entry site, IRES) as was originally shown for picornaviruses ([3, 57, 181, 189, 266, 292, 314] for reviews) and has since been described for a number of animal cellular and viral mRNAs [27, 257, 342, 343]; it may also occur in yeast [175]. A number of plant viral mRNAs are not capped [366 for review] and, therefore, must have a cap-independent ribosome entry site. Cap-independent translation initiation might still be dependent on ribosome association with the RNA 5' end and not involve a true IRES. For CPMV, three members of the potyvirus group, BYDV and STNV, cap-independent translation was shown to be conferred by viral sequences *in vitro* and partially also *in vivo*. In the potyviruses, the important RNA sequences are located upstream of the AUG codon that initiates translation of the polyprotein (TEV [53], PPV [280], PVY [218], TuMV [16]); in BYDV [353] and STNV [74, 334] sequences at the 5′ end and near the 3′ end of the RNA are required, and in CPMV, the sequence resides between the two first start codons on the M RNA (positions 161 and 512 [333, 350]). Competition experiments suggest that the function of the TEV and TuMV sequences depends on interaction with a cellular factor also involved in cap-dependent translation [16, 53]. The TuMV sequence is active independent of its orientation and, like the CPMV sequence, also downstream of a scanning-inhibiting secondary structure element [16, 333]; however, this structural element reduced the translation by a factor of five for the TuMV leader [16] and expression in the CPMV case was quite low and not different from other internal initiation events on the respective reporter RNA [333]. The effect of the PVY sequence was completely abolished by an antisense oligonucleotide for the first 16 nucleotides, suggesting that the important sequence is close to the RNA 5′ end [16]. The cap independence conferred by the TEV leader depends on sequences located more than 80 nucleotides downstream of the 5′ end [53]. Efficient, cap-independent initiation with BYDV and STNV RNA required sequences near the 5′ end and near the 3′ end [74, 334, 353].

The PPV, PVY and CPMV sequences were tested as internal entry sites between two ORFs. With the PPV sequence, only low levels of downstream ORF expression were observed in *in vitro* translation systems, which led to the conclusion that the PPV leader does not act as a true IRES [280]. Expression obtained with the PVY sequence was interpreted as proof for internal ribosome entry [218]; however, the data are difficult to evaluate because no quantitative comparison of downstream ORF translation with translation from a monocistronic RNA was made. The CPMV sequence allowed low levels of downstream ORF translation in the reticulocyte lysate [350], but was tested without success in an animal cell system [23]. In general, caution is required in extrapolating *in vitro* data to the *in vivo* situation. A sequence upstream of the TNV coat protein ORF was found to allow internal ribosome entry in wheat germ extract but not in tobacco protoplasts [240]. *In vitro* translation systems are known to accept also uncapped RNA, depending on the exact

translation conditions like ionic strength and translation factor concentration. The relatively long AUG-less regions that are tested for their IRES potential would direct all initiation capacity by spurious ribosome binding to the next downstream AUG codon and may, therefore, appear more efficient than other RNA sequences. The reported cases of internal ribosome entry on plant viral RNAs seem to need more substantiation.

The leader sequences of the como- and potyviruses are not as long or as complex as those of the distantly related picornaviruses that support internal initiation of translation. There is also no evidence so far that any of the plant viruses inhibits plant translation to create an advantage for translation of its own mRNAs. The picornaviruses inhibit cap-dependent translation by various mechanisms and in the absence of this inhibition viral mRNAs often compete only poorly with cellular RNAs [258]. Plant virus RNAs seem to compete with cellular mRNAs by reducing the requirements for or by increasing the affinity to initiation factors. The leader sequences of many plant viruses (but not all) have been found to increase translation in *in vitro* systems, in transfected plant protoplasts, and in transgenic plants by unknown mechanisms [53, 77, 91, 118, 119, 183, 184, 234, 255, 270, 308, 309, 311, 335, 367] (reviewed by Gallie, this issue). Translation of the uncapped STNV RNA requires eIF 4F at a lower concentration than other RNAs and is not affected by the presence of cap analogues [44], whereas translation of the capped, efficient AMV RNA 4 leader requires less eIF–4F and –4B than, for example, the β-globin RNA, but is clearly cap-dependent [102, 153]. The enhancing effect of the TMV Ω sequence in a yeast extract is independent of the cap-binding factor eIF–4E but still requires eIF–4A (the supposed RNA-unwinding factor) [7], although TMV RNA is naturally capped and Ω lacks strong secondary structure. The TMV Ω sequence may act differently to other translation enhancing plant viral RNA leaders since it is active in *E. coli* and almost all eukaryotic translation systems [119] (but not or only poorly in monocot cells [120]). The activity of Ω is end-dependent in eukaryotic translation systems [197], and two sequence motifs are crucial in plant protoplasts [126]; for the activity in *E. coli* the 5′ end with homology to a recently defined sequence element that can functionally replace a normal Shine-Dalgarno sequence may be responsible [121, 177]. No particular sequence motifs with enhancing effect have been found in other plant viral RNA leaders. For a more detailed review of these features see the contributions of D. Gallie (this volume).

It is also likely that some cellular mRNAs have leader sequences which modulate translation due to their particular translation factor requirements. For example, the untranslated regions of the barley α-amylase specifically enhanced translation in aleurone cells [129]. Other candidates for such leader sequences are the mRNAs translated under stress or developmental conditions that are detrimental to the translation of most other cellular mRNAs, and where discrimination occurs at the level of translation initiation ([127] for review).

In other eukaryotes, leader sequences and also 3'-untranslated sequences can modulate translation and RNA stability in concert with mRNA-specific proteins ([235, 316] for reviews); no example of such a mechanism has been described so far for plants.

Effect of the 3'-untranslated region

All the examples described above demonstrate flexibility in the efficiency and location of translation initiation. Somewhat surprisingly, the 3'-untranslated region (including the poly(A) tail) alone or in conjunction with the 5' leader can have a positive influence on translation initiation (reviewed by Gallie, this issue). These effects of the RNA 3' ends are probably commonplace but as they have been studied in only a few cases are treated here as exceptions. The mechanism for the enhancement is not known, but it is probably similar for the poly(A) tail and for the non-polyadenylated 3' ends of some RNA viruses [122, 125, 288]. In yeast, effects of the poly(A) tail are effective on the formation of the 80S ribosome [251, 290] and are mediated by the poly(A)-binding protein (PABP) [180, 250 for reviews]. The importance of the PABP was also observed in a pea *in vitro* translation system [304]. In plants, sequences far away from the RNA 5' end in conjunction with 5' leaders can cause cap-independence or at least reduce cap-dependence, as shown for STNV [334], TMV [370], BMV [191] and BYDV [353]; this suggests involvement in ribosome binding to the RNA. The stimulatory effects of 3'-UTRs seem to be quite variable in different assay systems and apparently are not always reproducible (e.g. the effects of the 3'-UTR of BMV described by Gallie and coworkers [128, 217] could not be reproduced by Lahser *et al.* [216]). The effects of the poly(A) tail in transfected plant protoplasts were much stronger in the presence of a 5' cap than in its absence. Exogenously added poly(A) stimulated translation of capped, non adenylated mRNAs in yeast [251] and inhibited that

of uncapped RNA in yeast [251] and in plant extracts [130], suggesting that poly(A) interacts with factors required for translation and that it may form a complex with the capped 5' end of the RNA, which is particularly efficient in ribosome binding (reviewed in detail by Gallie, this issue). In the presence of PABP, an oligo(A) tract in the mRNA leader is inhibitory for translation [81]. In *Xenopus* oocytes the poly(A) tail was found to enhance translation reinitiation [117], which means that in this case it acts after the first round of translation of an ORF on ribosomes that have translated the ORF already once. It is unclear how such a recycling of ribosomes could occur, but polysomal RNAs with a circular appearance have been observed in a variety of cases [64, 92, 171, 215], and sequence interactions between the ends of plant viral RNAs have been observed or suggested [74, 75, 103, 319, 334]. This interpretation has similarities with the ribosome shunt. It has been noted that the ribosome shunt *in vivo* might also serve for such a recycling since in the pararetroviral RNAs the shunt donor site is also present in the terminal redundant part at the 3' end of the RNA [114]. A ribosome recycling might help to increase the density of ribosomes on RNAs that can be successfully translated. In most present day translation studies, kinetics of polysome loading are not considered and only the amount of synthesized protein after the incubation period is measured. Also, in most of the studies of the effect of poly(A) tails or 3'-untranslated regions in yeast or plant systems, no discrimination between first initiation and re-initiation was made, and the kinetics of protein production with the different leader and 3'-untranslated regions still has to be investigated. Loading of mRNAs into large polysomes can be considerably slower than the transit time of a translating ribosome on such an mRNA [254], which suggests that it is not determined by primary binding of ribosomes to an RNA, but by rebinding of ribosomes that have translated already once. Reinitiation is also differently affected by initiation inhibitors and, therefore, may be functionally different from first initiation events [254]. While the effect of a cap structure on translation efficiency in transfected plant protoplasts was evident at the earliest measured time point, the full effect of the poly(A) tail was apparent only later [120]. Reinitiation may also be less or not cap-dependent, which may be one explanation why in some cases the 3'-UTR together with some 5' sequences conferred cap independence (see above). If reinitiation is partially involved in the mechanisms of translation enhancement by 3' regions, it is to be expected that these effects will differ with different reporter genes and possibly also with the precise linkage of the reporter ORF to the 3'-untranslated region, since in this case the mode of translation termination and subsequent ribosome migration should be important.

In other eukaryotes, 3'-UTRs of specific mRNAs have been found also to contain a number of other signals involved in RNA transport and localization, specific translation repression, and translation-dependent or independent RNA degradation ([79] for review). Strong differences in gene expression levels in plants have been described for reporter constructs with different 3'-untranslated regions [176], but the cause of the effect (translation, 3'-end processing, RNA stability, transport or localization) has not been elucidated.

Cotranslational effects

The translation process is in general precise and competing events such as misreading by 'wrong' tRNAs or shifts in the reading phase occur with a significant frequency only if decoding of the codon at the A site by a cognate tRNA or a termination factor is slow or disturbed [11]. In bacteria, translating ribosomes have been found to even skip longer stretches of the RNA completely [25, 172, 357]. This extreme is so far unknown in eukaryotes but smaller 'programmed errors' occur during translation elongation. The paradigm for such processes in eukaryotes is frameshifting and stop codon suppression in retroviruses and retrotransposons, which use both mechanisms to express their *pol* gene as a fusion protein with the upstream capsid proteins in a stoichiometrically controlled ratio ([156, 286] for reviews). A similar mechanism is also used by other viruses of mammals [40, 182] and yeast [341], but not by the plant pararetroviruses, which are relatives of the animal pararetroviruses [167, 286]. In these viruses, the *pol* ORF is either translated separately from the upstream, overlapping ORF corresponding to the retroviral *gag*-ORF [297] or is part of a long precursor ORF that also contains the *gag* functions [286]. Instead, frameshift events have been detected or proposed for pol gene expression of luteo- [85], enamo- [82], diantho- [195, 287], sobemo- [230] and carlaviruses [141]. Frameshift may also occur in the BWYV ST9-associated RNA [61] and in closteroviruses [4, 190]. In all cases except the last, frameshift occurs leftwards to the −1 reading phase (Table 3). The frameshift sites of BYDV-PAV [39, 85, 132], two strains of PLRV [214, 274], RCNMV [195] and CfMV [230] have been studied in greater detail. Like the frameshift signal of retroviruses and many retrotransposons they consist of a 'shifty' heptanucleotide sequence of the type X.XXY.YYZ (arranged in codons of the 0 frame) and a close downstream secondary structure element (Table 3). The sequences of a number of retroviral transframe-proteins have been determined and on this basis it has been suggested that the frameshift occurs when the peptidyl tRNA at the XXY codon at the P site and the aminoacyl RNA at the YYZ codon at the A site simultaneously slip backwards one nucleotide to the XXX and YYY codons, respectively [178]. Frameshifting does not necessarily lead to a unique transframe protein but rather to a small number of variants with slightly different amino acid sequences around the frameshift site [179, 356]. Frameshifting requires that the two tRNAs at the A and the P sites can stably interact with the new codons with mismatches only at the wobble position. It is caused or enhanced by ribosome pausing due to the downstream structural element [60, 313, 329, 330, 336] and can be influenced by the translation frequency [169]. The downstream element can be a simple hairpin structure or an elaborate pseudoknot [56, 301, 329, 330]. The nature of the codons (or their cognate tRNAs) is also important: of all possible YYZ codons, so far only AAC, UUU, UUA and recently AAU have been found [156, 243, 329].

That most of the studied frameshift events in plants follow the same mechanism was elucidated by determination of the sequence of the BYDV transframe protein [85] and by mutation analyses of the shifty heptanucleotides and the supposed downstream pause elements [132, 195, 214, 274, 363]. Frameshift sites were analyzed either in their natural contexts or by inserting the sites into a reporter ORF. Depending on the frameshift signal and the assay system, frameshift efficiencies of 1–30% were observed. High efficiencies were obtained with the Polish strain of PLRV and with CfMV in *in vitro* translation systems [214, 230], while the other signals direct frameshifting at only 1–4% efficiency in different *in vitro* systems or in plant protoplasts. Signals consisting of A and U nucleotides tend to produce higher efficiencies than those containing G or C [141, 274]. The BWYV ST9-associated RNA may contain a particularly efficient frameshift site, which has, however, not yet been identified [61].

A stop codon found immediately downstream of many of the shifty sites has a positive effect in some assay systems, probably because it contributes to ribosome pausing [39, 85]. In wheat germ extract, the BYDV frameshift event is greatly stimulated by a

Table 3. Frameshift. Sequences at proven or suggested sites in plant viral mRNAs. Virus acronyms and the family are given. The frameshift site is arranged in the 0 reading frame. For the −1 frameshift sites, the given sequence starts with the first nucleotide of the shifty site (X.XXY.YYN) and ends at the beginning of the downstream pause signal. Data are compiled from the indicated references; if no reference is given, data are from the compilation by Miller *et al.* [243].

	Frameshift site	Pause signal	Reference
−1 shifts			
Luteoviruses subgroup 1			
BYDV-PAV	G.GGU.UUU.UAG.AGG	hairpin or pseudoknot	243
SDV	G.GUU.UUU.UAG.AGGG	hairpin	
Luteoviruses subgroup 2			
PLRV-G	U.UUA.AAU.GGG.ACA	hairpin	274
PLRV	U.UUA.AAU.GGG.CAA	pseudoknot	132, 214
BWYV	G.GGA.AAC.GGG.AAG	pseudoknot	132, 349
BYDV-RPV	G.GGA.AAC.GGG.AAG	pseudoknot	
Enamovirus			
PEMV RNA2	U.UUU.UGG.UAG	hairpin	
PEMV RNA1	G.GGA.AAC.GGA.UUA.U	pseudoknot	
Dianthovirus			
RCNMV	G.GAU.UUU.UAG.GCG	hairpin	195
CRSV	G.GAU.UUU.UAA.GU	hairpin	287
Sobemovirus			
CfMV	U.UUA.AAC.UGC.CAG.CG	hairpin	230
SBMV	U.UUA.AAC.UGC.UUG.CG	hairpin	230
RYMV	U.UUA.AAC.UGC.CAG.GG	hairpin	230
Carlavirus			
PVM	U.AGA.AAA.UGA	none?	141
+1 shifts (proposed)			
Closterovirus			
BYV	CGG.GUU.UAG.CUC	pseudoknot	4
CTV	CGC.GUU.CGC	none?	190

sequence located more than 3 kb downstream of the site [353], and sequences 40–100 nucleotides upstream of the site may also be required [246].

The frameshift efficiency of a given signal is not only determined by the *cis*-acting sequences but also by the available tRNAs. In infected *Nicotiana benthamiana* plants, shift-promoting tRNA (classes) exist that decode UUU, UUA, UUC and the asparagine codon AAU but none that decode the asparagine codon AAC [195].

The supposed frameshift signal of PVM differs from those described above. The shifty sequence U.AGA.AAA.UGA allows only one basepair to be formed after the shift between RNA and tRNA at the P site, which is hardly sufficient to drive the shift. It was, therefore, suggested that in this case the shift occurs after peptidyl transfer and translocation when the last codon of the ORF (AAA) together with the peptidyl tRNA occupies the P site and the ribosome awaits recognition of the stop codon at the A site by the release factor [141]. This model was supported by mutagenesis analyses of the role of the stop codon and of the preceding A stretch. The frameshift signal was less effective than those described above (about 0.3% in the reticulocyte lysate) and the postulated frameshift product was not detectable in infected plants. It is noteworthy that the downstream ORF, which should be linked by the frameshift event to the upstream coat protein ORF, has its own translation start codon and is also translated separately (at least *in vitro*) by an unknown mechanism [141].

A number of yeast retrotransposons use a +1 frameshift to produce the correspondent of a *gag-pol* fusion protein [22]. In the case of Ty3, the sequence GCG.AGU.U is decoded as Ala-Val by reading the underlined codons. No slippage of the tRNA^{Ala} is possible but the frameshift efficiency depends on the presence of the 'rare' AGU codon and certain features of the tRNA that decodes the GCG codon or (more general) other special codons located upstream of a rare codon. Artificial frameshift sites can be constructed by combining codons that are decoded by such 'shifty' tRNAs with rare codons [261, 351]. A similar codon configuration may be present in a suggested +1 frameshift site between closterovirus ORFs 1a and 1b, which contains a rare CGG codon (CTV [190]) or a stop codon (BYV [4]) and, at least for CTV, lacks obvious secondary structure [190]. Frameshifting has still to be proven for these viruses.

The mammalian type C retroviruses use stop codon suppression instead of frameshifting to produce the *gag-pol* fusion protein [97, 156, 365]. Suppression occurs with an efficiency of about 5% *in vitro* and *in vivo* and is accomplished by misreading of an UAG termination codon by a glutaminyl tRNA [97]. A UGA stop codon at the same position is decoded as arginine, cysteine or tryptophan [97]. Since the stop codons of normal cellular genes are not suppressed detectably under the same conditions, it was concluded that a specific context is responsible for the effect in retroviral RNAs. The presence of a pseudoknot structure 8 nucleotides downstream was required for efficient readthrough *in vitro* [98, 360, 361], while *in vivo* only evidence for the requirement of a stem structure was obtained [96].

Stop codon suppression has been suggested as a mechanism for the expression of some seed storage proteins whose ORFs are interrupted by stop codons [93, 223, 352]. It is, however, not always clear whether these ORFs are on functional mRNAs or whether they represent silent pseudogenes. It has been pointed out that storage protein genes contain a high number of CAA and CAG codons (coding for glutamine) which could mutate into a stop codon by a simple C-to-T transition [93]. A stop codon in the hordein gene λ-hor1–14 was suppressed in bombarded barley endosperm with an efficiency of only 0.6% [93].

A large number of plant RNA viruses use stop codon suppression to produce components of RNA dependent RNA-polymerases or elongated coat proteins that are probably required for transmissibility by their respective vectors [243, 326, 366]. Unlike for retroviruses, no structure requirements for stop-codon suppression have been observed in plant systems. Rather readthrough is dependent on the immediate sequence context of the stop codon and on the presence of certain tRNA species that can decode such stop codons as sense codons (Table 4). Suppressible stop codons have been inserted into GUS ORFs and readthrough has been quantified in transfected plant protoplasts. The context of the stop codon of the tobacco mosaic virus 126 kDa ORF (CAA.UAG.CAA.UUA) allowed particularly efficient (around 5%) readthrough for all three possible stop codons [306, 307, 346]. The leakiness of this stop codon context was used to express angiotensin-I-converting enzyme inhibitor peptide as the C-terminal extension of a subfraction of TMV coat proteins from an accordingly engineered infectious TMV clone [148]. Mutagenesis analysis of the readthrough sequence defined (C/A)(A/C)A.UAG.CAR.YYA (with R=purine, Y=pyrimidine) as optimal consensus. This consensus sequence, like the UGA readthrough signals of TRV [150], PEBV [229], PCV and SBW-MV [163] and of sindbis virus RNA [219], fits to the statistical analysis of eukaryotic stop codon contexts: efficiently recognized stop codons normally avoid a C directly downstream but rather have a purine in this position [9, 42, 43, 236, 344]. In the light of this finding, the readthrough of the UAG.G signals of the carmo- and luteoviruses, and of the tombusviruses [159, 296] (Table 4) has to be functionally different and may require additional *cis*-acting sequences or specific tRNAs. The presence of a conserved CCCCA motif or repeated CCXXXX motifs downstream of the luteo- and carmovirus readthrough sites has been noted but the involvement in readthrough has not been experimentally tested so far [243]. The signals of TRV, Car-MV, MCDV, BYDV and BWYV were active in *in vitro* systems or in infected plants [12, 52, 100, 265] but were rather ineffective in transfected tobacco protoplasts [306].

The sequences of plant viral readthrough proteins produced *in vivo* have not been determined, but it is likely that the stop codons are decoded by particular tRNAs, as has been found in *in vitro* systems and *Xenopus* oocytes, where addition of tRNA^{Tyr} with a GΨA anticodon but not with a GUA anticodon greatly stimulated readthrough of the TMV signal [15, 20, 21, 368]. Interestingly, this tRNA is rare in young wheat leaves but is abundant in older tissue, suggesting that processes like readthrough (and possibly also frameshifting) are developmentally regulated. The

Table 4. Leaky termination codons of plant virus ORFs. The immediate sequence context of partially suppressed termination codons (underlined) and the protein resulting from translational suppression are listed. References to those viral sequences not further mentioned in this review can be found in the compilation by Miller *et al.* [243].

	Read-through stop codon	Read-through protein	References
Tobamovirus			
TMV	CAA.**UAG**.CAA.UUA	replicase	264, 306, 346
Tymovirus			
TYMV	CAA.**UAG**.CAA.UCA	polyprotein extension	247, 38
Furovirus			
BNYVV	CAA.**UAG**.CAA.UUA	coat protein extension	35, 373
Luteovirus			
BYDV	SCC.AAA.**UAG**.GUA.GAC	coat protein extension	243
PLRV	CCC.AAA.**UAG**.GUA.GAC	coat protein extension	
BWYV	CCC.AAA.**UAG**.GUA.GAC	coat protein extension	
SDV	GCU.AAA.**UAG**.GUA.GAC	coat protein extension	
BWYV-ST9	AAA.**UAG**.GGC	replicase	
Enamovirus			
PEMV	UCC.CUC.**UGA**.GGG.AC	coat protein extension	
Carmovirus			
CarMV	AAA.**UAG**.GGG	replicase (1. stop)	
	CAG.**UAG**.UUG	replicase (2. stop)	
MCMV	AAA.**UAG**.GGG	replicase	
CCFV	CGC.**UAG**.GGG	replicase	
MNSV	AAC.**UAG**.GGG	replicase	
TCV	CGC.**UAG**.GGG	replicase	
Necrovirus			
TNV	AAA.**UAG**.GGG	replicase	68
Tobravirus			
TRV	UUA.**UGA**.CGG.UUU	replicase	150
Tombusvirus			
TBSV	AAA.**UAG**.GGG	replicase	159, 296
Sobemovirus			
PCV RNA1	AAA.**UGA**.CGG	replicase	163
SBWMV RNA1	AAA.**UGA**.CGG	replicase	303
SBWMV RNA2	AGU.**UGA**.CGG	coat protein extension	

TMV readthrough site was also efficiently suppressed (20%) in transfected tobacco protoplasts by a bean tRNA^{leu} with an artificially introduced CUA anticodon [51] and in protoplasts and transgenic plants (up to 10%) by modified trp tRNAs [104]. The fact that such suppressor tRNAs can be overexpressed without strong phenotypic changes suggests that normal stop codons are not significantly suppressed by these tRNAs. It has, however, been noted that suppression is more efficient in protoplasts than in plants, suggesting that only

transgenic plants are regenerated which have a lower expression level of the suppressor tRNA [51].

The leaky UGA stop codon of TRV is suppressed by a tRNA^{Trp} with a CmCA anticodon [369]. One such tRNA originated from the chloroplast and was more efficient than that from the cytoplasm. Since the respective chloroplast and mitochondrial tRNAs are almost identical and TRV appears to be associated with mitochondria in infected cells it is possible that the virus uses the mitochondrial tRNA as suppressor *in vivo* [369]. At present it is unknown which tRNA suppresses the UAG.G stop codon of the luteoviruses.

Different suppressing tRNAs could be the reason that readthrough efficiencies vary with the assay system, but possible involvement of *cis*-active sequences distal to the readthrough sites may also account for the variation. For example, in *in vitro* translation experiments with the full-length TMV RNA, fourfold higher readthrough efficiencies have been obtained [264] than with the artificial constructs of Skuzeski *et al.* [306, 307], and it has to be determined whether this is an effect of more distal *cis*-acting sequences or of the different assay systems.

In other organisms, reading of UGA stop codons as selenocysteinyl codons is a special type of stop codon suppression ([95, 157] for reviews). Selenocysteine has been found in plant proteins [41] and a selenocysteine-tRNA recognizing UGA has been identified in *Beta vulgaris* [157]. However, it is unknown whether UGA misreading is the only way of selenocysteine incorporation into plant proteins and whether this process requires additional *cis*-active sequences on the mRNA, as was found in animal cells [30, 302] and *E. coli* [160].

For retroviruses, it is still a matter of discussion whether virus infection alters the frameshift or stop codon suppression capacity of the cell. Controversial results on induction of synthesis of shifty tRNAs or on differences in the levels of tRNA modifications with consequences for the frameshift efficiency have been presented ([156] for review). For plants, however, nothing comparable is known for tRNAs involved in stop codon suppression, although concentration differences in different tissues have been reported [21, 285], and the findings of Kim and Lommel [195] may indicate that the frameshift-prone tRNAs may also be represented differently in different plants.

Translation under special conditions

Many responses of plants to developmental and environmental signals or to stresses like wounding, heat shock, lack of water or oxygen have post-transcriptional components (for reviews see Gallie, this issue; [127, 143, 320]). Generally the translation machinery concentrates on a number of specific mRNAs while others are not translated anymore and are either stored as inactive RNAs or degraded. The pattern of polysome-associated mRNAs and overall translational activity can vary widely, but little is known about the fate of specific mRNAs and about the controlling *cis*- and *trans*-acting features. Translational

control is exerted mainly at the levels of initiation [29, 277] and elongation [10, 28, 99, 248, 305, 310, 348, 354]. Heat-shock mRNAs are more competitive under heat stress and this feature has been attributed to their leader sequences [131, 269, 293]. These sequences are, however, very heterogeneous and no common motifs have been discerned so far [187].

The association of mRNAs with architectural components of the cell or the precise location of an mRNA in the cytoplasm can influence the translational activity on that RNA [67, 149]. An alteration of such association or sequestration seems to be one level of translational control under conditions of stress or changes in developmental programs in plant cells [10, 13, 24, 70, 256, 277], but the molecular basis of this phenomenon is unclear.

Conclusions

A number of features that modulate translation have been detected in mRNAs translated in animal and yeast cells. Some of these are also found in RNAs translated in the cytoplasm of plant cells, predominantly on viral mRNAs. Quantitative control of the synthesis of two (or more) proteins from one mRNA is achieved by leaky scanning, frameshifting or stop codon suppression, or transactivated reinitiation on plant viral mRNAs. With the possible exception of stop codon suppression, it is not known whether any of these mechanisms is active also in the control of nuclear gene expression or is so plant, tissue, or development specific that it could be part of the host-range control of plant viruses. The 5' leader sequences and the 3'-untranslated region separately or in conjunction increase the cap-dependent or -independent affinity for ribosomes and probably can serve as a paradigm for cellular RNAs that are translated under special conditions. Short ORFs are found in the leaders of a number of nuclear and viral genes, but it is unknown whether these sORFs simply serve to reduce downstream translation or whether they regulate translation in response to cellular conditions. Further work is required to link these features to the translation control events observed in plant cells in response to a variety of stresses and developmental or environmental signals.

In particular, it is to be expected that further study of the influence of 5' leaders and 3'-UTRs on (cap-independent) translation, of the translation factors involved in this process in plants, and further analysis of the unusual translation mechanisms of the plant pararetroviruses will contribute essential knowledge to our understanding of eukaryotic translation processes.

Acknowledgements

We acknowledge the critical reading of the manuscript by Pat King and the steady help and interest of Ingo Potrykus. The authors are supported by biotechnology grants of the Swiss National Funds.

References

1. Abastado JP, Miller PF, Jackson BM, Hinnebusch AG: Suppression of ribosomal reinitiation at upstream open reading frames in amino-acid-starved cells forms the basis for GCN4 translational control. Mol Cell Biol 11: 486–496 (1991).
2. Acland P, Dixon M, Peters G, Dickson C: Subcellular fate of the Int-2 oncoprotein is determined by choice of initiation codon. Nature 343: 662–665 (1990).
3. Agol VI: The 5′-untranslated region of picornaviral genomes. Adv Virus Res 40: 103–180 (1991).
4. Agranovsky AA, Koonin EV, Boyko VP, Maiss E, Frötschl R, Lunina NA, Atabekov JG: Beet yellows closterovirus: complete genome structure and identification of a leader papain-like thiol protease. Virology 198: 311–324 (1994).
5. Ahlquist P, Dasgupta R, Shih DS, Zimmern D, Kaesberg P: Two-step binding of eukaryotic ribosomes to brome mosaic virus RNA 3. Nature 281: 277–282 (1979).
6. Ainley WM, Key JL: Development of a heat shock inducible expression cassette for plants: characterization of parameters for its use in transient expression assays. Plant Mol Biol 14: 949–967 (1990).
7. Altmann M, Blum S, Wilson TMA, Trachsel H: The 5′-leader sequence of tobacco mosaic virus RNA mediates initiation-factor–4E-independent, but still initiation-factor–4A-dependent translation in yeast extracts. Gene 91: 127–129 (1990).
8. Angenon G, Uotila J, Kurkela SA, Teeri TH, Botterman J, Van Montague M, Depicker A: Expression of dicistronic transcription units in transgenic tobacco. Mol Cell Biol 9: 5676–5684 (1989).
9. Angenon G, Van Montague M, Depicker A: Analysis of the stop codon context in plant nuclear genes. FEBS Lett 271: 144–149 (1990).
10. Apuya NR, Zimmermann JL: Heat shock gene expression is controlled primarily at the translational level in carrot cells and somatic embryos. Plant Cell 4: 657–665 (1992).
11. Atkins JF, Weiss RB, Thompson S, Gesteland RF: Towards a genetic dissection of the basis of triplet decoding, and its natural subversion: Programmed reading frame shifts and hops. Annu Rev Genet 25: 201–228 (1991).
12. Bahner I, Lamb J, Mayo MA, Hay RT: Expression of the genome of potato leafroll virus: readthrough of the coat protein termination condon in vivo. J Gen Virol 71: 2251–2256 (1990).
13. Bailey-Serres J, Freeling M: Hypoxic stress-induced changes in ribosomes of maize seedling roots. Plant Physiol 94: 1237–1243 (1990).
14. Baim SB, Sherman F: mRNA structures influencing translation in the yeast Saccharomyces cerevisiae. Mol Cell Biol 8: 1591–1601 (1988).
15. Barciszewski J, Barciszewska M, Suter B, Kubli E: Plant tRNA suppressors: in vivo readthrough properties and nucleotide sequence of yellow lupin seeds tRNA^{Tyr}. Plant Sci 40: 193–196 (1985).
16. Basso J, Dallaire P, Charest PJ, Devantier Y, Laliberte J-F: Evidence for an internal ribosome entry site within the 5′-untranslated region of turnip mosaic potyvirus RNA. J Gen Virol 75: 3157–3165 (1994).
17. Baughman GA, Howell SH: Cauliflower mosaic virus 35S RNA leader region inhibits translation of downstream genes. Virology 167: 125–135 (1988).
18. Baughman GA, Jacobs JD, Howell SH: Cauliflower mosaic virus gene VI produces a symptomatic phenotype in transgenic tobacco plants. Proc Natl Acad Sci USA 85: 733–737 (1988).
19. Beccera SP, Rose JA, Hardy M, Baroudy BM, Anderson CW: Direct mapping of adeno-associated virus capsid protein B and C: a possible AUC initiation codon. Proc Natl Acad Sci USA 76: 7919–7923 (1985).
20. Beier H, Barciszewska M, Krupp G, Mitnacht R, Gross HJ: UAG readthrough during TMV RNA translation: Isolation and sequence of two tRNAs^{Tyr} with suppressor activity from tobacco plants. EMBO J 3: 351–356 (1984).
21. Beier H, Barciszewska M, Sickinger H-D: The molecular basis for the differential translation of TMV RNA in tobacco protoplasts and wheat germ extracts. EMBO J 3: 1091–1096 (1984).
22. Belcourt MF, Farabaugh PJ: Ribosomal frameshifting in the yeast retrotransposon Ty: tRNA slippage on a 7 nucleotide minimal site. Cell 62: 339–352 (1990).
23. Belsham GJ, Lomonossoff GP: The mechanism of translation of cowpea mosaic virus middle component RNA: no evidence for internal initiation from experiments in an animal cell transient expression assay. J Gen Virol 72: 3109–3113 (1991).
24. Beltrán-Peña E, Ortiz-López A, Sánchez de Jiménez E: Synthesis of ribosomal proteins from stored mRNAs early in seed germination. Plant Mol Biol 28: 327–336 (1995).
25. Benhar I, Engelberg-Kulka H: Frameshifting of the E. coli trpR gene occurs by the bypassing of a segment of its coding sequence. Cell 72: 121–130 (1993).
26. Benkowski LA, Ravel JM, Browning KS: mRNA binding properties of wheat germ protein synthesis initiation factor 2. Biochem Biophys Res Comm 214: 1033–1039 (1995).
27. Berlioz C, Darlix J-L: An internal ribosome entry mechanism promotes translation of murine leukemia virus gag polyprotein precursors. J Virol 69: 2214–2222 (1995).
28. Berry JO, Carr JP, Klessig DF: mRNAs encoding ribulose-1,5-bisphosphate carboxylase remain bound to polysomes but are not translated in amaranth seedlings transferred to darkness. Proc Natl Acad Sci USA 85: 4190–4194 (1988).
29. Berry JO, Breiding DE, Klessig DF: Light-mediated control of translation initiation of ribulose-1,5-bisphosphate carboxylase in amaranth cotyledons. Plant Cell 2: 795–803 (1990).
30. Berry MJ, Banu L, Chen YY, Mandel SJ, Kieffer JD, Harney JW, Larsen PR: Recognition of UGA as a selenocysteine codon in type I deiodinase requires sequences in the 3′ untranslated region. Nature 353: 273–276 (1991).
31. Bevan M: Binary Agrobacterium vectors for plant transformation. Nucl Acids Res 12: 8711–8720 (1984).

32. Boeck R, Kolakofski D: Positions +5 and +6 can be major determinants of the efficiency of non-AUG initiation codons for protein synthesis. EMBO J 13: 3608–3617 (1994).

33. Bonetti B, Fu L, Moon J, Bedwell DM: The efficiency of translation termination is determined by a synergistic interplay between upstream and downstream sequences in *Saccharomyces cerevisiae*. J Mol Biol 251: 334–345 (1995).

34. Bonneville J-M, Sanfaçon H, Fütterer J, Hohn T: Posttranscriptional transactivation in cauliflower mosaic virus. Cell 59: 1135–1143 (1989).

35. Bouzoubaa S, Ziegler V, Beck D, Guilley H, Richards K, Jonard G: Nucleotide sequence of beet necrotic yellow vein virus RNA-2. J Gen Virol 67: 1689–1700 (1986).

36. Boyd L, Thummel CS: Selection of CUG and AUG initiator codons for *Drosophila* E74A translation depends on downstream sequences. Proc Natl Acad Sci USA 90: 9164–9167 (1993).

37. Boyer SK, Shotwell MA, Larkins BA: Evidence for the translational control of storage protein gene expression in oat seeds. J Biol Chem 267: 17449–17457 (1992).

38. Bransom KL, Weiland JJ, Tsai C-H, Dreher TW: coding density of the turnip yellow mosaic virus genome: roles of the overlapping coat protein and p206-readthrough coding regions. Virology 206: 403–412 (1995).

39. Brault V , Miller WA: Translational frameshifting mediated by a viral sequence in plant cells. Proc Natl Acad Sci USA 89: 2262–2266 (1992).

40. Brierley I, Rolley N, Jenner AJ, Inglis SC: Mutational analysis of the RNA pseudoknot component of a coronavirus ribosomal frameshifting signal. J Mol Biol 220: 889–902 (1991).

41. Brown TA, Shrift A: Identification of selenocysteine in the proteins of selenate-grown *Vigna radiata*. Plant Physiol 66: 758–761 (1980).

42. Brown CM, Stockwell PA, Trotman CNA, Tate WP: Sequence analysis suggests that tetra-nucleotides signal the termination of protein synthesis in eukaryotes. Nucl Acids Res 18: 6339–6345 (1990).

43. Brown CM, Dalphin ME, Stockwell PA, Tate WP: The translational termination signal database. Nucl Acids Res 21: 3119–3123 (1993).

44. Browning KS, Fletcher L, Ravel JM: Evidence that the requirements for ATP and wheat germ initiation factors 4A and 4F are affected by a region of satellite tobacco necrosis virus RNA that is 3′ to the ribosomal binding site. J Biol Chem 263: 9630–9634 (1988).

45. Buckingham RH: Codon context and protein synthesis – enhancements of the genetic code. Biochimie 76: 351–354 (1994).

46. Bugler B, Amalric F, Prats H: Alternative initiation of translation determines cytoplasmic or nuclear localization of basic fibroblast growth factor. Mol Cell Biol 11: 573–577 (1992).

47. Bulmer M: Coevolution of codon usage and transfer RNA abundance. Nature 325: 728–730 (1987).

48. Callis J, Fromm M, Walbot V: Expression of mRNA electroporated in plant and animal cells. Nucl Acids Res 15: 5823–5831 (1987).

49. Campbell WH, Gowri G: Codon usage in higher plants, green algae, and cyanobacteria. Plant Physiol 92: 1–11 (1990).

50. Cao JH, Geballe AP: Translational inhibition by a human cytomegalovirus upstream open reading frame despite inefficient utilization of its AUG codon. J Virol 69: 1030–1036 (1995).

51. Carneiro VTC, Pelletier G, Small I: Transfer RNA-mediated suppression of stop codons in protoplasts and transgenic plants. Plant Mol Biol 22: 681–690 (1993).

52. Carrington JC, Morris TJ: Characterization of cell-free translation products of carnation mottle virus genomic and subgenomic RNAs. Virology 144: 1–10 (1985).

53. Carrington JC, Freed DD: Cap-independent enhancement of translation by a plant potyvirus 5′-untranslated region. J Virol 64: 1590–1597 (1990).

54. Carroll R, Derse D: Translation of equine infectious anemia virus bicistronic tat-rev mRNA requires leaky ribosome scanning of the tat CTG initiation codon. J Virol 67: 1433–1440 (1993).

55. Cavener DR, Ray SC: Eukaryotic start and stop translation sites. Nucl Acids Res 19: 3185–3192 (1991).

56. Chamorro M, Parkin N, Varmus HE: An RNA pseudoknot and an optimal heptameric shift site are required for highly efficient ribosomal frameshifting on a retroviral messenger RNA. Proc Natl Acad Sci USA 89: 713–717 (1992).

57. Chen CY, Sarnow P: Initiation of protein synthesis by the eukaryotic translational apparatus on circular RNAs. Science 268: 415–417 (1995).

58. Chen G, Müller M, Potrykus I, Hohn T, Fütterer J: Rice tungro bacilliform virus: transcription and translation in protoplasts. Virology 204: 91–100 (1994).

59. Chen GFT, Inouye M: Role of the AGA/AGG codons, the rarest codons in global gene expression in *Escherichia coli*. Genes Devel 8: 2641-2652 (1994).

60. Chen X, Chamorro M, Lee SI, Shen LX, Hines JV, Tinoco Jr I, Varmus HE: Structural and functional studies of retroviral RNA pseudoknots involved in ribosomal frameshifting: nucleotides at the junction of the two stems are important for efficient ribosomal frameshifting. EMBO J 14: 842–852 (1995).

61. Chin L-S, Foster JL, Falk BW: The beet western yellows virus ST9-associated RNA shares structural and nucleotide sequence homology with carmo-like viruses. Virology 192: 473–482 (1993).

62. Chiorini JA, Boal TR, Miyamoto S, Safer B: A difference in the rate of ribosomal elongation balances the synthesis of eukaryotic translation initiation factor (eIF)-2 and eIF-2β. J Biol Chem 268: 13748–13755 (1993).

63. Cho HJ, Morikawa H, Murooka Y: Expression pattern of bacterial polycistronic genes in tobacco cells. J Ferment Bioeng 80: 111–117 (1995).

64. Christensen AK, Kahn LE, Bourne CM: Circular polysomes predominate on the rough endoplasmic reticulum of somatropes and mammotropes in the rat anterior pituitary. Am J Anat 178: 1–10 (1987).

65. Clements JM, Laz TM, Sherman F: Efficiency of translation initiation by non-AUG codons in *Saccharomyces cerevisiae*. Mol Cell Biol 8: 4533–4536 (1988).

66. Collins RF, Roberts M, Phoenix DA: Codon usage in *Escherichia coli* may modulate translation initiation. Biochem Soc Transact 23: 76 (1995).

67. Condeelis J: Elongation factor 1-alpha, translation and the cytoskeleton. Trends Biochem Sci 20: 169–170 (1995).

68. Coutts RHA, Rigden JE, Slabas AR, Lomonossoff GP: The complete nucleotide sequence of tobacco necrosis virus strain D. J Gen Virol 72: 1521–1529 (1991).

69. Craigen WJ, Lee CC, Caskey CT: Recent advances in peptide chain termination. Mol Microbiol 4: 861–865 (1990).

70. Crosby JS, Vayda ME: Stress-induced translational control in potato tubers may be mediated by polysome associated proteins. Plant Cell 3: 1013–1023 (1991).

71. Crowley KS, Reinhart GD, Johnson AE: The signal sequence moves through a ribosomal tunnel into a noncytoplasmic aqueous environment at the ER membrane early in translocation. Cell 73: 1101–1115 (1993).

72. Curran J, Kolakofsky D: Scanning independent ribosomal initiation of the sendai virus X protein. EMBO J 7: 2869–2874 (1988).

73. Damiani RD, Wessler SR: An upstream open reading frame represses expression of Lc, a member of the R/B family of maize transcriptional activators. Proc Natl Acad Sci USA 90: 8244–8248 (1993).

74. Danthinne X, Seurinck J, Meulewaeter F, Van Montagu M, Cornelissen M: The 3′ untranslated region of satellite tobacco necrosis virus RNA stimulates translation in vitro. Mol Cell Biol 13: 3340–3349 (1993).

75. Dasgupta R, Ahlquist P, Kaesberg P: Sequence of the 3′ untranslated region of brome mosaic virus coat protein messenger RNA. Virology 104: 339–346 (1980).

76. Dasso MC, Milburn SC, Hershey JWB, Jackson RJ: Selection of the 5′-proximal translation initiation site is influenced by mRNA and eIF-2 concentrations. Eur J Biochem 187: 361–371 (1990).

77. Datla RSS, Bekkaoui F, Hammerlindl JK, Pilate G, Dunstan DI, Crosby WL: Improved high-level constitutive foreign gene expression in plants using an AMV RNA4 untranslated leader sequence. Plant Sci 94: 139–149 (1993).

78. De Boer HA, Kastelein RA: Biased codon usage. In: Reznikoff W, Gold L (eds) Maximizing Gene Expression, pp. 225-285. Butterworths, Boston (1986).

79. Decker CJ, Parker P: Diversity of cytoplasmic functions for the 3′ untranslated region of eukaryotic transcripts. Curr Opin Cell Biol 7: 386–392 (1995).

80. Degnin CR, Schleiss MR, Cao J, Geballe AP: Translational inhibition mediated by a short upstream open reading frame in the human cytomegalovirus gpUL4 (gp48) transcript. J Virol 67: 5514–5521 (1993).

81. De Melo Neto OP, Standart N, Desa CM: Autoregulation of poly(A)-binding protein synthesis in vitro. Nucl Acids Res 23: 2198–2205 (1995).

82. Demler SA, De Zoeten GA: The nucleotide sequence and luteovirus-like nature of RNA 1 of an aphid non-transmissable strain of pea enation mosaic virus. J Gen Virol 72: 1819–1834 (1991).

83. De Tapia M, Himmelbach A, Hohn T: Molecular dissection of the cauliflower mosaic virus translation transactivator. EMBO J 12: 3305–3314 (1993).

84. Dewey RE, Wilson RF, Novitzky WP, Goode JH: The AAPT1 gene of soybean complements a cholinephosphotransferase-deficient mutant of yeast. Plant Cell 6: 1495–1507 (1994).

85. Di R, Dinesh-Kumar SP, Miller WA: Translational frameshifting by barley yellow dwarf virus RNA (PAV serotype) in Escherichia coli and in eukaryotic cell-free extracts. Mol Plant-Microbe Interact 6: 444–452 (1993).

86. Dinesh-Kumar SP, Brault V, Miller WA: Precise mapping and in vitro translation of a trifunctional subgenomic RNA of barley yellow dwarf virus. Virology 187: 711–722 (1992).

87. Dinesh-Kumar SP, Miller WA: Control of start codon choice on a plant viral RNA encoding overlapping genes. Plant Cell 5: 679–692 (1993).

88. Dixon L K, Hohn T: Initiation of translation of the cauliflower mosaic virus genome from a polycistronic mRNA: evidence from deletion mutagenesis. EMBO J 3: 2731-2736 (1984).

89. Donahue TF, Cigan AM, Pabich EK, Castilho Valavicius B: Mutations at a Zn(II) finger motif in the yeast eIF-2β gene alter ribosomal start-site selection during the scanning process. Cell 54: 621–632 (1988).

90. Dorris DR, Erickson FL, Hannig EM: Mutations in GCD11, the structural gene for eIF2-gamma in yeast, alter translational regulation of GCN4 and the selection of the start site for protein synthesis. EMBO J 14: 2239–2249 (1995).

91. Dowson Day MJ, Ashurst JL, Mathias SF, Watts JW: Plant viral leaders influence expression of a reporter gene in tobacco. Plant Mol Biol 23: 97–109 (1993).

92. Dubochet J, Morel C, Lebleu B, Herzberg M: Structure of globin mRNA and mRNA-protein particles: use of dark-field electron microscopy. Eur J Biochem 36: 465–472 (1973).

93. Entwistle J, Knudson S, Müller M, Cameron-Mills V: Amber codon suppression: the in vivo and in vitro analysis of two C-hordein genes from barley. Plant Mol Biol 17: 1217–1231 (1991).

94. Fajardo JE, Shatkin AJ: Translation of bicistronic viral mRNA in transfected cells: Regulation at the level of elongation. Proc Natl Acad Sci USA 87: 328–332 (1990).

95. Farabaugh PJ: Alternative readings of the genetic code. Cell 74: 591–596 (1993).

96. Felsenstein KM, Goff SP: Mutational analysis of the gag-pol junction of Moloney murine leukemia virus: Requirements for expression of the gag-pol fusion protein. J Virol 66: 6601–6608 (1992).

97. Feng Y-X, Copeland TD, Oroszlan S, Rein A, Levin JG: Identification of amino acids inserted during suppression of UAA and UGA termination codons at the gag-pol junction of Moloney murine leukemia virus. Proc Natl Acad Sci USA 87: 8860–8863 (1990).

98. Feng Y-X, Yuan H, Rein A, Levin JG: Bipartite signal for read-through suppression in murine leukemia virus mRNA: an eight-nucleotide purine-rich sequence immediately downstream of the gag termination codon followed by an RNA pseudoknot. J Virol 66: 5127–5132 (1992).

99. Fennoy SL, Bailey Serres J: Post-translational regulation of gene expression in oxygen-deprived roots of maize. Plant J 7: 287-295 (1995).

100. Filichkin SA, Lister RM, McGrath PF, Young MJ: In vivo expression and mutational analysis of the barley yellow dwarf virus readthrough gene. Virology 205: 290–299 (1994).

101. Filipowicz W, Haenni A-L: Binding of ribosomes to 5′-terminal leader sequences of eukaryotic messenger RNAs. Proc Natl Acad Sci USA 76: 3111–3115 (1979).

102. Fletcher L, Corbin SD, Browning KS, Ravel JM: The absence of a m7G cap on β-globin mRNA and alfalfa mosaic virus RNA 4 increases the amounts of initiation factor 4F required for translation. J Biol Chem 32: 19582–19587 (1990).

103. Florentz C, Brian JP, Giegé R: Possible functional role of viral tRNA-like structures. FEBS Lett 176: 295–300 (1984).

104. Franklin S, Lin TY, Folk WR: Construction and expression of nonsense suppressor tRNAs which function in plant cells. Plant J 2: 583–588 (1992).

105. French R, Jancke M, Ahlquist P: Bacterial genes inserted in an engineered RNA virus. Efficient expression in monocotyledonous plant cells. Science 231: 1294–1297 (1986).

106. Frolova L, Legoff X, Rasmussen HH, Cheperegin S, Drugeon G, Kress M, Arman I, Haenni AL, Celis JE, Philippe M, Justesen J, Kirilev L: A highly conserved eukaryotic pro-

182

tein family possessing properties of polypeptide chain release factor. Nature 372: 701–703 (1994).

107. Fujimoto H, Itoh K, Yamamoto M, Kyozuka J, Shimamoto K: Insect resistant rice generated by introduction of a modified δ-endotoxin gene of *Bacillus thuringiensis*. Bio/technology 11: 1151–1155 (1993).

108. Fütterer J, Gordon K, Bonneville JM, Sanfaçon H, Pisan B, Penswick J, Hohn T: The leading sequence of caulimovirus large RNA can be folded into a large stem-lop structure. Nucl Acids Res 16: 8377–8390 (1988).

109. Fütterer J, Gordon K, Pfeiffer P, Sanfaçon H, Pisan B, Bonneville JM, Hohn T: Differential inhibition of downstream gene expression by the cauliflower mosaic virus 35S RNA leader. Virus Genes 3: 45–55 (1989).

110. Fütterer J, Bonneville J-M, Gordon K, De Tapia M, Karlsson S, Hohn T: Expression from polycistronic cauliflower mosaic virus pregenomic RNA. In: Posttranscriptional Control of Gene Expression. NATO ASI Series H49, pp. 349–357 (1990).

111. Fütterer J, Gordon K, Sanfaçon H, Bonneville JM, Hohn T: Positive and negative control of translation by the leader sequence of cauliflower mosaic virus pregenomic 35S RNA. EMBO J 9: 1697–1707 (1990).

112. Fütterer J, Hohn T: Translation of a polycistronic mRNA in the presence of the cauliflower mosaic virus transactivator protein. EMBO J 10: 3887–3896 (1991).

113. Fütterer J, Hohn T: Role of an upstream open reading frame in the translation of polycistronic mRNAs in plant cells. Nucl Acids Res 20: 3851–3857 (1992).

114. Fütterer J, Kiss-László Z, Hohn T: Non-linear ribosome migration on cauliflower mosaic virus 35S RNA. Cell 73: 789–802 (1993).

115. Fütterer J, Potrykus I, Valles Brau MP, Dasgupta I, Hull R, Hohn T: Splicing in a plant pararetrovirus. Virology 198: 663–670 (1994).

116. Fütterer J, Potrykus I, Bao Y, Li L, Burns TM, Hull R, Hohn T: Position dependent ATT initiation during plant pararetrovirus rice tungro bacilliform virus translation. J Virol 70: 2999–3010 (1996).

117. Galili G, Kawata EE, Smith LD, Larkins BA: Role of the 3'-poly(A) sequence in translational regulation of mRNAs in Xenopus laevis oocytes. J Biol Chem 263: 5764–5770 (1988).

118. Gallie DR, Sleat DE, Watts JW, Turner P, Wilson TM: The 5'-leader sequence of tobacco mosaic virus RNA enhances the expression of foreign gene transcripts *in vitro* and *in vivo*. Nucl Acids Res 15: 3257–3273 (1987).

119. Gallie DR, Sleat DE, Watts JW, Turner P, Wilson TM: A comparison of eukaryotic viral 5'-leader sequences as enhancers of mRNA expression *in vivo*. Nucl Acids Res 15: 8693–8711 (1987).

120. Gallie DR, Lucas WJ, Walbot V: Visualizing mRNA expression in plant protoplasts: factors influencing efficient mRNA uptake and translation. Plant Cell 1: 301–311 (1989).

121. Gallie DR, Kado CI: A translational enhancer derived from tobacco mosaic virus is functionally equivalent to a Shine-Dalgarno sequence. Proc Natl Acad Sci USA 86: 129–132 (1989).

122. Gallie DR, Walbot V: RNA pseudoknot domain of tobacco mosaic virus can functionally substitute for a poly(A) tail in plant and animal cells. Genes Devel 4: 1149–1157 (1990).

123. Gallie DR: The cap and poly(A) tail function synergistically to regulate mRNA translational efficiency. Genes Devel 5: 2108–2116 (1991).

124. Gallie DR, Feder JN, Schimke RT, and Walbot V: Post-transcriptional regulation in higher eukaryotes: The role the reporter gene in controlling expression. Mol Gen Genet 228: 258–264 (1991).

125. Gallie DR, Feder JN, Schimke RT, Walbot V: Functional analysis of the tobacco mosaic virus tRNA-like structure in cytoplasmic gene regulation. Nucl Acids Res 19: 5031–5036 (1991).

126. Gallie DR, Walbot V: Identification of the motifs within the tobacco mosaic virus 5'-leader responsible for enhancing translation. Nucl Acids Res 20: 4631–4638 (1992).

127. Gallie DR: Posttranscriptional regulation of gene expression in plants. Annu Rev Plant Physiol Plant Mol Biol 44: 77–105 (1993).

128. Gallie DR, Kobayashi M: The role of the 3'-untranslated region of non-polyadenylated plant viral RNAs in regulating translational efficiency. Gene 142: 159–165 (1994).

129. Gallie DR, Young TE: The regulation of gene expression in transformed maize aleurone and endosperm protoplasts – analysis of promoter activity, intron enhancement, and mRNA untranslated regions on expression. Plant Physiol 106: 929–939 (1994).

130. Gallie DR, Tanguay R: Poly(A) binds to initiation factors and increases cap-dependent translation *in vitro*. J Biol Chem 269: 17166–17173 (1994).

131. Gallie DR, Caldwell C, Pitto L: Heat shock disrupts cap and poly(A) tail function during translation and increases mRNA stability of introduced reporter mRNA. Plant Physiol 108: 1703–1713 (1995).

132. Garcia A, Van Duin J, Pleij CWA: Differential response to frameshift signals in eukaryotic and prokaryotic translational systems. Nucl Acids Res 21: 401–406 (1993).

133. Geballe AP, Morris DR: Initiation codons within 5'-leaders of mRNAs as regulators of translation. Trend Biochem Sci 19: 159–164 (1994).

134. Godefroy-Colburn T, Ravelonandro M, Pinck L: Cap accessibility correlates with the initiation efficiency of alfalfa mosaic virus RNAs. Eur J Biochem 147: 549–552 (1985).

135. Goldman E, Rosenberg AH, Zubay G, Studier FW: Consecutive low-usage leucine codons block translation only when near the 5' end of a message in *Escherichia coli*. J Mol Biol 245: 467–473 (1995).

136. Goodall GJ, Filipowicz W: The AU-rich sequences in the introns of plant nuclear pre-mRNAs are required for splicing. Cell 58: 473–483 (1989).

137. Gordon K, Pfeiffer P, Fütterer J, Hohn T: *In vitro* expression of cauliflower mosaic virus genes. EMBO J 7: 309–317 (1988).

138. Gordon K, Fütterer J, Hohn T: Efficient initiation of translation at non-AUG triplets in plant cells. Plant J 2: 809–813 (1992).

139. Gowda S, Wu FC, Scholthof HB, Shepherd RJ: Gene VI of figwort mosaic virus (caulimovirus group) functions in posttranscriptional expression of genes on the full-length RNA transcript. Proc Natl Acad Sci USA 86: 9203–9207 (1989).

140. Gowda S, Scholthof HB, Wu FC, Shepherd RJ: Requirement of gene VII in cis for the expression of downstream genes on the major transcript of figwort mosaic virus. Virology 185: 867–871 (1991).

141. Gramstat A, Prüfer D, Rohde W: The nucleic acid-binding zinc finger protein of potato virus M is translated by internal initiation as well as by ribosomal frameshifting involving a shifty stop codon and a novel mechanism of P-site slippage. Nucl Acids Res 22: 3911–3917 (1994).

142. Grant CM, Hinnebusch AG: Effect of sequence context at the stop codons on efficiency of reinitiation in GCN4 translational control. Mol Cell Biol 14: 606–618 (1994).

143. Green PJ: Control of mRNA stability in higher plants. Plant Physiol 102: 1065–1070 (1993).

144. Grünert S, Jackson RJ: The immediate downstream codon strongly influences the efficiency of utilization of eukaryotic translation initiation codons. EMBO J 9: 3618–3630 (1994).

145. Gu Z, Harrod R, Rogers EJ, Lovett PS: Anti-peptidyl transferase leader peptides of attenuation-regulated chloramphenicol-resistance genes. Proc Natl Acad Sci USA 91: 5612–5616 (1994).

146. Guerineau F, Lucy A, Mullineaux P: Effect of two consensus sequences preceding the translation initiator codon on gene expression in plant protoplasts. Plant Mol Biol 18: 815–818 (1992).

147. Gultyaev AP, Van Batenburg FHD, Pleij CWA: The computer simulation of RNA folding pathways using a genetic algorithm. J Mol Biol 250: 37–51 (1995).

148. Hamamoto H, Sugiyama Y, Nakagawa N, Hashida E, Matsunaga Y, Takemoto S, Watanabe Y, Okada Y: A new tobacco mosaic virus vector and its use for the systemic production of angiotensin-I-converting enzyme inhibitor in transgenic tobacco and tomato. Bio/technology 11: 930–932 (1993).

149. Hamill D, Davies J, Drawbridge J, Suprenant KA: Polyribosome targeting to microtubules – enrichment of specific mRNAs in a reconstituted microtubule preparation from sea urchin embryos. J Cell Biol 127: 973–984 (1994).

150. Hamilton WDO, Boccara M, Robinson DJ, Baulcombe DC: The complete nucleotide sequence of tobacco rattle virus RNA-1. J Gen Virol 68: 2563–2575 (1987).

151. Hann SR, Sloan-Brown K, Spotts GD: Translational activation of the non-AUG-initiated c-myc 1 protein at high cell densities due to methionine deprivation. Genes Devel 6: 1229–1240 (1992).

152. Hann SR: Regulation and function of non-AUG-initiated proto-oncogenes. Biochimie 76: 880–886 (1994).

153. Hann LE, Gehrke L: mRNAs containing the unstructured 5′ leader sequence of alfalfa mosaic virus RNA 4 translate inefficiently in lysates from poliovirus-infected Hela cells. J Virol 69: 4986–4993 (1995).

154. Harrod AU, Lovett PS: Peptide inhibitors of peptidyltransferase alter the conformation of domains IV and V of large subunit rRNA: a model for nascent peptide control of translation. Proc Natl Acad Sci USA 92: 8650–8654 (1995).

155. Hatfield DL, Choi IS, Lee BJ, Jung J-E: Selenocysteyl-tRNAs recognize UGA in Beta vulgaris, a higher plant, and in Gliocladium virens, a filamentous fungus. Biochem Biophys Res Comm 184: 254-259 (1992).

156. Hatfield DL, Levin JG, Reim A, Oroszlan S: Translational suppression in retroviral gene expression. Adv Vir Res 41: 193–239 (1992).

157. Hatfield DL, Diamond A: UGA: a split personality in the universal genetic code. Trends Genet 9: 69–70 (1993).

158. Hay JM, Jones MC, Blackebrough ML, Dasgupta I, Davies JW, Hull R: An analysis of the sequence of an infectious clone of rice tungro bacilliform virus, a plant pararetrovirus. Nucl Acids Res 19: 2615–2621 (1991).

159. Hearne PQ, Knorr DA, Hillman BI, Morris TJ: The complete genome structure and synthesis of infectious RNA from clones of tomato bushy stunt virus. Virology 177: 141–151 (1990).

160. Heider J, Baron C, Böck A: Coding from a distance: dissection of the mRNA determinants required for the incorpora-tion of selenocysteine into protein. EMBO J 11: 3759–3766 (1992).

161. Hensgens LAM, Fornerod MWJ, Rueb S, Winkler AA, Van der Veen S, Schilperoort RA: Translation controls the expression level of a chimaeric reporter gene. Plant Mol Biol 20: 921–938 (1992).

162. Hershey JWB: Translational control in mammalian cells. Annu Rev Biochem 60: 717–755 (1991).

163. Herzog E, Guilley H, Manohar SK, Dollet M, Richards K, Fritsch C: Complete nucleotide sequence of peanut clump virus RNA 1 and relationships with other fungus-transmitted rod-shaped viruses. J Gen Virol 75: 3147–3155 (1994).

164. Herzog E, Guilley H, Fritsch C: Translation of the second gene of peanut clump virus RNA 2 occurs by leaky scanning in vitro. Virology 208: 215–225 (1995).

165. Hinnebusch AG: Translational control of GCN4 – an in vivo barometer of initiation-factor activity. Trends Biochem Sci 19: 409–414 (1994).

166. Hoekema A, Kastelein RA, Vasser M, De Boer HA: Codon replacement in the PGK1 gene of Saccharomyces cerevisiae: Experimental approach to study the role of biased codon usage in gene expression. Mol Cell Biol 7: 2914–2924 (1987).

167. Hohn T, Fütterer J: Pararetroviruses and retroviruses: a comparison of expression strategies. Semin Virol 2: 55–70 (1991).

168. Hohn T, Fütterer J: Transcriptional and translational control of gene expression in cauliflower mosaic virus. Curr Opin Genet Devel 2: 90–96 (1992).

169. Honigman A, Falk H, Mador N, Rosental T, Panet A: Translation frequency of the human T-cell leukemia virus (HTLV-2) gag gene modulates the frequency of ribosomal frameshifting. Virology 208: 312–318 (1995).

170. Horvath P, Suganuma A, Inaba M, Pan YB, Gupta KC: Multiple elements in the 5′-untranslated region downregulate c-sis messenger RNA translation. Cell Growth Diff 6: 1103–1110 (1995).

171. Hsu MT, Coca-Prodos M: Electron microscopic evidence for the circular form of RNA in the cytoplasm of eukaryotic cells. Nature 280: 339–340 (1979).

172. Huang WM, Ao S-Z, Casjens S, Orlandi R, Zeikus R, Weiss R, Winge D, Fang M: A persistent untranslated sequence within bacteriophage T4 DNA topoisomerase gene 60. Science 239: 1005–1012 (1988).

173. Hunter TR, Hunt T, Knowland J, Zimmern D: Messenger RNA for the coat protein of tobacco mosaic virus. Nature 260: 759–764 (1976).

174. Iida S, Mittelsten-Scheid O, Saul MW, Seipel K, Miyazaki C, Potrykus I: Expression of a downstream gene from a bicistronic transcription unit in transgenic tobacco plants. Gene 119: 199–205 (1992).

175. Iizuka N, Najita L, Franzusoff A, Sarnow P: Cap-dependent and cap-independent translation by internal initiation of mRNAs in cell extracts prepared from Saccharomyces cerevisiae. Mol Cell Biol 14: 7322–7330 (1994).

176. Ingelbrecht ILW, Herman LMF, Dekeyser RA, Van Montague MC, Depicker AG: Different 3′ end regions strongly influence the level of gene expression inplant cells. Plant Cell 1: 671–680 (1989).

177. Ivanov IG, Alexandrova RA, Dragulev BP, Abouhaidar MG: A second putative mRNA binding site on the Escherichia coli ribosome. Gene 160: 75–79 (1995).

178. Jacks T, Madhani HD, Masiarz FR, Varmus HE: Signals for ribosomal frameshifting in the Rous sarcoma virus gag-pol region. Cell 55: 447–458 (1988).

184

179. Jacks T, Power MD, Masiarz FR, Luciw PA, Barr PJ, Varmus HE: Characterization of ribosomal frameshifting in HIV-1 *gag-pol* expression. Nature 331: 280-283 (1988).

180. Jackson RJ, Standart N: Do the poly(A) tail and 3′ untranslated region control mRNA translation? Cell 62: 15–24 (1990).

181. Jang SK, Pestova TV, Hellen CUT Witherell GW, Wimmer E: Cap-independent translation of picornavirus RNAs: structure and function of the internal ribosome entry site. Enzyme 44: 292–309 (1990).

182. Jiang B, Monroe SS, Koonin EV, Stine SE, Glass RI: RNA sequence of astrovirus: distinctive genomic organization and putative retrovirus-like ribosomal frameshifting signal that directs viral replicase synthesis. Proc Natl Acad Sci USA 90: 10539–10543 (1993).

183. Jobling SA, Gehrke L: Enhanced translation of chimeric messenger RNAs containing a plant viral untranslated leader sequence. Nature 325: 622–625 (1987).

184. Jobling SA, Cuthbert CM, Rogers SG, Fraley RT, Gehrke L: *In vitro* transcription and translation efficiency of chimeric SP6 messenger RNAs devoid of 5′ vector nucleotides. Nucl Acids Res 16: 4483–4498 (1988).

185. Johansson HE, Belsham GJ, Sproat BS, Hentze MW: Target-specific arrest of mRNA translation by antisense 2′-*O*-alkyloligoribonucleotides. Nucl Acids Res 22: 4591–4598 (1994).

186. Joshi CP: An inspection of the domain between putative TATA box and translation start site in 79 plant genes. Nucl Acids Res 16: 6643–6653 (1987).

187. Joshi CP, Nguyen HT: 5′ untranslated leader sequences of eukaryotic mRNAs encoding heat shock induced proteins. Nucl Acids Res 23: 541–549 (1995).

188. Kaempffer R, Van Emmelo J, Fiers W: Specific binding of eukaryotic initiation factor 2 to stallite tobacco necrosis virus RNA at a 5′-terminal sequence comprising the ribosome binding site. Proc Natl Acad Sci USA 78: 1542–1546 (1981).

189. Kaminski A, Hunt SL, Gibbs CL, Jackson RJ: Internal initiation of mRNA translation in eukaryotes. Genet Engin 16: 115–155 (1994).

190. Karasev AV, Boyko VP, Gowda S, Nikolaeva OV, Hilf ME, Koonin EV, Niblett CL, Cline K, Gumpf DJ, Lee RF, Garnsey SM, Lewandowski, Dawson WO: Complete sequence of the citrus tristeza virus RNA genome. Virology 208: 511–520 (1995).

191. Karpova OV, Mavrodieva VA, Tomashevskaya OL, Rodionova NP, Atabekov JG: The 3′-untranslated region of brome mosaic virus RNA does not enhance translation of capped mRNAs *in vitro*. FEBS Lett 360: 281-285 (1995).

192. Kato T, Shirano Y, Kawazu T, Tada Y, Itoh E, Shibata D: A modified β-glucuronidase gene: Sensitive detection of plant promoter activities in suspension-cultured cells of tobacco and rice. Plant Mol Biol Rep 9: 333–339 (1991).

193. Kim J-K, Gamble Klein P, Mullet JE: Ribosomes pause at specific sites during synthesis of membrane-bound chloroplast reaction center protein D1. J Biol Chem 266: 14931–14938 (1991).

194. Kim J-K, Hollingsworth MJ: Localization of *in vivo* ribosome pause sites. Anal Biochem 206: 183–188 (1992).

195. Kim KH, Lommel SA: Identification and analysis of the site of −1 frameshifting in red clover necrotic mosaic virus. Virology 200: 574–582 (1994).

196. Kiss-László Z, Blanc S, Hohn T: Splicing of cauliflower mosaic virus is essential for viral infectivity. EMBO J 14: 3552–3562 (1995).

197. Konarska M, Filipowicz W, Domdey H, Gross HJ: Binding of ribosomes to linear and circular forms of the 5′-terminal leader fragment of tobacco mosaic virus. RNA Eur J Biochem 114: 221–227 (1981).

198. Kozak M: Point mutations define a sequence flanking the AUG initiator codon that modulates translation by eukaryotic ribosomes. Cell 44: 283–292 (1986).

199. Kozak M: Influences of mRNA secondary structure on initiation by eukaryotic ribosomes. Proc Natl Acad Sci USA 83: 2850–2854 (1986).

200. Kozak M: Effects of intercistronic length on the efficiency of reinitiation by eukaryotic ribosomes. Mol Cell Biol 7: 3438–3445 (1987).

201. Kozak M: Leader length and secondary structure modulate mRNA function under conditions of stress. Mol Cell Biol 8: 2737–2744 (1988).

202. Kozak M: The scanning model for translation: an update. J Cell Biol 108: 229–241 (1989).

203. Kozak M: Context effects and inefficient initiation at non-AUG codons in eukaryotic cell free translation systems. Mol Cell Biol 9: 5073–5080 (1989).

204. Kozak M: Circumstances and mechanisms of inhibition of translation by secondary structure in eukaryotic mRNAs. Mol Cell Biol 9: 5134–5142 (1989).

205. Kozak M: Downstream secondary structure facilitates recognition of initiator codons by eukaryotic ribosomes. Proc Natl Acad Sci USA 87: 8301–8305 (1990).

206. Kozak M: Effects of long 5′ leader sequences on initiation by eukaryotic ribosomes *in vitro*. Gene Exp 1: 117–125 (1991).

207. Kozak M: Structural features in eukaryotic mRNAs that modulate the initiation of translation. J Biol Chem 266: 19867–19870 (1991).

208. Kozak M: A consideration of alternative models for the initiation of translation in eukaryotes. Crit Rev Biochem Mol Biol 21: 385–402 (1992).

209. Kozak M: Regulation of translation in eukaryotic systems. Annu Rev Cell Biol 8: 197–225 (1992).

210. Kozak M: Determinants of translational fidelity and efficiency in vertebrate mRNAs. Biochimie 76: 815–821 (1994).

211. Kozak M: Adherence to the first-AUG rule when a second AUG codon follows closely upon the first. Proc Natl Acad Sci USA 92: 2662–2666 (1995).

212. Koziel MG, Beland GL, Bowman C, Carozzi NB, Crenshaw R, Crossland L, Dawson J, Desai N, Hill M, Kadwell S, Launis K, Lewis K, Maddox D, McPherson K, Meghji MR, Merlin E, Rhodes R, Warren GW, Wright M, Evola S: Field performance of elite transgenic maize plants expressing an insecticidal protein derived from Bacillus thuringiensis. Bio/technology 11: 194–200 (1993).

213. Kudlicki W, Kitaoka Y, Odom OW, Kramer G, Hardesty B: Elongation and folding of nascent ricin chains as peptidyl-tRNA on ribosomes – the effect of amino acid deletions on these processes. J Mol Biol 252: 203–212 (1995).

214. Kujawa AB, Drugeon G, Hulanicka D, Haenni A-L: Structural requirements for efficient translational frameshifting in the synthesis of the putative viral RNA-dependent RNA polymerase of potato leafroll virus. Nucl Acids Res 21: 2165–2171 (1993).

215. Ladhoff AM, Uerlings I, Rosenthal S: Electron microscopic evidence of circular molecules of 9-S globin mRNA from rabbit reticulocytes. Mol Biol Rep 7: 101–106 (1981).

216. Lahser FC, Marsh LE, Hall TC: Contributions of the brome mosaic virus RNA–3 3′-nontranslated region to replication and translation. J Virol 67: 3295–3303 (1993).

217. Leathers V, Tanguay R, Kobayashi M, Gallie DR: A phylogenetically conserved sequence within viral 3′ untranslated RNA pseudoknots regulates translation. Mol Cell Biol 13: 5331–5347 (1993).

218. Levis C, Astier-Manifacier S: The 5′ untranslated region of PVY RNA, even located in internal position, enables initiation of translation. Virus Genes 7: 367–379 (1993).

219. Li G, Rice CM: The signal for translational readthrough of a UGA codon in sindbis virus RNA involves a single cytidine residue immediately downstream of the termination codon. J Virol 67: 5062–5067 (1993).

220. Li YZ, Ma HM, Zhang Jl, Wang ZY, Hong MM: Effects of the first intron of rice waxy gene on the expression of foreign genes in rice and tobacco protoplasts. Plant Sci 108: 181–190 (1995).

221. Liebhaber SA, Cash F, Eshleman SS: Translation inhibition by an mRNA coding region secondary structure is determined by its proximity to the AUG initiation codon. J Mol Biol 226: 609–621 (1992).

222. Lim VI: Analysis of action of the wobble adenine on codon reading within the ribosome. J Mol Biol 252: 277–282 (1995).

223. Liu C-N, Rubinstein I: Transcriptional characterization of an α-zein gene cluster in maize. Plant Mol Biol 22: 323–336 (1993).

224. Lodish HF, Rose JK: Relative importance of 7-methylguanosine in ribosome binding and translation of VSV mRNA in wheat germ and reticulocyte cell-free systems. J Biol Chem 252: 1181–1188 (1977).

225. Lohmer S, Maddaloni M, Motto M, Salamini F, Thompson RD: Translation of the mRNA of the maize transcriptional activator opaque-2 is inhibited by upstream open reading frames present in the leader sequence. Plant Cell 5: 65–73 (1993).

226. Lovett PS: Nascent peptide regulation of translation. J Bact 176: 6415–6417 (1994).

227. Lütcke HA, Chow KC, Mickel FS, Moss KA, Kern HF, Scheele GA: Selection of AUG codons differs in plants and animals. EMBO J 6: 43–48 (1987).

228. Luukkonen BGM, Tan W, Schwartz S: Efficiency of reinitiation of translation on human immunodeficiency virus type 1 mRNAs is determined by the length of the upstream open reading frame and by the intercistronic distance. J Virol 69: 4086–4094 (1995).

229. Mac Farlane SA, Taylor SC, King DI, Hughes G, Davies JW: Pea early browning virus RNA1 encodes four polypeptides including a putative zinc-finger protein. Nucl Acids Res 17: 2245–2260 (1989).

230. Makinen K, Naess V, Tamm T, Truve E, Aaspollu A, Saarma M: The putative replicase of the cocksfoot mottle sobemovirus is translated as a part of the polyprotein by −1 ribosomal frameshift. Virology 207: 566–571 (1995).

231. Malkin LI, Rich A: Partial resistance of nascent polypeptide chains to proteolytic digestion due to ribosomal shielding. J Mol Biol 26: 329–346 (1967).

232. Maquat LE: When cells stop making sense: Effects of nonsense codons on RNA metabolism in vertebrate cells. RNA 1: 453–465 (1995).

233. Mayfield SP, Yohn CB, Cohen A, Danon A: Regulation of chloroplast gene expression. Annu Rev Plant Physiol Plant Mol Biol 46: 147–166 (1995).

234. Mazier M, Levis C, Chaybani R, Astier-Manifacier S, Tourneur J, Robaglia C: Enhancement of translational activity mediated by potyviral 5′-untranslated sequence in vivo but not in vitro. C R Acad Sci Ser III 317: 1065–1072 (1994).

235. McCarthy JEG, Kollmus H: Cytoplasmic mRNA-protein interactions in eukaryotic gene expression. Trends Biochem Sci 20: 191–197 (1995).

236. McCaughan KK, Brown CM, Dalphin ME, Berry MJ, Tate WP: Translational termination efficiency in mammals is influenced by the base following the stop codon. Proc Natl Acad Sci USA 92: 5431–5435 (1995).

237. Mehdi H, Ono E, Gupta KC: Initiation of translation at CUG, GUG and ACG codons in mammalian cells. Gene 91: 173–178 (1990).

238. Merrick WC: Mechanism and regulation of eukaryotic protein synthesis. Microbiol Rev 56: 291–315 (1992).

239. Merrick WC: Eukaryotic protein synthesis – an in vitro analysis. Biochimie 76: 822–830 (1994).

240. Meulewater F, Cornelissen M, Van Emmelo J: Subgenomic RNAs mediate expression of cistrons located internally on the genomic RNA of tobacco necrosis virus strain A. J Virol 66: 6419–6428 (1992).

241. Michelet B, Lukaszewicz M, Dupriez V, Boutry M: A plant plasma membrane proton-ATPase gene is regulated by development and environment and shows signs of a translational regulation. Plant Cell 6: 1375–1389 (1994).

242. Miller PF, Hinnebusch AG: Sequences that surround the stop codons of upstream open reading frames in GCN4 mRNA determine their distinct functions in translational control. Genes Devel 3: 1217–1225 (1989).

243. Miller WA, Dinesh-Kumar SP, Paul CP: Luteovirus gene expression. Crit Rev Plant Sci 14: 179–211 (1995).

244. Mirkov TE, Mathews DM, Du Plessis DH, Dodds JA: Nucleotide sequence and translation of satellite tobacco mosaic virus RNA. Virology 170: 139–146 (1989).

245. Moffat JG, Tate WP, Lovett PS: The leader peptides of attenuation-regulated chloramphenicol resistance genes inhibit translation termination. J Bact 176: 7115–7117 (1994).

246. Mohan BR, Dinesh-Kumar SP, Miller WA: Genes and cis-acting sequences involved in replication of barley yellow dwarf virus-PAV RNA. Virology 212: 186–195 (1995).

247. Morch MD, Boyer JC, Haenni AL: Overlapping open reading frames revealed by complete nucleotide sequencing of turnip yellow mosaic virus genomic RNA. Nucl Acids Res 16: 6157–6173 (1988).

248. Morelli JK, Shewmaker CK, Vayda ME: Biphasic stimulation of translational activity correlates with induction of translation elongation factor 1 subunit alpha upon wounding in potato tubers. Plant Physiol 106: 897–903 (1994).

249. Mottagui-Tabar S, Bjornsson A, Issaksson LA: The second to last amino acid in the nascent peptide as a codon context determinant. EMBO J 13: 249–257 (1994).

250. Munroe D, Jacobson A: Tales of poly(A): a review. Gene 91: 151–158 (1990).

251. Munroe D, Jacobson A: mRNA poly(A) tail, a 3′ enhancer of translation initiation. Mol Cell Biol 10: 3441–3455 (1990).

252. Murray EE, Lotzer J, Eberle M: Codon usage in plant genes. Nucl Acids Res 17: 477–493 (1989).

253. Murray EE, Rocheleau T, Eberle M, Stock C, Sekar V, Adang M: Analysis of unstable RNA transcripts of insecticidal crystal protein genes of Bacillus thuringiensis in transgenic plants and electroporated protoplasts. Plant Mol Biol 16: 1035–1050 (1991).

254. Nelson EM, Winkler MM: Regulation of mRNA entry into polysomes: parameters affect polysome size and the fraction of mRNA in polysomes. J Biol Chem 262: 11501–11506 (1987).

186

255. Nicolaisen M, Johansen E, Poulsen GB, Borkhardt B: The 5′ untranslated region of pea seedborne mosaic potyvirus RNA as a translational enhancer in pea and tobacco protoplasts. FEBS Lett 303: 169–172 (1992).

256. Nover L, Scharf K-D, Neumann D: Cytoplasmic heat shock granules are formed from precursor particles and are associated with a specific set of mRNAs. Mol Cell Biol 9: 1298–1308 (1989).

257. Oh S-K, Scott MP, Sarnow P: Homeotic gene antennapedia mRNA contains 5′-noncoding sequences that confer translation initiation by internal ribosome binding. Genes Devel 6: 1643–1653 (1992).

258. Ohlmann T, Rau M, Morley SJ, Pain VM: Proteolytic cleavage of initiation factor eIF-4-Gamma in the reticulocyte lysate inhibits translation of capped mRNAs but enhances that of uncapped mRNAs. Nucl Acids Res 23: 334–340 (1995).

259. Oliveira CC, McCarthy JEG: The relationship between eukaryotic translation and mRNA stability – a short upstream open reading frame strongly inhibits translational initiation and greatly accelerates mRNA degradation in the yeast *Saccharomyces cerevisiae*. J Biol Chem 270: 8936–8943 (1995).

260. Pain VM: Translational control during amino acid starvation. Biochimie 76: 718–728 (1994).

261. Pande S, Vimaladithan A, Zhao H, Farabaugh PJ: Pulling the ribosome out of frame by +1 at a programmed frameshift site by cognate binding of aminoacy-tRNA. Mol Cell Biol 15: 298–304 (1995).

262. Peabody DD: Translation initiation at non-AUG triplets in mammalian cells. J Biol Chem 264: 5031–5035 (1989).

263. Pease RJ, Leiper RJ, Harrison GB, Scott J: Studies on the translocation of the amino terminus of apolipoprotein B into the endoplasmic reticulum. J Biol Chem 270: 7261–7271 (1995).

264. Pelham HRB: Leaky UAG termination codon in tobacco virus RNA. Nature 272: 469–471 (1978).

265. Pelham HRB: Translation of tobacco rattle virus RNAs *in vitro*: four proteins from three RNAs. Virology 97: 256–265 (1979).

266. Pelletier J, Sonenberg N: Internal initiation of translation of eukaryotic mRNA directed by a sequence derived from poliovirus RNA. Nature 334: 320–325 (1988).

267. Perlak FJ, Fuchs RL, Dean DA, McPherson SL, Fischhoff DA: Modification of the coding sequence enhances plant expression of insect control protein genes. Proc Natl Acad Sci USA 88: 3324–3328 (1991).

268. Pinck M, Fritsch C, Ravelonandro M, Thivent C, Pinck L: Binding of ribosomes to the 5′ leader sequence ($N = 258$) of RNA 3 from alfalfa mosaic virus. Nucl Acids Res 9: 1087–1100 (1981).

269. Pitto L, Gallie DR, Walbot V: The role of the leader sequence during thermal repression of translation in maize, tobacco and carrot protoplasts Plant Physiol 100: 1827–1833 (1992).

270. Pooggin MM, Skryabin KG: The 5′-untranslated leader sequence of potato virus X RNA enhances the expression of a heterologous gene *in vivo*. Mol Gen Genet 234: 329–331 (1992).

271. Poole ES, Brown CM, Tate WR: The identity of the base following the stop codon determines the efficiency of *in vivo* translational termination in *Escherichia coli*. EMBO J 14: 151–158.

272. Potapov AP, Trianaalonso FJ, Nierhaus KH: Ribosomal decoding processes at codons in the A or P sites depend differently on 2′-OH groups. J Biol Chem 270: 17680–17684 (1995).

273. Prats H, Kaghad M, Prats AC, Klagsbrun M, Lelias JM, Liauzin P, Chalon P, Tauber JP, Amalric F, Smith JA, Caput D: High molecular mass forms of basic fibroblast growth factor are initiated by alternative CUG codons. Proc Natl Acad Sci USA 86: 1836–1840 (1989).

274. Prüfer D, Tacke E, Schmitz J, Kull B, Kaufmann A, Rohde W: Ribosomal frameshifting in plants: A novel signal directs the – 1 frameshift in the synthesis of the putative replicase of potato leafroll luteovirus. EMBO J 11: 1111–1117 (1992).

275. Putterill JJ, Gardner RC: Initiation of translation of the β-glucuronidase reporter gene at internal AUG codons in plant cells. Plant Sci 62: 199–205 (1989).

276. Quaedvlieg N, Dockx J, Rook F, Weisbeek P, Smeekens S: The homeobox gene ATH1 of *Arabidopsis* is derepressed in the photomorphogenic mutants cop1 and det1. Plant Cell 7: 117–129 (1995).

277. Reinbothe S, Reinbothe C, Parthier B: Methyl jasmonate represses translation initiation of a specific set of mRNAs in barley. Plant J 4: 459–467 (1993).

278. Rhoads RE: Cap recognition and the entry of mRNA into the protein synthesis cycle. Trends Biochem Sci 13: 52–56 (1988).

279. Rhoads RE: Regulation of eukaryotic protein synthesis by initiation factors. J Biol Chem 268: 3017–3020 (1993).

280. Riechmann JL, Lain S, Garcia JA: Identification of the initiation codon of plum pox potyvirus genomic RNA. Virology 185: 544–552 (1991).

281. Riis B, Rattan SIS, Clark BFC, Merrick WC: Eukaryotic protein elongation factors. Trends Biochem Sci 15: 420–424 (1990).

282. Rochaix J-D: Post-transcriptional steps in the expression of chloroplast genes. Annu Rev Cell Biol 8: 1–28 (1992).

283. Rochon DM, Johnston JC: Infectious transcripts from cloned cucumber necrosis virus cDNA: evidence for a bifunctional subgenomic RNA. Virology 181: 656–665 (1991).

284. Rogers SG, Fraley RT, Horsch RB, Levine AD, Flick JS, Brand LA, Fink CL, Mozer T, O'Connel K, Sanders PR: Evidence for ribosome scanning during translation initiation of mRNAs in transformed plant cells. Plan Mol Biol Rep 3: 111–116 (1985).

285. Rohde W, Gramstat A, Schmitz J, Tacke E, Prüfer D: Plant viruses as model systems for the study of non-canonical translation mechanisms in higher plants. J Gen Virol 75: 2141–2149 (1994).

286. Rothnie HM, Chapdelaine Y, Hohn T: Pararetroviruses and retroviruses: a comparative review of viral structure and gene expression strategies. Adv Virus Res 44: 1–67 (1994).

287. Ryabov EV, Generozov EV, Kendall TL, Lommel SA, Zavriev SK: Nucleotide sequence of carnation ringspot dianthovirus RNA-1. J Gen Virol 75: 243–247 (1994).

288. Ryabova LA, Torgashov AF, Kurnasov OV, Bubunenko MG, Spirin AS: The 3′-terminal untranslated region of alfalfa mosaic virus RNA 4 facilitates the RNA entry into translation in a cell-free system. FEBS Lett 326: 264–266 (1993).

289. Ryazanov AG, Rudkin BB, Spirin AS: Regulation of protein synthesis at the elongation stage. FEBS Lett 285: 170–175 (1991).

290. Sachs AB, Davies RW: The poly(A) binding protein is required for poly(A) shortening and 60S ribosomal subunit-dependent translation initiation. Celol 58: 857–867 (1989).

291. Saier MH: Differential codon usage – A safeguard against inappropriate expression of specialized genes. FEBS Lett 362: 1–4 (1995).

292. Scheper GC, Voorma HO, Thomas AAM: Basepairing with 18S ribosomal RNA in internal initiation of translation. FEBS Lett 352: 271–275 (1994).

293. Schöffl F, Rieping M, Baumann G, Bevan M, Angermüller S: The function of plant heat shock promoter elements in the regulated expression of chimaeric genes in transgenic tobacco. Mol Gen Genet 217: 246–253 (1989).

294. Scholthof HB, Gowda S, Wu FC, Shepherd RJ: The full-length transcript of caulimovirus is a polycistronic mRNA whose genes are transactivated by the product of gene VI. J Virol 66: 3131–3139 (1992).

295. Scholthof HB, Wu FC, Gowda S, Shepherd RJ: Regulation of caulimovirus gene expression and the involvement of cis-acting elements on both viral transcripts. Virology 190: 403–412 (1992).

296. Scholthof KGB, Scholthof HB, Jackson AO: The tomato bushy stunt virus replicase proteins are coordinately expressed and membrane associated. Virology 208: 365–369 (1995).

297. Schultze M, Hohn T, Jiricny J: The reverse transcriptase gene of cauliflower mosaic virus is translated separately form the capsid gene. EMBO J 9: 1177–1185 (1990).

298. Sedman SA, Gelembiuk GW, Mertz JE: Translation initiation at a downstream AUG occurs with increased efficiency when the upstream AUG is located very close to the 5' cap. J Virol 64: 453–457 (1990).

299. Sha YS, Broglio EP, Cannon JF, Schoelz JE: Expression of a plant viral polycistronic mRNA in yeast Saccharomyces cerevisiae mediated by a plant virus translational transactivator. Proc Natl Acad Sci USA 92: 8911–8915 (1995).

300. Shakin SH, Liebhaber SA: Destabilization of messenger RNA/complementary DNA duplexes by the elongating 80S ribosome. J Biol Chem 261: 16018–16025 (1986).

301. Shen LX, Tinoco I: The structure of an RNA pseudoknot that causes efficient frameshifting in mouse mammary tumor virus. J Mol Biol 247: 963–978 (1995).

302. Shen Q, Leonard JL, Newburger PE: Structure and function of the selenium translation element in the 3'-untranslated region of human cellular glutathione peroxidase mRNA. RNA 1: 519–525 (1995).

303. Shirako Y, Wilson TMA: Complete nucleotide sequence and organization of the bipartite RNA genome of soil-borne wheat mosaic virus. Virology 195: 16–32 (1993).

304. Sieliwanowicz B: The influence of poly(A)-binding proteins on translation of poly(A)+ RNA in a cell-free system from embryo axes of dry pea seeds. Biochim Biophys Acta 908: 54–59 (1987).

305. Skadsen RW, Scandalios JG: Translational control of photo-induced expression of the Cat2 catalase gene during leaf development in maize. Proc Natl Acad Sci USA 84: 2785–2789 (1987).

306. Skuzeski JM, Nichols LM, Gesteland RF: Analysis of leaky viral translation termination codons in vivo by transient expression of improved β-glucuronidase vectors. Plant Mol Biol 15: 65–79 (1990).

307. Skuzeski JM, Nichols LM, Gesteland RF, Atkins JF: The signal for a leaky UAG stop codon in several plant viruses includes the two downstream codons. J Mol Biol 218: 365–373 (1991).

308. Sleat DE, Gallie DR, Jefferson RA, Bevan MW, Turner, PC Wilson TMA: Characterization of the 5'-leader of tobacco mosaic virus RNA as a general enhancer of translation in vitro. Gene 60: 217–225 (1987).

309. Sleat DE, Hull R, Turner PC, Wilson TMA: Studies on the mechanism of translational enhancement by the 5'-leader sequences of tobacco mosaic virus RNA. Eur J Biochem 175: 75–86 (1988).

310. Slovin JP, Tobin EM: Synthesis and turnover of the light-harvesting chlorophyll a/b-protein in Lemna gibba grown with intermittent red light: possible translational control. Planta 154: 465–474 (1982).

311. Smirnyagina EV, Morozov SY, Radionova NP, Miroschnichenko NA, Solovyev AG, Fedorkin ON, Atabekov JG: Translational efficiency and competitive ability of mRNAs with 5'-untranslated αβ-leader of potato virus X RNA. Biochimie 73: 587–598 (1991).

312. Smith D, Yarus M: tRNA-tRNA interactions within cellular ribosomes. Proc Natl Acad Sci USA 86: 4397–4401 (1989).

313. Somogyi P, Jenner AJ, Brierley, I, Inglis SC: Ribosomal pausing during translation of an RNA pseudoknot. Mol Cell Biol 13: 6931–6940 (1993).

314. Sonenberg N: Picornavirus RNA translation continues to surprise. Trends Genet 7: 105–106 (1991).

315. Springer BA, Sligar SG: High-level expression of sperm whale myoglobin in Escherichia coli. Proc Natl Acad Sci USA 84: 8961–8965 (1987).

316. Standart N, Jackson RJ: Regulation of translation by specific protein mRNA interactions. Biochimie 76: 867–879 (1994).

317. Stansfield I, Jones KM, Kushnirov VV, Dagkesamanskaya AR, Poznyakowski AI, Paushkin SV, Nierras CR, Cox BS, Ter-Avanesyan MD, Tuite MF: The products of the SUP45 (eRF1) and SUP35 genes interact to mediate translation termination in Saccharomyces cerevisiae. EMBO J 14: 4365–4373 (1995).

318. Stansfield I, Jones KM, Tuite MF: The end in sight: terminating translation in eukaryotes. Trend Biochem Sci 20: 489–491 (1995).

319. Strazielle C, Benoit H, Hirth L: Particularités structurales de l'acide ribonucléique extrait du virus de la mosaïque jaune du navet. II. J Mol Biol 13: 735–748 (1965).

320. Sullivan ML, Green PJ: Post-transcriptional regulation of nuclear-encoded genes in higher plants: the roles of mRNA stability and translation. Plant Mol Biol 23: 1091–1104 (1993).

321. Suzuki N, Sugawara M, Kusano T: Rice dwarf phytoreovirus segment S12 transcript is tricistronic in vitro. Virology 191: 992–995 (1992).

322. Tacke E, Prüfer D, Salamini F, Rohde W: Characterization of a potato leafroll luteovirus subgenomic RNA: differential expression by internal translation initiation and UAG suppression. J Gen Virol 71: 2265–2272 (1990).

323. Tacke E, Kull B, Prüfer D, Reinold S, Schmitz J, Salamini F, Rohde W: PLRV expression in potato. In: Bills DD, Kung S-D (eds) Viral Pathogenesis and Disease Resistance. World Scientific Publishing, River Edge (1994).

324. Tahara SM, Dietlin TA, Dever TE, Merrick WC, Worrilow LM: Effect of eukaryotic initiation factor 4F on AUG selection in a bicistronic mRNA. J Biol Chem 266: 3594–3601 (1991).

325. Takahashi H, Shimamoto K, Ehara Y: Cauliflower mosaic virus gene VI causes growth suppression, development of necrotic spots and expression of defence-related genes in transgenic tobacco plants. Mol Gen Genet 216: 188–194 (1989).

326. Tamada T, Kusume T: Evidence that the 75K readthrough protein of beet necrotic yellow vein virus RNA-2 is essential for transmission by the fungus Polymyxa betae. J Gen Virol 72: 1497–1504 (1991).

188

327. Tanaka T, Nishihara M, Seki M, Sakamoto A, Tanaka K, Irifune K, Morikawa H: Successful expression in pollen of various plant species of *in vitro* synthesized mRNA introduced by particle bombardment. Plant Mol Biol 28: 337–341 (1995).

328. Taylor JL, Jones JDG, Sandler S, Mueller GM, Bedbrook J, Dunsmuir P: Optimizing the expression of chimeric genes in plant cells. Mol Gen Genet 210: 572–577 (1987).

329. Ten Dam EB, Pleij CWA, Bosch L: RNA pseudoknots: translational frameshifting and readthrough on viral RNAs. Virus Genes 4: 121–136 (1990).

330. Ten Dam EB, Verlaan PWG, Pleij CWA: Analysis of the role of the pseudoknot component in the SRV-1 gag-pro ribosomal frameshift signal – Loop lengths and stability of the stem regions. RNA 1: 146–154 (1995).

331. Thach RE: Cap recap: the involvement of eIF-4F in regulating gene expression. Cell 68: 177–180 (1992).

332. Thomas AAM, Scheper GC, Voorma HO: Hypothesis: is eukaryotic initiation factor 2 the scanning factor? New Biol 4: 404–407 (1992).

333. Thomas AAM, Ter Haar E, Wellink J, Voorma HO: Cowpea mosaic virus middle component RNA contains a sequence that allows internal binding of ribosomes and that requires eukaryotic initiation factor 4F for optimal translation. J Virol 65: 2953–2959 (1991).

334. Timmer RT, Benkowski LA, Schodin D, Lax SR, Metz AM, Ravel JM, Browning KS: The 5′ and 3′ untranslated regions of satellite tobacco necrosis virus RNA affect translational efficiency and dependence on a 5′ cap structure. J Biol Chem 268: 9504–9510 (1993).

335. Tomashevskaya OL, Solovyev AG, Karpova OV, Fedorkin ON, Rodionova P, Morozov, SY, Atabekov JG: Effects of sequence elements in the potato virus X RNA 5′-nontranslated αβ-leader on its translation enhancing activity. J Gen Virol 74: 2717–2724 (1993).

336. Tu C, Tzeng TW, Bruenn JA: Ribosomal movement impeded at a pseudoknot required for frameshifting. Proc Natl Acad Sci USA 89: 8636–8640 (1992).

337. Tulin EE, Tsutsumi K, Eijiri S: Continuously coupled transcription-translation system for the production of rice cytoplasmic aldolase. Biotech Bioeng 45: 511–516 (1995).

338. Tuite MF, Stansfield I: Translation – Knowing when to stop. Nature 372: 614–615 (1994).

339. Turner R, Bate N, Twell D, Foster GD: Analysis of a translational enhancer upstream from the coat protein open reading frame of potato virus S. Arch Virol 134: 321–333 (1994).

340. Tyc K, Konarska M, Gross HJ, Filipowicz W: Multiple ribosome binding to the 5′-terminal leader sequence of tobacco mosaic virus RNA. Assembly of an 80S ribosome-mRNA complex at an AUU codon. Eur J Biochem 140: 503–511 (1984).

341. Tzeng T-H, Tu C-L, Bruenn JA: Ribosomal frameshifting requires a pseudoknot in the *Saccharomyces cerevisiae* double-stranded RNA virus. J Virol 66: 999–1006 (1992).

342. Vagner S, Gensac M-C, Maret A, Bayard F, Amalric F, Prats H, Prats A-C: Alternative translation of human fibroblast growth factor 2 mRNA occurs by internal entry of ribosomes Mol Cell Biol 15: 35–44 (1995).

343. Vagner S, Waysbort A, Marenda M, Gensac MC, Amalric F, Prats AC: Alternative translation initiation of the Moloney murine leukemia virus mRNA controlled by internal ribosome entry involving the p57/PTB splicing factor. J Biol Chem 270: 20376–20383 (1995).

344. Valle RPC, Morch M-D: Stop making sense. Regulation at the level of termination in eukaryotic protein synthesis. FEBS Lett 235: 1–15 (1988).

345. Valle RPC, Haenni A-L: Peptide chain termination. In: Trachsel H (ed) Translation in Eukaryotes, pp. 177–189. CRC Press, Boca Raton, FL (1991).

346. Valle RPC, Drugeon G, Devignes-Morch MD, Legocki AB, Haenni A-L: Codon context effects in virus translational readthrough. A study *in vitro* of the determinants of TMV and Mo-MuLV amber suppression. FEBS Lett 306: 133–139 (1992).

347. Vancanneyt G, Rosahl S, Willmitzer L: Translatability of plant mRNAs strongly influences its accumulation in transgenic plants. Nucl Acids Res 18: 2917–2921 (1990).

348. Vayda ME, Shewmaker CK, Morelli JK: Translational arrest in hypoxic potato tubers is correlated with the aberrant association of elongation factor EF-1-alpha with polysomes. Plant Mol Biol 28: 751–757 (1995).

349. Veidt I, Bouzoubaa SE, Leiser R-M, Ziegler-Graf V, Guilley H, Richards K, Jonard G: Synthesis of full-length transcripts of beet western yellows virus RNA: messenger properties and biological activity in protoplasts. Virology 186: 192–200 (1992).

350. Verver J, Le Gall O, Van Kammen A, Wellink J: The sequence between nucleotides 161 and 512 of cowpea mosaic virus M RNA is able to support internal initiation of translation *in vitro*. J Gen Virol 72: 2339–2345 (1991).

351. Vimaladithan A, Farabaugh PJ: Special peptidyl-tRNA molecules can promote translational frameshift without slippage. Mol Cell Biol 14: 8107–8116 (1994).

352. Wandelt C, Feix G: Sequence of a maize 21 kd zein gene from maize containing an in-frame stop codon. Nucl Acids Res 17: 2354 (1989).

353. Wang S, Miller WA: A sequence located 4.5 to 5 kilobases from the 5′ end of the barley yellow dwarf virus (PAV) genome strongly stimulates translation of uncapped RNA. J Biol Chem 270: 13446–13452 (1995).

354. Webster C, Kim C-Y, Roberts JKM: Elongation and termination reactions of protein synthesis on maize root tip polysomes studied in a homologous cell-free system. Plant Physiol 96: 418–425 (1991).

355. Weiland J, Dreher TW: Infectious TYMV RNA from cloned cDNA: Effects *in vitro* and *in vivo* of point substitutions in the initiation codons of two extensively overlapping ORFs. Nucl Acids Res 17: 4675–4687 (1989).

356. Weiss R, Dunn DM, Shuh M, Atkins JF, Gesteland RF: *E. coli* ribosomes re-phase on retroviral frameshift signals at rates ranging from 2–50 percent. New Biol 1: 159–169 (1989).

357. Weiss RB, Huang WM, Dunn DM: A nascent peptide is required for ribosomal bypass of the coding gap in bacteriophage T4 gene 60. Cell 62: 117–126 (1990).

358. Werner M, Feller A, Messenguy F, Piérard A: The leader peptide of yeast gene CPA1 is essential for the translational repression of its expression. Cell 49: 805–813 (1987).

359. Wiedmann B, Sakai H, Davies TA, Wiedmann M: A protein complex required for signal-sequence-specific sorting and translocation. Nature 370: 434–440 (1994).

360. Wills NM, Gesteland RF, Atkins JF: Evidence that a downstream pseudoknot is required for translational readthrough of the Moloney murine leukemia virus gag stop codon. Proc Natl Acad Sci USA 88: 6991–6995 (1991).

361. Wills NM, Gesteland RF, Atkins JF: Pseudoknot-dependent read-through of retroviral gag termination codons: import-

ance of sequences in the spacer of loop 2. EMBO J 13: 4137–4144 (1994).

362. Wolin SL, Walter P: Ribosome pausing and stacking during translation of a eukaryotic mRNA. EMBO J 7: 3559–3569 (1988).

363. Xiong Z, Kim KH, Kendall TL, Lommel SA: Synthesis of the putative red clover necrotic mosaic virus RNA polymerase by ribosomal frameshifting *in vitro*. Virology 193: 213–221 (1993).

364. Yonath A, Leonard KR, Wittmann HG: A tunnel in the large ribosomal subunit revealed by three-dimensional image reconstruction. Science 236: 813–816 (1987).

365. Yoshinaka Y, Katoh I, Copeland TD, Oroszlan S: Translational readthrough of an amber termination codon during synthesis of feline leukemia virus protease. J Virol 55: 870–873 (1985).

366. Zaccomer B, Haenni A-L, Macaya G: The remarkable variety of plant virus genomes. J Gen Virol 76: 231–247 (1995).

367. Zelenina DA, Kulaeva OI, Smirnyagina EV, Solovyev AG, Miroshnichenko NA, Fedorkin ON, Rodionova NP, Morozov SY, Atabekov JG: Translation enhancing properties of the 5′-leader of potato virus X genomic RNA. FEBS Lett 296: 267–270 (1992).

368. Zerfass K, Beier H: Pseudouridine in the anticodon GΨA of plant cytoplasmic tRNA^{Tyr} is required for UAG and UAA suppression in the TMV specific context. Nucl Acids Res 20: 5911–5918 (1992).

369. Zerfass K, Beier H: The leaky UGA termination codon of tobacco rattle virus RNA is suppressed by tobacco chloroplast and cytoplasmic tRNAs^{Trp} with CmCA anticodon. EMBO J 11: 4167–4173 (1992).

370. Zeyenko VV, Ryabova LA, Gallie DR, Spirin AS: Enhancing effect of the 3′-untranslated region of tobacco mosaic virus RNA on protein synthesis *in vitro*. FEBS Lett 354: 271–273 (1994).

371. Zhang SP, Goldman E, Zubay G: Clustering of low usage codons and ribosomal movement. J Theor Biol 170: 339–3554 (1994).

372. Zhouraleva G, Frolova L, Le Goff X, Le Guellec R, Inge-Vechtomov S, Kisselev L, Philippe M: Termination of translation in eukaryotes is governed by two interacting polypeptide chain release factors, eRF1 and eRF3. EMBO J 14: 4065–4072 (1995).

373. Ziegler V, Richards K, Guilley H, Jonard G, Putz C: Cell-free translation of beet-necrotic yellow vein virus: read-through of the coat protein cistron. J Gen Virol 66: 2079–2087 (1985).

374. Zijlstra C, Hohn T: Cauliflower mosaic virus gene VI controls translation from dicistronic expression units in transgenic *Arabidopsis* plants. Plant Cell 4: 1471–1484 (1992).

375. Zimmer A, Zimmer AM, Reynolds K: Tissue specific expression of the retinoic acid receptor-beta-2: Regulation by short open reading frames in the 5′-noncoding region. J Cell Biol 127: 1111–1119 (1994).

Plant Molecular Biology **32**: 191–222, 1996.
© 1996 *Kluwer Academic Publishers.*

191

Molecular chaperones and protein folding in plants

Rebecca S. Boston[1], Paul V. Viitanen[2] & Elizabeth Vierling[3,*]
[1]*Department of Botany, North Carolina State University, Raleigh, NC 27695, USA;* [2]*Molecular Biology Division, Central Research and Development Department, E.I. DuPont de Nemours and Company, Experimental Station, Wilmington, DE 19880–0402, USA;* [3]*Department of Biochemistry, University of Arizona, Tucson, AZ 85721, USA* (*author for correspondence)*

Key words: heat shock proteins, foldases, BiP, protein transport, protein disulfide isomerase, calnexin

Contents

Abstract	191
Introduction	192
Cytosolic protein folding activities	193
HSP70 or stress seventy chaperone family	193
HSP90 and signal transduction	195
HSP100 and thermotolerance	196
TriC: the TCP1-containing cytosolic 'chaperonin'	197
The small HSPs: potential chaperones	198
Cytosolic immunophilins: peptidyl-prolyl *cis-trans* isomerases	199
Co-chaperones	199
ER chaperones and foldases	200
Protein folding in the endoplasmic reticulum	200
BiP, the binding protein	201
Role of BiP in assembly and packaging of seed storage proteins	201
GRP94	202
Foldases in the ER: PDI and PPI	203
Calnexin	203
Calreticulin	204
Small heat-shock proteins	204
Chloroplast-localized chaperones and foldases	204
The higher-plant chloroplast chaperonins	204
The Rubisco connection	205
Chloroplast cpn60 (ch-cpn60)	206
Chloroplast cpn10 (ch-cpn10)	207
Functional properties of the purified chloroplast chaperonins	208
Remaining questions about chloroplast chaperonins	209
Other chloroplast chaperones and foldases	209
Chaperone activities in mitochondria	211
Summary and future prospects	212
Acknowledgements	212
References	212

Abstract

Protein folding *in vivo* is mediated by an array of proteins that act either as 'foldases' or 'molecular chaperones'. Foldases include protein disulfide isomerase and peptidyl prolyl isomerase, which catalyze the rearrangement of disulfide bonds or isomerization of peptide bonds around Pro residues, respectively. Molecular chaperones are a diverse group of proteins, but they share the property that they bind substrate proteins that are in unstable, non-native structural states. The best understood chaperone systems are HSP70/DnaK and HSP60/GroE, but considerable data support a chaperone role for other proteins, including HSP100, HSP90, small HSPs and calnexin. Recent research indicates that many, if not all, cellular proteins interact with chaperones and/or foldases during their lifetime in the cell. Different chaperone and foldase systems are required for synthesis, targeting, maturation and degradation of proteins in all cellular compartments. Thus, these diverse proteins affect an exceptionally broad array of cellular processes required for both normal cell function and survival of stress conditions. This review summarizes our current understanding of how these proteins function in plants, with a major focus on those systems where the most detailed mechanistic data are available, or where features of the chaperone/foldase system or substrate proteins are unique to plants.

Introduction

In the past ten years a significant body of literature has accumulated demonstrating that *in vivo* protein folding is mediated by an array of proteins that act either as 'foldases' or 'molecular chaperones'. Numerous excellent reviews detail both the historical background and recent progress on protein folding *in vivo* involving these proteins [99, 100, 121, 165, 189, 216]. Foldases include enzymes such as protein disulfide isomerase (PDI) [88] and the 'immunophilins' or peptidyl prolyl isomerase (PPI or rotamase) [270] that have demonstrated catalytic activities which increase the rate of protein folding [101, 311]. As a group the molecular chaperones are quite diverse, but they share the property that they bind substrate proteins that are in unstable, non-native structural states. They are distinct from foldases in that they are not strictly catalysts of protein folding. Chaperone interactions with non-native proteins serve a variety of functions that are specific to different chaperones and include: facilitating folding of nascent proteins as they exit the ribosome, promoting folding of proteins to their final native state, holding substrates in an unstructured form that is competent for membrane transport, maintaining proteins in specific conformations, preventing aggregation of unfolded proteins, and promoting renaturation of aggregated proteins. The latter two functions are particularly important for cells experiencing high temperature and other stresses. It is therefore not surprising that many molecular chaperones were first identified in one or more organisms as heat shock proteins (HSPs) [51, 99, 121, 185, 231]. However, it is now recognized that the same or closely related proteins are frequently essential components of normal cells. Given the range of processes affected by protein folding and therefore by foldase and chaperone activities, these proteins are rapidly being implicated as important players in virtually all aspects of cell biology.

Major classes of molecular chaperones and foldases in eukaryotes are listed in Table 1 along with 'co-chaperones', which are accessory proteins known to mediate activity of specific chaperones [99, 246]. The bacterial counterpart of each protein is also listed, and underscores the evolutionary conservation of the protein folding processes. The best understood chaperone systems are HSP70/DnaK and HSP60/GroE (see [113] for review). For both these systems detailed genetic experiments as well as mechanistic studies *in vitro* have been performed to characterize their activities, and for mammalian heat-shock cognate 70 and bacterial GroE three-dimensional structural data are available [24, 81]. Probably the least well understood classes are the HSP100 and small HSPs (smHSPs). Although genetic evidence implicates HSP100 in protein disaggregation [230], such an activity has not yet been demonstrated in a reconstituted system. In contrast, significant data have documented chaperone activity *in vitro* for the smHSPs [137, 138, 170], but as yet these activities have not been confirmed by studies *in vivo*. The foldases have well-described enzymatic activities [88, 270, 275] but their major functions *in vivo*, particularly with regard to substrates, need further investigation.

Almost all of the chaperones and foldases listed in Table 1 have now been identified in plant cells. The goal of this review is to summarize our current understanding of how these proteins function in plants, with a major focus on those systems where the most detailed mechanistic data are available, or where features of the chaperone/foldase system or substrate proteins are unique to plants. Previous reviews should be consulted for background information on chaperones and the stress response in plants [31, 121, 216, 231]. It is important to note, as evident from Table 1, that in eukaryotes, chaperones as well as some foldases are found in essentially every cellular compartment, which is not surprising given the ubiquitous need for protein folding. It should also be pointed out that multiple chaperones can interact with a single substrate along a protein folding pathway [51, 99, 121]. To emphasize the importance of chaperones and foldases for specific cellular processes, as well as the proposed functional interactions between different chaperones, data are reviewed with reference to different intracellular compartments.

Table 1. Major classes of foldases and molecular chaperones in eukaryotes

Protein class[a]	Intracellular location	Prokaryotic homologue
Chaperones		
Clp proteins		
HSP100(ClpB)	cytoplasm, mitochondrion	ClpB
ClpA/C	chloroplast	ClpA/C
HSP90		HtpG
HSP80/90	cytoplasm	
GRP94	ER	
HSP70		DnaK
HSP/HSC70	cytoplasm/nucleus, chloroplast, mitochondrion	
BiP/GRP78	ER	
Chaperonins		GroEL
HSP60/Cpn60	chloroplast, mitochondrion	
TRIC/TCP-1	cytoplasm	TF55 [b]
Calnexin	ER	
Nucleoplasmin	Nucleus	
Small HSPs	cytoplasm, mitochondrion, chloroplast, ER[c]	IBPA/B
Co-chaperones[d]		
DnaJ/HSP40[e] (Hsp70)	cytoplasm, mitochondrion, ER	DnaJ
GrpE (HSP70)	mitochondrion	GrpE
Cpn10 (Cpn60)	mitochondrion, chloroplast	GroES
Foldases		
Protein disulphide isomerase	ER	DsbA
Peptidyl prolyl isomerase (or immunophilins)		
Cyclophilin	cytoplasm, ER, mitochondria	PPIa,b
FK506 binding protein	cytoplasm, ER	

[a] Only the major members of each class are listed, in many cases multiple alternative names have appeared in the literature, see [100] for additional nomenclature.
[b] In archaebacteria.
[c] There is no specific nomenclature for the small HSPs found in different cellular compartments; they are referred to as small HSPs with a molecular weight value.
[d] The major chaperone with which each protein interacts appears in parentheses.
[e] Eukaryotic DnaJ homologues are diverse and have multiple names, see [54, 282] for review.

Cytosolic protein folding activities

HSP70 or stress seventy chaperone family

As the major site of protein synthesis, the cytosol contains significant levels of chaperones required for normal cellular function. Heat shock protein 70 (HSP70), also known as 'stress seventy' or heat shock cognate 70 (HSC70), is a prominent chaperone found in the cytosol of virtually all eukaryotic cells [51, 121]. HSP70s are among the most highly conserved proteins known, with ca. 50% identical residues between the *Escherichia coli* homologue DnaK and the eukaryotic HSP70s [20]. HSP70 function is essential for viability of eukaryotic cells, which has been demonstrated most directly in genetic experiments in the yeast *Saccharomyces cerevisiae* (see [50] for review). HSP70 proteins also increase in abundance during heat shock (the origin of the name), and the increase can be due to more vigorous transcription of the same genes expressed in the absence of stress, or to induction of other genes [51, 63, 69].

HSP70 proteins have two major domains, an amino-terminal ATPase domain of ca. 45 kDa, and a carboxyl-terminal peptide-binding domain of ca. 25 kDa. The structure of the amino-terminal domain, which is the most highly conserved region, has been solved at atomic resolution for bovine brain HSP70 [81]. Interestingly, the tertiary structure of the ATP binding fold resembles that of actin and hexokinase [21, 82]. ATP binding and hydrolysis seem to modulate the affinity of HSP70 for substrate [228]. The peptide-binding domain has not yet been solved at the three dimensional level, but it has been modeled to resemble the peptide-binding cleft of MHC I proteins [248], a structure consistent with its demonstrated peptide-binding ability [316]. The ATPase activity of HSP70 is stimulated by peptides [83] and by unfolded proteins [289], and this stimulation is lost when the carboxyl-terminal domain is removed. The most variable region of the protein is adjacent to the carboxyl terminus. This region may dictate some degree of substrate specificity, and/or interactions with specific co-chaperones in the DnaJ and GrpE family.

Experiments using the ER form of HSP70, BiP, have defined in general the peptide-binding preferences of HSP70 [16, 84, 153]. The binding site is thought to accommodate 7–8 residue long peptides in an extended conformation. Peptides with certain alternating aliphatic and aromatic residues are preferred by BiP. Although members of the HSP70 family share general peptide-binding characteristics, recent evidence indicates that they may have different peptide-binding specificities [86, 107]. Most importantly, it should be recognized that there is no consensus binding motif on the substrate. Instead, common structural elements allow HSP70 to bind to a wide range of non-native proteins.

Current models propose that HSP70 binds to nascent polypeptides thereby preventing improper folding before synthesis of the polypeptide chain is complete [90, 113]. The binding of unfolded polypeptides to HSP70 in the cytoplasm also appears to serve a number of more specific functions (see [50, 121, 185, 246] for review) such as uncoating of clathrin-coated vesicles [147] and maintaining precursors in a form competent for transport into organelles [50]. HSP70 has also been demonstrated to concentrate in the nucleus and nucleolus during stress, where it may be involved in pre-ribosome assembly [235]. HSP70 is also found in the steroid hormone receptor complex, and may be required for assembly of this complex (see section below on HSP90).

Numerous genes encoding cytosolic HSP70 proteins have been isolated from plants, including maize [11, 250], petunia [330], potato (EMBL accession number Z11983), soybean [249], carrot [184], spinach [155], *Arabidopsis* [334], tomato [183] and pea [63], and from the alga *Chlamydomonas* [220]. It is typical of a single plant species to have multiple genes for cytosolic HSP70s all of which encode proteins that are greater than 90% identical to each other at the amino acid level. Even *S. cerevisiae* has multiple cytosolic HSP70 genes (see [49] for review), and it remains unknown to what degree these proteins are functionally distinct. Observations of differential regulation of specific HSP70 genes in plants supports the hypothesis that this diversity is not just gene redundancy, but further understanding of the biochemical distinctions between these subtly different HSP70 isoforms is required.

There are considerable descriptive data concerning HSP70 expression in a variety of plants, including studies during heat and other stresses, as well as investigations of tissue-specific and developmental regulation [29, 47, 63, 68, 313]. These data are consistent with the ability of HSP70s to interact with diverse substrates and therefore take part in many different cellular processes. Plant cytosolic HSP70s presumably participate in folding of newly synthesized proteins, facilitate transport of proteins into organelles and perform other functions as has been demonstrated for their counterparts in other eukaryotes. However, there have been only limited mechanistic investigations of plant cytosolic HSP70s. Evidence has been obtained that cytosolic HSP70 from pea is capable of uncoating clathrin-coated vesicles [151], as has been demonstrated for mammalian HSP70. Miernyk and colleagues [210] showed that HSP70 in a wheat germ cell-free translation extract associated with newly translated proteins in an ATP-dependent fashion. In addition, HSP70 increased efficiency of precursor protein translocation into maize microsomal membranes as demonstrated by immunodepletion and add-back experiments. The same researchers have studied properties and activities of recombinant tomato cytosolic HSP70 and a site-directed mutant in which the proposed ATP-binding site was deleted [211]. They report that the normal wild-type protein has ATPase activity that can be stimulated by DnaJ and GrpE, similar to the *E. coli* HSP70 homologue DnaK [182], while the mutant is completely inactive. The mutant protein also fails to bind to ATP agarose and to stimulate protein translocation into microsomes. Stimulation of purified maize cytosolic HSP70 ATPase activity by cations, peptides and co-chaperones has also been partially characterized [343]. Two reports consider the oligomeric structure of HSP70, which appears to be variable, depending on ATP concentration. Anderson *et al.* [5] report HSP70 exists as monomers, dimers and higher-molecular-weight oligomers in protein fractions that contain both cytosolic and ER forms of HSP70. Whether the oligomeric complexes contain HSP70s alone or HSP70s in association with co-chaperones (see below) or target proteins was not determined. Addition of ATP to oligomers causes an increase in monomers and dimers, while ATP added to dimers leads to both monomer and oligomer formation. Miernyk and Hayman [211] also see three HSP70 forms, and find that mutant HSP70 lacking the ATP-binding site cannot form dimers. The significance of these multiple HSP70 forms remains unknown.

HSP90 and signal transduction

The HSP90 family of proteins, like the HSP70 family of proteins, are highly conserved among both prokaryotes and eukaryotes, sharing at least 40% amino acid identity [111, 185]. Similar also to HSP70, one or more HSP90 family members are typically constitutive, abundant cellular proteins, and yeast cells require at least one HSP90 for viability [4]. Both biochemical and genetic evidence support a role for HSP90 in modulating cellular signal transduction through effects on protein conformation/folding ([222, 242, 261] and references therein). In the mid-1980s it was recognized that HSP90 was a component a high-molecular-weight steroid receptor complex in mammalian cells, and the assembly and composition of this complex has been studied extensively *in vitro* in reticulocyte lysate extracts. In the current model HSP90 binding to steroid receptors, such as the glucocorticoid and estrogen receptors, is required for the receptors to be in a high-affinity hormone-binding state. Subsequent release of HSP90 must then occur before the receptor can bind DNA and activate the transcriptional reponse to hormone. HSP90 has also been implicated in regulation of signal transduction by certain protein kinases [261]. Genetic evidence for these functions was first obtained in a heterologous system in which mammalian steroid hormone receptor or kinase activities were assayed in *S. cerevisiae* [238, 335]. Reduced HSP90 levels eliminated efficient hormone induction of reporter gene transcription and suppressed the toxicity of high level expression of the *v-src* tyrosine protein kinase. However, it should be noted that not all tyrosine protein kinases expressed in yeast are affected

by HSP90 levels, and no endogenous yeast substrates of HSP90 have yet been identified.

Confirmation of a signalling role for HSP90 has been obtained more recently in other organisms. Mutations in *Drosophila* Hsp83 (an HSP90 familiy member) were found to reduce receptor tyrosine kinase activity of the *sevenless* gene [53], and in *Schizosaccharomyces pombe* Hsp90 was found to physically and genetically interact with the Wee1 tyrosine kinase [2]. In mammalian cells, certain benzoquinone ansamycins, drugs that were thought to inhibit tyrosine kinases, act by binding to HSP90, and presumably inhibit functional interactions of HSP90 in signal tranduction complexes [325]. Interestingly, other chaperones and foldases are believed to be involved in the HSP90 protein complexes modulating receptor and kinase activity [19]. HSP70 and certain immunophilins (see below) are also components of the steroid receptor/HSP90 complex [34, 242]. Furthermore, the HSP70 co-chaperone DnaJ was identified in yeast as required for HSP90 function in assays of heterologous receptor and kinase activity [150]. As seen for other processes involving chaperones, in many cases modulation of signal transduction appears to require the activity of more than one chaperone.

The majority of information about HSP90 function concerns its chaperone activity in normal cells. Whether it performs the same and/or additional roles during heat stress needs further investigation. Based on studies *in vitro*, Buchner and colleagues [138, 140, 141, 326] have proposed that HSP90 also prevents thermal denaturation and aggregation of protein substrates. They have shown that HSP90 *in vitro* will bind to early unfolding intermediates of model substrates and thereby prevent their irreversible inactivation. Unlike HSP70 activity, the reaction is nucleotide-independent and presumably depends on the rate constants of HSP90 binding and release relative to the rate constants of substrate denaturation. To date there is no direct evidence for this type of activity *in vivo* and it is not known how this activity may be related to HSP90 function in signal transduction.

There is very little information from higher plants on this important class of cytosolic chaperones. Genes encoding HSP90s along with their patterns of expression at the mRNA level have been reported in a number of plant species including maize [197], tomato [158], *Pharbitis nil* [79], *Brassica napus* [160] and *Arabidopsis* [46, 288, 336]. The cloned genes show various patterns of tissue-specific expression and only some of the genes are heat regulated. In *Arabidopsis* expression of certain HSP90 genes was high in pollen cells, anther primordia and around shoot apical meristems [336]. Pollen expression was also observed in maize from the pre-meiotic to the binucleate phase [197]. Tomato HSP90 mRNA was found to be abundant in shoot and root apicies [158]. In *Brassica* HSP90 was also induced by cold treatment [160]. The functional significance of these mRNA expression patterns remains unknown.

Essentially no published studies have examined the HSP90 proteins in plant cells, and no potential substrates for HSP90s have been identified in plants. However, recent studies have demonstrated that wheat germ extracts, like rabbit reticulocyte extracts, will assemble glucocorticoid receptor/HSP90 complexes in which the receptor is converted to a steroid-binding form ([134] and P. Krishna, personal communication). Similar to the mammalian complexes, the receptor/HSP90 complexes contain HSP70 and at least one immunophilin, providing evidence for the conservation of multiple chaperone interactions in HSP90 function (P. Krishna, personal communication). It will be of great interest to ultimately determine in what ways HSP90 chaperone function may be coupled to signal transduction and gene activation in plant cells.

HSP100 and thermotolerance

In plants, the synthesis during heat shock of a protein or proteins in the 100 kDa size range was recognized a number of years ago. Recent cloning of the corresponding gene from soybean [172], *Arabidopsis* [268] and maize (J. Nieto-Sotolo, personal communication) revealed that the encoded proteins are highly homologous to yeast Hsp104 [232]. To date, no homologous proteins have been cloned from any other higher eukaryotes. However, these HSP100 proteins belong to a larger class of proteins, the Clp proteins, which have been studied in bacteria and are now believed to be a new class of molecular chaperones [124, 269, 284]. Three related Clp proteins have been described in bacteria, ClpA, B and C, all of which are ATPases containing two non-homologous, conserved ATP-binding motifs. Another group of Clp proteins has also been identified, but diverges from the former groups in that it has only a single ATP-binding domain. The best characterized member of this second group is ClpX [331]. The eukaryotic (yeast and plant) HSP100 proteins are most similar to the ClpB subgroup [172, 232, 268]. Homologues of some of the other family members are also found in plants, but appear to be organelle-localized and are discussed in subsequent sections of this review.

The HSP100/Clp family of chaperones are interesting in that rather than preventing protein aggregation and misfolding, as is attributed to other chaperones, evidence indicates that these proteins function in protein disaggregation and/or protein degradation [124]. Bacterial ClpA and ClpX have been shown to facilitate the proteolytic activity of the *E. coli* protease ClpP, but with different substrate specificities [320]. Surprisingly, ClpB does not appear to have similar ClpP protease stimulating activity. It is still a formal possibility that ClpB stimulates a protease other than ClpP, but there is no evidence for such an activity. However, both ClpB and yeast Hsp104 have a peptide-stimulated ATPase activity ([333], S. Lindquist, personal communication).

Insight into the role of the ClpB subclass has come primarily from studies of Hsp104 in yeast. Hsp104 is not required for growth at any temperature in yeast, but it is highly induced by temperature and other stresses, and it has been shown genetically to be required for yeast thermotolerance [263] as well as tolerance to other stresses [264]. The potential chaperone activity of yeast Hsp104 was recently demonstrated through the ability of wild-type, but not *hsp104* mutant cells, to promote reactivation of heat-inactivated bacterial luciferase *in vivo* [230]. Furthermore, although both wild-type and *hsp104* mutant cells accumulated intracellular protein aggregates during heat stress, only wild-type cells were able to clear these aggregates from the cells. The ATPase activity of Hsp104 is essential for these functions, as shown by experiments with Hsp104 ATPase defective mutants. In assays of RNA splicing activity *in vitro*, addition of Hsp104 was found to reactivate heat-inactivated splicing extracts [305]. Thus, Hsp104 has been proposed to be directly involved in resolubilization of protein aggregates, potentially in both the nucleus and cytoplasm.

A chaperone activity for HSP100 involving protein aggregates has also been suggested by the observation that Hsp104 is required for propagation of the yeast non-Mendelian factor [*psi*+] [41]. The yeast [*psi*+] factor is believed to be a prion-like protein that is encoded by the SUP35 gene and has homology to EF-1α. Moderate levels of Hsp104 are required for the maintenance and propagation of [*psi*+], while high levels of Hsp104 will cure cells of [*psi*+]. As prions are believed to be normal cellular proteins with an alternative and self-propagating (or templating) structure, Lindquist and colleagues suggest that

Hsp104 is necessary to form and/or maintain the proper self-propagating structure of [*psi*+], while too much Hsp104 interferes with the template activity of [*psi*+] [41]. It will be of great interest to investigate at a more mechanistic level, the interaction of Hsp104 with protein substrates such as the SUP35 gene product. This has been difficult because the proposed substrates are large insoluble complexes, and no system has been devised to measure Hsp104 activities *in vitro* (S. Lindquist, personal communication).

Although to date there have been no biochemical or mechanistic studies of the cytosolic HSP100 proteins in plants, the *Arabidopsis* and soybean homologues have both been shown to substitute for the function of Hsp104 in the development of thermotolerance in yeast [172, 268]. The *Arabidopsis* protein is not expressed in normal vegetative tissues, but accumulates to high levels during heat stress, consistent with a more specific role in stress responses, such as resolubilization of abnormal protein aggregates. Interestingly, the *Arabidopsis* protein also accumulates during seed development and is present in dry seeds (E. Vierling and L. Larson, unpublished). It could be speculated that this developmental pattern of expression reflects a need for protein resolubilization/renaturation during the rehydration of the seed. Whether this HSP100 protein is involved in other stress tolerances as seen in yeast has not been adequately explored, but an antigenically related protein has been reported to be induced by abscisic acid in rice [168]. In total, these data indicate that the plant protein is capable of similar functions as the yeast protein and further suggest that HSP100 proteins may be critical for the development of thermotolerance in plants. It will certainly also be of interest to determine if plant HSP100 plays a role in any non-Mendelian inheritance phenomena, similar to the effects of yeast HSP104 on the propagation of yeast prion-like proteins.

TriC: the TCP1-containing cytosolic 'chaperonin'

In the eukaryotic cytosol the protein complex known as TRiC, the tcp1 ring complex (TCP-1 is *t* complex polypeptide 1), is believed to be a distantly related, structural and functional homologue of the well described chaperonin known as GroEL in *E. coli*, which in turn has close homologues in both chloroplasts and mitochondria of higher cells (see below) [72, 125, 149, 329]. As isolated from mammalian cells, eukaryotic TriC is a large (ca. 900 kDa) hetero-oligomeric complex containing multiple polypeptides between 55–60 kDa [89, 178, 252]. The first identified subunit of this complex was shown to be identical to mouse TCP-1, which was originally described as an abundant testis protein encoded in the mouse *t*-complex. Genes encoding at least six additional subunits of this complex, with between 27–35% identity to TCP1 have been cloned from mammalian cells [162]. The primary structure of these proteins shows limited identity with GroEL, but TCP1 is nearly 40% identical to TF55, a protein with ATPase activity that is induced to high levels during heat stress in certain archaebacteria [236]. However, the quaternary structure of the archaebacterial TF55 complex and mammalian TriC are very similar to GroE, in that they comprise two stacked rings of 8 or 9 subunits [91, 149, 237, 310].

Studies both *in vivo* and *in vitro* support a chaperone function for eukaryotic TriC. Although it is possible that TRiC is not required for folding of all proteins, excellent data support the conclusion that TRiC is essential for proper folding of actin and tubulin. Yaffe and colleagues [337] demonstrated that tubulin expressed in rabbit reticulocyte lysates is associated in an unfolded form with a TCP1-containing complex. Addition of Mg-ATP promoted release and correct folding of tubulin. Cowan and colleagues similarly showed that reticulocyte lysate TRiC could fold recombinant actin in an ATP-dependent fashion [91]. These researchers have continued more mechanistic studies of both the actin and tubulin refolding reactions [92, 205]. Hartl and colleagues have proposed that TRiC may play a more general role in folding of many newly synthesized proteins [90]. They showed that TRiC was a component of a high molecular mass complex required for correct folding and activation of luciferase during translation in reticulocyte lysates. In addition to these *in vitro* studies, several studies in yeast provide genetic evidence for a role of TCP1 like proteins in actin and microtubule function [39, 295, 302].

Very little is known about TRiC and protein folding in plants. Studies indicating that TRiC was involved in phytochrome folding [217] were later found to be in error [218, 229]. Thus, to date other than the identification of several DNA sequences encoding polypeptides with homology to TCP1 [71, 215] the nature and function of this complex in plants remain to be investigated. It will certainly be interesting to know if TRiC is important for plant cytoskeletal dynamics, or what other proteins in the cytosol are dependent on TRiC for folding.

The small HSPs: potential chaperones

The small heat-shock proteins, or smHSPs, were the first HSPs cloned from higher plants, yet they remain the most enigmatic and have only recently been suggested to act as a type of molecular chaperone [138]. These proteins are typically between 16 and ca. 30 kDa in size, and comprise five and possibly six gene families that are evolutionarily conserved among both monocots and dicots [317, 318]. The gene families encode proteins that are targeted to different intracellular locations, including the cytosol (class I and II), chloroplast, mitochondrion, endoplasmic reticulum and perhaps other endomembrane compartments. Although the smHSPs are not very highly conserved (as little as 30% amino acid similarity between some families), all smHSPs share a signature carboxyl-terminal domain of ca. 100 amino acids found in other eukaryotic smHSPs and in the α-crystallin structural proteins of the eye lens. This domain distinguishes them from other small proteins induced by high temperature. The smHSPs from many different organisms are found in high molecular weight complexes *in vivo*, between 200 to 800 kDa, that appear to be homo-oligomers [7, 318]. Recombinant pea HSP18.1 (a class I cytosolic smHSPs) and pea HSP17.7 (a class II cytosolic smHSP) expressed in *E. coli* form soluble, 12 subunit homo-oligomers [170], which are similar in size to smHSP structures observed *in vivo*.

Like HSP100, the smHSPs are not detected in the absence of stress in leaf or root tissues, but the cytosolic proteins can accumulate to over 1% of total cell protein during heat stress [64, 128]. However, a number of studies have also documented expression of certain cytosolic smHSPs in reproductive organs during various stages of plant development (see [318] for review). In general these expression data suggest a specialized role for these putative chaperones, as distinct from participation in constitutively required protein folding processes. A complete review of our current knowledge of the evolution, structure and expression of the smHSPs can be found in Waters *et al.* [318].

Evidence that the smHSPs act as chaperones has been derived entirely from *in vitro* studies using recombinant smHSP proteins. Recombinant smHSPs from mammals and plants promote renaturation of proteins diluted from denaturant, prevent heat-induced protein aggregation, and facilitate reactivation of heat-inactivated model substrates [126, 139, 170]. The plant cytosolic class I and II smHSPs act in substoichiometric to stoichiometric amounts in these reactions [170].

Similar activities have been reported for isolated vertebrate α-crystallins [126, 209]. In contrast to chaperones of the HSP70 and chaperonin classes, activities of the smHSPs do not appear to be stimulated by nucleotides.

Under temperature conditions at which pea HSP18.1 (a class I plant smHSP) prevents aggregation of substrate proteins (34–45 °C), the target proteins bind to the smHSP as observed on non-denaturing gels or size-exclusion chromatography (G.J. Lee and E. Vierling, in prep.). Several non-native proteins may be bound to the surface of one smHSP oligomer, and substrate protein can be shown to be in a highly protease-susceptible conformation (G.J. Lee and E. Vierling, in prep.). Similarly, when plant cell extracts containing smHSPs are heated, immunodetectable smHSP shifts to a higher molecular weight on non-denaturing gels, while smHSPs heated alone show no such size shift [143]. A model consistent with these *in vitro* data is that smHSPs capture unfolding polypeptides by hydrophobic interactions and thus keep them in a state competent to refold. The stable smHSP substrate complexes have not however, been effectively disassociated *in vitro* without concommitant disassociation of the smHSP oligomers themselves. It is plausible that, *in vivo*, substrate release may be triggered by synergistic interactions with other chaperone proteins and lead to functional refolding. Alternatively smHSP/substrate interactions may facilitate substrate degradation by proteolytic systems. Thus the smHSPs would act as a highly effective first defense against irreversible protein damage/unfolding.

Although the idea that smHSPs act as molecular chaperones in plants was first suggested over 6 years ago [144], evidence that the smHSPs act as chaperones *in vivo* remains to be acquired. It should be noted that while there is considerable evidence that smHSPs are capable of conferring increased cellular thermotolerance in mammalian cells [7], only correlative data are available in higher plants. The most recent evidence in this vein comes from Schöffl and colleagues [171]. These workers produced transgenic *Arabidopsis* plants that constitutively express an active heat-shock-transcription factor and report that these plants are significantly more thermotolerant than wild-type plants. The plants constitutively express at least one of the cytosolic smHSPs, consistent with a role for these proteins in thermotolerance. However, expression of this transcription factor most likely affects multiple genes, including those encoding the HSP70, 90 and 100 classes of chaperones, and it is difficult to make conclusions about the role of any one induced component. Nonetheless, these plants are an exciting demonstration of the overall relevance of the heat-shock response to thermotolerance.

Cytosolic immunophilins: peptidyl-prolyl cis/trans isomerases

A major group of protein foldases is the peptidyl-prolyl cis/trans isomerases (PPIases) also known as rotamases, which catalyze isomerization of peptide bonds around Pro residues, a normally slow step in protein folding [275, 311]. PPIases are found in the cytosol of both prokaryotic and eukaryotic cells, as well as in a number of eukaryotic organelles. There are two structurally distinct classes of PPIases that are inhibited by different immunosuppressant drugs: the cyclophilins (CyP) which bind cyclosporin A (CsA) and the FK506-binding proteins (FKBPs) which bind the compound FK506 [311]. Because of these drug-binding activities these PPIases have also been termed immunophilins [275]. In mammalian cells, the CyP-CsA or FKBP-FK506 complexes have been shown to bind and inactivate the Ca^{2+}-dependent phosphatase calcineurin, and thereby block major cellular signalling pathways required for activation of T lymphocytes. A similar effect on calcineurin has also been observed in yeast [52]. The inactivation of calcineurin is apparently unrelated to the PPIase activity of these proteins. The immunophilins are also components of the HSP90 steroid hormone receptor complex described above. In mammals, the HSP56 component of this complex was identified as an FKBP and a 40 kDa component was determined to be a CsA-binding protein [242]. Since HSP56 is incorporated into the HSP90 complex even in the presence of FK506, it appears that complex formation does not require PPI activity. The FK506-binding site is still accessible in the complex, indicating that the complex has PPI activity. The role of immunophilins in the steroid hormone receptor complex remains undefined, although recent data suggest that HSP56 may be involved in nuclear localization of the complex [55]. Thus, although there is considerable data demonstrating the foldase activity of PPIases, their roles in vivo are poorly defined, and may well extend beyond proline isomerization.

Both cytosolic FKBPs and cyclophilins have been identified in plants, and several of the corresponding genes have been cloned. Gasser and colleagues [93] reported the first cloning of 18 kDa cyclophilin homologues from tomato, maize and Brassica napus,

and found that in the latter two species the genes are members of small gene families. The deduced amino acid sequences are 74% identical to human CyP. When expressed in E. coli the tomato 18 kDa cyclophilin showed PPIase activity and CsA binding. Five cytosolic cyclophilin genes have now been cloned from Arabidopsis and expression of the genes shows organ specificity ([116], C. Gasser, personal communication). In Vicia faba multiple CsA- and FK506-binding proteins were purified by affinity chromatography using these drugs [190]. Cell fractionation indicated some of the affinity purified proteins were cytosolic, while others were most likely chloroplast or mitochondrial proteins. The homologue of the HSP56/FKBP found associated with HSP90 was also recently cloned from Arabidopsis, but its direct association with HSP90 in plant cells in vivo has not yet been confirmed (C. Gasser, personal communication).

There is only indirect evidence addressing the functions of these foldases in plant cells. FK506 and CsA were recently shown to inhibit the Ca^{2+}-dependent inactivation of K^+ channels in Vicia faba guard cells [192]. This effect might be due to interference with a signal transduction pathway involving phosphorylation and calcineurin as shown for T lymphocyte activation. However, to date the target of the immunosuppressant/immunophilin complexes in plants is unknown, and in fact no calcineurin homologue has been identified in plants [3]. Other attempts to use these drugs to interfere in a range of different plant cell signalling pathways have been unsuccessful (C. Gasser, personal communicaton). An FKBP is associated with HSP90 glucocorticoid receptor complexes assembled in wheat germ extracts (P. Krishna, personal communication). This finding again demonstrates the evolutionary conservation of these protein interactions, but does not further illuminate their function.

Co-chaperones

As indicated in Table 1, chaperones in the HSP70 class are known to interact with accessory proteins or 'co-chaperones' that modulate their activity. These co-chaperones, DnaJ and GrpE, were first defined in E. coli. DnaJ stimulates the rate of ATP hydrolysis of DnaK, the E. coli HSP70 homologue, while GrpE acts as a nucleotide exchange factor for DnaK [182]. In eukaryotes, DnaJ homologues have also been implicated in determining more specifically how HSP70s interact with substrate proteins [246]. The structure and potential functions of eukaryotic DnaJ homologues have been summarized in three recent reviews [54, 246, 282]. Surprisingly, although homologues of GrpE have been identified in eukaryotic organelles, a cytosolic counterpart has not yet been isolated. Hartl and colleagues have recently isolated an HSP70 binding protein (HIP) that they propose may take the place of GrpE in the eukaryotic cytosol [127].

As mentioned in the discussion of HSP70s above, plant cytosolic HSP70 ATPase activity is increased in the presence of *E. coli* DnaJ. Thus, at least part of this regulatory interaction is conserved between prokaryotes and eukaryotes [343]. Genes encoding cytosolic DnaJ proteins have also been isolated from several plant species [15, 243, 342, 344]. The homologue isolated from *Atriplex numularia*, called ANJ1, was found to complement a yeast mutation in MAS5, a protein required for import into yeast mitochondria, implicating ANJ1 in a similar activity in plants. The function of ANJ1 was further shown to be dependent on the presence of a functional CAQQ isoprenylation motif at the carboxyl-terminus [123]. Recombinant ANJ1 could be farnesylated *in vitro*, and *in vivo* ANJ1 was associated with cellular membranes. In total these data implicate ANJ1 as a co-chaperone for HSP70 involved in transport of proteins across membranes in plants.

ER chaperones and foldases

Protein folding in the endoplasmic reticulum

Nascent secretory proteins as well as those that will reside in any of the compartments of the extensive plant endomembrane system enter the ER during translocation and must be folded into their native structures within the lumen of the ER. In fact, the ER lumen is dedicated primarily to protein folding as proteins enter it in an unfolded conformation and leave it fully folded [257]. The protein secretory pathway in higher plants has been the subject of other reviews and will not be covered in detail here ([12, 227, 304], Bar-Peled *et al.*, this volume). Rather, we will focus on protein-protein interactions that occur between molecular chaperones and newly synthesized proteins within the ER.

General characteristics of the ER translocation and folding pathway have been obtained in studies of yeast, animal and plant systems (reviewed in [256]). In each of these systems, the folding process begins with chaperones associating with the polypeptide chain as it emerges into the ER lumen. Although the exact sequence of folding events is not yet understood, a number of protein participants have been identified (Table 1).

BiP (binding protein, or GRP78, glucose-regulated protein), calnexin, calreticulin, GRP94 (endoplasmin or glucose-regulated protein 94), and PDI (protein disulfide isomerase) have all been implicated in protein-protein interactions in the ER. These and other proteins are important for translocation and/or folding in yeast and mammalian systems. Counterparts for several of these ER molecular chaperones have been discovered in higher plants (reviewed in [12] and [304]). The presence of such similar proteins in different kingdoms provides evidence for a common pathway of folding and translocation and evolutionarily conserved modes of action for these ER chaperones. Several processes in plants, such as protein body formation and secretion of cell wall precursors, however, do not have obvious counterparts in mammalian systems. Thus, an understanding of protein folding and assembly in these processes may offer new insights into chaperone function in the ER.

Chaperone function is generally examined by trapping the interactions between chaperones and target proteins. Agents that increase the tendency of proteins to misfold lead to increases in the level of several ER chaperones and thus facilitate detection of the chaperone-target protein complexes [169]. The most common means of inducing ER chaperones in plants is incubation of cells or tissue with tunicamycin, an antibiotic that inhibits N-linked glycosylation [56, 61, 85]. The lack of glycosylation apparently hinders proper protein folding and slows assembly of oligomers [32, 70]. Azetidine-2-carboxylic acid (AZC), a proline analog, was also used successfully to induce ER chaperone levels in maize (R.L. Wrobel and R.S. Boston, unpublished results). An alternative to these relatively non-specific ways of generating altered proteins was used by Vitale and colleagues who generated assembly-defective phaseolin mutants by recombinant DNA technology [233]. A number of other treatments, including addition of plant growth regulators and exposure to plant pathogens have resulted in increased accumulation of chaperone mRNAs [59, 146, 312]. These latter treatments cause increased traffic through the ER. Whether chaperone induction is a general cellular response to this traffic or a more specialized stress response is not known. The following sections of this review focus on participation of molecular chaperones in protein-protein interactions that appear to facilitate proper protein folding and assembly in the ER.

BiP, the binding protein

The binding protein (BiP), a member of the HSP70 family of proteins, is the best studied of the ER resident chaperones in plants (for reviews see [102, 234]). This soluble protein recognizes no consensus binding site on target proteins but instead interacts more tightly with non-native protein domains than with native regions of target proteins as described above [17, 25, 100]. The model for BiP action presumes a function to prevent aggregation of misfolded proteins or protein domains. BiP binds polypeptides to stabilize partially folded intermediates, then releases and rebinds to sites that remain in non-native conformations [154].

Of the ER chaperones that have been identified in higher plants, BiP is most likely involved in one of the earliest stages of protein folding. In yeast, kar2 (BiP⁻) mutants and cross-linking experiments have been used to show that BiP is associated with nascent polypeptides, and is required in the ER for translocation and proper folding of newly synthesized proteins [265, 283, 306]. Support for a similar role for plant BiP was obtained by partial complementation of a temperature sensitive yeast kar2 mutant by a tobacco BiP cDNA [61]. Although the complementation data do not directly address the role of BiP in plants, they do show that amino acid residues important for BiP function in yeast have been evolutionarily conserved between yeast and higher plants.

A striking difference between BiP genes in higher plants and other organisms is their organization as multigene families. The discovery of mutations in the KAR2 (BiP) gene from yeast led to the demonstration that KAR2 was a single, essential gene in that organism [226, 253]. Similarly, spinach appears to have only one BiP gene [6]. In contrast, tobacco, maize and soybean BiPs are encoded by several genes ([61, 146]; R.L. Wrobel and R.S. Boston, unpublished results). It will be interesting to investigate the biological significance of differences in gene number with respect to gene regulation and possible redundancy of function.

A number of studies have addressed the regulation of plant BiP gene expression at the level of RNA and protein accumulation (for review see [102, 234]). BiP RNA levels increased in response to environmental stimuli as well as to agents that perturb protein folding. In contrast, protein levels reflected increases in RNA levels in some instances but were not dramatically changed in others [6, 59, 146]. Interpretation of such data is confounded by lack of gene-specific probes for multigene family members and by our lack of knowledge of the functional forms of the protein. Thus, further experimentation will be required to determine the important signals affecting BiP gene expression.

Reversal of the binding of BiP to substrates by ATP has been used both as a diagnostic assay for molecular chaperone activity and as a basis for affinity purification [62, 103, 323]. Interestingly, maize BiP ATPase activity is stimulated by low concentrations of calcium but inhibited by high concentrations of either magnesium or calcium [343]. These results contrast with those for bovine BiP which does not show an increase in ATPase activity with calcium nor is it inhibited by high concentrations of magnesium [327]. Whether these differences reflect different catalytic mechanisms for plant and animal BiPs remains to be determined.

Interestingly, the ATPase activity of maize BiP can be increased by the co-chaperones DnaJ and GrpE from yeast [343]. Likewise, BiP ATPase activity is stimulated by synthetic peptides. Although the stimulation of BiP ATPase activity resembled that seen for cytosolic HSP70s, differences in target recognition were observed. Further investigation of functional differences in chaperones from various subcellular compartments should improve our understanding of the biological importance of these proteins in plants.

Role of BiP in assembly and packaging of seed storage proteins

Most of the research related to ER chaperone function in plants has involved studies of seeds, in which the majority of protein synthesis is devoted to production of storage proteins. Storage proteins must not only be folded correctly at the level of individual polypeptides but, depending on the plant, must also be tightly packaged in an ordered arrangement within specialized storage bodies, assembled into higher-ordered structures, covalently modified, and/or transported to storage vacuoles beyond the ER [227].

BiP has been shown to associate with nascent prolamine storage proteins during protein body formation in rice. Li et al. [181] showed that BiP co-fractionated with polysomes from immature rice seed in sucrose density gradients. Prior treatment with puromycin or removal of RNA with RNase resulted in a reduction in the amount of BiP in complexes and in the release of nascent prolamine chains. These data indicate that BiP was associated with the nascent prolamin chains in the ER lumen. The association of BiP with nascent polypeptides in rice protein bodies is similar to the association of BiP with nascent polypeptides in yeast

[265, 283, 306]. However, in yeast, translocation can occur post-translationally as well as co-translationally whereas in plants the translocation process appears to be strictly co-translational.

In addition to its association with nascent polypeptides, BiP has also been shown to bind full-length polypeptides. Gillikin *et al.* [103] used antibodies against either BiP or the soybean seed storage protein β-conglycinin to co-immunoprecipitate BiP and β-conglycinin. These authors estimated that levels of free β-conglycinin were 1000-fold higher than those of BiP-β-conglycinin complexes. This ratio is not surprising given the propensity of BiP to undergo only transient interactions with proteins in a native conformation (reviewed in [234]). Vitale and colleagues detected complexes of BiP and phaseolin in bean cotyledons treated with tunicamycin [56]. In subsequent studies, these investigators examined the role of BiP during the folding and assembly of trimeric phaseolin in developing bean cotyledons [303]. Pulse-labeling of cotyledons followed by immunoprecipitation with anti-BiP or anti-phaseolin antibodies revealed association of BiP with monomers of phaseolin but not with fully assembled trimers. Successful assembly of monomers into trimers abolished detectable BiP-phaseolin complexes presumably by masking exposed BiP-binding sites during trimer assembly [303]. A similar stable association of BiP with unassembled protein subunits was observed in tobacco protoplasts expressing an assembly-defective bean phaseolin mutant [233]. Association of BiP with normal, assembly-competent phaseolin was not observed in the tobacco protoplast system. In contrast, in rice endosperm, Li *et al.* [181] were able to co-immunoprecipitate BiP and full-length prolamines with anti-prolamine antisera. The significance of the different stabilities of complexes between BiP and normal seed proteins in bean and rice is not clear. Rice prolamincs do not appear to assemble into discrete higher ordered multimers, yet they do undergo intermolecular interactions with other storage proteins within protein bodies [227]. It is possible that the BiP-storage protein complexes in rice may be more stable than are ones from beans. Alternatively, rice prolamines may fold more slowly or interact with a different subset of chaperones than do the bean phaseolins. In any case, taken together, these results are consistent with a role for BiP in folding and assembly of newly synthesized seed storage proteins.

Interaction of BiP with seed proteins in common bean and soybean was abolished by addition of ATP prior to immunoprecipitation [56, 103, 303]. Likewise, mature prolamine-BiP complexes in rice protein bodies were also sensitive to ATP treatment [181]. Surprisingly, association of BiP with nascent prolamine polypeptides in rice was ATP-insensitive even in the presence of 5 mM MgATP [181]. Insensitivity of BiP-peptide complexes to ATP levels of 10 μm has also been reported in experiments with mammalian BiP *in vitro* [25]. However, such complexes were disrupted by 100 μm ATP. The significance of the ATP-insensitive complexes to folding of nascent proteins remains unknown.

Differential ATP sensitivity of complexes between BiP and nascent or mature prolamines is consistent with two different points of interaction occurring between these proteins during prolamine synthesis and assembly within protein bodies [181]. More in-depth discussion of possible roles for chaperones in determining whether proteins are assembled into higher-ordered complexes in protein bodies or transported from the ER is provided by Okita and Rogers [227].

In maize, high levels of BiP are associated with ER and ER-derived protein bodies in the endosperm mutants *floury-2* (*fl2*), *Defective endosperm* (De*-B30) and *Mucronate* (*Mc*) [22, 85, 196]. In the *fl2* mutant, the signal sequence processing site of a major storage protein has been mutated such that it cannot be recognized by processing machinery in maize microsomes [45]. This mutant protein remains associated with ER membranes (J.W. Gillikin, F. Zhang and R.S. Boston, unpublished results). Coincident with accumulation of this mutant zein, BiP is induced and is localized at the periphery of protein bodies [341]. Similarly, in transgenic *Arabidopsis* expressing a truncated maize γ-zein storage protein, BiP co-localized with the maize protein in reticular/amorphous structures but was found at the same site as normal γ-zeins only at very early stages of γ-zein accumulation in the ER [98]. Such data lead us to suggest that the increase in BiP in these cases is triggered by accumulation of misfolded proteins or improperly assembled protein complexes. Direct proof, however, remains to be obtained.

GRP94

GRP94 or endoplasmin is the ER luminal member of the HSP90 protein family [138]. Like Hsp90, GRP94 can associate with target proteins, but little information is available about its physiological role. Recent work by Melnick *et al.* [206] showed that GRP94 in mammalian cells could be found in association with BiP.

Based on differential binding to oxidized and reduced immunoglobulin molecules, these authors proposed that BiP and GRP94 act sequentially [207]. BiP would bind to an early reduced form of an intermediate in the folding pathway and dissociate rapidly, but GRP94 would bind to a fully oxidized intermediate and dissociate much more slowly [207].

In plants, GRP94 has been identified in barley, bean, maize, periwinkle and tobacco [23, 56, 59, 60, 276, 312]. GRP94 is induced by stimuli that induce BiP in maize and bean ([56]; R.L. Wrobel and R.S. Boston, unpublished results). Little effect of heat shock has been observed on GRP94/endoplasmin expression but increased protein traffic through the secretory system is coincident with increased GRP94 in cell cultures and in barley infected with powdery mildew [312]. Whether increases in the amount of protein trafficking itself or other stimuli are responsible for the increase in GRP94 expression remains to be determined. Functional analysis of GRP94 interactions with target proteins or other chaperones in plants have not yet been reported.

Foldases in the ER: PDI and PPI

Protein disulfide isomerase (PDI) is generally accepted to have a physiological role as a catalyst of disulfide bond formation in the oxidizing environment of the ER [257]. By promoting formation and rearrangement of disulfides, PDI acts to stabilize the tertiary and quaternary structure of proteins and thus facilitates protein folding. Both PDI genes and proteins have been characterized from plants [48, 110, 179, 251, 278, 279, 281]. Direct evidence supporting the role of PDI in protein folding was obtained in translation-processing studies with γ-gliadin storage protein from wheat. Bulleid and Freedman [27] analyzed co-translational disulfide bond formation in a truncated form of γ-gliadin synthesized in the presence of dog pancreas microsomes. Microsomes depleted of lumenal content by alkaline lysis were defective in disulfide formation. Control microsomes permitted disulfide bond formation as did depleted microsomes to which PDI had been restored.

PDI is induced in alfalfa and tobacco cells treated with tunicamycin [59, 281]. PDI is also induced by treatment of barley aleurone cells with the plant hormone GA_3 [59]. In the latter case, a large increase occurs in protein traffic through the secretory system. These data are consistent with findings from animal systems in which PDI levels increase in proportion to accumulation of S-S bonded proteins [87].

PDI has multiple domains that may be involved in additional activities [88]. Homology to phosphatidylinositol-specific lipase C, β subunit of prolyl 4-hydroxylase, thioredoxin, a subunit of the microsomal triglyceride transfer protein complex, the glycosylation site-binding protein of the oligosaccharide transferase complex and ERp72, an ER lumenal protein with three copies of the active-site sequences of PDI, have all been reported. Characterization of an alfalfa cDNA clone revealed sequence homology to mammalian ERp72 and induction of the corresponding RNA in tunicamycin-treated suspension cultures similar to the induction of mammalian ERp72 by tunicamycin treatment of animal cells [280]. Surprisingly, the deduced amino acid sequence of the corresponding protein lacked the COOH-terminal K/HDEL consensus sequence associated with BiP, GRP94 and mammalian ERp72. At present, the functional importance of these homologies to protein folding in higher plants is still mostly speculative.

PPIases (immunophilins) also appear to be present in the ER of eukaryotes, as evidenced by data from yeast and mammalian cells [142, 224]. However, work on these proteins has only just begun in plants. A small family of genes encoding an 18 kDa FKBP with a putative signal peptide has been cloned from *Arabidopsis*. The FKBP18 from fava bean has been shown to be inhibited by both FK506 and rapamycin (S. Luan, personal communication).

Calnexin

The first report of a plant calnexin resulted from a serendipitous discovery of an *Arabidopsis* calnexin cDNA clone in a search for genes encoding chloroplast envelope proteins [129]. The *Arabidopsis* calnexin has a single membrane-spanning domain as judged by tryptic digestion of native protein in plant microsomes. In maize, RNA gel blot and immunoblot analyses indicated an increase in calnexin RNA and protein in the *fl2*, *Mc* and De*-B30 mutants that have been previously shown to have high levels of BiP (R.L. Wrobel and R.S. Boston, unpublished results).

The biological function of calnexin is not entirely clear in any system. Calnexin associates with partially deglucosylated glycoproteins in mammalian microsomes (reviewed in [328]). Recently, Kim and Arvan used DTT to slow the protein folding process in thyrocytes [148]. Subsequent differences in the amounts of protein that could be co-immunoprecipitated with calnexin and BiP led them to propose that calnexin and

204

BiP act sequentially during folding in a 'precursor-product' relationship. Although these results present an alternative explanation to the concept that calnexin associates with polypeptides after BiP has acted in the translocation process, it should be noted that DTT affects levels of other molecular chaperones as well as the kinetics of protein folding [59, 169].

Recent reports demonstrating the presence of calnexin outside the ER provide perhaps the most intriguing possibilities for calnexin function [57, 322]. Delmer and colleagues identified a 65 kDa plasma membrane polypeptide that interacts with callose synthase in cotton fibers [58]. Subsequent cDNA cloning and sequence homology searches with the corresponding gene revealed identity with calnexin [57]. This surprising finding is consistent with a report that calnexin can be found in the CD3 complex of the T cell receptor at the surface of mammalian thyrocytes [322]. In the latter case, the calnexin ER retention signal may be masked by protein-protein interactions with CD3.

Masking of a lumenal ER retention signal has been previously proposed for a secreted maize auxin-binding protein [145, 221]. Calnexin may provide yet another example of such a masking phenomenon. Certainly, identification of proteins associated with cotton calnexin and characterization of the transport of calnexin to the plasma membrane will expand our understanding of calnexin function.

Calreticulin

Calreticulin is a Ca^{2+}-binding protein of ca. 55 kDa whose function is presumed to be calcium storage in the ER [156]. Previous work in plants showed the presence of immunological cross-reacting material in pear pollen, accumulation of a calreticulin-like protein in spinach leaves and pea, and the presence of two different calreticulin cDNA clones from barley [4, 35, 115, 208]. Recently, Denecke and colleagues [59] used co-immunoprecipitation with anti-calreticulin antibody to show that calreticulin associates with a 75 kDa polypeptide in tobacco leaf protoplasts . The association was disrupted upon treatment with ATP. Although the identity of the 75 kDa polypeptide was not established, its molecular weight, together with its release upon ATP treatment, are suggestive that the protein is BiP. Co-immunoprecipitation of smaller polypeptides with calreticulin was also observed when protoplasts were incubated with tunicamycin or given a heat-shock. Such results are consistent with a chaperone role for

calreticulin and provide an exciting opportunity for further defining the role of this protein *in vivo*.

Small heat-shock proteins

Emerging evidence that smHSPs may represent a new class of chaperones, as discussed above relative to the cytosolic forms of these proteins, suggests that related smHSPs found in the plant endomembrane system may also have protein folding activities. An ER-localized smHSP was first described in pea and soybean [117]. The protein has a typical signal peptide that is removed *in vitro* and *in vivo*, as well as a predicted ER retention signal. There is no evidence that these proteins are expressed in the absence of heat stress in leaves or roots, nor are they expressed in developing seeds under normal growth conditions. Absence of these proteins from normal tissues suggests they do not participate in normal protein trafficking through this membrane system as do the chaperones just discussed. Rather they may play a more specialized role under stress conditions.

Proteins with homology to the pea and soybean ER smHSPs have now also been cloned from *Arabidopsis* [118] and potato [296]. Interestingly, the potato protein was isolated in a screen for proteins induced during cold storage and represents the only documented instance of ER-smHSP expression in the absence of heat stress. Another cDNA encoding a smHSP that has a signal peptide but lacks an ER retention signal and does not appear to be a member of the same ER-localized protein class has been cloned from soybean [163]. Whether proteins similar to this latter protein are produced by other plants, as well as the exact intracellular localization of this protein remian unknown.

It should be noted that to date, only higher plants have been demonstrated to contain endomembrane-localized smHSPs; these proteins do not appear to be made by non-plant eukaryotes. Whether they replace the function of another eukaryotic chaperone during stress, or have a function specialized to plant substrates or processes warrants further investigation.

Chloroplast-localized chaperones and foldases

The higher-plant chloroplast chaperonins

The GroE-related chaperonins [120] are ubiquitous molecular chaperones that are constitutively expressed in eubacteria and those eukaryotic organelles that evolved from procaryotes (e.g. chloroplasts and mito-chondria). From bacteria [33, 120] and fungi [40, 254], to higher plants [14, 120] and animals [194, 239], two distinct sequence-related family members are recognized: chaperonin 60 (cpn60) and chaperonin 10 (cpn10). Together these proteins facilitate the folding, assembly, and translocation of numerous other proteins [97, 100, 121]. Genetic studies have clearly established that both chaperonin components are essential for viability in yeast [40, 255] and bacteria [78], and this is likely true for higher eukaryotes as well. Considering that most of our mechanistic insight into the chaperonins has come from studies on the GroEL (cpn60) and GroES (cpn10) proteins of *E. coli*, it seems relevant to summarize some of the more important properties of the bacterial prototypes (reviewed in [97, 100, 121]).

GroEL, the crystal structure of which was recently reported [24], is a tetradecamer of two stacked rings of seven identical ca. 60 kDa subunits [122]. Through conformational changes mediated by its weak K^+-dependent ATPase activity [122, 300], GroEL oscillates between states of high and low affinity for non-native proteins [97, 100, 121, 136, 202, 291]. In its high-affinity state (i.e. in the absence of adenine nucleotides), GroEL forms stable complexes with a variety of unfolded proteins [298] that bind within its central cavity [167]. Its preferred substrates are 'molten-globule'-like intermediates that are rich in secondary structure, but lack the organized tertiary structure of the native state [97, 121, 201]. While the formation of a stable complex with GroEL suppresses undesirable 'off-path' reactions such as aggregation [26], it also interferes with productive folding events [164]. In general, for efficient folding to resume, the bound protein must be released from GroEL, if only moment-arily, and occasionally this can be accomplished by the mere addition of adenine nucleotides [97, 121]. However, release per se does not ensure proper folding, and under 'non-permissive' conditions [271] where unassisted spontaneous folding is not observed, the GroEL-assisted folding reaction strictly requires both ATP hydrolysis and GroES [97, 121, 201, 271].

The co-chaperonin, GroES, is a single toroid of seven identical ca. 10 kDa subunits. It, too, physic-ally interacts with GroEL, but only in the presence of adenine nucleotides [33]. The complexes that are formed with ADP alone are stable, asymmetric, 'bullet-shaped' particles [167, 262] that consist of 1 mol of $GroEL_{14}$, 1 mol of $GroES_7$, and 7 mol of tightly bound ADP [136, 167, 290]; the GroES is restricted to only one end of the GroEL cylinder. Within the asymmet-ric complex, the ATPase activity of one entire GroEL ring is completely inhibited [136, 290], while the oth-er hydrolyzes ATP with enhanced cooperativity [109] and altered kinetic properties [290]. However, this is an oversimplification of an highly dynamic system. After a single round of ATP hydrolysis (by the unin-hibited GroEL toroid), the originally bound GroES and ADP are released from the asymmetric complex [291]. Moreover, although the subject remains controversial [77], the cycle of breakdown and reformation of asym-metric complexes may proceed through symmetrical 'football-shaped' intermediates that consist of 1 mol of $GroES_{14}$ and 2 mol of $GroES_7$ [8, 272, 291]. It has been suggested that the role of GroES is to coordin-ate all of the subunits in one ring of GroEL to convert to their low affinity state in synchrony [291]. This ensures the transient but complete release of the bound protein substrate, affording it an opportunity to fold unhindered in solution [291, 321]. According to this model, those molecules that fail to achieve the native state and instead partition to kinetically trapped mis-folded species are recaptured by GroEL, and are effi-ciently recycled during subsequent rounds of release and rebinding [136, 291, 321].

While the folding reactions of other chaperonins are likely to be very similar to that of GroEL and GroES, the higher-plant chloroplast homologues pos-sess several intriguing structural and functional prop-erties that are not shared by other family members (recently reviewed in [94]). Before examining these in detail, however, we will briefly summarize the histor-ical events that led to the discovery of these proteins.

The Rubisco connection

Nearly a decade before it was shown that purified *E. coli* chaperonins could assist in protein folding [104], the 'Rubisco large subunit binding protein', now known as the chloroplast cpn60 (ch-cpn60) [120], was caught in the act of doing just that (reviewed in [73, 75, 76, 94, 95, 112]). The salient observation was that prior to their assembly into the Rubisco holoenzyme, the chloroplast-encoded Rubisco large subunits transi-ently bound to another plastid protein [10]. The latter was a huge oligomer consisting of ca. 60 kDa subunits, with a native molecular mass >600 kDa. Based on the precursor-product relationship between Rubisco large subunits associated with the binding protein, and those incorporated into the Rubisco holoenzyme, it was sug-gested that the assembly of the plant Rubisco required the assistance of another protein [10]. As noted elsewhere [73], at a time when it was widely assumed that all proteins fold spontaneously and there was little precedence for molecular chaperones, these observations were not fully appreciated.

Over the ensuing years, many attempts have been made to elucidate the role of the ch-cpn60 in the assembly of the higher plant Rubisco (a hexadecamer of eight large and eight small subunits), using isolated chloroplasts [18, 30, 213, 259, 260] and chloroplast extracts [130–132]. While some of these experiments are extremely difficult to interpret due to the complicated nature of the target protein and biological systems under study, it is still generally believed that the chaperonins are obligatory for plant Rubisco assembly [73, 75, 76, 95, 112, 258]. Perhaps the best evidence for this is that MgATP is able to discharge Rubisco large subunits from the ch-cpn60, and stimulate their incorporation into the Rubisco holoenzyme [18, 213], and that these events are blocked by anti-ch-cpn60 antibodies [30]. Other indirect observations that support this notion include: (1) the inability of the plant Rubisco to correctly assemble in *E. coli* [95]; (2) the tendency of Rubisco large subunits to aggregate *in vitro* [308]; and (3) the requirement for GroEL and GroES in the assembly of various prokaryotic Rubiscos during expression in *E. coli* [95, 105].

Ironically, while the mechanism of plant Rubisco assembly still remains obscure, a simpler dimeric Rubisco from *Rhodospirillum rubrum* has provided an excellent *in vitro* substrate for studying chaperonin mechanism [104, 300]. Based on results with this protein, it now seems likely that the chloroplast chaperonins facilitate the folding, not the assembly, of Rubisco large subunits. While they might also participate in the folding of Rubisco small subunits [75, 96], there is currently no evidence that chaperonins directly mediate the formation of quaternary structure.

Chloroplast cpn60 (ch-cpn60)

The ch-cpn60 is synthesized as a nuclear-encoded precursor that is subsequently imported into chloroplasts [94, 119, 120]. The protein is constitutively expressed [75, 119, 219, 301], although its levels increase slightly during heat-shock [114, 301]. Similar to GroEL, the native ch-cpn60 is a cylindrical 14-mer comprised of two stacked rings with sevenfold symmetry [244, 294, 301], and exhibits a weak intrinsic ATPase activity [244, 301]. Surprisingly, however, the chloroplast GroEL homologue appears to be a hetero-oligomeric protein. Native tetradecamers purified from pea (*Pisum sativum*) [119, 219, 301], barley (*Hordeum vulgare*) [219], and wheat (*Triticum aestivum*) [219] all consist of roughly stoichiometric amounts of two distinct ca. 60 kDa proteins, that are referred to as the α and β subunits. The latter are distinguishable during SDS-PAGE [119], are immunologically distinct species [219], and yield different protease digestion patterns [219]. Historically, it was the purification of the pea ch-cpn60 that enabled the production of antibodies [119] that were used to clone the α subunits of wheat and castor bean (*Ricinus communis*) [120]. Analysis of the cloned plant genes revealed substantial amino acid sequence homology to GroEL and led to the identification of a new family of proteins, christened the chaperonins [120].

The next major breakthrough resulted from the isolation of cDNAs encoding the α and β subunits of the *Brassica napus* ch-cpn60, and the recognition that their predicted amino acid sequences were highly divergent [200]. Unexpectedly, the two isoforms were only about 50% identical, no more similar to each other than they were to GroEL. Similar findings have been reported for *Arabidopsis thaliana* [200] and pea [65, 301]. While multiple cpn60 genes have also been identified in mitochondria [212, 292] and certain bacteria including *Mycobacterium tuberculosis* [157], *Bradyrhizobium japonicum* [80], and *Synechocystis* [173], apart from chloroplasts, there are no reports of purified cpn60 14-mers that contain similar amounts of two divergent subunits [119, 219, 301]. Indeed, given this observation and arguments based on 7-fold symmetry, it is tempting to speculate that ch-cpn60 tetradecamers are comprised of one ring of α subunits and one ring of β subunits. However, while such a structure is intuitively satisfying and analogous to that proposed for the double-ring Archaeosome, a TCP1-like chaperonin [245] (see above), the subunit composition and organization of ch-cpn60 tetradecamers has not been unequivocally determined (however, see below). Thus, the possible existence of homo-oligomers (e.g. α_{14} and β_{14}) and/or unrestricted hetero-oligomers (e.g. $\alpha_x\beta_{14-x}$) cannot currently be excluded.

A number of studies have examined the stability of ch-cpn60 tetradecamers (ch-cpn60$_{14}$) *in vitro*. In the presence of MgATP, the latter dissociate into lower molecular weight species [18, 119, 219, 260], including α and β monomers [219]. This phenomenon is enhanced at lower temperatures and requires ATP hydrolysis [18, 119]. That substantial dissociation could occur at physiological levels of ATP initially cast doubt that the ch-cpn60 was a 14-mer *in vivo* [119]. However, it was soon recognized that a dynamic equilibrium exists between monomers and tetradecamers, and that extreme dilution *in vitro* favors dissociation [219, 260]. Paradoxically, Lissin has recently shown that the reassembly of urea-dissociated ch-cpn60$_{14}$ from its monomeric subunits also requires adenine nucleotides [187]. This reaction is highly cooperative and is only observed at protomer concentrations exceeding 15 μM; the physiological concentration of cpn60 subunits in the chloroplast stroma is ca. 150 μM [219]. Based on these observations, it was proposed that adenine nucleotides are required for the formation of assembly-competent ch-cpn60 monomers, which in turn, are in equilibrium with 14-mers [187]. This effect of adenine nucleotides appears to be universal, since similar results have been obtained with GroEL [187, 188] and yeast mt-cpn60 [187].

The ability to reconstitute tetradecamers *in vitro* provides a powerful strategy for probing the subunit organization of the ch-cpn60. With this approach, it has recently been shown that purified β subunits can self-assemble into functional 14-mers that are active in protein folding [65]. As anticipated, the β-assembly reaction requires adenine nucleotides, is highly dependent on protein concentration, and is potentiated by GroES homologues. In contrast, purified α subunits only incorporate into 14-mers in the presence of β subunits, and this reaction also depends on adenine nucleotides. That the α subunits greatly stimulate the assembly of β subunits, and the resultant particles contain similar amounts of both isoforms, strongly supports the notion that native ch-cpn60 tetradecamers consist of one ring each of α and β subunits. These observations complement previous *in vivo* studies in which the α and β subunits of *Brassica napus* were expressed in *E. coli* [43, 44], both together and individually. In summary, the available evidence suggests that the majority of the α and β subunits reside in shared macromolecular complexes. However, from the results of the *in vitro* reconstitution experiments [65], it is also possible that plastids contain a smaller population of pure β 14-mers.

Chloroplast cpn10 (ch-cpn10)

The ch-cpn10 was initially identified through its ability to form a stable complex with GroEL [14, 94, 194]. As already noted, the two *E. coli* chaperonins only bind to each other in the presence of ATP or ADP [33, 300], and this highly conserved interaction has

proven quite useful for 'fishing out' hitherto unknown GroES homologues from crude cell-free extracts [14, 194, 254]. A pea chloroplast protein that was identified in this manner was able to assist GroEL during protein folding, and exhibited substantial N-terminal sequence homology to GroES [14]. However, its subunit molecular mass was nearly twice that of bacterial or mt-cpn10. The structural basis for this apparent anomaly was provided by the fortuitous cloning of the homologous protein from spinach [14]. Remarkably, the cloned spinach gene comprised a single open reading frame which encoded two complete GroES-like sequences, that were fused head-to-tail to form a single protein. The two 'halves' of this molecule are connected by a short stretch of amino acid residues that are ideally suited for domain linkage [9]. Despite these unusual characteristics, ch-cpn10 subunits form toroidal structures that superficially resemble GroES in the electron microscope [9].

The ch-cpn10 is synthesized as a nuclear-encoded precursor with an N-terminal transit peptide that is cleaved upon import [14]. Like the ch-cpn60, it is constitutively expressed, although its levels may increase slightly during thermal stress [301]. The predicted cleavage site for the cloned spinach precursor is in excellent agreement with a partial N- terminal sequence that was obtained from the purified pea protein [14]; the mature ch-cpn10 subunit is ca. 21 kDa. Importantly, both halves of the binary co-chaperonin are highly conserved at a number of amino acid residues that are thought to be important for function [14], and each possesses a polypeptide segment analogous to the so-called 'mobile loop' region of GroES ([159] and S. Landry, personal communication). The latter plays a critical role in the interaction between GroEL and GroES [166].

The binary organization of the ch-cpn10 presumably reflects a special adaptation of plants that has occurred in response to their possession of two divergent cpn60 isoforms. In support of this notion, the unique ca. 21 kDa co-chaperonin is widely distributed throughout the plant kingdom and is restricted to photosynthetic eukaryotes [9, 301]. Immunologically detectable levels of this protein are present in plant species that reflect an evolutionary divergence of at least 4×10^8 years [9]. In contrast, the cpn10 of cyanobacteria is of the normal 10 kDa variety, despite the fact that these photosynthetic organisms also possess multiple genes for cpn60 [173]. It appears that the binary ch-cpn10 resulted from either gene duplication or gene fusion after the endosymbiotic event that gave rise to chloroplasts. While plastids might also contain a more conventional GroES homologue, no such species have yet been identified.

The dual conservation of key residues in the binary ch-cpn10 immediately suggested that both halves of the molecule were active and may perform different functions [14]. One plausible scenario is that the two cpn10 domains interact differentially with the divergent α and β subunits. Perhaps one of the ch-cpn60 isoforms has evolved to accommodate the folding of the higher plant Rubisco or some other chloroplast protein(s) that requires 'special' folding assistance. That the chaperonins of a particular organism or organelle may have become specialized in response to their environment or the proteins that they normally encounter has received little attention. However, there are hints in the literature suggesting that this might be the case [80, 157, 173, 212, 292]. Other notable examples include the stringent requirement of the mt-cpn60 for a co-chaperonin of mitochondrial origin [66, 254, 299], and the obligatory role of the Gp31 protein (a bacteriophage T4-encoded GroES homologue [297]) in the GroEL-mediated folding of the major T4 coat protein (S. M. van der Vies, personal communication). Although the exact relationship between the two ch-cpn60 isoforms and the two halves of the binary ch-cpn10 remains to be elucidated, a recent study has shown that the latter can indeed function autonomously. Thus, when expressed individually in *E. coli*, both the N-terminal and C-terminal domains of the spinach ch-cpn10 are able to complement GroES-deficient mutants [9]. Unfortunately, for reasons that are not yet understood, neither half molecule is functionally active in assays with either GroEL [9] or the ch-cpn60 *in vitro* ([9] and P. V. Viitanen, unpublished observations).

Functional properties of the purified chloroplast chaperonins

Most of our mechanistic understanding of the chloroplast chaperonins has come from recent studies with the purified proteins [9, 14, 94, 301]. Needless to say, these studies were greatly facilitated by the cloning [14] and overexpression [9] of the spinach ch-cpn10. Prior to its discovery, it was shown that the ch-cpn60 could assist in the folding of prokaryotic Rubisco, under conditions *in vitro* where spontaneous folding does not occur [104]. Similar to the situation with GroEL, the recovery of active Rubisco from the ch-cpn60-Rubisco binary complex required both ATP and GroES, but the yield of folded Rubisco with this heterologous system was only 25% of that obtained with GroEL and GroES. This initially suggested that for full activity, the ch-cpn60 required its own co-chaperonin [104]. However, more recent experiments with Rubisco and mitochondrial malate dehydrogenase have shown that the ch-cpn60 functions equally well with bacterial, mitochondrial, or chloroplast cpn10 [301]. Thus, the unique binary ch-cpn10 is not obligatory for the folding reaction mediated by the ch-pn60, at least with regard to the two model proteins that have been studied *in vitro*.

As implied above, the available evidence suggests that the folding reactions mediated by GroEL and the ch-cpn60 are mechanistically similar. Both involve the formation of a stable binary complex with an unfolded protein substrate, and a subsequent discharge reaction that requires ATP hydrolysis [301] and the participation of cpn10 [104, 301]. Moreover, the $t_{1/2}$ for Rubisco refolding at 25 °C is virtually identical for GroEL and its chloroplast counterpart (i.e. ca. 3 min). During the chaperonin cycle, GroEL and GroES physically interact with each other and, depending on the adenine nucleotide present, form characteristic complexes that are distinguishable in the electron microscope (i.e. 'bullets' versus 'footballs'). That similar complexes are observed with the purified chloroplast chaperonins [301], is further testimony to the extreme conservation of mechanism that has apparently occurred in the evolution from bacteria to higher plants. Despite these similarities, however, the folding reaction mediated by the ch-cpn60 does not appear to require K^+ ions [301]. Indeed, preliminary experiments suggest that in contrast to GroEL [290, 300] and the mt-cpn60 of both yeast [254] and mammals [66, 299], the ATPase activity of the ch-cpn60 is not K^+-dependent [301]. However, at present the possible existence of tightly bound K^+ ions in the purified ch-cpnbo preparation cannot be excluded.

In addition to interacting with its cognate partner, the ch-cpn10 is also able to service bacterial and mitochondrial cpn60. For example, it can form stable complexes with GroEL [9, 14], inhibit GroEL's ATPase activity [9], and assist GroEL in the folding of prokaryotic Rubisco [9, 14]. Thus, despite their obvious structural differences, GroES and the binary ch-cpn10 are functionally interchangeable *in vitro* by a number of criteria. In contrast to GroES, however, the ch-cpn10 can also partially interact with the yeast mt-cpn60. This conclusion is supported not only by *in vitro* protein folding assays [254], but also by functional complementation experiments in yeast (S. Rospert, A.

A. Gatenby, and P. V. Viitanen, unpublished observations). Presumably, one of the two ch-cpn10 domains shares some subtle homology with the yeast mt-cpn10, that is absent in GroES.

Like other GroEL homologues [97, 100, 121, 298], the ch-cpn60 forms stable complexes with a variety of proteins that are structurally unrelated in their native states. In addition to its well known interactions with Rubisco large and small subunits [73, 75, 76, 95, 112], it stably associates with various proteins that are imported into isolated chloroplasts [96, 193, 195, 293]. Although it is not known whether the ch-cpn60 can assist in the folding of all of these proteins, the results certainly suggest that it has a rather broad substrate specificity. Moreover, there are a number of cases where the interaction between the ch-cpn60 and the imported protein appears to be an obligatory part of the import/assembly pathway [10, 96, 195, 293]. For example, the disappearance of the ch-cpn60-Rieske FeS protein complex during in vitro import experiments correlates with the appearance of properly folded Rieske FeS protein in thylakoid membranes [195]. Interestingly, two of the imported proteins that were found to interact with the ch-cpn60, namely, the Rieske FeS protein [195] and ferredoxin-NADP+ reductase [293], also transiently bound to the chloroplast Hsp70. This suggests that the folding of these and perhaps other chloroplast proteins requires the sequential assistance of multiple molecular chaperones. A similar conclusion was reached from reconstitution experiments in vitro with the multisubunit coupling factor CF_1 core [36].

Remaining questions about chloroplast chaperonins

The chloroplast chaperonins appear to play a prominent role in plastid protein folding. Consistent with this notion are the drastic phenotypic alterations (ranging to lethality at the extreme) that were observed in transgenic tobacco plants expressing low levels of the β ch-cpn60 [340]. Ironically, these antisense plants accumulated normal or slightly elevated levels of active Rubisco. While this result does not rule out a possible important role for β subunits in the folding of the plant Rubisco, similar studies should be conducted for the α subunits. This seems particularly relevant in light of the in vitro assembly experiments [65], where β subunits were strictly required for the incorporation of α subunits into 14-mers. Indeed, it would be very interesting to examine the oligomeric state and overall content of α subunits in the β-antisense plants, and vice versa.

In summary, further insight into the chloroplast chaperonins will require a much better understanding of the organization of ch-cpn60 tetradecamers, and how the divergent α and β subunits interact with the two halves of the binary ch-cpn10. Or, are these intriguing coincidences just another example of one of nature's red herrings? While both domains of the ch-cpn10 are capable of autonomous function when expressed individually in E. coli [9], recent experiments suggest that this might not be the case when they are linked together in the intact molecule [13].

Other chloroplast chaperones and foldases

Members of most of the other chaperone/foldase protein families have been identified in chloroplasts of higher plants, as well as in some algal groups and cyanobacteria. HSP70 family members were first reported in pea chloroplasts by Marshall et al. [198] and in the chloroplasts of Euglena gracilis by Ajir-Shapira et al. [1]. It is now well-established that chloroplasts contain a major soluble HSP70 homologue that is more similar to the bacterial DnaK than to the eukaryotic cytosolic HSP70s [67, 199, 285, 314]. In higher plants the protein is nuclear-encoded, but, not surprisingly, it is chloroplast-encoded in certain algae [247, 266, 315]. The chloroplast HSP70s also have biochemical properties similar to their bacterial counterparts with regard to autophosphorylation and ATPase activity [177]. Additional HSP70 proteins are associated with the chloroplast envelope [155, 198], but appear to be more eukaryotic-like.

The function of the stromal, DnaK-like HSP70 is likely to be related to the functions of its counterparts in bacteria and mitochondria, and the reader should consult several excellent reviews that deal with these subjects [240, 246, 286]. There has been limited direct investigation of the chaperone functions of HSP70 in chloroplasts. One study reported that stromal HSP70 was required for insertion of light-harvesting chlorophyll-binding proteins (LHCP) into the chloroplast membrane [338]. However, a subsequent study showed that although a stromal factor is required for LHCP integration, the factor was not HSP70 [339]. In fact, it has recently been established that this stromal factor is composed at least in part of a homologue of SRP54, a signal recognition particle component [180], which itself may have some type of chaperone activity. Other data, as mentioned above, have provided some evidence that stromal HSP70 interacts with proteins newly imported into the chloroplast [195, 293], as is the case for yeast and *Neurospora* mitochondrial HSP70 [286].

Perhaps the most exciting development concerning plastid HSP70 comes from evidence for a role for HSP70 chaperone activity in protein translocation at the chloroplast envelope as reviewed by Gray and Row [108]. Two groups have found an envelope HSP70 associated with precursor proteins arrested during translocation [274, 309]. Partial amino acid sequence data indicate the protein is similar to eukaryotic HSP70s (D. Schnell, personal communication). No analogous HSP70 homologues, distinct from the matrix-localized HSP70, have been implicated in protein transport into mitochondria. It is currently hypothesized that this chloroplast HSP70 protein may be required for ATP dependent precursor unfolding at the chloroplast envelope, and that this activity accounts in part for the ATP requirement of protein translocation.

A nuclear-encoded HSP100 homologue which localizes to chloroplasts has been defined in pea, tomato and *Arabidopsis* [106, 214, 277, 284]. HSP100 homologues have also been sequenced from the chloroplast genome of the chromophytic alga *Heterosigma* and from the cyanobacterium *Synechococcus* (A. Clarke, personal communication). These proteins are all members of the ClpC class of HSP100 proteins, distinct from the cytosolic ClpB types discussed above. Interestingly, recent data implicate an essential role for the ClpC protein, potentially similar to that of ClpA in *E. coli*. In higher plants, ClpC is constitutively expressed [268, 277] and in *Synechococcus* attempts to disrupt the single-copy gene have been unsuccessful, implying lethality of the transformants (A. Clarke, personal communication). It has also proven difficult to generate antisense transgenic *Arabidopsis* plants with less than 30% of wt ClpC levels (J. Shanklin, personal communication). Shanklin and colleagues have now demonstrated that recombinant pea ClpC facilitates the degradation of casein by *E. coli* ClpP in an ATP-dependent manner [277]. The chloroplast genome encodes the proteolytic ClpP subunit, and the same authors reported purified tobacco ClpP would hydrolyze peptide substrates. ClpP appears to be an essential chloroplast gene, in the sense that parasitic plants, which have lost the vast majority of protein coding genes from their chloroplast genome, still retain the ClpP gene [332]. Recently nuclear encoded ClpP homologues have also been identified, but the significance of having both nuclear- and chloroplast-encoded copies is unknown [267, 277]. Drawing on the model developed in studies of *E. coli* ClpA/ClpP interactions,

chloroplast ClpC appears to act as an ATP-dependent chaperone to present proteins for degradation to chloroplast ClpP. Chloroplast ClpC/ClpP thus represents the first energy-dependent proteolytic system identified in chloroplasts. Interestingly, there is no evidence for an enhanced role of the chloroplast ClpC/ClpP protease during heat stress, as the transcript level for ClpC actually declines during heat treatment [268].

A chloroplast-localized member of the smHSP family has been cloned from a number of higher plant species [38, 139, 225, 324], and observed in several others [42]. The protein is nuclear-encoded and has a transit peptide which is cleaved following chloroplast import to yield a mature protein on the order of 21 kDa in size. As a member of the smHSP family of proteins, it has the conserved carboxyl-terminal heat shock domain. A unique feature in the amino-terminal half of the protein, compared to other plant smHSPs, is a 22 amino acid domain predicted to form an amphipathic α-helix, in which the hydrophobic side is dominated by Met residues and all residues on the hydrophilic side have been conserved between monocot and dicot proteins [38]. A *Chenopodium* smHSP which lacks this domain and was originally reported to be chloroplast-localized [152] is now known to be localized to the mitochondrion (K. Kloppstech, personal communication). This domain would appear to be an important structural feature of the higher plant proteins, although its function remains unknown.

As has been described for the cytosolic smHSPs, the chloroplast protein is not detected in normal plants grown under controlled, non-stress conditions. The protein accumulates during heat stress, comprising up to 0.02% of total cell protein under some heat stress conditions, and is stable during recovery with a half-life of >50 h [37]. Given the absence of the chloroplast smHSP under normal conditions, it cannot be involved in folding processes required for normal development and maintenance of plastids, as are HSP70 and Cpn60. Rather, it is likely to be involved in functions unique to the stressed condition. To date these proteins have not been tested in any direct functional assays. Based on their homology to eukaryotic cytosolic smHSPs from both plants and other organisms, which are now proposed to be a type of molecular chaperone (see above), it is reasonable to propose that these proteins also have molecular chaperone activity. If so, it would seem likely that they have a restricted set of substrates, considering the low level to which they accumulate in the chloroplast. Kloppstech, Ohad and colleagues have proposed the protein functions to protect photosystem II [161], but interactions with specific PSII components have not been directly demonstrated. Critical functional studies of these proteins remain to be performed.

Immunophilins have also been described in chloroplasts. Nuclear-encoded chloroplast cyclophilins have been cloned from *Arabidopsis thaliana* [186] and *Vicia faba* [191]. The protein is localized to the chloroplast stroma and the mRNA appears to be greatly enriched in leaves compared to roots. Although PPIase activity might be proposed to play a role in protein import into chloroplasts, import of proteins into isolated pea chloroplasts was unaffected by high concentrations of CsA [186]. In *V. faba* high-temperature stress caused an increase in chloroplast CyP mRNA levels, implicating a role for PPIase in chloroplast stress responses [191]. Chloroplast-localized FK506 proteins, which have been identified by affinity chromatography [190], have not yet been further characterized.

Given the prokaryotic/endosymbiotic nature of the chloroplast, it will not be surprising to eventually identify members of the HSP90 chaperone family, as well as homologues of the GrpE and DnaJ co-chaperones, in these organelles.

Chaperone activities in mitochondria

Chaperone activities in plant mitochondria are much less well characterized than those in other plant cell compartments. Chaperones are clearly important for the biogenesis and continued function of mitochondria, and for a recent general review of this topic the reader should consult Martinus *et al.* [203]. We present here only a brief overview of progress on the plant mitochondrial chaperones. To date, with the exception of cpn10, no co-chaperones or foldases have been identified in plant mitochondria.

HSP70 proteins are known to be localized to plant mitochondria, but there have been few attempts to investigate their function. Mitochondrial HSP70 proteins are nuclear-encoded and have been characterized in several plant species, including pea [319], tomato [223] and *Arabidopsis*, which has two mitochondrial HSP70 genes (W. Zolotor and E. Vierling, unpublished observations). As seen for cytosolic HSP70s, at least one mitochondrial HSP70 gene is transcribed even in the absence of stress, but increased transcription also occurs during heat stress. In spinach the mitochondrial HSP70 was found to be associated in part with the inner mitochondrial membrane [307]. The role of these proteins in plant mitochondria can only be extrapolated from work in yeast and *Neurospora* indicating they function as chaperones during transport and folding of proteins in mitochondria [203, 286].

Research on plant mitochondrial cpn60 and cpn10 has also been limited, particularly with regard to mechanism. However, both chaperonin components have been identified in plant mitochondria [28, 241], and what is known about their expression and interaction with other proteins has recently been reviewed [94].

A member of the HSP100/Clp protein family, designated Hsp78, has been identified in yeast mitochondria [176]. It is most closely related to *E. coli* ClpB as are the plant cytosolic HSP100 proteins, and was also found to be induced during heat stress. Deletion of the gene was not lethal, nor did it impair growth under respiratory or heat stress conditions. Recently, Langer and colleagues obtained evidence for an interaction of HSP78 with the mitochondrial HSP70 protein encoded by the *SSC1* gene [273]. HSP78 overexpression suppressed a ts defect in *SSC1* and HSP78 was found associated with newly imported proteins in the ts mutant. They suggest the protein acts as a chaperone and is part of a salvage pathway under conditions of limiting HSP70. Such a protein has not yet been found associated with plant mitochondria, but as for other chaperones, it is likely to be conserved and may well soon be identified.

Although many workers had reported smHSPs associated with mitochondria, definitive evidence for organelle-localized smHSPs was only recently obtained by Lenne working with *Pisum sativum* [174, 175]. Homologous genes have now also been cloned from soybean [163] and *Chenopodium* (K. Kloppstech, personal communication). The proteins have amino terminal, mitochondrial targeting sequences and the mature regions contain the defined carboxyl-terminal heat shock domain but overall are less than 40% identical to the other classes of plant smHSPs. They are not present in control plants, but are highly induced by heat stress. How these proteins function as chaperones and the nature of their potential substrates requires further investigation.

Immunophilins from plant mitochondria have also not yet been described. However, cyclophilins have been described in yeast and *Neurospora crassa* mitochondria, indicating they are very possibly present in plants. Recent data from these model systems suggests they are involved in mitochondrial protein import [204].

Summary and future prospects

Research on molecular chaperones and protein folding has increased exponentially in the past ten years. It is now well-established that many, if not all, cellular proteins interact with chaperones and/or foldases during their lifetime in the cell. As we have outlined above, different chaperone and foldase systems are required for synthesis, targeting, maturation and degradation of proteins in all cellular compartments. Thus, these diverse proteins affect an exceptionally broad array of cellular processes required for both normal cell function and survival of stress conditions. In addition to the proteins we have discussed, other proteins have also been classified as chaperones [74, 76], including nucleoplasmin, presequence binding protein and SRP54 in eukaryotes, and trigger factor, SecB and PapD [133] in prokaryotes. Propeptides have also been termed 'intramolecular chaperones' because they can be required for correct folding of the mature protein [135].

Many important questions remain to be answered concerning the molecular mechanism of chaperone and foldase action. A crucial question is how different chaperones recognize their substrates and the nature of substrate binding. The role of ATP and stoichiometry of ATP hydolysis relative to protein folding also remain unresolved. Perhaps most importantly, we have little more than hypotheses regarding the identity of critical chaperone substrates within cells. Addressing this latter question will no doubt require new genetic approaches as well as biochemistry. For example, it has only been through genetic experiments that the interaction of HSP90 with signal transduction components has been confirmed in eukaryotes.

Although studies of chaperones in model systems such as *E. coli* and yeast have provided a basic roadmap for the analysis of these proteins in all organisms, significant questions concerning protein folding in plants can only be addressed by further studies directly with plants. We know already that there are a number of unique features of several plant chaperone systems. For example, only chloroplasts have two divergent cpn60 subunits and a 'double' cpn10 co-chaperone. The smHSPs, only recently described as being chaperones, show diversity and intracellular targets unique to plants. Other distinctive properties of plant chaperones and foldases will certainly be discovered, and basic information about known chaperones and foldases is still lacking in plants. Additional studies of protein folding in specialized plant processes and organelles are also fruitful areas for new discoveries. As discussed above, work with chloroplast cpn60 was instrumental in demonstrating the folding activity of this class of chaperones. Research on the deposition of plant storage proteins has also provided some of the best data on the role of the ER chaperones *in vivo*. It will certainly be of interest to know how plant chaperones and foldases may participate in other processes in plants such as cell wall deposition, transport through plasmodesmata and signal transduction to name only a few. The next decade promises continued advances in this fascinating area of cell biology.

Acknowledgements

We would like to thank our many colleagues who communicated unpublished results and manuscripts in press: A. Clarke, C. Gasser, A. Gatenby, P. Krishna, S. Lindquist, S. Luan, J. Miernyk, J. Nieto-Sotolo, T. Okita, N. Raikhel, D. Schnell, J. Shanklin and A. Vitale. We also thank those individuals who critically read all or part of this manuscript: R. Azpiroz, S. Erickson-Viitanen, C. Gasser, J. Gillikin, A. Mehta and J. Shanklin. We gratefully acknowledge financial support for research performed in our laboratories as granted from the National Science Foundation (R.B.), the NC Agricultural Research Service (R.B.), the National Institutes of Health (E.V.), the National Research Intitiative Competitive Grants Program of the US Department of Agriculture (E.V.), State of Arizona Hatch Funds (E.V.) and The American Cancer Society (E.V.).

References

1. Ajir-Shapira D, Leustek T, Dalie B, Weissbach H, Brot N: Hsp70 proteins, similar to *Escherichia coli* DnaK, in chloroplasts and mitochondria of *Euglena gracilis*. Proc Natl Acad Sci 87: 1749–1752 (1990).
2. Aligue R, Akhavan-Niak H, Russell P: A role for Hsp90 in cell cycle control: wee1 tyrosine kinase activity requires interaction with Hsp90. EMBO J 13: 6099–6106 (1994).
3. Allen GJ, Sanders D: Calcineurin, a type 2B protein phosphatase, modulates the Ca^{2+}-permeable slow vacuolar ion channel of stomatal guard cells. Plant Cell 7: 1473–1483 (1995).
4. Allen NS, Tiwari SC: Localization of calreticulin, a major calcium binding protein, in plant cells. Plant Physiol 96 (suppl): 42 (1991).
5. Anderson JV, Haskell DW, Guy CL: Differential influence of ATP on native spinach 70-kilodalton heat-shock cognates. Plant Physiol 104: 1371–1380 (1994).

6. Anderson JV, Li Q, Haskell DW, Guy CL: Structural organization of the spinach endoplasmic reticulum-luminal 70-kDa heat shock cognate gene and expression of 70-kDa heat-shock genes during cold acclimation. Plant Physiol 104: 1359–1370 (1994).

7. Arrigo A-P, Landry J: Expression and function of the low-molecular weight heat shock proteins. In: Morimoto R, Tissieres A, Georgopolous C (eds), The Biology of Heat Shock Proteins and Molecular Chaperones, pp. 335–373. Cold Spring Harbor Laboratory Press, Cold Spring Harbor, NY 1994.

8. Azem A, Kessel M, Goloubinoff P: Characterization of a functional GroEL$_{14}$(GroES$_7$)$_2$ chaperonin hetero-oligomer. Science 265: 653–656 (1994).

9. Baneyx F, Bertsch U, Kalbach CE, Van der Vies SM, Soll J, Gatenby AA: Spinach chloroplast cpn21 co-chaperonin possesses two functional domains fused together in a toroidal structure and exhibits nucleotide-dependent binding to plastid chaperonin 60. J Biol Chem 270: 10695–10702 (1995).

10. Barraclough R, Ellis RJ: Protein synthesis in chloroplasts. IX. Assembly of newly-synthesized large subunits into ribulose bisphosphate carboxylase in isolated intact pea chloroplasts. Biochim Biophys Acta 608: 19–31 (1980).

11. Bates EEM, Vergne P, Dumas C: Analysis of the cytosolic hsp70 gene family in *Zea mays*. Plant Mol Biol 25: 909–916 (1994).

12. Bednarek SY, Raikhel NV: Intracellular trafficking of secretory proteins. Plant Mol Biol 20: 133–150 (1992).

13. Bertsch U, Soll J: Functional analysis of isolated cpn10 domains and conserved amino acid residues in spinach chloroplast co-chaperonin by site-directed mutagenesis. Plant Mol Biol 29: 1039–1055 (1995).

14. Bertsch U, Soll J, Seetharam R, Viitanen PV: Identification, characterization, and DNA sequence of a functional 'double' groES- like chaperonin from chloroplasts of higher plants. Proc Natl Acad Sci USA 89: 8696–8700 (1992).

15. Bessoule J-J: Occurrence and sequence of a DnaJ protein in plant (*Allium porrum*) epidermal cells. FEBS Lett 323: 51–54 (1993).

16. Blond-Elguindi S, Cwirla SE, Dower WJ, Lipshutz RJ, Sprang SR, Sambrook JF, Gething M-JH: Affinity panning of a library of peptides displayed on bacteriophages reveals the binding specificity of BiP. Cell 75: 717–728 (1993).

17. Blond-Elguindi S, Fourie AM, Sambrook JF, Gething M-JH: Peptide-dependent stimulation of the ATPase activity of the molecular chaperone BiP is the result of conversion of oligomers to active monomers. J Biol Chem 268: 12730–12735 (1993).

18. Bloom MV, Milos P, Roy H: Light-dependent assembly of ribulose–1,5-bisphosphate carboxylase. Proc Natl Acad Sci USA 80: 1013–1017 (1983).

19. Bohen SP, Kralli A, Yamamoto KR: Hold 'em and fold 'em: chaperones and signal transduction. Science 268: 1303–1304 (1995).

20. Boorstein WR, Ziegelhoffer T, Craig EA: Molecular evolution of the HSP70 multigene family. J Mol Evol 38: 1–17 (1994).

21. Bork P, Sander C, and Valencia A: An ATPase domain common to prokaryotic cell cycle proteins, sugar kinases, actin, and hsp70 heat shock proteins. Proc Natl Acad Sci USA 89: 7290–7294 (1992).

22. Boston RS, Fontes EBP, Shank BB, Wrobel RL: Increased expression of the maize immunoglobulin binding protein homologue b–70 in three zein regulatory mutants. Plant Cell 3: 497–505 (1991).

23. Boston RS, Gillikin JW, Wrobel RL: Coordinate induction of three luminal ER-stress proteins in maize endosperm mutants. J Cell Biochem 19A: 143 (1995).

24. Braig K, Otwinowski Z, Hegde R, Boisvert DC, Joachimiak A, Horwich AL, Sigler PB: The crystal structure of the bacterial chaperonin GroEL at 2.8 å. Nature 371: 578–586 (1994).

25. Brot N, Redfield B, Qiu N-H, Chen G-J, Vidal V, Carlino A, Weissbach H: Similarity of nucleotide interactions of BiP and GTP-binding proteins. Proc Natl Acad Sci USA 91: 12120–12124 (1994).

26. Buchner J, Schmidt M, Fuchs M, Jaenicke R, Rudolph R, Schmid FX, Kiebhaber T: GroE facilitates refolding of citrate synthase by suppressing aggregation. Biochemistry 30: 1586–1591 (1991).

27. Bulleid N, Freedman R: Defective co-translational formation of disulfide bonds in protein disulphide isomerase-deficient microsomes. Nature 335: 649–651 (1988).

28. Burt WJE, Leaver CJ: Identification of a chaperonin–10 homologue in plant mitochondria. FEBS Lett 339: 139–141 (1994).

29. Cabané M, Calvet P, Vincens P, Boudet AM: Characterization of chilling-acclimation-related proteins in soybean and identification of one as a member of the heat shock protein (HSP 70) family. Planta 190: 346–353 (1993).

30. Cannon S, Wang P, Roy H: Inhibition of ribulose bisphosphate carboxylase by antibody to a binding protein. J Cell Biol 103: 1327–1335 (1986).

31. Carazo JM, Marco S, Abella G, Carrascosa JL, Secilla J-P, Muyal M: Electron microscopy study of GroEL chaperonin: different views of the aggregate appear as a function of cell growth temperature. J Struct Biol 106: 211–220 (1991).

32. Ceriotti A, Pedrazzini E, Bielli A, Giovinazzo G, Bollini R, Vitale A: Assembly and intracellular transport of phaseolin, the major storage protein of *Phaseolus vulgaris*. J Plant Physiol 145: 648–653 (1995).

33. Chandrasekhar GN, Tilly K, Woolford C, Hendrix R, Georgopoulos C: Purification and properties of the groES morphogenetic protein of *Escherichia coli*. J Biol Chem 261: 12414–12419 (1986).

34. Chang H-CJ, Lindquist S: Conservation of Hsp90 macromolecular complexes in *Saccharomyces cerevisiae*. J Biol Chem 269: 24983–24988 (1994).

35. Chen F, Hayes PM, Mulrooney DM, Pan A: Identification and characterization of cDNA clones encoding plant calreticulin in barley. Plant Cell 6: 835–843 (1994).

36. Chen GG, Jagendorf AT: Chloroplast molecular chaperone-assisted refolding and reconstitution of an active multisubunit coupling factor CF$_1$ core. Proc Natl Acad Sci USA 91: 11497–11501 (1994).

37. Chen Q, Lauzon LM, DeRocher AE, Vierling E: Accumulation, stability, and localization of a major chloroplast heat-shock protein. J Cell Biol 110: 1873–1883 (1990).

38. Chen Q, Vierling E: Analysis of conserved domains identifies a unique structural feature of a chloroplast heat shock protein. Mol Gen Genet 226: 425–431 (1991).

39. Chen X, Sullivan DS, Huffaker TC: Two yeast genes with similarity to TCP-1 are required for microtubule and actin function *in vivo*. Proc Natl Acad Sci USA 91: 9111–9115 (1994).

40. Cheng MY, Hartl F-U, Martin J, Pollock RA, Kalousek F, Neupert W, Hallberg EM, Hallberg RL, Horwich AL: The mitochondrial chaperonin hsp60 is essential for assembly of protein imported into yeast mitochondria. Nature 337: 620–625 (1989).

214

41. Chernoff YO, Lindquist SL, Ono B, Inge-Vechtomov SG, Liebman SW: Role of the chaperone protein Hsp104 in propagation of the yeast prion-like factor [*psi*$^+$]. Science 268: 880–884 (1995).

42. Clarke AK, Critchley C: Characterisation of chloroplast heat shock proteins in young leaves of C$_4$ monocotyledons. Physiol Plant 92: 118–130 (1994).

43. Cloney LP, Bekkaoui DR, Wood MG, Hemmingsen SM: Assessment of plant chaperonin-60 gene function in *Escherichia coli*. J Biol Chem 267: 23333–23336 (1992).

44. Cloney LP, Wu HB, Hemmingsen SM: Expression of plant chaperonin–60 genes in *Escherichia coli*. J Biol Chem 267: 23327–23332 (1992).

45. Coleman CE, Lopes MA, Gillikin JW, Boston RS, Larkins BA: A defective signal peptide in the maize high-lysine mutant floury 2. Proc Natl Acad Sci USA 92: 6828–6831 (1995).

46. Conner TW, LaFayette PR, Nagao RT, Key JL: Sequence and expression of a HSP83 from *Arabidopsis thaliana*. Plant Physiol 94: 1689–1695 (1990).

47. Cordewener JHG, Hause G, Görgen E, Busink R, Hause B, Dons HJM, van Lammeren AAM, van Lookeren Campagne MM, Pechan P: Changes in synthesis and localization of members of the 70-kDa class of heat-shock proteins accompany the induction of embryogenesis in *Brassica napus* L. microspores. Planta 196: 747–755 (1995).

48. Coughlan SJ, Hastings C, Winfrey JR: Molecular characterization of plant endoplasmic reticulum: identification of protein disulfide-isomerase as the major reticuloplasmin. Eur J Biochem 235: 215–244 (1996).

49. Craig EA: Regulation and function of the HSP70 multigene family of *Saccharomyces cerevisiae*. In: Morimoto RI, Tissières A, Georgopoulos C (eds), Stress Proteins in Biology and Medicine, pp. 301–321. Cold Spring Harbor Laboratory Press, Cold Spring Harbor, NY (1990).

50. Craig EA, Baxter BK, Becker J, Halladay J, Ziegelhoffer T: Cytosolic hsp70s of *Saccharomyces cerevisiae*: roles in protein synthesis, protein translocation, proteolysis, and regulation. In Morimoto RI, Tissières A, Georgopoulos C (eds) The Biology of Heat Shock Proteins and Molecular Chaperones, pp. 31–52. Cold Spring Harbor Press, Cold Spring Harbor, NY (1994).

51. Craig EA, Gambill BD, Nelson RJ: Heat shock proteins: molecular chaperones of protein biogenesis. Microbiol Rev 57: 402–414 (1993).

52. Cunningham KW, Fink GR: Calcineurin-dependent growth control in *Saccharomyces cerevisiae* mutants lacking PMC1, a homolog of plasma membrane Ca^{2+} ATPases. J Cell Biol 124: 351–363 (1994).

53. Cutforth T, Rubin GM: Mutations in *Hsp83* and *cdc37* impair signaling by the sevenless receptor tyrosine kinase in *Drosophila*. Cell 77: 1027–1036 (1994).

54. Cyr DM, Langer T, Douglas MG: DnaJ-like proteins: molecular chaperones and specific regulators of Hsp70. Trends Biochem Sci 19: 176–181 (1994).

55. Czar MJ, Lyons RH, Welsh MJ, Renoir JM, Pratt WB: Evidence that the FK506-binding immunophilin heat shock protein 56 is required for trafficking of the glucocorticoid receptor from the cytoplasm to the nucleus. Mol Endocrinol 9: 1549–1560 (1995).

56. D'Amico L, Valsasina B, Daminati MG, Fabrini MS, Nitti G, Bollini R, Ceriotti A, Vitale A: Bean homologoues of the mammalian glucose regulated proteins: induction by tunica-mycin and interaction with newly-synthesized storage proteins in the endoplasmic reticulum. Plant J 2: 443–455 (1992).

57. Delmer DP, Amor Y: Cellulose biosynthesis. Plant Cell 7: 987–1000 (1995).

58. Delmer DP, Volokita M, Solomon M, Fritz U, Delphendahl W, Herth W: A monoclonal antibody recognizes a 65 kDa higher plant membrane polypeptide which undergoes cation-dependent association with callose synthase *in vitro* and co-localizes with sites of high callose deposition *in vivo*. Protoplasma 176: 33–42 (1993).

59. Denecke J, Carlsson LE, Vidal S, Höglund A-S, Ek B, Van Zeijl MJ, Sinjorgo KMC, Palva ET: The tobacco homologue of mammalian calreticulin is present in protein complexes *in vivo*. Plant Cell 7: 391–406 (1995).

60. Denecke J, Ek B, Caspers M, Sinjorgo KMC, Palva ET: Analysis of sorting signals responsible for the accumulation of soluble reticuloplasmins in the plant endoplasmic reticulum. J Exp Bot (Suppl.) 44: 213–221 (1993).

61. Denecke J, Goldman MHS, Demolder J, Seurink J, Botterman J: The tobacco luminal binding protein is encoded by a multigene family. Plant Cell 1025–1035 (1991).

62. Denecke J, Vitale A: The use of protoplasts to study protein synthesis and transport by the plant endomembrane system. In: Galbraith D, Bourque D, Bohnert D (eds) Methods in Cell Biology Part B, pp. 335–348. Academic Press, New York (1995).

63. DeRocher A, Vierling E: Cytoplasmic HSP70 homologues of pea: differential expression in vegetative and embryonic organs. Plant Mol Biol 27: 441–456 (1995).

64. DeRocher AE, Helm KW, Lauzon LM, Vierling E: Expression of a conserved family of cytoplasmic low molecular weight heat shock proteins during heat stress and recovery. Plant Physiol 96: 1038–1047 (1991).

65. Dickson R, Howard R, Alldrick SP, Ellis RJ, Viitanen PV: Reconstitution of higher plant chloroplast chaperonin 60 tetradecamers active in protein folding. J Biol Chem, submitted (1996).

66. Dickson R, Larsen B, Viitanen PV, Tormey MB, Geske J, Strange R, Bemis LT: Cloning, expression, and purification of a functional nonacetylated mammalian mitochondrial chaperonin 10. J Biol Chem 269: 26858–26864 (1994).

67. Domoney C, Ellis N, Turner L, Casey R: A developmentally regulated early-embryogenesis protein in pea (*Pisum sativum* L.) is related to the heat-shock protein (HSP70) gene family. Planta 184: 350–355 (1991).

68. Duck N, McCormick S, Winter J: Heat shock protein hsp70 cognate gene expression in vegetative and reproductive organs of *Lycopersicon esculentum*. Proc Natl Acad Sci 86: 3674–3678 (1989).

69. Dure L, III: A repeating 11-mer amino acid motif and plant desiccation. Plant J 3: 363–369 (1993).

70. Dwek RA: Glycobiology: More functions for oligosaccharides. Science 269: 1234–1235 (1995).

71. Ehmann B, Krenz M, Mummert E, Schäfer E: Two Tcp-1-related but highly divergent gene families exist in oat encoding proteins of assumed chaperone function. FEBS Lett 336: 313–316 (1993).

72. Ellis J: Protein folding: cytosolic chaperonin confirmed. Nature 358: 191–192 (1992).

73. Ellis RJ: Molecular chaperones: the plant connection. Science 250: 954–959 (1990).

74. Ellis RJ: Chaperone function: cracking the second half of the genetic code. Plant J 1: 9–13 (1991).

75. Ellis RJ, van der Vies SM: The Rubisco subunit binding protein. Photosyn Res 16: 101–115 (1988).

76. Ellis RJ, van der Vies SM: Molecular chaperones. Annu Rev Biochem 60: 321–347 (1991).

77. Engel A, Hayer-Hartl MK, Goldie KN, Pfeifer G, Hegerl R, Müller S, Da Silva ACR, Baumeister W, Hartl FU: Functional significance of symmetrical versus asymmetrical GroEL-GroES chaperonin complexes. Science 269: 832–836 (1995).

78. Fayet O, Ziegelhoffer T, Georgopoulos C: The groES and groEL heat shcok gene products of *Escherichia coli* are essential for bacterial growth at all temperatures. J Bact 171: 1379–1385 (1989).

79. Felsheim RF, Das A: Structure and expression of a heat-shock protein 83 gene of *Pharbitis nil*. Plant Physiol 100: 1764–1771 (1992).

80. Fischer HM, Babst M, Kaspar T, Acuña G, Arigoni F, Hennecke H: One member of a *groESL*-like chaperonin multigene family in *Bradyrhizobium japonicum* is co-regulated with symbiotic nitrogen fixation genes. EMBO J 12: 2901–2912 (1993).

81. Flaherty KM, DeLuca-Falherty C, McKay DB: Three-dimensional structure of the ATPase fragment of a 70K heat-shock cognate protein. Nature 346: 623–628 (1990).

82. Flaherty KM, McKay DB, Kabsch W, Holmes KC: Similarity of the three-dimensional structures of actin and the ATPase fragment of a 70- kDa heat shock cognate protein. Proc Natl Acad Sci USA 88: 5041–5045 (1991).

83. Flynn GC, Chappell TG, Rothman JE: Peptide binding and release by proteins implicated as catalysts of protein assembly. Science 245: 285–390 (1989).

84. Flynn GC, Pohl J, Flocco MT, Rothman JE: Peptide-binding specificity of the molecular chaperone BiP. Nature 353: 726–730 (1991).

85. Fontes EBP, Shank BB, Wrobel RL, Moose SP, OBrian GR, Wurtzel ET, Boston RS: Characterization of an immunoglobulin binding protein homologue in the maize *floury-2* endosperm mutant. Plant Cell 3: 483–496 (1991).

86. Fourie AM, Sambrook JF, Gething M-JH: Common and divergent peptide binding specificities of hsp70 molecular chaperones. J Biol Chem 269: 30470–30478 (1994).

87. Freedman RB: Protein disulphide isomerase: multiple roles in the modification of nascent secretory proteins. Cell 57: 1069–1072 (1989).

88. Freedman RB, Hirst TR, Tuite MF: Protein disulphide isomerase: building bridges in protein folding. Trends Biochem Sci 19: 331–336 (1994).

89. Frydman J, Nimmesgern E, Erdjument-Bromage H, Wall JS, Tempst P, Hartl F-U: Function in protein folding of TRiC, a cytosolic ring complex containing TCP-1 and structurally related subunits. EMBO J 11: 4767–4778 (1992).

90. Frydman J, Nimmesgern E, Ohtsuka K, Hartl FU: Folding of nascent polypeptide chains in a high molecular mass assembly with molecular chaperones. Nature 370: 111–117 (1994).

91. Gao Y, Thomas JO, Chow RL, Lee G-H, Cowan NJ: A cytoplasmic chaperonin that catalyzes β-actin folding. Cell 69: 1043–1050 (1992).

92. Gao Y, Vainberg IE, Chow RL, Cowan NJ: Two cofactors and cytoplasmic chaperonin are required for the folding of α- and β-tubulin. Mol Cell Biol 13: 2478–2485 (1993).

93. Gasser CS, Gunning DA, Budelier KA, Brown SM: Structure and expression of cytosolic cyclophilin/peptidyl-prolyl *cis-trans* isomerase of higher plants and production of active tomato cyclophilin in *Escherichia coli*. Proc Natl Acad Sci USA 87: 9519–9523 (1990).

94. Gatenby AA: The chaperonins of photosynthetic organisms. In Ellis RJ (ed) The chapronins, Academic Press, pp. 65–90 (1996).

95. Gatenby AA, Ellis RJ: Chaperone function: the assembly of ribulose bisphosphate carboxylase-oxygenase. Annu Rev Cell Biol 6: 125–149 (1990).

96. Gatenby AA, Lubben TH, Ahlquist P, Keegstra K: Imported large subunits of ribulose bisphosphate carboxylase/oxygenase, but not imported β-ATP synthase sununits, are assembled into holoenzyme in isolated chloroplasts. EMBO J 7: 1307–1314 (1988).

97. Gatenby AA, Viitanen PV: Structural and functional aspects of chaperonin-mediated protein folding. Annu Rev Plant Physiol Plant Mol Biol 45: 469–491 (1994).

98. Geli MI, Torrent M, Ludevid D: Two structural domains mediate two sequential events in gamma-zein targeting: protein endoplasmic reticulum retention and protein body formation. Plant Cell 6: 1911–1922 (1994).

99. Georgopoulos C, Welch WJ: Role of the major heat shock proteins as molecular chaperones. Annu Rev Cell Biol 9: 601–634 (1993).

100. Gething M-J, Sambrook J: Protein folding in the cell. Nature 355: 33–45 (1992).

101. Gilbert HF: Protein chaperones and protein folding. Curr Opin Biotechnol 5: 534–539 (1994).

102. Gillikin JW, Boston RS: Plant BiPs. In: Gething MJ (ed) Guidebook to Molecular Chaperones and Protein Folding Catalysts. Sambrook and Tooze Publications at Oxford University Press, New York, in press (1996).

103. Gillikin JW, Fontes EPB, Boston RS: Protein-protein interactions in the endoplasmic reticulum. In: Galbraith D, Bourque D, Bohnert D (eds) Methods in Cell Biology Part B, pp. 309–323. Academic Press, New York (1995).

104. Goloubinoff P, Christeller JT, Gatenby AA, Lorimer GH: Reconstitution of active dimeric ribulose bisphosphate carboxylase from an unfolded state depends on two chaperonin proteins and Mg-ATP. Nature 342: 884–889 (1989).

105. Goloubinoff P, Gatenby AA, Lorimer GH: GroE heat-shock proteins promote assembly of foreign prokaryotic ribulose bisphosphate carboxylase oligomers in *Escherichia coli*. Nature 337: 44–47 (1989).

106. Gottesman S, Squires C, Pichersky E, Carrington M, Hobbs M, Mattick JS, Dalrymple B, Kuramitsu H, Shiroza T, Foster T, Clark WP, Ross B, Squires CL, Maurizi MR: Conservation of the regulatory subunit for the Clp ATP-dependent protease in prokaryotes and eukaryotes. Proc Natl Acad Sci 87: 3513–3517 (1990).

107. Gragerov A, Gottesman ME: Different peptide binding specificities of hsp70 family members. J Mol Biol 241: 133–135 (1994).

108. Gray JC, Row PE: Protein translocation across chloroplast envelope membranes. Trends Cell Biol 5: 243–247 (1995).

109. Gray TE, Fersht AR: Cooperativity in ATP hydrolysis by GroEL is increased by GroES. FEBS Lett 292: 254–258 (1991).

110. Grynberg A, Nicolas J, Drapron R: Some characteristics of protein disulfide isomerase (E.C. 5.3.4.1) from wheat (*Triticum vulgare*) embryo. Biochimie 60: 547–551 (1978).

111. Gupta RS: Phylogenetic analysis of the 90 kD heat shock family of protein sequences and an examination of the relationship among animals, plants, and fungi species. Mol Biol Evol 12: 1063–1073 (1995).

112. Gutteridge S, Gatenby AA: Rubisco synthesis, assembly, mechanism, and regulation. Plant Cell 7: 809–819 (1995).

216

113. Hartl FU, Martin J: Molecular chaperones in cellular protein folding. Curr Opin Struct Biol 5: 92–102 (1995).

114. Hartman DJ, Dougan D, Hoogenraad NJ, Hoj PB: Heat shock proteins of barley mitochondria and chloroplasts: Identification of organellar hsp 10 and 12: putative chaperonin 10 homologues. FEBS Lett 305: 147–150 (1992).

115. Hassan A-M, Wesson C, Trumble WR: Calreticulin is the major Ca^{++} storage protein in the endoplasmic reticulum of the pea plant (*Pisum sativum*). Biochem Biophys Res Commun 211: 54–59 (1995).

116. Hayman GT, Miernyk JA: The nucleotide and deduced amino acid sequences of a peptidyl-prolyl *cis-trans* isomerase from *Arabidopsis thaliana*. Biochim Biophys Acta 1219: 536–538 (1994).

117. Helm KW, LaFayette PR, Nagao RT, Key JL, Vierling E: Localization of small heat shock proteins to the higher plant endomembrane system. Mol Cell Biol 13: 238–247 (1993).

118. Helm KW, Schmeits J, Vierling E: An endomembrane-localized small heat-shock protein from *Arabidopsis thaliana*. Plant Physiol 107: 287–288 (1995).

119. Hemmingsen SM, Ellis RJ: Purification and properties of ribulosebisphosphate carboxylase large subunit binding protein. Plant Physiol 80: 269–276 (1986).

120. Hemmingsen SM, Woolford C, Van der Vies SM, Tilly K, Dennis DT, Georgopoulos CP, Hendrix R, Ellis RJ: Homologous plant and bacterial proteins chaperone oligomeric protein assembly. Nature 333: 330–334 (1988).

121. Hendrick JP, Hartl F-U: Molecular chaperone functions of heat-shock proteins. Annu Rev Biochem 62: 349–384 (1993).

122. Hendrix RW: Purification and properties of groE, a host protein involved in bacteriophage lambda assembly. J Mol Biol 129: 375–392 (1979).

123. Holdridge C, Dorsett D: Repression of *hsp70* heat shock gene transcription by the suppressor of hairy-wing protein of *Drosophila melanogaster*. Mol Cell Biol 11: 1894–1900 (1991).

124. Horwich AL: Molecular chaperones: resurrection or destruction. Curr Biol 5: 455–458 (1995).

125. Horwich AL, Willison KR: Protein folding in the cell: functions of two families of molecular chaperone, hsp 60 and TF55-TCP1. Phil Trans R Soc Lond [Biol] 339: 313–326 (1993).

126. Horwitz J: α-Crystallin can function as a molecular chaperone. Proc Natl Acad Sci USA 89: 10449–10453 (1992).

127. Höhfeld J, Minami Y, Hartl F-U: Hip, a novel cochaperone involved in the eukaryotic Hsc70/Hsp40 reaction cycle. Cell 83: 589–598 (1995).

128. Hsieh M-H, Chen J-T, Jinn T-L, Chen Y-M, Lin C-Y: A class of soybean low molecular weight heat shock proteins. Immunological study and quantitation. Plant Physiol 99: 1279–1284 (1992).

129. Huang L, Franklin AE, Hoffman NE: Primary structure and characterization of an *Arabidopsis thaliana* calnexin-like protein. J Biol Chem 268: 6560–6566 (1993).

130. Hubbs A, Roy H: Synthesis and assembly of large subunits into ribulose bisphosphate carboxylase/oxygenase in chloroplast extracts. Plant Physiol 100: 272–281 (1992).

131. Hubbs AE, Roy H: Assembly of *in vitro* synthesized large subunits into ribulose-bisphosphate carboxylase/oxygenase: formation and discharge of an L8-like species. J Biol Chem 268: 13519–13525 (1993).

132. Hubbs AE, Roy H: Assembly of *in vitro*-synthesized large subunits into ribulose bisphosphate carboxylase/oxygenase is sensitive to Cl$^-$, requires ATP, and does not proceed when

large subunits are synthesized at temperatures >32 °C. Plant Physiol 101: 523–533 (1993).

133. Hultgren SJ, Jacob-Dubuisson F, Jones CH, Bränden C-I: PapD and superfamily of periplasmic immunoglobulin-like pilus chaperones. Adv Protein Chem 44: 99–123 (1993).

134. Hutchison KA, Stancato LF, Owens-Grillo JK, Johnson JL, Krishna P, Toft DO, Pratt WB: The 23-kDa acidic protein in reticulocyte lysate is the weakly bound component of the hsp foldosome that is required for assembly of the glucocorticoid receptor into a functional heterocomplex with hsp90. J Biol Chem 270: 18841–18847 (1995).

135. Inouye M: Intramolecular chaperone: the role of the propeptide in protein folding. Enzyme 45: 314–321 (1991).

136. Jackson GS, Staniforth RA, Halsall DJ, Atkinson T, Holbrook JJ, Clarke AR, Burston SG: Binding and hydrolysis of nucleotides in the chaperonin catalytic cycle: Implications for the mechanism of assisted protein folding. Biochemistry 32: 2554–2563 (1993).

137. Jaenicke R, Creighton TE: Protein folding: Junior chaperones. Curr Biol 3: 234–235 (1993).

138. Jakob U, Buchner J: Assisting spontaneity: The role of Hsp90 and small Hsps as molecular chaperones. Trends Biochem Sci 19: 205–211 (1994).

139. Jakob U, Gaestel M, Engel K, Buchner J: Small heat shock proteins are molecular chaperones. J Biol Chem 268: 1517–1520 (1993).

140. Jakob U, Lilie H, Meyer I, Buchner J: Transient interaction of Hsp90 with early unfolding intermediates of citrate synthase. Implications for heat shock *in vivo*. J Biol Chem 270: 7288–7294 (1995).

141. Jakob U, Meyer I, Bügl H, André S, Bardwell JCA, Buchner J: Structural organization of procaryotic and eucaryotic Hsp90. Influence of divalent cations on structure and function. J Biol Chem 270: 14412–14419 (1995).

142. Jin Y-J, Albers MW, Lane WS, Bierer BE, Schreiber SL, Burakoff SJ: Molecular cloning of a membrane-associated human FK506- and rapamycin-binding protein, FKBP-13. Proc Natl Acad Sci USA 88: 6677–6681 (1991).

143. Jinn T-L, Chen Y-M, Lin C-Y: Chracterization and physiological function of class I low molecular weight heat shock protein complexes in soybean. Plant Physiol 108: 693–701 (1995).

144. Jinn TL, Yeh YC, Chen YM, Lin C-Y: Stabilization of soluble proteins *in vitro* by heat shock proteins-enriched ammonium sulfate fraction from soybean seedlings. Plant Cell Physiol 30: 463–469 (1989).

145. Jones AM, Herman EM: KDEL-containing auxin binding protein is secreted to the plasma membrane and cell wall. Plant Physiol 101: 595–606 (1993).

146. Kalinski A, Rowley DL, Loer DS, Foley C, Buta G, Herman EM: Binding-protein expression is subject to temporal, developmental and stress-induced regulation in terminally differentiated soybean organs. Planta 195: 611–621 (1995).

147. Keith B, Dong X, Ausubel FM, Fink GR: Differential induction of 3-deoxy-D-*arabino*-heptulosonate 7-phosphate synthase genes in *Arabidopsis thaliana* by wounding and pathogenic attack. Proc Natl Acad Sci USA 88: 8821–8825 (1991).

148. Kim PS, Arvan P: Calnexin and BiP act as sequential molecular chaperones during thyroglobulin folding in the endoplasmic reticulum. J Cell Biol 128: 29–38 (1995).

149. Kim S, Willison KR, Horwich AL: Cystosolic chaperonin subunits have a conserved ATPase domain but diverged polypeptide-binding domains. Trends Biochem Sci 19: 543–548 (1994).

50. Kimura Y, Yahara I, Lindquist S: Role of the protein chaperone YDJ1 in establishing Hsp90-mediated signal transduction pathways. Science 268: 1362–1365 (1995).

51. Kirsch T, Beevers L: Uncoating of clathrin-coated vesicles by uncoating ATPase from developing peas. Plant Physiol 103: 205–212 (1993).

152. Knack G, Liu Z, Kloppstech K: Low molecular mass heat-shock proteins of a light-resistant photoautotrophic cell culture. Eur J Cell Biol 59: 166–175 (1992).

153. Knarr G, Gething MJ, Modrow S, Buchner J: BiP binding sequences in antibodies. J Biol Chem 270:27589–27594 (1995).

154. Knittler MR, Haas IG: Interaction of BiP with newly synthesized immunoglobulin light chain molecules: Cycles of sequential binding and release. EMBO J 11: 1573–1581 (1992).

155. Ko K, Bornemisza O, Kourtz L, Ko ZW, Plaxton WC, Cashmore AR: Isolation and characterization of a cDNA clone encoding a cognate 70-kDa heat shock protein of the chloroplast envelope. J Biol Chem 267: 2986–2993 (1992).

156. Koch GLE: In: Rothblatt J, Novick P, Stevens T (eds), Guidebook to the Secretory Pathway, pp. 82–83. Sambrook and Tooze Publications at Oxford University Press, New York (1994).

157. Kong TH, Coates ARM, Butcher PD, Hickman CJ, Shinnick TM: *Mycobacterium tuberculosis* expresses two chaperonin–60 homologues. Proc Natl Acad Sci USA 90: 2608–2612 (1993).

158. Koning AJ, Rose R, Comai L: Developmental expression of tomato heat-shock cognate protein 80. Plant Physiol 100: 801–811 (1992).

159. Koonin EV, van der Vies SM: Conserved sequence motifs in bacterial and bacteriophage chaperonins. Trends Biochem Sci 20: 14–15 (1995).

160. Krishna P, Sacco M, Cherutti JF, Hill S: Cold-induced accumulation of hsp90 transcripts in *Brassica napus*. Plant Physiol 107: 915–923 (1995).

161. Kruse E, Kloppstech K: Heat shock proteins in plants: an approach to understanding the function of plastid heat shock proteins. In: Barber J (ed), The Photosystems: Structure, Function and Molecular Biology, pp. 409–442. Elsevier Sicence Publishers, Amsterdam (1992).

162. Kubota H, Hynes G, Carne A, Ashworth A, Willison K: Identification of six *Tcp-1*-related genes encoding divergent subunits of the TCP-1-containing chaperonin. Curr Biol 4: 89–99 (1994).

163. LaFayette PR, Nagao RT, O'Grady K, Vierling E, Key JL: Molecular characterization of cDNAs encoding low-molecular-weight heat shock proteins of soybean organelles. Plant Mol Biol 30: 159–169 (1996).

164. Laminet AA, Ziegelhoffer T, Georgopoulos C, Plückthun A: The *Escherichia coli* heat shock proteins GroEL and GroES modulate the folding of the β-lactamase precursor. EMBO J 9: 2315–2319 (1990).

165. Landry SJ, Gierasch LM: Polypeptide interactions with molecular chaperones and their relationship to *in vivo* protein folding. Annu Rev Biophys Biomol Struct 23: 645–669 (1994).

166. Landry SJ, Zeilstra-Ryalls J, Fayet O, Georgopoulos C, Gierasch LM: Characterization of a functionally important mobile domain of GroES. Nature 364: 255–258 (1993).

167. Langer T, Pfeifer G, Martin J, Baumeister W, Hartl F-U: Chaperonin-mediated protein folding: GroES binds to one end of the GroEL cylinder, which accommodates the protein substrate within its central cavity. EMBO J 11: 4757–4765 (1992).

168. Lata Singla S, Grover A: Antibodies raised against yeast HSP 104 cross-react with a heat and abscisic acid-regulated polypeptide in rice. Plant Mol Biol 22: 1177–1180 (1993).

169. Lee AS: Coordinated regulation of a set of genes by glucose and calcium ionophores in mammalian cells. Trends Biochem Sci 12: 20–23 (1987).

170. Lee GJ, Pokala N, Vierling E: Structure and *in vitro* molecular chaperone activity of cytosolic small heat shock proteins from pea. J Biol Chem 270: 10432–10438 (1995).

171. Lee JH, Hübel A, Schöffl F: Derepression of the activity of genetically engineered heat shock factor causes constitutive synthesis of heat shock proteins and increased thermotolerance in transgenic *Arabidopsis*. Plant J 8: 603–609 (1995).

172. Lee Y-RJ, Nagao RT, Key JL: A soybean 101-kD heat shock protein complements a yeast *HSP104* deletion mutant in acquiring thermotolerance. Plant Cell 6: 1889–1897 (1994).

173. Lehel C, Los D, Wada H, Györgyei J, Horváth I, Kovács E, Murata N, Vigh L: A second *groEL*-like gene, organized in a *groESL* operon is present in the genome of *Synechocystis* sp. PCC 6803. J Biol Chem 268: 1799–1804 (1993).

174. Lenne C: Sequence and expression of the mRNA encoding HSP22, the mitochondrial small heat-shock protein in pea leaves. Biochem J 311: 805–813 (1995).

175. Lenne C, Douce R: A low molecular mass heat-shock protein is localized to higher plant mitochondria. Plant Physiol 105: 1255–1261 (1994).

176. Leonhardt SA, Fearon K, Danese PN, Mason TL: *HSP78* encodes a yeast mitochondrial heat shock protein in the Clp family of ATP-dependent proteases. Mol Cell Biol 13: 6304–6313 (1993).

177. Leustek T, Amir-Shapira D, Toledo H, Brot N, Weissbach H: Autophosphorylation of 70 kDa heat shock proteins. Cell Mol Biol 38: 1–10 (1992).

178. Lewis VA, Hynes GM, Zheng D, Saibil H, Willison K: T-complex polypeptide–1 is a subunit of a heteromeric particle in the eukaryotic cytosol. Nature 358: 249–252 (1992).

179. Li C-P, Larkins BA: Expression of protein disulfide isomerase is elevated in the endosperm of the maize *floury–2* mutant. Plant Mol Biol, in press (1996).

180. Li X, Henry R, Yuan J, Cline K, Hoffman NE: A chloroplast homologue of the signal recognition particle subunit SRP54 is involved in the posttranslational integration of a protein into thylakoid membranes. Proc Natl Acad Sci USA 92: 3789–3793 (1995).

181. Li X, Wu Y, Zhang D-Z, Gillikin JW, Boston RS, Franceschi VR, Okita TW: Rice prolamine protein body biogenesis: a BiP-mediated process. Science 262: 1054–1056 (1993).

182. Liberek K, Marszalek J, Ang D, Georgopoulos C, Zylicz M: *Escherichia coli* DnaJ and GrpE heat shock proteins jointly stimulate ATPase activity of DnaK. Proc Natl Acad Sci USA 88: 2874–2878 (1991).

183. Lin T-Y, Duck NB, Winter J, Folk WR: Sequences of two hsc 70 cDNAs from *Lycopersicon esculentum*. Plant Mol Biol 16: 475–478 (1991).

184. Lin X, Chern M, Zimmerman JL: Cloning and characterization of a carrot hsp70 gene. Plant Mol Biol 17: 1245–1249 (1991).

185. Lindquist S, Craig EA: The heat shock proteins. Annu Rev Genet 22: 631–677 (1988).

186. Lippuner V, Chou IT, Scott SV, Ettinger WF, Theg SM, Gasser CS: Cloning and characterization of chloroplast and cytosolic

218

forms of cyclophilin from *Arabidopsis thaliana*. J Biol Chem 269: 7863–7868 (1994).

187. Lissin NM: In vitro dissociation and self-assembly of three chaperonin 60s: the role of ATP. FEBS Lett 361: 55–60 (1995).

188. Lissin NM, Venyaminov SY, Girshovich AS: (Mg-ATP)-dependent self-assembly of molecular chaperone GroEL. Nature 348: 339–342 (1990).

189. Lorimer GH: Role of accessory proteins in protein folding. Curr Biol 2: 26–34 (1992).

190. Luan S, Albers MW, Schreiber SL: Light-regulated, tissue-specific immunophilins in a higher plant. Proc Natl Acad Sci USA 91: 984–988 (1994).

191. Luan S, Lane WS, Schreiber SL: pCyP B: a chloroplast-localized, heat-shock-responsive cyclophilin from fava bean. Plant Cell 6: 885–892 (1994).

192. Luan S, Li W, Rusnak F, Assmann SM, Schreiber SL: Immunosuppressants implicate protein phosphatase regulation of K^+ channels in guard cells. Proc Natl Acad Sci USA 90: 2202–2206 (1993).

193. Lubben TH, Donaldson GK, Viitanen PV, Gatenby AA: Several proteins imported into chloroplasts form stable complexes with the GroEL-related chloroplast molecular chaperone. Plant Cell 1: 1223–1230 (1989).

194. Lubben TH, Gatenby AA, Donaldson GK, Lorimer GH, Viitanen PV: Identification of a groES-like chaperonin in mitochondria that facilitates protein folding. Proc Natl Acad Sci USA 87: 7683–7687 (1990).

195. Madueno F, Napier JA, Gray JC: Rieske iron-sulfur protein associates with both cpn60 and hsp70 in the chloroplast stroma. Plant Cell 5: 1865–1876 (1993).

196. Marocco A, Santucci A, Cerioli S, Motto M, Di Fonzo N, Thompson R, Salamini F: Three high-lysine mutations control the level of ATP-binding HSP70-like proteins in the maize endosperm. Plant Cell 3: 507–515 (1991).

197. Marrs KA, Casey ES, Capitant SA, Bouchard RA, Dietrich PS, Mettler IJ, Sinibaldi RM: Characterization of two maize hsp90 heat shock protein genes: expression during heat shock, embryogenesis, and pollen development. Devel Genet 14: 27–41 (1993).

198. Marshall JS, DeRocher AE, Keegstra K, Vierling E: Identification of heat shock protein hsp70 homologues in chloroplasts. Proc Natl Acad Sci USA 87: 374–378 (1990).

199. Marshall JS, Keegstra K: Isolation and characterization of a cDNA clone encoding the major Hsp70 of the pea chloroplastic stroma. Plant Physiol 100: 1048–1054 (1992).

200. Martel R, Cloney LP, Pelcher LE, Hemmingsen SM: Unique composition of plastid chaperonin–60: α and β polypeptide-encoding genes are highly divergent. Gene 94: 181–187 (1990).

201. Martin J, Langer T, Boteva R, Schramel A, Horwich AL, Hartl F-U: Chaperonin-mediated protein folding at the surface of groEL through a 'molten globule'-like intermediate. Nature 352: 36–42 (1991).

202. Martin J, Mayhew M, Langer T, Hartl FU: The reaction cycle of GroEL and GroES in chaperonin-assisted protein folding. Nature 366: 228–233 (1993).

203. Martinus RD, Ryan MT, Naylor DJ, Herd SM, Hoogenraad NJ, Hoj PB: Role of chaperones in the biogenesis and maintenance of the mitochondrion. FASEB J 9: 371–378 (1995).

204. Matouschek A, Rospert S, Schmid K, Glick BS, Schatz G: Cyclophilin catalyzes protein folding in yeast mitochondria. Proc Natl Acad Sci USA 92: 6319–6323 (1995).

205. Melki R, Cowan NJ: Facilitated folding of actins and tubulins occurs via a nucleotide-dependent interaction between cytoplasmic chaperonin and distinctive folding intermediates. Mol Cell Biol 14: 2895–2904 (1994).

206. Melnick J, Aviel S, Argon Y: The endoplasmic reticulum stress protein GRP94, in addition to BiP, associates with unassembled immunoglobulin chains. J Biol Chem 267: 21303–21306 (1992).

207. Melnick J, Dul JL, Argon Y: Sequential interaction of the chaperones BiP and GRP94 with immunoglobulin chains in the endoplasmic reticulum. Nature 370: 373–375 (1994).

208. Menegazzi P, Guzzo F, Baldan B, Mariani P, Treves S: Purification of calreticulin-like protein(s) from spinach leaves. Biochem Biophys Res Commun 190: 1130–1135 (1993).

209. Merck KB, Groenen PJTA, Voorter CEM, de Haard-Hoekman WA, Horwitz J, Bloemendal H, de Jong WW: Structural and functional similarities of bovine α-crystallin and mouse small heat-shock protein. A family of chaperones. J Biol Chem 268: 1046–1052 (1993).

210. Miernyk JA, Duck NB, Shatters RG, Jr., Folk WR: The 70-kilodalton heat shock cognate can act as a molecular chaperone during the membrane translocation of a plant secretory protein precursor. Plant Cell 4: 821–829 (1992).

211. Miernyk JA, Hayman GT: ATPase activity and molecular chaperone function of the stress70 proteins. Plant Physiol 110: 419–424 (1996).

212. Miller SG, Leclerc RF, Erdos RF: Identification and characterization of a testis-specific isoform of a chaperonin in a moth, *Heliothis virescens*. J Mol Biol 214: 407–422 (1990).

213. Milos P, Roy H: ATP-released L subunits participate in the assembly of ribulose bisphosphate carboxylase. J Cell Biochem 24: 153–162 (1984).

214. Moore T, Keegstra K: Characterization of a cDNA clone encoding a chloroplast-targeted Clp homologue. Plant Mol Biol 21: 525–537 (1993).

215. Mori M, Murata K, Kubota H, Yamamoto A, Matsushiro A, Morita T: Cloning of a cDNA encoding the *Tcp-1* (*t* complex polypeptide) homologue of *Arabidopsis thaliana*. Gene 122: 381–382 (1992).

216. Morimoto R, Tissières A, Georgopolous C: The biology of heat shock proteins and molecular chaperones. Cold Spring Harbor Laboratory Press, Cold Spring Harbor, NY (1994).

217. Mummert E, Grimm R, Speth V, Eckerskorn C, Schiltz E, Gatenby AA, Schäfer E: A TCP1-related molecular chaperone from plants refolds phytochrome to its photoreversible form. Nature 363: 644–648 (1993).

218. Mummert E, Grimm R, Speth V, Eckerskorn C, Schiltz E, Gatenby AA, Schäfer E: A TCP1-related molecular chaperone from plants refolds phytochrome to its photoreversible form – Correction. Nature 372: 709 (1994).

219. Musgrove JE, Johnson RA, Ellis RJ: Dissociation of the ribulosebisphosphate-carboxylase large-subunit binding protein into dissimilar subunits. Eur J Biochem 163: 529–534 (1987).

220. Müller FW, Igloi GL, Beck CF: Structure of a gene encoding heat-shock protein HSP70 from the unicellular alga *Chlamydomonas reinhardtii*. Gene 111: 165–173 (1992).

221. Napier RM, Fowke LC, Hawes C, Lewis M, Pelham HRB: Immunological evidence that plants use both HDEL and KDEL for targeting proteins to the endoplasmic reticulum. J Cell Sci 102: 261–271 (1992).

222. Nathan DF, Lindquist S: Mutational analysis of Hsp90 function: interactions with a steroid receptor and a protein kinase. Mol Cell Biol 15: 3917–3925 (1995).

3. Neumann D, Emmermann M, Thierfelder JM, Nieden U, Clericus M, Braun HP, Nover L, Schmitz UK: HSP68-a DnaK-like heat-stress protein of plant mitochondria. Planta 190: 32–43 (1993).

24. Nielsen JB, Foor F, Siekierka JJ, Hsu M-J, Ramadan N, Morin N, Shafiee A, Dahl AM, Brizuela L, Chrebet G, Bostian KA, Parent SA: Yeast FKBP-13 is a membrane-associated FK506-binding protein encoded by the nonessential gene *FKB2*. Proc Natl Acad Sci USA 89: 7471–7475 (1992).

225. Nieto-Sotelo J, Vierling E, Ho T-HD: Cloning, sequence analysis, and expression of a cDNA encoding a plastid-localized heat shock protein in maize. Plant Physiol 93: 1321–1328 (1990).

226. Normington K, Kohno K, Kozutsumi Y, Gething M-J, Sambrook J: *S. cerevisiae* encodes an essential protein homologous in sequence and function to mammalian BiP. Cell 57: 1223–1236 (1989).

227. Okita TW, Rogers JC: Compartmentation of proteins in the endomembrane system of plant cells. Annu Rev Plant Physiol Plant Mol Biol, in press (1996).

228. Palleros DR, Reid KL, Shi L, Welch WJ, Fink AL: ATP-induced protein-Hsp70 complex dissociation requires K$^+$ but not ATP hydrolysis. Nature 365: 664–666 (1993).

229. Parker W, Wells TA, Meza-Keuthen S, Kim I-S, Song P-S: Purification and characterization of a 60-kDa protein from oat, formerly known as a TCP1-related chaperone. J Protein Chem 14: 53–58 (1995).

230. Parsell DA, Kowal AS, Singer MA, Lindquist S: Protein disaggregation mediated by heat-shock protein Hsp104. Nature 372: 475–478 (1994).

231. Parsell DA, Lindquist S: The function of heat-shock proteins in stress tolerance: degradation and reactivation of proteins. Annu Rev Genet 27: 437–496 (1993).

232. Parsell DA, Sanchez Y, Stitzel JD, Lindquist S: Hsp104 is a highly conserved protein with two essential nucleotide-binding sites. Nature 353: 270–273 (1991).

233. Pedrazzini E, Giovinazzo G, Bollini R, Ceriotti A, Vitale A: Binding of BiP to an assembly-defective protein in plant cells. Plant J 5: 103–110 (1994).

234. Pedrazzini E, Vitale A: The binding protein, BiP, and the synthesis of secretory proteins. Plant Physiol Biochem 34: 207–216 (1996).

235. Pelham HRB: Function of the hsp70 protein family: an overview. In: Morimoto RI, Tissieres A, Georgopoulos C (eds) Stress Proteins in Biology and Medicine, pp. 287–299. Cold Spring Harbor Laboratory Press, Cold Spring Harbor, NY (1990).

236. Phipps BM, Hoffmann A, Stetter KO, Baumeister W: A novel ATPase complex selectively accumulated upon heat shock is a major cellular component of thermophilic archaebacteria. EMBO J 10: 1711–1722 (1991).

237. Phipps BM, Typke D, Hegerl R, Volker S, Hoffmann A, Stetter KO, Baumeister W: Structure of a molecular chaperone from a thermophilic archaebacterium. Nature 361: 475–477 (1993).

238. Picard D, Khursheed B, Garabedian MJ, Fortin MG, Lindquist S, Yamamoto KR: Reduced levels of hsp90 compromise steroid receptor action *in vivo*. Nature 348: 166–168 (1990).

239. Picketts DJ, Mayanil CSK, Gupta RS: Molecular cloning of a Chinese hamster mitochondrial protein related to the 'chaperonin' family of bacterial and plant proteins. J Biol Chem 264: 12001–12008 (1989).

240. Polissi A, Goffin L, Georgopoulos C: The *Escherichia coli* heat shock response and bacteriophage lambda development. FEMS Microbiol Rev 17: 159–169 (1995).

241. Prasad TK, Hallberg RL: Identification and metabolic characterization of the *Zea mays* mitochondrial homologue of the *Escherichia coli* GroEL protein. Plant Mol Biol 12: 609–618 (1989).

242. Pratt WB: The role of heat shock proteins in regulating the function, folding, and trafficking of the glucocorticoid receptor. J Biol Chem 268: 21455–21458 (1993).

243. Preisig-Müller R, Kindl H: Plant dnaJ homologue: molecular cloning, bacterial expression, and expression analysis in tissues of cucumber seedlings. Arch Biochem Biophys 305: 30–37 (1993).

244. Pushkin AV, Tsuprun VL, Solovjeva NA, Shubin VV, Evstigneeva ZG, Kretovich WL: High molecular weight pea leaf protein similar to the groE protein of *Escherichia coli*. Biochim Biophys Acta 704: 379–384 (1982).

245. Quaite-Randall E, Trent JD, Josephs, R., Joachimiak A: Conformational cycle of the Archaeosome, a TCP1-like chaperonin from *Sulfolobus shibatae*. J Biol Chem 270: 28818–28823 (1995).

246. Rassow J, Voos W, Pfanner N: Partner proteins determine multiple functions of Hsp70. Trends Cell Biol 5: 207–212 (1995).

247. Reith M, Munholland J: An hsp70 homologue is encoded on the plastid genome of the red alga, *Porphyra umbilicalis*. FEBS Lett 294: 116–120 (1991).

248. Rippmann F, Taylor WR, Rothbard JB, Green NM: A hypothetical model for the peptide binding domain of hsp70 based on the peptide binding domain of HLA. EMBO J 10: 1053–1059 (1991).

249. Roberts JK, Key JL: Isolation and characterization of a soybean hsp70 gene. Plant Mol Biol 16: 671–683 (1991).

250. Rochester DE, Winter JA, Shah DM: The structure and expression of maize genes encoding the major heat shock proteins, hsp70. EMBO J 5: 451–458 (1986).

251. Roden LT, Miflin BJ, Freedman RB: Protein disulphide-isomerase is located in the endoplasmic reticulum of developing wheat endosperm. FEBS Lett 138: 121–124 (1982).

252. Rommelaere H, Van Troys M, Gao Y, Melki R, Cowan NJ, Vandekerckhove J, Ampe C: Eukaryotic cytosolic chaperonin contains t-complex polypeptide 1 and seven related subunits. Proc Natl Acad Sci USA 90: 11975–11979 (1993).

253. Rose MD, Misra LM, Vogel JP: *KAR2*, a karyogamy gene, is the yeast homologue of the mammalian BIP/GRP78 gene. Cell 57: 1211–1221 (1989).

254. Rospert S, Glick BS, Jenö P, Schatz G, Todd MJ, Lorimer GH, Viitanen PV: Identification and functional analysis of chaperonin 10, the groES homologue from yeast mitochondria. Proc Natl Acad Sci USA 90: 10967–10971 (1993).

255. Rospert S, Junne T, Glick BS, Schatz G: Cloning and disruption of the gene encoding yeast mitochondrial chaperonin 10, the homolog of *E. coli* groES. FEBS Lett 335: 358–360 (1993).

256. Rothblatt J, Novick P, Stevens T: Guidebook to the Secretory Pathway. Sambrook and Tooze Publications at Oxford University Press, New York (1994).

257. Rowling PJE: Folding, assembly and posttranslational modification of proteins within the lumen of the endoplasmic reticulum. Subcell Biochem 21: 41–80 (1993).

258. Roy H: Rubisco assembly: a model system for studying the mechanism of chaperonin action. Plant Cell 1: 1035–1042 (1989).

259. Roy H, Bloom M, Milos P, Monroe M: Studies on the assembly of large subunits of ribulose bisphosphate

220

carboxylase in isolated pea chloroplasts. J Cell Biol 94: 20–27 (1982).

260. Roy H, Hubbs A, Cannon S: Stability and dissociation of the large subunit of RuBisCo binding protein complexes *in vitro* and in organello. Plant Physiol 86: 50–53 (1988).

261. Rutherford SL, Zuker CS: Protein folding and the regulation of signaling pathways. Cell 79: 1129–1132 (1994).

262. Saibil HR, Zheng D, Roseman AM, Hunter AS, Watson GMF, Chen S, Auf der Mauer A, O'Hara BP, Wood SP, Mann NH, Barnett LK, Ellis RJ: ATP induces large quaternary rearrangements in a cage-like chaperonin structure. Curr Biol 3: 265–273 (1993).

263. Sanchez Y, and Lindquist SL: HSP104 required for induced thermotolerance. Science 248: 1112–1115 (1990).

264. Sanchez Y, Taulien J, Borkovich KA, Lindquist S: Hsp104 is required for tolerance to many forms of stress. EMBO J 11: 2357–2364 (1992).

265. Sanders SL, Whitfield KM, Vogel JP, Rose MD, Schekman RW: Sec61p and BiP directly facilitate polypeptide translocation into the ER. Cell 69: 353–365 (1992).

266. Scaramuzzi CD, Stokes HW, Hiller RG: Heat shock Hsp70 protein is chloroplast-encoded in the chromophytic alga *Pavlova lutherii*. Plant Mol Biol 18: 467–476 (1992).

267. Schaller A, Ryan CA: Cloning of a tomato cDNA (Gen-Bank L388581) encoding the proteolytic subunit of a Clp-like energy dependent proteolysis. Plant Physiol 108: 1341(1995).

268. Schirmer EC, Lindquist S, Vierling E: An *Arabidopsis* heat shock protein complements a thermotolerance defect in yeast. Plant Cell 6: 1899–1909 (1994).

269. Schirmer EC, Lindquist SL: The HSP100 family. In: Gething MJ (ed), Guidebook to Molecular Chaperones and Protein Folding Catalysts. Oxford University Press, in press (1966).

270. Schmid FX, Mayr LM, Mucke M, Schonbrunner ER: Prolyl isomerases: role in protein folding. Adv Protein Chem 44: 25–66 (1993).

271. Schmidt M, Buchner J, Todd MJ, Lorimer GH, Viitanen PV: On the role of groES in the chaperonin-assisted folding reaction. Three case studies. J Biol Chem 269: 10304–10311 (1994).

272. Schmidt M, Rutkat K, Rachel R, Pfeifer G, Jaenicke R, Viitanen P, Lorimer G, Buchner J: Symmetric complexes of GroE chaperonins as part of the functional cycle. Science 265: 656–659 (1994).

273. Schmitt M, Neupert W, Langer T: Hsp78, a Clp homologue within mitochondria, can substitute for chaperone functions of mt-hsp70. EMBO J 14: 3434–3444 (1995).

274. Schnell DJ, Kessler F, Blobel G: Isolation of components of the chloroplast protein import machinery. Science 266: 1007–1012 (1994).

275. Schreiber SL: Chemistry and biology of the immunophilins and their immunosuppressive ligands. Science 251: 283–287 (1991).

276. Schroder G, Beck M, Eichel J, Vetter H, Schroder J: HSP90 homologue from Madagascar periwinkle (*Catharanthus roseus*): cDNA sequence, regulation of protein expression and location in the endoplasmic reticulum. Plant Mol Biol 23: 583–598 (1993).

277. Shanklin J, DeWitt ND, Flanagan JM: The stroma of higher plant plastids contain ClpP and ClpC, functional homologues of *E. coli* ClpP and ClpA: a two component ATP-dependent protease. Plant Cell 7: 1713–1722 (1995).

278. Shimoni Y, Segal G, Zhu X, Galili G: Nucleotide sequence of a wheat cDNA encoding protein disuphide isomerase. Plant Physiol 107: 281(1995).

279. Shorrosh BS, Dixon RA: Molecular cloning of a putative plant endomembrane protein resembling vertebrate protein disulfide-isomerase and a phosphatidylinositol-specific phospholipase C. Proc Natl Acad Sci USA 88: 10941–10945 (1991).

280. Shorrosh BS, Dixon RA: Molecular characterization and expression of an alfalfa protein with sequence similarity to mammalian ERp72, a glucose-regulated endoplasmic reticulum protein containing active site sequences of protein disphide isomerase. Plant J 2: 51–58 (1992).

281. Shorrosh BS, Subramaniam J, Schubert KR, Dixon RA: Expression and localization of plant protein disulfide isomerase. Plant Physiol 103: 719–726 (1993).

282. Silver PA, Way JC: Eukaryotic dnaJ homologues and the specificity of Hsp70 activity. Cell 74: 5–6 (1993).

283. Simons JF, Ferro-Novick S, Rose MD, Helenius A: BiP/Kar2p serves as a molecular chaperone during carboxypeptidase Y folding in yeast. J Cell Biol 130: 41–49 (1995).

284. Squires C, Squires CL: The Clp proteins: proteolysis regulators or molecular chaperones. J Bact 174: 1081–1085 (1992).

285. Strsalka K, Tsugeki R, Nishimura M: Heat shock induces synthesis of plastid-associated hsp70 in etiolated and greening pumpkin seedlings. Fol Histochem Cytobiol 32: 45–49 (1994).

286. Stuart RA, Cyr DM, Craig EA, Neupert W: Mitochondrial molecular chaperones: their role in protein translocation. Trends Biochem Sci 19: 87–92 (1994).

287.

288. Takahashi T, Naito S, Komeda Y: Isolation and analysis of the expression of two genes for the 81-kilodalton heat-shock proteins from *Arabidopsis*. Plant Physiol 99: 383–390 (1992).

289. Takenaka IM, Hightower LE: Transforming growth factor-$\beta 1$ rapidly induces Hsp70 and Hsp90 molecular chaperones in cultured chicken embryo cells. J Cell Physiol 152: 568–577 (1992).

290. Todd MJ, Viitanen PV, Lorimer GH: Hydrolysis of adenosine 5′- triphosphate by *Escherichia coli* GroEL: Effects of GroES and potassium ion. Biochemistry 32: 8560–8567 (1993).

291. Todd MJ, Viitanen PV, Lorimer GH: Dynamics of the chaperonin ATPase cycle: implications for facilitated protein folding. Science 265: 659–666 (1994).

292. Tsugeki R, Mori H, Nishimura M: Purification, cDNA cloning and Northern-blot analysis of mitochondrial chaperonin 60 from pumpkin cotyledons. Eur J Biochem 209: 453–458 (1992).

293. Tsugeki R, Nishimura M: Interaction of homologues of Hsp70 and Cpn60 with ferredoxin-NADP$^+$ reductase upon its import into chloroplasts. FEBS Lett 320: 198–202 (1993).

294. Tsuprun VL, Boekema EJ, Samsonidze TG, Pushkin A: Electron microscopy of the complexes of ribulose-1,5-bisphosphate carboxylase (Rubisco) and Rubisco subunit-binding protein from pea leaves. FEBS Lett 289: 205–209 (1991).

295. Ursic D, Culbertson MR: The yeast homologue to mouse Tcp-1 affects microtubule-mediated processes. Mol Cell Biol 11: 2629–2640 (1991).

296. Van Berkel J, Salamini F, Gebhardt C: Transcripts accumulating during cold storage of potato (*Solanum tuberosum* L.) tubers are sequence related to stress-responsive genes. Plant Physiol 104: 445–452 (1994).

297. van der Vies SM, Gatenby AA, Georgopoulos C: Bacteriophage T4 encodes a co-chaperonin that can substitute for *Escherichia coli* GroES in protein folding. Nature 368: 654–656 (1994).

98. Viitanen PV, Gatenby AA, Lorimer GH: Purified chaperonin 60 (groEL) interacts with the nonnative states of a multitude of *Escherichia coli* proteins. Protein Sci 1: 363–369 (1992).

299. Viitanen PV, Lorimer GH, Seetharam R, Gupta RS, Oppenheim J, Thomas JO, Cowan NJ: Mammalian mitochondrial chaperonin 60 functions as a single toroidal ring. J Biol Chem 267: 695–698 (1992).

300. Viitanen PV, Lubben TH, Reed J, Goloubinoff P, O'Keefe DP, Lorimer GH: Chaperonin-Facilitated refolding of ribulosebisphosphate carboxylase and ATP hydrolysis by chaperonin 60 (groEL) are K$^+$ dependent. Biochemistry 29: 5665–5671 (1990).

301. Viitanen PV, Schmidt M, Buchner J, Suzuki T, Vierling E, Dickson R, Lorimer GH, Gatenby A, Soll J: Functional characterization of the higher plant chloroplast chaperonins. J Biol Chem 270: 18158–18164 (1995).

302. Vinh DB-N, Drubin DG: A yeast TCP-1-like protein is required for actin function *in vivo*. Proc Natl Acad Sci USA 91: 9116–9120 (1994).

303. Vitale A, Bielli A, Ceriotti A: The binding protein associates with monomeric phaseolin. Plant Physiol 107: 1411–1418 (1996).

304. Vitale A, Ceriotti A, Denecke J: The role of the endoplasmic reticulum in protein synthesis, modification and intracellular transport. J Exp Bot 44: 1417–1444 (1993).

305. Vogel JL, Parsell DA, Lindquist S: Heat-shock proteins Hsp104 and Hsp70 reactivate mRNA splicing after heat inactivation. Curr Biol 5: 306–317 (1995).

306. Vogel JP, Misra LM, Rose MD: Loss of BiP/GRP78 function blocks translocation of secretory proteins in yeast. J Cell Biol 110: 1885–1895 (1990).

307. Von Stedingk EM, Glaser E: The molecular chaperone mhsp72 is partially associated with the inner mitochondrial membrane both in normal and heat stressed *Spinacia oleracea*. Biochem Mol Biol Int 35: 1307–1314 (1995).

308. Voordouw G, van der Vies SM, Bouwmeister PP: Dissociation of ribulose-1,5-bisphosphate carboxylased oxygenase from spinach by urea. Eur J Biochem 141: 313–318 (1984).

309. Waegemann K, Soll J: Characterization of the protein import apparatus in isolated outer envelopes of chloroplasts. Plant J 1: 149–158 (1991).

310. Waldmann T, Nimmesgern E, Nitsch M, Peters J, Pfeifer G, Müller S, Kellermann J, Engel A, Hartl F-U, Baumeister W: The thermosome of *Thermoplasma acidophilum* and its relationship to the eukaryotic chaperonin TRiC. Eur J Biochem 227: 848–856 (1995).

311. Walsh CT, Zydowsky LD, McKeon FD: Cyclosporin A, the cyclophilin class of peptidylprolyl isomerases, and blockade of T cell signal transduction. J Biol Chem 267: 13115–13118 (1992).

312. Walther-Larsen H, Brandt J, Collinge DB, Thordal-Christensen H: A pathogen-induced gene of barley encodes a HSP90 homologue showing striking similarity to vertebrate forms resident in the endoplasmic reticulum. Plant Mol Biol 21: 1097–1108 (1993).

313. Wang C, Lin B-L: The disappearance of an hsc70 species in mung bean seed during germination: purification and characterization of the protein. Plant Mol Biol 21: 317–329 (1993).

314. Wang H, Goffreda M, Leustek T: Characteristics of an Hsp70 homologue localized in higher plant chloroplasts that is similar to DnaK, the Hsp70 of prokaryotes. Plant Physiol 102: 843–850 (1993).

315. Wang S, Liu X-Q: The plastid genome of *Cryptomonas* Phi encodes an hsp70-like protein, a histone-like protein, and an acyl carrier protein. Proc Natl Acad Sci USA 88: 10783–10787 (1991).

316. Wang TF, Chang J, Wang C: Identification of the peptide binding domain of Hsc70. J Biol Chem 268: 26049–26051 (1993).

317. Waters E: The molecular evolution of the small heat shock proteins in plants. Genetics 141: 785–795 (1995).

318. Waters ER, Lee GJ, Vierling E: Evolution, structure and function of the small heat shock proteins in plants. J Exp Bot 47: 325–338 (1996).

319. Watts FZ, Walters AJ, Moore AL: Characterisation of PHSP1, a cDNA encoding a mitochondrial HSP70 from *Pisum sativum*. Plant Mol Biol 18: 23–32 (1992).

320. Wawrzynow A, Wojtkowiak D, Marszalek J, Banecki B, Jonsen M, Graves B, Georgopoulos C, Zylicz M: The ClpX heat-shock protein of *Escherichia coli*, the ATP-dependent substrate specificity component of the ClpP-ClpX protease, is a novel molecular chaperone. EMBO J 14: 1867–1877 (1995).

321. Weissman JS, Kashi Y, Fenton WA, Horwich AL: GroEL-mediated protein folding proceeds by multiple rounds of binding and release of nonnative forms. Cell 78: 693–702 (1994).

322. Wiest DL, Burgess WH, McKean D, Kearse KP, Singer A: The molecular chaperone calnexin is expressed on the surface of immature thymocytes in association with clonotype-independent CD3 complexes. EMBO J 14: 3425–3433 (1995).

323. Welch WJ, Feramisco JR: Rapid purification of mammalian 70,000-dalton stress proteins: affinity of the proteins for nucleotides. Mol Cell Biol 5: 1229–1237 (1985).

324. Weng J, Wang Z-F, Nguyen HT: Nucleotide sequence of a *Triticum aestivum* cDNA clone which is homologous to the 26 kDa chloroplast-localized heat shock protein gene of maize. Plant Mol Biol 17: 255–258 (1991).

325. Whitesell L, Mimnaugh EG, De Costa B, Myers CE, Neckers LM: Inhibition of heat shock protein HSP90-pp60^{v-src} heteroprotein complex formation by benzoquinone ansamycins: essential role for stress proteins in oncogenic transformation. Proc Natl Acad Sci USA 91: 8324–8328 (1994).

326. Wiech H, Buchner J, Zimmermann R, Jakob U: Hsp90 chaperones protein folding *in vitro*. Nature 358: 169–170 (1992).

327. Wilbanks SM, DeLuca-Flaherty C, McKay DB: Structural basis of the 70-kilodalton heat shock cognate protein ATP hydrolytic activity. J Biol Chem 269: 12893–12898 (1994).

328. Williams DB: Calnexin: a molecular chaperone with a taste for carbohydrate. Biochem Cell Biol 73: 123–132 (1995).

329. Willison KR, Kubota H: The structure, function, and genetics of the chaperonin containing TCP-1 (CCT) in eukaryotic cytosol. In: Morimoto RI, Tissières A, Georgopoulos C (eds) The Biology of Heat Shock Proteins and Molecular Chaperones, pp. 299–312. Cold Spring Harbor Laboratory Press, Cold Spring Harbor, NY (1994).

330. Winter J, Wright R, Duck N, Gasser C, Fraley R, Shah D: The inhibition of petunia hsp70 mRNA processing during CdCl$_2$ stress. Mol Gen Genet 211: 215–319 (1988).

331. Wojtkowiak D, Georgopoulos C, Zylicz M: Isolation and characterization of ClpX, a new ATP-dependent specificity component of the Clp protease of *Escherichia coli*. J Biol Chem 268: 22609–22617 (1993).

332. Wolfe KH, Morden CW, Palmer JD: Function and evolution of a minimal plastid genome from a nonphotosynthetic parasitic plant. Proc Natl Acad Sci USA 89: 10648–10652 (1992).

333. Woo KM, Kim KI, Goldberg AL, Ha DB, Chung CH: The heat-shock protein ClpB in *Escherichia coli* is a protein-activated ATPase. J Biol Chem 267: 20429–20434 (1992).

222

334. Wu CH, Casper T, Browse J, Lindquist S, Somerville C: Characterization of an HSP70 cognate gene family in *Arabidopsis*. Plant Physiol 88: 731–740 (1988).

335. Xu Y, Lindquist S: Heat-shock protein hsp90 governs the activity of pp60^{v-src} kinase. Proc Natl Acad Sci USA 90: 7074–7078 (1993).

336. Yabe N, Takahashi T, Komeda Y: Analysis of tissue-specific expression of *Arabidopsis thaliana* HSP90-family gene *HSP81*. Plant Cell Physiol 35: 1207–1219 (1994).

337. Yaffe MB, Farr GW, Miklos D, Horwich AL, Sternlicht ML, Sternlicht H: TCP1 complex is a molecular chaperone in tubulin biogenesis. Nature 358: 245–248 (1992).

338. Yalovsky S, Paulsen H, Michaeli D, Chitnis PR, Nechushtai R: Involvement of a chloroplast HSP70 heat shock protein in the integration of a protein (light-harvesting complex protein precursor) into the thylakoid membrane. Proc Natl Acad Sci USA 89: 5616–5619 (1992).

339. Yuan J, Henry R, Cline K: Stromal factor plays an essential role in protein integration into thylakoids that cannot be replaced by unfolding or by heat shock protein Hsp70. Proc Natl Acad Sci USA 90: 8552–8556 (1993).

340. Zabaleta E, Oropeza A, Assad N, Mandel A, Salerno G, Herrera- Estrella L: Antisense expression of chaperonin 60β in transgenic tobacco plants leads to abnormal phenotypes and altered distribution of photoassimilates. Plant J 6: 425–432 (1994).

341. Zhang F, Boston RS: Increases in binding protein (BiP) accompany changes in protein body morphology in three high lysine mutants of maize. Protoplasma 171: 142–152 (1992).

342. Zhou R, Kroczynska B, Hayman GT, Miernyk JA: *AtJ2*, an arabidopsis homologue of *Escherichia coli dna*J. Plant Physiol 108: 821–822 (1995).

343. Zhou R, Miernyk JA: ATPase activities of the maize stress70 molecular chaperone proteins. J Biol Chem, in press (1996).

344. Zhu J-K, Shi J, Bressan RA, Hasegawa PM: Expression of an *Atriplex nummularia* gene encoding a protein homologous to the bacterial molecular chaperone DnaJ. Plant Cell 5: 341–349 (1993).

Plant Molecular Biology **32:** 223–249, 1996.
© 1996 *Kluwer Academic Publishers.*

Transport of proteins in eukaryotic cells: more questions ahead

Maor Bar-Peled, Diane C. Bassham & Natasha V. Raikhel*
*MSU-DOE Plant Research Laboratory, Michigan State University, East Lansing, MI 48824–1312, USA (*author for correspondence)*

Key words: secretory pathway, nucleus, chloroplasts, mitochondria, peroxisomes, vacuole, endoplasmic reticulum, Golgi, protein trafficking

Contents

Abstract   223
Introduction   224
Transport to chloroplasts   224
Transport within the chloroplast   225
Transport to mitochondria   226
Transport to peroxisomes   227
Transport to the nucleus   227
The secretory pathway   228
   Endoplasmic reticulum: the secretory
    pathway entry site   229
     The signal sequence: signal peptide
      (SP), signal anchor (SA)   229
     The tail anchor   230
   Signal sequence recognition, targeting
    and insertion across the ER   230
   ER localization   231
   Transport of secretory proteins from
    the ER-to-Golgi network   233
The Golgi complex   235
   Possible mechanisms for Golgi
    retention   235
   Transport along the Golgi complex   236

Transport from the Golgi apparatus to
   the plasma membrane   236
Transport of proteins to the cell
   surface in plants   236
Transport to the vacuole/lysosome   237
   Transport of proteins to the
    mammalian lysosome   237
   Transport to the yeast vacuole   238
   Transport of proteins to plant vacuoles   238
   Vacuolar sorting signals in soluble
    plant proteins   238
   Components of the plant vacuolar
    sorting machinery   239
   Are there multiple mechanisms for the
    transport of proteins to the vacuole?   239
   Transport of membrane proteins to the
    vacuole   240
   Alternative route to the vacuole   240
Perspectives   240
Acknowledgements   241
References   241

Abstract

Some newly synthesized proteins contain signals that direct their transport to their final location within or outside of the cell. Targeting signals are recognized by specific protein receptors located either in the cytoplasm or in the membrane of the target organelle. Specific membrane protein complexes are involved in insertion and translocation of polypeptides across the membranes. Often, additional targeting signals are required for a polypeptide to be further transported to its site of function. In this review, we will describe the trafficking of proteins to various cellular organelles (nucleus, chloroplasts, mitochondria, peroxisomes) with emphasis on transport to and through the secretory pathway.

Introduction

Most proteins in eukaryotic cells are synthesized on cytosolic ribosomes. Some of these newly synthesized proteins remain soluble in the cytoplasm; others are further routed to various subcellular organelles, such as the mitochondria, chloroplasts, peroxisomes, secretory system, and nucleus, or to the plasma membrane. Alternatively, some proteins are secreted out of the cell. Targeting to these different subcellular compartments is often achieved by the synthesis of precursor proteins (preproteins) containing targeting signals. It is now generally accepted that targeting signals are recognized by specific protein receptors located either in the cytoplasm or in the membrane of the target organelle. Both insertion into and translocation of preproteins across membranes is mediated by a set of membrane proteins located in the translocation 'channels' through which the polypeptide chains move. The translocated protein is folded and assembled through interaction with molecular chaperones. This general mechanism enables proteins to be targeted to, and subsequently translocated across, the correct intracellular membrane, while cellular membrane impermeability to other solutes is maintained. Often, secondary targeting sequence information on the polypeptide ensures further movement to a different subcompartment within the organelle (e.g. thylakoids in the chloroplast; the intermembrane space in the mitochondrion) or to various compartments (e.g. the vacuole and the Golgi) of the secretory pathway.

The molecular dissection of the protein transport machinery relies on *in vitro* or semi *in vitro* systems to identify signals or targeting receptors and to elucidate transport mechanisms. Such assays have initiated the study of transport of proteins across the endoplasmic reticulum (ER) [24], as well as the import of proteins into chloroplasts [104], mitochondria [154], peroxisomes [246], and nuclei [182]. In this review we describe the signals directing a polypeptide to a particular location in a cell, the cellular components recognizing the targeting signals, and attempts to understand the mechanisms through which the targeting event is achieved. Emphasis is placed on recent advances in the molecular mechanisms of transport through the secretory system. We provide a brief overview of discoveries related to trafficking in other organelles, as well as relevant citations for further detailed reading.

Transport to chloroplasts

Chloroplasts have three distinct membranes (outer, inner, and thylakoid membranes) which define three aqueous compartments (intermembrane space, stroma, and thylakoid lumen; see Fig. 1). Since many chloroplast proteins are encoded in the nucleus, specific transport mechanisms exist that control their traffic and import from the cytosol to their final destination in the chloroplast. These nuclear-encoded chloroplast proteins are synthesized with an N-terminal presequence (the transit peptide) which targets polypeptides to the chloroplast and further allows their translocation across the chloroplast envelope into the stroma (see for review [51]). The mechanism of transport of a precursor protein to the chloroplast and its docking at the envelope has not been fully elucidated. However, this process may be mediated by cytosolic factors [271]. Some presequences of mitochondria and chloroplast precursor proteins have similar characteristics, and mistargeting between these organelles have been observed *in vivo* and *in vitro* [29, 113], further supporting the possibility of involvement of cytosolic factors chaperoning precursors to the correct organelle. Furthermore, distribution mechanisms could exist to direct equal amounts of precursors to each chloroplast. To date, there is no consensus as to whether the transit peptide interacts initially with receptor proteins or with the chloroplast lipid bilayer, itself; however, it is possible that both interactions occur.

A few outer envelope membrane proteins are synthesized in the cytosol without a cleavable presequence [145]; however, at least two, OEP86 and OEP75, are synthesized with cleavable transit peptides [106, 263]. The putative transit peptide of OEP86 is very long (ca. 150 amino acids) and relatively negatively charged in contrast to transit peptides that direct proteins to the stroma; the latter is much shorter (40 to 80 amino acids), consisting of hydroxylated amino acid and lacking acidic amino acid residues. Furthermore, the peptidase removing the signal peptide from OEP86 protein has not been identified yet, and it is unlikely to be a stromal peptidase [106]. The putative transit peptide of OEP75 has characteristics similar to stromal presequences [263].

Various laboratories are currently isolating proteins involved in translocation across the chloroplast with the goal of a precise understanding of the specific molecular events that direct import. Thus far, biochemical identification of a complex [272] and chemical cross-linking strategies have led to the isolation of the two

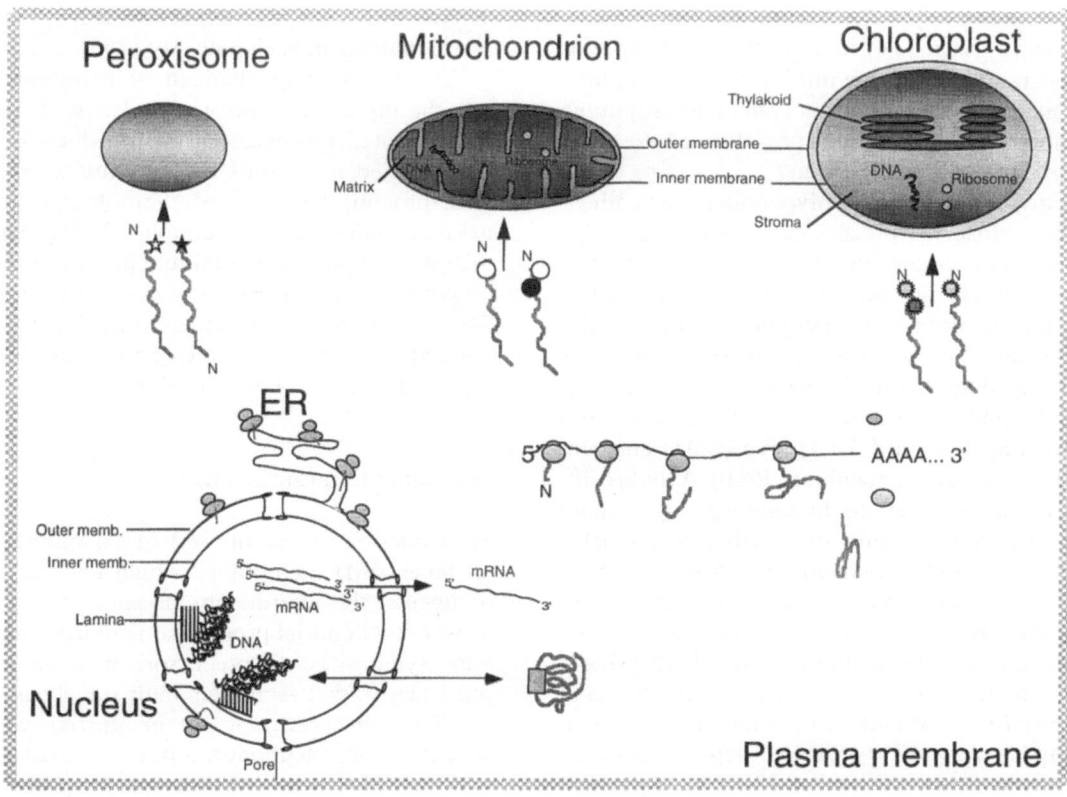

Figure 1. Various signal sequences direct proteins to sub-cellular organelles. Targeting to different compartments is achieved by the synthesis of proteins containing targeting signals: to the nucleus (■); to the chloroplast (⊗); to mitochondrion (◯); to peroxisome (★ or ⋆). These targeting signals are recognized by specific protein receptors, located either in the cytoplasm or in the membrane of the target organelle. Often, secondary targeting sequence information on the polypeptide ensures further movement to a different compartment within the organelle: e.g. thylakoids (✦); intermembrane space in mitochondria (●) or to various subcompartments (e.g. vacuole and Golgi, see Fig. 2) of the secretory pathway. The targeting signals directing proteins to mitochondria and chloroplasts are often cleaved from the preproteins by specific peptidases whereas signals directing nuclear targeting are not cleaved.

above-mentioned outer envelope membrane proteins, OEP75 and OEP86 [206, 232, 272]; in addition, a 34 kDa GTP-binding protein [125] and an Hsp70-related protein [235] were purified as components of the protein import apparatus.

The import of chloroplast precursors occurs at contact sites between the outer and inner envelope membranes and the precursor proteins span both envelope membranes while in transit [231]. The hypothetical model for protein translocation across the chloroplast envelope is as follows (see for review [81]): The initial steps are thought to involve the interaction/association of the precursor proteins, via their transit peptides, with either lipids embedded into the outer membrane and/or with associated proteins. The unfolded or partially-unfolded precursor forms can interact with components of the import apparatus described above. The import

of stromal precursor proteins across the chloroplast envelope is a GTPase-dependent process. However, the import of outer membrane precursor proteins requires ATP [106] suggesting different translocation apparatuses across the outer chloroplast envelope.

Transport within the chloroplast

Many imported chloroplast polypeptides are further sorted to the inner envelope, thylakoid membrane, or thylakoid lumen. The targeting sequences for inner membrane proteins have been studied for only few proteins. The N-terminal presequence of the phosphate translocator contains stromal targeting information, and the N-terminal region of the mature protein possibly mediates the insertion of the protein into the inner membrane [130]. If outer and inner membrane proteins are translocated via the same protein-import apparatus, it raises an intriguing question; how are inner membrane proteins that contain hydrophobic transmembrane region(s) able to pass through the outer membrane without being arrested?

Some thylakoid lumenal polypeptides are synthesized in the cytosol as precursors with bipartite signals at their amino-termini [221] (see Fig. 1). The first signal targets the precursor to the chloroplast envelope and facilitates polypeptide import into the stroma. During this process, the primary signal is proteolyticaly removed in the stroma [15], exposing the second signal. The second signal, the thylakoid lumen-targeting domain, directs polypeptide translocation across the thylakoid membrane [260]. A thylakoid-specific protease cleaves the lumen-targeting domain from the polypeptide during, or shortly after, translocation across the thylakoid membrane [86].

There are at least two pathways for soluble protein transport into thylakoids [40, 114] and, apparently, a different one for the membrane thylakoid protein insertion. The translocation of some soluble lumenal proteins requires ATP and stromal factors [39, 114], and resembles the prokaryotic Sec-dependent translocation mechanisms. Yuan et al. [283] have purified from the stroma of pea chloroplasts a protein that is able to support transport of certain soluble precursors into isolated thylakoids. This stromal protein (CPSecAp, 110 kDa) is active as a homodimer (200–250 kDa), requires ATP and GTP, but is inhibited by azide, similar to SecA, which is part of the bacterial protein translocation machinery. Furthermore, CPSecAp neither supports the transport of the membrane thylakoid protein (light-harvesting chlorophyll a/b protein, LHCP), nor transports two other lumenal proteins, OE17 and OE23 [283]. Moreover, Robinson's group have isolated a SecY homologue from Arabidopsis plants and have demonstrated that cpSecYp is nuclear-encoded as a 58 kDa precursor that is processed to the mature protein, 46 kDa, which is located in the thylakoid membrane [137].

For other thylakoid lumenal proteins (e.g. OE17 and OE23), the translocation process is unusual, requiring only the pH gradient across the thylakoid membrane [39, 46]. Chaddock et al. [34] have reported that a specific twin-arginine motif, located before the hydrophobic region in the presequence of some thylakoid soluble proteins, directs the precursor across the thylakoid membrane in a pH-dependent manner. Existence of two distinct thylakoid-targeting pathways

in vivo has been also demonstrated by isolation of maize mutants in each pathway [265].

The insertion mechanism of membrane proteins into the thylakoid is not yet understood. It has been demonstrated, however, that the translocation of LHCP into isolated thylakoids requires the stromal chloroplast protein, 54CP [146], homologous to the signal recognition particle subunit, SRP54. The 54 kDa SRP subunit plays a pivotal role in the translocation of preproteins into the ER. The chloroplast homologue, 54CP, binds to LHCP and is required for thylakoid integration. The precise role of 54CP in LHCP targeting still remains to be established.

Transport to mitochondria

Mitochondria are composed of an outer (OM) and an inner (IM) membrane, which enclose two compartments: the intermembrane space and the matrix. Most mitochondrial proteins are nuclear-encoded proteins synthesized as precursors with an N-terminal presequence that facilitates their translocation across the OM. Some precursors are recognized by a cytosolic chaperone or mitochondrial import stimulating factor (MSF), which in an ATP-dependent process, directs these precursors in their partially folded state to the OM import subunits [84, 85]. Other precursors with a basic amino acid presequence domain seem to interact directly with the cytosolic domain of OM import subunits [95, 160, 167, 175].

The mitochondrial outer membrane (MOM) contains surface-bound proteins that, together with membrane-embedded components, facilitate the translocation of precursor proteins across the MOM. The MOM complex proteins, as well as the mitochondrial assembly (MAS) complex proteins, have been isolated from Neurospora crassa and from yeast. Each complex consists of several subunits [91, 105, 167, 217]. These complexes are associated with other proteins, which are thought to form insertion channels for translocation of mitochondrial precursor proteins (see for review [133, 168]).

The translocation apparatus consists of two channels, outer and inner. Some precursors destined for the intermembrane mitochondrial space are synthesized without a cleavable presequence (see for review [242]) and cross the OM with the apparent involvement of the IM proteins [71]. However, many precursors are synthesized with cleavable presequences. Precytochrome b2 and c2 presequences carry two targeting signals

227

and are processed in two steps. The initial N-terminal hydrophilic domain of the presequence (the matrix-targeting signal), is cleaved off in the matrix by a mitochondrial processing peptidase [83]. The remaining C-terminal hydrophobic portion of the presequence (a sorting signal to the intermembrane space), is removed by the Imp1p-Imp2p protease complex, located peripherally on the external face of the IM [189]. The import of proteins across the IM into the matrix occurs where the two membranes are closely apposed. The driving force for import seems to be a combination of the inner membrane potential and the binding and release of chaperones in the matrix (see for review [210]). A novel genetic screen in yeast for mitochondrial protein import mutants has led to isolation of several genes encoding essential mitochondrial IM (MIM) proteins [25, 63, 153].

In plants, Perryman et al. [207] have identified a mitochondrial OM protein, MOM42, involved in the import of precursor protein. The plant MOM42p is possibly a homologue of the Neurospora and yeast protein; an antibody to MOM42 inhibits import of bound precursor into the organelle [207]. In addition, a mitochondrial processing peptidase has been cloned from potato [62]. As plants contain plastids and mitochondria the specificity of targeting to the correct organelle may require additional check point factors to prevent mistargeting.

Transport to peroxisomes

Peroxisomes are ubiquitous intracellular organelles containing enzymes involved in a variety of oxidizing processes. Unlike chloroplasts and mitochondria, peroxisomes are enclosed by a single membrane. Membrane or soluble polypeptides destined for the peroxisome are synthesized in the cytosol on free polysomes prior to their import into the organelle. At least two distinct peroxisomal targeting signals, PTS1 and PTS2, are involved in the import of proteins into the peroxisomal matrix. PTS1s comprise a non-cleavable C-terminal tripeptide SKL, or closely related variants of this tripeptide [79], whereas PTS2s consist of a cleavable N-terminal leader sequence with some similarities to mitochondrial targeting sequences. The transport of some PTS1-containing proteins requires cytosolic factors, including members of the 70 kDa heat shock protein family [274]. A candidate for a PTS1 receptor was first identified in methylotrophic yeast through a screen for peroxisome assembly

mutants [80]. One mutant, pas8, was unable to import a PTS1-containing polypeptide but was fully competent to import a PTS2-containing polypeptide. The PAS8p shares homology with the yeast mitochondrial import receptor, Mas70p. Terlecky et al. [259] have characterized the PAS8p (68 kDa) protein, finding it to be associated with the cytoplasmic side of the peroxisomal membrane. Furthermore, PAS8p contains a tetratricopeptide repeat region that binds to the PST1 targeting sequences with a very high affinity of 460 nM. A functional homologue of the PAS8 gene has recently been cloned from human cells [57], suggesting evolutionarily conserved transport.

Plant cells possess several classes of peroxisomes, and chimeric proteins fused to PTS-like sequences are correctly directed into the various types of peroxisomes [194, 266]. Studies aimed at identifying peroxisomal OM proteins [43] and developing an in vivo peroxisome import assay in plants [10] are initial steps toward understanding the specific molecular events related to import into this organelle.

Transport to the nucleus

The nuclear envelope is composed of three primary structures (see Fig. 1): the outer and inner membranes including the intermembrane space, termed the perinuclear compartment, the nuclear pore complexes (NPCs), and the nuclear lamina (see for review [72]). NPCs are inserted in the nuclear membrane at the junction of the outer and inner membranes to form a pore membrane domain. These pores are anchored to the inner membrane, presumably by the nuclear lamina. The outer nuclear membrane is continuous with the endoplasmic reticulum (ER) and associated with ribosomes. It has been assumed that the nuclear envelope is structurally and functionally distinct from the ER, however, a recent study has suggested that the nuclear envelope may retain various ER functions, such as translocation of proteins [254]. Polypeptides destined for the nucleus are synthesized with short basic amino-acid sequences, nuclear localization signals (NLSs) that direct their import into the nucleus (for a comprehensive review describing different NLSs and the mechanisms of nuclear import in yeast, animals, and plants, see [101]). Transport across the nucleus has several unique features. Unlike cleavable targeting signals which are removed during the translocation of soluble proteins into the chloroplast (see for review [124]), mitochondria [93], peroxisomes [138], and secretory pathway (see for review [20]), nuclear localization signals are not cleaved from nuclear proteins, thus allowing them to shuttle in and out of the nucleus.

A second interesting feature of proteins destined for the nucleus is that some are incorporated after their synthesis into other parts of the cell (i.e. the ER or the plasma membrane); alternatively these proteins may remain 'inactive' in the cytoplasm and, upon induction, may be released from their location or be modified to expose their NLS and be imported to the nucleus [199, 275]. In plants, the shuttling of some nuclear proteins in and out of the nucleus is regulated by light (see for reviews [55, 101]). A third feature is the observation that small polypeptides, lacking NLSs, may diffuse through the pore complex, although the *in vivo* significance of this is unclear. Moreover, the same pore that mediates the shuttling of proteins is also thought to function in the transport of completely different macromolecules (i.e. RNA).

The import of nuclear proteins via the nuclear pore complex seems to be a two-step process. The binding of NLS-containing protein domains to the NPC requires soluble cytosolic factors and energy (ATP and GTP) in mammalian cells [4], yeast [230], and possibly in plant cells (Hicks and Raikhel, unpublished results). After the interaction of NLS-binding proteins with the NPC, nuclear proteins are actively translocated through the pore into the nuclear interior. Both steps require multiple cytosolic factors. To date, several soluble cytosolic factors which bind NLSs and stimulate import have been purified from vertebrates: the 54 and 56 kDa proteins [3] (also termed importin-α [74, 76, 78], m-importin [116], and karyopherin α1 and α2 [215]), and a 97 kDa protein (p97), which serves as an adaptor between the NLS-binding proteins and the NPC [2, 37] (also termed importin-β [74, 76, 78] and karyopherinβ [215]. Translocation across the pore requires, in addition to these factors, two more cystolic proteins: the GTP-binding protein Ran/TC4 [162, 169] and a small 15 kDa protein [170]. The role of Ran and the 15 kDa protein is still unclear. A cDNA homologous to karyopherin α has been cloned from *Arabidopsis* plants, and the gene product (58 kDa) is present in all plant tissues (Smith and Raikhel, unpublished).

Our understanding of the mechanisms and signals that target proteins to the inner nuclear membrane is still unclear. It has been suggested that proteins are embedded in the ER and then move by diffusion or NLS-mediated 'dragging' from the ER membrane to inner nuclear membrane [251]. It is also proposed that the movement is active, requiring ATP, and occurs along the edge of the pore (see for review [279]). In addition, targeting to the inner membrane may be effected by binding to nuclear proteins (i.e. lamina, chromatin) or other inner membrane proteins [279]. It has been reported that the amino-terminal domain of an inner nuclear membrane protein, the lamin B receptor, can redirect a reporter protein to this membrane [247, 251].

The secretory pathway

The secretory pathway is a major site for the biosynthesis, folding, assembly, and modification of numerous soluble and membrane proteins as well as for the synthesis and transport of lipids, polysaccharides, and glycoproteins. This pathway is composed of the ER, Golgi apparatus, plasma membrane, short-lived small vesicles; also part of this system are vacuoles (in yeast and plant cells) and lysosomes (in mammalian cells) (see Fig. 2). Each compartment possesses a distinct morphological structure and differs in its lumenal, lipid, and protein contents, enabling efficient synthesis and modification of the molecules that pass through it. This pathway mediates the proper delivery and sorting of proteins to a variety of subcellular compartments. Proteins targeted to the various compartments of the pathway start their journey after translocation to the ER, in a signal sequence-mediated process and are subsequently transported to the Golgi complex. Additional sequences are required for targeting proteins to their final compartmental destination. When these targeting sequences are lacking, soluble proteins are routed directly to the plasma membrane and secreted. A unique feature of the secretory pathway is that proteins are transported in small membrane-bound vesicles [200], unlike the other trafficking systems already described. These cargo vesicles, 50–150 nm in diameter, pinch off from a donor membrane and are targeted, docked, and fused through the action of many cytosolic factors to the next acceptor membrane (see for review [214]). This mode of trafficking correctly sorts not only soluble proteins but also membrane components. Moreover, this process is not unidirectional: transport can occur from the ER to cell surface (anterograde transport) as well as in reverse (retrograde transport). Any forward flow of vesicles must be matched by a reverse flow of equal magnitude if each compartment is to maintain its identity.

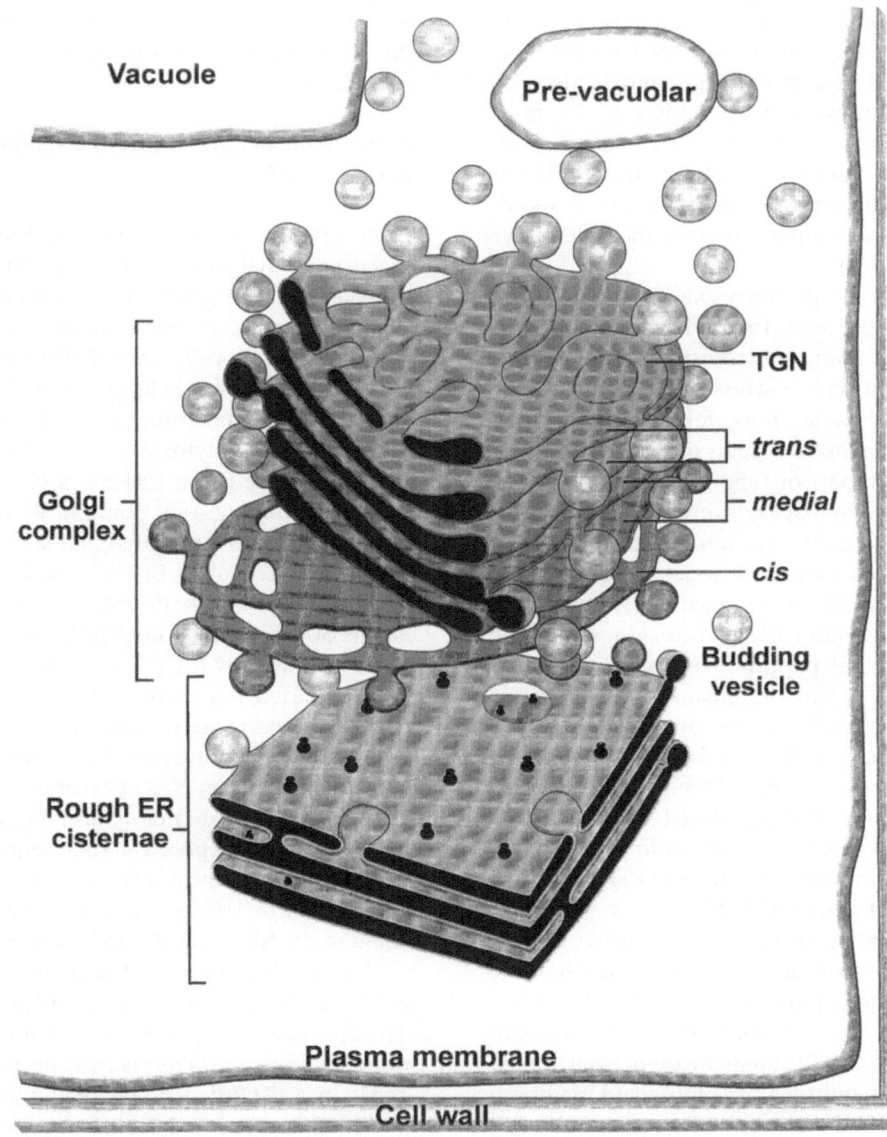

Figure 2. Transport of proteins along the secretory pathway is mediated by small vesicles that pinch off from a donor membrane and are targeted, docked, and fused, via many cytosolic factors that coat the vesicles, at the next acceptor membrane. Most proteins targeted to the various compartments of the pathway are initially translocated into the ER, a signal sequence-mediated process, and are subsequently transported to the Golgi complex, the plasma membrane, or the vacuoles. Additional targeting sequences are required for their final compartmental destination (to ER, Golgi, or vacuole). Without these targeting sequences soluble proteins are directed to the plasma membrane and secreted.

Endoplasmic reticulum: the secretory pathway entry site

The signal sequence: signal peptide (SP), signal anchor (SA)

All signal sequences direct nascent soluble and membrane proteins into the ER. Two such ER-targeting sig-nals (signal sequences) have been identified. Those that are cleaved from the nascent chain (either from membrane or soluble proteins) by a specific ER-signal peptidase upon or during membrane insertion are termed signal peptides (SP). The other group, signal anchors (SA), are uncleaved signal sequences that direct proteins to the ER membrane and subsequently serve to anchor the proteins in the membrane. SPs are usually present at the NH_2 terminus of the nascent chain; membrane proteins possessing such signals also contain stop-transfer sequences (a region of 18 to 25 hydrophobic residues followed by a cluster of positively charged amino acids), which function to prevent complete translocation of the nascent chain across the membrane and to anchor the protein to the lipid bilayer (see for review [103]).

Integral membrane proteins exhibit various transmembrane topologies (see for review [103]). Type I proteins, which contain one transmembrane domain and possess cytoplasmic carboxyl-terminus are synthesized with a cleavable SP as well as a stop transfer. Signal anchor proteins also span the membrane once and can be either type I or type II proteins; the latter exhibit amino-terminal cytoplasmic orientation. SPs consist of three domains: an amino-terminal, positively charged region (1 to 5 residues long); a central hydrophobic region (7 to 15 residues); and a polar carboxyl-terminal region (3 to 7 residues) (see for review [268]). Signal peptides vary considerably in length between different organisms, and no precise sequence conservation has been found. Various types of SPs can redirect engineered foreign proteins into the ER of plant cells. These proteins include HSA, a naturally occurring secretory protein [243]; GUS, PAT, and NPTII, three bacterial cytoplasmic proteins [52]; bacterial chitinase, a secretory protein [151]; and ALB, a cytoplasmic seed albumin [115].

Integral membrane proteins are inserted into the lipid bilayer either co-translationally during their biosynthesis or post-translationally after they have been fully synthesized in the cytoplasm. Co-translational insertion, probably the dominant mode of insertion into or across the ER membrane of soluble and membrane proteins, starts from the N-terminal portion of the preprotein. The presentation and position of SP and SA into the ER translocation site are different [186].

The tail anchor

Some integral membrane proteins are not synthesized with signal sequences, instead they possess a hydrophobic region near the C-terminus that orients them with the amino-terminus in the cytosol. This group of proteins (termed tail anchor [136]), includes the v- and t-SNAREs and has been implicated in vesicular transport; they are fully synthesized in the cytosol prior to their translocation into the ER (post-translational translocation mode). Two proteins of this class have been tested and found to insert efficiently into the ER post-translationally in a ribosome-independent manner [118, 135].

Signal sequence recognition, targeting and insertion across the ER

Both types of signal sequences (SP or SA) are recognized by a cytosolic component, the signal recognition particle (SRP). The initial steps in co-translational translocation in mammalian cells start during polypeptide elongation. The nascent chains must be at least 60 amino acid residues long before the amino-terminal signal sequence emerging from the ribosome becomes accessible to the cytosolic SRP subunit, SRP54 in its nucleotide-free form. SRP54, a GTP-binding protein with a methionine-rich domain, is involved in recognition of various types of signal sequences [152]. Part of the SRP complex, SRP54 consists of six different proteins (SRP 9, 14, 19, 54, 68, and 72 kDa) as well as one 300 nucleotide long RNA molecule, termed 7S RNA (see for review [273]). Wiedmann et al. [278] have reported that another protein complex, a nascent chain-associated complex (NAC), may also play a role in the targeting process or function as a quality-control checkpoint; NAC prevents binding of the SRP to ribosomes that do not exhibit a nascent chain structure with fully exposed signal sequences. The SRP, once bound to the nascent chain, arrests translation and targets the ribosome-linked nascent polypeptide chain to the SRP receptor (SR) on the ER membrane.

SR is a heterodimer, ER-membrane protein comprised of SRα and SRβ subunits (see for review [273]). The SRP receptor appears to act both as a GTP-loading protein and as a GTPase activating protein (GAP) [42, 165]. In a GTP-dependent reaction, the SRP is displaced from both the signal sequence and the ribosome. The efficient transfer of the nascent chain from SRP54 to the membrane translocation site requires the presence of the Sec61p complex, a major component of the protein conducting channel of the ER-translocation machinery. This has been demonstrated by various approaches such as electrophysiology [244], cross-linking [171], and measurement of fluorescent probe incorporation into the translocating polypeptide chain [47]. The mechanisms through which the nascent chain is transferred as it moves from the SRP into the protein-conducting channel, and the opening of the channel, are not yet understood. The Sec61p complex consists of three membrane proteins (α, β, and γ); two of these resemble the proteins SecYp and SecEp, which function in the export of proteins in prokaryotes [94]. Sec61p is tightly associated with membrane-bound ribosomes [123], suggesting that the nascent polypeptide is transferred directly from the ribosome into the protein-conducting channel [75]. Translocation can be fully reproduced *in vitro* using proteoliposomes containing only the SRP receptor, the Sec61p complex, and the translocation chain-associated membrane (TRAM) protein [77].

T. Rapoport's group has reported that the complex of SRP-nascent chain ribosomes is bound to Sec61p in a loose manner and that the nascent chain contacts Sec61α [121]. It is postulated that during this event the signal sequence contacts either the Sec61p complex, or the ER phospholipids, or both. Subsequently, the signal sequence is recognized by the Sec61p complex, and the nascent chain is inserted into the translocation site where its signal sequence interacts with the TRAM protein. The ribosomes become firmly bound to the Sec61p complex and the translocation channel opens toward the lumen of the ER. Next, the signal sequence is cleaved by the signal peptidase, the TRAM protein disengages, and the nascent chain adopts a transmembrane structure. According to this study [121], co-translational protein translocation through the ER is comprised of two checkpoints. First, the recognition event by SRP, if successful, results in the movement of the nascent chain to the second stage where an independent assessment by membrane components is made; only when this assessment is completed is the substrate transported across the ER membrane. Both type I and II membrane proteins containing SA have been found in close proximity to Sec61p during insertion to the ER [102]. It is not known how the membrane protein is transferred from the protein conducting channel to its location within the lipid bilayer.

In yeast, several temperature-sensitive lethal mutants (sec61, sec62, sec63, sec65, and sss1) have been isolated that fail to properly localize a signal peptide-bearing cytosolic protein to the ER lumen [56, 64, 65, 253]. Protein translocation occurs both co- and post-translationally but, surprisingly, disruption of genes that encode the SRP and SRP receptor subunits is not lethal in yeast [164]. A complex of proteins is required for protein post-translational translocation across the ER of yeast: this complex includes the ER lumenal chaperone protein, Kar2p (the mammalian homologue is called binding protein, BiP); Sec63p (an integral ER membrane protein having a lumenal domain similar to the *Escherichia coli* DnaJ protein); Sec71p; Sec72p, and Sbh1p (a homologue of the mam-

malian Sec61β) [31, 32, 66, 202]. The driving force for translocation of proteins through the channel is still unclear.

In addition, some integral type II membrane proteins, lacking a signal sequence but possessing a hydrophobic region near their C-termini, are post-translationally inserted into the ER [150]. Insertion is accomplished in the presence of ATP but in the absence of ribosomes, Sec61p complex, and SRP [135], suggesting alternative insertion mechanisms across the ER.

An understanding of the ER-targeting machinery in mammalian, yeast, and bacteria systems has revealed that plants may utilize a similar pathway. A cDNA homologous to Sec61 has been cloned from algae [172]; a Sec61-β homologue has been cloned from *Arabidopsis*, a Sec61-γ also from rice [94], and an SSS1 (suppressor of sec61 in yeast) homologue from rice [64]. The SRP-like complex mediates the co-translational insertion of nascent polypeptide chains across microsomes prepared from wheat germ [211] and maize [33]. More recently, a homologue of the 54 kDa subunit of SRP has been cloned from *Arabidopsis* [149] and tomato [132]. The tomato gene product has been shown to bind to the 7S RNA and SRP19 protein subunit of SRP [132]. These observations strongly suggest the evolutionary conservation of components of the protein translocation complex. An interesting question is whether ribosomes bind randomly to any translocation sites or to specific regions of the ER. For example, are nuclear membrane proteins inserted at different sites to proteins that will be secreted. In plants there is some evidence that mRNA can be targeted to distinct ER regions (see for review [193]).

ER localization

The ER is an extensive membrane-bound organelle composed of a network of cisternae stretching through the cytoplasm that is subdivided into two structurally distinct domains: the rough (RER), which is covered with ribosomes, and the smooth (SER). Although the cisternae of RER and SER can be connected, and similar proteins are found in both compartments, it is evident that certain proteins, such as the Sec61p complex, are located exclusively in the RER [12]. Furthermore, the Sec61p complex has been found in the outer nuclear membrane that also contains ribosomes and is contiguous with the RER [254]. Beyond its function as the entry site for most secretory pathway proteins, the ER is the permanent compartment of proteins involved in lipid and oligosaccharide biosynthesis, and creates the environment for folding, assembly, maturation, as well as degradation of 'visitor' soluble and membrane proteins (see for review [264]). Proteins traveling from the ER to the Golgi exit the ER at different rates, suggesting the effect of ER factors on this process. It has been demonstrated recently that quality control involved in the assembly, folding, and maturation of glycoproteins in the ER, is responsible for their retention in this compartment until they are correctly folded and assembled [96, 257]. Several mechanisms existing within the ER function to retain specific 'home' proteins that reside in that environment, while simultaneously allowing other soluble and membrane proteins to exit the ER on their way to the next secretory pathway compartment, the Golgi complex.

Lumenal ER resident proteins such as BiP and calreticulin and some type II integral membrane ER proteins such as Sec20p (in which the carboxyl terminus is inside the ER), possess a carboxy-terminal sorting domain consisting of the four amino acid KDEL or closely related tetrapeptide variants [204]. These KDEL-like sequences, when fused to normally secreted reporter proteins, retain them in the ER of plant, yeast, and mammalian cells [174, 204, 264]. Other ER signals have been identified on the cytoplasmatic domain of some type I and type II ER integral membrane proteins, such as dibasic amino acid (double lysine, KKXX, and KxKxx at the COOH terminal, or the double arginine motifs in the NH_2 terminal [70, 117, 234, 261]. Although these signals maintain proteins in the ER, some ER proteins can acquire a carbohydrate modification, known to occur in post-ER compartments (i.e. the Golgi complex), suggesting that they must have once escaped from the ER [50, 70, 163, 208]. Thus, it seems that ER localization is more complex than just a signal for retention: resident ER proteins can be selectively retained, but those that 'slip' through this first filter are selectively retrieved from one or more downstream compartments by a continuous retrieval mechanism [250]. Proteins that carry the KDEL-like signal are recognized and bound by a receptor, Erd2p, in the Golgi [112, 143, 144]. Upon ligand binding, the KDEL receptor moves to the ER [144], where the retrieved protein dissociates presumably, in a pH-dependent manner [280]. It has been shown that KDEL receptor-mediated retrograde *trans* occurs from as far away as the *trans* Golgi network (TGN) to the ER [163]. This observation is further supported by the distribution of the KDEL receptor,

Erd2p, throughout the Golgi stack [82]. Erd2p has been isolated from mutated yeast cells secreting BiP [237], and later from mammalian cells [112, 143] and plant cells [139]. Interestingly, the *Arabidopsis ERD2* gene can complement the yeast *erd2* mutant [139], whereas the mammalian counterpart cannot [143]. The Erd2p is thought to contain seven membrane-spanning domains, and mutational analysis has revealed that the seventh transmembrane domain is important for the receptor targeting/recycling functions [262]. The mRNA level of the plant *ERD2* gene [14] is upregulated in *Arabidopsis thaliana* upon treatment with tunicamycin (affecting the folding of proteins) or cold temperature (affecting the transport of vesicles) and redistributing Erd2p from *cis*-Golgi to the *trans* side of the Golgi stack [82]. This suggests that a specific signal transduction mechanism exists between the secretory pathway and the nucleus that monitors the folding of cargo proteins and the transport of vesicles from the ER.

The mechanism for ER retention/retrieval is conserved between plant, mammalian, and yeast cells. Although several reporter proteins carrying ER signals are typically retained within the ER of plants, certain engineered plant proteins tagged with the KDEL-like sequence are not localized to the ER. Phytohemagglutinin, a vacuolar storage protein engineered by the addition of a carboxy-terminal KDEL peptide, is retarded in the ER but not retained there [99]. Similarly, addition of a K/HDEL to phaseolin, a vacuolar protein, is not sufficient for retention in the ER. Furthermore, SH-EP, an endogenous vacuolar endopeptidase, possesses a KDEL sequence at its C-terminus that is cleaved from the mature protein by vacuolar protease [192]. These vacuolar-localized proteins possess vacuolar sorting information (see for reviews [38, 68]). Thus, the vacuolar sorting machinery can successfully compete against the ER-KDEL-retrieval machinery. In addition, the native auxin-binding protein also possesses a KDEL sequence; however, ca. 5% of this protein is found at the plasma membrane [120]. Thus, the KDEL sequence may not be presented to the ER retention machinery suggesting that other mechanisms exist to secrete a portion of a protein or to regulate its distribution.

The exact mechanisms for the recognition, retention, and targeting of these 'escaped' ER-resident proteins with their receptor(s) and their subsequent release are still unclear. Moreover, it is difficult to imagine a retention mechanism that operates entirely by salvaging lost proteins. That some ER-resident proteins do not carry a clear ER-retention signal (e.g. type II ER-integral membrane protein Sec12p) suggests that other mechanisms, in addition to those mentioned (here and below), exist for the proper retention/retrieval of resident ER membrane proteins.

The ER quality control system ensures that every newly synthesized protein entering the ER must be retained in that compartment until it matures, a specific period for each polypeptide. The ER is the only compartment in the secretory pathway where a high concentration of chaperones exists: these include BiP, calnexin, calreticulin, GRP94, and HSP74. Thus, the ER provides an environment optimal for protein folding and assembly [87]. These chaperones assist the folding and oligomeric assembly of glycoproteins, mediate partial ER retention of incompletely folded and oligomerized proteins, and prevent premature degradation [96, 216, 276]. It has been observed that certain wheat and rice storage proteins lacking a KDEL-like motif are retained in the ER and assembled into dense protein bodies within that compartment before being exported. This assembly process could be due to the aggregation of hydrophobic proteins or may perhaps depend on specific cellular factors or other retention signals. Such factors can retain the newly arrived polypeptides in the ER until the protein body is completely formed. It has been reported that BiP molecules, ER-resident proteins involved in assembly and quality control, form a complex with the nascent chain of prolamine [147, 193]. The N-terminal region (PQQPFPQ) of γ-gliadin is a signal for the retention and accumulation of this wheat storage protein in dense protein bodies in the ER [5].

Transport of secretory proteins from the ER-to-Golgi network

The movement of soluble and integral proteins from the ER to the Golgi apparatus is the first vesicle-mediated step in the process of protein secretion. Several reports have demonstrated that soluble and transmembrane proteins are concentrated and sorted from resident ER proteins before leaving the ER at specialized exit regions [7, 8, 166]. Transport between the membrane-bound compartments of the secretory pathway is thought to involve the formation of coated vesicle intermediates that bud from one membrane (donor) compartment and are targeted and subsequently fuse with an acceptor membrane compartment (for comprehensive reviews [214, 223]; and Fig. 3). Vesicular transport from the ER to the Golgi has been extens-

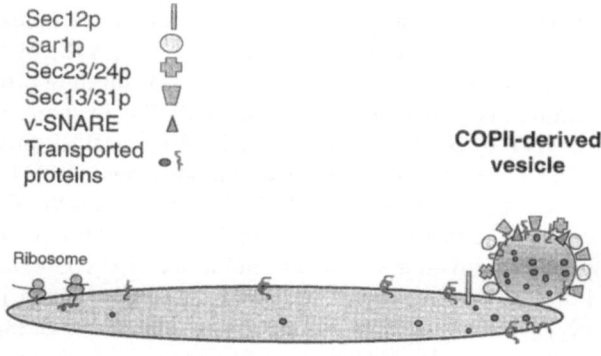

Sec12p

Sar1p

Sec23/24p

Sec13/31p

v-SNARE

Transported proteins

COPII-derived vesicle

Ribosome

ER cisterna

Figure 3. Model of COPII-vesicles budding from ER. The *in vitro* budding of one type of vesicle from the ER requires the integral membrane protein Sec12p, GTP-binding protein Sar1p, and the cytosolic factors Sec23/Sec24p and Sec13/Sec31p that coat the bud to form COPII coated vesicles. The COPII vesicles also contain v-SNARE proteins that are required for targeting and fusion with the Golgi membrane. The position of the coat proteins do not imply specific association.

ively studied in the yeast *Saccharomyces cerevisiae*; moreover, a large collection of mutants capable of blocking transport at this stage are now available. More than 25 genes have been implicated in this process [214].

Our understanding of the function of each protein involved in vesicle-mediated transport derives from a combination of the following methods: *in vitro* assays that reconstitute ER to Golgi protein transport in semi-intact cells or in microsomal preparations [6, 225]; use of temperature-sensitive yeast mutants; depletion or addition of cellular proteins under study; and use of antibodies toward specific functional proteins [67, 148]. Assays that measure vesicle budding, targeting, docking, and fusion have allowed the characterization of each protein and the required co-factors for each stage [220]. It has been observed that ER to Golgi protein transport in yeast mutants (such as Sec12, Sec13, Sec16, or Sec23 proteins) is blocked; these proteins are probably involved in the formation of ER-derived transport vesicles [214]. Cells depleted of Ypt1p, Sec22, Sec18, Sec17, Bos1, or Sed5 proteins are deficient in protein transport from the ER to the Golgi membrane, resulting in the accumulation of ca. 50 to 60 nm vesicles and enlarged ER membranes [18, 90, 122, 236, 240].

The formation and release of vesicles from ER membranes requires the function of Sec12p and COPII coat structure composed of Sar1p, Sec23/24p complex, and Sec13/31p complex [11, 13, 54, 100, 178, 191, 213, 220, 228]. Sar1p (21 kDa) is a GTP-binding protein that is activated after its bound GDP is exchanged for GTP by Sec12p [13]. Sec12p is an integral ER protein whose cytosolic domain accelerates the GDP/GTP exchange of Sar1p [13, 54]. Sec23p/24p is a 400 kDa complex consisting of two Sec23p (85 kDa) polypeptides and an associated protein, Sec24p (105 kDa) [100]. The hydrolysis of GTP by purified Sar1p is very slow, but the addition of monomeric Sec23p enhances Sar1p GTPase activity over 10-fold, suggesting that Sec23 acts as a GTPase-activating protein (GAP) [282]. Sec24p has no effect on GAP activity but is necessary for vesicle formation [282]. Sec13p (33 kDa) is a hydrophilic protein [213] that interacts with Sec31 (150 kDa) protein to form a large complex (>700 kDa) [228]. The function of the Sec13p complex is still unclear, although both proteins share several repeated domains [213].

Using a refined *in vitro* system assay containing the purified cytosolic factors Sar1p, Sec23/24p complex, and Sec13/31p complex at saturating amounts in combination with washed yeast membranes, the budding and release of competent vesicles from the ER have been reconstituted [228]. Membrane-bound, activated Sar1p recruits and initiates the assembly of the COPII coat structure, composed of the Sec23/24p and Sec13/31p complexes, resulting in the formation of an ER-derived vesicle [12]. Sar1p hydrolyzes GTP in a reaction stimulated by Sec23p [282], presumably leading to the dissociation of inactive cytosolic Sar1p-GDP from the coated vesicle. COPII (Sec23/24p, Sec13/31p coat structure) initiates the budding of the transit vesicle from the ER.

The formation of buds from the ER requires GTP. Vesicles released from the ER in the presence of purified factors and GTP are functional, in that they can be targeted and fused with the Golgi membrane and deliver their cargo proteins. However, ER vesicles prepared in the presence of GTP analogs are nonfunctional, as they cannot be fused with Golgi. These nonfunctional vesicles contain higher levels of Sar1p relative to functional vesicles [12]. GTP hydrolysis is not required for the formation of vesicles because in the presence of either non-hydrolyzable analogues of GTP (GTPγS and GMP-PNP), or the GDP analog (GDPγS), the ER-derived vesicles are released [12, 190]. Triggered by a mechanism yet to be elucidated, possibly the shedding of the inactive Sar1p from the coat, the coat proteins dissociate from the membrane vesicle, exposing pro-

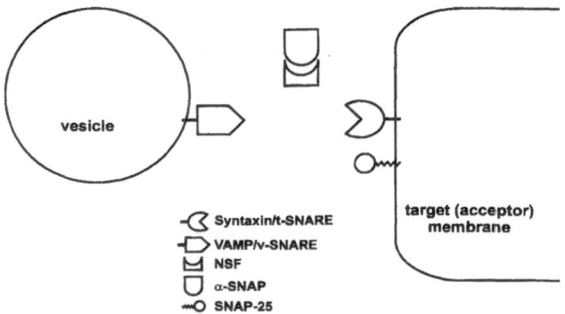

Figure 4. Schematic model for targeting and fusion of a vesicle with its cognate target membrane. The SNARE model suggests that integral membrane proteins on the vesicles (v-SNAREs; VAMP) interact specifically with proteins on the target membrane (t-SNAREs; syntaxin, SNAP-25) and with cytosolic factors, α-SNAP and NSF. The SNARE complex is disassembled upon ATP hydrolysis leading to fusion of the vesicle with the target membrane. (Reproduced with permission of Trends in Plant Science.)

teins such as, Bos1p (27 kDa), Sec22p (24 kDa), Sly1p (75 kDa), Bet1p, and Emp24p [12, 219, 229] that are required for vesicle targeting and fusion.

The stable attachment of the transport vesicles to the Golgi membrane requires the function of Ypt1p and Sec18 proteins, whereas fusion between the vesicle and the target membrane requires ATP and Ca^{2+} [220]. The ER-Golgi docking complex contains at least three species of v-SNARE proteins: Sec22p, Bos1p, and a newly identified ca. 23 kDa protein, Ykt6p ([248]; for details, see below). With the help of Ypt1 (mammalian homologue, Rab1p), a small GTP-binding protein, these proteins interact with their cognate *cis*-Golgi localized Sed5p [9], a t-SNARE protein [248]. Rothman's group [248] has demonstrated the various docking and fusion steps involved in vesicle fusion to Golgi. The exact protein(s) on the ER-derived transport vesicle that interact(s) specifically and only with its cognate docking protein(s) on the *cis*-Golgi membrane is (are) still unknown. Furthermore, the regulation of budding (regarding vesicle selection of cargo protein and the amounts of cargo attached or loaded) is not yet understood. Likewise, the process of vesicle pinching from the membrane and subsequent attachment and fusion with its next compartment requires further investigation. It is very likely, however, that GTP-binding proteins are involved in these regulatory processes.

Functional mammalian and plant gene homologues to *SAR1*, *SEC12*, *YPT1* have been identified [53, 134, 203], as well as mammalian *SEC23* and *SEC13* [196, 239]. In plants, the mRNA levels of the secretory pathway genes (a*SAR1*, a*SEC12*, a*SEC23*, a*YPT1*) [14, 49] and their proteins are generally low compared to constitutively expressed housekeeping genes such as actin (Bar-Peled and Raikhel, in preparation); moreover, regulation of these genes is tissue-specific [14, 49]. Although the secretory pathway is active in all plant tissues, there is evidence suggesting that roots, and especially root tips, exhibit higher secretory activity than do leaves [14]. In *Arabidopsis*, aSar1p levels were found to be higher in roots and in suspension cultures compared with leaves, whereas aSec12p exists in these tissues at hardly detectable levels (Bar-Peled and Raikhel, in preparation). Overexpression of sense and antisense genes in concert with dominant-negative mutants is currently being used to further dissect the roles of aSar1p, aSec12p, aSec13p, aSec23p, and aErd2p in ER-Golgi transport and to understand the differences between various plant tissues (Bar-Peled and Raikhel, in preparation).

Transport between the ER and Golgi is highly complex, involving additional coat proteins, COPI, that participate in vesicle transport early in the secretory pathway. The coatomer, COPI, is composed of seven polypeptides: Ret1p (mammalian homologue α-COP, 160 kDa), Sec26p (β-COP, 110 kDa), Sec27p (β'-COP, 102 kDa), Sec21p (γ-COP, 98 kDa), δ-COP (61 kDa), ε-COP (31 kDa), ζ-COP (20 kDa), and an abundant 20 kDa GTP-binding protein, ARF, an ADP-ribosylation factor [238]. Ample evidence supports the role of COPI in mediating vesicle and cargo transport between the ER and Golgi [205]; *sec21*, *sec26*, and *sec27* yeast mutants fail to transport soluble secretory proteins between the ER and Golgi [60, 111], whereas the *ret1-1* mutant fails to retrieve ER membrane proteins possessing the KKXX signal [140]. The cytoplasmic KKXX signal binds specifically to some COPI protein subunits: α-COPI, β'-COPI, and ε-COPI, *in vitro* [44]. The transit of a tagged KKXX protein in wild-type cells proceeds from the ER to the Golgi and subsequently back to the ER, whereas in *ret1-1* cells, the KKXX protein continues transport to the cell surface. Thus, it may be that some subunits of COPI function in the sorting or retrieval of ER proteins that escape to the Golgi [227].

The involvement of COPI in vesicle transport between the ER and Golgi has been further substantiated by the following analyses. In its GTP-binding domain, mutated Arf1p triggered the accumulation of a normally secreted membrane protein in the pre-Golgi compartment [48]; in addition, ca. 70% of β-COP is localized on the *cis*-side of the Golgi complex [195]. Bednarek *et al.* [21] provided further confirm-

ation of the role of COPI in ER-to-Golgi transport utilizing *in vitro* experiments and demonstrating that COPI-derived vesicles bud from the ER. Some COPI proteins share strikingly similar functions with COPII proteins, such as the GTPases Arf1p and Sar1p and the repeat domains of Ret1p (α-COP), Sec27p (β'-COP) and Sec13/31p. The type of cargo transported via COPI or COPII is yet unknown, as is the stage at which ER-derived vesicles are coated with COPI.

The Golgi complex

Possible mechanisms for Golgi retention

The Golgi complex is composed of several distinct membrane structures: the *cis*-Golgi, *medial*-Golgi, and *trans*-Golgi cisternae (see Fig. 2). A stack of closely apposed and flattened cisternae is the central feature of the Golgi stack. Protease treatment causes the cisternae to separate [41] suggesting that adjacent cisternae are held together by intercisternal proteins. Indeed, some intercisternal components bind short cytoplasmic tails of resident *medial*-Golgi membrane enzymes via ionic interaction [245].

Localization of membrane proteins in the Golgi depends on their membrane-spanning domain and cytoplasmic domain. Replacing the membrane-spanning domain of a reporter protein with that of a Golgi protein results in a hybrid molecule that localizes to the the correct cisterna of the Golgi apparatus [184]. Two models for retention have been proposed (see for review [185]). According to the 'kin recognition' hypothesis, homodimeric proteins may form large hetero-oligomers with other residents of the same Golgi compartment by interaction via their transmembrane and/or stalk regions. The size of these kin oligomers would prevent their introduction into Golgi transport vesicles [184]. By replacing the cytoplasmic regions on several *medial*- and *trans*-Golgi membrane enzymes with that of an ER-localized protein, Nilsson *et al.* [183] have demonstrated that each of these chimeric proteins is retained in the ER. Moreover, when only one *medial*-Golgi protein was engineered, it caused relocation of the other endogenous *medial*-Golgi protein to the ER. An alternative model [28] suggests that retention in the Golgi is based on the length of the transmembrane polypeptide as a function of its interaction with the different lipid composition in the compartments of the secretory pathway [173]. Because the plasma membrane is thought to be thicker than the

Golgi membrane, polypeptides with shorter transmembrane regions would be retained in the Golgi, whereas longer ones would be transported further [157].

A role for cytoplasmic domains in localization of integral membrane proteins to the TGN has been demonstrated for several proteins. For example, furin, a type I membrane protein, that cycles between the TGN and the plasma membrane contains two regions within the cytoplasmic domain which contribute to its localization [270].

Transport along the Golgi complex

Transport of cargo proteins along the Golgi complex is mediated by small vesicles budding from one cisterna and fusing with the next. Each cisterna contains a specific complement of enzymes that can sequentially modify proteins being transported through the Golgi (see for reviews [142, 255]). Subsequently, cargo proteins reach the *trans*-Golgi network (TGN) and are transported either to the plasma membrane or the vacuole. Vesicular transport between adjacent Golgi stacks is mediated by COPI; from extensive investigation in mammalian systems, a clear model of vesicular transport has now emerged (see for review [222]). COPI coat assembly is a regulated process: transport is initiated by the attachment of the small-molecular-weight GTP-binding protein, Arf1p, to the donor compartment via interaction with a membrane-bound GTP exchange factor and a membrane associated Arf1p receptor (yet unknown) [97, 98, 222]. Bound Arf1p-GTP stimulates recruitment of the coatomer, a 7 subunit protein complex, from the cytosol and budding occurs when the coatomer binds [88, 201]. Finally, the hydrolysis of ARF-GTP, presumably at the target membrane, results in coat disassembly, permitting the vesicle to fuse with its target membrane [61, 256] possibly via the SNAREs model (see below). Transport between Golgi stacks also appears to be regulated by the small GTP-binding protein Rab6, either inhibiting anterograde transport or as a positive regulator of retrograde transport [156].

In plants, ARF homologues have been described and found to be very abundant in all tissues relative to the ER to Golgi genes [14, 218], yet their function in plants remains to be elucidated. A functional Rab6 homologue has been identified in *Arabidopsis* and is able to complement a yeast *rab6* mutant [22].

The *trans*-Golgi network (TGN) is thought to be the compartment where proteins are sorted to the plasma membrane for secretion or to the vacuole. For soluble proteins, secretion to the exterior of the cell does not require positive sorting information other than the signal sequence, and secretion is therefore considered to be a default pathway.

Transport from the Golgi apparatus to the plasma membrane

Vesicles carrying plasma membrane and secreted proteins bud from the TGN and are targeted to the cell surface, where they fuse with the plasma membrane. The type of coat on these transport vesicles is still unknown. In yeast, components of the transport machinery acting between the Golgi apparatus and the plasma membrane have been isolated by genetic and biochemical means. A screen for secretion-defective (*sec*) mutants has allowed a number of genes to be identified which are required specifically in this late stage of the secretory pathway [187, 188]: *SEC1*, *2*, *3*, *4*, *5*, *6*, *8*, *9*, *10*, and *15*. Three of the gene products, Sec6p, Sec8p and Sec15p, are part of a large, multi-subunit complex located both in the cytoplasm and associated with the plasma membrane [258]. This complex contains five additional, yet unidentified, polypeptides. The complex has been localized by immunofluorescence analysis to small bud tips, sites of rapid cell growth in yeast. It may, therefore, be involved in either the docking of vesicles at the membrane or in the fusion reaction between the vesicle and plasma membrane. As the complex is not found to be stably associated with transport vesicles, it is unlikely to form part of a vesicle coat such as coatomer. The Sec6/8/15 complex is disrupted in *sec3*, *sec5* and *sec10* mutants with specific proteins missing from the complex in each case, which implies an interaction between these gene products [258].

Studies in various organisms, in particular mammalian neuronal cells and yeast, have led to the proposal of a general model for the docking of vesicles with their target membrane (SNARE or SNAP-receptor hypothesis). A protein in the vesicle membrane (v-SNARE; synaptobrevin) interacts with proteins in the target membrane (t-SNAREs; syntaxin, SNAP-25) along with some soluble factors (α-SNAP and NSF) to enable the vesicle to dock at the target membrane and to permit fusion to occur [249]. Different isoforms of the v-SNAREs and t-SNAREs reside in various cell membranes and vesicles, and thus could provide specificity for the docking reaction. For example, the t-SNARE Sed5p is localized to the *cis*-Golgi membrane [9], whereas Sso1p is at the plasma membrane [1]. The exact role of the SNAREs is still in debate [30], and the mechanism by which membrane fusion occurs is still unknown.

In neurons, vesicles dock at the pre-synaptic membrane and a signal is required before fusion and release of the vesicle contents occurs (regulated secretion). In yeast, however, docking and fusion are apparently not regulated (constitutive secretion). Despite this difference, homologues of neuronal proteins have been found which are required for constitutive secretion. Two yeast plasma membrane proteins, Sso1p and Sso2p, are syntaxin homologues identified as high copy suppressors of *sec1* mutations [1]. Synaptobrevin homologues have also been cloned from yeast (*SNC1* and *SNC2*), localized to post-Golgi vesicles and demonstrated to function in secretion [212]. Sec9p was identified as the yeast counterpart of SNAP-25 and shown to interact with both *SSO* and *SNC* gene products, indicating that constitutive and regulated secretion both occur via similar mechanisms involving analogous gene products [27, 45]. Sec4p is a small GTP-binding protein of the rab family, and a *sec4* mutant can be suppressed by overproduction of Sec9p, indicating that Sec4p may act to regulate the formation or activity of the SNARE complex [27]. Sec1p may also be involved in the regulation of vesicle docking and fusion; a homologue of this protein participates in vesicle fusion in mammalian cells [209].

Transport of proteins to the cell surface in plants

While a number of components involved in Golgi-to-plasma membrane transport have now been identified in other systems, very little is known about this phase of transport in plant cells. As the proteins required for this process appear to be conserved in divergent organisms, it may be expected that similar machineries and mechanisms exist in plants. In fact, several plant proteins have been expressed in yeast and correctly targeted to the plasma membrane (see for review [17]). One interesting system in which this process is being studied is the peribacteroid membrane (PBM) of legume root nodules. This membrane is derived from the plasma membrane after *Rhizobium* is endocytosed. Small GTP-binding proteins are involved in the biogenesis of the PBM, and cDNA clones encoding plant homologues of Rab1p and Rab7p have been isolated [36]. Both genes are induced during nodulation, and plants expressing these genes in an antisense orientation are defective in nodulation, indicating a requirement for the gene products in this process. A phosphatidylinositol 3-kinase is also induced during

nodule formation in soybean and is thought to play a role in the formation of the PBM [109].

For some proteins, post-translational modifications can affect their final location. Nodulin-24 is a nodule specific protein of the PBM [35]. This protein is synthesized as a soluble protein with a signal peptide that directs its transport into the ER lumen and is then cleaved. Mature nodulin-24 is found in the PBM and can only be removed using detergent, as would be expected for an integral membrane protein. The mature protein is also significantly larger than the newly synthesized nodulin-24; thus, it has been suggested that this is due to the attachment of membrane lipids.

A sucrose binding protein (SBP) which appears to follow an unusual route for localization has been identified in soybean on the outer face of the plasma membrane. Extraction of the protein with sodium carbonate or urea causes the dissociation of some of the SBP from the membrane, whereas detergent is needed for complete extraction. It has, therefore, been suggested that SBP is a peripheral membrane protein with a proportion of the protein tethered to the plasma membrane [197]. SBP is synthesized on free ribosomes *in vivo* and is not transported into microsomes *in vitro*, indicating that it may not be secreted via the ER and Golgi but rather by an unusual pathway that remains to be elucidated [198].

Plant cells also secrete many soluble proteins. No sorting signal is required for secretion, and a soluble protein containing a signal peptide with no additional signal is transported extracellularly. Secretion is therefore considered to be a default pathway.

Transport to the vacuole/lysosome

Transport of proteins to the mammalian lysosome

The lysosome is considered to be the mammalian equivalent of the vacuole and transport to this organelle has been extensively studied. Most soluble lysosomal proteins contain a carbohydrate targeting signal consisting of phosphorylated mannose residues. This signal is recognized by membrane-bound mannose-6-phosphate receptors which target them to lysosomes (see for review [131, 267]). Two mannose-6-phosphate receptors have been identified, a cation-dependent and a cation-independent receptor. However, some proteins are transported by a mannose-6-phosphate independent pathway, not yet well characterized [267].

Transport to the yeast vacuole

Transport of proteins to the yeast vacuole does not require a carbohydrate modification; rather, peptide sequences are responsible for targeting. The sorting information for the vacuolar proteins carboxypeptidase Y (CPY) [119] and proteinase A [129] resides in an N-terminal propeptide which is cleaved in the vacuole to yield the active mature form. Over forty yeast mutants have been isolated which are defective in vacuolar protein sorting (*vps* mutants), and the identification of the disrupted genes in these mutants is leading to an understanding of the mechanisms involved in the sorting pathway. Recently, the *VPS10* gene was shown to encode a sorting receptor for CPY, which was not required for the sorting of any other vacuolar proteins tested [155]. Other *VPS* genes are required for the sorting of multiple vacuolar proteins. For example, Vps1p is an 80 kDa GTP-binding protein required for the sorting of both CPY and proteinase A [224]. Small GTP-binding proteins are involved in many transport steps throughout the cell, and *VPS21* encodes a member of this family which is homologous to the mammalian *RAB5* gene and is important for vacuolar transport [110]. Vps15p is a protein kinase which forms a complex with, and activates, Vps34p, a phosphatidylinositol 3-kinase [252]. These proteins may thus be involved in the regulation of vesicle trafficking to the vacuole.

In addition to the better characterized pathway of transport to the vacuole via the Golgi complex, an additional route of transport exists in yeast. At least two proteins (aminopeptidase I and (-mannosidase) are known to be delivered post-translationally to the vacuole, directly from the cytosol, by a mechanism which is independent of the secretory pathway. Five different mutants have been identified which are defective in the transport of aminopeptidase I to the vacuole but which have no effect on delivery via the secretory pathway [89], and are thus likely to encode components of this alternative transport system.

Transport of proteins to plant vacuoles

The plant vacuole is a large organelle and consists of a matrix surrounded by a single bilayer membrane (the tonoplast) which, in mature cells, may occupy up to ninety percent of the total cell volume. At various stages of development, different vacuole types have been observed, each containing a discrete complement of proteins. The vacuole has multiple functions, including the maintenance of turgor, cell growth, digestion of macromolecules and storage of materials such as proteins and secondary metabolites (see for review [26]). Proteins are thought to be transported to the vacuole from the TGN by clathrin-coated vesicles [92].

Vacuolar sorting signals in soluble plant proteins

At the TGN, some proteins are sorted from the pathway of secretion to be transported to the vacuole. In the case of soluble proteins, this requires positive sorting information in addition to the signal sequence for entry into the ER. Three different types of sorting signal have been identified for transport of proteins to the plant vacuole: N-terminal propeptides (NTPP), C-terminal propeptides (CTPP) or regions of the mature protein. In the case of NTPPs and CTPPs, the sorting signal is removed upon deposition of the protein in the vacuole. It should be noted that plant vacuolar targeting signals do not appear to function in yeast [68].

Sweet potato sporamin contains an NTPP, the deletion of which results in its secretion in transformed tobacco cells [159]. This indicates that the NTPP is required for vacuolar targeting. Another protein containing an NTPP is aleurain, a vacuolar thiol protease from barley. Regions of the aleurain NTPP which are able to redirect the normally secreted reporter protein endoproteinase B to the vacuole have been identified [108] and thus must act as vacuolar sorting signals. A common motif is present in the NTPPs of the sporamin and aleurain precursors, and in the potential NTPPs of some other vacuolar proteins; site-directed mutagenesis of a conserved isoleucine residue in this motif of prosporamin abolishes sorting to the vacuole [38, 176, 177].

Other soluble proteins contain vacuolar targeting information at their C-terminus. Barley lectin is synthesized with a 15 amino acid glycosylated CTPP. Deletion of the CTPP causes barley lectin to be secreted from tobacco cells [23], indicating that the CTPP is necessary for vacuolar targeting. When the CTPP is added to the C-terminus of cucumber chitinase, a secreted protein, this enzyme is found in the vacuole, demonstrating that the CTPP is also sufficient for sorting to the vacuole [19]. Similar experiments showed that the seven amino acid CTPP of basic tobacco chitinase is both necessary and sufficient for vacuolar targeting [181]. Extensive site-directed mutagenesis analysis of both the barley lectin [59] and tobacco chitinase [180] CTPPs demonstrated that there is no specific peptide sequence required for a vacuolar sorting signal and that a wide variety of sequences are able to function in vacuolar targeting in both cases. In the case of barley lectin, as few as three amino acids of the CTPP are sufficient for targeting, whereas six to seven amino acids of the tobacco chitinase CTPP are required. However, the addition of two or more glycines to the barley lectin CTPP, or the movement of the glycosylation site closer to the C-terminus, causes secretion, presumably due to steric hindrance. This implies that some component of the sorting machinery recognizes the CTPP from the C-terminus. Other proteins with CTPPs containing vacuolar targeting information include the tobacco proteins AP24 and β-1,3-glucanase [161].

It has also been demonstrated that vacuolar sorting information may reside within regions of the mature protein rather than in a cleavable propeptide. A domain of phytohemagglutinin (PHA), consisting of thirty amino acids predicted to be exposed on the surface of the protein, was able to redirect 50% of a normally secreted reporter protein to the plant vacuole [269]. Field bean legumin also appears to contain vacuolar sorting information within the mature polypeptide, and large regions of the protein were found necessary for transport to the vacuole [226].

Despite the presence of a vacuolar targeting signal, it has been observed that some root cells in bean contain PHA in the cell wall. In root meristem cells, PHA is found in the vacuole as expected, but in elongating and differentiating cells, the same protein is present only in the cell wall [128]. This may be due to the lack of recognition of the vacuolar targeting signal by the sorting apparatus in these cells, leading to cell type-specific alternate targeting of PHA. A similar observation has been made for protease inhibitors in transgenic tomato plants, which are normally vacuolar but in root cells are also found in the cell wall [179]. This raises the possibility that other proteins may be targeted to an alternate location in different tissue and cell types.

Components of the plant vacuolar sorting machinery

The vacuolar targeting signals of several soluble plant proteins are now relatively well defined; however, few proteins involved in the targeting process have been identified. A protein (BP-80) has been isolated from pea clathrin-coated vesicles by affinity chromatography using a peptide corresponding to the proaleurain NTPP. It has been proposed that the binding protein, a glycoprotein whose binding domain is located in the lumen of the vesicles, is a vacuolar targeting receptor [126]. BP-80 also binds to the prosporamin NTPP and the 2S albumin C-terminal targeting signal as well as

to the proaleurain NTPP, although no similarities could be identified between the binding motifs in these proteins. However, no binding has been observed to the probarley lectin CTPP. BP-80 thus appears to have a broad binding specificity, but is probably not involved in the targeting of all vacuolar proteins [127]. The existence of a receptor for CTPP-containing proteins has been implicated by the observation that tobacco chitinase, although correctly targeted to the vacuole at low levels of expression, is secreted when expressed at a high level in tobacco protoplasts [180]. This indicates that the sorting system is saturable, and thus is likely to be receptor-mediated. Further evidence for a CTPP receptor is the ability of a CTPP to redirect reporter proteins to the vacuole [19, 158].

An *Arabidopsis* syntaxin homologue has been identified which may be involved in vacuolar protein transport [16]. The yeast Pep12 protein is a member of the syntaxin family (see above), and a yeast *pep12* mutant is defective in the sorting of proteins to the vacuole. A cDNA was isolated from *Arabidopsis* by functional complementation of the *pep12* mutant and encodes a protein (aPep12p) homologous to Pep12p and other syntaxins: aPep12p may, therefore, be involved in the docking of transport vesicles at the vacuolar membrane or prevacuolar compartment.

Other genes putatively involved in vacuolar protein sorting have been isolated by homology to yeast genes. A PI 3-kinase gene has been isolated from *Arabidopsis* showing some sequence homology to the yeast *VPS34* gene [277], and a chimeric gene containing regions of the plant gene and regions of the yeast gene can complement the yeast *vps34* mutant. An *Arabidopsis* gene showing homology to the yeast *VPS1* gene has also been identified [58] although it has not been demonstrated to function in vacuolar sorting. In addition, small GTP-binding proteins have been found to be associated with vesicles carrying vacuolar proteins in pumpkin cotyledon cells, and their function in protein transport has been proposed [241].

Are there multiple mechanisms for the transport of proteins to the vacuole?

The presence of multiple types of targeting signals for the transport of soluble proteins to the plant vacuole raises the question of whether the mechanisms for transporting these proteins differs. When sporamin (containing an NTPP) and barley lectin (containing a CTPP) are co-expressed in tobacco, both proteins are transported to the same vacuole, as shown by electron microscopic immunolocalization [233]. The sorting signals of barley lectin and sporamin have been exchanged and are still able to function in vacuolar targeting in tobacco, indicating that these signals are interchangeable [158]. However, the vacuolar transport of proteins containing the barley lectin CTPP is sensitive to the fungal metabolite wortmannin, whereas transport of proteins containing the sporamin NTPP is relatively insensitive to this inhibitor [158]. This implies that there is more than one mechanism for transport to the vacuole in tobacco cells: this is also supported by the selective binding of the potential receptor protein BP-80 to some targeting signals and not to others ([127]; see above). One interesting question which remains to be answered is whether proteins containing different types of targeting signals are transported to the vacuole in the same vesicle, or whether distinct populations of vesicles exist for different proteins.

Transport of membrane proteins to the vacuole

The delivery of membrane proteins to the tonoplast has been studied in the case of α-TIP, an integral membrane protein with six membrane-spanning domains that transports water across the tonoplast. The transport inhibitors, brefeldin A and monensin, block the transport of soluble proteins to the plant vacuole; however, neither inhibitor prevents the transport of α-TIP to the tonoplast, suggesting that soluble and membrane proteins are transported via different mechanisms [73]. Regions of α-TIP involved in sorting to the tonoplast have been studied by fusion to the reporter protein phosphinotricine acetyltransferase [107]. A portion of α-TIP containing the sixth transmembrane domain and the cytoplasmic tail is able to redirect the reporter to the tonoplast in tobacco, and α-TIP lacking the cytoplasmic tail is still found in the tonoplast. Thus, it has been concluded that either the sixth transmembrane domain contains targeting information, or that the tonoplast is the default destination for membrane proteins in the plant secretory pathway. To resolve this issue, the sorting of more vacuolar membrane proteins needs to be studied, and several tonoplast proteins have now been cloned which could be used for this purpose (see for review [17]).

Alternative route to the vacuole

An unusual route to the vacuole has been described for the prolamin seed storage proteins in wheat [69]. This pathway is independent of the Golgi apparatus and

may occur by an autophagy-like process [141]. The prolamins initially assemble into protein bodies in the ER. Electron microscopy studies have demonstrated that protein bodies then become detached from the ER cisternae and are surrounded by small vesicles which apparently fuse to form small vacuoles. This alternative route of transport to the vacuole may also be used by seed storage proteins of other cereal species [69].

Perspectives

Work in diverse organisms (from prokaryotes to higher eukaryotes) suggests that basic trafficking mechanisms involving the recognition of targeting signals followed by translocation across a membrane are evolutionarily conserved. The targeting signals for transport into the ER, from the chloroplast stroma to the thylakoid lumen, and export to the periplasmic space in *E. coli* all contain similar structural features and in some cases are functionally interchangeable. Homologous proteins involved in the transport process can also be found in these various systems. Although higher organisms acquired the basic prokaryotic transport machinery, various components were adapted to assist and regulate this process. Plant cells have additional sorting requirements when compared with other organisms. They contain both mitochondria and plastids and thus must maintain the fidelity of targeting to each of these organelles. Some cells also contain glyoxisomes, specialized types of peroxisomes, which contain a specific complement of proteins, and cells therefore need to differentiate between proteins destined for these organelles.

The secretory pathway is unique in that, after transport into the ER, no further membrane translocation seems to be required for proteins to reach their final destination. Rather, transport is mediated by small vesicles that bud from one compartment and fuse with the next. Although in many cases much is known about the proteins that are required for vesicle budding and docking, very little is understood about the regulation of these processes. For example, thus far only three cytosolic factors have been shown to be necessary for a vesicle to bud from the ER *in vitro*; moreover, it has been shown that this process can occur even in the absence of passenger proteins [281]. This suggests that regulation of budding does not occur in the *in vitro* system, although this is unlikely to be the case *in vivo*. Therefore, we can envision that factors exist which may determine when a bud forms. It is not known if a signal transduction mechanism exists for detecting whether a bud contains cargo proteins before a vesicle can form *in vivo*. Alternatively, vesicles may form at the same rate irrespective of the amount of material being transported through the secretory pathway. In plants, it seems likely that the capacity of the pathway alters in response to different secretory loads and conditions and, therefore, signaling pathways must exist to ensure its regulation. In fact, genes encoding components of the transport machinery have been shown to be upregulated in response to environmental conditions [14]. Many components have been identified from yeast genetic screens which are involved in vesicle transport *in vivo*, but which are not essential for the process *in vitro*, and are thus candidates for regulatory proteins.

The number of proteins identified in plants which are involved in transport through the secretory pathway is relatively small, and it is, therefore, not clear how conserved the transport mechanisms are between plants and other organisms. However, homologues of some genes involved in vesicular transport in other organisms have been isolated from plants and thus at least the basic machinery for vesicle trafficking may be similar. Transport to the vacuole appears to be one example where sorting signals differ between mammals, yeast and plants and it is likely that the receptors which recognize these signals are also different. This is clear for targeting to the mammalian lysosome by the mannose-6-phosphate-dependent pathway; there is probably no equivalent carbohydrate-dependent mechanism in yeast or plant cells.

One question which remains unanswered is at which point sorting to the vacuole occurs within the secretory pathway. It is generally assumed that secreted proteins are sorted from vacuolar proteins at the TGN; however, it has not been demonstrated in plants that this is the case. For example, are secreted and vacuolar proteins transported in the same vesicles between the ER and Golgi, or do separate vesicle populations already exist for proteins destined for different locations? Furthermore, are NTPP- and CTPP-containing proteins transported to the vacuole in the same vesicle?

Another interesting question concerns the transport machinery itself: are newly synthesized components of this machinery, such as SNAREs, transported to their site of function in an inactive form? If not, potential vesicle targeting problems can be imagined, with vesicles carrying the itinerant SNAREs fusing with incorrect target membranes. A similar question arises when recycling of transport components is considered: are v-SNAREs recycled back from their target membrane as inactive proteins?

The identification of components of targeting machineries in plant cells allows us to address various issues regarding the plant-specific regulation and adaptation of these machineries. For example, vesicle transport through the secretory pathway in mammalian cells is blocked at low temperatures; in plants, however, the pathway must remain functional at lower temperatures, and the transport components must reflect this difference. There are also some indications that there are differences in the secretory pathway activity between various plant tissues, in particular between roots and leaves, and tissue-specific regulation of the genes involved must, therefore, occur.

The unique role of sub-compartments of the plant secretory pathway in processes such as cell wall synthesis and plant defense emphasizes the need for a more complete understanding of this system and its plant-specific functions. As all of the components involved in these processes become known, this will allow the complete reconstitution of transport *in vitro*. It should then also become possible to address new questions concerning the mechanisms and regulation of transport, and the means by which signaling between different sub-compartments occurs to co-ordinate the different stages of protein synthesis, transport and function.

Acknowledgements

We thank Drs Sebastian Bednarek, Maarten Chrispeels, John Froehlich, Glenn Hicks and Ken Keegstra for their helpful comments and suggestions on the manuscript, and other members of the Raikhel group for fruitful debates and discussions. Research was supported by grants from the National Science Foundation (MCB-9507030), Department of Energy (DE-AC02–76ERO-1338) and Michigan State Research Excellence Funds to N.V.R.

References

1. Aalto MK, Ronne H, Keranen S: Yeast syntaxins Sso1p and Sso2p belong to a family of related membrane proteins that function in vesicular transport. EMBO J 12: 4095–4104 (1993).
2. Adam EJH, Adam SA: Identification of cytosolic factors required for nuclear location sequence-mediated binding to the nuclear envelope. J Cell Biol 125: 547–555 (1994).

242

3. Adam SA, Gerace L: Cytosolic proteins that specifically bind nuclear location signals are receptors for nuclear import. Cell 66: 837–847 (1991).

4. Adam SA, Sterne-Marr R, Gerace L: Nuclear protein import in permeabilized mammalian cells requires soluble cytosolic factors. J Cell Biol 111: 807–816 (1990).

5. Altschuler Y, Rosenberg N, Harel R, Galili G: The N- and C-terminal regions regulate the transport of wheat γ-gliadin through the endoplasmic reticulum in *Xenopus* oocytes. Plant Cell 5: 443–450 (1993).

6. Baker D, Hicke L, Rexach M, Schleyer M, Schekman R: Reconstitution of SEC gene product-dependent intercompartmental protein transport. Cell 54: 335–344 (1988).

7. Balch WE, Farquhar MG: Beyond bulk flow. Trends Cell Biol 5: 16–19 (1995).

8. Balch WE, McCaffery JM, Plutner H, Farquhar MG: Vesicular stomatitis virus glycoprotein is sorted and concentrated during export from the endoplasmic reticulum. Cell 76: 841–852 (1994).

9. Banfield DK, Lewis MJ, Rabouille C, Warren G, Pelham HRB: Localization of Sed5, a putative vesicle targeting molecule, to the *cis*-Golgi network involves both its transmembrane and cytoplasmic domains. J Cell Biol 127: 357–371 (1994).

10. Banjoko A, Trelease RN: Development and application of an in vivo plant peroxisome import system. Plant Physiol 107: 1201–1208 (1995).

11. Barlowe C, D'Enfert C, Schekman R: Purification and characterization of Sar1p, a small GTP-binding protein required for transport vesicle formation from the endoplasmic reticulum. J Biol Chem 268: 873–879 (1993).

12. Barlowe C, Orci L, Yeung T, Hosobuchi M, Hamamoto S, Salama N, Rexach MF, Ravazzola M, Amherdt M, Schekman R: COPII: a membrane coat formed by Sec proteins that drive vesicle budding from the endoplasmic reticulum. Cell 77: 895–908 (1994).

13. Barlowe C, Schekman R: SEC12 encodes a guanine nucleotide exchange factor essential for transport vesicle budding from the ER. Nature 365: 347–349 (1993).

14. Bar-Peled M, Conceicao AS, Frigerio L, Raikhel NV: Expression and Regulation of a ERD2, a gene encoding the KDEL receptor homolog in plants and other genes encoding proteins involved in ER-Golgi vesicular trafficking. Plant Cell 7: 667–676 (1995).

15. Bassham DC, Bartling D, Mould RM, Dunbar B, Weisbeek P, Herrmann RG, Robinson C: Transport of proteins into chloroplasts: delineation of envelope 'transit' and thylakoid 'transfer' signals within the presequences of three imported thylakoid lumen proteins. J Biol Chem 266: 23606–23610 (1991).

16. Bassham DC, Gal S, Conceição AS, Raikhel NV: An *Arabidopsis* syntaxin homologue isolated by functional complementation of a yeast *pep12* mutant. Proc Natl Acad Sci USA 92: 7262–7266 (1995).

17. Bassham DC, Raikhel NV: Transport proteins in the plasma membrane and the secretory system: more questions than answers. Trends Plant Sci, in press (1996).

18. Becker J, Tan T-J, Trepte H-H, Gallwitz D: Mutational analysis of the putative effector domain of the GTP-binding Ypt1 protein in yeast suggests specific regulation by a novel GAP activity. EMBO J 10: 785–792 (1991).

19. Bednarek SY, Raikhel NV: The barley lectin carboxyl-terminal propeptidis a vacuolar protein-sorting determinant in plants. Plant Cell 3: 1195–1206 (1991).

20. Bednarek SY, Raikhel NV: Intracellular trafficking of secretory proteins. Plant Mol Biol 20: 133–150 (1992).

21. Bednarek SY, Ravazzola M, Hosobuchi M, Amherdt M, Perrelet A, Schekman R, Orci L: COPI- and COPII-coated vesicles bud directly from the endoplasmic reticulum in yeast. Cell, in press (1996).

22. Bednarek SY, Reynolds TL, Schroeder M, Grabowski R, Hengst L, Gallwitz D, Raikhel NV: A small GTP-binding protein from *Arabidopsis thaliana* functionally complements the yeast *YPT6* null mutant. Plant Physiol 104: 591–596 (1994).

23. Bednarek SY, Wilkins TA, Dombrowski JE, Raikhel NV: A carboxyl-terminal propeptide is necessary for proper sorting of barley lectin to vacuoles of tobacco. Plant Cell 2: 1145–1155 (1990).

24. Blobel G, Dobberstein B: Transfer of proteins across membranes. II. Reconstitution of functional rough microsomes from heterologous components. J Cell Biol 67: 852–862 (1975).

25. Blom J, Kübrich M, Rassow J, Voos W, Dekker PJT, Maarse AC, Meijer M, Pfanner N: The essential yeast protein MIM44 (encoded by *MPI1*) is involved in an early step of preprotein translocation across the mitochondrial inner membrane. Mol Cell Biol 13: 7346–7371 (1993).

26. Boller T, Wiemken A: Dynamics of vacuolar compartmentation. Annu Rev Plant Physiol 37: 137–164 (1986).

27. Brennwald P, Kearns B, Champion K, Keränen S, Bankaitis V, Novick P: Sec9 is a SNAP–25-like component of a yeast SNARE complex that may be the effector of Sec4 function in exocytosis. Cell 79: 245–258 (1994).

28. Bretscher MS, Munro S: Cholesterol and the Golgi apparatus. Science 261: 1280–1281 (1993).

29. Brink S, Flugge UI, Chaumont F, Boutry M, Emmerman M, Schmitz U, Becker K, Pfanner N: Preproteins of chloroplast envelope inner membrane contain targeting information for receptor-dependent import into fungal mitochondria. J Biol Chem 2696: 16478–16485 (1994).

30. Broadie K, Prokop A, Bellen HJ, O'Kane CJ, Schulze KL, Sweeney ST: Syntaxin and synaptobrevin function downstream of vesicle docking in *Drosophila*. Neuron 15: 663–673 (1995).

31. Brodsky JL, Goeckeler J, Schekman R: BiP and Sec63p are required for both co- and post-translational protein translocation into the yeast endoplasmic reticulum. Proc Natl Acad Sci USA 92: 9643–9646 (1995).

32. Brodsky JL, Schekman R: A Sec63p-BiP complex from yeast is required for protein translocation in a reconstituted proteoliposomes. J Cell Biol 123: 1355–1363 (1993).

33. Campos N, Palau J, Torrent M, Ludevid MD: Diversity of 7 SL RNA from the signal recognition particle of maize endosperm. Nucl Acids Res 17: 1573–1588 (1989).

34. Chaddock AM, Mant A, Karnauchov I, Brink S, Hermann RG, Klosgen RB, Robinson C: A new type of signal peptide: central role of a twin-arginine motif in transfer signals for the ΔpH-dependent thylakoidal protein translocase. EMBO J 14: 2715–2722 (1995).

35. Cheon CI, Hong Z, Verma DPS: Nodulin–24 follows a novel pathway for integration into the peribacteroid membrane in soybean root nodules. J Biol Chem 269: 6598–6602 (1994).

36. Cheon CI, Lee NG, Siddique AB, Bal AK, Verma DPS: Roles of plant homologues of Rab1p and Rab7p in the biogenesis of the peribacteroid membrane, a subcellular compartment formed de novo during root nodule symbiosis. EMBO J 12: 4125–4135 (1993).

37. Chi NC, Adam EJH, Adam SA: Sequence and characterization of cytoplasmic nuclear protein import factor p97. J Cell Biol 130: 265–274 (1995).

38. Chrispeels MJ, Raikhel NV: Short peptide domains target proteins to plant vacuoles. Cell 68: 613–616 (1992).

39. Cline K, Ettinger WF, Theg SM: Protein-specific energy requirements for protein transport across or into thylakoid membranes. Two lumenal proteins are transported in the absence of ATP. J Biol Chem 267: 2688–2696 (1992).

40. Cline K, Henry R, Li C, Yuan J: Multiple pathways for protein transport into or across the thylakoid membrane. EMBO J 12: 4105–4114 (1993).

41. Cluett EB, Brown WJ: Adhesion of Golgi cisternae by proteinaceous interactions: intercisternal bridges as putative adhesive structures. J Cell Sci 103: 773–784 (1992).

42. Connolly T, Gilmore R: GTP hydrolysis by complexes of the signal recognition particle and signal recognition particle receptor. J Cell Biol 123: 799–807 (1993).

43. Corpas FJ, Bunkelmann J, Trelease RN: Identification and immunochemical characterization of a family of peroxisomes membrane proteins (PMPs) in oilseed glyoxysomes. Eur J Cell Biol 656: 280–290 (1994).

44. Cosson P, Letourneur F: Coatomer interaction with di-lysine endoplasmic reticulum retention motifs. Science 263: 1629–1631 (1994).

45. Couve A, Gerst JE: Yeast Snc proteins complex with Sec9. J Biol Chem 269: 23391–23394 (1994).

46. Creighton AM, Hulford A, Mant A, Robinson D, Robinson C: A monomeric, tightly folded stromal intermediate on the delta pH-dependent thylakoidal protein transport pathway. J Biol Chem 270: 1663–1669 (1995).

47. Crowley KS, Rienhart GD, Johnson AE: The signal sequence moves through a ribosomal tunnel into a noncytoplasmic aqueous environment at the ER membrane early in translocation. Cell 73: 1101–115 (1993).

48. Dascher C, Balch WE: Dominant inhibitory mutants of ARF1 block endoplasmic reticulum to Golgi transport and trigger disassembly of the Golgi apparatus. J Biol Chem 269: 1437–1448 (1994).

49. Davies C: Cloning and characterization of a tomato GTPase-like gene related to yeast and *Arabidopsis* genes involved in vesicular transport. Plant Mol Biol 24: 525–531 (1994).

50. Dean N, Pelham HRB: Recycling of proteins from the Golgi compartment to the ER in yeast. J Cell Biol 111: 369–377 (1990).

51. de Boer AD, Weisbeek PJ: Biochim Biophys Acta 1071: 221–253 (1991).

52. Denecke J, Botterman J, Deblaere R: Protein secretion in plant cells can occur via a default pathway. Plant Cell 2: 51–59 (1990).

53. D'Enfert C, Gensse M, Gaillardin C: Fission yeast and a plant have functional homologues of the Sar1 and Sec12 proteins involved in ER to Golgi traffic in budding yeast. EMBO J 11: 4205–4211 (1992).

54. D'Enfert C, Wuesthube LJ, Lila T, Schekman R: Sec12p-dependent membrane binding of the small GTP-binding protein Sar1p promotes formation of transport vesicles from the ER. J Cell Biol 114: 663–670 (1991).

55. Deng X-W: Fresh view on light signal transduction in plants. Cell 76: 423–426 (1994).

56. Deshaies RJ, Schekman R: A yeast mutant defective at an early stage in import of secretory protein precursors into the endoplasmic reticulum. J Cell Biol 105: 633–645 (1987).

57. Dodt G, Braverman N, Wong C, Moser A, Moser HW, Watkins P, Valle D, Gould SJ: Mutations in the PTS1 receptor gene, *PXR1*, define complementation group 2 of the peroxisome biogenesis disorders. Nature Genet 9: 115–125 (1995).

58. Dombrowski JE, Raikhel NV: Isolation of a cDNA encoding a novel GTP-binding protein of *Arabidopsis thaliana*. Plant Mol Biol 28: 1121–1126 (1995).

59. Dombrowski JE, Schroeder MR, Bednarek SY, Raikhel NV: Determination of the functional elements within the vacuolar targeting signal of barley lectin. Plant Cell 5: 587–596 (1993).

60. Duden R, Hosobuchi M, Hamamoto S, Winey M, Byers B, Schekman R: Yeast β- and β'-coat proteins (COP): Two coatomer subunits essential for endoplasmic reticulum-to-Golgi protein traffic. J Biol Chem 269: 24486–24495 (1994).

61. Elazar Z, Orci L, Ostermann J, Amherdt M, Tanigawa G, Rothman JE: ADP-ribosylation factor and coatomer couple fusion to vesicle budding. J Cell Biol 124: 415–424 (1994).

62. Emmermann M, Schmitz UK: Two cDNA clones encoding isoforms of the β-subunit of the general mitochondrial processing peptidase from potato. Plant Physiol 107: 1467–1468 (1995).

63. Emtage JLT, Jensen RE: *MAS6* encodes an essential inner membrane component of the yeast mitochondrial protein import pathway. J Cell Biol 122: 1003–1012 (1993).

64. Esnault Y, Blondel M-O, Deshaies RJ, Schekman R, Képès F: The yeast SSS1 gene is essential for secretory protein translocation and encodes a conserved protein of the endoplasmic reticulum. EMBO J 12: 4083–4093 (1993).

65. Esnault Y, Feldheim D, Blondel MO, Schekman R, Kepes F: SSS1 encodes a stabilizing component of the Sec61 subcomplex of the yeast protein translocation apparatus. J Biol Chem 269: 27478–27485 (1994).

66. Feldheim D, Schekman R: Sec72p contributes to the selective recognition of signal peptides by the secretory polypeptide translocation complex. J Cell Biol 126: 935–943 (1994).

67. Franzusoff A, Lauzé E, Howell KE: Immunoisolation of Sec7p-coated transport vesicles from the yeast secretory pathway. Nature 355: 173–175 (1992).

68. Gal S, Raikhel NV: Protein sorting in the endomembrane system of plant cells. Curr Opin Cell Biol 5: 636–640 (1993).

69. Galili G, Altschuler Y, Levanony H: Assembly and transport of seed storage proteins. Trends in Cell Biol 3: 437–442 (1993).

70. Gaynor EC, Te Heesen S, Graham TR, Aebi M, Emr SD: Signal-mediated retrieval of a membrane protein from the Golgi to the ER in yeast. J Cell Biol 127: 653–665 (1994).

71. Glick BS, Beasley EM, Schatz G: Protein sorting in mitochondria. Trends Biochem Sci 17: 453–459 (1992).

72. Goldberg MW, Allen TD: Structural and functional organization of the nuclear envelope. Curr Opin Cell Biol 7: 301–309 (1995).

73. Gomez L, Chrispeels MJ: Tonoplast and soluble vacuolar proteins are targeted by different mechanisms. Plant Cell 5: 1113–1124 (1993).

74. Görlich D, Kostka S, Kraft R, Dingwall C, Laskey RA, Hartmann E, Prehn S: Two different subunits of importin cooperate to recognize nuclear localization signals and bind them to the nuclear envelop. Curr Opin Cell Biol 5: 383–392 (1995).

75. Görlich D, Prehn S, Hartmann E, Kalies K-U, Rapoport TA: A mammalian homologue of SEC61p and SECYp is associated with ribosomes and nascent polypeptide during translocation. Cell 71: 489–503 (1992).

244

76. Görlich D, Prehn S, Laskey RA, Hartmann E: Isolation of a protein that is essential for the first step of nuclear protein import. Cell 79: 767–78 (1994).

77. Görlich D, Rapoport TA: Protein translocation into proteoliposomes reconstituted from purified components of the endoplasmic reticulum membrane. Cell 75: 615–630 (1993).

78. Görlich D, Vogel F, Mills AD, Hartmann E, Laskey RA: Distinct functions for the two importin subunits in nuclear protein import. Nature 377: 246–248 (1995).

79. Gould SJ, Keller GA, Hosken N, Wilkinson J, Subramani S: A conserved tripeptide sorts proteins to peroxisomes. J Cell Biol 108: 1657–1664 (1989).

80. Gould SJ, McCollum D, Spong AP, Heyman JA, Subramani S: Development of the yeast *Pichia pastoris* as a model organism for a genetic and molecular analysis of peroxisome assembly. Yeast 8: 613–628 (1992).

81. Gray JC, Row PE: Protein translocation across chloroplast envelope membranes. Trends Biochem Sci 5: 243–247 (1995).

82. Griffiths G, Ericsson M, Krijnse-locker J, Nilsson T, Goud B, Söling H-D, Tang BL, Wong SH, Hong W: Localization of the Lys, Asp, Glu, Leu tetrapeptide receptor to the Golgi complex and the intermediate compartment in mammalian cells. J Cell Biol 127: 1557–1574 (1994).

83. Gruhler A, Ono H, Guiard B, Neupert W, Stuart RA: A novel intermediate on the import pathway of cytochrome b_2 into mitochondria: evidence for conservative sorting. EMBO J 14: 1349–135 (1995).

84. Hachiya N, Komiya T, Alam R, Iwahasi J, Sakaguchi M, Omura T, Mihara, K: MSF, a novel cytoplasmic chaperone which functions in precursor targeting to mitochondria. EMBO J 13: 5146–5154 (1994).

85. Hachiya N, Mihara K, Suda K, Horst M, Schatz G, Lithgow T: Reconstitution of the initial steps of mitochondrial protein import. Nature 376: 705–708 (1995).

86. Hageman J, Robinson C, Smeekens S, Weisbeek P: A thylakoid processing protease is required for complete maturation of the lumen protein plastocyanin. Nature 324: 567–569 (1986).

87. Hammond C, Helenius A: Quality control in the secretory pathway. Curr Opin Cell Biol 7: 523–529 (1995).

88. Hara-Kuge S, Kuge O, Orci L, Amherdt M, Ravazzola M, Wieland FT, Rothman JE: *En bloc* incorporation of coatomer subunits during the assembly of COP-coated vesicles. J Cell Biol 124: 883–892 (1994).

89. Harding TM, Morano KA, Scott SV, Klionsky DJ: Isolation and characterization of yeast mutants in the cytoplasm to vacuole protein targeting pathway. J Cell Biol 131: 591–602 (1995).

90. Hardwick KG, Pelham HRB: *SED5* encodes a 39-kDa membrane protein required for vesicular transport between the ER and Golgi complex. J Cell Biol 119: 513–521 (1992).

91. Harkness TAA, Nargang FE, Van der Klei I, Neupert W, Lill R: A crucial role of the mitochondrial protein import receptor MOM19 for the biogenesis of mitochondria. J Cell Biol 124: 637–648 (1994).

92. Harley SM, Beevers L: Coated vesicles are involved in the transport of storage proteins during seed development in *Pisum sativum* L. Plant Physiol 91: 674–678 (1989).

93. Hartl F-U, Neupert W: Protein sorting to mitochondria: evolutionary conservations of folding and assembly. Science 247: 930–938 (1990).

94. Hartmann E, Sommer T, Prehn S, Görlich D, Jentsch S, Rapoport TA: Evolutionary conservation of components of the protein translocation complex. Nature 367: 654–657 (1994).

95. Haucke V, Lithgow V, Rospert S, Hahne K, Schatz G: The yeast mitochondrial protein import receptor Mas20p binds precursor proteins through electrostatic interaction with the positively charged presequence. J Biol Chem 270: 5565–22570 (1995).

96. Hebert DN, Foellmer B, Helinius A: Glucose trimming and reglucosylation determine glycoprotein association with calnexin in the endoplasmic reticulum. Cell 81: 425–433 (1995).

97. Helms JB, Palmer DJ, Rothman JE: Two distinct populations of ARF bound to Golgi membranes. J Cell Biol 121: 751–760 (1993).

98. Helms JB, Rothman JE: Inhibition by Brefeldin A of Golgi membrane enzyme that catalyzes exchange of guanine nucleotide bound to ARF. Nature 360: 352–354 (1992).

99. Herman EM, Tague BW, Hoffman LM, Kjemtrup SE, Chrispeels MJ: Retention of phytohemagglutinin with carboxyterminal tetrapeptide KDEL in the nuclear envelope and the endoplasmic reticulum. Planta 182: 305–312 (1990).

100. Hicke L, Yoshihisa T, Schekman R: Sec23p and a novel 105 kD protein function as a multimeric complex to promote vesicle budding and protein transport from the ER. Mol Cell Biol 3: 667–676 (1992).

101. Hicks GR, Raikhel NV: Protein import into the nucleus: an integrated view. Annu Rev Cell Dev Biol 11: 155–188 (1995).

102. High S, Andersen SSL, Görlich D, Hartmann E, Prehn S, Rapoport TA, Dobberstein B: Sec61p is adjacent to nascent type I and type II signal-anchor proteins during their membrane insertion. J Cell Biol 121: 743–750 (1993).

103. High S, Dobberstein B: Mechanisms determining the transmembrane disposition of proteins. Curr Opin Cell Biol 4: 581–586 (1992).

104. Highfield PE, Ellis RJ: Synthesis and transport of the small subunit of chloroplast ribulose bisphosphate carboxylase. Nature 271: 420–424 (1978).

105. Hines V, Schatz G: Precursor binding to yeast mitochondria. J Biol Chem 268: 449–545 (1993).

106. Hirsch S, Muckel E, Heemeyer F, Von Heijne G, Soll J: A receptor component of the chloroplast protein translocation machinery. Science 266: 1989–1992 (1994).

107. Höfte H, Chrispeels MJ: Protein sorting to the vacuolar membrane. Plant Cell 4: 995–1004 (1992).

108. Holwerda BC, Padgett HS, Rogers JC: Proaleurain vacuolar targeting is mediated by short contiguous peptide interactions. Plant Cell 4: 307–318 (1992).

109. Hong Z, Verma DPS: A phosphatidylinositol 3-kinase is induced during soybean nodule organogenesis and is associated with membrane proliferation. Proc Natl Acad Sci USA 91: 9617–9621 (1994).

110. Horazdovsky BF, Busch GR, Emr SD: *VPS21* encodes a rab5-like GTP-binding protein that is required for the sorting of yeast vacuolar proteins. EMBO J 13: 1297–1309 (1994).

111. Hosobuchi M, Kreis T, Schekman R: SEC21 is a gene required for ER to Golgi protein transport that encodes a subunit of a yeast coatomer. Nature 360: 603–605 (1992).

112. Hsu VW, Shah N, Klausner RD: A brefeldin A-like phenotype is induced by the overexpression of a human ERD-2- like protein, ELP–1. Cell 69: 625–635 (1992).

113. Huang J, Hack E, Thornburg RW, Myers AM: A yeast mitochondrial leader peptide functions in vivo as a dual targeting signal for both chloroplasts and mitochondria. Plant Cell 2: 1249–1260 (1990).

114. Hulford A, Hazell L, Mould RM, Robinson C: Two distinct mechanisms for the translocation of proteins across the thylakoid membrane, one requiring the presence of stromal protein factor and nucleotide triphosphates. J Biol Chem 269: 3251–3256 (1994).

115. Hunt DC, Chrispeels MJ: The signal peptide of a vacuolar protein is necessary and sufficient for the efficient secretion of a cytosolic protein. Plant Physiol 96: 18–25 (1991).

116. Imamoto N, Shimamoto T, Takao T, Tachibana T, Kose S, Matsubae M, Sekimoto T, Shimonishi Y, Yoneda Y: In vivo evidence for involvement of a 58 kDa component of nuclear pore-targeting complex in nuclear protein import. EMBO J 14: 3617–3625 (1995).

117. Jackson MR, Nilsson T, Peterson PA: Retrieval of transmembrane proteins to the endoplasmic reticulum. J Cell Biol 121: 317–333 (1993).

118. Jäntti J, Keränen S, Toikkanen J, Kuismanen E, Ehnholm C, Söderlund H, Olkkonen VM: Membrane insertion and intracellular transport of yeast syntaxin Sso2p in mammalian cells. J Cell Sci 107: 3623–3633 (1994).

119. Johnson LM, Bankaitis VA, Emr SD: Distinct sequence determinants direct intracellular sorting and modification of a yeast vacuolar protease. Cell 48: 875–885 (1987).

120. Jones AM, Herman EM: KDEL- containing auxin-binding protein is secreted to the plasma membrane and cell wall. Plant Physiol 101: 595–606 (1993).

121. Jungnickel B, Rapoport TA: A posttargeting signal sequence recognition event in the endoplasmic reticulum membrane. Cell 82: 261–270 (1995).

122. Kaiser C, Schekman R: A distinct sets of SEC genes govern transport vesicle formation and fusion early in the secretory pathway. Cell 61: 727–733 (1990).

123. Kalies KU, Gorlich D, Rapoport TA: Binding of ribosomes to the rough endoplasmic reticulum mediated by the Sec61p-complex. J Cell Biol 126: 925–934 (1994).

124. Keegstra K: Transport and routing of proteins into chloroplasts. Cell 56: 247–253 (1989).

125. Kessler F, Blobel B, Patel HA, Schnell DJ: Identification of two GTP-binding proteins in the chloroplast protein import machinery. Science 266: 1035–1039 (1994).

126. Kirsch T, Paris N, Butler JM, Beevers L, Rogers JC: Purification and initial characterization of a potential plant vacuolar targeting receptor. Proc Natl Acad Sci USA 91: 3403–3407 (1994).

127. Kirsch T, Saalbach G, Raikhel NV, Beevers L: Specificity of a potential vacuolar targeting receptor for vacuolar targeting information. Mol Biol Cell 6: 104a (1995).

128. Kjemtrup S, Borksenious O, Raikhel NV, Chrispeels MJ: Targeting and release of phytohemagglutinin from the roots of bean seedlings. Plant Physiol 109: 603–610 (1995).

129. Klionsky DJ, Banta LM, Emr SD: Intracellular sorting and processing of a yeast vacuolar hydrolase: proteinase A propeptide contains vacuolar targeting information. Mol Cell Biol 8: 2105–2116 (1988).

130. Knight JS, Gray JC: The N-terminal hydrophobic region of the mature phosphate translocator is sufficient for targeting to the chloroplast inner envelope membrane. Plant Cell 7: 1421–1432 (1995).

131. Kornfeld S: Structure and function of the mannose 6-phosphate/insulin-like growth factor II receptor. Annu Rev Biochem 61: 307–330 (1992).

132. Krolkiewicz S, Sänger HL, Niesbach-Klösgen U: Structural and functional characterization of the signal recognition particle-specific 54 kDa (SRP54) of tomato. Mol Gen Genet 245: 565–576 (1994).

133. Kubrich M, Dietmeier K, Pfanner N: Genetic and biochemical dissection of the mitochondrial protein-import machinery. Curr Genet 27: 393–403 (1995).

134. Kuge O, Dascher C, Orci L, Rowe T, Amherdt M, Plutner H, Ravazzola M, Rothman JE, Balch WE: Sar1 promotes vesicle budding from the endoplasmic reticulum but not Golgi compartments. J Cell Biol 125: 51–65 (1994).

135. Kutay U, Ahnert-Hilger G, Hartmann E, Wiedenmann B, Rapoport TA: Transport route for synaptobrevin via a novel pathway of insertion into endoplasmic reticulum membrane. EMBO J 14: 217–223 (1995).

136. Kutay U, Hartmann E, Rapoport TA: A class of membrane proteins with a C-terminal anchor. Trends Biochem Sci 3: 72–75 (1993).

137. Laidler V, Chaddock AM, Knott TG, Walker D, Robinson C: A SecY homolog in Arabidopsis thaliana. Sequence of full-length cDNA clone and import of the precursor protein into chloroplast. J Biol Chem 270: 17664–17667 (1995).

138. Lazzarow PB, Fujiki Y: Biogenesis of peroxisomes. Annu Rev Cell Biol 1: 489–530 (1985).

139. Lee H-I, Gal S, Newman TC, Raikhel NV: The Arabidopsis endoplasmic reticulum retention receptor functions in yeast. Proc Natl Acad Sci USA 90: 11433–11437 (1993).

140. Letourneur F, Gaynor EC, Hennecke S, Demouliere C, Duden R, Emr SD, Riezman H, Cosson P: Coatomer is essential for retrieval of dilysine-tagged proteins to the ER. Cell 79: 1199–1207 (1994).

141. Levanony H, Rubin R, Altschuler Y, Galili G: Evidence for a novel route of wheat storage proteins to vacuoles. J Cell Biol 119: 1117–1128 (1992).

142. Levy S, Staehelin LA: Synthesis, assembly and function of plant cell wall macromolecules. Curr Opin Cell Biol 4: 856–862 (1992).

143. Lewis MJ, Pelham HRB: A human homologue of the yeast HDEL receptor. Nature 348: 162–163 (1990).

144. Lewis MJ, Pelham HRB: Ligand-induced redistribution of a human KDEL receptor from the Golgi complex to the endoplasmic reticulum. Cell 68: 353–364 (1992).

145. Li H-m, Moore T, Keegstra K: Targeting of proteins to the outer envelope membrane uses a different pathway than transport into chloroplasts. Plant Cell 3: 709–717 (1991).

146. Li X, Henry R, Yuan J, Cline K, Hoffman NE: A chloroplast homolog of the signal recognition subunit SRP54 is involved in the posttranslational integration of a protein into thylakoid membrane. Proc Natl Acad USA 92: 3788–3793 (1995).

147. Li X, Wu Y, Zhang D-Z, Gillikin JW, Boston RS, Franceschi VR, Okita TW: Rice prolamine protein body biogenesis: a BiP-mediated process. Science 262: 1054–1056 (1993).

148. Lian JP, Ferro-Novick S: Bos1p, an integral membrane protein of the endoplasmic reticulum to the Golgi transport vesicles, is required for their fusion competence. Cell 73: 735–745 (1993).

149. Lindstrom JT, Chu B, Belanger F: Isolation and characterization of an Arabidopsis thaliana gene for the 54 kDa subunit of the signal recognition particle. Plant Mol Biol 23: 1265–1272 (1993).

150. Linstendt AD, Foguet M, Renz M, Seelig HP, Glick BS, Hauri H-P: A C-terminally-anchored Golgi protein is inserted into the endoplasmic reticulum and then transported to the Golgi apparatus. Proc Natl Acad Sci USA 92: 5102–5105 (1995).

246

151. Lund P, Lee RY, Dunsmuir P: Bacterial chitinase is modified and secreted in transgenic tobacco. Plant Physiol 91: 130–135 (1989).

152. Lütcke H, High S, Römisch K, Ashford AJ, Dobberstein B: The methionine-rich domain of the 54 kDa subunit of signal recognition particle is sufficient for the interaction with signal sequences. EMBO J 11: 1543–1551 (1992).

153. Maarse AC, Blom J, Grivell LA, Meijer M: MPI1, an essential gene encoding a mitochondrial membrane protein, is possibly involved in protein import into yeast mitochondria. EMBO J 11: 3619–3628 (1992).

154. Maccecchini ML, Rudin Y, Blobel G, Schatz G: Import of proteins into mitochondria: precursor forms of the extra mitochondrially made F_1-ATPase subunits in yeast. Proc Natl Acad Sci USA 76: 343–347 (1979).

155. Marcusson EG, Horazdovsky BF, Cereghino JL, Gharakhanian E, Emr SD: The sorting receptor for yeast vacuolar carboxypeptidase Y is encoded by the VPS10 gene. Cell 77: 579–586 (1994).

156. Martinez O, Schmidt A, Salaméro J, Hoflack B, Roa M, Goud B: The small GTP-binding protein rab6 functions in intra-Golgi transport. J Cell Biol 127: 1575–1588 (1994).

157. Masibay AS, Balaji PV, Boeggeman EE, Qasba PK: Mutational analysis of the Golgi retention signal of bovine β–1,4,-galactosyltransferase. J Biol Chem 268: 9908–9916 (1993).

158. Matsuoka K, Bassham DC, Raikhel NV, Nakamura K: Different sensitivity to wortmannin of two vacuolar sorting signals indicates the presence of distinct sorting machineries in tobacco cells. J Cell Biol 130: 1307–1318 (1995).

159. Matsuoka K, Nakamura K: Propeptide of a precursor to a plant vacuolar protein required for vacuolar targeting. Proc Natl Acad Sci USA 88: 834–838 (1991).

160. Mayer A, Nargang FE, Neupert W, Lill R: MOM22 is a receptor for mitochondrial targeting sequences and cooperates with MOM19. EMBO J 14: 4204–4211 (1995).

161. Melchers LS, Sela-Buurlage MB, Vloemans SA, Woloshuk CP, Van Roekel JSC, Pen J, Van den Elzen PJM, Cornelissen BJC: Extracellular targeting of the vacuolar tobacco proteins AP24, chitinase and β–1,3-glucanase in transgenic plants. Plant Mol Biol 21: 583–593 (1993).

162. Melchior F, Paschal B, Evans J, Gerace L: Inhibition of nuclear protein import by nonhydrolyzable analogues of GTP and identification of the small GTPase Ran/TC4 as an essential transport factor. J Cell Biol 123: 1649–1659 (1993).

163. Miesenböck G, Rothman JE: The capacity to retrieve escaped ER proteins extends to the trans-most cisterna of the Golgi stack. J Cell Biol 129: 309–319 (1995).

164. Miller JD, Tajima S, Lauffer L, Walter P: The β subunit of the signal recognition particle receptor is a transmembrane GTPase that anchors thealpha subunit, a peripheral membrane GTPase, tothe endoplasmic reticulum membrane. J Cell Biol 128: 273–282 (1995).

165. Miller JD, Wilhelm H, Gierasch L, Gilmore R, Walter P: GTP binding and hydrolysis by the signal recognition particle during initiation of protein translocation. Nature 366: 351–354 (1993).

166. Mizuno M, Singer J: A soluble secretory protein is first concentrated in the endoplasmic reticulum before transfer to the Golgi apparatus. Proc Natl Acad Sci USA 90: 5732–5736 (1993).

167. Moczko M, Ehmann B, Gärtner F, Hönlinger A, Sächfer E, Pfanner, N: Deletion of the receptor MOM19 strongly impairs import of cleavable preproteins into Saccharomyces cerevisiae mitochondria. J Biol Chem 269: 9045–9051 (1994).

168. Moore AL, Wood CK, Watts, FZ: Protein import into plant mitochondria. Annu Rev Plant Physiol Plant Mol Biol 45: 545–575 (1994).

169. Moore MS, Blobel G: The GTP-binding protein Ran/TC4 is required for protein import into the nucleus. Nature 365: 661–663 (1993).

170. Moore MS, Blobel G: Purification of a Ran-interacting protein that is required for protein import into the nucleus. Proc Natl Acad Sci USA 91: 10212–10216 (1994).

171. Mothes W, Prehn S, Rapoport TA: Systematic probing of the environment of a translocating secretory protein during translocation through the ER membrane. EMBO J 13: 3973–3982 (1994).

172. Muller SB, Rensing SA, Martin WF, Maier UG: cDNA cloning of a Sec61 homologue from the cryptomonad alga Pyrenomonas salina. Curr Genet 26: 410–414 (1994).

173. Munro S: An investigation of the role of transmembrane domains in Golgi protein retention. EMBO J 14: 4695–4704 (1995).

174. Munro S, Pelham HRB: C-terminal signal prevents the secretion of luminal ER proteins. Cell 48: 899–907 (1987).

175. Murakami H, Blobel G, Pain D: Signal sequence region of mitochondrial precursor proteins bind to mitochondrial import receptor. Proc Natl Acad Sci USA 90: 3358–3362 (1993).

176. Nakamura K, Matsuoka K: Protein targeting to the vacuole in plant cells. Plant Physiol 101: 1–5 (1993).

177. Nakamura K, Matsuoka K, Mukumoto F, Watanabe N: Processing and transport to the vacuole of a precursor to sweet potato sporamin in transformed tobacco cell line BY–2. J Exp Bot 44: 331–338 (1993).

178. Nakano A, Muramatsu M: A novel GTP-binding protein, Sar1p, is involved in transport from the endoplasmic reticulum to the Golgi apparatus. J Cell Biol 109: 2677–2691 (1989).

179. Narváez-Vásquez J, Franceschi VR, Ryan CA: Proteinase-inhibitor synthesis in tomato plants: evidence for extracellular deposition in roots through the secretory pathway. Planta 189: 257–266 (1993).

180. Neuhaus J-M, Pietrzak M, Boller T: Mutation analysis of the C-terminal vacuolar targeting peptide of tobacco chitinase: low specificity of the sorting system, and gradual transition between intracellular retention and secretion into the extracellular space. Plant J 5: 45–54 (1994).

181. Neuhaus J-M, Sticher L, Meins FJr, Boller T: A short C-terminal sequence is necessary and sufficient for the targeting of chitinases to the plant vacuole. Proc Natl Acad Sci USA 88: 10362–10366 (1991).

182. Newmeyer DD, Finlay DR, Forbes DJ: In vitro transport of a fluorescent nuclear protein and exclusion of non-nuclear proteins. J Cell Biol 103: 2091–2102 (1986).

183. Nilsson T, Hoe MH, Slusarewics P, Rabouille C, Watson R, Hunte F, Watzele G, Berger EG, Warren G: Kin recognition between medial Golgi enzymes in HeLa cells. EMBO J 10: 3567–3575 (1994).

184. Nilsson T, Slusarewics P, Hoe MH, Warren G: Kin recognition: a model for the retention of Golgi enzymes. FEBS Lett 330: 1–4 (1993).

185. Nilsson T, Warren G: Retention and retrieval in the endoplasmic reticulum and the Golgi apparatus. Curr Opin Cell Biol 6: 517–521 (1994).

186. Nilsson I, Whitley P, Von Heijne G: The COOH-terminal ends of internal signal and signal-anchor sequences are positioned

differently in the ER translocase. J Cell Biol 126: 1127–1132 (1994).

187. Novick P, Ferro S, Schekman R: Order of events in the yeast secretory pathway. Cell 25: 461–469 (1981).

188. Novick P, Field C, Schekman R: Identification of 23 complementation groups required for post-translational events in the yeast secretory pathway. Cell 21: 205–215 (1980).

189. Nunnari J, Fox TD, Walter P: A mitochondrial protease with two catalytic subunits of nonoverlapping specificities. Science 262: 1997–2004 (1993).

190. Oka T, Nakano A: Inhibition of GTP hydrolysis by Sar1p causes accumulation of vesicles that are a functional intermediate of the ER-to-Golgi transport in yeast. J Cell Biol 124: 425–434 (1994).

191. Oka T, Nishikawa S, Nakano A: Reconstitution of GTP-binding Sar1 protein function in ER to Golgi transport. J Cell Biol 114: 671–679 (1991).

192. Okamoto T, Nakayama H, Seta K, Isobe T, Minamikawa T: Posttranslational processing of a carboxy-terminal propeptide containing a KDEL sequence of plant vacuolar cysteine endopeptidase (SH-EP). FEBS Lett 351: 31–34 (1994).

193. Okita TW, Li X, Roberts MW: Targeting of mRNAs to domains of the endoplasmic reticulum. Trends Biochem Sci 4: 91–96 (1994).

194. Olsen LJ, Ettinger WF, Damsz B, Matsudaira K, Webb MA, Harada JJ: Targeting of glyoxysomal proteins to peroxisomes in leaves and roots of a higher plant. Plant Cell 5: 941–952 (1993).

195. Oprins A, Duden R, Kreis TE, Geuze HJ, Slot JW: β-COP localizes mainly to the *cis*-Golgi side in exocrine pancreas. J Cell Biol 121: 49–59 (1993).

196. Orci L, Ravazzola M, Meda P, Holcomb C, Moore H-P, Hicke L, Schekman R: Mammalian Sec23p homologue is restricted to the endoplasmic reticulum transitional cytoplasm. Proc Natl Acad Sci USA 88: 8611–8615 (1991).

197. Overvoorde PJ, Grimes HD: Topographical analysis of the plasma membrane-associated sucrose binding protein from soybean. J Biol Chem 269: 15154–15161 (1994).

198. Overvoorde PJ, Grime HD: Targeting and transport of the sucrose binding protein to the soybean plasma membrane. J Cell Biochem Suppl 19A: A3–214 (1995).

199. Pahl HL, Baeuerle PA: A novel signal transduction pathway from the endoplasmic reticulum to the nucleus is mediated by transcription factor NF-B. EMBO J 14: 2580–2588 (1995).

200. Palade GE: Intracellular aspects of the process of protein secretion. Science 187: 347–358 (1975).

201. Palmer DJ, Helms JB, Beckers CJM, Orci L, Rothman JE: Binding of coatomer to Golgi membranes requires ADP-ribosylation factor. J Biol Chem 268: 12083–12089 (1993).

202. Panzner S, Dreier L, Hartmann E, Kostka S, Rapoport TA: Posttranslational protein transport in yeast reconstituted with a purified complex of Sec proteins and Kar2p. Cell 81: 561–570 (1995).

203. Park YS, Song O, Kwak JM, Hong SW, Lee HH, Nam HG: Functional complementation of a yeast vesicular transport mutation ypt1–1 by a *Brassica napus* cDNA clone encoding a small GTP-binding protein. Plant Mol Biol 26: 1725–1735 (1994).

204. Pelham HRB: Evidence that luminal ER proteins are sorted from secreted proteins in a post-ER compartment. EMBO J 7: 913–918 (1988).

205. Pepperkok R, Scheel J, Horstmann H, Hauri H-P, Griffiths G, Kreis TE: β-COP is essential for biosynthetic membrane transport from the endoplasmic reticulum to the Golgi complex in vivo. Cell 74: 71–82 (1993).

206. Perry SE, Keegstra K: Envelope membrane proteins that interact with chloroplastic precursor proteins. Plant Cell 6: 93–105 (1994).

207. Perryman RA, Mooney B, Harmey MA: Identification of a 42-kDa plant mitochondrial outer membrane protein, MOM42, involved in the import of precursor proteins into plant mitochondria. Arch Biochem Biophys 316: 659–664 (1995).

208. Peter F, Van PN, Söling HD: Different sorting of Lys-Asp-Glu-Leu proteins in rat liver. J Biol Chem 267: 10631–10637 (1992).

209. Pevsner J, Hsu S-C, Scheller RH: n-Sec1: a neural-specific syntaxin-binding protein. Proc Natl Acad Sci USA 91: 1445–1449 (1994).

210. Pfanner N, Meijer M: Pulling in the proteins. Curr Biol 5: 132–135 (1995).

211. Prehn S, Wiedmann M, Rapoport TA, Zweib C: Protein translocation across wheat germ microsomal membranes requires an SRP-like component. EMBO J 6: 2093–2097 (1987).

212. Protopopov V, Govindan B, Novick P, Gerst JE: Homologs of the synaptobrevin/VAMP family of synaptic vesicle proteins function on the late secretory pathway in *S. cerevisiae*. Cell 74: 855–861 (1993).

213. Pryer NK, Salama NR, Schekman RW, Kaiser CA: Cytosolic Sec13p is required in cytoplasmic form for the ER to Golgi *in vitro*. J Cell Biol 120: 867–875 (1993).

214. Pryer NK, Wuestehube LJ, Schekman R: Vesicle-mediated protein sorting. Annu Rev Biochem 61: 471–516 (1992).

215. Radu A, Blobel G, Moore MS: Identification of a protein complex that is required for nuclear protein import and mediates docking of import substrate to distinct nucleoporins. Proc Natl Acad Sci USA 92: 1767–1773 (1995).

216. Rajagopalan S, Xu Y, Brenner MB: Retention of unassembled components of integral membrane proteins by calnexin. Science 263: 387–390 (1994).

217. Ramage L, Junne T, Hahne K, Lithgow T, Schatz G: Functional cooperation of mitochondrial protein import receptors in yeast. EMBO J 12: 4115–4123 (1993).

218. Regad F, Bardet C, Tremousaygue D, Moisan A, Lescure B, Axelos M: cDNA cloning and expression of an *Arabidopsis* GTP-binding protein of the ARF family. FEBS Lett 316: 133–136 (1993).

219. Rexach MF, Latterich M, Schekman RW: Characteristics of encoplasmic reticulum-derived transport vesicles. J Cell Biol 126: 1133–1148 (1994).

220. Rexach M, Schekman R: Distinct biochemical requirements for the budding, targeting and fusion of ER-derived transport vesicles. J Cell Biol 114: 219–229 (1991).

221. Robinson C, Cai D, Hulford A, Brock IW, Michl D, Hazell L, Schmidt I, Herrmann RG, Klosgen RB: The presequence of a chimeric construct dictates which of two mechanisms are utilized for translocation across the thylakoid membrane: evidence for the existence of two distinct translocation systems. EMBO J 13: 279–285 (1994).

222. Rothman JE: Mechanisms of intracellular protein transport. Nature 372: 55–63 (1994).

223. Rothman JE, Orci L: Molecular dissection of the secretory pathway. Nature 355: 409–415 (1992).

224. Rothman JH, Raymond CK, Gilbert T, O'Hara PJ, Stevens TH: A putative GTP-binding protein homologous to interferon-inducible Mx proteins performs an essential function in yeast protein sorting. Cell 61: 1063–1074 (1990).

225. Ruohola H, Kastan Kabcenell A, Ferro-Novick S: Reconstitution of protein transport from the endoplasmic reticulum to the Golgi complex in yeast: the acceptor Golgi compartment is defective in the sec23 mutant. J Cell Biol 107: 1465–1476 (1988).

226. Saalbach G, Jung R, Kunze G, Saalbach I, Adler K, Müntz K: Different legumin protein domains act as vacuolar targeting signals. Plant Cell 3: 695–708 (1991).

227. Salama NR, Schekman R: The role of coat proteins in the biosynthesis of secretory proteins. Curr Opin Cell Biol 7: 536–543 (1995).

228. Salama NR, Yeung T, Schekman R: The sec13p complex and reconstitution of vesicles budding from the ER with purified cytosolic proteins. EMBO J 12: 4073–4082 (1993).

229. Schimmöller F, Singer-Krüger B, Schröder S, Krüger U, Barlowe C, Riezman H: The absence of Emp24p, a component of ER-derived COPII-coated vesicles, causes a defect in transport of selected proteins to the Golgi. EMBO J 14: 1329–1339 (1995).

230. Schlenstedt G, Hurt E, Doye V, Silver PA: Reconstitution of nuclear protein transport with semi-intact yeast cells. J Cell Biol 123: 785–798 (1993).

231. Schnell DJ, Blobel, G: Identification of intermediates in the pathway of protein import into chloroplasts and their localization to envelope contact sites. J Cell Biol 120: 103–115 (1993).

232. Schnell DJ, Kessler F, Blobel G: Isolation of components of the chloroplast protein import. Science 266: 1007–1012 (1994).

233. Schroeder MR, Borksenious ON, Matsuoka K, Nakamura K, Raikhel NV: Colocalization of barley lectin and sporamin in vacuoles of transgenic tobacco plants. Plant Physiol 101: 451–458 (1993).

234. Schutze MP, Peterson PA, Jackson MR: An N-terminal double-arginine motif maintains type II membrane proteins in the endoplasmic reticulum. EMBO J 13: 1696–1705 (1994).

235. Seedorf M, Waegemann K, Soll J: A constituent of the chloroplast import complex represents a new type of GTP-binding protein. Plant J 7235: 401–411 (1995).

236. Segev N, Mulholland J, Botstein D: The yeast GTP-binding Ypt1 protein and a mammalian counterpart are associated with the secretion machinery. Cell 52: 915–924 (1988).

237. Semenza JC, Hardwick KG, Dean N, Pelham HRB: ERD2, a yeast gene required for the receptor-mediated retrieval of luminal ER proteins from the secretory pathway. Cell 61: 1349–1357 (1990).

238. Serafini T, Orci L, Amehrdt M, Brunner M, Kahn RA, Rothman JE: ADP-ribosylation factor is a subunit of the coat of Golgi derived COP-coated vesicles: a novel role for a GTP binding protein. Cell 67: 239–253 (1991).

239. Shaywitz DA, Orci L, Ravazzola M, Swaroop A, Kaiser CA: Human SEC13Rp functions in yeast and is located on transport vesicles budding from the endoplasmic reticulum. J Cell Biol 128: 769–777 (1995).

240. Shim J, Newman A, Ferro-Novick S: The BOS1 gene encodes an essential 27 kD putative membrane protein that is required for vesicular transport from the ER to the Golgi complex in yeast. J Cell Biol 113: 55–64 (1991).

241. Shimada T, Nishimura M, Hara-Nishimura I: Small GTP-binding proteins are associated with the vesicles that are targeted to vacuoles in developing pumpkin cotyledons. Plant Cell Physiol 35: 995–1001 (1994).

242. Shore GC, McBride HM, Millar DG, Steenaart AE, Nguyen M: Import and insertion of proteins into the mitochondrial outer membrane. Eur J Biochem 227: 9–18 (1995).

243. Sijmons PC, Dekker BMM, Schrammeijer B, Verwoerd TC, Van den Elzen PJM, Hoekema A: Production of correctly processed human serum albumin in transgenic plants. Bio/technology 8: 217–221 (1990).

244. Simon SM, Blobel G: A protein-conducting channel in the endoplasmic reticulum. Cell 65: 371–380 (1991).

245. Slusarewics P, Nilsson T, Hui N, Watson R, Warren G: Isolation of a matrix that binds medial Golgi enzymes. J Cell Biol 124: 405–413 (1994).

246. Small GH, Imanaka T, Shio H, Lazzarow PB: Efficient association of in vitro translation products with purified, stable Candida tropicalis peroxisomes. Mol Cell Biol 7: 1848–1855 (1987).

247. Smith S, Blobel G: The first membrane spanning region of the lamin B receptor is sufficient for sorting to the inner nuclear membrane. J Cell Biol 120: 631–637 (1993).

248. Søgaard M, Tani K, Ye RR, Geromanos S, Tempst P, Kirchhausen T, Rothman JE, Söllner T: A Rab protein is required for the assembly of SNARE complexes in the docking of transport vesicles. Cell 78: 937–948 (1994).

249. Söllner T, Whiteheart SW, Brunner M, Erdjument-Bromage H, Geromanos S, Tempst P, Rothman JE: SNAP receptors implicated in vesicle targeting and fusion. Nature 362: 318–324 (1993).

250. Sönnichsen B, Füllekrug J, Nguyen Van P, Diekmann W, Robinson DG, Mieskus G: Retention and retrieval: both mechanisms cooperate to maintain calreticulin in the endoplasmic reticulum. J Cell Sci 107: 2705–2717 (1994).

251. Soullam B, Worman HJ: Signals and structural features involved in integral membrane protein targeting to the inner nuclear membrane. J Cell Biol 130: 15–27 (1995).

252. Stack JH, Herman PK, Schu PV, Emr SD: A membrane-associated complex containing the Vps15 protein kinase and the Vps34 PI 3-kinase is essential for protein sorting to the yeast lysosome-like vacuole. EMBO J 12: 2195–2204 (1993).

253. Stirling CJ, Rothblatt J, Hosobuchi M, Deshaies R, Schekman R: Protein translocation mutants defective in the insertion of integral membrane proteins into the endoplasmic reticulum. Mol Biol Cell 3: 129–142 (1992).

254. Strambio-de-Castillia C, Blobel G, Rout MP: Isolation and characterization of nuclear envelopes from the yeast Saccharomyces. J Cell Biol 131: 19–31 (1995).

255. Sturm A: N-Glycosylation of plant proteins. In: Montreuil J, Schachter H, Vliegenthart JFG (eds) Glycoproteins, pp. 521–541. Elsevier, Amsterdam (1995).

256. Tanigawa g, Orci L, Amherd M, Ravazzola M, Helms JB, Rothman JE: Hydrolysis fo bound GTP by ARF protein triggers uncoatinf of Golgi-derived COP-coated vesicles. J Cell Biol 123: 1365–1371 (1993).

257. Tatu U, Hammond C, Helenius A: Folding and oligomerization of influenza hemagglutinin in the ER and the intermediate compartment. EMBO J 14: 1340–1348 (1995).

258. TerBush DR, Novick P: Sec6, Sec8, and Sec15 are components of a multisubunit complex which localizes to small bud tips in Saccharomyces cerevisiae. J Cell Biol 130: 299–312 (1995).

259. Terlecky SR, Nuttley WM, McCollum D, Sock E, Subramani S: The Pichia pastoris peroxisomal protein PAS8p is the receptor for the C-terminal tripeptide peroxisomal targeting signal. EMBO J 14: 3627–3634 (1995).

260. Theg SM, Scott SV: Protein import into chloroplasts. Trends Cell Biol 3: 186–190 (1993).

261. Townsley FM, Pelham HRB: The KKXX signal mediates retrieval of membrane proteins from the Golgi to the ER in yeast. Eur J Cell Biol 64: 211–216 (1994).

262. Townsley FM, Wilson D, Pelham HRB: Mutational analysis of the human KDEL receptor: distinct structural requirements for Golgi retention, ligand binding and retrograde transport. EMBO J 12: 2821–2829 (1993).

263. Tranel PJ, Froehlich J, Goyal A, Keegstra K: A component of the chloroplastic import apparatus is targeted to the outer envelope membrane via a novel pathway. EMBO J 14: 2436–2446 (1995).

264. Vitale A, Ceriotti A, Denecke J: The role of endoplasmic reticulum in protein synthesis, modification and intracellular transport. J Exp Bot 44: 1417–1444 (1993).

265. Voelker R, Barkan A: Two nuclear mutations disrupt distinct pathways for targeting proteins to the chloroplast thylakoid. EMBO J 14: 3905–3914 (1995).

266. Volokita M: The carboxy-terminal end of glycolate oxidase directs a foreign protein into tobacco leaf peroxisomes. Plant J 1: 361–366 (1991).

267. Von Figura K, Hasilik A: Lysosomal enzymes and their receptors. Ann Rev Biochem 55: 167–193 (1986).

268. Von Heijne G: Membrane protein assembly: rules of the game. BioEssays 17: 25–30 (1994).

269. Von Schaewen A, Chrispeels MJ: Identification of vacuolar sorting information in phytohemagglutinin, an unprocessed vacuolar protein. J Exp Bot 44: 339–342 (1993).

270. Voorhees P, Deignan E, Van Donselaar E, Humphrey J, Marks MS, Peters PJ, Bonifacino JS: An acidic sequence within the cytoplasmic domain of furin functions as a determinant of trans-Golgi network localization and internalization from the cell surface. EMBO J 14: 4961–4975 (1995).

271. Waegemann K, Paulsen H, Soll J: Translocation of proteins into isolated chloroplasts require cytosolic factor to obtain import competence. FEBS Lett 261: 89–92 (1990).

272. Waegemann K, Soll J: Characterization of the protein import apparatus in isolated outer envelopes of chloroplasts. Plant J 1: 149–158 (1991).

273. Walter P, Johnson AE: Signal sequence recognition and protein targeting to the endoplasmic reticulum membrane. Annu Rev Cell Biol 10: 87–119 (1994).

274. Walton PA, Wendland M, Subramani S, Rachubinski RA, Welch WJ: Involvement of 70-kD heat shock proteins in peroxisomal import. J Cell Biol 125: 1037–1046 (1994).

275. Wang X, Sato R, Brown MS, Hua X, Goldstein JL: SREBP–1, a membrane-bound transcription factor released by sterol-regulated proteolysis. Cell 77: 53–62 (1994).

276. Ware FF, Vassilakos A, Peterson PA, Jackson MR, Lehrman MA, Williams DB: The molecular chaperone calnexin binds $Glc_1Man_9GlcNAc_2$ oligosaccharides as an initial step in recognizing unfolded glycoproteins. J Biol Chem 270: 4697–4704 (1995).

277. Welters P, Takegawa K, Emr SD, Chrispeels MJ: AtVPS34, a phosphatidylinositol 3-kinase of Arabidopsis thaliana, is an essential protein with homology to a calcium-dependent lipid binding domain. Proc Natl Acad Sci USA 91: 11398–11402 (1994).

278. Wiedmann B, Sakai H, Davis TA, Wiedmann M: A protein complex required for signal sequence-specific sorting and translocation. Nature 370: 434–440 (1994).

279. Wiese C, Wilson KL: Nuclear membrane dynamics. Curr Opin Cell Biol 5: 387–394 (1993).

280. Wilson D, Lewis MJ, Pelham HRB: pH-dependent binding of KDEL to its receptor in vitro. J Biol Chem 268: 7465–7468 (1993).

281. Yeung T, Barlowe C, Schekman R: Uncoupled packaging of targeting and cargo molecules during transport vesicle budding from the endoplasmic reticulum. J Biol Chem 270: 30567–30570 (1995).

282. Yoshihisa T, Barlowe C, Schekman R: Requirement of a GTP-ase activating protein in vesicle budding from the endoplasmic reticulum. Science 259: 1466–1468 (1993).

283. Yuan J, Henry R, McCaffery M, Cline K: SecA homolog in protein transport within chloroplast: evidence for endosymbiont-derived sorting. Science 266: 765–798 (1994).

Plant Molecular Biology **32**: 251–273, 1996.
© 1996 *Kluwer Academic Publishers.*

Plasmodesmal cell-to-cell transport of proteins and nucleic acids

Laurel A. Mezitt & William J. Lucas*
Section of Plant Biology, Division of Biological Sciences, University of California, Davis, CA 95616, USA
(* *author for correspondence*)

Key words: cell-to-cell transport, plasmodesmata, macromolecular trafficking, supracellular control proteins, developmental domains, viral movement proteins

Contents

Abstract   251
Introduction   252
Plasmodesmata and the symplasmic
    concept   252
Plasmodesmal structure in relation to
    function   252
Plasmodesmata and cell-to-cell
    trafficking of viral proteins   254
MP-mediated trafficking of vRNA/DNA
    through plasmodesmata   255
Molecular determinants of viral protein
    trafficking through plasmodesmata   256
Role of macromolecular trafficking in
    plant development   257
    KNOTTED1, a maize transcription
        factor, functions
        non-cell-autonomously   257

KN1 can taffic through
    plasmodesmata!   260
KN1 can selectively traffic its own
    RNA through plasmodesmata   260
KN1 has multiple homologues in
    maize and other plant species   260
A new class of plant genes: the
    supracellular control proteins (SCPs)   261
Role of SCPs in floral morphogenesis   262
    Are SCPs involved in floral organ
        determination?   263
SCPs and coordination of development at
    the whole-plant level   267
Concluding remarks   270
Acknowledgements   270
References   270

Abstract

The complexity associated with post-translational processing, in terms of protein sorting and delivery is now well understood. Although such studies have been focused almost exclusively on the fate of proteins within the cell in which they are synthesized, recent studies indicate that it is time to broaden this focus to incorporate the concept of intercellular targeting of proteins. Direct evidence is now available that viral and endogenous proteins can be synthesized in a particular cell and subsequently transported into neighboring (or more distant) cells. Plasmodesmata, plasma membrane-lined cytoplasmic pores, are thought to establish the intercellular pathway responsible for this cell-to-cell trafficking of macromolecules (proteins and nucleic acids). These recent findings establish a new paradigm for understanding the manner in which higher plants exert control over developmental processes. We discuss the concept that programming of plant development involves supracellular control achieved by plasmodesmal trafficking of informational molecules, herein defined as supracellular control proteins (SCPs). This novel concept may explain why, in plants, cell fate is determined by position rather than cell lineage. Finally, the circulation of long-distance SCPs, within the phloem, may provide the mechanism by which the plant signals to the shoot apical meristem that it is time to switch to the reproductive phase of its development.

Introduction

In plant biology it is usual to contemplate the fate of a newly synthesized protein in terms of it being targeted to a specific region, organelle, or compartment (including the cell wall) associated with the cell in which this post-transcriptional event took place. Recent studies on both viral and endogenous proteins suggests that this viewpoint is in need of revision, as it has been demonstrated that proteins synthesized in a particular cell can be transported to neighboring (or more distant) cells located within the same tissue. Furthermore, a large class of newly identified proteins are synthesized in one cell type (companion cells) prior to their entry into the phloem long-distance transport system for delivery to distant organs! Plasmodesmata, plasma membrane-lined cytoplasmic pores, are thought to establish the intercellular pathway through which such macromolecular (protein and also nucleic acid) trafficking takes place. We here review the experimental basis for this paradigm and then proceed to advance new roles for this novel macromolecular trafficking pathway in terms of the orchestration of developmental processes.

Plasmodesmata and the symplasmic concept

To understand the fundamental role of plasmodesmata in plant biology, it is first necessary to briefly describe their formation and distribution throughout the plant. Cell initials, located within the shoot apical meristem, produce the cells that undergo further division and differentiation to form stems, axillary meristems and leaves (Fig. 1, panels A–C). During this process, primary plasmodesmata are formed across each developing cell plate to establish cytoplasmic continuity between cells in the same layer; secondary plasmodesmata are produced across the existing walls of the L1, L2 and L3 (Fig. 1, panel B), thereby ensuring cytoplasmic continuity between all cells of the developing tissues (for a comprehensive review of this process, see [53]). As individual cells expand, the degree to which their cytoplasm remains connected to neighboring cells depends upon their developmental fate. For example, mesophyll cells maintain intimate cytoplasmic contact either by modifying the existing primary plasmodesmata through the addition of secondary branches, or by the *de novo* formation of highly branched secondary plasmodesmata [19, 22]. At other locations within the mature leaf, plasmodesmal densities may be reduced

to low levels, with the actual pattern being a characteristic of the individual species [32, 98]. A typical example is illustrated in Fig. 1 (panels D and E). It is important to stress that in the entire body of the plant there is only one cell type that, at maturity, becomes completely isolated from its neighboring cells, and this is the guard cell of the stomatal complex (Fig. 1, panel E). This plant-wide cytoplasmic continuum, established by plasmodesmata, is termed the symplasm.

Plasmodesmal structure in relation to function

Ultrastructural studies have established that the plasma membrane of each plasmodesma exists as a cylindrical extension of the two adjoining cells. The central region of unmodified primary plasmodesmata is occupied by an appressed form of the endoplasmic reticulum (ER) which is connected to the ER in the neighboring cell(s) (Fig. 1, panel F). Thus, plasmodesmata establish both a cytoplasmic and an ER continuum that extends from the shoot apical meristem all the way down the plant axis. Clearly, regulation of this symplasmic pathway must be of paramount importance in terms of exerting developmental or physiological control at specific sites within the plant body.

Control over the type of molecules and ions that can pass between cells is established at two levels, but each appears to involve the same basic substructural elements. Detailed analyses of plasmodesmal substructure have been provided by Tilney *et al.* [95], Ding *et al.* [20], Lucas *et al.* [53] and Botha *et al.* [3]. Figure 1 (panels F and G) illustrates the basic structural model of the primary plasmodesma. Globular proteins (ca. 3 nm in diameter) appear to be embedded in the inner and outer leaflets of the plasma membrane and appressed ER, respectively [20]. These integral membrane proteins appear to be interlinked by another class of proteins which span across the cytoplasmic annulus that is located between the plasma membrane and the appressed ER. The physical arrangement of the proteins in this supramolecular complex divides the cytoplasmic annulus into 8–10 microchannels, each creating a tortuous intercellular path with a diameter of approx. 2–3 nm (Fig. 1, panels F and G).

Microinjection studies have established that the size exclusion limit (SEL) of these microchannels is in the range of 800–1000 Da [74, 94], which is consistent with a diameter of ca. 3 nm (see Fig. 1, Panel G). Hence, the creation of these microchannels within the cytoplasmic annulus appears to establish the first level

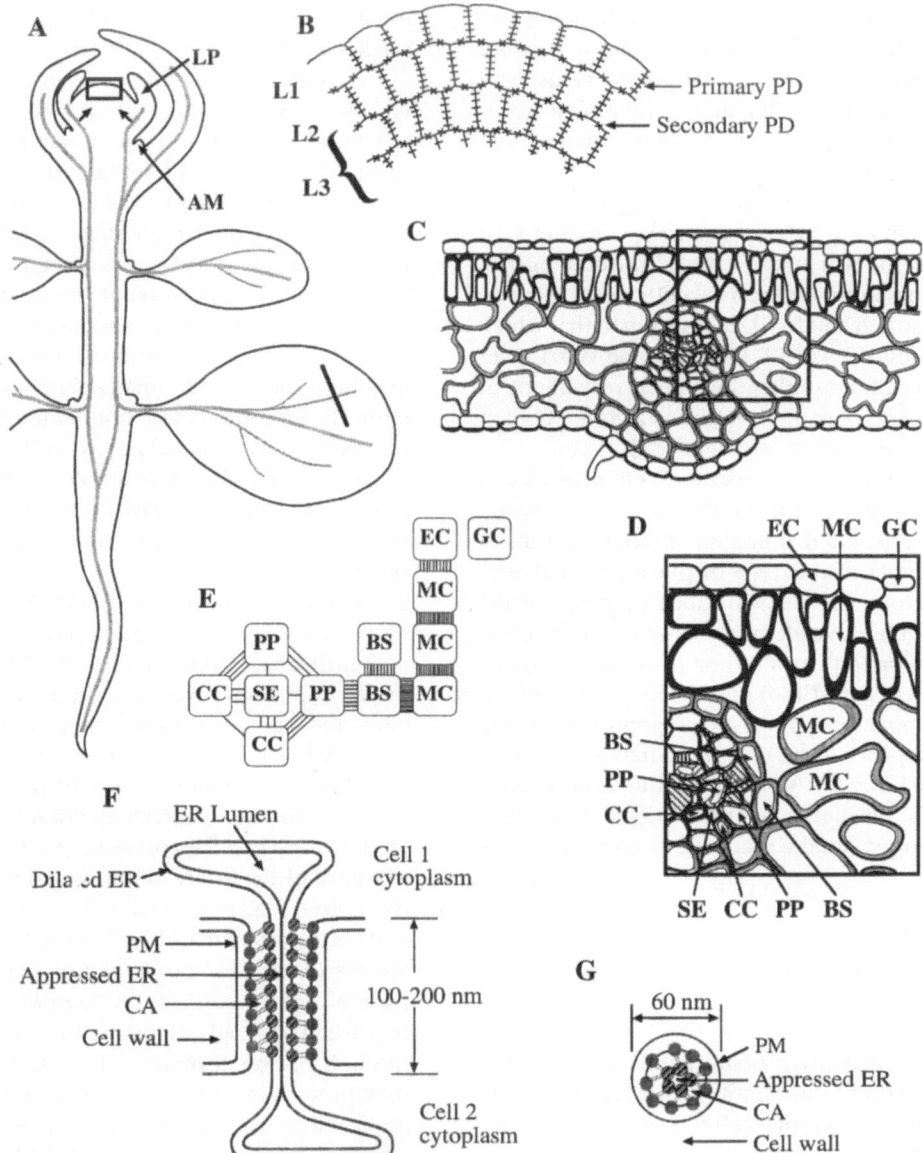

Figure 1. Schematic representation of the supracellular nature of higher plants. A. General arrangement of the plant axis, involving the shoot apical meristem (boxed region represents the central zone of the meristem), leaf primordia (LP), axillary meristems (AM), expanding leaves, the stem and mature leaves, and the root, with its associated root apical meristem. Vascular (phloem and xylem) tissues (in grey) interconnect all regions of the plant. Nutrients essential for growth of the meristem are delivered, via the phloem, into the region beneath the meristem proper. Transport of these nutrients, and communication signals delivered via the phloem, are thought to move from this region to the apex via the symplasm; this route is indicated by arrows. B. Central zone of the shoot apical meristem (boxed region in panel A), comprised of three layers L1, L2 and L3, illustrating the arrangement of the cells and the presence of primary and secondary plasmodesmata that interconnect cells within and between layers, respectively. C. Transverse section through the mature region of a leaf (solid line on leaf in panel A). D. Higher magnification of boxed region in panel C indicating the cell types associated with the vascular tissue and surrounding mesophyll; BS, bundle sheath cell; CC, companion cell; EC, epidermal cell; GC, guard cell; MC, mesophyll cell; PP, phloem parenchyma cell; SE, sieve element. E. Diagrammatic representation of the spatial distribution of plasmodesmata between the cell types illustrated in panel D. Relative density of plasmodesmata indicated by the number of lines connecting specific cell types. F. Diagrammatic representation of a longitudinal section through a primary plasmodesma. Globular integral membrane proteins (3 nm diameter) are located in the inner and outer leaflets of the plasma membrane and appressed ER, respectively, and these proteins are thought to be interconnected by special linking proteins [20]; CA, cytoplasmic annulus, PM, plasma membrane. G. Transverse section of a primary plasmodesma. Note that the CA is divided into a number of microchannels with diameters of ca. 2–3 nm. (Redrawn from [54] with permission.)

of control over what can move through the symplasmic route. Intercellular exchange of ions and small metabolites (sugars, organic acids, amino acids, and hormones that are all smaller than 1 kDa) can occur by diffusion. However, it is important to stress that this pathway operates under metabolic and physiological control, in that plasmodesmal SEL is regulated both by transcellular gradients in turgor pressure [70] and the cytosolic level of ATP [14]. Introduction of kinase inhibitors into mesophyll cells results in an increase in plasmodesmal SEL equivalent to that which occurs under reduced ATP (T. Fujiwara and W.J. Lucas, unpublished results), implicating the involvement of phosphorylation/dephosphorylation as a component of a plasmodesmal regulatory element [13].

Physiological (or wound-induced) changes in cell turgor can give rise to the establishment of special symplasmically isolated domains. Perhaps the most important example of this type of plasmodesmal control is found along the long-distance pathway of the phloem, where the sieve element-companion cell complex appears to operate as a turgor-restricted symplasmic domain. Here the SEL of plasmodesmata interconnecting the companion cells to the neighboring phloem parenchyma appears to be down-regulated to a value below 400 Da [71, 99]. Clearly, modulation of plasmodesmal SEL must represent a fundamental aspect of the manner in which the plant exerts control over its symplasm.

Plasmodesmata and cell-to-cell trafficking of viral proteins

In the absence of plasmodesmal proteins, the cytoplasmic annulus would be sufficiently wide (ca. 10 nm) to permit the uncontrolled intercellular exchange of proteins and mRNA. Thus, the microchannels in the cytoplasmic annulus must have evolved to regulate the cell-to-cell exchange of both small and large molecules. Until very recently, it was generally accepted that plasmodesmata allowed only the exchange of small metabolites and ions [74]. The only exception appeared to involve cell-to-cell movement of plant viruses, which was though to take place through what appeared to be modified plasmodesmata [55, 74]. Interestingly, it was studies on plant viruses that led to the discovery that plasmodesmata exert a second level of control over molecular exchange between cells. There is now an expanding body of experimental evidence that demonstrates, unequivocally, that plasmodesmata

mediate the cell-to-cell trafficking of both proteins and nucleic acids.

With the advances made by molecular virology, it is now clear that most plant viruses encode a nonstructural protein(s) that is essential for viral movement from the site of replication into the surrounding, uninfected cells [13, 17, 55, 63]. Details of the mechanism(s) by which these viral proteins potentiate transport of the viral infectious material have come from the integration of a number of experimental approaches. Initial studies focused on mutational analysis of these putative viral movement proteins (MPs), and this approach established that whereas viral replication (in protoplast) was unaffected by the absence of functional MP, the local cell-to-cell spread of the virus was inhibited [55]. For example, pioneering studies on tobacco mosaic virus (TMV) established that a 30 kDa protein was essential for the establishment of infection [16, 65]. Cytological studies indicated that during TMV infection this putative 30 kDa MP was localized to mesophyll plasmodesmata [96].

Further support for the role of the TMV MP in cell-to-cell movement was provided by studies on transgenic tobacco plants expressing a cDNA encoding the TMV MP. These transgenic plants could be infected by TMV that contained mutant forms of the MP. Of equal importance, microinjection studies performed on these TMV MP-expressing plants revealed that the presence of the MP resulted in a significant increase in mesophyll plasmodesmal SEL. from control values of 850 Da to greater than 10 kDa [109]. Collectively, these studies (infection, mutational analysis, immunolocalization and microinjection) supported the hypothesis that the TMV MP mediates the movement of TMV through plasmodesmata. The additional finding that mutations in the TMV coat protein block virion formation, but not infectivity [15, 25, 93], in conjunction with the discovery that the TMV MP has the capacity to bind to nucleic acids [109, 13, 72], provided support for the hypothesis that the TMV MP is capable of facilitating the cell-to-cell transport of unencapsidated vRNA [25, 26].

Direct evidence that a viral MP has the capacity to interact with plasmodesmata was gained through microinjection experiments in which the MP was introduced into the cytoplasm of mesophyll cells in control plants. Here, the viral MP was first expressed in *Escherichia coli*, then extracted from inclusion bodies and fluorescently labeled with fluorescein isothiocyanate (FITC-MP). Experiments of this type were performed on the 35 kDa MP of red clover necrotic mosaic virus (RCNMV), the 30 kDa MP of TMV, the 30 kDa MP of cucumber mosaic virus (CMV) (all single-stranded RNA viruses), and on the 33 kDa MP (BL1) of the bipartite, single-stranded DNA virus, bean dwarf mosaic virus (BDMV). Microinjection of the BDMV FITC-MP into bean mesophyll cells (a natural host for BDMV) resulted in the almost immediate spread of fluorescence into neighboring cells [68]. The FITC-BDMV MP continued to move out in a radial pattern away from the injected cell and, within ca. 2–4 min, a fluorescence pattern became established and remained stable for a considerable period (Fig. 2, panels A and B [68]). Identical responses were obtained when the FITC-MP of RCNMV, TMV or CMV was injected into mesophyll cells located within mature leaves of their natural host plants [21, 31, 102]. Control experiments using mutant forms of each MP that were incapable of supporting viral infection always resulted in the retention of the FITC-MP within the injected mesophyll cell. Collectively, these results provide unambiguous experimental proof that viral MPs can interact with the supramolecular complex of proteins within plasmodesmata to mediate their cell-to-cell transport.

In comparison to these direct FITC-MP microinjection experiments, a somewhat different response was recorded when 9.4 kDa FITC-dextran was injected into transgenic tobacco plants expressing the TMV MP. In these experiments, the 9.4 kDa FITC-dextran continued to move out into the surrounding cells to the point where it could no longer be detected, even with the extreme sensitivity of the Hamamatsu VIM analytical detection system used in these studies [109]. Clearly, the difference between these two experimental systems is that in the transgenic tobacco plants the TMV MP was being expressed to high levels in all mesophyll cells, whereas in the direct injection experiments performed on cowpea, bean and tobacco plants, only a very small quantity of MP was introduced into one cell. Hence, the restricted movement of the RCNMV, BDMV and CMV FITC-MP likely reflects the situation in which the concentration of MP eventually falls below a threshold level required for effective plasmodesmal interaction.

Coinjection into tobacco mesophyll cells of Escherichia coli-synthesized TMV MP with FITC-dextrans of known molecular weight resulted in the expected increase in plasmodesmal SEL. Indeed, in these experiments 20 kDa FITC-dextran moved from cell to cell [102]. Experiments performed with the MP of RCNMV, CMV and BDMV confirmed that these viral MPs also have the capacity to interact with meso-

phyll plasmodesmata to increase the SEL to greater than 10 but less than 20 kDa [21, 31, 68]. Thus, it appears that an increase in SEL may be essential for macromolecular transport through plasmodesmata.

MP-mediated trafficking of vRNA/DNA through plasmodesmata

The ability of the viral MP to mediate cell-to-cell transport of vRNA/DNA has also been established using microinjection procedures. In these experiments, RNA/DNA was fluorescently labeled with the nucleic acid dye, TOTO-1 [33]. Introduction of TOTO-labeled RNA or DNA alone always resulted in retention of fluorescence in the injected cell. However, coinjection of TOTO-labeled single-stranded RNA with, for example, RCNMV MP, resulted in rapid movement of fluorescence from the target cell into the surrounding mesophyll [31]. The specificity of this MP-RNA interaction was established by the fact that the RCNMV MP was not able to potentiate cell-to-cell transport of either single- or double-stranded DNA. In contrast, the BDMV MP was found to be competent in potentiating the cell-to-cell transport of any form of double-stranded DNA, but not of single-stranded DNA nor single- or double-stranded RNA [68]. We shall return to this non-sequence-specific aspect of viral MP-mediated trafficking of nucleic acids when we discuss cell-to-cell transport of endogenous mRNA.

Although there is now ample evidence that virally encoded proteins can be synthesized in the cytosol of one cell and subsequently move through plasmodesmata into many of the surrounding cells, the argument can be made that these trafficking events are orchestrated by the virus and have little, if anything, to do with cell-to-cell transport of endogenous proteins. Such a position could be defended if the time course for the transport of injected FITC-MP or TOTO-RNA/DNA plus MP was similar to that observed for the cell-to-cell spread of viral infection. The difference between the two events, being cell-to-cell transport within seconds when fluorescently labeled probes are injected and hours to invade the neighboring cells when virus particles are microinjected or inoculated [18, 24], provides strong support for the hypothesis that the viral MP 'hitches a ride' on a fully functional endogenous plasmodesmal macromolecular trafficking system (Fig. 2, panel C). An expanding body of both direct and indirect evidence is now accumulating in support of this novel hypothesis. The important ques-

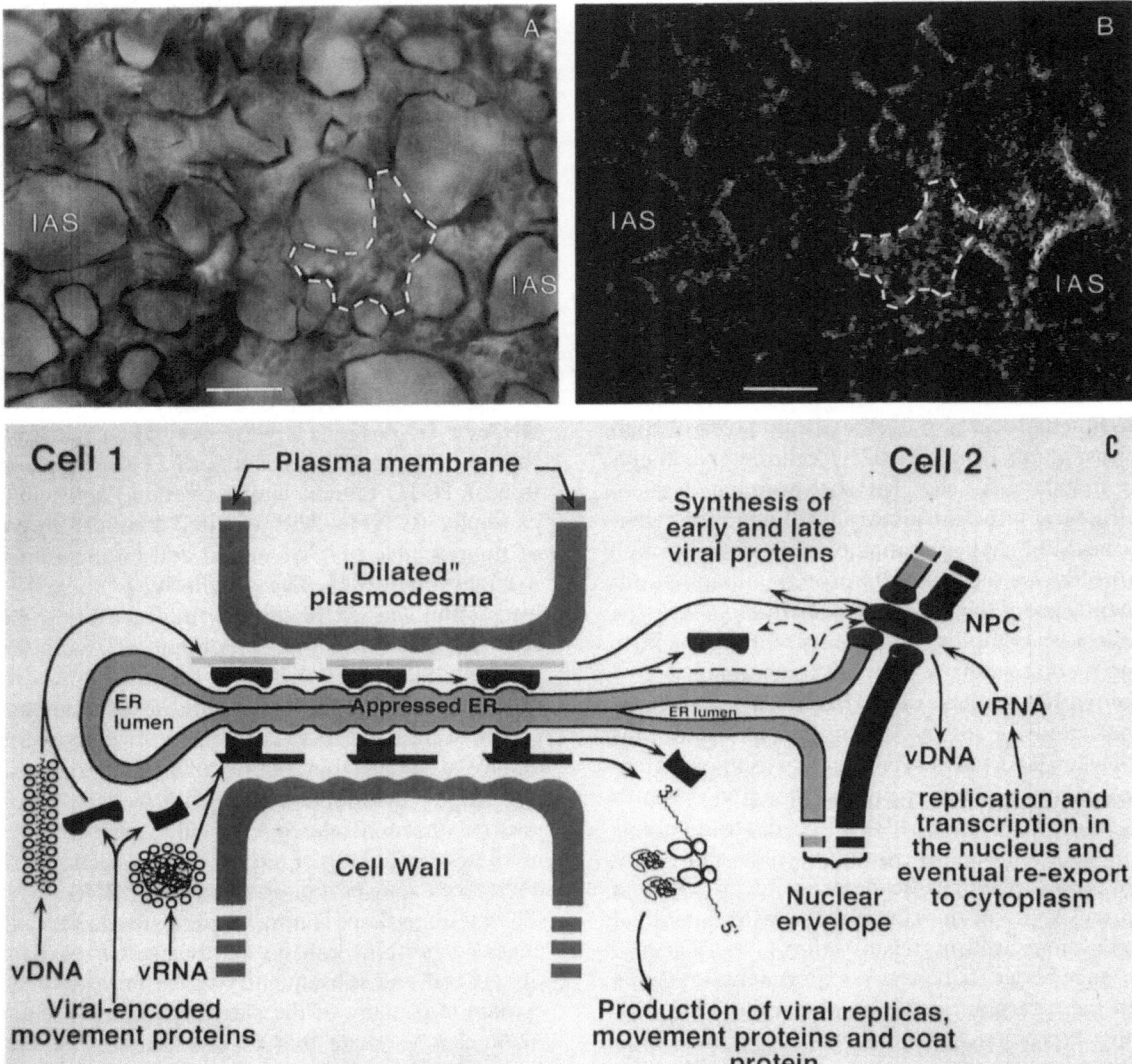

Figure 2. Viral MPs mediate in the plasmodesmal trafficking of infectious vRNA or vDNA. A. Bright-field image of bean mesophyll tissue in which FITC-labeled BDMV MP was injected (target cell outlined by white broken line). B. Steady-state fluorescence image acquired 4 min after FITC-labeled BDMV MP was injected into the target mesophyll cell indicated in panel A. Scale bars = 20 μm; IAS, intercellular air spaces. C. Model of the mechanisms by which vDNA and vRNA, synthesized in an infected plant cell, may be transported through plasmodesmata into surrounding uninfected cells. Viral-encoded MPs are targeted to, and interact with, the substructural components of the plasmodesma resulting in the dilation of the micro-channels and transport of the MP into the neighboring cell. When the MP is bound to its respective viral genomic infectious material (vDNA or vRNA) this complex is also transported through the plasmodesma. During this transport step, the infectious material may be unwound, either by an interaction between the vDNA/vRNA and the endogenous plasmodesmal proteins (structural and macromolecular trafficking components), the MP, or both. (Panels A and B from Noueiry *et al.* [68], with permission; panel C from Lucas and Gilbertson [55], with permission.)

tion that needs to be resolved is how and when did the viral MP acquire the ability to potentiate plasmodesmal transport of its vRNA/DNA?

Molecular determinants of viral protein trafficking through plasmodesmata

The sorting of newly synthesized proteins to the correct cellular, or supracellular compartment involves both molecular chaperones and specific targeting sequences that are embedded in the protein (see paper by Bar-Pekd, Basham and Raikhel, this issue). Here it is interesting to note that such targeting, or transit peptide sequences appear to be reasonably conserved across evolutionary boundaries. For example, although there are differences in the nuclear localization sequences (NLS) required for protein import into the nucleus in animal and plant systems, each NLS appears to retain function in the heterologous condition. Initial studies on viral MPs sought to identify equivalent simple sequence motifs responsible for plasmodesmal trafficking [49, 64]. Analysis of a large number of MPs, over a range of plant viral groups, has failed to reveal the existence of such a conserved motif. The underlying basis for this situation likely reflects the greater degree of complexity associated with the regulation of endogenous plasmodesmal macromolecular trafficking and the presence of specialized symplasmic domains [54].

Mutations engineered within specific regions of the viral MP do cause a loss of function, in that such mutant forms no longer appear capable of interacting with plasmodesmata to traffic from cell to cell (see [55] and references therein). However, although each MP has an identifiable region required for plasmodesmal targeting and SEL increase (Fig. 3), insufficient information is currently available concerning the tertiary structure of these MPs to permit the identification of possible common structural domains. Irrespective of this situation, there is now ample evidence that molecular interactions between the viral MP and host factors, including the plasmodesmal supramolecular complex, determine both host range and the tissues within the host that can be infected [55].

Numerous viruses enter the plant through insect vectors that feed on the phloem. Of these, some can egress, via plasmodesmal trafficking, from the phloem long-distance transport pathway all the way to cells in the periphery of the infected tissue or organ. Other viruses are limited to the vascular, or phloem tissues, a condition consistent with the hypothesis that the MPs of such viruses lack the specific molecular determinant(s) for trafficking across the plasmodesmata located at the bundle sheath-mesophyll boundary [55, 78]. Additional support for the hypothesis that plasmodesmata form special symplasmic boundaries was also provided by recent infectivity studies performed on TMV [23, 67] and RCNMV. Infectious RCNMV clones expressing various mutant forms of the MP were inoculated onto the same type of leaf tissue used for

FITC-MP and TOTO-labeled RCNMV RNA microinjection experiments [31]. Detailed immunolocalization studies of these infected plants revealed that, whereas specific mutations in the RCNMV MP did not interfere with the ability of the RCNMV MP to engage in the trafficking of its vRNA through mesophyll and bundle sheath plasmodesmata, viral entry into the companion cell-sieve element complex could be blocked (Y. Wang et al., unpublished results).

Collectively, results of this nature support the hypothesis that plasmodesmata have the capacity to divide the symplasm into specific domains and that macromolecular trafficking from one such domain into another is highly regulated by plasmodesmal protein-MP (endogenous protein) interactions. The importance of this capacity of plasmodesmata, in terms of establishing symplasmic domains, will now be examined from the viewpoint of exerting supracellular control over plant development.

Role of macromolecular trafficking in plant development

It has long been appreciated that cell fate in developing plant tissues is determined by position rather than by lineage [37]. Thus, the orchestration of development must involve the capacity to engage in the exchange of positional information, which directs transcriptional/post-transcriptional events essential for the imprinting of cell fate. Given the increasing body of experimental evidence that now supports the concept that plasmodesmata mediate in the cell-to-cell transport (exchange) of proteins and nucleic acids, it seems highly likely that the delivery of this putative positional information will be achieved by macromolecular trafficking through plasmodesmata [53, 56]. This system would form a link between hormone action and cell fate. An emerging body of direct and indirect experimental evidence provides strong support for this hypothesis.

KNOTTED1, a maize transcription factor, functions non-cell-autonomously

Dominant mutations at the *knotted1 (kn1)* locus of maize result in the formation of knot-like outgrowths and ectopic ligule fringes on leaf blades [85]. Detailed mutant analysis revealed that this phenotype was caused by ectopic expression of *kn1* around the lateral veins of the malformed leaves. In wild-type plants,

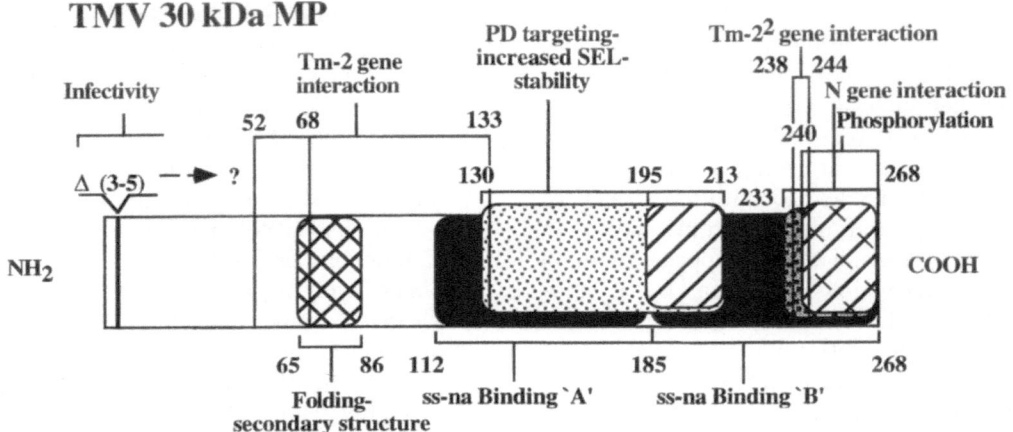

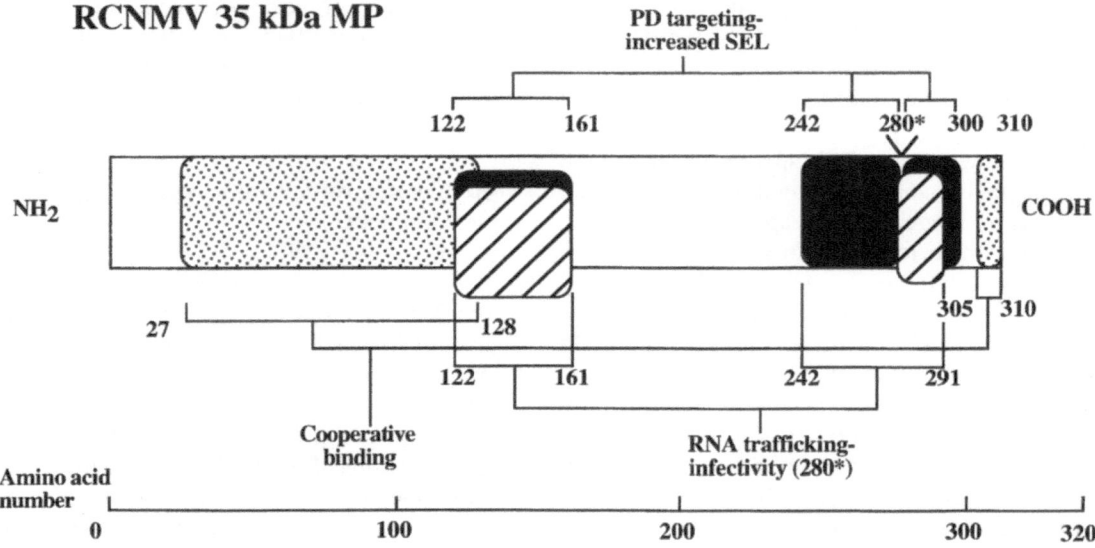

Figure 3. Functional domains identified within the TMV and RCNMV MPs. Boxed, shaded, and hatched regions denote amino acid regions associated with the indicated function(s). Numbers correspond to the specific amino acids involved in the function(s) or establishing the boundaries of the functional domains. PD, plasmodesmata; SEL, size exclusion limit; ss-na, single-stranded nucleic acid. (From Lucas and Gilbertson [55], with permission.)

KNOTTED1 (*KN1*) is expressed only in the meristem and developing vascular bundles [87], where it is thought to be involved in the maintenance of the indeterminate state [44, 86, 87]. As soon as cells on the periphery of the meristem become designated to form organs, they cease to express *KN1* (Fig. 4, panel A [44]). Thus, the formation of outgrowths on leaves in the *kn1* mutants seems to be caused by the dedifferentiation of cells in the leaf blade, due to the presence of KN1 [86].

Analysis of sectors of wild-type tissue in *kn1* mutant maize plants revealed that the ectopic expression of *kn1* in only the middle mesophyll-bundle sheath layer of the leaf was sufficient to produce the knotted phenotype. However, the knots and ectopic ligules formed from all layers of the leaf, including the normally determinate cells of the epidermis. Thus, KN1 appears to be able to act non-cell-autonomously: expression in one set of cells leads to a change of growth pattern in the surrounding cells [85]. As KN1 has been identified as a homeobox containing transcription factor [100], the movement of KN1 across cellular boundaries could account for its observed non-cell-autonomous effects [44].

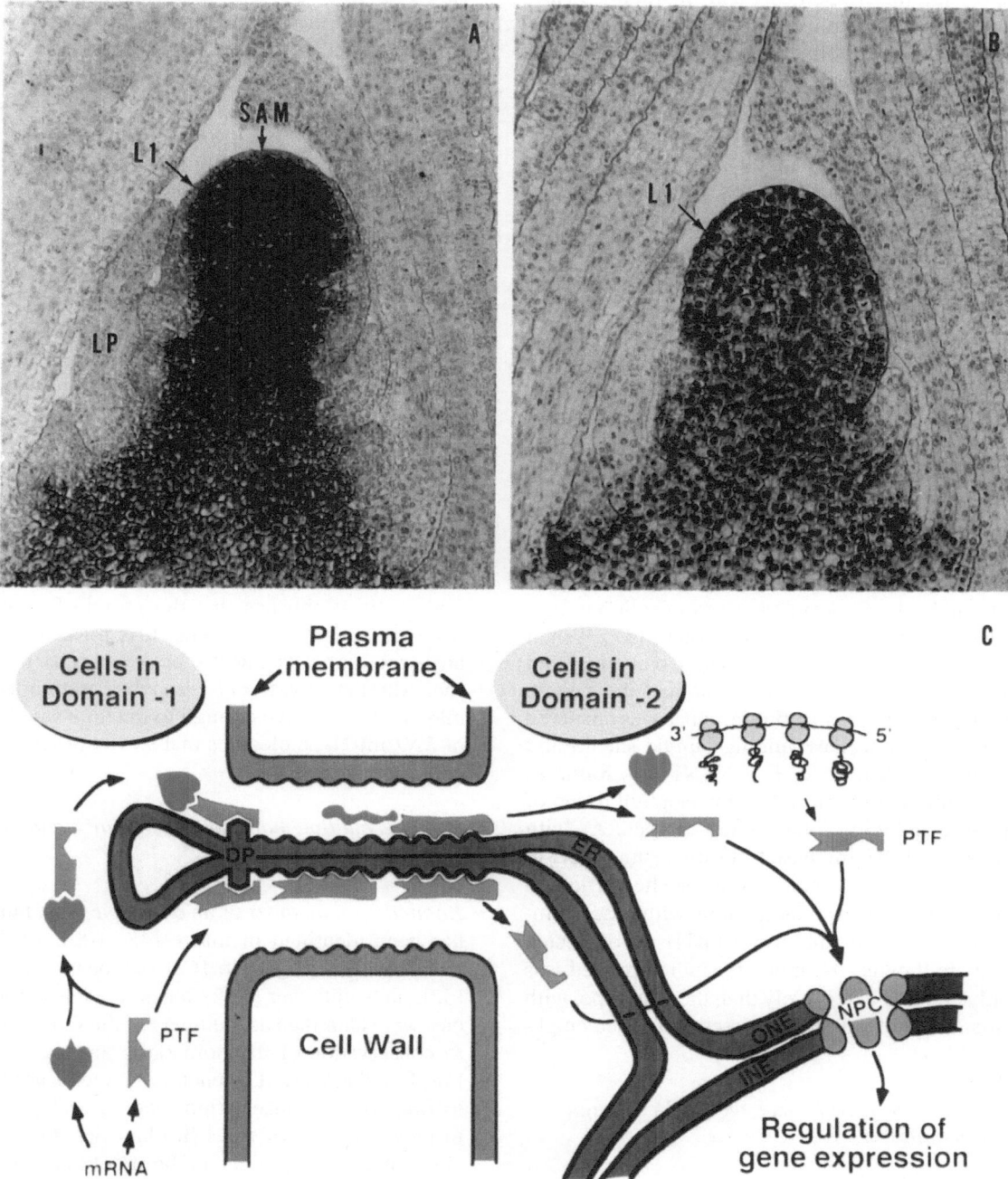

Figure 4. Plant-encoded transcription factors, such as KNOTTED1, traffic through plasmodesmata to mediate supracellular control over developmental processes. A and B. Serial sections of a maize vegetative shoot apex, processed for *in situ* hybridization for *KN1* mRNA and immunolocalization of KN1, respectively, reveal the presence of KN1 in L1 cells in which *KN1* mRNA is not detected. C. Proposed involvement of plasmodesmata in cell-to-cell trafficking of macromolecules. An mRNA encoding a plant transcription factor (PTF), such as *KN1*, is transcribed in cells located in domain-1. The encoded protein has a plasmodesmal localization motif that allows it to bind to a putative plasmodesmal docking protein (DP) positioned at the terminal region of a microchannel. Binding may result in a relaxation in the crosslinking proteins associated with this microchannel, thereby causing an increase in SEL and allowing the PTF access to the internal translocation machinery. Chaperones may be required to deliver the PTF to the DP and to renature it once it is transported along the length of the microchannel and released into the cytoplasm of cells located in domain-2 (or domain-3). The presence of a nuclear localization sequence allows the PTF to enter the nucleus, via the nuclear pore complex (NPC). (Note that we will shortly define PTFs that can traffic through plasmodesmata as supracellular control proteins, or SCPs.) SAM, shoot apical meristem; LP, leaf primordia; L1, outermost layer of SAM; ONE, outer nuclear envelope; INE, inner nuclear envelope. (Reproduced from Lucas *et al.* [57] (A and B) and Lucas [56] (C), with permission.)

Support for this hypothesis can be found in the meristem of wild-type maize plants, where immunolocalization studies established the presence of KN1 in cells in which *KN1* mRNA could not be detected by *in situ* hybridization analysis [44, 57]. Figure 4 depicts the expression pattern of *KN1* mRNA (panel A) and the cellular location of KN1 (panel B) in the maize vegetative meristem. Whereas *KN1* mRNA is not detected in the L1 layer of the meristem, KN1 clearly accumulates in the nuclei of the L1 cells. Hence, in both the *kn1* leaves and in the wild-type meristem, KN1 appears to exert control over transcription in cells where it does not seem to be translated. The logical conclusion that arises from these observations is that KN1 has a capacity equivalent to viral MPs, in that it can interact with plasmodesmata to traffic from the vascular cells, through the mesophyll to the epidermal cells, or from the L2 to the L1 cells in the meristem (Fig. 4, panel C).

KN1 can traffic through plasmodesmata!

To directly demonstrate that KN1 has the ability to traffic through plasmodesmata, microinjection experiments were performed with this protein [57]. When *E. coli*-expressed KN1 was FITC-labeled and injected into maize mesophyll cells that adjoined the bundle sheath cells of a lateral vein, fluorescence was observed to rapidly move into surrounding bundle sheath and mesophyll cells. Likewise, FITC-KN1 was found to traffic very efficiently through the mesophyll tissue of other plants, including tobacco and cowpea. As with the viral MPs, KN1 increased plasmodesmal SEL to greater than 20 kDa. To demonstrate that the trafficking of KN1 involves specific interactions with other components in the cell, mutant forms of KN1 were injected into mesophyll cells. All mutants studied moved less efficiently and less extensively than the wild type, with one being incapable of trafficking even into the neighboring cell [57].

KN1 can selectively traffic its own mRNA through plasmodesmata

Given the similarities between the trafficking of KN1 and that of viral MPs, experiments were undertaken to determine whether KN1 was also capable of trafficking RNA. Coinjection into tobacco mesophyll cells of unlabeled KN1 with TOTO-labeled *KN1* mRNA revealed that KN1 could indeed mediate the cell-to-cell movement of its own RNA! When *KN1* mRNA was injected in the absence of KN1, no trafficking

occurred. Interestingly, KN1 was not able to mediate the trafficking of either CMV RNA (a nucleic acid which has previously been shown to traffic when coinjected with its CMV MP [21]) or the *KN1* antisense RNA. In contrast, the CMV MP could mediate trafficking of both its own vRNA and the *KN1* mRNA [57]. Thus, KN1 has a unique property not yet identified in any viral MP, in that it has the ability to bind to a specific RNA sequence, thereby allowing it to engage in selective trafficking of its own mRNA. A virus can usurp control over the plant's translational machinery [103] to the point where it can amplify its vRNA to a level where it is the most abundant RNA species in the cytoplasm of the infected cell. Therefore, viral MPs may have lost specificity of RNA binding, as it may no longer be of value, in evolutionary terms.

Whether the ability of KN1 to traffic its own RNA is an integral part of the protein's function remains to be determined. As mentioned above, *in situ* hybridization studies did not detect the presence of *KN1* mRNA in the L1 cells of maize vegetative or inflorescence meristems (see Fig. 4). The presence of KN1 in these same cells establishes that the protein has trafficked from the underlying L2 cells. It is possible that KN1 does not traffic its mRNA between cells in this tissue. Alternatively, the *in situ* hybridization technique may not be sensitive enough to detect a small number of *KN1* mRNA molecules that were trafficked into L1 cells by KN1.

KN1 has multiple homologues in maize and other plant species

Knotted1 is a member of an extensive gene family that has been identified in maize [48, 100], *Arabidopsis* [52], tobacco [28], rice [62], soybean [59], tomato [52], and sunflower [52]. Genes in each of these species were identified as members of the same family due to conservation of the homeobox and ELK domains [48] found in KN1. The encoded proteins are all likely to function as transcription factors, and they appear to play a variety of roles in plant development. The OSH1 protein, expressed in the shoot tips of rice seedlings, is a very closely related homologue of KN1 [48, 62]. When rice protoplasts are transformed with plasmids carrying the *OSH1* gene, they produce plants which exhibit a leaf phenotype closely resembling that of knotted maize plants [62]. A closely related *KN1* homologue in *Arabidopsis*, *KNAT1*, also shows parallels with KN1 in its function in that it is expressed in the vegetative meristem and is excluded from the cells that form organ primordia [52]. Furthermore, ectopic expression of *KNAT1*, driven by the viral 35S promoter, resulted in altered rosette and cauline leaf morphology. Interestingly, the same wrinkled, lobed leaf phenotype could be produced by ectopic expression of either *KNAT1* or *KN1* [52].

Transgenic tobacco expressing a 35S-*KN1* construct also exhibits aberrant leaf morphology. Here, moderate expression of *KN1* results in wrinkled, misshapen leaves, whereas high expression leads to the formation of shoots that arise from the leaf surface [86]. A similar range of phenotypes can be produced by the ectopic expression of *KNAT1* [52] or *OSH1* [46] in transgenic tobacco lines. Thus, the functions of KN1 and its homologues appear to be highly conserved. Considering the ability of KN1 to traffic in a range of plant species and tissue types, plasmodesmal transport and non-cell-autonomous control over development may well be common facets of the functions of KN1 homologues.

A new class of plant genes: the supracellular control proteins (SCPs)

Members of the *KN1* gene family may represent a new class of plant proteins whose site(s) of action can extend beyond the cell in which synthesis took place. We here define such molecules that can both traffic through plasmodesmata and control gene expression (at a site distant from the point of synthesis) as supracellular control proteins (SCPs). Control over gene expression by an SCP is envisaged to occur at one of three levels. At the DNA level, an SCP could directly control transcription, or it could activate (or deactivate) a transcription factor. Such modification to transcription factor activity could occur through dimerization, phosphorylation (or dephosphorylation), or control over the location of the transcription factor within the cell. At the RNA level, an SCP could control gene expression by directing the intercellular trafficking of mRNA, by controlling translation, or modifying the rate of mRNA degradation. As above, these functions could occur directly, or indirectly, through the activation (or deactivation) of a second protein. At the protein level, an SCP could activate, or deactivate, the function of a protein in any one of the above-mentioned ways. We further propose that an SCP may often require a cofactor(s) to mediate its function(s). An SCP cofactor is defined as a molecule that is not itself trafficked, but can control either the trafficking or the supracellular activity of an SCP.

Various functions that may be incorporated into an individual SCP or SCP cofactor are presented in Fig. 5. Every SCP must contain a domain that allows it to interact with plasmodesmata in order to traffic from cell to cell. This domain may necessarily include a second domain that alters plasmodesmal SEL, as we have not yet encountered proteins which move through plasmodesmata without an increase in SEL. (It will be of great interest to determine the evolutionary relationship between these SCP domains and the functionally equivalent domains for viral MPs.) The remainder of the SCP could contain a range of combinations of the other functional domains, as illustrated in Fig. 5; examples are presented for KN1 and some putative SCPs that will be discussed in detail below. SCP cofactors could be made up of some of the same domains as SCPs, but these would not include the plasmodesmal trafficking domain. Numerous functions can be envisaged for an SCP cofactor. One important role may be associated with the intracellular localization of an SCP. For example, an SCP such as KN1 must be able to both traffic from cell to cell *and* control transcription. One set of cofactors might be involved in the delivery of KN1 to the plasmodesma for trafficking, while another set may mediate in the delivery of KN1 to the nucleus for transcriptional regulation.

Control over nuclear localization has been shown to play an important role in *Drosophila* development, where the ratio of nuclear to cytoplasmic DORSAL protein determines dorsal-ventral polarity [89]. Work with the COP1 transcription factor of *Arabidopsis* has indicated that control over nuclear localization may also play a role in plant development [101]. Thus, in a given cell, the relative affinity of KN1 for plasmodesmal trafficking versus nuclear import could be determined by the relative levels of cofactors which chaperone KN1 along the cytoskeleton on a plasmodesmal-directed route (Fig. 5, putative cofactor N1) and those which chaperone KN1 along a nuclear-directed route. Alternatively, nuclear versus plasmodesmal localization of KN1 could depend on regulation of an SCP cofactor that contains an essential chaperone motif. Such a cofactor might be a nuclear-directed chaperone that is activated by phosphorylation (Fig. 5, putative cofactor N2). When the activated cofactor is present in a cell, KN1 would be localized to the nucleus; when the cofactor is absent or dephosphorylated, KN1 would traffic to neighboring cells.

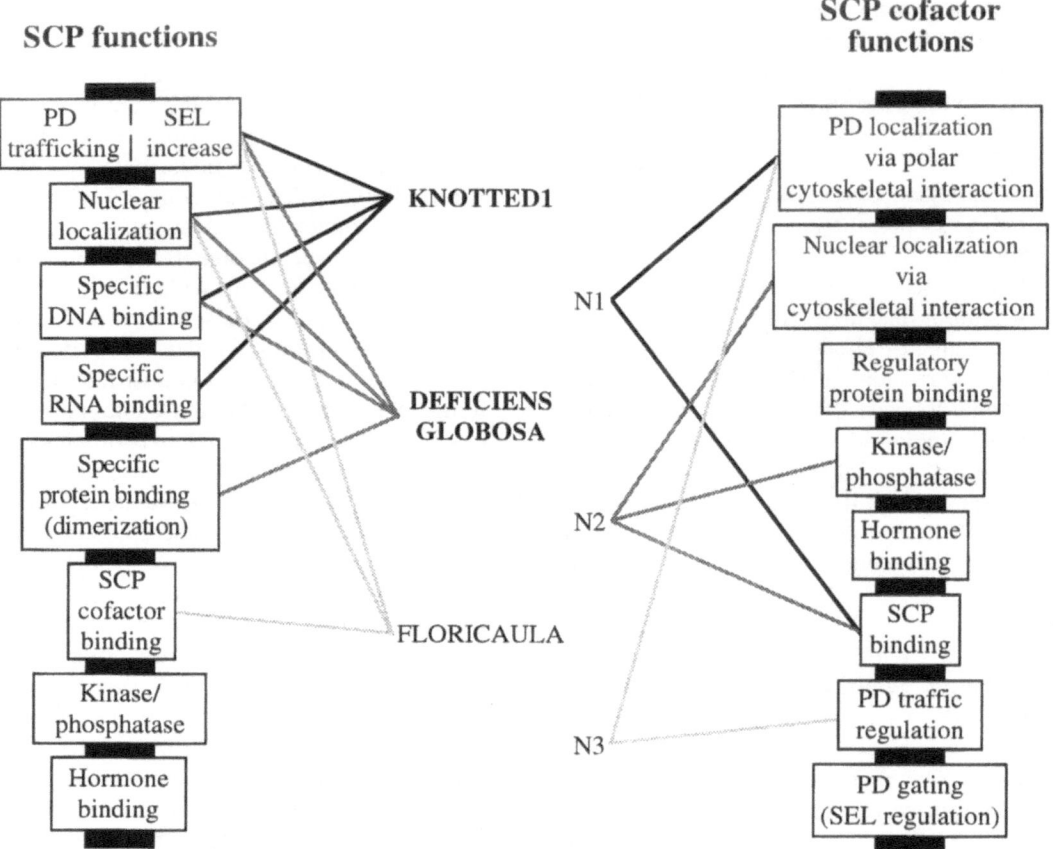

Figure 5. Functional domains contributing to the evolution of SCPs and SCP cofactors. A wide range of functions for SCPs and SCP cofactors may have evolved through genetic rearrangement of DNA segments encoding peptides with the functions shown. The insertion of peptides capable of regulatory interactions or specific nucleic acid binding, into proteins capable of intercellular trafficking would create an SCP. Further evolution of these SCPs would then give the plant a novel means of coordinating development within tissues. Specific mutations within plasmodesmal trafficking domains may have allowed some proteins to interact with plasmodesmata without trafficking themselves. A peptide with this capability (or the capability to bind trafficking proteins) that subsequently gained a regulatory domain would create an SCP cofactor which may have led to tighter control over the timing of intercellular trafficking through a specific tissue. The identified SCPs (bold type), putative SCPs, and SCP cofactors (plain type) may have evolved from the juxtaposition of the functional domains indicated by the connecting lines.

Another role for SCP cofactors could be in mediating general control over plasmodesmal trafficking. For instance, a specific cofactor might bind tightly to a putative plasmodesmal docking protein (see Fig. 4, panel C) and prevent all macromolecular trafficking (Fig. 5, putative cofactor N3). Expression of such a cofactor in a cell would isolate it from macromolecular signaling within its developmental domain. This cell (or a file of such cells) might then proceed to divide to form a supracellular domain of its (their) own. An opposite role could be played by a cofactor which activates plasmodesmal trafficking of macromolecules. SCPs translated in a specific cell might

accumulate while the plasmodesmal binding sites are blocked. Upon expression of the activating cofactor, the plasmodesmata would become competent to traffic, allowing the release of a wave of SCPs from that cell, or group of cells.

Role of SCPs in floral morphogenesis

Several examples of non-cell-autonomous control over development in the floral meristem have been documented. Work with tomato periclinal chimeras has shown that the inner layers of the meristem can determine the meristem size and organ numbers of the developing flower when the L1 has a different genotype from the L2 and L3, or when the L1 and L2 have a different genotype from the L3 [92]. The genotype of the L2 and L3 can also direct the formation of normal petals when the L1 carries the *lateral suppressor* mutation (*lateral suppressor* mutants fail to form petals due to the lack of initiation of second whorl organs (see Fig. 7)) [91]. Furthermore, studies of a camellia periclinal chimera have demonstrated that the L1 layer of the floral meristem can direct the development of reproductive organs in a flower that contains L2 and L3 layers from a variety that does not form stamens or carpels [90]. These examples all support the hypothesis that a control factor, produced by one layer of the meristem, can traffic into another layer to direct development.

Work with *Antirrhinum majus* chimeras provides strong evidence that such non-cell-autonomous control over floral development involves macromolecules rather than small signal molecules. *Antirrhinum* plants with loss of function mutations at the *floricaula* (*flo*) locus produce secondary inflorescences in place of flowers [9]. Thus, the plant requires wild-type FLO expression to create a floral meristem. Chimeric plants produced by the reversion of transposon-induced mutations at the *flo* locus have recently been described [8, 36]. As illustrated in Fig. 6 (panel A), restoration of the wild-type *FLO* gene in the L1 layer alone allows the formation of almost normal flowers. This demonstrates that a mobile signal can pass from the L1 layer to the inner cells of the meristem. Interestingly, a reversion in the L3 is not as effective as a reversion in the L1 in terms of restoring a normal phenotype. Revertants in L3 do produce flowers, but these flowers are severely malformed. An L2 revertant shows a phenotype that is intermediate between the phenotypes of the L1 and L3 revertants. These findings suggest that the signal acting between cell layers moves in a polar manner. Because polar movement over these distances cannot be a result of diffusion, it is unlikely that the signal is a small, diffusible molecule. This leads us to the question of whether FLO could be an SCP?

It has been suggested that FLO may be a transcription factor [9], and its *Arabidopsis* homolog, LEAFY, has been shown to localize to the nucleus of cells in the floral meristem [51, 106]. Hence, following translation in the L1 layer, FLO might undergo plasmodesmal trafficking into the underlying L2 and L3 cells, where it could enter the nucleus to direct down-stream genes involved in control over normal flower development (Fig. 6, panel B). This trafficking could be regulated

in a polar manner, making trafficking from the L3 to the L2 and L1 much less efficient. Polar trafficking of a macromolecule could be accomplished in different ways. An SCP cofactor could chaperone FLO along cytoskeletal elements toward plasmodesmata located across a specific wall of the cell. Alternatively, plasmodesmata along the various walls of a cell could have different affinities for specific SCPs. If FLO does not itself have the ability to traffic through plasmodesmata, it might act upstream of an SCP which would traffic in the manner described for FLO (Fig. 6, panel B).

Although the expression of *FLO* mRNA in the L1 layer alone is sufficient to direct flower development, *FLO* is normally expressed in all layers of the floral meristem. This fact makes the trafficking of FLO, or a downstream SCP, seem redundant, because in a wild-type plant the developmental program initiated by the activity of FLO should be activated autonomously in all cells of the floral meristem. However, normal development depends on an intricate network of interacting proteins, hormones, and intracellular signals that must be perfectly synchronized in space and time. The plant ultimately regulates this network by modulating the specific concentration of each of the interacting molecules in each cell. We propose that the intercellular trafficking of SCPs provides the plant with an additional 'fail-safe' method of regulating the cell-to-cell distribution and concentration of specific proteins. Within developing domains of the plant, such as the floral meristem, important control molecules must be turned on in a coordinated fashion within the entire domain. Giving key molecules the ability to traffic, from cell to cell, would ensure that each and every cell has the same concentration of the control factor and will thus be fully competent to develop coordinately with its surrounding cells.

Are SCPs involved in floral organ determination?

The floral meristem develops into a flower by producing four concentric whorls of organs: sepals, petals, stamens, and carpels. Genetic studies have begun to dissect the manner in which a plant establishes the identity of the floral meristem and of each type of floral organ. In Table 1 we list genes from *Arabidopsis thaliana* and *Antirrhinum majus* that are involved in floral meristem development [see 58, 69, 110 for topical reviews]. The functions of these genes have been deduced from the analysis of mutants, and in most cases the mRNA expression pattern of the gene corresponds to its proposed site of function. A model for

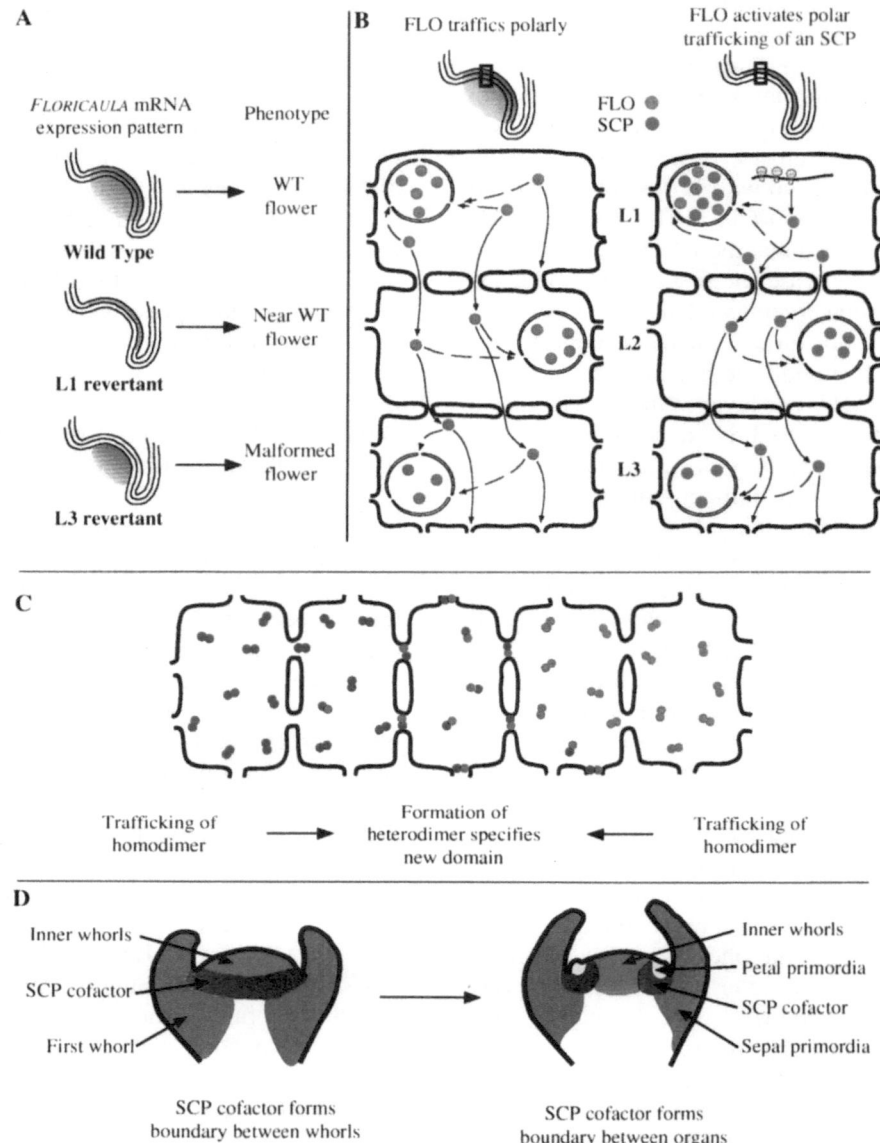

Figure 6. Models for the functions of SCPs and SCP cofactors in the floral meristem. A. Illustration of *FLORICAULA* mRNA expression patterns and the resulting floral phenotypes in wild-type *Antirrhinum* and in periclinal chimeras formed by the reversion of transposon-induced mutations in *floricaula* mutants. B. Two models for the polar intercellular communication that leads to restoration of normal flower development in Ll revertants. Either FLO itself traffics into the L2 and L3 layers, to direct development, or FLO activates an unidentified downstream *SCP* which then traffics to the inner layers. Solid arrows indicate a strong interaction with an SCP cofactor which directs trafficking in a polar manner; dashed arrows indicate a weaker interaction with an SCP cofactor that directs nuclear localization. C. Specification of a new supracellular domain, or a boundary between domains, by the formation of a heterodimer from two SCPs trafficking towards each other. Dimer formation between SCPs may be a factor affecting both the DNA binding specificity and the plasmodesmal trafficking ability of MADS-box proteins. When intercellular trafficking results in altered concentrations of MADS-box proteins in specific cells, new heterodimers may form and they could then alter the developmental pathway of the cells or isolate the cells from their neighbors by shutting down plasmodesmal trafficking. D. Isolation of supracellular domains by a boundary-forming SCP cofactor. Proper development of whorls, and organs within whorls, may depend on barriers that eliminate trafficking between supracellular domains. Such barriers could be produced by cofactors which bind to plasmodesmal docking proteins to prevent all macromolecular trafficking between domains.

Outer whorls   Inner whorls

Whorl 1   Whorl 2   Whorl 3   Whorl 4

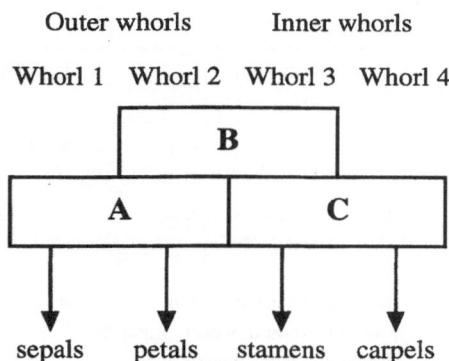

sepals   petals   stamens   carpels

Figure 7. Schematic presentation of the ABC model of floral organ determination. Within the developing floral meristem, organ identity genes fall into three categories, designated A, B, and C. Expression of these genes in the overlapping pattern shown gives rise to four unique domains within the meristem, each of which develops into a whorl of organs of the indicated type. Regulatory interactions between genes in the different classes appear to play a role in maintaining the patterns of expression. Such interactions, and the initial establishment of the four domains, may involve intercellular trafficking of macromolecules, via plasmodesmata.

floral organ specification has been developed, and a simplified version is presented in Fig. 7. This model proposes three genetic functions, designated A, B, and C, within the developing floral meristem; organ primordia expressing functions A, A+B, B+C, and C give rise to sepals, petals, stamens, and carpels, respectively.

Regulatory interactions maintaining the boundaries of expression for some of the floral identity genes have been identified. For instance, antagonistic interactions between class A and class C genes are thought to maintain the boundary between the outer and inner whorls. Thus, the domain of *AP1* expression seems to be limited to the outer two whorls by the expression of *AG*, as *AP1* RNA expression extends into whorls three and four in *ag* mutants [35]. Conversely, *AP2* seems to downregulate the expression of *AG* in the outer two whorls; i.e., *AG* mRNA expression expands into the outer whorls in *ap2* mutants [27]. This second interaction may involve an unidentified gene(s) in addition to *AP2*, because *AP2* is expressed in all four whorls but only limits *AG* expression in whorls one and two [45]. The inner boundary of *B* gene expression seems to be maintained by the expression of *SUP*, because *sup* mutants show an invasion (expansion) of *B* gene activity into the fourth whorl [5]. Thus, it is clear that different organ identity genes can regulate each other. However, the exact nature of these regulatory interactions remains to be established. Likewise, it is still

unclear how the patterns of expression of the organ identity genes are initially established to allow the correct timing and position of organ initiation.

Pattern formation within the floral meristem could be explained on the basis of intercellular trafficking of transcription factors, i.e., SCPs. Microinjection experiments performed with FITC-DEF A and FITC- GLO have established that both proteins have the capacity to traffic from cell to cell and to increase plasmodesmal SEL (W.J. Lucas *et al.*, unpublished results). Molecular studies on these two proteins have established that they are MADS box containing transcription factors that bind DNA as a heterodimer to regulate petal and stamen development and to positively regulate their own expression [83, 97, 112]. Many additional genes involved in floral development contain the highly conserved MADS domain (Table 1), which indicates that they too most likely act as transcription factors and are candidate SCPs. Microinjection studies will confirm whether the encoded proteins can traffic through plasmodesmata.

All of the floral MADS-box genes (Table 1) contain a conserved region, designated the K-domain, which is thought to be involved in dimer formation. The possibility that each of the MADS-box proteins may be able to form multiple heterodimers, as well as homodimers, suggests that they may 'mix and match' to form a wide range of transcription factors, each with specific targets. The particular MADS- box dimers active in a given nucleus would be determined by which proteins are present, their relative concentrations, and the specific binding constants for a given dimer pair. If this ability to 'mix and match' were combined with controlled plasmodesmal trafficking and nuclear localization, a complex, dynamic, and highly regulatable matrix of transcriptional control elements could establish a pattern of supracellular domains in the floral meristem. These supracellular domains could give rise to specific whorls in the meristem and to individual organs within each whorl.

Two models for the establishment of supracellular domains within the floral meristem are depicted in Fig. 6. In the first model (panel C), the mRNA for two MADS-box genes is translated in different regions of the meristem and the putative SCPs traffic towards each other. When the SCPs meet, they form a heterodimer which has a higher binding affinity than either homodimer. The presence of the heterodimer in a limited number of cells could establish either a new domain, or a boundary between two domains. The former would occur if the heterodimeric transcrip-

Table 1. Genes involved in floral meristem development.

Gene[a]		Sequence information	Function	mRNA expression information	References[c]
LEAFY (FLORICAULA)	LFY (FLO)	Putative transcription factor	Floral meristem identity	Throughout early meristem, Early whorl 1, Whorls 2–4,	81, **105**, 107 **(9)**
APETALA1 (SQUAMOSA)	AP1 (SQUA)	MADS box K domain	Floral meristem identity, Sepal and petal determination, Class A	Throughout early meristem, Whorls 1 and 2, (Bracts, Whorls 1, 2, and 4)	4, 40, **60** **(38)**
CAULIFLOWER	CAL	MADS box K domain	Floral meristem identity, Redundancy with AP1 function	Throughout early meristem, Low expression in whorls 1 and 2	4, **47**
APETALA2	AP2	Putative transcription factor	Floral meristem identity, Sepal and petal determination, Class A	Throughout early meristem, All whorls	11, 40. **45**, 50
UNUSUAL FLORAL ORGANS (FIMBRIATA)	UFO (FIM)	No homology to known sequences	Floral meristem identity, Mediation between meristem identity and organ identity genes	Center of early meristem, Between whorl 1 and inner whorls, Junctions of petas with adjacent whorls	39, 51. 108 **(84)**
AGAMOUS-LIKE 2	AGL2	MADS box K domain	Mediation between meristem identity and organ identity genes?	Throughout meristem, All whorls	**30**
AGAMOUS-LIKE 4	AGL4	MADS box K domain	Mediation between meristem identity and organ identity genes?	Throughout meristem, Transient expression in whorl 1, Whorls 2–4	**79**
AGAMOUS-LIKE 9	AGL9	MADS box	Mediation between meristem	Whorls 2–4	**75**
APETALA3 (DEFICIENS)	AP3 (DEFA)	MADS box K domain	Petal and stamen determination, Class B	Whorls 2 and 3, (Transient expression in center of meristem, Whorls 2, and 3)	6, **42**, 43 (83, **88**, 112)
PISTILLATA (GLOBOSA)	PI (GLO)	MADS box K domain	Petal and stamen determination, Class B	Whorls 3 and 4 meristem, Whorls 2 and 3, (Whorls 2 and 3)	6, **34** **(97)**
AGAMOUS (PLENA)	AG (PLE)	MADS box K domain	Stamen and carpel identity, Determinacy of floral meristem, Class C	Whorls 3 and 4	6, **111** **(7)**
SUPERMAN	SUP	Zinc finger	Restriction of B function in whorl 4	Inner edge of whorl 3	5, 82, **76**

[a] All information applies to *Arabidopsis thaliana* except information in parentheses, which applies to *Antirrhinum majus.*
[b] mRNA expression patterns listed apply only to early through intermediate floral meristems. Some genes show additional domains of expression.
[c] Citations in bold refer to papers in which the sequences of the genes were first reported.

tion factor activates a new developmental pathway, and the latter would occur if the presence of the heterodimer leads to downregulation of plasmodesmal trafficking. Both the size and the position of such a new domain of cells would be determined by the amount of each protein initially translated, the timing of trafficking (proteins may traffic in waves that are released either by plasmodesmal activators or by attainment of a threshold concentration), and the rate of cell-to-cell trafficking. The developmental course taken by the new domain would depend on the specific matrix of control elements present in the cells. Some genes in the floral meristem are expressed in complementary regions and could function in the manner just described. For example, AG might interact in this way with AP1 to establish a boundary between whorls two and three.

A second model for pattern formation in the meristem involves the activity of SPC cofactors (Fig. 6, panel D). If cofactors exist that can shut down plasmodesmal trafficking, they could be responsible for the separation of domains destined to develop along different pathways. For instance, such cofactors might be necessary to separate cells designated to different whorls or to different organs within a whorl. Expression of such cofactors may be driven by the formation of heterodimers (as described for the situation depicted in Fig. 6, panel C) or by specific combinations of transcription factors found at certain positions in the meristem. Possible candidates for this type of cofactor are UFO and FIM (Table 1). These highly homologous proteins show no sequence similarity to any known proteins, and they are thought to act between the meristem identity and organ identity genes. It has been hypothesized that they regulate transcription of the organ identity genes because reduced mRNA levels of such genes has been shown in *ufo* and *fim* mutants [51, 84]. There is contradictory evidence, however, that shows *ufo* mutants with wild-type expression of organ identity genes [108]. The expression pattern of *fim* mRNA in the *Antirrhinum* meristem (and *ufo* mRNA in *Arabidopsis*) is consistent with a role in boundary formation between sepals and petals [39, 84]. Similarly, the phenotypes of both mutants include chimeric organs and disorganized whorls, consistent with the notion that developmental programs were not correctly compartmentalized. Experiments involving the coinjection of UFO or FIM with known SCPs would provide an important test as to whether the functions of these proteins include downregulation of trafficking through plasmodesmata.

Another protein that could play a role in boundary formation is SUP. As described above, *SUP* expression seems to be involved in defining the inner boundary of *AP3* and *PI* expression. Interestingly, *SUP* is expressed in a narrow domain that overlaps completely with the innermost 3–4 cells of the *AP3* expression domain [76]. If AP3 is capable of intercellular trafficking, via plasmodesmata, as is its homologue DEF A (Table 1), the presence of SUP in a cell may prevent AP3 trafficking. If this were the case, *sup* mutants would not be able to prevent AP3 trafficking, thereby allowing AP3 activity to intrude into whorl four to direct the formation of extra stamens.

Definitive tests of the SCP models presented in Fig. 6 will require the use of a combination of experimental techniques using cellular, molecular and genetic approaches. Confirmation of the basic tenets of the SCP model have already been established by studies on KN1 [57]. Further confirmation would strengthen the foundation for incorporation of the supracellular concept into working models of plant development [53, 56].

SCPs and coordination of development at the whole-plant level

The final aspect of cell-to-cell transport of proteins and nucleic acids that we will address relates to macromolecular trafficking around the plant via the phloem. Clearly, as just discussed, the development of plant organs involves intricate networks of control molecules. But the development of such organs in one part of the plant must also be coordinated with the activities of the plant as a whole. An excellent example of this type of whole-plant integration is the control over floral evocation. Extensive studies on numerous species have aimed to establish the conditions which allow a given plant to make the transition to flowering (reviewed in [1, 2]). Factors which influence this transition include: the specific genotype of the plant, day length, temperature, water availability, nutritional stress, extent of vegetative growth, and exposure to specific hormones. Induction of this floral transition can depend on signals from the leaves, roots, or shoot apex, each of which can sense different environmental stimuli. The only interspecies generalization that can be made is that the transition to flowering can be activated by multiple combinations of environmental conditions and/or internal events.

Insight into the possible mechanism(s) by which the plant instructs the shoot apical meristem to shift its developmental fate, to become an inflorescence meristem, may be gained from recent studies performed on the long-distance transport system of the phloem. Analysis of the phloem sap, at various positions along the transport path, has revealed that it contains hundreds of proteins ranging from low to high molecular mass [29, 66, 77, 80]. These proteins are moving through the sieve tubes of the higher plant phloem which is comprised of enucleate sieve elements [73]. Thus, these phloem-mobile proteins must be synthesized in the surrounding nucleate cells prior to their entry into the long-distance transport stream. Strong support for this hypothesis was provided by ^{35}S-methionine labeling experiments which demonstrated that ^{35}S-labeled proteins were confined to the companion cells and the associated sieve elements [29]. As the abundant proteins range from 20 to 60 kDa, it would appear that their entry into the sieve elements would likely occur via plasmodesmal trafficking.

Several phloem proteins have been cloned [41, 80], and microinjection studies performed on a 13 kDa abundant rice phloem protein [41] revealed that it can mediate its cell-to-cell transport (Ishiwatari *et al.*, unpublished results). As anticipated from other such studies, an increase in plasmodesmal SEL also occurred in association with the transport of this protein out of the injected cells. Preliminary studies on another phloem protein has yielded similar results. The point to note here is that due to the relatively inaccessible nature of the companion cells to microinjection, the above-mentioned studies were performed on mesophyll cells. Thus, although it is clear that these phloem proteins behave in essence like KN1 and the viral MPs, the question must be asked as to why, *in vivo*, these proteins are restricted to the companion cells (presumed sites of synthesis) and the sieve elements?

A plausible answer to this question would be that SCP cofactors (or a specialized sub-set of plasmodesmal cofactors) are involved in the delivery of these proteins from the ribosomes to the special plasmodesmata that interconnect the companion cells to the sieve elements [53]. (Microinjection of these phloem proteins into mesophyll cells, where the special cofactors would be absent, appears to allow their diffusion within the cytoplasm and eventual interaction with plasmodesmal binding sites.) Thus, induction of gene expression within the companion cells of a mature leaf can result in the synthesis of a protein that is then selectively trafficked across the companion cell-sieve element plasmodesmata for delivery to a specific location within the plant. Control over entry and egress of such a long-distance signaling protein (long range SCP) would be as described for general plasmodesmal macromolecular trafficking. Finally, as viruses appear to be capable of using the phloem for the entry and transport of viral ribonucleoprotein complexes (MP-vRNA), the possibility cannot be discounted that the plant also transports SCP-mRNA complexes via this communication (delivery) system.

These new findings allow the development of a plausible scheme for the transition to flowering in higher plants (Fig. 8). Organ identity and the perception of environmental stimuli result in the production of specific signals (hormones and SCPs) in various parts of the plant. Some of these signals are transported throughout the plant by means of the xylem and the phloem, where they influence the signal balance in different tissues. For the explicit case in which floral induction involves the delivery of signals via the phloem, the following may occur. The leaf will be receiving hormonal signals from the root, as well as signals produced in the leaf in response to the light regime (day length). When these signal elements reach a critical balance, within the companion cells of the phloem, the genes involved in the production of the flowering signal become derepressed. Once this signal reaches a critical threshold required for plasmodesmal trafficking it will become loaded into the phloem and subsequently transported to the meristem.

In order to respond to the signal from the leaf, the meristem must be in a competent state. The meristem reaches this state by the maintenance of a specific balance of signals for a critical period of time, in a scheme analogous to that described for the leaf. The competence of the meristem is likely to involve two factors. The first is the ability of the meristem to communicate the signal from the newly differentiated phloem to the upper layers of the meristem (see Fig. 1, panel A); the second is the ability of the cells in the upper layers of the meristem to alter their developmental pathways in response to the incoming signal.

A florigenic signal that travels through the phloem has never been conclusively identified [2], making it likely that such a signal is present in very low abundance. Thus, it does not seem possible that a small diffusible signal molecule would travel directly from the phloem, through the many cells of the meristem, to affect development of cells in the L1, L2 and L3 layers. Rather, the phloem signal more likely acts directly on the companion cells into which it is unloaded, where it could elicit the production of a second signal. This secondary signal might be an SCP or an SCP-mRNA complex. As illustrated in Fig. 9 (panel A), the distance over which an SCP can operate may be too short to permit its efficient delivery to the true shoot apical meristem. This problem would be resolved if the SCP traffics through the intervening tissue as an SCP-mRNA complex (Fig. 9, panel B). With such a system, the transcription of relatively few mRNA molecules in the companion cell could rapidly propagate a signal all the way to the central zone of the meristem. Thus, the competence of the meristem to communicate the signal from the phloem to the upper layers might involve the presence of cofactors which allow the trafficking of a florigenic SCP and its mRNA.

Once the florigenic signal from the phloem reaches the upper layers of the meristem, these cells must be competent to respond to the signal. If the signal is an SCP, this competence might involve the presence of an SCP cofactor. Such a cofactor could chaperone the SCP into the nucleus or modify the SCP in a way that would activate its florigenic function. Alternatively, the SCP might function in the cytosol to activate a control factor that is produced in cells poised to enter the florigenic pathway. Thus, if conditions throughout

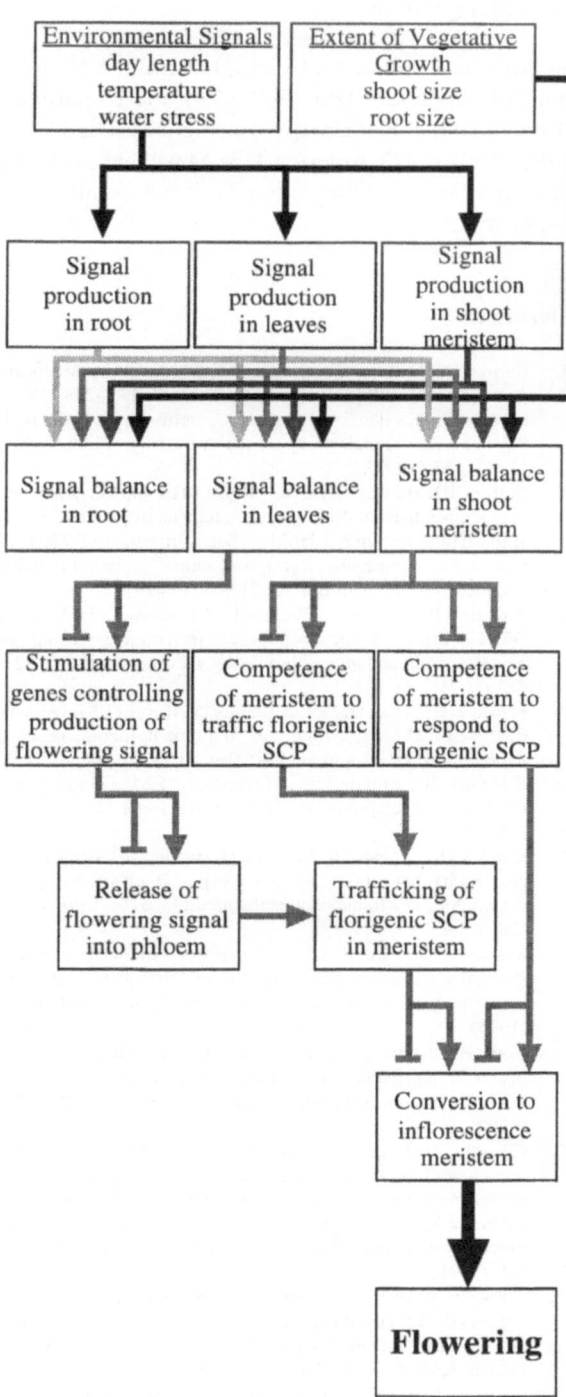

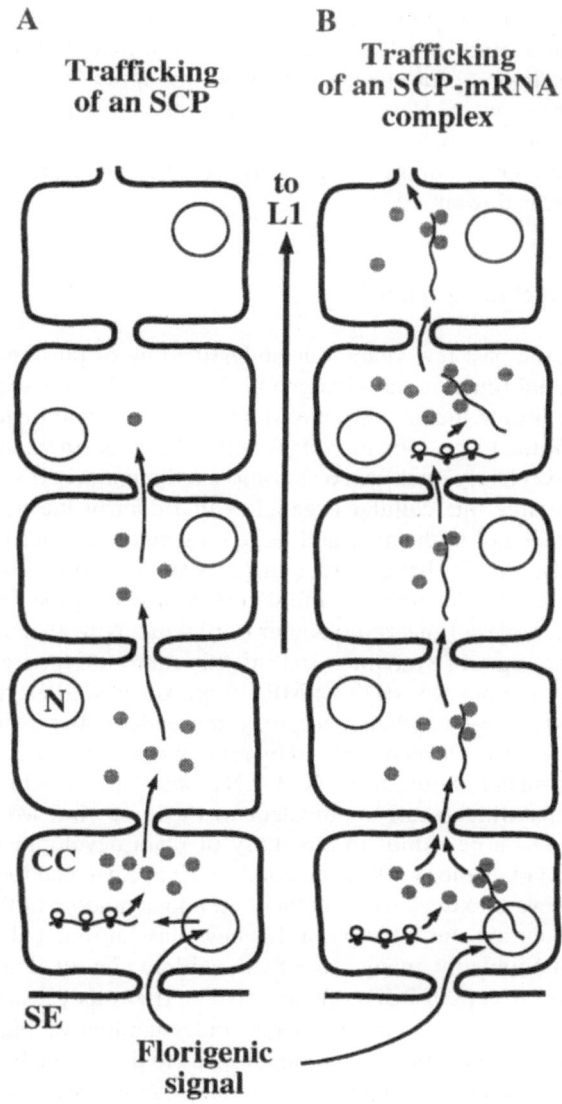

A

**Trafficking
of an SCP**

B

**Trafficking
of an SCP-mRNA
complex**

to
L1

N

CC

SE

**Florigenic
signal**

Figure 8. Schematic representation of the complex matrix of input signals involved in orchestrating control over the shoot apical meristem, in terms of its induction to become an inflorescence meristem (see text for details).

Figure 9. Alternate models for the manner in which a phloem-borne florigenic signal (see Fig. 8) may act to elicit the synthesis and cell-to-cell transport of a meristem SCP responsible for the conversion of the shoot apical meristem to an inflorescence meristem. A. Florigenic signal enters the companion cells (CC) from the sieve elements (SE) where it acts to induce expression of a *FLORIGENIC SCP*. The SCP synthesized in the CC would then traffic, via plasmodesmata, towards the central zone of the shoot apical meristem. The efficacy of this signalling event will be directly proportional to the number of cells present in the pathway. B. As in Panel A, except that the florigenic SCP travels from cell to cell via an SCP-mRNA complex. In this situation, in each cell, SCP bound to its mRNA would be in equilibrium with free SCP. As the SCP and SCP-mRNA complex move from cell to cell, free SCP would gradually be degraded. This would alter the equilibrium and expose the mRNA to the cytoplasm, thus allowing ribosomes to bind and translate new SCP. As the concentration of free SCP again builds it could displace the SCP mRNA from the ribosomes and mediate further cell-to-cell trafficking. Such a mechanism would ensure that the florigenic SCP has the capability to pass from the phloem through the overlying layers of cells to the central zone of the shoot apical meristem (see Fig. 1, Panel A).

the plant are conducive to flowering, different organs could send signals through the phloem that elicit the trafficking of a florigenic SCP to the upper layers of the meristem. Here, cells that are competent to respond to the SCP will alter their developmental programs to produce an inflorescence meristem that subsequently forms flowers.

Concluding remarks

In the past few years, our understanding of plasmodesmal function has changed radically. No longer seen as simple 'holes' allowing small molecules and metabolites to pass from cell to cell, plasmodesmata are now clearly established as complex, dynamic, and regulatable intercellular organelles that control the trafficking of both small and large molecules. Studies on plant viruses have yielded initial insights into the true nature of plasmodesmal function by providing the first examples of macromolecular trafficking. Now that the paradigm of systemic viral infection through the trafficking of vRNA(DNA)-MP complexes is well established, we can begin to study the endogenous plant system so cleverly utilized by plant viruses. The recent characterization of an SCP, KN1, that has trafficking capabilities nearly equivalent to those of viral MPs, opens a new door to the study of plant development and physiology. Over the coming decade, we can look forward to the development of an understanding of the evolution and function of this new class of molecules. Undoubtedly, many more SCPs will be identified and the concept of SCP cofactors will be rigorously tested. Parallel advancements in the understanding of plasmodesmal structure should allow us to determine how SCPs and their putative cofactors interact with plasmodesmal proteins to potentiate spatial and temporal control over macromolecular trafficking within the plant. These discoveries will allow us to gain insights into the role of plasmodesmata in intercellular communication, the establishment of supracellular domains within tissues, and the coordination of developmental events involving communication via hormones and long-distance SCPs. A seminal outcome from these endeavors would be the discovery that the elusive florigen is indeed a long-distance SCP.

Acknowledgements

This work was supported by grants from the National Science Foundation (IBN–9406974), the Department of Energy Division of Energy Biosciences (DE-FG03–94ER20134) and JT America. L.A.M was supported by a National Science Foundation Training Grant (BIR–94–14106).

References

1. Bernier G: The control of floral evocation and morphogenesis. Annu Rev Plant Physiol Plant Mol Biol 39: 175–219 (1988).
2. Bernier G, Havelange A, Houssa C, Petitjean A, Lejeune P: Physiological signals that induce flowering. Plant Cell 5: 1147–1155 (1993).
3. Botha CEJ, Hartley BJ, Cross RHM: The ultrastructure and computer-enhanced digital image analysis of plasmodesmata at the Kranz mesophyll-bundle sheath interface of *Themeda triandra* var. imberbis (Retz) A. Camus in conventionally-fixed leaf blades. Ann Bot 72: 255–261 (1993).
4. Bowman JL, Alvarez J, Weigel D, Meyerowitz EM, Smyth DR: Control of flower development in *Arabidopsis thaliana* by *APETALA1* and interacting genes. Development 119: 721–743 (1993).
5. Bowman JL, Sakai H, Jack T, Weigel D, Mayer U, Meyerowitz EM: SUPERMAN, a regulator of floral homeotic genes in *Arabidopsis*. Development 114: 599–615 (1992).
6. Bowman JL, Smyth DR, Meyerowitz EM: Genes directing flower development in *Arabidopsis*. Plant Cell 1: 37–52 (1989).
7. Bradley D, Carpenter R, Sommer H, Hartley N, Coen E: Complementary floral homeotic phenotypes result from opposite orientations of a transposon at the *plena* locus of *Antirrhinum*. Cell 72: 85–95 (1993).
8. Carpenter R, Coen ES: Transposon induced chimeras show that *floricaula*, a meristem identity gene, acts non-autonomously between cell layers. Development 121: 19–26 (1995).
9. Coen ES, Romero JM, Doyle S, Elliot R, Murphy G, Carpenter R: *floricaula*: a homeotic gene required for flower development in *Antirrhinum majus*. Cell 63: 1311–1322 (1990).
10. Citovsky V, Knorr D, Schuster G, Zambryski P: The P30 movement protein of tobacco mosaic virus is a single-strand nucleic acid binding protein. Cell 60: 637–647 (1990).
11. Citovsky V, Zambryski P: How do plant virus nucleic acids move through intercellular connections? BioEssays 13: 373–379 (1991).
12. Citovsky V, Wong ML, Shaw AL, Venkataram Prasad BV, Zambryski P: Visualization and characterization of tobacco mosaic virus movement protein binding to single-stranded nucleic acids. Plant Cell 4: 397–411 (1992).
13. Citovsky V: Probing plasmodesmal transport with plant viruses. Plant Physiol 102: 1071–1076 (1993).
14. Cleland RE, Fujiwara T, Lucas WJ: Plasmodesmal-mediated cell-to-cell transport in wheat roots is modulated by anaerobic stress. Protoplasma 178: 81–85 (1994).
15. Dawson WO, Bubrick P, Grantham GL: Modifications of the tobacco mosaic virus coat protein gene affecting replication,

movement and symptomatology. Phytopathology 78: 783–789 (1988).

16. Deom CM, Oliver MJ, Beachy RN: The 30-kilodalton gene product of tobacco mosaic virus potentiates virus movement. Science 237: 389–394 (1987).

17. Deom CM, Lapidot M, Beachy RN: Plant virus movement proteins. Cell 69: 221–224 (1992).

18. Derrick PM, Barker H, Oparka KJ: Increase in plasmodesmatal permeability during cell-to-cell spread of tobacco rattle virus from individually inoculated cells. Plant Cell 4: 1405–1412 (1992).

19. Ding B, Haudenshield JS, Hull RJ, Wolf S, Beachy RN, Lucas WJ: Secondary plasmodesmata are specific sites of localization of the tobacco mosaic virus movement protein in transgenic tobacco plants. Plant Cell 4: 915–928 (1992).

20. Ding B, Turgeon R, Parthasarathy MV: Substructure of freeze substituted plasmodesmata. Protoplasma 169: 28–41 (1992).

21. Ding B, Li Q, Nguyen L, Palukaitis P, Lucas WJ: Cucumber mosaic virus 3a protein potentiates cell-to-cell trafficking of CMV RNA in tobacco plants. Virology 207: 345–353 (1995).

22. Ding B, Lucas WJ: Secondary Plasmodesmata: Biogenesis, Special Functions and Evolution. In: Smallwood M, Knox P, Bowles D (eds) Membranes: Specialized Functions in Plant Cells. Bios Scientific Publishers, Oxford (in press).

23. Ding XS, Shintaku MH, Arnold SA, Nelson RS: Accumulation of mild and severe strains of tobacco mosaic virus in minor veins of tobacco. Mol Plant-Microbe Interact 8: 32–40 (1995).

24. Dolja VV, McBride HJ, Carrington JC: Tagging of plant potyvirus replication and movement by insertion of Beta-glucuronidase into the viral polyprotein. Proc Natl Acad Sci USA 89: 10208–10212 (1992).

25. Dorokhov YL, Alexandrova NM, Miroschnichenko NA, Atabekov JG: Isolation and analysis of virus-specific ribonucleoprotein of tobacco mosaic virus-infected tobacco. Virology 127: 237–252 (1983).

26. Dorokhov YL, Alexandrova NM, Miroschnichenko NA, Atabekov JG: The informosome-like virus-specific ribonucleoprotein (vRNP) may be involved in the transport of tobacco mosaic virus infection. Virology 137: 127–134 (1994).

27. Drews GN, Bowman JL, Meyerowitz EM: Negative regulation of the *Arabidopsis* homeotic gene *AGAMOUS* by the *APETALA2* product. Cell 65: 991–1002 (1991).

28. Feng XH, Kung SD: Identification of differentially expressed members of tobacco homeobox families by differential PCR. Biochem Biophys Res Comm 198: 1012–1019 (1994).

29. Fisher DB, Wu Y, Ku MSB: Turnover of soluble proteins in the wheat sieve tube. Plant Physiol 100: 1433–1441 (1992).

30. Flanagan CA, Ma H: Spatially and temporally regulated expression of the MADS-box gene *AGL2* in wild-type and mutant *Arabidopsis* flowers. Plant Mol Biol 26: 581–595 (1994).

31. Fujiwara T, Giesman-Cookmeyer D, Ding B, Lommel SA, Lucas WJ: Cell-to-cell trafficking of macromolecules through plasmodesmata potentiated by the red clover necrotic mosaic virus movement protein. Plant Cell 5: 1783–1794 (1993).

32. Gamalei YV: Structure and function of leaf minor veins in trees and herbs. A taxonomic review. Trees 3: 96–110 (1989).

33. Glazer AN, Rye HS: Stable dye-DNA intercalation complexes as reagents for high-sensitivity fluorescence detection. Nature 359: 859–861 (1992).

34. Goto K, Meyerowitz EM: Function and regulation of the *Arabidopsis* floral homeotic gene *PISTILLATA*. Genes Devel 8: 1548–1560 (1994).

35. Gustafson-Brown C, Savidge B, Yanofsky MF: Regulation of the *Arabidopsis* floral homeotic gene *APETALA1*. Cell 76: 131–143 (1994).

36. Hantke SS, Carpenter R, Coen ES: Expression of *floricaula* in single cell layers of periclinal chimeras activates downstream homeotic genes in all layers of floral meristems. Development 121: 27–35 (1995).

37. Huala E, Sussex IM: Determination and cell interactions in reproductive meristems. Plant Cell 5: 1157–1165 (1993).

38. Huijser P, Klein J, Lonnig WE, Meijer H, Saedler H, Sommer H: Bracteomania, an inflorescence anomaly, is caused by the loss of function of the MADS-box gene *squamosa* in *Antirrhinum majus*. EMBO J 11: 1239–1249 (1992).

39. Ingram GC, Goodrich J, Wilkinson MD, Simon R, Haughn GW, Coen ES: Parallels between *UNUSUAL FLORAL ORGANS* and *FIMBRIATA*, genes controlling flower development in *Arabidopsis* and *Antirrhinum*. Plant Cell 7: 1501–1510 (1995).

40. Irish VF, Sussex IM: Function of the *apetala-1* gene during *Arabidopsis* floral development. Plant Cell 2: 741–753 (1990).

41. Ishiwatari Y, Honda C, Kawashima I, Nakamura S, Hirano H, Mori S, Fujiwara T, Hayashi H, Chino M: Thioredoxin h is one of the major proteins in rice phloem sap. Planta 195: 456–463 (1995).

42. Jack T, Brockman LL, Meyerowitz EM: The homeotic gene *APETALA3* of *Arabidopsis thaliana* encodes a MADS box and is expressed in petals and stamens. Cell 68: 683–697 (1992).

43. Jack T, Fox GL, Meyerowitz EM: *Arabidopsis* homeotic gene *APETALA3* ectopic expression: transcriptional and posttranscriptional regulation determine floral organ identity. Cell 76: 703–716 (1994).

44. Jackson D, Veit B, Hake S: Expression of maize *KNOTTED1* related homeobox genes in the shoot apical meristem predicts patterns of morphogenesis in the vegetative shoot. Development 120: 405–413 (1994).

45. Jofuku KD, den Boer BGW, Montague MV, Okamuro JK: Control of *Arabidopsis* flower and seed development by the homeotic gene *APETALA2*. Plant Cell 6: 1211–1225 (1994).

46. Kano-Murakami Y, Yanai T, Tagiri A, Matsuoka M: A rice homeotic gene, *OSH1*, causes unusual phenotypes in transgenic tobacco. FEBS Let 334: 365–368 (1993).

47. Kempin SA, Savidge B, Yanofsky MF: Molecular basis of the *cauliflower* phenotype in *Arabidopsis*. Science 267: 522–525 (1995).

48. Kerstetter R, Vollbrecht E, Lowe B, Veit B, Yamaguchi J, Hake S: Sequence analysis and expression patterns divide the maize *knotted1*-like homeobox genes into two classes. Plant Cell 6: 1877–1887 (1994).

49. Koonin EV, Mushegian AR, Ryabov EV, Dolja VV: Diverse groups of plant RNA and DNA viruses share related movement proteins that may possess chaperone-like activity. J Gen Virol 72: 2895–2903 (1991).

50. Kunst L, Klenz JE, Martinez-Zapater J, Haughn GW: *AP2* gene determines the identity of perianth organs in flowers of *Arabidopsis thaliana*. Plant Cell 1: 1195–1208 (1989).

51. Levin JZ, Meyerowitz EM: *UFO*: an *Arabidopsis* gene involved in both floral meristem and floral organ development. Plant Cell 7: 529–548 (1995).

52. Lincoln C, Long J, Yamaguchi J, Serikawa K, Hake S: A *knotted1*-like homeobox gene in *Arabidopsis* is expressed in the vegetative meristem and dramatically alters leaf morpho-

logy when overexpressed in transgenic plants. Plant Cell 6: 1859–1876 (1994).

53. Lucas WJ, Ding B, van der Schoot C: Plasmodesmata and the supracellular nature of plants. New Phytol 125: 435–476 (1993).

54. Lucas WJ, Wolf S: Plasmodesmata: the intercellular organelles of green plants. Trends Cell Biol 3: 308–315 (1993).

55. Lucas WJ, Gilbertson RL: Plasmodesmata in relation to viral movement within leaf tissues. Annu Rev Phytopath 32: 387–411 (1994).

56. Lucas WJ: Plasmodesmata: intercellular channels for macromolecular transport in plants. Curr Opin Cell Biol 7: 673–680 (1995).

57. Lucas WJ, Bouche-Pillon S, Jackson DP, Nguyen L, Baker L, Ding B, Hake S: Selective trafficking of KNOTTED1 homeodomain protein and its mRNA through plasmodesmata. Science 270: 1980–1983 (1995).

58. Ma H: The unfolding drama of flower development: recent results from genetic and molecular analyses. Genes Dev 8: 745–756 (1994).

59. Ma H, McMullen MD, Finer JJ: Identification of a homeobox-containing gene with enhanced expression during soybean (Glycine max L.) somatic embryo development. Plant Mol Biol 24: 465–473 (1994).

60. Mandel MA, Gustafson-Brown C, Savidge B, Yanofsky MF: Molecular characterization of the Arabidopsis floral homeotic gene APETALA1. Nature 360: 273–277 (1992).

61. Mandel MA, Yanofsky MF: A gene triggering flower formation in Arabidopsis. Nature 377: 522–524 (1995).

62. Matsuoka M, Ichikawa H, Saito A, Tada Y, Fujimura T, Kano-Murakami Y: Expression of a rice homeobox gene causes altered morphology of transgenic plants. Plant Cell 5: 1039–1048 (1993).

63. Maule AJ: Virus movement in infected plants. Crit Rev Plant Sci 9: 457–473 (1991).

64. Melcher U: Similarities between putative transport proteins of plant viruses. J Gen Virol 71: 1009–1018 (1990).

65. Meshi T, Watanabe Y, Saito T, Sugimoto A, Maeda T, Okada Y: Function of the 30 kd protein of tobacco mosaic virus: involvement in cell-to-cell movement and dispensability for replication. EMBO J 6: 2557–2563 (1987).

66. Nakamura S, Hayashi H, Mori S, Chino M: Protein phosphorylation in the sieve tubes of rice plants. Plant Cell Physiol 34: 927–933 (1993).

67. Nelson RS, Li G, Hodgson RAJ, Beachy RN, Shintaku MH: Impeded phloem-dependent accumulation of the masked strain of tobacco mosaic virus. Mol Plant-Microbe Interact 6: 45–54 (1993).

68. Noueiry AO, Lucas WJ, Gilbertson RL: Two proteins of a plant DNA virus coordinate nuclear and plasmodesmal transport. Cell 76: 925–932 (1994).

69. Okamuro JK, den Boer BGW, Jofuku KD: Regulation of Arabidopsis flower development. Plant Cell 5: 1183–1193 (1993).

70. Oparka KJ, Prior DAM: Direct evidence for pressure-generated closure of plasmodesmata. Plant J 2: 741–750 (1992).

71. Oparka KJ, Duckett CM, Prior DAM, Fisher DB: Real-time imaging of phloem unloading in the root tip of Arabidopsis. Plant J 6: 759–766 (1994).

72. Osman TAM, Hayes RJ, Buck KW: Cooperative binding of the red clover necrotic mosaic virus movement protein to single-stranded nucleic acids. J Gen Virol 73: 223–227 (1992).

73. Parthasarathy MV: Sieve-element structure. In: Zimmermann MH, Milburn JA (eds) Encyclopedia of Plant Physiology, New Series: Transport in Plants I. Phloem Transport, vol 1, pp. 3–38. Springer-Verlag, Berlin, pp. 3–38 (1975).

74. Robards AW, Lucas WJ: Plasmodesmata. Annu Rev Plant Physiol Plant Mol Biol 41: 369–419 (1990).

75. Rounsley SD, Ditta GS, Yanofsky MF: Diverse roles for MADS box genes in Arabidopsis development. Plant Cell 7: 1259–1269 (1995).

76. Sakai H, Medrano LJ, Meyerowitz EM: Role of SUPERMAN in maintaining Arabidopsis floral whorl boundaries. Nature 378: 199–203 (1955).

77. Sakuth T, Schobert C, Pecsvaradi A, Eichholz A, Komor E, Orlich G: Specific proteins in the sieve tube exudate of Ricinus communis L. seedlings: separation, characterization and in vivo labelling. Planta 191: 207–213 (1993).

78. Sanger M, Passmore B, Falk BW, Bruening G, Ding B, Lucas WJ: Symptom severity of beet western yellows virus strain ST9 is conferred by the ST9-associated RNA and is not associated with virus release from the phloem. Virology 200: 48–55 (1994).

79. Savidge B, Rounsley SD, Yanofsky MF: Temporal relationship between the transcription of two Arabidopsis MADS box genes and the floral organ identity genes. Plant Cell 7: 721–733 (1995).

80. Schobert C, Grossmann P, Gottachalk M, Komor E, Pecsvaradi A, zur Nieden U: Sieve-tube exudate from Ricinus communis L. seedlings contains ubiquitin and chaperones. Planta 196: 205–210 (1995).

81. Schultz EA, Haughn GW: LEAFY, a homeotic gene that regulates inflorescence development in Arabidopsis. Plant Cell 3: 771–781 (1991).

82. Schultz EA, Pickett FB, Haughn GW: The FLO10 gene product regulates the expression domain of homeotic genes AP3 and PI in Arabidopsis flowers. Plant Cell 3: 1221–1237 (1991).

83. Schwarz-Sommer Z, Hue I, Huijser P, Flor PJ, Hansen R, Tetens F, Lonnig WE, Saedler H, Sommer H: Characterization of the Antirrhinum floral homeotic MADS-box gene deficiens: evidence for DNA binding and autoregulation of its persistent expression throughout flower development. EMBO J 11: 251–263 (1992).

84. Simon R, Carpenter R, Doyle S, Coen E: Fimbriata controls flower development by mediating between meristem and organ identity genes. Cell 78: 99–107 (1994).

85. Sinha N, Hake S: Mutant characters of Knotted maize leaves are determined in the innermost tissue layers. Dev Biol 141: 203–210 (1990).

86. Sinha NR, Williams RE, Hake S: Overexpression of the maize homeobox gene, KNOTTED–1, causes a switch from determinate to indeterminate cell fates. Genes Devel 7: 787–795 (1993).

87. Smith LG, Green B, Veit B, Hake S: A dominant mutation in the maize homeobox gene, Knotted-1, causes its ectopic expression in leaf cells with altered fates. Development 116: 21–30 (1992).

88. Sommer H, Beltran JP, Huijser P, Pape H, Lonnig WE, Saedler H, Schwarz-Sommer Z: Deficiens, a homeotic gene involved in the control of flower morphogenesis in Antirrhinum majus: the protein shows homology to transcription factors. EMBO J 9: 605–613 (1990).

89. St Johnston D, Nusslein-Volhard C: The origin of pattern and polarity in the Drosophila embryo. Cell 68: 201–219 (1992).

90. Stewart RN, Meyer FG, Dermen H: Camellia + 'Daisy Eagleson', a graft chimera of *Camellia sasanqua* and *C. japonica*. Am J Bot 59: 515–524 (1972).

91. Szymkowiak EJ, Sussex IM: Effect of *lateral suppressor* on petal initiation in tomato. Plant J 4: 1–7 (1993).

92. Szymkowiak EJ, Sussex IM: The internal meristem layer (L3) determines floral meristem size and carpel number in tomato periclinal chimeras. Plant Cell 4: 1089–1100 (1992).

93. Takamatsu N, Ishiakwa M, Meshi T, Okada Y: Expression of bacterial chloramphenicol acetyltransferase gene in tobacco plants mediated by TMV-RNA. EMBO J 6: 307–311 (1987).

94. Terry BR, Robards AW: Hydrodynamic radius alone governs the mobility of molecules through plasmodesmata. Planta 171: 145–157 (1987).

95. Tilney LG, Cooke TJ, Connelly PS, Tilney MS: The structure of plasmodesmata as revealed by plasmolysis, detergent extraction, and protease digestion. J Cell Biol 112: 739–747 (1991).

96. Tomenius K, Clapham D, Meshi T: Localization by immunogold cytochemistry of the virus-coded 30K protein in plasmodesmata of leaves infected with tobacco mosaic virus. Virology 160: 363–371 (1987).

97. Trobner W, Ramirez L, Motte P, Hue I, Huijser P, Lonnig WE, Saedler H, Sommer H, Schwarz-Sommer Z: *GLOBOSA*: a homeotic gene which interacts with *DEFICIENS* in the control of *Antirrhinum* floral organogenesis. EMBO J 11: 4693–4704 (1992).

98. van Bel AJE: Strategies of phloem loading. Annu Rev Plant Physiol Plant Mol Biol 44: 253–281 (1993).

99. van Bel AJE, van Rijen HVM: Microelectrode-recorded development of the symplasmic autonomy of the sieve element/companion cell complex in the stem phloem of *Lupinus luteus* L. Planta 192: 165–175 (1994).

100. Vollbrecht E, Veit B, Sinha N, Hake S: The developmental gene *Knotted–1* is a member of a maize homeobox gene family. Nature 350: 241–243 (1991).

101. von Arnim AG, Deng XW: Light inactivation of *Arabidopsis* photomorphogenic repressor COP1 involves a cell-specific regulation of its nucleocytoplasmic partitioning. Cell 79: 1035–1045 (1994).

102. Waigmann E, Lucas WJ, Citovsky V, Zambryski P: Direct functional assay for tobacco mosaic virus cell-to-cell movement protein and identification of a domain involved in increasing plasmodesmal permeability. Proc Natl Acad Sci USA 91: 1433–1437 (1994).

103. Wang D, Maule AJ: Inhibition of host gene expression associated with plant virus infection. Science 267: 229–231 (1995).

104. Weigel D: The APETALA2 domain is related to a novel type of DNA binding domain. Plant Cell 7: 388–389 (1995).

105. Weigel D, Alvarez J, Smyth DR, Yanofsky MF, Meyerowitz EM: *LEAFY* controls floral meristem identity in *Arabidopsis*. Cell 69: 843–859 (1994).

106. Weigel D, Meyerowitz EM: Activation of floral homeotic genes in *Arabidopsis*. Science 261: 1723–1726 (1993).

107. Weigel D, Nilsson O: A developmental switch sufficient for flower initiation in diverse plants. Nature 377: 495–500 (1995).

108. Wilkinson MD, Haughn GW: *UNUSUAL FLORAL ORGANS* controls meristem identity and organ primordia fate in *Arabidopsis*. Plant Cell 7: 1485–1499 (1995).

109. Wolf S, Deom CM, Beachy RN, Lucas WJ: Movement protein of tobacco mosaic virus modifies plasmodesmatal size exclusion limit. Science 246: 377–379 (1989).

110. Yanofsky MF: Floral meristems to floral organs: genes controlling early events in *Arabidopsis* flower development. Annu Rev Plant Physiol Plant Mol Biol 46: 167–188 (1995).

111. Yanofsky MF, Ma H, Bowman JL, Drews GN, Feldmann KA, Meyerowitz EM: The protein encoded by the *Arabidopsis* homeotic gene *agamous* resembles transcription factors. Nature 346: 35–39 (1990).

112. Zachgo S, Silva EA, Motte P, Trobner W, Saedler H, Schwarz-Sommer Z: Functional analysis of the *Antirrhinum* floral homeotic *DEFICIENS* gene in vivo and in vitro by using a temperature-sensitive mutant. Development 121: 2861–2875 (1995).

Plant Molecular Biology **32**: 275–302, 1996.
© 1996 *Kluwer Academic Publishers.*

Proteolysis in plants: mechanisms and functions

Richard D. Vierstra
Department of Horticulture, University of Wisconsin-Madison, Madison, WI 53706, USA

Key words: Arabidopsis, biotechnology, ClpAP protease, protein degradation, 20S and 26S proteasome, ubiquitin

Contents

Abstract	275	Functions of proteolysis	287
Introduction	276	Removal of abnormal/damaged	
Mechanisms for degrading proteins	276	proteins	288
General features	276	N-end rule pathway	289
Degradation of proteins in the		Supply of amino acids	289
cytoplasm and nucleus	277	Control of enzymatic pathways	290
Ubiquitin-dependent proteolytic		Control of various cell regulators	291
pathway	277	Phytochrome A	292
20S and 26S Proteasomes	280	Other regulatory proteins in plants	292
Specificity of the ubiquitin pathway	282	Timing of the cell cycle	293
Ubiquitin-independent pathways	285	Programmed cell death	293
Protein degradation in organelles	285	Applications in biotechnology	294
Vacuoles	285	Conclusions	295
Chloroplasts	286	Acknowledgements	296
Other organelles	287	References	296

Abstract

Proteolysis is essential for many aspects of plant physiology and development. It is responsible for cellular housekeeping and the stress response by removing abnormal/misfolded proteins, for supplying amino acids needed to make new proteins, for assisting in the maturation of zymogens and peptide hormones by limited cleavages, for controlling metabolism, homeosis, and development by reducing the abundance of key enzymes and regulatory proteins, and for the programmed cell death of specific plant organs or cells. It also has potential biotechnological ramifications in attempts to improve crop plants by modifying protein levels. Accumulating evidence indicates that protein degradation in plants is a complex process involving a multitude of proteolytic pathways with each cellular compartment likely to have one or more. Many of these have homologous pathways in bacteria and animals. Examples include the chloroplast ClpAP protease, vacuolar cathepsins, the KEX2-like proteases of the secretory system, and the ubiquitin/26S proteasome system in the nucleus and cytoplasm. The ubiquitin-dependent pathway requires that proteins targeted for degradation become conjugated with chains of multiple ubiquitins; these chains then serve as recognition signals for selective degradation by the 26S proteasome, a 1.5 MDa multisubunit protease complex. The ubiquitin pathway is particularly important for developmental regulation by selectively removing various cell-cycle effectors, transcription factors, and cell receptors such as phytochrome A. From insights into this and other proteolytic pathways, the use of phosphorylation/dephosphorylation and/or the addition of amino acid tags to selectively mark proteins for degradation have become recurring themes.

Introduction

The ultimate post-transcriptional control of gene expression involves the proteolytic breakdown of the encoded protein back to its constituent amino acids. Here, protein degradation not only represents an important recycling system for amino acids but also represents the final step in what can be a complex cascade of regulatory events controlling gene function [for reviews see 25, 34, 95, 207]. In the past decade, it has become increasingly obvious that the ability of cells to switch from one developmental state to another or to adapt to new environmental conditions often requires the rapid dismantlement of existing regulatory networks, a process frequently dependent on proteolysis. Examples range from the control of metabolism and cell specification to the progression of the cell cycle and the initiation of various signal transduction pathways. Moreover, both the speed (proteins can have half-lives <5 min) and irreversibility of proteolysis provide advantages to cellular regulation not offered by other mechanisms. With respect to the various post-translational regulatory processes, it should be emphasized that none are more influential or pervasive than protein breakdown in determining the final concentration of active proteins.

The purpose of this chapter is to describe our present understanding of how plant proteins are degraded and to illustrate the various ways that proteolysis can be used as a regulatory mechanism. This review will not focus solely on plants but will also include important paradigms from animal and bacterial systems that have relevance to plant protein breakdown [34, 75, 95, 143]. As will be seen, proteolysis is an intricate process, involving a multitude of pathways to select and catabolize target molecules. It can range from single cleavages that activate (or inactivate) proteins to the total digestion of the polypeptide. From the initial analysis of the few proteolytic pathways identified in plants, it is clear that the level of complexity required to degrade proteins may eventually rival that required to initially synthesize them [25, 207]. For example, we estimate that the ubiquitin-dependent proteolytic pathway alone may involve over 100 genes in *Arabidopsis thaliana* (or 0.5% of the coding region); more than 45 of which have been identified to date [207, R.D. Vierstra, unpublished]. Given that over 10 000 proteins can exist simultaneously in any given plant cell, it is also not surprising that cells have evolved highly sophisticated mechanisms for select-

ive target recognition, thus avoiding the indiscriminate breakdown of other proteins.

Mechanisms for degrading proteins

General features

At first glance, protein degradation would appear to simply involve a protease (or set of proteases) digesting a protein. However, several general observations argued early on that the process cannot be that simple [44, 77, 103]. First, most *in vivo* proteolysis requires energy. Because peptide bond hydrolysis is an exergonic reaction, and because most purified proteases are energy-independent, this requirement presupposed that energy-dependent steps must exist and that they may control proteolysis. Second, protein degradation is fast, so rapid in fact that detecting partial breakdown products is often difficult. This rapidity implies that once the proteolytic machinery finds a suitable target, it uses multiple protease activities to completely digest the target before another target is chosen. The efficient removal of partial cleavage fragments may be physiologically essential as these peptides could interfere with a multitude of protein/protein interactions should they accumulate. However, the failure to detect partial breakdown products also represents a major technical barrier in understanding how proteins catabolized and which proteases are responsible [207].

A third feature is that although most proteases are specific to certain amino acids sequences (e.g. trypsin cleaves after Arg and Lys residues) and/or sites within the protein (i.e. internal peptide bonds [endopeptidases] or terminal peptide bonds [exopeptidases including N-terminal aminopeptidase or C-terminus carboxypeptidases]), they are not typically restricted to specific proteins [12]. In fact, this lack of specificity has created an unusual nomenclature for proteases. Unlike other enzymes which are generally classified according to their substrates or products, proteases are typically classified based on the essential elements within their active sites. Examples include cysteine-, serine-, aspartic acid- and metalloproteases which require the aforementioned components in their cleavage reactions [12]. Because of this broad specificity, most proteases must be regulated or compartmentalized to avoid random breakdown of all intracellular proteins.

The fourth feature is that proteolysis is highly selective. Even in the same cellular milieu, protein half-lives can range from minutes to weeks [25, 77, 207]. Moreover, the turnover rate of individual proteins can vary dramatically depending on the conformational state and location of the protein, or on the developmental and physiological state of the cell. For example, the half-life of the plant morphogenic photoreceptor, phytochrome A, can vary by ca. 100-fold depending on whether its in the Pr or Pfr forms [208]. This selectivity implies that proteolytic mechanisms exist that individually recognize appropriate targets and that these mechanisms can be regulated. While it was originally thought that the overall physico-chemical properties (e.g. molecular mass, isoelectric point, thermal stability) of proteins govern their half-lives [see 44, 77], it is now clear that the essential determinants are often contained within small, sometimes conserved domains [50, 73, 117, 153, 199, 205, 218]. Several domains that confer a short half-life are functionally transferable to other proteins, thus raising the intriguing possibility that protein half-lives can be rationally re-engineered [73, 117, 199, 205, 218].

Nonetheless, it should be emphasized that not all proteins are continually susceptible to degradation. In fact, the half-life of total protein can be quite long (ca. 4–7 days in non-stressed plants [44, 103, 148, J. Walker and R.D. Vierstra, unpublished]) indicating that only a small percentage of proteins (<10%) undergoes rapid breakdown at any given moment and that most proteins actually turn over very slowly. But, these short-lived proteins are often responsible for rate-limiting steps in metabolic pathways or act as critical regulators [34, 77, 103, 207]. By helping control the levels of these proteins, degradation can have a profound, but energy cost-effective, influence on cell biology.

In the past decade, substantial progress has been made toward our understanding of protein degradation in both prokaryotes and eukaryotes [for reviews see 34, 77, 143, 207]. Evidence has emerged that several distinct pathways exist in plants with each cellular compartment having one or more (Fig. 1). The types of pathways are consistent with the evolutionary origin of each compartment; for example, chloroplasts and mitochondria appear to use pathways similar to those found in prokaryotes whereas the cytoplasm and nucleus have pathways in common with other eukaryotes [25, 207]. Examination of several of these proteolytic pathways has allowed us to answer, at least in a rudimentary way, some of the fundamental questions concerning protein turnover: What is the nature of the energy requirement? Where does proteolysis occur? How does proteolysis occur so rapidly and completely? And, how is proteolysis so exquisitely selective?

Degradation of protein in the cytoplasm and nucleus

Ubiquitin-dependent proteolytic pathway

Our first insights into how cytoplasmic and nuclear proteins are degraded were made ca. 15 years ago by Hershko and colleagues with the discovery of a major proteolytic pathway involving the small protein, ubiquitin [for reviews see 34, 95, 97]. In this pathway, short-lived proteins are broken down by a multi-subunit protease called the 26S proteasome following their conjugation with multiple molecules of ubiquitin. The pathway was characterized initially using rabbit reticulocytes and subsequently has been shown to exist in a wide range of other eukaryotes, including humans, yeast (*Saccharomyces cerevisiae*), *Drosophila*, *Caenorhabditis elegans*, *Arabidopsis thaliana*, wheat, and various other plant species [34, 75, 109, 207]. Most, if not all, of the major steps have been elucidated from studies with these organisms and are illustrated in Fig. 3.

Given ubiquitin's central position in the proteolytic pathway, understanding its unusual structure and mode of synthesis has been helpful in determining how the pathway functions mechanistically. As the name implies, ubiquitin is indeed ubiquitous, being present in species from all kingdoms. It is arguably the most conserved protein yet identified; its 76-amino acid sequence is identical among all higher-plant species analyzed to date and differs by only one residue to that in the alga *Chlamydomonas reinhardtii*, by two residues to yeast ubiquitin, and by three residues from the invariant sequence present in animals [26]. Sequence homologues have also been found in an archaebacterium [217] and a eubacterium [54]. Whether the ubiquitin pathway is widely distributed in these prokaryotic kingdoms is unclear; for example, *Escherichia coli* does not appear to contain ubiquitin [143].

X-ray crystallographic structures of plant and animal ubiquitins show that the molecule consists of a compact globular domain with a flexible, protruding C-terminus (Fig. 2). The compact structure is stabilized by extensive hydrogen bonding that accounts for ubiquitin's unusual resistance to acid, base, and heat denaturation and its ability to rapidly refold to its native conformation once unfolded [20, 209]. As will be seen below, the exposed C-terminal Gly-76 participates in a number of essential reactions in the path-

278

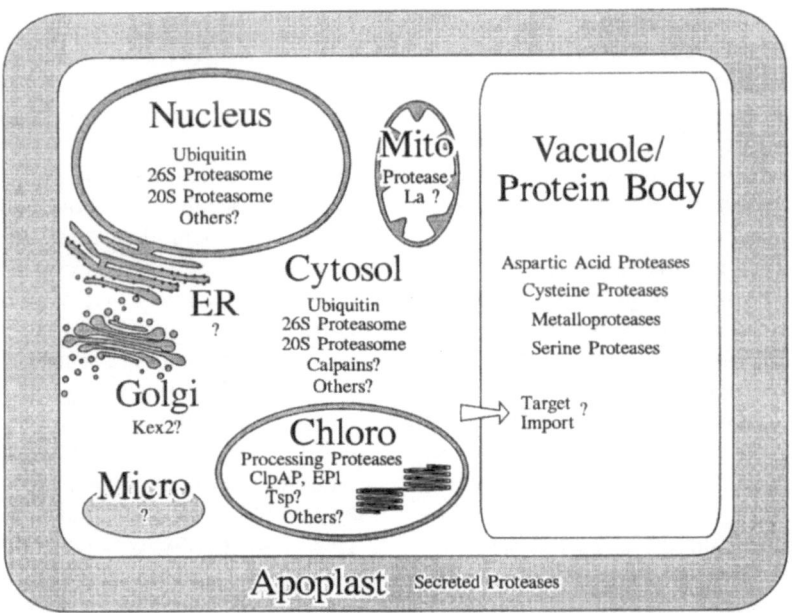

Figure 1. Cellular location of various plant proteases and proteolytic pathways in plant cells. Diagrammatic representation of the subcellular compartments within a typical plant cell along with the proteases that have been identified within the compartment. Details of each protease or pathway are described in the text. Question marks denote proteolytic activities or pathways that have been detected but not yet confirmed or proposed but not yet detected in plants. Chloro, chloroplasts; ER, endoplasmic reticulum; Micro, microbodies; Mito, mitochondria.

way. Its substitution or removal [which occurs rapidly in plant extracts by endogenous proteases cleaving at Arg-74 (Fig. 2)] renders the molecule completely inactive [209]. In plants, as in other eukaryotes, ubiquitin is primarily localized in the cytoplasm and nucleus with trace amounts in the vacuole and membrane fractions [15]. It is not present within plastids despite early reports to the contrary [see 15].

Ubiquitin is also unusual among eukaryotic proteins in that it is encoded by complex multi-gene families that synthesize ubiquitin as a natural protein fusion (Fig. 3) [26, 109]. While the reason behind this organization is unclear, its conservation among organisms as diverse as unicellular eukaryotes, angiosperms and mammals, suggests that it serves an important role in ubiquitin synthesis and/or function. In plants, a number of ubiquitin fusion genes have been described; the best characterized family is from *Arabidopsis* where 14 different ubiquitin genes exist (*AtUBQ1–14* [22, 27, 28]). In each case, functional ubiquitin monomers are released from the fusion protein by a unique group of proteases, designated ubiquitin C-terminal hydrolases (or ubiquitin proteases), that specifically cleave the α-amino peptide bond that follows the C-terminal Gly-76 of each ubiquitin moiety [26, 109, 97]. Processing is

rapid in plants and may occur co-translationally, thus preventing unprocessed ubiquitin fusions from accumulating [63, 100]. The α-amino hydrolases are constrained to having ubiquitin sequence at the N-terminal side of the cleavage site, but are unaffected by sequence at the C-terminal side of the cleavage site, provided that proline is not the first residue [205]. The uncommon specificity of these hydrolases has allowed synthetic ubiquitin fusions to be exploited as a novel method to express proteins *in vivo* with N-termini besides methionine [7, 55, 100, 205].

In one type of ubiquitin gene fusion, tandem arrays of ubiquitin coding regions are fused, thus directing the synthesis of a polyubiquitin precursor [26, 109] (Fig. 3). Polyubiquitin genes containing 6, 5, 4 and 3 ubiquitin repeats are present in *Arabidopsis* [22, 28]; two seven-repeat and two six-repeat genes are found in maize [33] and sunflower [17], respectively; two four-repeat and one six-repeat genes exist in flax [2]; and an astonishing 52 ubiquitin-repeat gene was detected in *Trypanosoma* [191]. Although the nucleotide sequences do vary among the ubiquitin coding repeats, each encodes the canonical ubiquitin amino acid sequence. The last repeat of each polyubiquitin gene almost always encodes extensions of one to sever-

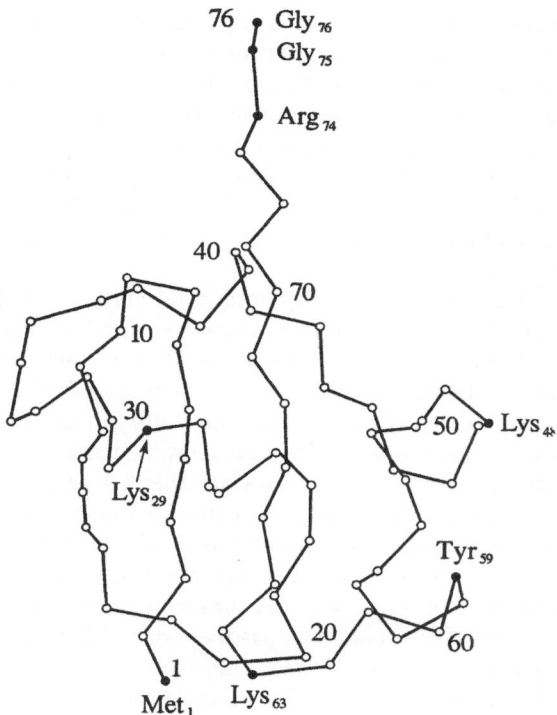

Figure 2. Three-dimensional structure of plant ubiquitin as determined by X-ray crystallography [210]. The lysine residues at positions 29, 48, and 63 which can participate in forming multiubiquitin chains are indicated. Tyr_{59} is typically modified with ^{125}I to track ubiquitin *in vitro* reactions.

al additional amino acids before the termination codon; these extra residues presumably prevent the ubiquitin pathway from using these polyubiquitin proteins until they are processed into monomers [26, 109].

In another type of gene fusion, single ubiquitin coding regions are appended to the $5'$ end of those encoding one of two unrelated ribosomal subunits, thus expressing ubiquitins with long C-terminal extensions [26, 27, 109] (Fig. 2). In plants, these extensions are either 52 or 79–82 amino acids long and are 70–85% identical to counterparts in animals and yeast [27]. Only after removal of the ubiquitin moiety, do these extension polypeptides associate with the ribosome [27, 60]. Because these ribosomal subunits are naturally expressed only as ubiquitin fusions and because they express poorly in an unfused form, it has been speculated that the ubiquitin moiety assists in the translation and/or stability of the subunits prior to their integration into the ribosome [60].

Loci predicted to encode ubiquitin-like proteins have also been found in *Arabidopsis*, yeast, several

animals, and Baculoviridae viruses [see 28, 87, 109, 132 and references therein]. Some are organized similarly to the polyubiquitin genes whereas others have a single ubiquitin-coding region with a C-terminal extension. Several of the non-plant versions are expressed and functional. In fact, a human ubiquitin-like protein, whose expression is enhanced by γ-interferon, can become conjugated to other proteins [132]. A majority of the *Arabidopsis* loci are not transcribed and thus likely represent pseudogenes [28]. However, a cDNA derived from one of these ubiquitin-like loci has been detected in an *Arabidopsis* cDNA library suggesting that some are functional genes [28].

Following synthesis of ubiquitin monomers, the first step in the ubiquitin-dependent proteolytic pathway is the covalent ligation of ubiquitin to proteins destined for breakdown (Fig. 3). This post-translational modification is accomplished by an enzymatic cascade involving ubiquitin-activating enzymes (or E1s), ubiquitin carrier or conjugating enzymes (or E2s), and sometimes ubiquitin-protein ligases (or E3s) [for reviews see 34, 95, 97, 109, 207]. In the first step, an E1 directs the ATP-dependent formation of a high energy thiol-ester intermediate, created by linking the C-terminus Gly-76 of ubiquitin to one of its cysteines [90, 91]. The activated ubiquitin is then transferred from the E1 to a specific cysteine in an E2 via transesterification. Finally, the E2 either ligates the ubiquitin directly to the target protein or transfers the activated ubiquitin to an associated E3 via another transesterification step [170]; the E3, in turn, transfers ubiquitin to the target protein. Studies with a number of eukaryotic species indicate that conjugation is hierarchial [34, 97, 207]. In yeast and *Arabidopsis*, for example, only one or two related E1s execute ubiquitin activation [53, 92, 144], but a multitude of E2s and E3s assist in the transfer of E1-bound ubiquitin to various targets [34, 109, 207]. Ubiquitin is linked to the target protein via an isopeptide bond between the C-terminal Gly-76 of ubiquitin and free lysyl ϵ-amino groups within the target. Structural studies of several ubiquitinated proteins have led to the notion that ubiquitin attachment to the target is often not restricted to contextually specific lysine(s), but in fact can be quite promiscuous [101, 199, K. Lohman and R.D. Vierstra, unpublished].

In a few cases, a single ubiquitin is appended to the target [95, 109]. These monoubiquitinated proteins appear to be metabolically stable, suggesting that adding a single ubiquitin moiety does not commit a protein to degradation but may serve to alter protein structure or function, possibly in a manner analogous to protein phosphorylation. However in most cases, the conjugation cascade modifies the target protein with multiple ubiquitins [95, 109]. Although this modification could occur by attaching single ubiquitins to different lysine residues within the protein, it most often occurs by attaching one or more chains of ubiquitin monomers. These chains subsequently provide a strong signal for degradation [29]. Multiubiquitin chains consist of ubiquitins linked together through ε-amino isopeptide bonds between the C-terminal Gly-76 of one ubiquitin and lysine residues in the adjacent ubiquitin. Lys-48 appear to be the most common residue involved in this intermolecular connection [29, 201]; through interactions among neighboring ubiquitins, these Lys-48-linked chains assemble into compact polymers with 2-fold symmetry [37]. In addition to Lys-48, several studies have implicated Lys-29 and Lys-63 in chain assembly [6, 112, 184]. All three of these lysines are found on the surface of ubiquitin's three-dimensional structure (Fig. 2).

How multiubiquitin chains are generated is unclear but two mechanisms are possible. The chains could be assembled directly on the target by reiterative rounds of ubiquitination or they could be preassembled as free chains and then attached *en masse* to the target in a single step. While current opinion favors the former route, three lines of evidence support the latter as a possible mechanism. First, several E2s have been identified in mammals and plants that can assemble multiubiquitin chains *in vitro* [31, 200]. One family encoded by wheat *TaUBC7* and *Arabidopsis AtUBC7/13/14* genes forms such chains using Lys-48 as the exclusive linkage [200, 203]. Second, free multiubiquitin chains can be detected in a variety of eukaryotes including several plant species and are often the most abundant ubiquitin conjugates present in cell extracts [201]. Third, free chains are as kinetically competent as ubiquitin monomers in ubiquitin conjugation reactions [31, 201]. Collectively, these data imply that the ubiquitin pathway can synthesize free multiubiquitin chains and use them directly in conjugation reactions *in vivo*.

Once a protein is tagged with one or more multiubiquitin chains, it has two possible fates. The ubiquitin moieties can be removed by one of a group of ubiquitin C-terminal hydrolases that specifically cleaves ubiquitins linked via isopeptide bonds [34, 97] (Fig. 3). These ε-amino hydrolases are potentially distinct from the α-amino hydrolases responsible for processing ubiquitin translational fusions [see above]. While some likely help recycle functional ubiquitins during tar-

get degradation by removing the residual peptide fragments from ubiquitin's C-terminus, others can deubiquitinate intact proteins. Recent observations that specific hydrolases are intimately involved in cell division [159] and certain aspects of development (e.g. eye cell fate in *Drosophila* [102]) and that there are 15 or more distinct types of hydrolase proteins in yeast [97] imply that deubiquitination may have important regulatory functions.

Ubiquitin C-terminal hydrolases fall into two broad classes [97]. One class of relatively small proteins (ca. 20 kDa) appear to remove small molecules (e.g. peptides, lysine, and glutathione) from ubiquitin's C-terminus. Surprisingly, one of these comprises up to 5% of the total protein in animal neuronal tissue [215]. Hydrolases in the other class are much larger (50–300 kDa) and cleave ubiquitin from a range of proteins. Some prefer ubiquitins linked via an α-amino linkages, ε-amino linkages, or can accommodate both. Members that prefer ε-amino linkages include yeast DOA4 [159] and mammalian isopeptidase T [88], both of which may function in the disassembly of multiubiquitin chains. In addition to their larger size, the second class is defined by the presence of two conserved motifs, one containing an essential cysteine and the other containing two essential histidines that are necessary for catalysis [102]. Little is know about ubiquitin C-terminal hydrolases in plants. Activities corresponding to both α- and ε-amino hydrolases have been detected in wheat germ [188]. Recently, several *Arabidopsis* genes have been identified that encode proteins structurally related to the large-size class of hydrolases including the presence of the conserved Cys and His boxes [N. Yan, T. Falbel and R.D. Vierstra, unpublished].

A second fate of ubiquitin conjugates is that they can be degraded by the 26S proteasome, a 1.5-MDa ATP-dependent proteolytic complex specific for such intermediates [95, 161, 193] (Fig. 3). The 26S proteasome degrades the target protein into amino acids and short peptides but releases the ubiquitin moieties in free, functional forms. In this way, ubiquitin serves as a reusable recognition signal for protein breakdown.

20S and 26S Proteasomes

The 26S proteasome contains ca. 30 polypeptides that dissociate in the absence of ATP into two subcomplexes of 20S and 19S, both of which are approximately 700 kDa in size [161, 164, 193] (Fig. 4). The 20S particle (known as the 20S proteasome, multicatalytic protease, or macropain) contains the catalytic core of the protease and is ATP-*in*dependent. It is present in both the nucleus and cytoplasm of animals [95, 193] and plants [25, 171, 193] with related species also found in some archaebacteria [135] and eubacteria [192]. Its distinctive hollow cylinder shape, which can be detected in plant extracts by electron microscopy [156, 171, 193], is created by the assembly of four stacked rings, each of which contains seven polypeptides (Fig. 3).

In the archaebacterium, *Thermoplasma acidophilum*, the subunit composition of the 20S particle is simple, the two outside rings are formed by identical α-polypeptides and the two inside rings are formed by identical β-polypeptides [135] (Fig. 4). Its composition in animals and plants is more heterogeneous, involving as many as 14 different α-like and 14 different β-like polypeptides that range in size from 22 to 35 kDa [156, 193]. Some β-type subunits are made as larger precursors that require proteolytic removal of an N-terminal extension prior to integration into the complex. In yeast, missense mutations within three different β-type subunits (PRE1, 2 and 3) leads to a slow growth phenotype, hypersensitivity to stress, and a failure to degrade ubiquitin conjugates, whereas complete disruptions of the corresponding genes are lethal [93, 176, 193]. Genes encoding two α-like and one β-like subunits have been identified in *Arabidopsis* that display greater than 50% amino acid sequence identity to counterparts in yeast and various animals [67, 69, 181]. An *Arabidopsis* line bearing a chromosomal deletion of one of the α-like subunits is phenotypically normal suggesting that a multi-gene family exists or that this subunit is non-essential [181]. In spinach, some of the 20S polypeptides may be glycosylated [171].

At least five types of protease activities are associated with the 20S complex, including chymotrypsin-like, trypsin-like, and peptidyl-glutamyl bond hydrolyzing activities [161, 164, 193]. One or more protease activities reside in members of the β-subunit family which use a novel active-site involving the N-terminal threonine [175]. Several selective chemical inhibitors of the mammalian 20S complex have been discovered that are effective both *in vitro* and *in vivo* [58, 165]. One of these, lactacystin, acts by covalently binding to the active-site threonine in one or more of the β-like subunits [58]. Whether these inhibitors are also effective in plants is currently under investigation [J. Walker and R.D. Vierstra, unpublished]. While the 20S complex can degrade unfolded proteins completely in the absence of ATP, it has difficulty with native proteins

implying that the 19S complex assists as an 'unfoldase' [51]. The cleavage patterns of the purified 20S proteasome is, for the most part, non-specific and typically generates peptide fragments 6–9 residues long [213]. *In vivo*, most of these peptides would then be completely degraded to amino acids by cytosolic peptidases. However, in a special case involving the presentation of foreign antigens in mammals, these peptides are transported to the endoplasmic reticulum (ER) and ultimately to the cell surface where they are presented to the immune system by MHC class I molecules [58, 165]. In mammals, the subunit composition of the rings can be altered by γ-interferon, suggesting that the catalytic specificity of the 20S proteasome can be modified by developmental or environmental cues [64].

The crystal structure of the 20S proteasome from *Thermoplasma* was recently solved to 0.34 nm [135]. It shows the complex to contain 3 cavities (Fig. 4). The central cavity is created by association of the two equatorial, β-subunit-containing rings and harbors the active-site threonines [135, 175]. The outside cavities are positioned at the interface between the α- and β-subunit-containing rings and form a narrow channel restricting access to the central lumen. In this way, the site of proteolysis is spatially isolated from the rest of the intracellular milieu in a structure that would allow only unfolded proteins to enter and amino acids and small peptides to exit [76].

The 19S particle binds to one or both ends of the 20S proteasome (Fig. 4). Because it imparts both ATP and ubiquitin dependence to the 26S particle, it is often referred to as the regulatory complex [161, 193] or the 700-kDa proteasome activator (PA700 [45]). Presumably, this complex assists in the recognition of ubiquitinated substrates, unfolds them, and then facilitates entry of the unfolded substrates into the lumen of the 20S proteasome. Electron micrographs of the 19S complex from rat and spinach show an identical V-shaped structure, the interior of which could be the site of protein unfolding [193] (Fig. 4).

The 19S complex contains approximately 15 subunits ranging in size from 35 to 110 kDa [161, 193]. At least five subunits have been identified as members of a newly recognized ATPase family, suggesting that they couple protein unfolding to ATP hydrolysis in a similar manner to chaperonins [45, 161]. Deletions of several of these in yeast arrest cell division [70, 80]. Another subunit, DOA4 in yeast, was recently shown to be a ubiquitin C-terminal hydrolase [159]. This hydrolase appears to regenerate free ubiquitins during the final stages of conjugate digestion as its deletion res-

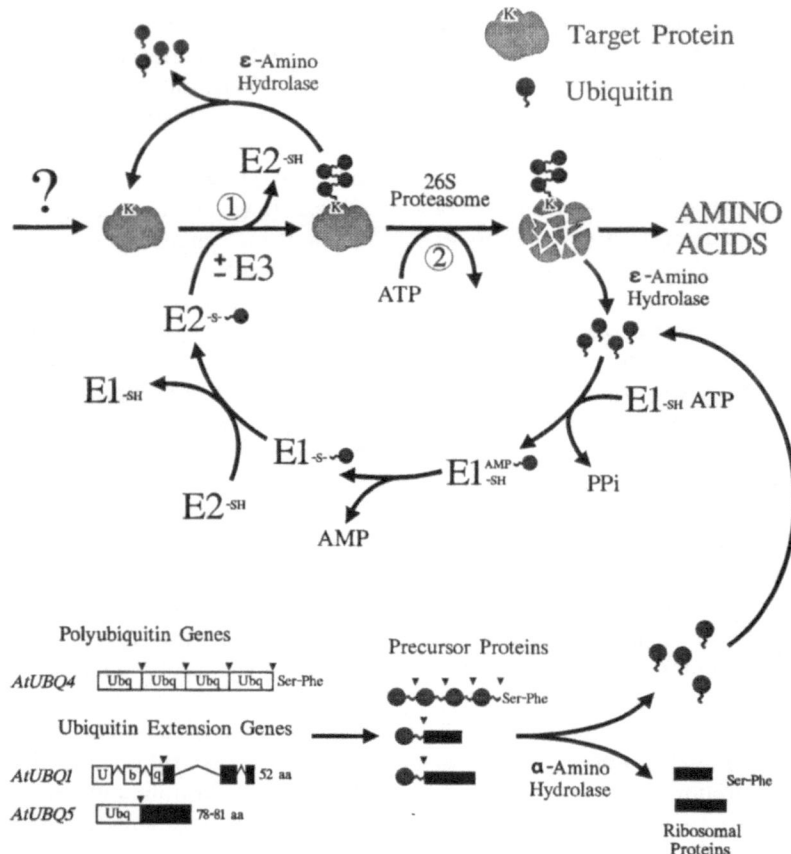

Figure 3. Pathway for ubiquitin-dependent proteolysis. The pathway begins with synthesis of ubiquitin fusion proteins, either polyubiquitin or ubiquitin extension, followed by their processing by α-amino ubiquitin C-terminal hydrolases to release ubiquitin monomers. Several of these ubiquitin monomers are then ligated to a protein targeted for degradation using an ATP-dependent reaction sequence involving E1s, E2s, and possibly E3s. The ubiquitinated protein is either disassembled by ε-amino ubiquitin C-terminal hydrolases or degraded to amino acids and peptides by the ATP-dependent 26S proteasome with the concomitant release of free ubiquitin. The ubiquitin genes provided as examples are *AtUBQ1, 4* and *5* isolated from *Arabidopsis thaliana* [22, 27]. K, lysine involved in ubiquitin attachment; Ubq, ubiquitin.

ults in the accumulation of ubiquitin chains linked to small peptides. As expected, the 19S complex also contains a subunit that binds ubiquitin [48, 202]. The gene encoding this 50-kDa polypeptide, designated MBP1 (for *m*ultiubiquitin *b*inding *p*rotein), was first isolated from an *Arabidopsis* cDNA library using free multiubiquitin chains as probes [202]. Sequence analysis subsequently showed MBP1 to be a member of a highly conserved gene family present in a wide variety of other eukaryotes including *Caenorhabditis*, *Drosophila*, man, yeast, rice, and castor beans [202]. The plant 19S particle has been isolated and visualized by electron microscopy [193], but only the MBP1 subunit has been characterized to date (see below).

Specificity of the ubiquitin pathway

Within the ubiquitin pathway, two important recognition events occur that determine specificity; the first selects appropriate substrates for ubiquitination, and the second identifies ubiquitin conjugates for breakdown by the 26S proteasome (Fig. 3). The first recognition event encompasses step(s) in the conjugation cascade involving E2s and/or E3s [34, 109, 207].

E2s are a heterogenous family of enzymes, generally ranging in size from 14 to 35 kDa, with distinct substrate specificities and E3 requirements [109, 207]. Structurally, all E2s share a common 150-amino acid core domain that has a pocket containing the essential cysteine required for forming the E2-thiol ester intermediate [36]. Even though there can be substan-

Proteasome

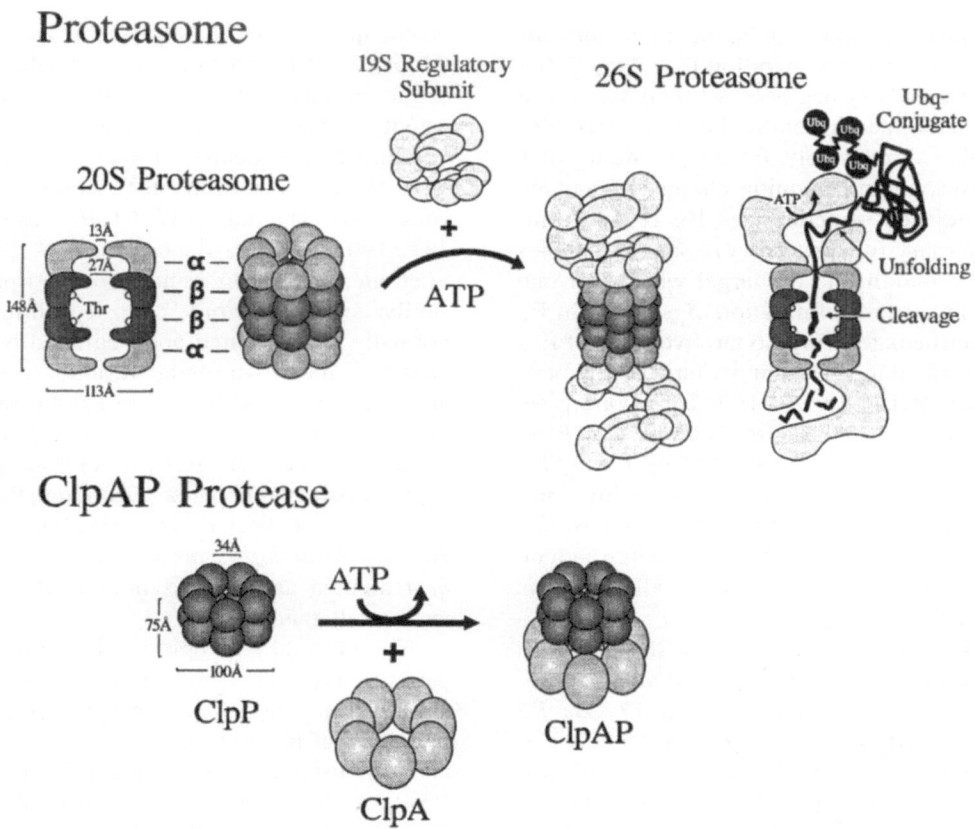

Figure 4. Proposed structures of the 20S and 26S proteasomes, involved in ubiquitin-dependent proteolysis, and the ClpAP protease. Structure of the 20S proteasome was determined by X-ray crystallography of that derived from the bacterium *Thermoplasma* [135]. Proposed structures of the 19S regulatory subunit of the 26S proteasome and ClpAP were created from electron microscopic images of the particles prepared from spinach and rat [193] or *E. coli* [61, 143], respectively. Details of the proteolytic complexes are described within the text. Thr, active-site threonine; Ubq, ubiquitin.

tial amino acid sequence dissimilarity within the core of different E2s (of up to 70%), they fold into a similar three-dimensional structure [36, 38]. Certain E2s also contain additional sequences within the core or extending beyond the N- or C-termini [109, 145]. Several of these additions allow E2s to interact directly with specific substrates (at least *in vitro*) suggesting that they play a role in substrate binding in the absence of an E3 [189, 200]. In fact, it has been shown that either transfer of natural C-terminal extension from one E2 core to another or addition of a synthetic protein-binding domain to an E2 core can be sufficient to confer E3 independence and appropriate substrate recognition [81, 189].

In yeast, genetic analysis has identified twelve E2 genes encoding eleven structurally different proteins (the exception being the closely related *ScUBC4* and

5 genes [34, 109]). A number of E2 genes have been discovered in plants as well. In *Arabidopsis*, seventeen E2 genes have been characterized to date encoding six different E2 types (*AtUBC1-17* [13, 68, 72, 190, 203]); four of these have counterparts in yeast. A cDNA encoding a seventh plant E2 type was recently isolated from tomato with homology to a 25-kDa mammalian E2 [31], but unrelated to any of those in yeast [S. van Nocker and R.D. Vierstra, unpublished]. Mutations in many of the yeast E2 genes lead to distinct phenotypes suggesting that the corresponding proteins ubiquitinate different substrates [34, 109]. For one target, the MATα2 repressor, multiple E2s work in concert suggesting that target specificity can be further expanded by various permutations of E2s forming heteromeric complexes [30].

In most cases, E3s appear to be the main elements responsible for substrate recognition [34, 205]. Little is known of these enzymes because their large size (100 to >300 kDa) and instability have impeded biochemical studies. Presumably, E3s have binding sites for both a corresponding ubiquitin-charged E2 and the target (or an adjacent structure [see 10, 169]). Formation of the ternary (or quaternary) complex enables transfer of the ubiquitin to the target which, for one class of E3s, involves the formation of a ubiquitin-E3 thiol-ester intermediate [170]. So far, five types of E3s have been described in yeast and mammals with evidence that others exist [10, 79, 169, 205]. One type, exemplified by the 225 kDa ScUBR1 protein from yeast, functions with the E2 encoded by the *RAD6* (or *ScUBC2*) gene and is responsible for ubiquitinating proteins based on the nature of their N-terminal residue (see below [205]). Its likely cognates include the rabbit E3s, E3α and β, that specifically interacts with a 14-kDa E2 [34, 95]).

Another E3 type, exemplified by the 100-kDa human E6-AP protein (and its possible yeast homologue ScUFD4 [112]) interacts with a 17-kDa E2 encoded by the *UBCH4/5* gene family [169, 170]. It recognizes a variety of intracellular substrates, including the tumor suppressor p53 when p53 is bound to the papillomavirus protein E6 [169]. Both E6-AP and ScUFD4 have a consensus ca. 30-residue C-terminal sequence, defined as the *hect* domain (for *h*omology to *E*6-AP *C-t*erminus), that includes a contextually conserved cysteine essential for ubiquitin transfer [104, 112, 170]. This *hect* signature was detected in a number of other proteins with previously unknown functions; subsequent biochemical studies on several of these showed that they have 'E3-like' activity [104]. Beyond the *hect* domain, the proteins share little sequence homology, suggesting that the rest of each molecule is involved in E2 specificity and/or the recognition of distinct substrates. Yeast RAD18 protein, required in postreplicative DNA repair, may represent a third type of E3 activity [10]. RAD18 associates with both the RAD6 E2 and single-stranded DNA and may help RAD6 ubiquitinate specific chromatin-associated proteins during the repair of single-stranded DNA gaps. This mechanism of *trans*-conjugation for RAD18 has been observed for ScUBR1 (and maybe E6-AP) as well [111, 169], indicating that E3s can also interact indirectly with their targets through association with other factors (nucleic acid or protein).

Several lines of evidence indicate that E3 counterparts to ScUBR1 and E6-AP exist in plants. First, biochemical activities similar to ScUBR1 and E6-AP can be detected in partially purified wheat germ extracts as factors that are essential for the conjugating activity of the wheat E2s TaUBC1 and TaUBC8 [71, 72]; based on sequence homology and enzymatic analyses, these E2s are probable functional homologues of yeast RAD6 and human UBCH4/5, respectively [169, 189, 190]. Second, Bachmair *et al.* [9] have identified an *Arabidopsis* mutant (*prt1-1*) phenotypically similar to yeast *Scubr1⁻*; i.e. it stabilizes substrates normally ubiquitinated and degraded because of the nature of their N-terminal residue. And third, several structural homologues to E6-AP (and ScUFD4) have been recently discovered in collections of randomly sequenced *Arabidopsis* cDNAs [P. Bates and R.D. Vierstra, unpublished]. Although their E3 activities have not yet been demonstrated biochemically, the encoded *Arabidopsis* proteins contain the consensus *hect* domain and the essential cysteine in their C-terminal domains.

The other key recognition event in the ubiquitin pathway involves association of multiubiquitinated proteins with the 26S proteasome (Fig. 3). One of the essential binding proteins is likely to be MBP1, recently discovered as an integral component of the 19S regulatory complex (see above [202]). Although *Arabidopsis* MBP1 can bind ubiquitin monomers, it prefers multiubiquitin chains containing four or more ubiquitins [202]: a binding specificity that may explain the need for multiubiquitination prior to target degradation by the 26S proteasome [29]. It is possible that binding of chains to MBP1 not only promotes recognition of conjugates by the 26S proteasome, but also serves to tether ubiquitinated proteins to the 26S proteasome until the entire target is unfolded and degraded.

How MBP1 binds multiubiquitin chains is clear. That *Arabidopsis* MBP1 and its yeast and human homologues can recognize ubiquitin chains even following SDS-denaturation and adhesion to nitrocellulose indicates that the primary sequence of MBP1 and/or a highly stable secondary structure is probably involved [48, 202]. In solution, free MBP1 can act as a potent and specific inhibitor of ubiquitin-dependent proteolysis *in vitro*, presumably by competing for conjugates with 26S proteasome-associated MBP1 [49]. One model for binding of MBP1 to multiubiquitin chains proposes that a hydrophobic patch found twice within many MBP1 homologues associates with a repeated hydrophobic patch on the surface of multiubiquitin chains [14, 202]. Nonetheless, it is likely that MBP1 does not work alone but in concert with other 19S subunits. van Nocker *et al.* [204] have demonstrated recently that deletion of the yeast counterpart is not lethal and only impairs degradation of a subset of ubiquitin pathway targets.

Ubiquitin-independent pathways

In addition to the ubiquitin-dependent system, the cytoplasmic and nuclear compartments of plants and animals likely have other proteolytic pathways. Two may involve the 20S and 26S proteasomes by themselves. For example, degradation of ornithine decarboxylase (ODC) in animals is dependent on the 26S proteasome but independent of ubiquitin [150]. Targeting requires association of ODC with antizyme, a small protein that appears to facilitate docking of ODC with the 26S complex [150]. This precedent shows that the 26S proteasome can recognize short-lived proteins by signals other than ubiquitination. Ubiquitin-independent recognition conceivably could occur through direct interaction of substrates with the proteasome, indirect associations of substrates through other factors (e.g. antizyme), or interactions of substrates by a more generalized, ubiquitin-independent tagging mechanism (e.g. phosphorylation, methylation).

Another proteolytic system could involve the Ca^{2+}-activated neutral protease, calpain, found in a variety of vertebrates, invertebrates, and fungi [42]. While the exact functions of calpain is still unclear, it may be involved in the complete degradation of mature proteins as well as the limited proteolysis of preproteins. Two isoforms of this protease are known that differ in their Ca^{2+} sensitivity. Both exist as a heterodimer between one of two distinct 80-kDa catalytic subunits that contain a cysteine-protease domain linked to a calmodulin-like Ca^{2+}-binding domain and a common 30-kDa regulatory subunit that contains another calmodulin-like domain [42]. Ca^{2+} is required not only for activity but also for autoproteolytic activation of each subunit's proenzyme precursor. Although the presence of calpains has not been unequivocally demonstrated in plants, Ca^{2+}-activated proteases have been detected [162]. As of yet, no plants genes have been isolated with convincing homology to calpain subunits from either animals or fungi (R.D. Vierstra, unpublished).

Protein degradation in organelles

Vacuoles

Vacuoles are the largest organelles in plants, occupying as much as 90% of the total cell volume. Like yeast vacuoles, they contain a variety of hydrolytic activities including a number of proteases, hence they are often called the 'lytic compartments' of the plant cell [44, 139] (Fig. 1). In fact, these vacuolar proteases account for most of the proteolytic activity measured in plant extracts and thus, are likely responsible for one of the main technical problems associated with trying to purify plant proteins intact. A wide range of proteases have been detected including, endo- and exoproteases, amino- and carboxyl peptidases, and aspartic acid-, cysteine- and metallo- and serine proteases; some are commercially important, including papain (papaya), ficin (fig), and bromelain (pineapple) [12]. Most vacuolar proteases perform optimally at acidic pH (pH 3–6), a condition that exists within the vacuole *in vivo*. None have been discovered that are energy-dependent.

Despite the myriad of proteases present, the role of vacuoles in general protein breakdown is still unresolved. By analogy with animal lysosomes, it was originally proposed that plant vacuoles are responsible for degrading most cellular proteins, including those from the cytoplasm and chloroplasts which were thought to enter the vacuole primarily via autophagy [139]. However, (1) the subsequent identification of plant proteolytic pathways outside of the vacuole [25, 207], (2) the ability of yeast defective in major vacuolar proteases to degrade normally protein from other compartments [113, 194], and (3) the ability of plants to degrade intracellular proteins even when most vacuolar protease activities are inhibited [148] have led researchers to question this role and, in fact, suggests that vacuolar proteases contribute little to total protein breakdown in a typical plant cell.

Nevertheless, collective evidence does not rule out more specialized proteolytic roles for plant vacuoles. In fact, recent data suggests that plants contain two types of vacuoles, one with an acidic pH like the lysosome [221]. One well documented proteolytic function involves protein bodies, a specialized form of the vacuole responsible for the storage and mobilization of protein reserves during seed germination [59, 216]. During seed maturation, specific storage proteins are synthesized on rough endoplasmic reticulum (ER) and subsequently transported into these membrane-bound vesicles. The vesicles likely arise from fragmentation of larger vacuoles. Proteins within these protein bodies are stable until germination, at which time specific proteases are synthesized *de novo* and transported into the vesicles to initiate proteolysis [16, 59, 216]. Some of the storage protein-degrading proteases are related to the cathepsin class of cysteine proteases, found in mammalian lysosomes [16, 99, 124]. This type of storage and mobilization is not restricted to seeds but can also be observed in leaves, seed pods, and seedling hypocotyls. Here, a small family of vegetative storage proteins are synthesized and sequestered in vacuoles during periods of high nitrogen availability and are subsequently degraded when the tissue becomes nitrogen-limited or senescent or when the stored amino acids are needed by sink tissues [185].

In addition to degrading storage proteins, vacuolar proteolysis likely serve other functions. Vacuolar proteases (and other hydrolytic activities) may help plants defend against pathogens, parasites, and herbivores by attacking the invader once the plant cell is lysed [18, 141]. Consistent with this role is the vacuolar location of various proteinaceous inhibitors to animal and fungal proteases [166]. Vacuolar proteases could act during the final stages of plant senescence by degrading any remaining cytoplasmic and organellar substrates after rupture of the vacuolar membrane. They may assist in the proteolytic processing of vacuolar zymogens. And finally, vacuolar proteases may help supply free amino acids during times of rapid growth, starvation, or stress [185, 194]. In addition to storage proteins, this breakdown could involve a variety of other vacuolar and cytosolic proteins.

If cytosolic targets are involved, stress-enhanced vacuolar degradation would necessitate active transport of proteins into the organelle. Although such a transport system has not yet been demonstrated in plants, an animal paradigm exists involving lysosomes, which are responsible for enhanced protein degradation during nutrient deprivation [50]. The pathway employs a cognate of the heat-shock 70 kDa protein family, PRP73, whose abundance in the cytosol increases about twenty-fold during starvation [32]. PRP73 binds to a group of cytosolic proteins that all bear a consensus motif, KFERQ, and then facilitates the ATP-dependent transport of the bound proteins into the lysosome where degradation commences [50]. About 20–30% of cytosolic protein in mammalian cells have this motif and thus are enlisted by this accelerated lysosomal degradation. It would be appealing to evoke a similar degradation system for plant vacuoles, especially during nitrogen deprivation and leaf and flower senescence.

Chloroplasts

Chloroplasts are protein-rich compartments in plants, containing up to 50% of the total cellular protein in photosynthetic tissue. As a result, much attention has been given to understanding how chloroplast proteins are degraded, especially during leaf senescence when much of the protein lost is of chloroplastic origin [43]. Chloroplast proteins were originally proposed to be degraded by vacuolar proteases [43] and more recently by the ubiquitin pathway [see 15, 207]. However, it now appears that neither mechanism is involved and that chloroplasts have variety of internal proteases, some of which require ATP [130, 137, 180] (Fig. 1).

One important chloroplast protease is related to the bacterial ATP-dependent protease Clp, first identified in *E. coli* [143]. Bacterial Clp is composed of two types of subunits: ClpP, a 21 kDa serine protease, and ClpA, a 81 kDa ATPase that uses ATP hydrolysis to activate ClpP and unfold protein substrates [143, 214]. Protein degradation requires the ATP-dependent assembly of two heptameric rings of ClpP subunits with a single ring likely composed of seven ClpA subunits [61, 119] (Fig. 4). Both the 20S proteasome and the GroEL chaperonin (involved in protein folding) also have a similar hollow barrel structure, suggesting that protein unfolding/folding is fostered by this three-dimensional arrangement [212].

Homologues to both *E. coli* ClpP and ClpA were first discovered in a variety of plants (including *Arabidopsis*, pea, tobacco, tomato, rice, and wheat) by DNA sequence homology [82, 86, 142, 147, 180] and since have been confirmed by functional assays of the encoded proteins [180]. Whereas, the plant ClpP protein is encoded by the chloroplast genome, the ClpA protein is encoded by the nuclear genome, synthesized in the cytoplasm, and transported in the chloroplast where it assembles with ClpP. In this way, the nucleus has the potential to tightly regulate chloroplast proteolysis by controlling the synthesis of ClpA regulatory subunit. Additional regulation may be accomplished by using alternate ClpA subunits. Recent evidence indicates that several types of ClpA proteins are present in bacteria [83], each of which may be differentially regulated and/or recognize distinct groups of substrates.

Besides ClpAP, chloroplasts have a variety of other proteases, including several neutral proteases [131], a prolyl endopeptidase [126], a stroma-located metalloprotease EP1 that may be involved in degrading Rubisco [23], and two proteases required for the removal of transit peptides from imported proteins [155]. None of the proteases for which clones are available appear to be encoded by the chloroplast genome, further underscoring the notion that most plastid proteolysis is accomplished with the help of nuclear-encoded enzymes. However, a maternally-inherited mutant has been characterized from *Oenothera* that blocks processing of chloroplast precursor proteins, indicating that the chloroplast does encode essential functions in this proteolytic process [110]. Another prominent bacterial protease that may be present in chloroplasts is protease La (encoded by the *LON* gene); a protease essential for degrading abnormal proteins in *E. coli* [75, 143]. La is an ATP-dependent serine protease assembled as an oligomer of identical 105-kDa subunits. Recently, a maize gene encoding a homologue of protease La was identified (W. Rapp and S. Barakat, pers. comm.). While the location of the encode protein remains to be determined, the initial translation product contains an N-terminal leader similar to the chloroplast transit sequence.

Other organelles

Currently little is known about how other plant organelles degrade proteins (Fig. 1). In animal cells, there is evidence that mitochondria, microbodies, and the ER and its connected secretory system have proteolytic pathways associated with the maturation and maintenance of each compartment [75, 123, 211]. In each case, degradation is highly selective. For example, rat mitochondria degrade 30–50% of total cellular protein within a hour of synthesis, with an even faster rate if the substrates are abnormal [47]. Animal mitochondria contain several protease activities; one is nuclear-encoded and appears related to the *E. coli* protease La [75].

Following translation, a substantial number of proteins enter the ER where they are extensively modified, assembled into complexes, and sorted *en route* to the Golgi, vacuole, various membranes, or apoplastic space. Proteolysis likely serves three main roles in the ER and its connected secretory system [123]. One is to remove unwanted normal proteins. A second is to dispose of improperly folded or assembled proteins. In animal, such dysfunctional proteins are retained in the ER (possibly by association of improper protein conformations with BiP, an ER-resident chaperonin [116]), where they are rapidly degraded by internal proteases

(half-lives of 10–60 min) [123]. Similar retention and removal of abnormal proteins likely occurs in plant ER as well. For example, several maize mutants defective in endosperm development not only inhibit accumulation of the zein storage protein family, presumably by increased proteolysis, but also concomitantly increase the levels of the plant BiP chaperonin [19].

A third proteolytic role of the ER is to assist in protein maturation by proteolytically processing larger precursors. One example of this in plants may involve systemin, an 18-amino acid peptide hormone synthesized during defense response in tomato [169]. The active peptide is generated by proteolytic processing an inactive 200-amino acid precursor. Processing likely occurs in the secretory system, possibly by a subtilisin-type serine protease related to the yeast KEX2 protease [167]. KEX2 protease is an integral membrane protein of the Golgi and is responsible for generating a number of peptide hormones, including the yeast α-factor mating pheromone, by cleaving at the carboxyl side of dibasic residues, Arg-Lys and Arg-Arg [11]. Plant genes encoding KEX2 homologues have recently been isolated from *Alnus*, *Arabidopsis*, and melon [163, 219].

Proteolysis is also required for the function of microbodies, especially when they differentiate into the various specialized forms during plant development. For example, the shift of glyoxysomes to peroxisomes in cotyledons requires both the import of enzymes involved in photorespiration and the selective removal of enzymes previously involved in lipid β-oxidation [56]. This proteolysis could be triggered by the import of specific proteases or the selective export of the unwanted proteins. The ubiquitin system has been implicated, at least in part, based on the recent discovery that genes essential for peroxisome biogenesis in the yeasts, *S. cerevisiae* and *Pichia pastoris,* encode E2s integrally bound to the cytoplasmic face of the peroxisomal membrane [40, 211].

Functions of proteolysis

A picture is emerging, especially over the last decade, that protein degradation is an essential component in many aspects of plant growth, development, and environmental responses. Although most data on function have been obtained with organisms better suited to genetic manipulation (e.g. *E. coli*, yeast, and *Drosophila* [34, 97, 102, 109, 143, 151]), the conservation of many proteolytic pathways implies that the conclusions drawn from these models systems pertain to plants as well. Based on several lines of evidence, the ubiquitin pathway likely plays a particularly pervasive role in plants. First, expression studies with a number of plant species, using immunoblot analysis for protein or northern blot analysis or GUS-reporter histochemistry for mRNA, show that ubiquitin and many other pathway enzymes are present in most, if not all, plants cells [22, 27, 69, 92, 195]. Second, the pathway is important to normal plant development based on the severe morphological abnormalities induced in tobacco when a functionally impaired mutant of ubiquitin (Lys-48 to Arg) is expressed [8]. Third, a number of plant proteins are probable ubiquitin-pathway targets, judging from the innumerable conjugates that can be detected in extracts from various plant species [15, 22, 201, 209]. However, as in all other organisms, the identity of most of these targets remains to be discovered.

Removal of abnormal/damaged proteins

One of the first recognized functions of proteolysis is its role in cellular housekeeping. Abnormal proteins continually arise by a variety of mechanisms including mutations, biosynthetic errors, spontaneous denaturation and free radical-induced damage and can be accelerated by environmental stresses such as heat shock, desiccation, high-fluence light, disease, nutrient deprivation, and exposure to heavy metals or amino acid analogues [65, 143, 207]. Accumulation of aberrant proteins not only squanders valuable supplies of plant nitrogen but also can disrupt the integrity of subcellular compartments and various macromolecular structures should they accumulate to sufficient levels. Removal of abnormal proteins is especially important to plants where cell division rates are much too slow to passively reduce intracellular concentrations. In many situations, damaged proteins can be repaired or refolded with chaperonins helping to restore native conformations [65]. However, for some proteins or in some situations where the levels of abnormal protein become too high (e.g. heat shock), proteolysis is an important solution. The chloroplast encoded 32-kDa D1 protein, one protein in the core of the photosystem II reaction center complex, is one such example. To maintain photosynthetic electron flow under high light stress, chloroplasts are continually replacing damaged D1 with functional counterparts [140].

Each major compartment in plants must have mechanisms for degrading abnormal proteins (Fig. 1). In the cytoplasm and nucleus, the ubiquitin pathway is an important route [109]. Consistent with this role, perturbations in the ubiquitin pathway heighten the sensitivity of tobacco and yeast to amino acid analogues and other conditions that exacerbate protein denaturation [8, 109]. In yeast, the ScUBC4/5 E2s are essential for removing abnormal proteins [109]. Their plant homologues include the wheat TaUBC8 and *Arabidopsis* AtUBC8-12 E2s [71, 72]. TaUBC8 works with a wheat E3 designated E3γ, at least *in vitro*, which may be a plant homologue of human E6-AP [71, 169]. The TaUBC8/E3γ pair is responsible for most of the ubiquitin-conjugating activity in wheat germ extracts, implying that this E2/E3 pair could play a prominent ubiquitinating role *in vivo* [71]. It is unclear which proteases remove abnormal chloroplast proteins. By analogy with bacteria, ClpAP (and possibly protease La) is a probable candidate given its essential role in removing abnormal proteins in *E. coli* [75, 143].

A variety of conditions the exacerbate protein denaturation also activate proteolytic pathways. For example, genes encoding ubiquitin and a number of conjugating enzymes are activated in plants, yeast, and animals by desiccation, heat shock, heavy metals, and infection [33, 66, 109, 207]. In bacteria, both the La and ClpAP proteases genes are heat shock-inducible [74, 143] whereas, in *Arabidopsis*, the ClpA gene is drought-inducible [122]. It should be noted that the stress activation of specific proteolytic genes may not be universal among plant species; although heat shock induction of ubiquitin genes is strong in maize [33], potato [62] and tobacco [66], it occurs weakly, if at all, in *Arabidopsis* [22]. Whether this implies that the ubiquitin system is not involved in the heat shock response in some plants (e.g. *Arabidopsis*) or is already at sufficient levels in these plants to handle the stress is unknown. Activation of proteolytic pathways is also intimately tied to activation of other stress-related proteins (e.g. chaperonins) involved in damage control [65]. Even in the absence of stress, chaperonin synthesis in bacteria, animals, and plants can be enhanced either by inactivating proteolytic pathways that remove abnormal proteins or by overloading proteolytic pathways with denatured protein [3, 74, 75, 109, 127]. Enhanced accumulation of BiP chaperonin in the ER of maize endosperm mutants defective in zein synthesis is a prime example in plants [19].

Recognition of incorrectly folded proteins probably involves general properties of the target in ways that allows many types of denatured proteins to be detected by chaperonins or proteolytic pathways regardless of their amino acid sequence. The most likely determinant would be an increased exposure of hydrophobic surfaces normally buried in native conformations [65]. This exposure could either provide binding sites for chaperonins involved in refolding/unfolding, recognition sites for proteolysis, or compel proteins to spontaneously aggregate. Aggregation could provide a convenient mechanism to sequester abnormal proteins prior to catabolism [143, 207].

For those aberrant proteins translated from truncated mRNAs, Sauer and coworkers recently discovered that *E. coli* has developed a highly sophisticated mechanism for their removal [117]. Because the termination codon is essential for releasing the nascent polypeptide from ribosome-bound tRNA once translation is complete, incomplete polypeptides remain bound to and thus stall ribosomes. This stalling triggers the association of the 10Sa RNA, a tRNA-like RNA containing a charged Ala, with the ribosome. The ribosome subsequently transfers the nascent polypeptide chain onto the bound Ala moiety. After the truncated mRNA is released, the ribosome switches to translating an internal sequence within the 10Sa RNA encoding the nonapeptide NDENYALAA, which is then followed by a termination codon. Once the chimeric polypeptide is released, the *E. coli* tail-specific protease, Tsp, recognizes the C-terminal addition and rapidly degrades the tagged protein [117, 182]. This translational switch serves two purposes: it facilitates release of defective mRNAs and marks potentially harmful proteins for breakdown. Tsp protease, which can also degrade normal proteins, has homologues in various Gram-positive and Gram-negative bacteria and animals [4, 117, 182]. In the cyanobacterium, *Synechocystis*, a Tsp homologue is responsible for proteolytically processing of the photosynthetic D1 protein [4]. An intriguing possibility is that a Tsp relative also exists in chloroplasts.

N-end rule pathway

Another group of aberrant proteins are those that are improperly processed or become mis-localized. In many cases, these errors generate polypeptides with N-termini atypical of proteins normally localized to the compartment. For example, whereas the N-termini of most cytoplasm and nuclear proteins are either Met, Gly, Ala, Ser, Thr, or Val and/or are N-acetylated [5], those that are incorrectly processed or from the ER and other organelles will not likely have these termini [5, 7]. Varshavsky and co-workers have discovered in yeast, mammalian cells, and *E. coli*, that these foreign N-termini are universally exploited as recognition sites for an N-end rule pathway that removes such unwanted proteins [7, 205]. However, this degradation scheme is not limited to aberrant proteins as the levels of several normal proteins are regulated by this method as well [136, 205].

In the cytoplasm and nucleus of animals and yeast, the N-end rule pathway involves components within the ubiquitin system and contains a hierarchy for amino acid recognition [205]. In yeast, recognition of inappropriate N-termini is accomplished by the RAD6 (or ScUBC2) E2 working in concert with the E3α, ScUBR1 [205]. RAD6/ScUBR1 directly bind to polypeptides with Arg, Lys, or bulky hydrophobic N-termini and hence these amino acids are called primary destabilizing residues within the N-end rule. Asn and Gln are called tertiary destabilizing amino acids because they first must be deamidated by an N-terminal deamidase to generate secondary destabilizing residues, Asp and Glu. Arg is added to these acidic N-termini by an arginyl transferase, using Arg-tRNA as the donor, to then generate the primary destabilizing Arg N-terminus that is finally recognized by ScUBR1.

Components of the N-end rule pathway have been identified in plants. The *Arabidopsis* E2s encoded by the *AtUBC1-3* genes appear to be the RAD6 counterparts based on amino acid sequence similarities and an analogous requirement for E3α *in vitro* [72, 190]. The *Arabidopsis* mutant (*prt1-1*) that is incapable of degrading proteins with Phe N-termini may affect the plant counterpart to ScUBR1 [9].

In *E. coli*, the N-end rule pathway is executed by ClpAP [197]. Its N-end rule hierarchy appears to be more restricted than those in yeast and animals and includes just bulky hydrophobic residues (e.g. Phe, Leu, Tyr, Trp) and Arg and Lys as destabilizing residues [197, 205]. Only bulky hydrophobic residues are primary destabilizing residues; they are recognized directly by ClpAP and degraded. Lys or Arg are secondary destabilizing residues. Leu is added to these N-terminal residues by a Leu/Phe transferase to generate N-termini with the primary destabilizing residues Leu. Given the presence of ClpAP homologues in chloroplasts [82, 180], it is likely that chloroplasts have the accompanying N-end rule pathway as well. In this way, chloroplast ClpAP could assist in the removal of incorrectly processed cytoplasmic precursors, as well as in the removal of misfolded proteins.

Supply of amino acids

Even though plants can synthesis all amino acids *de novo*, a substantial portion of new proteins are derived from recycled amino acids [44, 103, 207]. These amino acids can be generated from the proteolytic housekeeping of abnormal and unwanted proteins, or they can be derived from specialized versions of normal proteins whose sole purpose is to store amino acids. Both seed storage proteins and vegetative storage proteins are examples of this latter group [59, 185, 216]. Degradation of vegetative storage proteins, and possibly other proteins, are accelerated during nitrogen deprivation suggesting that plants have signalling pathway(s) linking the supply of free amino acids to the rate of intracellular proteolysis [185]. By analogy with animal and bacterial starvation responses, these pathways could monitor levels of uncharged tRNAs as an indirect measure of low amino acid supplies [174].

Extracellular proteolysis can also provide an important source of amino acids in plants. This process not only necessitates the export of proteases into the apoplast but also the existence of amino acid and peptide transporters for import of proteolytic products. Both of these components have been detected in a variety of plants [25, 59, 186]. For example, extracellular proteolysis of endosperm storage proteins in cereal seeds provides most of the amino acids used by the developing embryo [59]. This degradation is accomplished by secretion of proteases from scutella epithelia and aleurone layers. During barley seed germination, a complex of proteases are secreted, including two cysteine proteases, EPA and EPB, with specificity toward the hordein storage proteins of the endosperm [124]. Amino acid and peptide transporters, needed to import proteolytic products, have been detected in plants as well; one from *Arabidopsis* is a member of a membrane-transport protein family also found in yeast and animals [see 186]. In a special case of extracellular proteolysis, carnivorous plants have developed a highly sophisticated morphology to trap, digest, and absorb insects as an additional source of amino acids and nitrogen [114].

Control of enzymatic pathways

One of the central functions of proteolysis is to help regulate metabolism by directly controlling the levels of key enzymes. Global use of this strategy first became evident from a survey of enzymes responsible for catalyzing either the first or rate-limiting step in various metabolic cascades; all had short half-lives [44, 77, 103]. In some cases, the short half-life is constitutive, whereas in others, it is developmental or environmentally regulated or induced by limiting substrate or excess product. Such instability allows cells to control the metabolic flux through a pathway simply by attenuating synthesis of a crucial enzyme and then allowing degradation to rapidly reduce its levels [77]. The advantage of degradation over other methods of metabolic regulation is its speed (half-lives of minutes) and that elimination of the protein negates any possibility that the enzyme can be reactivated inappropriately. Examples of such short-lived plants proteins include: NADPH protochlorophyllide oxidoreductase, fructose bisphosphatase, ATP sulfurylase, HMG-CoA reductase, ornithine decarboxylase, squalene synthetase, and phenylalanine ammonia lyase which catalyze important or committed steps in chlorophyll production, carbon, and sulfur metabolism, and sterol, spermine, isoprenoid, and lignin biosynthesis, respectively [25, 44, 103, 207]. Such proteolytic regulation can also encompass whole metabolic pathways. For example, the entire β-oxidation pathway is removed from the cotyledons of dicotyledonous plants during the transition from glyoxylate- to photosynthate-dependent growth [56].

Nitrate reductase (NR), which catalyzes the first step in the conversion of nitrate to ammonia, is one of the best studied plant examples of metabolic regulation by proteolysis [41]. By a combination of mechanisms, NR activity increases in the presence of nitrate and light and decreases in their absence. Darkness or removal of nitrate not only down-regulates NR gene transcription but also induces the rapid, but reversible inactivation of the NR protein. In spinach and maize, NR is inactivated by phosphorylation of the protein [41, 115]. This inactivation is then followed by rapid degradation of the NR polypeptide. Although a connection between phosphorylation and NR proteolysis has not yet been made, it is conceivable that the added phosphates aid in the recognition of inactive NR by specific proteases [41]. As NR is a cytosolic enzyme, participation of the ubiquitin pathway is plausible. A region of ca. 50 amino acids near the N-terminus of tobacco NR appears to be required for its inactivation and/or proteolysis [154].

Protein breakdown also controls the levels of ACC synthase, the enzyme which catalyzes the first step in ethylene biosynthesis, the conversion of *S*-adenosyl-L-methionine to 1-aminocyclopropane-1-carboxylic acid (ACC). This cytoplasmic protein has an extremely short half life (20–120 min depending of the tissue or plant species [118]). Although the mechanism of ACC breakdown is unknown, it does appear to be energy-dependent as uncouplers of oxidative phosphorylation block protein loss [120]. Studies with both protein kinase and phosphatase inhibitors suggest that ACC synthase degradation, like that of NR, is regulated by phosphorylation/dephosphorylation of the protein [183].

With respect to multi-subunit complexes, proteolysis also assists in correcting the inappropriate stoichiometry of subunits and in maintaining correct enzyme/cofactor ratios [207]. Rubisco is an example of a multi-subunit enzyme whose stoichiometry is corrected in this way. In the absence of the chloroplast-encoded large subunit, unassembled nuclear-encoded small subunit is rapidly degraded upon import into the organelle [172]. Similarly, chlorophyll a/b-binding proteins and plastocyanin are rapidly catabolized in the absence of their respective cofactors, chlorophyll and Cu^{2+} [145, 149]. The stoichiometric accumulation of a number of other chloroplast and mitochondrial enzyme complexes may be corrected in a similar fashion to help overcome a lack of precise coordination between the organelle and nuclear genomes. How the proteolytic machineries recognize unassembled subunits is unclear, but because many incompletely assembled complexes are conformationally unstable, they may be detected by the same general features as are abnormal proteins.

Control of various cell regulators

In addition to controlling many metabolic pathways, short-lived proteins also play crucial roles in various regulatory processes, including signal reception and transduction, homeosis, transcription, and cell growth and division [34, 95, 97, 205, 207]. In animals and yeast, a growing number of such regulators appears to be specifically degraded by the ubiquitin/26S proteasome pathway. Examples include transcription factors cJUN, cFOS, MOS, GCN4, the p53 tumor suppressor, V(D)J recombination activator protein RAG2, components of the NFκB transcriptional complex, the yeast G-α protein Gpa1, and the MATα2 repressor involved in yeast mating-type switch [34, 97, 125, 129, 136, 153, 199, 205]. The yeast STE2 plasma membrane receptor is also conjugated with ubiquitin upon α-factor induction, but degradation appears to take place in the lysosome/vacuole [96]. Various other cell surface receptors are ubiquitinated upon ligand engagement,

including the platelet derived growth factor receptor, T-cell antigen receptor, and the immunoglobulin E receptor, but whether this ubiquitination targets the receptors for degradation has not yet been established [34, 101]. In addition to those regulatory proteins for which direct experimental evidence exists, a whole host of other regulatory proteins are predicted to be short-lived based on the phenotypic consequences of specific proteolytic defects. For example, mutations in specific components of the ubiquitin pathway affect DNA repair, peroxisomal biogenesis and protein translocation into the ER in yeast [40, 97, 109, 211], or alter eye cell fate, neuronal development, and cell proliferation in *Drosophila* [102, 138, 151]. These restricted phenotypes suggest that the mutations block the degradation of specific short-lived, but as yet unidentified, regulatory proteins.

In several situations, the chain of events responsible for regulatory protein degradation by the ubiquitin system has been partially deciphered and the domains within the target proteins responsible for their short half-life have been defined. Degradation of the yeast MATα2 repressor requires several E2s (ScUBC4/5, 6, and 7) and involves at least two different recognition domains within the repressor protein [30]. Mammalian c-JUN requires a 27 amino acid sequence near the C-terminus of the protein for ubiquitination, but any lysine(s) can serve as the attachment site [199]. Howley and coworkers have shown that p53 degradation can be accelerated in Hela cells by trans-conjugation. In this case, a third component (E6) provided by the various 'high risk' papillomaviruses promotes contact of p53 with the E6-AP E3 leading to the enhanced ubiquitination of p53 [169].

An intriguing example of proteolytic regulation is the activation of NF-κB, a human transcriptional activator involved in the defense response [158, 198]. It is synthesized as a 105-kDa precursor (p105) that is processed into a 50-kDa mature form (p50) by proteolytically removing the C-terminal half of the molecule. p50 resides in the cytoplasm under non-stressed conditions as a ternary complex with p65 (or RelB) and IκB, an inhibitory protein that masks the nuclear localization signal of the p50/p65 heterodimer. Activation by a number of defense signals, such as tumor necrosis factor α, triggers the selective destruction of the IκB subunit; the rest of the NF-κB complex (p50/p65) then enters the nucleus to transcriptionally activate a number of defense related genes. Both the processing of precursor p105 and the removal of IκB require ubiquitin conjugation and the 26S proteasome [158, 198].

Thus, the ubiquitin pathway can not only selectively remove polypeptides from a multi-subunit complex but can also selectively degrade a limited part of a single polypeptide. Initial studies indicate that selective ubiquitination of IκB requires phosphorylation of the inhibitor at either one of two adjacent serines [21, 198].

Phytochrome A

The best studied short-lived regulatory protein in plants is phytochrome A, a member of a morphogenic photoreceptor family involved in light perception [208]. Phytochromes are dimeric cytoplasmic proteins, with each subunit consisting of a linear tetrapyrrole chromophore covalently linked to a ca. 120-kDa polypeptide. They regulate photomorphogenesis by switching between two photointerconvertible forms, a red light-absorbing from Pr that is biologically inactive, and a far-red light-absorbing from Pfr that is biologically active. Upon conversion to Pfr, phytochromes initiate a diverse array of physiological and developmental responses that allow plants to optimize reception of photosynthetic light and to coordinate their life cycle with daylength. All other members of the phytochrome family are expressed at low levels and are stable both as Pr and Pfr. However, phytochrome A is unique in that it is highly expressed in young seedlings and while stable as Pr ($t_{1/2} > 100$ hr), it is rapidly degraded once converted to Pfr ($t_{1/2}$ ca. 1 hr) [208]. The short half-life of phytochrome A Pfr helps seedlings adapt to continually fluctuating light conditions as they grow through the soil by removing previous light signals stored as Pfr.

Shanklin *et al.* [178] investigated the possible involvement of the ubiquitin pathway in phytochrome A degradation in oat and found that soon after Pfr formation, the chromoprotein becomes rapidly ubiquitination *in vivo*. Like other targets of the ubiquitin pathway subsequently examined, ladders of ubiquitin-phytochrome conjugates were evident following SDS-PAGE, consistent with the addition of multiubiquitin chains of various lengths [178, 179]. Although a direct link between ubiquitination and Pfr degradation remains to be made, various kinetic analyses present a strong case for ubiquitin's involvement: (1) Pfr-induced ubiquitination and degradation could be observed in a variety of plant species, both monocot and dicot [106]; (2) the levels of phytochrome ubiquitin conjugates directly correlated with the extent of degradation [105, 106, 178]; and (3) ubiquitin-

phytochrome conjugates were turned over much more rapidly than Pfr, consistent with the kinetics expected for a degradation intermediate [105, 106]. One interesting aspect of Pfr degradation is that both ubiquitination and breakdown follow a rapid ($t_{1/2}$ ca. 2 s at 25 °C), energy-dependent aggregation of Pfr in the cytoplasm [208]. The function of this aggregation is unknown but it may help quickly inactivate excess Pfr by sequestering it in a form that is amenable to slower proteolytic destruction.

Selective ubiquitination of phytochrome A could involve specific structural differences and/or differential aggregation between the Pr and Pfr forms [208]. The extreme C-terminus of phytochrome A appears to be required as its removal stabilizes the chromoprotein as Pfr [R.C. Clough and R.D. Vierstra, unpublished]. Interestingly, this sequence is highly divergent between phytochrome A and the other more stable phytochromes [208]. Initial studies have also implicated an internal domain as a multiubiquitin chain attachment site (residues 742–790) [179]. However, substitution of the invariant lysines in this domain to arginines does not affect the rate of Pfr degradation, suggesting that ubiquitin attachment may not be restricted to this region [K. Lohman and R.D. Vierstra, unpublished]. Like most other natural substrates of the ubiquitin pathway, the E2/E3 pair involved in Pfr recognition has not been identified.

Other regulatory proteins in plants

The levels of a variety of other plant regulatory proteins are also likely to be controlled by proteolysis but at present, few have been analyzed at this level [207]. One potential example is the *Arabidopsis* homeodomain protein, SHOOTMERISTEMLESS (STM), required for shoot meristem formation [134]. The STM transcript is expressed in the meristem and then rapidly disappears as leaf primordia emerge implying that the protein is only needed within a restricted window of differentiation. A similar fate may befall APETALA3, a floral homeodomain protein essential for early specification of petal and stamens [107]. A number of proteins whose expression is induced by auxin have extremely short half-lives [1]. These data in combination with the discovery that one auxin-insensitive mutant, *axr1*, affects a protein with some amino acid sequence similarity to E1s [128], suggests that the ubiquitin pathway and auxin responses are intertwined. Proteolysis of a key regulatory protein also may be involved in the wound response of tomato as the response can be also be induced *in vivo* by the aminopeptidase inhibitor bestatin [168].

Timing of the cell cycle

To ensure correct progression through mitosis and meiosis, cells have adopted elaborate timing mechanisms and checkpoints to coordinate DNA replication, chromosome pairing and segregation, and cell division [108, 152]. Recent work on the cell cycle in fission and budding yeasts, various metazoans, and mammals has shown that correct traversal requires the timed proteolytic removal and replacement of key regulatory proteins [34, 97, 152]. In fact, proteolysis regulates a number of important checkpoints including: entry in S phase (DNA replication) from G1, progression through S, entry in M phase (mitosis), completion of anaphase, and the exit from mitosis [152].

Two types of proteins that must be degraded for correct progression through the cell cycle bind to and alter the activity of the cyclin-dependent kinase (CDK), the master switch of the cycle [152]. They are the S-phase and M-phase cyclins, which are positive regulators originally discovered because their oscillating levels coincided with the cell cycle [152], and a CDK inhibitor (p40^{SIC1} in yeast [173] or p27 in mammals [157]). Initial degradation of CDK-inhibitor derepresses CDK which in turn initiates passage across the G1 to S boundary. Sequential accumulation of various cyclins then promotes various CDK activities. These activations not only initiate specific steps in G1, S, G2 and M phases but also promotes the rapid degradation of the associated cyclins [152]. In this way, a wave of cyclins is created, signaling that one checkpoint in the cell cycle has been completed and that the next step can be attempted. In addition to cyclins and the CDK inhibitor, data suggest that a third protein must also be degraded for cells to initiate anaphase and may be required for the release of sister chromatids [98]. Finally, proteolytic loss of all cyclins at the end of mitosis prevents inappropriate entry into another round of cell division until the cell is ready [152].

In yeast and animal cells, removal of the CDK inhibitor and various cyclins (both S- and M-phase-specific) requires the ubiquitin-dependent proteolytic pathway [73, 157, 173, 177]. In fact, it was observed long before the discovery of cyclin and CDK inhibitor degradation, that one consistent phenotype of mutants in E1 and several 26S proteasome subunits is cell cycle arrest [70, 80, 109]. In yeast, degradation of the CDK inhibitor p40^{SIC1} requires the E2, encoded by the

CDC34 (or *ScUBC3*) gene [173], whereas degradation of the different cyclins require the E2s either encoded by the *CDC34* or the *ScUBC9* genes [177, 218]. Other factors (possibly E3s) are also involved in cyclin ubiquitination and appear to assemble into a large complex [121]. The S- and M-phase cyclins are degraded by different mechanisms [177, 218]. M-phase cyclins contain a specific targeting signal (called the destruction box) for ubiquitination; transfer of this box to other proteins is sufficient to induce rapid breakdown of the recipient protein in a cell cycle-specific manner [73]. Preliminary data suggest that degradation of S-phase cyclins requires their CDK-dependent phosphorylation [218].

Given that homologous mechanisms of cell cycle control exist in fungi and animals, we expect that the plant cell cycle is also regulated by protein degradation. Plant counterparts to cdc28 kinase and cyclins have been isolated [108]. However, no components specifically involved in the ubiquitination of cell-cycle factors have yet been detected. Transcripts encoding ubiquitin and several 20S proteasome subunits accumulate in proliferating *Arabidopsis* and tobacco cell cultures [66, 69]. Whether this activation indicates a direct role of the pathway during cell division or an indirect role by providing constituents essential to actively growing cells in not known.

Programmed cell death

One of the natural consequences in the development of multi-cellular organisms is the timed disintegration of specific cells [57]. This can be confined to single cells or small regions or can occur on a massive scale and involve whole organs. Examples in plants include leaf, flower and ovary senescence, fruit ripening, xylem and periderm maturation, petiole abscission, programmed abortion of organ primordia in unisex flowers, tapetum and stomium degeneration in anthers, valve dehiscence in seed pods, and the hypersensitive response during pathogen invasion [25, 85, 207]. A number of genetic and pharmacological studies indicated early on that cell disintegration is a highly controlled, complex process initiated by both intrinsic and extrinsic signals. Programmed cell death generally involves the activation of both nucleases and proteases to efficiently degrade the resident nucleic acids and proteins [25, 57, 146]. Presumably, this catabolism economizes the loss of nitrogen and carbon by exporting them to areas of growth or storage. This recycling is most evident during leaf senescence where up to 70% of the total leaf protein can be retrieved [43, 196].

Leaf, flower, ovary, and fruit senescence has been intensively studied because they represent important agriculture problems. Onset of senescence is controlled, in part, by cytokinins and gibberellins and involves activation of a number of proteases [43, 84, 94, 196]. Several of these proteases have been isolated, including *Arabidopsis* SAG2 and SAG12 [94, 133] and pea TPP [84], which are cysteine proteases whose transcripts dramatically increase in senescing leaves and ovaries, respectively. ClpAP mRNA levels also rise during *Arabidopsis* leaf senescence implicating ClpAP in the senescence-induced loss of chloroplast proteins [S.-S. Gan and R. Amasino, unpublished]. The ubiquitin pathway may also be involved. Both ubiquitin and ubiquitin conjugates have been shown to increase in daylily flower senescence [39] whereas perturbation of the ubiquitin pathway accelerates leaf senescence in tobacco [8].

Xylogenesis has been recently studied as an example of a programmed cell death process restricted to individual plant cells. One model system involves the hormone-induced *in vitro* differentiation of *Zinnia elegans* mesophyll cells into tracheids. During *Zinnia* xylogenesis, the accumulation of several proteases has been observed [46, 220]. Ubiquitin has also been implicated in xylogenesis. In intact *Coleus blumei* stems, increased levels of immunodetectable ubiquitin was associated with regions of newly differentiated xylem [35]. Using promoter-GUS fusions, Thoma *et al.* [195] found enhanced expression of a number of E2 genes in differentiating *Arabidopsis* vascular tissue. And finally, marked abnormalities in tobacco vascular tissue can be induced by expressing a non-functional ubiquitin [8].

Applications in biotechnology

In addition to its role in cell physiology, proteolysis also has important ramifications in attempts to improve crop plants through genetic engineering [100]. Here, manipulations of proteolysis can not only enhance the accumulation of foreign proteins intended to confer beneficial traits but also may be used to repress accumulation of unwanted endogenous proteins that interfere with important agronomic processes. In fact, several strategies to enhance or repress protein accumulation by proteolytic approaches have been developed in the past few years [see 81, 100, 205].

One of the obvious problems when attempting to ectopically express proteins to high levels is that plants, like other organisms, often recognize foreign proteins and degrade them rapidly. Thus, even when all other transcriptional and post-transcriptional processes are optimized, proteolysis can still be a major barrier to adequate accumulation [100]. As a result, new emphasis has been placed on understanding the factors that regulate protein stability and on developing methods to interfere with the responsible proteases. In several cases, protein turnover is controlled by short amino acid sequences (e.g. cyclin destruction box, N-terminal residue, KFERQ [73, 50, 205, 218]). Increased stability (and thus increased accumulation) can be engineered, in some cases, simply by removing these instability domains. For example, removal of the destruction box from M-phase cyclins has been shown to stabilize cyclins and permanently induce repetitive rounds of the cell cycle in *Xenopus* oocyte extracts [73]. The recent identification of essential proteolytic pathways also offers the potential to use pharmacological and genetic strategies to inactivate the interfering proteases. In this regard, the newly discovered inhibitors of the 20S and 26S proteasome may be useful in preventing plant senescence processes that require these proteolytic complexes [49, 58, 165]. A genetic approach has been successful in enhancing protein production in *E. coli* and yeast with the disruption of the ClpAP and La proteases [75, 143] or vacuolar proteases [113] and the ubiquitin pathway [109, 205], respectively. However, comparable mutants in plants are not yet available.

In an unusual case, ubiquitin itself has been exploited to augment protein accumulation. This approach, first developed by Butt and colleagues [24, 55], involves expressing proteins as translational fusions to the C-terminus of ubiquitin. It was based on the knowledge that (1) poorly expressed proteins often can be stabilized by fusion with a highly stable protein, (2) ubiquitin is a highly stable protein, and that (3) natural ubiquitin C-terminal fusions exist. In yeast and *E. coli*, expression of ubiquitin-fused proteins was dramatically enhanced; the accumulation of some recalcitrant proteins were increased over 200-fold [24, 55]. Furthermore in several cases, the protein products were found to be more intact and active than their nonfused counterparts. Whereas the engineered products remained as ubiquitin fusions when expressed in *E. coli* [24], in yeast they were correctly processed following the C-terminal Gly-76 of ubiquitin by endogenous ubiquitin α-amino hydrolases to release both the fused protein and ubiquitin in intact forms [55, 205] (Fig. 3).

A similar fusion strategy has been shown to work in plants as well, and can result in significant increases in protein accumulation [63, 100]. As in yeast, these fusions are rapidly processed (possibly cotranslationally) to yield non-fused products. Likewise, the plant hydrolases appear capable of processing many types of synthetic ubiquitin fusions provided that Pro is not the first residue following ubiquitin Gly[76] (D. Hondred, J. Walker, and R.D. Vierstra, unpublished). In addition to enhancing protein accumulation, this method also permits the synthesis of proteins/peptides with N-termini besides Met.

Proteolysis can also be used as a strategy to remove unwanted proteins. One approach involves re-engineering proteins to enhance their recognition by proteolytic pathways. In several cases, the addition of instability domains has been exploited to convert otherwise stable proteins into ones that are rapidly degraded [73, 117, 199, 205, 218]. In a specialized case, Dohmen et al. [52] appended a temperature-sensitive domain for degradation thus creating a fusion protein with a turnover rate that could be environmentally accelerated. An alternative approach could involve re-engineering proteolytic pathways to rapidly degrade stable, unmodified proteins. Gosink and Vierstra [81] recently demonstrated the potential of this approach with respect to the ubiquitin pathway. They showed that E2 target recognition could be redefined in vitro simply by engineering E2s with appropriate protein-binding domains fused to their C-termini. The binding domains could originate from naturally occurring proteins that interact with the target or be artificial peptides selected by their binding affinity [81]. Addition of these binding domains facilitated in vitro ubiquitination of the target which could then lead to the ATP-dependent degradation of a normally stable protein. If successful in vivo, this targeted proteolytic approach could provide benefits not offered by the genetic methods commonly used to attenuate protein accumulation (e.g. antisense and gene silencing). These include: (1) its catalytic nature, (2) the fact that neither the target protein nor its corresponding genes need be altered, (3) the ability to selectively recognize individual proteins or a whole family of related proteins by choosing the appropriate recognition site for E2 binding, (4) and the ability to target proteins not encoded by the host cell [81]. The latter benefit could allow plant cells to be 'preimmunized' against pathogenic invasion by targeting key pathogen proteins for destruction.

Conclusions

As can be seen from the wealth of data accumulating rapidly over the past few years, the field of protein degradation in plants is entering an exciting period. We are beginning to realize that proteolysis is more complex than previously understood, pervades a multitude of cellular processes, and provides cells with a number of creative strategies for regulation. Multiple degradative systems are now known to operate within plants, with several delineated by compartmental boundaries. The constellation of identified plant proteases is growing exponentially. In some cases, we now understand why proteolysis requires energy and where it is consumed, how proteolysis is sequestered from other cellular processes, why proteolysis does not generate large peptide intermediates, and how proteins are recognized and targeted for degradation.

Converging information from bacterial, animal, and plant systems show that the organization of proteolytic systems and their methods for detecting protein targets are frequently conserved, thus allowing information derived from one kingdom to be potentially useful in another. Examples include the detection of bacterial ClpAP [82, 180] and La protease [75] homologues in plants, respectively, and the detection of ubiquitin and the 20S proteasome in some archaebacteria [135, 217] and eubacteria [38, 192]. Three fascinating themes have emerged. One is the use of peptide tags as signals for degradation; these include the addition of ubiquitin to internal Lys residues [34, 95], N-terminal arginylation in the N-end rule [205], and C-terminal addition of the AANDENYALAA tag specific for the bacterial Tsp protease [117]. The second is the use of oligomeric barrel-shaped structures in the folding/unfolding and/or degradation of proteins [61, 143, 212] (Fig. 4). Presumably, this arrangement protects other cytosolic constituents from processes occuring inside. The third theme is that proteolysis and phosphorylation may be intimately connected with phosphorylation providing signals for protein activation/inactivation as well as degradation [21, 41, 125, 129, 153, 183, 198, 218]. In addition to degrading proteins in their entirety, proteolytic systems can also partially degrade polypeptides or can selectively remove individual subunits from oligomeric proteins complex. In some cases, these subunits function as inhibitors or activators thus providing another level of metabolic regulation. From an understanding of the mechanisms of protein breakdown, several biotechnological applications have emerged, especially the ability to manipulate protein half-lives in vivo

[100]. Clearly, as more is known of this fascinating process, its applications to agriculture and medicine could be profound.

Acknowledgements

I thank Drs R. Amasino, J. Callis, D. Hondred, W. Rapp and S. Gan for providing information in advance of publication. I am also grateful to various members of my laboratory, past and present, for generating the data discussed herein and for their thoughtful editorial review of the manuscript. This work was supported by grants from the U.S. Department of Agriculture NRICGP (94-37301-03347), the U.S. Department of Energy (DF-EG02-88ER13968), the Consortium for Plant Biotechnology Research (92-34190-6941), the Research Division of the UW College of Agriculture and Life Sciences (Hatch 2858), and the DOE/NSF/USDA Collaboration Program on Research in Plant Biology (BIR 92-20331).

References

1. Abel S, Oeller PW, Theologis A: Early auxin-induced genes encode short-lived nuclear proteins. Proc Natl Acad Sci USA 91: 326–330 (1994).
2. Agarwal ML, Cullis CA: The ubiquitin-encoding multigene family of flax, *Linum usitatissimum*. Gene 99: 69–75 (1991).
3. Anathan J, Goldberg AL, Vollemy R: Abnormal proteins serve as eukaryotic signals and trigger the activation of heat shock genes. Science 232: 522–524 (1986).
4. Anbudurai PR, Mor TS, Ohad I, Shestakov SV, Pakrasi HB: The *ctpA* gene encodes the C-terminal processing protease for the D1 protein of the photosystem II reaction center complex. Proc Natl Acad Sci USA 91: 8082–8086 (1994).
5. Arfin S Bradshaw R: Cotranslational processing and protein turnover in eukaryotic cells. Biochemistry 27: 7979–7990 (1988).
6. Arnason T, Ellison MJ: Stress resistance in *Saccharomyces cerevisiae* is strongly correlated with assembly of a novel type of multiubiquitin chain. Mol Cell Biol 14: 7876–7883 (1994).
7. Bachmair A, Finley D, Varshavsky A: *In vivo* half-life of a protein is a function of its amino-terminal residue. Science 234: 179–186 (1986).
8. Bachmair A, Becker F, Masterson V, Schell J: Pertubation of the ubiquitin system causes leaf curling, vascular tissue alterations, and necrotic lesions in a higher plant. EMBO J 9: 4543–4549 (1991).
9. Bachmair A, Becker F, Schell J: Use of a reporter transgene to generate *Arabidopsis* mutants in ubiquitin-dependent proteolysis. Proc Natl Acad Sci USA 90: 418–421 (1993)
10. Bailly V, Lamb J, Sung P, Prakash S, Prakash L: Specific complex formation between yeast Rad6 and Rad18 proteins: a potential mechanism for targeting Rad6 ubiquitin-conjugating activity to DNA damage sites. Genes Devel 8: 811–820 (1994).
11. Barr PJ: Mammalian subtilisins: the long sought dibasic processing endoproteases. Cell 66: 1–3 (1991).
12. Barret AJ: The classes of proteolytic enzymes. In: Dalling MJ (ed) Plant ProteolyticEnzymes, pp. 1–16. CRC Press, Boca Raton, FL (1986).
13. Bartling D, Rehling P, Weiler EW: Functional expression and molecular characterization of AtUBC2–1, a novel ubiquitin conjugating enzyme (E2) from *Arabidopsis thaliana*. Plant Mol Biol 23: 387–396 (1993).
14. Beal R, Deveraux Q, Xia G, Rechsteiner M, Pickart C: Surface hydrophobic residues of multiubiquitin chains essential for proteolytic targeting. Proc Natl Acad Sci USA 93: 861–866 (1996).
15. Beers E, Moreno TN, Callis JA: Subcellular localization of ubiquitin and ubiquitinated proteins in *Arabidopsis*. J Biol Chem 267: 15432–15439 (1992).
16. Bethke PC, Hillmer S, Jones RL: Isolation of intact protein storage vacuoles from barley aleurone. Plant Physiol 110: 521–529 (1996).
17. Binet M-N, Weil J-H, Tessier L-H: Structure and expression of sunflower ubiquitin genes. Plant Mol Biol 17: 395–407 (1991).
18. Boller T: Roles of proteolytic enzymes in interactions of plants with other organisms. In: Dalling MJ (ed) Plant Proteolytic Enzymes, pp. 67–96. CRC Press, Boca Raton, FL (1986).
19. Boston RS, Fontes EBP, Shank BB, Wrobel RL: Increased expression of the maize immunoglobulin binding protein homolog b–70 in three zein regulatory mutants. Plant Cell 3: 497–505 (1991).
20. Briggs MS, Roder H: Early hydrogen-bonding events in the folding reaction of ubiquitin. Proc Natl Acad Sci USA 89: 2017–2021 (1992).
21. Brown K, Gerstberger S, Carlson L, Franzoso G, Siebenlist U: Control of IκB-α by site-specific, signal-induced phosphorylation. Science 267: 1485–1488 (1995).
22. Burke TJ, Callis JA, Vierstra RD: Characterization of a polyubiquitin gene in *Arabidopsis thaliana*. Mol Gen Genet 213: 435–443 (1988).
23. Bushnell T, Bushnell D, Jagendorf AT: A purified zinc protease of pea chloroplasts, EP1, degrades the large subunit of ribulose–1,5-bisphosphate carboxylase/oxygenase. Plant Physiol 103: 585–591 (1993).
24. Butt TR, Jannalagadda S, Monia B, Sternberg E, Marsh JA, Stadel JM, Ecker DJ, Crooke ST: Ubiquitin fusion augments the yield of cloned gene products in *Escherichia coli*. Proc Natl Acad Sci USA 86: 2540–2544 (1989).
25. Callis, JA: Regulation of protein degradation. Plant Cell 7: 845–857 (1995).
26. Callis J, Vierstra RD: Ubiquitin and ubiquitin genes in higher plants. Oxford Surv Plant Mol Cell Biol 6: 1–30 (1989).
27. Callis JA, Raasch JA, Vierstra RD: Ubiquitin extension proteins in *Arabidopsis thaliana*: structure, localization, and expression of their promoters in transgenic tobacco. J Biol Chem 265: 12486–12493 (1990).
28. Callis JA, Carpenter TB, Sun CW, Vierstra RD: Structure and evolution of genes encoding polyubiquitin and ubiquitin-like proteins in *Arabidopsis thaliana* ecotype Columbia. Genetics 139: 921–939 (1995).
29. Chau V, Tobias JW, Bachmair A, Marriott D, Ecker DJ, Gonda DK, Varshavsky A: A multiubiquitin chain is confined to

specific lysine in a targeted short-lived protein. Science 243: 1576–1583 (1989).

30. Chen P, Johnson P, Sommer T, Jentsch S, Hochstrasser M: Multiple ubiquitin-conjugating enzymes participate in the *in vivo* degradation of the yeast MATα2 repressor. Cell 74: 357–369 (1993).

31. Chen Z, Niles EG, Pickart CM: Isolation of a cDNA encoding a mammalian multiubiquitinating enzyme ($E2_{25k}$) and overexpression of the functional enzyme in *Escherichia coli*. J Biol Chem 266: 15698–15704 (1991).

32. Chiang H-L, Terlecky SR, Plant CP, Dice JF: A role for a 70-kilodalton heat shock protein in lysosomal degradation of intracellular proteins. Science 246: 382–385 (1989).

33. Christensen AH, Sharrock RA, Quail PH: Maize polyubiquitin genes: structure, thermal perturbation of expression and transcript splicing, and promoter activity following transfer to protoplasts by electroporation. Plant Mol Biol 18: 675–689 (1992).

34. Ciechanover, A: The ubiquitin-proteasome proteolytic pathway. Cell 79: 13–21 (1994).

35. Collins BA, Reed PD, Rubinstein B: Ubiquitinated proteins in differentiating vascular tissue of *Coleus blumei*. Plant Physiol 102: 125 (1993).

36. Cook WJ, Jeffrey LC, Sullivan ML, Vierstra RD: Threedimensional structure of a ubiquitin conjugating enzyme (E2). J Biol Chem 267: 15116–15121 (1992).

37. Cook WJ, Jeffrey LC, Carson M, Chen Z, Pickart CM: Structure of a diubiquitin conjugate and a model for interaction with ubiquitin conjugating enzyme (E2). J Biol Chem 267: 16467–16471 (1992).

38. Cook WJ, Jeffrey LC, Xu Y, Chau V: Tertiary structures of class I ubiquitin-conjugating enzymes are highly conserved: crystal structure of yeast Ubc4. Biochemistry 32: 13809–13817 (1993).

39. Courtney SE, Rider CC, Stead AD: Changes in protein ubiquitination and the expression of ubiquitin-encoding transcripts in daylily petals during floral development and senescence. Physiol Plant 91: 196–204 (1994).

40. Crane DI, Kalish JU, Gould SJ: The *Pichia pastoris PAS4* gene encodes a ubiquitin-conjugating enzyme required for peroxisome assembly. J Biol Chem 269: 21835–21844 (1994).

41. Crawford NM: Nitrate: nutrient and signal for plant growth. Plant Cell 7: 859–868 (1995).

42. Croall DE, DeMartino GN: Calcium-activated neutral protease (calpain) system: structure, function, and regulation. Physiol Rev 71: 813–847 (1991).

43. Dalling MJ, Nettleton AM: Chloroplast senescence and proteolytic enzymes. In: Dalling MJ (ed) Plant Proteolytic Enzymes, pp. 125–53. CRC Press, Boca Raton, FL (1986).

44. Davies DD: Physiolocial aspects of protein turnover. In: Coulter D, Partier B (eds) Encyclopedia of Plant Physiology, vol 14A, pp. 189–228. Springer-Verlag, Berlin (1982).

45. DeMartino GN, Moomaw CR, Zagnitko OP, Proske RJ, Ma C-P, Afendis SJ, Swaffield JC, Slaughter CA: PA700, an ATP dependent activator of the 20S proteasome, is an ATPase containing multiple members of a nucleotide-binding protein family. J Biol Chem 269: 20878–20884 (1994).

46. Demura T, Fukuda H: Novel vascular cell-specific genes whose expression is regulated temporally and spatially during vascular system development. Plant Cell 6: 967–981 (1994).

47. Desautels M, Goldberg AL: Liver mitochondria contain an ATP-dependent vanadate-sensitive pathway for the degrad-

ation of proteins. Proc Natl Acad Sci USA 79: 18691873 (1982).

48. Deveraux Q, Ustrell V, Pickart C, Rechsteiner M: A 26S protease subunit that binds ubiquitin conjugates. J Biol Chem 269: 7059–7061 (1994).

49. Deveraux Q, van Nocker S, Mahaffey D, Vierstra RD, Rechsteiner M: Inhibition of ubiquitin-mediated proteolysis by the *Arabidopsis* 26S protease subunit S5a. J Biol Chem 270: 29660–29663 (1995).

50. Dice JF: Molular determinants of protein half-lives in eukaryotic cells. FASEB J 1: 349–357 (1987).

51. Dick LR, Aldrich C, Jameson SC, Moomaw CR, Pramanik BC, Doyle CK, DeMartino GN, Bevan MJ, Forman JM, Slaughter CA: Proteolytic processing of ovalbumin and-β-galactosidase by the proteasome to yield antigenic peptides. J Immunol 152: 3884–3894 (1994).

52. Dohmen RJ, Wu P, Varshavsky A: Heat-inducible degron: a method for constructing temperature-sensitive mutants. Science 263: 1273–1276 (1994).

53. Dohmen RJ, Stappen R, McGrath JP, Forrova H, Kolarov J, Goffeau A, Varshavsky A: An essential yeast gene encoding a homolog of ubiquitin-activating enzyme. J Biol Chem 270: 18099–18109 (1995).

54. Durner J, Boger P: Ubiquitin in the prokaryote *Anabaena variabilis*. J Biol Chem 270: 3720–3725 (1995).

55. Ecker DJ, Stadel JM, Butt TR, Marsh JA, Monia BP, Powers DA, Gorman JA, Clark PE, Warren F, Shatzman A, Crooke ST: Increasing gene expression in yeast by fusion to ubiquitin. J Biol Chem. 264:7715–7719 (1989).

56. Eising R, Gerhardt B: Catalase degradation in sunflower cotyledons during peroxisome transition from glyoxysomal to leaf peroxisomal function. Plant Physiol 84: 225–232 (1987).

57. Ellis RE, Yuan J, Horwitz HR: Mechanisms and functions of cell death. Annu Rev Cell Biol 7: 663–698 (1991).

58. Fenteany G, Standaert RF, Lane WS, Choi S, Corey EJ, Schreiber SL: Inhibition of proteasome activities and subunit-specific amino-terminal threonine modification by lactacystin. Science 268: 726–731 (1995).

59. Fincher GB: Molular and cellular biology associated with endosperm mobilization in germinating cereal grains. Annu Rev Plant Physiol Plant Mol Biol 40: 305–346 (1989).

60. Finley D, Bartel B, Varshavsky A: The tails of ubiquitin precursors are ribosomal proteins whose fusion to ubiquitin facilitates ribosome biogenesis. Nature 338: 394–401 (1989).

61. Flannagan JM, Wall JS, Capel MS, Schneider DK, Shanklin JS: Scanning transmission electron microscopy and smallangle scattering provide evidence that native *Escherichia coli* ClpP is a tetradecamer with an axial pore. Biochemistry 34: 10910–10917 (1995).

62. Garbarino JE, Rockhold DR, Belknap WR: Expression of stress-responsive ubiquitin genes in potato tubers. Plant Mol Biol 20: 235–244 (1992).

63. Garbarino JE, Oosumi T, Belknap WR: Isolation of a polyubiquitin promoter and its expression in transgenic potato plants. Plant Physiol 109: 1371–1378 (1995).

64. Gaszynska M, Rock KL, Goldberg AL: γ-Interferon and expression of MHC genes regulate peptide hydrolysis by proteasomes. Nature 365: 264–267 (1993).

65. Gatenby AA, Viitanen PV: Structural and functional aspects of chaperonin-mediated protein folding. Annu Rev Plant Physiol Plant Mol Biol 45: 469–491 (1994).

66. Genschik P, Parmentier Y, Durr A, Marbach J, Criqui M-C, Jamet E, Fleck J: Ubiquitin genes are differentially regulated in protoplast-derived cultures of *Nicotiana sylvestris* and in

298

response to various stresses. Plant Mol Biol 20: 897–910 (1992).

67. Genschik P, Philipps G, Gigot C, Fleck J: Cloning and sequence analysis of a cDNA clone from *Arabidopsis thaliana* homologous to a proteasome α subunit from *Drosophila*. FEBS Lett 309: 311–315 (1992).

68. Genschik P, Durr A, Fleck J: Differential expression of several E2-type ubiquitin-carrier protein genes at different developmental stages of *Arabidopsis thaliana* and *Nicotiana sylvestris*. Mol Gen Genet 244: 548–556 (1994).

69. Genschik P, Jamet E, Philipps, Parmentier Y, Gigot C, Fleck J: Molular characterization of a β-type proteasome subunit from *Arabidopsis thaliana* co-expressed at high level with an α-type proteasome subunit early in the cell cycle. Plant J 6: 537–546 (1994).

70. Ghislain M, Udvardy A, Mann C: *S. cerevisiae* 26S proteasome mutants arrest cell division in G2/metaphase. Nature 366: 358–362 (1993).

71. Girod P-A, Vierstra RD: A major ubiquitin conjugation system in wheat germ extracts involves a 15-kDa ubiquitin conjugating enzyme (E2) homologous to the yeast *UBC4/UBC5* gene products. J Biol Chem 268: 955–960 (1993).

72. Girod P-A, Carpenter TB, van Nocker S, Sullivan ML, Vierstra RD: Homologs of the essential ubiquitin conjugating enzymes UBC1, 4, and 5 in yeast are encoded by a multigene family in *Arabidopsis thaliana*. Plant J 3: 545–552 (1993).

73. Glotzner M, Murray A, Kirschner MW: Cyclin is degraded by the ubiquitin pathway. Nature 349: 132–138 (1991).

74. Goff SA, Goldberg AL: Production of abnormal proteins in *E. coli* stimulates transcription of *lon* and other heat shock genes. Cell 41: 587–595 (1985).

75. Goldberg AL: The mechanism and functions of ATP-dependent proteases in bacterial and animal cells. Eur J Biochem 203: 9–23 (1992).

76. Goldberg AL: Functions of the proteasome: the lysis at the end of the tunnel. Science 268: 522–523 (1995).

77. Goldberg AL, St John AC: Intracellular protein degradation in mammalian and bacterial cells: part 2. Annu Rev Biochem 45: 747–803 (1976).

78. Goldberg Al, Rock KL: Proteolysis, proteasomes and antigen presentation. Nature 357: 375–379 (1992).

79. Gonen H, Stancovski I, Shkedy D, Hadari T, Bercovich B, Bengal E, Mesilati S, AbuHatoum O, Schwartz AL, Ciechanover A: Isolation, characterization, and partial purification of a novel ubiquitin-protein ligase, E3. J Biol Chem 271: 302–310 (1996).

80. Gordon C, McGirk D, Dillon P, Rosen C, Hastie ND: Defective mitosis due to a mutation in the gene for a fission yeast 26S proteasome subunit. Nature 366: 355–357 (1993).

81. Gosink M, Vierstra RD: Redirecting the specificity of ubiquitination through modification of ubiquitin conjugating enzymes (E2s). Proc Natl Acad Sci USA 92: 9117–9121 (1995).

82. Gottesman S, Squires C, Pichersky E, Carington M, Hobbs M, Mattick JS, Dalrymple B, Kuramitsu H, Shiroza T, Foster T, Clark WP, Ross B, Squires CL, Maurizi MR: Conservation of the regulatory subunit for the Clp ATP-dependent protease in prokaryotes and eukaryotes. Proc Natl Acad Sci USA 87: 3513–3517 (1990).

83. Gottesman S, Clark WP, Crecy-Lagard VD, Maurizi MR: ClpX, an alternative subunit for the ATP-dependent Clp protease of *Escherichia coli*: sequence and *in vivo* activities. J Biol Chem 268: 22618–11626 (1993).

84. Granell A, Harris N, Pisabarro AG, Carbonell J: Temporal and spatial expression of a thiolprotease gene during pea ovary senescence and its regulation by gibberellin. Plant J 2: 907–915 (1992).

85. Greenberg JT, Ausubel FM: *Arabidopsis* mutants compromised for the control of cellular damage during pathogenesis and aging. Plant J 4: 327–341 (1993).

86. Grey JC, Hird Sm, Dyer TA: Nucleotide sequence of a wheat chloroplast gene encoding the proteolytic subunit of an ATP-dependent protease. Plant Mol Biol 15: 947–950 (1990).

87. Guarino LA, Smith G, Dong W: Ubiquitin is attached to membranes of Bacuolvirus particles by a novel type of phospholipid anchor. Cell (1995).

88. Hadir T, Warms JVB, Rose IA, Hershko A: A ubiquitin C-terminal isopeptidase that acts on polyubiquitin chains. J Biol Chem 267: 719–727 (1992).

89. Hatfield PM, Vierstra RD: Ubiquitin-dependent proteolytic pathway in wheat germ: Isolation of multiple forms of ubiquitin-activating enzyme. Biochemistry 28: 735–742 (1989).

90. Hatfield PM, Callis J, Vierstra RD: Cloning of ubiquitin activating enzyme from wheat and expression of a functional protein in *Escherichia coli*. J Biol Chem 265: 15813–15817 (1990).

91. Hatfield PM, Vierstra RD: Multiple forms of ubiquitin-activating enzyme (E1) from wheat: identification of an essential cysteine by *in vitro* mutagenesis. J Biol Chem 267: 14799–14803 (1992).

92. Hatfield PM, Gosink MM, Carpenter TB, Vierstra RD: The ubiquitin-activating enzyme (E1) gene family in *Arabidopsis thaliana*. Plant J (submitted).

93. Heinemeyer W, Gruhler A, Möhrle V, Mahé Y, Wolf DH: PRE2, highly homologous to the human major histocompatibility complex-linked *Ring10* gene, codes for a yeast proteasome unit necessary for chymotryptic activity and degradation of ubiquitinated proteins. J Biol Chem 268: 5115–5120 (1993).

94. Hensel LL, Grbic V, Baumgarten DA, Bleecker AB: Developmental and age-related processes that influence the longevity and senescence of photosynthetic tissues in *Arabidopsis*. Plant Cell 5: 553–564 (1993).

95. Hershko A, Ciechanover A: The ubiquitin system for protein degradation. Annu Rev Biochem 61: 761–807 (1992).

96. Hicke L, Riezman H: Ubiquitination of a yeast plasma membrane receptor signals its ligand-stimulated endocytosis. Cell 84: 277–287 (1996).

97. Hochstrasser M: Ubiquitin, proteasomes, and the regulation of intracellular protein degradation. Curr Opin Cell Biol 7: 215–223 (1995).

98. Holloway SL, Glotzer M, King RW, Murray AW: Anaphase is initiated by proteolysis rather than by inactivation of maturation-promoting factor. Cell 73: 1393–1402 (1993).

99. Holwerda BC, Padgett HS, Rogers JC: Proaleurain vacuolar targeting is mediated by short contiguous peptide interactions. Plant Cell 4: 307–318 (1992).

100. Hondred D, Vierstra RD: Novel applications of the ubiquitin-dependent proteolytic pathway in plant genetic engineering. Curr Opin Biotechnol. 3: 147–151 (1992).

101. Hou D, Cenciarelli D, Jensen JP, Nguyen HB, Weissman AM: Activation-dependent ubiquitination of a T cell antigen receptor subunit on multiple intracellular lysines. J Biol Chem 269: 14244–14247 (1994).

102. Huang Y, Baker RT, Fischer-Vize JA: Control of cell fate by a deubiquitinating enzyme encoded by the *fat facets* gene. Science 270: 1828–1831 (1995).

103. Huffaker RC, Peterson LW: Protein turnover in plants and possible means of its regulation. Annu Rev Plant Physiol 25: 363–392 (1974).

104. Huibregtse JM, Scheffner M, Beaudenon S, Howley PM: A family of proteins structurally and functionally related to the E6-AP ubiquitin-protein ligase. Proc Natl Acad Sci USA 92: 2563–2567 (1995).

105. Jabben M, Shanklin J, Vierstra RD: Ubiquitin-phytochrome conjugates: pool dynamics during *in vivo* phytochrome degradation. J Biol Chem 264: 4998–5005 (1989).

106. Jabben M, Shanklin J, Vierstra RD: Red light-induced accumulation of ubiquitin-phytochrome conjugates in both moncots and dicots. Plant Physiol 90: 380–384 (1989).

107. Jack T, Fox GL, Meyerowitz EM: *Arabidopsis* homeotic gene *APETALA3* ectopic expression: transcriptional and posttranscriptional regulation determine floral organ identity. Cell 76: 703–716 (1994).

108. Jacobs TW: Cell cycle control. Annu Rev Plant Physiol Plant Mol Biol 46: 317–339 (1995).

109. Jentsch S: The ubiquitin-conjugation system. Annu Rev Genet 26: 179–207 (1992).

110. Johnson EM, Schnabelrauch LS, Sears BB: A plastome mutation affects processing of both chloroplast and nuclear DNA-encoded plastid proteins. Mol Gen Genet 225: 106–112 (1991).

111. Johnson ES, Gonda DK, Varshavsky A: *Cis-trans* recognition and subunit-specific degradation of short-lived proteins. Nature 346: 287–291 (1990).

112. Johnson ES, Ma PC, Ota IM, Varshavsky A: A proteolytic pathway that recognizes ubiquitin as a degradation signal. J Biol Chem 270: 17442–17456 (1995).

113. Jones EW: Three proteolytic systems in the yeast *Saccharomyces cerevisiae*. J Biol Chem 266: 7963–7966 (1991).

114. Juniper BE, Robins RJ, Joel DM (eds): The Carnivorous Plants. Academic Press, London (1989).

115. Kaiser WM, Huber SC: Post-translational regulation of nitrate reductase in higher plants. Plant Physiol 106: 817–821 (1994).

116. Kassenbrock CK, Garcia PD, Walter P, Kelly RB: Heavy-chain binding protein recognizes aberrant polypeptides translocated *in vitro*. Nature 333: 90–93 (1988).

117. Keiler KC, Waller PRH, Sauer RT: Role of peptide tagging system in degradation of proteins synthesized from damaged messenger RNA. Science 271: 990–993 (1996).

118. Kende H: Ethylene biosynthesis. Annu Rev Plant Physiol Plant Mol Biol 44: 283–307 (1993).

119. Kessel M, Maurizi MR, Kim B, Kocsis E, Trus BL, Singh K, Steven AC: Homology in structural organization between *E. coli* ClpAP protease and the eukaryotic 26S proteasome. J Mol Biol 250: 587–594 (1995).

120. Kim WT, Yang SF: Turnover of 1-aminocyclopropane–1-carboxylic acid synthase protein in wounded tomato fruit tissue. Plant Physiol. 100: 1126–1131 (1992).

121. King RW, Peters J-M, Tugendreich S, Rolfe M, Hieter P, Kirschner MW: A 20S complex containing CDC27 and CDC16 catalyzes the mitosis-specific conjugation of ubiquitin to cyclin B. Cell 81: 279–288 (1995).

122. Kiyosue T, Yamaguchi-Shinozaki K, Shinozaki K: Characterization of cDNA for a dehydration-inducible gene that encodes a ClpA, B-like protein in *Arabidopsis thaliana* L. Biochem Biophys Res Commun 196: 1214–1220 (1993).

123. Klausner RD, Sitia R: Protein degradation in the endoplasmic reticulum. Cell 62: 611–614 (1990).

124. Koehler SM, Ho T-HD: Hormonal regulation, processing and secretion of cysteine proteinases in barley aleurone layers. Plant Cell 2: 769–783 (1990).

125. Kornitzer D, Raboy B, Kulka RG, Fink GR: Regulated degradation of the transcription factor Gcn4. EMBO J 13: 6021–6030 (1994).

126. Kuwabara T: Characterization of a prolyl endopeptidase from spinach thylakoids. FEBS Lett 300: 127–130 (1992).

127. Lee Y-R, Nagao RT, Lin C-Y, Key JL: Induction and regulation of heat-shock gene expression by an amino acid analog in soybean seedlings. Plant Physiol 110: 241–248 (1996).

128. Leyser HMO, Lincoln CA, Timpte C, Lammer D, Turner J, Estelle M: *Arabidopsis* auxin resistance gene *AXR1* encodes a protein related to ubiquitin-activating enzyme E1. Nature 364: 161–164 (1993).

129. Lin W-C, Desiderio S: Regulation of V(D)J recombination activator protein RAG–2 by phosphorylation. Science 260: 953–958 (1993).

130. Liu X-Q, Jagendorf AT: ATP-dependent proteolysis in pea chloroplasts. FEBS Lett 166: 248–252 (1984).

131. Liu X-Q, Jagendorf AT: Neutral peptidases in the stroma of pea chloroplasts. Plant Physiol 81: 603–608 (1986)

132. Loeb KR, Haas AL: The interferon-inducible 15 kDa ubiquitin homolog conjugates to intracellular proteins. J Biol Chem 267: 7806–7813 (1992).

133. Lohman KN, Gan S, John MC, Amasino RM: Molular analysis of natural leaf senescence in *Arabidopsis thaliana*. Physiol Plant 92: 322–328 (1994).

134. Long JA, Moan EI, Medford JI, MK Barton: A member of the KNOTTED class of homeodomain proteins encoded by the *STM* gene of *Arabidopsis*. Nature 379: 66–69 (1996).

135. Löwe J, Stock D, Jap F, Zwickl P, Baumeister W, Huber R: Crystal structure of the 20S proteasome from the archaeon *T. acidophilum* at 3.4 Å resolution. Science 268: 533–539 (1995).

136. Madura K, Varshavsky A: Degradation of Gα by the N-End Rule pathway. Science 265: 1454–1458 (1994).

137. Malek L, Bogorad L, Ayers AR, Goldberg AL: Newly synthesized proteins are degraded by an ATP-stimulated proteolytic process in isolated pea chloroplasts. FEBS Lett 166: 253–257 (1984).

138. Mansfield, E, Hersperger E, Biggs J, Shearn A: Genetic and molecular analysis of *hyperplastic discs*, a gene whose product is required for regulation of cell proliferation in *Drosophila melanogaster* imaginal discs and germ cells. Devel Biol 165: 507–526 (1994).

139. Matile PH: Protein degradation. In: Coulter D, Partier B (eds) Encyclopedia of Plant Physiology, New Series, Vol. 14A, pp. 169–188, Springer-Verlag, Berlin (1982).

140. Mattoo AK, Hoffman-Falk H, Marder JB, Edelman M: Regulation of protein metabolism; coupling of photosynthetic electron transport to the *in vivo* degradation of the rapidly metabolized 32-kilodalton protein in the chloroplast membranes. Proc Natl Acad Sci USA 81: 1380–1384 (1984).

141. Mauch F, Staehelin LA: Functional implications of the subcellular localization of ethylene-induced chitinase and β–1,3 glucanase in bean leaves. Plant Cell 1: 447–457 (1989).

142. Maurizi MR, Clark WP, Kim SH, Gottesman S: ClpP represents a unique family of serine proteases. J Biol Chem 265: 12456–12552 (1990).

143. Maurizi MR: Protease and protein degradation in *Escherichia coli*. Experientia 48: 178–201 (1992).

144. McGrath JP, Jentsch S, Varshavsky A: *UBA1*: an essential yeast gene encoding-ubiquitin-activating enzyme. EMBO J 10: 227–236 (1991).

145. Merchant S, Bogorad L: Rapid degradation of apoplastocyanin in Cu(II)-deficient cells of *Chlamydomonas reinhardtii*. J Biol Chem 261: 15850–15853 (1986).

146. Mittler R, Lam E: *In situ* detection of nDNA fragmentation during the differentiation of trachery elements in higher plants. Plant Physiol. 108: 489–493 (1995).

147. Moore T, Keegstra K: Characterization of a cDNA clone encoding a chloroplast-targeted Clp homologue. Plant Mol Biol 21: 525–537 (1993).

148. Moriyasu Y: Examination of the contribution of vacuolar proteases to intracellular protein degradation in *Chara corallina*. Plant Physiol 109: 1309–1315 (1995).

149. Mullet JE, Gamble-Klein P, Klein RR: Chlorophyll regulates accumulation of the plastid-encoded chlorophyll apoproteins CP43 ad D1 by increasing apoprotein stability. Proc Natl Acad Sci USA 87: 4038–4042 (1990).

150. Murakami Y, Matsufuji S, Kameji T, Hayashi S-I, Igarashi K, Tamura T, Tanaka K, Ichihara A: Ornithine decarboxylase is degraded by the 26S proteasome without ubiquitination. Nature 360: 597–599 (1992).

151. Muralidhar MG, Thomas JB: The *Drosophila bendless* gene encodes a neural protein related to ubiquitin-conjugating enzymes. Neuron 11: 253–266 (1993).

152. Murray A: Cyclin ubiquitination: the destructive end of mitosis. Cell 81: 149–152 (1995).

153. Nishizawa M, Furuno N, Okazaki K, Tanaka H, Ogawa Y, Sagata N: Degradation of Mos by the N-terminal proline (Pro$_2$)-dependent ubiquitin pathway on fertilization of *Xenopus* eggs: possible significance of natural selection for Pro$_2$ in Mos. EMBO J 12: 4021–4027 (1993).

154. Nussaume L, Vincentz M, Meyer C, Boutin J-P, Caboche M: Post-transcriptional regulation of nitrate reductase by light is abolished by an N-terminal deletion. Plant Cell 7: 611–621 (1995).

155. Oblong JE, Lamppa GK: Identification of two structurally related proteins involved in proteolytic processing of precursors targeted to the chloroplast. EMBO J 11: 4401–4409 (1992).

156. Ozaki M, Fujinami K, Tanaka D, Amemiya Y, Sato T, Ogura N, Nakagawa H: Purification and initial characterization of the proteasome from the higher plant *Spinacia oleracea*. J Biol Chem 267: 21678–21684 (1992).

157. Pagano M, Tam SW, Theodras AM, Beer-Romero, P, Del Sal G, Chau V, Yew PR, Draetta GF, Rolfe M: Role of the ubiquitin proteasome pathway in regulating abundance of the cyclin-dependent kinase inhibitor p27. Science 269: 682–685 (1995).

158. Palmobella VJ, Rando OJ, Goldberg AL, Maniatis T: The ubiquitin-proteasome pathway is required for processing the NF-κB1 precursor protein and the activation of NF-κB. Cell 78: 773–785 (1994).

159. Papa FR, Hochstrasser M: The yeast *DOA4* gene encodes a deubiquitinating enzyme related to the product of the human *tre-2* oncogene. Nature 366: 313–319 (1993).

160. Pearce G, Strydom D, Johnson S, Ryan CA: A polypeptide from tomato leaves induces wound-inducible proteinase inhibitor proteins. Science 253: 895–898 (1991).

161. Rechsteiner M, Hoffman L, Dubiel W: The multicatalytic and 26S proteases. J Biol Chem 268: 6065–6068 (1993).

162. Reddy, ASN, Safadi F, Beyette JR, Mykles DL: Calcium-dependent proteinase activity in root cultures of *Arabidopsis*. Biochem Biophys Res Commun 199: 1089–1095 (1994).

163. Ribeiro A, Akkermans AD, van Kammen A, Bisseling T, Pawlowski K: A nodule-specific gene encoding a subtilisin-like protease is expressed in early stages of actinorhizal nodule development. Plant Cell 7: 785–794 (1995).

164. Rivett AJ: Proteasomes: multicatalytic proteinase complexes. Biochem J 291: 1–10 (1993).

165. Rock KL, Gramm C, Rothstein L, Clark K, Stein R, Dick L, Hwang D, Goldberg AL: Inhibitors of the proteasome block degradation of most cell proteins and the generation of peptides presented on MHC class I molecules. Cell 78: 761–771 (1994).

166. Ryan CA: Proteolytic enzymes and their inhibitors in plants. Annu Rev Plant Physiol 24: 173–196 (1973).

167. Schaller A, Ryan CA: Identification of a 50 kDa systemin-binding protein in tomato plasma membranes having Kex2-like properties. Proc Natl Acad Sci USA 91: 11802–11806 (1994).

168. Schaller A, Bergey DR, Ryan CE: Induction of wound response genes in tomato leaves by bestatin, an inhibitor of amino peptidases. Plant Cell 7: 1893–1898 (1995).

169. Scheffner M, Huibregtse JM, Vierstra RD, Howley PM: The HPV-16 E6 and E6-AP complex functions as a ubiquitin-protein ligase in ubiquitination of p53. Cell 75: 495–505 (1993).

170. Scheffner M, Nuber U, Hulbregtse JM: Protein ubiquitination involving an E1-E2-E3 enzyme ubiquitin-thioester cascade. Nature 363: 81–83 (1995).

171. Schliephacke M, Kremp A, Schmid, H-P, Kohler K, Kull U: Prosomes (proteasomes) of higher plants. Eur J Cell Biol 55: 114–121 (1991).

172. Schmidt GW, Mishkind ML: Rapid degradation of unassembled ribulose 1,5-bisphosphate carboxylase small subunit in chloroplasts. Proc Natl Acad Sci USA 80: 2632–2636 (1983).

173. Schwob E, Böhm T, Mendenhall MD, Nasmyth K: The B-type cyclin kinase inhibitor p40^{SIC1} controls the G1 to S transition in *S. cerevisiae*. Cell 79: 233–244 (1994).

174. Scornik OA: Role of protein degradation in the regulation of cellular protein content and amino acid pools. FASEB J 43: 1283–1288 (1984).

175. Seemüller E, Lupas A, Stock D, Löwe J, Huber R, Baumeister W: Proteasome from *Thermoplasma acidophilum*: a threonine protease. Science 268: 579–582 (1995).

176. Seufert W, Jentsch S: *In vivo* function of the proteasome in the ubiquitin pathway. EMBO J 11: 3077–3080 (1992).

177. Seufert W, Futcher B, Jentsch S: Role of a ubiquitin-conjugating enzyme in degradation of S- and M-phase cyclins. Nature 373: 78–81 (1995).

178. Shanklin J, Jabben M, Vierstra RD: Red light-induced formation of ubiquitin-phytochrome conjugates: identification of possible intermediates of phytochrome degradation. Proc Natl Acad Sci USA 84: 359–363 (1987).

179. Shanklin J, Jabben M, Vierstra RD: Partial purification and peptide mapping of ubiquitin-phytochrome conjugates from oat. Biochemistry 28: 6028–6034 (1989).

180. Shanklin J, DeWitt ND, Flanagan JM: The stroma of higher plant plastids contain ClpP and ClpC, functional homologues of *Escherichia coli* ClpP and ClpA: an archetypal two component ATP-dependent protease. Plant Cell 7: 1713–1722 (1995).

181. Shirley BW, Goodman HM: An *Arabidopsis* gene homolgous to mammalian and insect genes encoding the largest proteasome subunit. Mol Gen Genet 241: 586–594 (1993).

182. Silber KR, Keiler KC, Sauer RT: Tsp: a tail-specific protease that selectively degrades proteins with non-polar C-termini. Proc Natl Acad Sci USA 89: 295–299 (1992).

183. Spanu P, Grosskopf DG, Felix G, Boller T: The apparent turnover of 1-aminocyclopropane–1-carboxylate synthase in tomato cells is regulated by protein phosphorylation and dephosphorylation. Plant Physiol 106: 529–535 (1994).

184. Spence J, Sadis S, Haas AL, Finley D: A ubiquitin mutant with specific defects in DNA repair and multiubiquitination. Mol Cell Biol 15: 1265–1273 (1995).

185. Staswick PE: Storage proteins of vegetative plant tissues. Annu Rev Plant Physiol Plant Mol Biol 45: 303–322 (1994).

186. Steiner H-Y, Song W, Zhang L, Naider F, Becker JM, Stacey G: An *Arabidopsis* peptide transporter is a member of a new class of membrane transport proteins. Plant Cell 6: 1289–1299 (1994).

187. Sullivan ML, Vierstra RD: A ubiquitin carrier protein from wheat germ is structurally and functionally similar to the yeast DNA repair enzyme encoded by *RAD6*. Proc Natl Acad Sci USA 86: 9861–9865 (1989).

188. Sullivan ML, Callis J, Vierstra RD: HPLC resolution of ubiquitin pathway from wheat germ. Plant Physiol 94: 710–716 (1990).

189. Sullivan ML, Vierstra RD: Cloning of a 16 kDa ubiquitin carrier protein (E2) from wheat and *Arabidopsis thaliana*: identification of functional domains by *in vitro* mutagenesis. J Biol Chem 266: 23878–23885 (1991).

190. Sullivan ML, Carpenter T, Vierstra RD: Homologues of wheat ubiquitin-conjugating enzymes *Ta*UBC1 and *Ta*UBC4 are encoded by small multigene families in *Arabidopsis thaliana*. Plant Mol Biol 24: 651–661 (1994).

191. Swindle J, Ajioka J, Eisen H, Sanwal B, Jacquemot C, Browder Z, Buck G: The genomic organization and transcription of the ubiquitin genes of *Trypanosoma cruzi*. EMBO J 7: 1121–1127 (1988).

192. Tamura T, Nagy I, Lupas A, Lottspeich F, Cejka Z, Schoofs G, Tanaka K, De Mot R, Baumeister W: The first characterization of a eubacterial proteasome: the 20S complex of *Rhodococcus*. Curr Biol 5: 766–774 (1995).

193. Tanahashi N, Tsurumi C, Tamura T, Tanaka K: Molular structures of 20S and 26S proteasomes. Enzyme Protein 47: 241–251 (1993).

194. Teichert U, Mechlers B, Müller H, Wolf DH: Lysosomal (Vacuolar) proteinases of yeast are essential catalysts for protein degradation, differentiation, and cell survival. J Biol Chem 264: 16037–16045 (1989).

195. Thoma S, Sullivan ML, Vierstra RD: Members of gene families encoding the ubiquitin-conjugating enzymes, *At*UBC1–3 and *At*UBC4–6, from *Arabidopsis* are differentially expressed. Plant Mol Biol, in press (1986).

196. Thomas H, Stoddart JL: Leaf senescence. Annu Rev Plant Physiol 31: 83–111 (1980).

197. Tobias JW, Shrader TE, Rocap G, Varshavsky A: The N-end rule in bacteria. Science 254: 1374–1377 (1991).

198. Traenckner EB-M, Wilk S, Baeuerle PA: A proteasome inhibitor prevents activation of NF-κB and stabilizes a newly phosphorylated form of IκB that is still bound to NF-κB. EMBO J 13: 5433–5441 (1994).

199. Treier M, Staszewski LM, Bohmann D: Ubiquitin-dependent c-jun degradation *in vivo* is mediated by the δ domain. Cell 78: 787–798 (1994).

200. van Nocker S, Vierstra RD: Cloning and characterization of a 20-kilodalton ubiquitin-carrier protein (E2) from wheat that catalyzes multi ubiquitin-chain formation *in vitro*. Proc Natl Acad Sci USA 88: 10297–10301 (1991).

201. van Nocker S, Vierstra RD: Multiubiquitin chains linked through lysine–48 are abundant *in vivo* and competent intermediates in the ubiquitin-dependent proteolytic pathway. J Biol Chem 268: 24766–24773 (1993).

202. van Nocker S, Deveraux Q, Rechsteiner M, Vierstra RD: *Arabidopsis MBP1* gene encodes a conserved ubiquitin recognition component of the 26S proteasome. Proc Natl Acad Sci USA 93: 856–860 (1996).

203. van Nocker S, Walker JM, Vierstra RD: A multigene family in *Arabidopsis thaliana* encodes constitutively expressed E2s capable of forming multiubiquitin chains *in vitro*. J Biol Chem, in press (1996).

204. van Nocker S, Saddis S, Rubin D, Glickman M, Fu H, Coux O, Wefes I, Finley D, Vierstra RD: The multiubiquitin chain-binding MCB1 is a component of the 26S proteasome in *Saccharomyces cerevisiae* and plays a nonessential substrate-specific role in protein turnover. Mol Cell Biol, In press (1996).

205. Varshavsky A: The N-end rule. Cell 69: 725–735 (1992).

206. Vierstra RD: Demonstration of ATP-dependent, ubiquitin conjugating activities in higher plants. Plant Physiol 84: 332–336 (1987).

207. Vierstra RD: Protein degradation in plants. Annu Rev Plant Physiol Plant Mol Biol 44: 385–410 (1993).

208. Vierstra RD: Phytochrome degradation. In: Kendrick RE, Kronenberg GHM (eds.) Photomorphogenesis in Plants, pp. 141–162. Martinus Nijhoff, Dordrecht, Netherlands (1994).

209. Vierstra RD, Langan SM, Haas AL: Purification and initial characterization of ubiquitin from the higher plant, *Avena sativa*. J Biol Chem 260:12015–12021 (1985).

210. Vijay-Kumar S, Buggs CE, Wilkinson KD, Vierstra RD, Hatfield PM, Cook WJ: Comparison of the three-dimensional structures of human, yeast, and oat ubiquitin. J Biol Chem 262: 6396–6399 (1987).

211. Weibel FF, Kunau WH: The Pas2 protein essential for peroxisome biogenesis is related to ubiquitin-conjugating enzymes. Nature 359: 73–76 (1992).

212. Weissman JS, Sigler PB, Horwich AL: From the cradle to the grave: ring complexes in the life of a protein. Science 268: 523–524 (1995).

213. Wensel T, Eckerskorn C, Lottspeich F, Baumeister W: Existence of a molecular ruler in proteasomes suggested by analysis of degradation products. FEBS Lett 349: 205–209 (1994).

214. Wickner S, Gottesman S, Skowyra D, Hoskins J, McKenney K, Maurizi MR: A molecular chaperone, ClpA, functions like DnaK and DnaJ. Proc Natl Acad Sci USA 91: 12218–12222 (1994).

215. Wilkinson KD, Lee K, Deshpande S, Duerksen-Hughes P, Boss JM, Pohl J: The neuronspecific protein PGP9.5 is a ubiquitin carboxyl-terminal hydrolase. Science 246: 670–673 (1989).

216. Wilson KA: Role of proteolytic enzymes in the mobilization of protein reserves in the germinating dicot seed. In: Dalling MJ (ed) Plant Proteolytic Enzymes, pp. 19–48. CRC Press, Boca Raton, FL (1986).

217. Wolf S, Lottspeich F, Baumeister W: Ubiquitin found in the archaebacteria *Thermoplasma acidophilum*. FEBS Lett 326: 42–44 (1993).

302

218. Yaglom J, Linskens MHK, Sadis S, Rubin DM, Futcher B, Finley D: p34^{Cdc28}-mediated control on Cln3 cyclin degradation. Mol Cell Biol 15: 731–741 (1995).

219. Yamagata H, Masuzawa T, Nagaoka Y, Ohnishi T, Iwasaki T: Cucumisin, a serine protease from melon fruits, shares structural homology with subtilisin and is generated from a larger precursor. J Biol Chem 169: 32725–32731 (1994).

220. Ye Z-H, Varner J: Gene expression patterns associated with *in vitro* tracheary element formation in isolated single mesophyll cells of *Zinnia elegans*. Plant Physiol 103: 805–813 (1993).

221. Paris N, Stanley CM, Jones RL, Rogers JC: Plant cells contain two functionally vacuolar compartments. Cell 85: 563–572 (1996).

Plant Molecular Biology **32:** 303–314, 1996.
© 1996 *Kluwer Academic Publishers.*

Regulation of gene expression in plant mitochondria

Stefan Binder[1], Anita Marchfelder[2] & Axel Brennicke[1,*]
[1]*Allgemeine Botanik, Universität Ulm, Albert-Einstein-Allee 11, D-89069 Ulm, Germany (*author for correspondence);* [2]*Institut für Genbiologische Forschung GmbH, Ihnestrasse 63, D-14195 Berlin, Germany*

Key words: promoters, RNA processing, splicing, ribosomal regulation

Contents

Abstract	303	Intron splicing	309
Introduction	303	RNA editing	310
Size variation of transcription units	304	Ribosomal regulation	310
Multiple promoters in the plant mitochondrial genome	304	Protein turnover and complex assembly	310
Promoter structures	304	Developmental regulation of gene expression	311
Different promoter types in plant mitochondria	306	Aberrant genes in mutants: nuclear genes influencing post-transcriptional processes	311
Regulation of transcriptional activity	306		
5′ Processing of mRNAs and rRNAs	306	Acknowledgements	312
3′ Processing of mRNAs and rRNAs	307	References	312
Processing of tRNAs	308		

Abstract

Many genes in plant mitochondria have been analyzed in the past 15 years and regulatory processes controlling gene expression can now be investigated. *In vitro* systems capable of initiating transcription faithfully at promoter sites have been developed for both monocot and dicot plants and will allow the identification of the interacting nucleic acid elements and proteins which specify and guide transcriptional activities. Mitochondrial activity, although required in all plant tissues, is capable of adapting to specific requirements by regulated gene expression. Investigation of the factors governing the quality and quantity of distinct RNAs will define the extent of interorganelle regulatory interference in mitochondrial gene expression.

Introduction

The mitochondrial genome is the remnant of an ancient endosymbiotic event in which the ancestor of present-day mitochondria successfully became specialized in chemical energy conversion, particularly respiration [17, 18]. The common functions of mitochondria in plants, animals and fungi are reflected by similarities in their genomes. A common, basic set of genes is encoded in all mitochondrial genomes, although their sizes differ greatly, with plant mitochondrial DNAs being by far the most complex [59]. This overview describes the recent developments and achievements in understanding plant mitochondrial genome expression. In doing so, it emphasizes the observed differences between individual species. The recent development of *in vitro* sytems for transcription initiation and tRNA processing has yielded intriguing details of transcriptional and post-transcriptional regulation in plant mitochondria.

Size variation of transcription units

Most of the coding regions in plant mitochondrial genomes are separated by several kilobases of non-coding regions [8, 19, 35, 37, 52, 65]. This organization implies that most of the essential coding information is expressed as monocistronic transcripts from individual promoters, for example the *atp6, atp9, cytb, cox1, cox2* and *cox3* genes in most plants. When transcripts are analysed by Northern blot hybridization, the mRNAs are found to be generally much larger than the actual coding regions of the genes and to include extended non-coding 5′ and 3′ transcribed regions of several hundred nucleotides. The frequent genomic recombinations in plant mitochondrial genomes can place genes that are far apart in one genome into close proximity in another species.

Genes located closely together almost invariably show co-transcription [38, 52], such as the 18S-5S-RNA genes [41, 77] or the *nad3-rps12* transcription units [22]. The transcribed spacers between, for example, the 18S-5S rRNA genes vary among plant species, ranging from just over 100 nt to more than 500 nt; sequences apparently can become integrated or lost by intra- or intermolecular recombination [41, 77]. One of the longest multicistronic transcription units has been observed in the *18S-5S-nad5* arrangement in *Oenothera* mitochondria, where the two ribosomal RNAs and the first two exons of the *nad5* gene are co-transcribed [77]. Even larger transcription units will most likely be found upon more detailed investigation of higher plants, particularly in relatively small genomes and consequently with a denser gene population such as the *Brassica hirta* genome with only 216 kb total complexity [57]. The liverwort *Marchantia* mitochondrial genome, with only 187 kb and a gene complement very similar to higher plants, has to a large extent retained prokaryotic ribosomal protein gene clusters, which are expected to be transcribed into large polycistronic precursors [55, 56]. Other genes in this genome overlap by several nucleotides, raising other problems of transcription and translation.

The stably transcribed fraction of higher-plant mitochondrial genomes has been estimated to be about one-third of the genomic complexity, for example in Brassicaceae and in Curcurbitaceae [39]. This RNA population is equivalent to the entire sequence information common to all plant mitochondrial genomes. The remaining 70% of the plant mitochondrial genome appears to be non-essential sequence information derived partly from chloroplast DNA, partly from nuc-

lear sequences and in part from the duplication of mitochondrial sequences degenerated to various degrees. These sequences are sometimes transcribed but the transcripts observed in run-on experiments are mostly not detected in the steady-state transcript population [15]. Exceptions are some tRNA genes of chloroplast origin, which are transcribed and processed to yield functional tRNAs for mitochondrial translation [30, 46, 76].

Multiple promoters in the plant mitochondrial genome

The presence of numerous transcription units implies a corresponding multitude of individual promoters of transcription and the presence of several promoters per genome has been confirmed by direct experimental investigation. These approaches have utilized the absence of cap structures at primary 5′ ends of mRNAs in plant mitochondria, which allows specific labelling of primary transcripts by *in vitro* capping [2, 10, 13]. In *Oenothera* mitochondria, a minimum of 15 promoters dispersed throughout the genome has been estimated to yield stable transcripts detectable by this selective approach [2].

For some genes, multiple promoters have been identified which are actively engaged in transcription and may be used for differential regulation of the respective gene activity. The *cox2* gene in maize, for example, is transcribed from a closely stacked set of promoter sequence, and the *atp9* gene in this plant is preceded by at least six active promoters spaced over several hundred nucleotides upstream of the coding region [48, 54, 70]. The transcripts arising from such multiple promoters have different sizes and thus complicate transcription patterns of individual genes, even in monocistronic units [50].

Transcription patterns of individual genes are further complicated by the presence of several gene copies in a given genome. In Northern blots, detectable transcripts of different size may arise from duplicated genes with differing 5′ or 3′ regions adjacent to the respective open reading frame that leads to variable extensions at either end.

Promoter structures

Specific labelling of RNA 5′ termini by *in vitro* capping has allowed the distinction between 5′ termini derived

wheat   RAaaNNG<u>CRTA</u>tAR<u>t</u>Ragt

maize   RARAANTRA<u>CRTA</u><u>T</u>

dicots   TNNNNN_TAA_{AT}^AN<u>CRTA</u>AGA<u>G</u>A

Figure 1. Plant mitochondrial promoter sequences have been deduced from a combination of primary transcript termini inspection and *in vitro* analyses. Among monocots an A/T-rich region and the tetranucleotide 5′-CRTA-3′ motif are the only features conserved between the potential promoter sequences usually surrounding transcription initiation sites. Extensive primary sequence variation is observed between individual promoters within any monocot species. In dicot mitochondria, a larger region of 9 nt is conserved, including the CRTA motif at the 5′ end containing the transcription initiation site. The A/T-rich region is located just upstream of the CRTA tetranucleotid. Besides these promoters of the CRTA type, other transcription initiation sites identified in plant mitochondria show no discernable sequence similarity to this consensus or to each other.

from *de novo* initiation and 5′ processing events. A number of 5′ termini identified by primer extension and S1 protection methods indeed turned out to be genuine transcription initiation sites. This observation shows that 5′ processing is not obligatory in plant mitochondria and that no further modification is necessary at primary 5′ termini [6]. Inspection of the sequences at these genuine transcription initiation sites reveals several conserved sequence features surrounding the first transcribed nucleotide (Fig. 1). In maize, for example, a region of 11 nt including the first two transcribed nucleotides shows conserved nucleotide identities at several positions, with a tetranucleotide CRTA well maintained [40, 48, 60, 61]. Only this tetranucleotide is apparent at transcription initiation sites in different monocots, including most of the promoter regions in wheat mitochondria [13, 27]. The CRTA motif is also conserved in dicot promoters; however, when monocot and dicot sequences are compared, structural differences become apparent between the two plant groups (Fig. 1). In dicot plants, a region of 9 nt is found to be highly conserved [2, 3, 4, 6, 10]. Plant-specific differences of promoter structure can be identified in an *in vitro* transcription assay, in which some heterologous promoters are less efficiently recognized than homologous sequences. In a pea mitochondrial transcription system, promoters from *Oenothera* were recognized less efficiently than some promoters from the more closely related soybean [4].

The recently developed *in vitro* transcription systems for several plants now allow for the detailed determination of the influence of individual nucleotide

identities upon promoter activity. The first successfully established *in vitro* transcription system for wheat mitochondria was primed with a *cox2* promoter and confirmed that the consensus sequence is indeed competent to programme transcription initiation [27]. Following the protocol established for wheat mitochondria, *in vitro* transcription initiation was also obtained in maize mitochondrial extracts, which were used to study the promoter of the atp1 gene in more detail [60, 61].

Deletion studies, point mutagenesis and linker-scanning experiments confirmed that in addition to the conserved tetranucleotide CRTA, other upstream sequence features extending to a region of about 17 nt are important for a fully competent promoter. Included in this sequence is a conserved purine-rich stretch about 15 nt upstream of the transcription initiation site. This arrangement has also been observed in comparisons of wheat mitochondrial promoters. The conserved tetranucleotide and the upstream purine-rich region are thus separated by about one helical turn of the DNA and the presence of both features may be important for docking the transcription initiation proteins to the DNA. Mutagenesis experiments with the maize *cox3* and *atp6* promoters extended the general validity of the *atp1* maize promoter studies to other genes [69, 70].

For dicot plants, an *in vitro* transcription system developed with pea mitochondrial lysates confirmed the competence of the conserved sequences of dicot plants around and upstream of transcription initiation sites to promote transcription [4]. As in monocot mitochondria, an upstream A-rich region is conserved and deletion studies show that the substitution of this region by other sequences results in reduced transcription initiation activity. In the pea *in vitro* transcription system, heterologous promoters from soybean and *Oenothera* were also tested [4]. As expected from the high conservation of the nucleotide sequences at these promoter regions in different dicot species, most of the promoters from these plants were correctly recognized, albeit with much lower efficiency with *Oenothera* templates. Species-specific factors thus most likely determine recognition efficiencies, although inspection of the primary sequences does not reveal any obvious correlating variation in sequence motifs. *In vitro* analysis of several soybean promoter regions in the pea extract showed that only those with the conserved sequence motif were recognized and that a promoter region lacking the conserved nonanucleotide motif of dicots could not be actively engaged by these mitochondrial fractions [4, 10].

Different promoter types in plant mitochondria

The *in vitro* results confirm the parallel presence of different types of promoter structures in plant mitochondria. Some of the promoters for protein-encoding genes differ from the consensus nonanucleotide type and the ribosomal rRNA genes appear to be transcribed from a different type of recognition sequence in both monocots and dicots [6, 51]. In maize, the 18S rRNA upstream sequence did not prime transcription *in vitro* [70]. The pea mitochondrial lysate, likewise, did not recognize the potential *Oenothera* ribosomal 26S RNA promoter region in order to initiate transcription (Stefan Binder, unpublished observations). Inspection of the sequences surrounding these promoters did not reveal any consensus motif and thus different types of recognition mechanism must be assumed, most likely with individual transcription cofactors specific for each sequence. In *Zea perennis,* such a non-conserved promoter region was dependent on a nuclear gene product encoded by the MCT gene, confirming that indeed distinct transcription cofactors act individually on these alternative promoter sequences [54]. Distinct recognition proteins of course do not preclude the simultaneous presence of different types of RNA polymerase in plant mitochondria, in analogy to the situation found in chloroplasts [21], where at least two types of RNA polymerases are actively involved in the transcription of chloroplast genes.

It is still unclear whether the use of different RNA polymerases and/or different cofactors in both organelles of a plant cell is related to the presence of two genetic systems, i.e. whether isosorting activities or factors are shared between the two organelles. In animal and fungal cells, which only have the additional genetic system of the mitochondria, one unique type of promoter appears to be active in mitochondria [12, 70]. Common to mitochondria in animals, fungi and plants appears to be the positioning of the conserved sequence block and the first transcribed nucleotide. The conserved sequence motif generally appears to include the first transcribed nucleotide(s), resulting in the conservation of the first nucleotide of all mRNAs in yeast mitochondria and the first two nucleotides in plant mitochondria [19, 70].

Regulation of transcriptional activity

The observation of different promoter activities in *in vitro* transcription systems has been confirmed by run-on experiments in organelles. Isolated mitochondria incorporate added nucleotides into RNA molecules by elongation of preinitiated incomplete transcripts. In these transcripts, as well as in transcripts elongated in mitochondrial lysates, the relative activity of transcription can be compared for individual genes. In one study of maize mitochondria, the two large ribosomal RNAs and the ribosomal protein *rps12* were found to be the most actively transcribed genes [16]. *Atp1, atp6, atp9, cob* and *cox3* follow in order of descending transcriptional activity. The results of another investigation differ somewhat by suggesting less clear distinctions between individual genes [49]. However, it is undisputed that the quantities of the transcripts synthesized in these experiments reflect individual promoter activities, and thus show that regulatory control of gene expression in plant mitochondria is exercised at least partially at the promoter level [19, 49]. Individual promoter activities most likely result from the effect of different promoter structures on the affinity for the transcription initiation complex, leading to the specific constitutive transcript levels. While the higher transcriptional activity of the ribosomal RNA genes reflects the higher abundance of these RNAs in the steady state RNA, transcriptional activity of protein-encoding genes does not correlate to the amount of the respective transcripts in the mitochondrial *in vivo* RNA population. Additional extensive control of mRNA levels (Fig. 2) must regulate post-transcriptionally the amounts of transcripts available for translation in plant mitochondria [19]. RNA stability thus appears to play a major role in regulating steady-state levels of RNA.

5' Processing of mRNAs and rRNAs

Several of the 5' termini mapped for mRNAs as well as rRNAs in plant mitochondria cannot be labelled by *in vitro* capping and may thus be derived from 5' processing events. The inability to cap an RNA terminus *in vitro*, however, does not necessarily imply that this RNA is derived by processing cleavage. Some *atp6* mRNA termini in maize mitochondria, for example, were found to initiate transcription *in vitro* although they could not be cap-labelled [32, 50, 69, 70]. Many other 5' termini do appear to be derived from genuine processing events and the sequences surrounding these termini have been investigated in an attempt to identify potential processing signals. Including the terminal sequences of the rRNA genes, two sequence blocks were found to be conserved in these regions and to

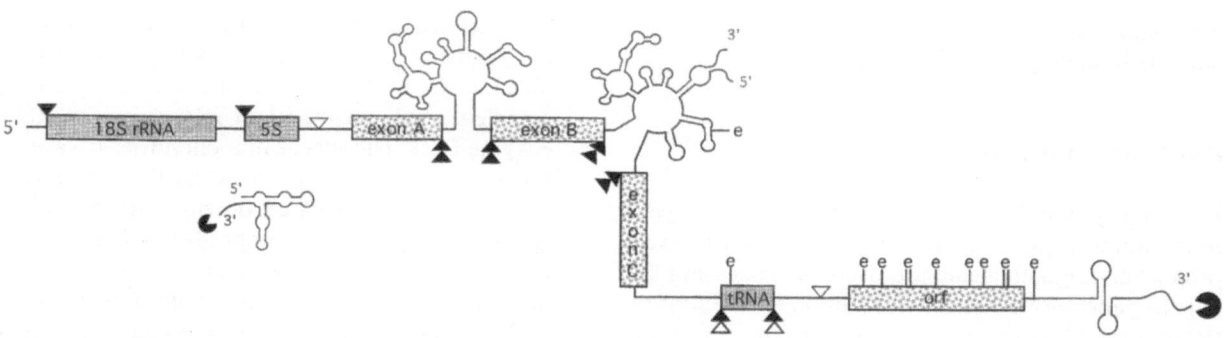

Figure 2. The different processing steps of RNA maturation in plant mitochondria are potential levels of post-transcriptional regulation in the organelle. Endonucleolytic cleavages occur at the 5′ termini of rRNAs (black triangles) and at the 5′ and 3′ termini of tRNAs (black and white triangle doublets) and as steps in the maturation of mRNA precursors (open triangles). Exonucleolytic maturation is required at the 3′ termini of the rRNAs and of mRNAs (open mouth symbol). Complex secondary stem-loop structures at the 3′ termini of mRNAs are potential signals for additional endonucleolytic digestion, for attenuation of exonucleolytic 3′ to 5′ digestion or for transcription termination. Intron excision (filled triangle doublet) involves *cis*-splicing as well as *trans*-splicing introns, the latter mediating the connection of independent mRNAs molecules into a single mature open reading frame. RNA editing (e) is required for the maturation of open reading frames as well as tRNAs and introns. Some RNA-editing sites in apparently non-coding regions may be connected to other regulatory processes.

act as potential processing signals [40, 64]. Confirmation of these deduced signal motifs awaits experimental evidence, for example by *in vitro* processing assays. 5′ processing sites, at least of rRNAs, will most likely be derived from endonucleolytic cleavages at these conserved signal regions and should thus generate detectable products in *in vitro* assays. The 18S rRNA has been found to be synthesized from upstream promoters and thus requires 5′ processing removal of 25–120 nt in different plants. In wheat, a 1 nt distant tRNA gene is used as a processing signal to mature simultaneously the tRNAs 3′ and the 18S rRNA 5′ ends [19, 41].

3′ Processing of mRNAs and rRNAs

mRNAs are not polyadenylated in plant mitochondria (or in chloroplasts) and thus require other means of sequence stabilization. Extensive 3′ processing is most likely required, since additional non-coding regions are usually included in transcripts of individual coding regions (Fig. 2). These 3′ trailer regions can extend up to several hundred nucleotides. Multiple 3′ termini are observed only exceptionally and the unique ends seen in individual transcription units suggest specific regulatory features for these regions.

mRNA 3′ termini map to one or very few scattered nucleotides in S-1 protection experiments. Primary inspection of sequences surrounding these termini identified conserved nucleotides just upstream of the 3′ mRNA terminus in the *cox2* and *atp1* genes of

Oenothera [66]. This region of some 15 identical nucleotides could be folded into a double stem-loop structure. Many of the sequences upstream of other mapped 3′ termini show an analogous ability to fold into either single or double stem-loop structures, for example in maize, rice, pea, petunia and sorghum. This secondary structure folding at a stable mRNA 3′ terminus is most likely associated with the formation and/or stabilization of this RNA end, but transcription termination may also be induced by such secondary structures [19, 66]. For some chloroplast transcripts, such foldback structures have been found to be recognized by specific RNA-binding proteins that protect the 3′ terminus against exonucleolytic degradation [68], and one can assume similar functions for the 3′ structures in mitochondria. This is supported by transcript analysis of *cyt b* genes in mitochondria of rice and in an alloplasmic line of wheat [31, 63], which showed two *cyt b* genes in both plants differing in their 3′ regions. Transcripts are only found for the genes containing a double stem-loop structure near the 3′ transcript termini; transcripts for the genes without these structures are not stably maintained. Since both copies appear to be transcribed in run-on experiments, at least in rice, these structures are associated with transcript stability. However, experimental data testing these assumptions is still required and await the development of the respective *in vitro* systems. Of particular interest will be a comparison of 3′ end generation/stabilization between chloroplasts and mitochondria to identify common and distinct features and components in the regulatory control mechanisms.

Processing of tRNAs

Particularly in animal mitochondria, but also in yeast mitochondria, tRNA genes are interspersed between protein-coding genes and are used as processing signals to generate smaller mRNAs upon excision of the tRNA molecules from large polycistronic transcripts [12]. In plant mitochondria, tRNA internal sequences have been proposed as candidates for transcription signals in analogy to RNA polymerase III transcription in the eukaryotic nucleus. Although this possibility has not been formally excluded, the presence of longer tRNA precursors and the recent identification of a promoter upstream of the tRNA^{Phe} gene in *Oenothera* mitochondria show that upstream promoters can be used for transcription of tRNA genes in mitochondria of plants. This tRNA promoter contains the conserved nonanucleotide sequence found at mRNA transcription initiation sites and thus appears to be recognized by the same transcription machinery as mRNAs [3]. The observation of tRNA precursors including upstream and downstream sequences of the actual coding regions implies that tRNAs in plant mitochondria have to be excised from larger precursor molecules (Figs. 2 and 3).

In vitro processing systems developed for wheat mitochondrial lysates [26] and *Oenothera* [45], potato [43] and pea [44] mitochondrial extracts confirm that plant mitochondria have the ability to excise tRNA molecules from longer transcripts. Processing activities have been identified in these mitochondrial lysates including an RNase P-like enzyme, which creates endonucleolytically the precise 5' tRNA end, a 3' endonucleolytic processing activity and an enzyme which is able to add the genomically non-encoded CCA triplet to the 3' tRNA terminus (Fig. 3). Micrococcal nuclease experiments suggest that the RNase P-like enzyme of plant mitochondria contains an RNA moiety, which is neccessary for activity, in analogy to RNase P enzymes characterized in pro- and eukaryotic genetic systems. A candidate sequence for this RNA moiety has tentatively been assigned, but verification of its participation in RNase P activity is still required [43]. Partial purification of the RNase P activity showed that it can be separated from both the CCA-adding enzyme and the 3' processing activity [43]. Biochemical characterization suggests that slightly degenerated tRNA structures

are accepted as substrates by the RNase P enzyme, although highly degenerated tRNA features still recognized by the *Escherichia coli* RNase P enzyme are not sufficient to direct the plant mitochondrial RNase P enzyme [42]. The wheat mitochondrial RNase P was found to correctly recognize tRNA-like structures with irregular D-loops located downstream of the rRNA genes in this plant [28], suggesting that, together with the tRNA^{fMet} located just one nucleotide upstream of the 5' terminus of the 18S RNA, both tRNAs and tRNA-like structures may also act as processing signals of precursor RNAs in plant mitochondria, similar to the tRNA punctuation model of animal mitochondria. This tRNA-mediated processing model is suggested by the endonucleolytic activity of the 3' processing enzyme in plant, animal and yeast mitochondria.

It will be of general interest to identify the structures and locations of the genes encoding the protein and the potential RNA moieties of the RNase P enzyme and of the 3' endonuclease in plant mitochondria, to determine the evolutionary origin of these molecules particularly with respect to the chloroplast system in higher plants. It this context, it should be noted that in the alga *Porphyra purpurea* an RNase P RNA subunit gene has been identified in the chloroplast genome [62]. At least in this alga, as well as in yeast mitochondria, the RNase P RNA moiety does not require import into the respective organelle, since it is encoded by a resident gene. Unfortunately, the low similarities of the RNase P RNA and the protein subunits prohibit their identification in distantly related species by heterologous hybridization techniques. Further investigation of the tRNA enzymes thus requires biochemical purification from the mitochondrial lysates.

The presence of such tRNA-processing activities in plant mitochondria suggests that tRNA genes of chloroplast origin integrated into the mitochondrial genome and transcribed from 'normal' mitochondrial promoters located upstream of the integrated chloroplast sequences should be excised from the precursor transcripts. *In vitro* processing analyses in wheat and potato mitochondrial extracts have shown that tRNAs of chloroplast origin are indeed correctly recognized and processed by the RNase P and the 3' processing activity [26, 42, 43].

Since not all of the required tRNAs are encoded in the mitochondrial genome in plants, additional tRNAs derived from nuclear-encoded genes are translocated from the cytoplasm [46]. At present it is still unclear whether these tRNAs are translocated into mitochondria as mature molecules or whether they pass the

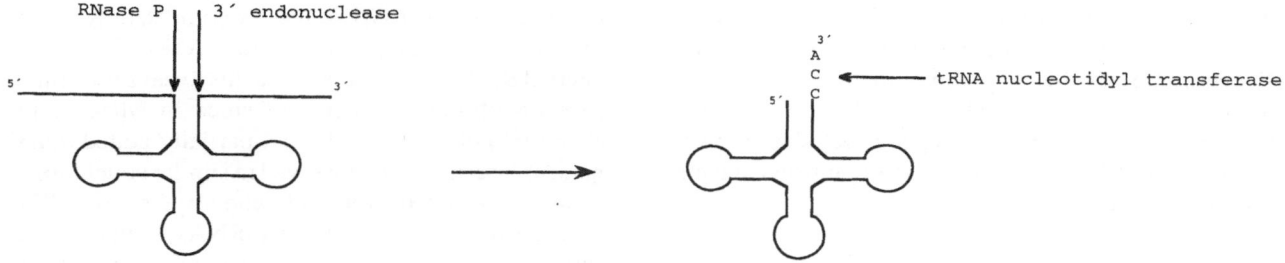

Figure 3. tRNA processing in plant mitochondria involves endonucleolytic cleavage by distinct enzymes at the 5′ terminus (RNase P) and at the 3′ terminus (3′ endonuclease). The tRNA nucleotidyltransferase adds the CCA terminus to the 3′ end of the maturing tRNA.

membrane as unprocessed precursors and mature in the mitochondrial compartment. In the latter case, the mitochondrial processing machinery would most likely be able to recognize these tRNA structures and transform them correctly into functional tRNA molecules.

Mitochondrial tRNA processing in plants can thus regulate the tRNAs of mitochondrial, chloroplast and potentially also of nuclear origin available for translation in the organelle. The processing of individual tRNA genes is controlled by different features of the respective tRNA as shown by the varying efficiency of excision. The processing of tRNA^{Phe}, for example, is regulated by the RNA-editing status of the precursor. Unedited precursors are not recognized by the RNase P and are not processed, at least *in vitro* [44]; edited precursor molecules on the other hand are rapidly converted to yield mature tRNA molecules.

Intron splicing

In higher plant mitochondria, numerous introns of the class II intron type have been identified in protein-coding genes (Fig. 2). Intron content varies from species to species, indicating that introns have been lost and gained independently during the evolution of plant species. These observations suggest that the plant mitochondrial group II introns are similar to this class of introns in other systems such as yeast mitochondria, where they behave as mobile elements that can integrate and excise on the genomic level.

Similar to the *psbA* gene in *Chlamydomonas* chloroplasts and to the *rps12* gene in higher-plant chloroplasts, some of the group II introns in plant mitochondria are physically disrupted with the adjacent exons encoded in far-distant regions of the respective genome (Fig. 2). These interrupted and consequently *trans*-splicing introns have been identified in the *nad1*,

nad2 and *nad5* genes in virtually all flowering plants tested [5,8, 11, 29, 33, 78]. The distant exons are transcribed (and most likely regulated) independently into different RNA molecules. These must be assembled in a *trans*-splicing reaction to yield the continuous open reading frame. Most intriguing is the situation of the penultimate intron of the *nad1* gene, which contains the only open reading frame with similarity to maturase proteins identified in plant mitochondria. This intron is found to be *trans*-spliced at different locations in wheat and petunia mitochondrial mRNAs [8]. While in petunia the maturase open reading frame is present in the upstream half of the *trans*-splicing intron, it is located in the downstream half of this intron in wheat mitochondria [8]. In broad bean and *Oenothera* mitochondria, this intron is not *trans*-spliced at all and the adjacent exons and the internal maturase-related open reading frame are located *in cis* [73, 78]. Extended base pairing has been observed at the locations of the intron interruptions in most of the *trans*-splicing introns, which suggests that stabilization of the secondary and tertiary intron structures may be involved in allowing *trans*-splicing processes and the correct assembly of the mature fully spliced mRNAs. All interruptions of these trans-splicing introns occur in domain 4 of the group II introns, which is the most variable domain in this intron class [8].

In several *cis*- as well as in some *trans*-splicing introns RNA editing events have been observed in conserved essential regions. Thus, these introns may require to be edited before correct excision can occur [5, 78]. RNA editing has been shown to be necessary for autocatalytic activity of a recombined intron, in which plant intron sequences were inserted into a yeast mitochondrial intron context [7]. Thus RNA editing may regulate intron splicing in plant mitochondria. Again *in vitro* systems are needed to analyse the intron excision processes in plant mitochondria in more detail, particularly with respect to the requirements of *trans*-splicing in these organelles. It will be interesting to determine whether in analogy to splicing of yeast mitochondrial introns, nuclear-encoded cofactors are required that may also be engaged in regulating post-transcriptional RNA maturation at the intron excision level in plant mitochondria.

RNA editing

In plant mitochondria, RNA editing modifies virtually all coding regions within mRNAs by C-to-U and also by U-to-C modifications and is required to generate correct open reading frames. RNA editing thus offers a unique level of regulatory control in plant mitochondria. Such potential control levels are most obvious in the generation of AUG initiation codons, for example in the *nad1* transcript in wheat mitochondria [8], in the creation or elimination of stop codons, in the intron sequences were RNA editing appears to be required for excision, and for tRNA processing (Fig. 2). The topic of RNA editing in plant mitochondria and chloroplasts is extensively dealt with in the article by Maier *et al.* (this volume).

Ribosomal regulation

Extensive gene-specific regulation of translation initiation has been observed in yeast mitochondria, where nuclear genes encode proteins that are required for efficient translation of individual mRNAs. In plant mitochondria, such translation-regulating proteins have as yet not been identified, although indirect evidence suggests potentially involved features in the untranslated 5' leader sequences of transcripts. Comparison of these leader sequences has identified three sequence blocks conserved upstream of, in particular, the *cox2, atp6* and *orf25* genes in different plant species, including both monocots and dicots [58]. These sequence blocks are not present upstream of other genes, suggesting that they may be involved in binding gene-specific proteins involved in regulating translation. The presence of introns apparently has little influence on an mRNA being incorporated into the ribosome, as suggested by a study of *cox2* transcripts in polysomes versus total mitochondrial RNA [71]. A similar distribution of intron-containing mRNAs and processed mRNAs for this gene was found in maize mitochondria in both polysomal and total RNA, suggesting that

intron-containing RNA is just as competent for ribosome entry as is the processed mRNA. Reading frame internal RNA editing likewise seems to have little influence on ribosome selectivity, since in wheat mitochondrial polysomes fully and partially edited *nad3-rps12* transcripts were observed in similar stoichiometric amounts as in the total mitochondrial mRNA [23].

Ribosomal recognition of mRNAs in plant mitochondria may involve sequences upstream of the translation initiation codon, since all transcripts contain long leader sequences. Searches for potential ribosome-binding sites, sequences potentially base-pairing with the 3' end of the 18S rRNA, have led to ambiguous results. Such potential base-pairing sequences can be found upstream of the AUG initiation codons of many genes, but not in all transcripts. Because the 3' terminus of the 18S rRNA in plant mitochondria is generally not complementary to the Shine-Dalgarno sequence of the *E. coli* translation system, different mechanisms may be involved in ribosome entry. *In vitro* translation systems for plant mitochondria have at present not been established. This difficulty seems to be a general feature of organelle protein synthesis since in animals and fungi it has also been impossible so far to obtain competent *in vitro* translation. A total wheat cell-free system, however, has been found to be capable of recognizing plant mitochondrial mRNAs and to synthesize at least some of the encoded mitochondrial polypeptides with apparently correct protein sizes [25]. However, extensive aberrant protein synthesis also occurs as expected in such heterologous experiments. Translation in heterologous systems using fully edited mRNAs should be feasible for plant mitochondrial transcripts, since at least in land plants the universal genetic code is used. The genetic code system appears to be different in some algae, for example in mitochondria of the red alga *Condrus crispus* the codon UGA is decoded as tryptophan [9].

Protein turnover and complex assembly

The above-mentioned investigation showing that both edited and unedited mRNAs can be translated indicates that different proteins can be made from a single gene. One can safely presume, however, that the protein made from unedited mRNAs is generally non-functional, since RNA editing in many genes has been found to change evolutionarily highly conserved amino acids, which from similarity comparisons are required for functionality. This is confirmed by protein sequence analysis of, for example, the *nad9* polypeptide isolated from complex I of the respiratory chain [20]. The complex contains only proteins made from edited mRNAs, although more than 50% of the mRNA population is unedited in several positions. Extensive regulation of complex assembly is thus expected to influence protein turnover, as only correct proteins, i.e. those polypeptides specified by completely edited mRNAs, are incorporated into a functional complex such as the ribosome or the respiratory chain components. Incorrect proteins synthesized from unedited mRNAs presumably cannot be integrated into the complex and are rapidly degraded. An effective protein degradation mechanism most likely is present in plant mitochondria which recognizes and attacks proteins not assembled into their respective functional contexts.

Developmental regulation of gene expression

Tissue and/or developmental regulation of plant mitochondrial gene expression has been investigated in different morphogenetic stages in wheat leaves [74]. In this analysis, Topping and Leaver compared the quantities of mitochondrial genes and the corresponding transcripts in different sectors from the meristematic part at the bottom to the mature tip of a wheat leaf. The copy number of mitochondrial genes per cell decreases about 5- to 10-fold from the basal section of the leaf upwards. The abundance of the corresponding transcripts was found to decrease gene-specifically between 2.5- and 20-fold. Although the overall decrease in mitochondrial genes and transcripts is probably related to the increasing activity of chloroplast chromosome amplification and transcription, these results nevertheless suggest some gene-specific regulation of mRNA abundance during development. Unlike the chloroplast, where different developmental stages of the entire organelle can be observed and corresponding massive changes in gene expression occur, mitochondria are expected to be essential in most of their functions in all stages of cell differentiation in a plant and thus to vary only quantitatively between individual cell types.

A crucial tissue-specific function of plant mitochondrial gene expression unique to a development-specific cell differentiation in plants has been deduced from the effect of mitochondrial mutations that lead to a cytoplasmic male-sterile phenotype (cms). In these mutants, all vegetative tissues appear to be normal in their physiological and biochemical mitochondri-al functions although the mitochondrial genome contains abnormal rearrangements [1, 14, 34, 36, 47, 72]. Only pollen development is interrupted at different stages depending on the type of mitochondrial mutation, which suggests that more than the normal vegetative mitochondrial function is required for pollen development. These observations have prompted a study using *in situ* hybridization of the expression of several mitochondrial and nuclear genes encoding mitochondrial proteins in individual cell layers and types contributing to pollen maturation [67]. It was found that the amounts of transcripts for four mitochondria-encoded genes, *atpA, atp9, cob* and *rrn26*, significantly increase in young meiotic cells, while nuclear-encoded genes for mitochondria proteins show little differences in their transcription rates. This observation confirms that the mitochondrial genes are specifically upregulated in the meiocyte and tapetal cells. A crucial function of mitochondria during pollen development is also supported by the distinct association of mitochondria with the nucleus. Mitochondria appear to surround the entire nuclear envelope during these stages of development, suggesting a requirement for mitochondrial functions in the nucleus. Clarification of the mechanisms of enhanced transcription within the mitochondrion will shed light on the specific functions of these organelles during pollen development and on the mechanisms by which the cms phenotype is established in plants.

Aberrant genes in mutants: nuclear genes influencing post-transcriptional processes

Commercial application of cms phenotypes for hybrid seed production depends on nuclear inherited restorer genes, which allow seed production by compensating for the mitochondrially caused disturbance of normal pollen development. The nuclear restorer functions identified in plant breeders' efforts towards establishing cms systems for commercial use are promising candidates for nuclear-encoded genes involved in gene- and/or tissue-specific regulation of mitochondrial gene expression. In the maize T-cytoplasm, rearrangement of the mitochondrial genome has created a new open reading frame [14, 36]. Expression of this chimeric gene disturbs pollen maturation while normal growth of the somatic tissues in a plant is not affected. Among the several restorer genes known to reduce or completely eliminate the cms phenotype, some specifically affect the transcription pattern of this gene. One of

312

the restorers introduces new processing sites into the mutant *T-urf13* transcripts and leads to almost complete reduction of the corresponding polypeptide synthesis [36].

The open reading frame of the mitochondrial complex IV gene *cox1* is extended by a recombination event in some sorghum cytoplasms and leads to a cms phenotype upon high expression. As yet unknown nuclear genes in certain nucleus-cytoplasm combinations revert the cms phenotype and compensate for the expression of this aberrant protein [1].

In sunflower mitochondria, a recombined novel open reading frame correlated to a cms phenotype is likewise influenced by specific nuclear gene expression. Comparison of northern blot analysis and run-on experiments has shown that the nuclear restorer genes act at the post-transcriptional level and alter transcript stability. The presence of these nuclear genes destabilizes only this mitochondrial transcript in a tissue-specific manner and acts in the tapetal and early meiotic cells during anther development [34, 47].

Besides nuclear restorer genes in cms systems, a nuclear influence acting gene-specifically within the mitochondrial compartment can also be identified in crosses between different plant lines possessing various nuclear backgrounds [24, 75]. When, for example, mitochondria from the maize relative teosinte are introduced into maize nuclear backgrounds, protein synthesis patterns are altered within the mitochondria [53]; synthesis of a novel polypeptide is initiated. It remains to be investigated whether this stimulation of gene expression occurs at the mRNA level or as a translational control. Both avenues of approach, the cms restorer system and crosses between different types of mitochondria and nuclear backgrounds, promise major advances at the molecular level towards identifying interorganelle regulatory proteins that control mitochondrial gene expression by a specific modulation. With advances in rapid nuclear gene identification, the identity of these regulatory genes can be expected to be unravelled in the very near future. This will allow access to the regulatory cascades in plant mitochondrial gene expression.

Acknowledgements

We are grateful to Evelyn Laible-Schmid for expert help in the assembly of this text. Experimental work in the laboratory is supported by grants from the Deutsche Forschungsgemeinschaft, a Landesforschungsschwerpunkt from the Forschungsministerium of Baden-Württemberg, the Bundesministerium für Bildung and Forschung and the Fond der Chemischen Industrie.

Note added in proof: The recent identification of a nuclear restorer gene in maize (rf2) coding for a mitochondrial aldehyde dehydrogenase suggests that metabolic enzymes influence the effects of mitochondrial gene expression either directly through gain of function or indirectly by their metabolic activity. [Cui X, Wise RP, Schnable PS: The rf2 nuclear restorer gene of male sterile T-cytoplasm maize. Science 272: 1334–1336 (1996)].

References

1. Bailey-Serres J, Hanson DK, Fox TD, Leaver CJ: Mitochondrial genome rearrangement leads to extension and relocation of the cytochrome c oxidase subunit I gene in sorghum. Cell 47: 567–576 (1986).
2. Binder S, Brennicke A: Transcription initiation sites in *Oenothera* mitochondria. J Biol Chem 268: 7849–7855 (1993).
3. Binder S, Brennicke A: A tRNA gene transcription initiation site is similar to mRNA and rRNA promoters in plant mitochondria. Nucl Acids Res 21: 5012–5019 (1993).
4. Binder S, Hatzack F, Brennicke A: A novel pea mitochondrial *in vitro* transcription system recognizes homologous and heterologous mRNA and tRNA promoters. J Biol Chem 270: 22182–22189 (1995).
5. Binder S, Marchfelder A, Brennicke A, Wissinger B: RNA editing in intron sequences may be required for *trans*-splicing of *nad2* transcripts in *Oenothera* mitochondria. J Biol Chem 267: 7615–7623 (1992).
6. Binder S, Thalheim C, Brennicke A: Transcription of potato 26S rRNA is initiated at its mature 5' end. Curr Genet 26: 519–523 (1994).
7. Boerner GC, Moerl M, Wissinger B, Brennicke A, Schmelzer C: RNA editing of a group II intron in *Oenothera* as a prerequisite for splicing. Mol Gen Genet 246: 739–744 (1995).
8. Bonen L: The mitochondrial genome: so simple yet so complex. Curr Opinion Genet Devel 1: 515–522 (1991).
9. Boyen C, Leblanc C, Bonnard G, Grienenberger JM, Kloareg B: Nucleotide sequence of the *cox3* gene from *Chondrus crispus*: evidence that UGA encodes tryptophan and evolutionary implications. Nucl Acids Res 22: 1400–1403 (1994).
10. Brown GG, Auchincloss AH, Covello PS, Gray MW, Menassa R, Singh M: Characterization of transcription initiation sites on the soybean mitochondrial genome allows identification of a transcription-associated sequence motif. Mol Gen Genet 228: 345–355 (1991).
11. Chapdelaine Y, Bonen L: The wheat mitochondrial gene for subunit I of the NADH dehydrogenase complex: a *trans*-splicing model for this gene-in-pieces. Cell 65: 465–472 (1991).
12. Clayton DA: Transcription of the mammalian mitochondrial genome. Annu Rev Biochem 53: 573–594 (1984).
13. Covello PS, Gray MW: Sequence analysis of wheat mitochondrial transcripts capped *in vitro*: definitive identification of transcription initiation sites. Curr Genet 20: 245–251 (1991).

14. Dewey RE, Levings III CS, Timothy DH: Novel recombinations in the maize mitochondrial genome produce a unique transcriptional unit in the texas male-sterile cytoplasm. Cell 44: 439–449 (1986).

15. Fejes E, Masters BS, McCarty DM, Hauswirth WW: Sequence and transcriptional analysis of a chloroplast insert in the mitochondrial genome of *Zea mays*. Curr Genet 13: 509–515 (1988).

16. Finnegan PM, Brown GG: Transcriptional and post-transcriptional regulation of RNA levels in maize mitochondria. Plant Cell 2: 71–83 (1990).

17. Gray MW: Origin and evolution of mitochondrial DNA. Annu Rev Cell Biol 5: 25–30 (1989).

18. Gray MW: The evolutionary origins of organelles. Trends Genet 5: 294–299 (1989).

19. Gray MW, Hanic-Joyce PJ, Covello PS: Transcription, processing and editing in plant mitochondria. Annu Rev Plant Physiol Plant Mol Biol 43: 145–175 (1992).

20. Grohmann L, Thieck O, Herz U, Schröder W, Brennicke A: Translation of *nad9* mRNAs in mitochondria from *Solanum tuberosum* is restricted to completely edited transcripts. Nucl Acids Res 22: 3304–3311 (1994).

21. Gruissem W, Barkan A, Deng XW, Stern DB: Transcriptional and post-transcriptional control of plastid mRNA levels in higher plants. Trends Genet 4: 258–263 (1988).

22. Gualberto JM, Bonnard G, Lamattina L, Grienenberger JM: Expression of the wheat mitochondrial *nad3-rps12* transcription unit: correlation between editing and mRNA maturation. Plant Cell 3: 1109–1120 (1991).

23. Gualberto JM, Wintz H, Weil JH, Grienenberger JM: The genes coding for subunit 3 of NADH dehydrogenase and for ribosomal protein S12 are present in the wheat and maize mitochondrial genomes and are co-transcribed. Mol Gen Genet 215: 118–127 (1988).

24. Gupta D, Abbott AG: Higher plant mitochondrial DNA expression. 2. Influence of nuclear background on the transcription of a mitochondrial open reading frame, ORF 25. Theor Appl Genet 82: 723–728 (1991).

25. Hack E, Hendrick CA, Al-Janabi SM, Crane VC, Girton LE: Translation in a wheat germ cell-free system of RNA from mitochondria of the normal and texas male-sterile cytoplasms of maize (*Zea mays* L.) Curr Genet 25: 73–79 (1994).

26. Hanic-Joyce PJ, Gray MW: Processing of transfer RNA precursors in a wheat mitochondrial extract. J Biol Chem 265: 13782–13791 (1990).

27. Hanic-Joyce PJ, Gray MW: Accurate transcription of a plant mitochondrial gene *in vitro*. Mol Cell Biol 11: 2035–2039 (1991).

28. Hanic-Joyce PJ, Spencer, DF, Gray MW: *In vitro* processing of transcripts containing novel tRNA-like sequences ('t-elements') encoded by wheat mitochondrial DNA. Plant Mol Biol 15: 551–559 (1990).

29. Haouazine N, Takvorian A, Jubier MF, Michel F, Lejeune B: The *nad6* gene and the exon d of *nad1* are co-transcribed in wheat mitochondria. Curr Genet 24: 533–538 (1993).

30. Joyce PBM, Gray MW: Chloroplast-like transfer RNA genes expressed in wheat mitochondria. Nucl Acids Res 17: 5461–5476 (1989).

31. Kaleikau EK, André CP, Walbot V: Structure and expression of the rice mitochondrial apocytochrome b gene (cob-1) and pseudogene (cob-2). Curr Genet 22: 463–470 (1992).

32. Kennell JC, Pring DR: Initiation and processing of *atp6*, T-*urf13* and ORF*221* transcripts from mitochondria of T cytoplasm maize. Mol Gen Genet 216: 16–24 (1989).

33. Knoop V, Schuster W, Wissinger B, Brennicke A: *Trans* splicing integrates an exon of 22 nucleotides into the nad5 mRNA in higher plant mitochondria. EMBO J 10: 3483–3493 (1991).

34. Laver HK, Reynolds SJ, Monéger F, Leaver CJ: Mitochondrial genome organization and expression associated with cytoplasmic male sterility in sunflower (*Helianthus annuus*). Plant J 1: 185–193 (1991).

35. Leaver CJ, Gray MW: Mitochondrial genome organization and expression in higher plants. Annu Rev Plant Physiol 33: 373–402 (1982).

36. Levings III CS: The Texas cytoplasm of maize: cytoplasmic male sterility and disease susceptibility. Science 250: 942–947 (1990).

37. Lonsdale DM: The plant mitochondrial genome. Biochem Plants 15: 229–295 (1989).

38. Liu AW, Narayanan KK, André CP, Kaleikau EK, Walbot V: Co-transcription of *orf*25 and *cox*III in rice mitochondria. Curr Genet 21: 507–513 (1992).

39. Makaroff CA, Palmer JD: Extensive mitochondrial specific transcription of the *Brassica campestris* mitochondrial genome. Nucl Acids Res 15: 5141–5156 (1987).

40. Maloney AP, Traynor PL, Levings III CS, Walbot V: Identification in maize mitochondrial 26S rRNA of a short 5'-end sequence possibly involved in transcription initiation and processing. Curr Genet 15: 207–212 (1989).

41. Maloney AP, Walbot V: Structural analysis of mature and dicistronic transcripts from the 18S and 5S ribosomal RNA genes of maize mitochondria. J Mol Biol 213: 633–649 (1990).

42. Marchfelder A, Brennicke A: Plant mitochondrial RNase P and *E. coli* RNase P have different substrate specificities. Biochem Mol Biol Int 29: 621–634 (1993).

43. Marchfelder A, Brennicke A: Characterization and partial purification of tRNA processing activities from potato mitochondria. Plant Physiol 105: 1247–1254 (1994).

44. Marchfelder A, Brennicke A, Binder S: RNA editing is required for efficient excision of tRNA^Phe from precursors in plant mitochondria. J Biol Chem 271: 1898–1903 (1996).

45. Marchfelder A, Schuster W, Brennicke A: *In vitro* processing of mitochondrial and plastid derived tRNA precursors in a plant mitochondrial extract. Nucl Acids Res 18: 1401–1406 (1990).

46. Marechal-Drouard L, Weil JH, Dietrich A: Transfer RNAs and transfer RNA genes in plants. Annu Rev Plant Mol Biol 44: 13–32 (1993).

47. Monéger F, Smart CJ, Leaver CJ: Nuclear restoration of cytoplasmic male sterility in sunflower is associated with the tissue-specific regulation a novel mitochondrial gene. EMBO J 13: 8–17 (1994).

48. Mulligan RM, Lau GT, Walbot V: Numerous transcription initiation sites exist for the maize mitochondrial genes for subunit 9 of the ATP synthase and subunit 3 of cytochrome oxidase. Proc Natl Acad Sci USA 85: 7998–8002 (1988).

49. Mulligan RM, Leon P, Walbot V: Transcriptional and posttranscriptional regulation of maize mitochondrial gene expression. Mol Cell Biol 11: 533–543 (1991).

50. Mulligan RM, Maloney AP, Walbot V: RNA processing and multiple transcription initiation sites result in transcript size heterogeneity in maize mitochondria. Mol Gen Genet 211: 373–380 (1988).

51. Nakazono M, Tsutsumi N, Sugiura M, Hirai A: A small repeated sequence contains the transcription initiation sites for both *trnfM* and *rrn26* in rice mitochondria. Plant Mol Biol 28: 343–346 (1995).

314

52. Newton KJ: Plant mitochondrial genomes: organization, expression and variation. Ann Rev Plant Physiol Plant Mol Biol 39: 503–532 (1988).

53. Newton KJ, Walbot V: Maize mitochondria synthesize organ-specific polypeptides. Proc Natl Acad Sci USA 83: 6879–6883 (1985).

54. Newton KJ, Winberg B, Yamato K, Lupold S, Stern DB: Evidence for a novel mitochondrial promoter preceding the *cox2* gene of perennial teosintes. EMBO J 14: 585–593 (1995).

55. Nozato N, Oda K, Yamato K, Ohta E, Takemura M, Akashi K, Fukuzawa H, Ohyama K: Cotranscriptional expression of mitochondrial genes for subunits of NADH dehydrogenase, *nad5, nad4, nad2,* in *Marchantia polymorpha.* Mol Gen Genet 237: 343–350 (1993).

56. Oda K, Yamato K, Ohta E, Nakamura Y, Takemura M, Nozato N, Akashi K, Kanegae T, Ogura Y, Kohchi T, Ohyama K: Gene organization deduced from the complete sequence of liverwort *Marchantia polymorpha* mitochondrial DNA: a primitive form of plant mitochondrial genome. J Mol Biol 223: 1–7 (1992).

57. Palmer JD, Herbon LA: Unicircular structure of the *Brassica hirta* mitochondrial genome. Curr Genet 11: 565–570 (1987).

58. Pring DR, Mullen JA, Kempken F: Conserved sequence blocks 5′ to start codons of plant mitochondrial genes. Plant Mol Biol 19: 313–317 (1992).

59. Quetier F, Vedel F: Heterogeneous population of mitochondrial DNA molecules in higher plants. Nature 268: 365–368 (1977).

60. Rapp WD, Shelley Lupold D, Mack S, Stern DB: Architecture of the maize mitochondrial *atp1* promoter as determined by linker-scanning and point mutagenesis. Mol Cell Biol 13: 7232–7238 (1993).

61. Rapp WD, Stern DB: A conserved 11 nucleotide sequence contains an essential promoter element of the maize mitochondrial *atp1* gene. EMBO J 11: 1065–1073 (1992).

62. Reith M: Molecular biology of Rhodophyte and Chromophyte plastids. Annu Rev Plant Physiol Plant Mol Biol 46: 549–575 (1995).

63. Saalaoui E, Litvak S, Araya A: The apocytochrome b from an alloplasmic line of wheat (*T. aestivum,* cytoplasm-*T. timopheevi*) exists in two differently expressed forms. Plant Sci 66: 237–246 (1990).

64. Schuster W, Brennicke A: Conserved sequence elements at putative processing sites in plant mitochondria. Curr Genet 14: 187–192 (1989).

65. Schuster W, Brennicke A: The plant mitochondrial genome: structure, information content, RNA editing and gene transfer. Annu Rev Plant Physiol Plant Mol Biol 45: 61–78 (1994).

66. Schuster W, Hiesel R, Isaac PG, Leaver CJ, Brennicke A: Transcript termini in messenger RNAs in higher plant mitochondria. Nucl Acids Res 14: 5943–5954 (1986).

67. Smart CJ, Monéger F, Leaver CJ: Cell-specific regulation of gene expression in mitochondria during anther development in sunflower. Plant Cell 6: 811–825 (1994).

68. Stern DB, Gruissem W: Control of plastid gene expression: 3′ inverted repeats act as mRNA processing and stablizing elements, but do not terminate transcription. Cell 15: 1145–1157 (1987).

69. Stern DB, Rapp WD: *In vitro* analysis of plant mitochondrial transcription. In: Brennicke A, Kück U (eds) Plant Mitochondria, pp. 181–192. VCH Weinheim (1993).

70. Tracy RL, Stern DB: Mitochondrial transcription initiation: promoter structures and RNA polymerases. Curr Genet 28: 205–216 (1995).

71. Yang AJ, Mulligan RM: Distribution of maize mitochondrial transcripts in polysomal RNA: evidence for non-selectivity in recruitment of mRNAs. Curr Genet 23: 532–536 (1993).

72. Young EG, Hanson MR: A fused mitochondrial gene associated with cytoplasmic male sterility is developmentally regulated. Cell 50: 41–49 (1987).

73. Thomson MC, Macfarlane JL, Beagley CT, Wolstenholme DR: RNA editing of *mat-r* transcripts in maize and soybean increases similarity of the encoded protein to fungal and bryophyte group II intron maturases: evidence that *mat-r* encodes a functional protein. Nucl Acids Res 22: 5745–5752 (1994).

74. Topping JF, Leaver CJ: Mitochondrial gene expression during wheat leaf development. Planta 182: 399–407 (1990).

75. Wang J, Barth J, Abbott, AG: Higher plant mitochondrial DNA expression. 1. Variant expression of plant mitochondrial open reading frame, ORF25, in B37N and B73N maize lines. Theor Appl Genet 82: 765–770 (1991).

76. Wintz H, Grienenberger JM, Weil JH, Lonsdale DM: Location and nucleotide sequence of two tRNA genes and a tRNA pseudo-gene in the maize mitochondrial genome: evidence for the transcription of a chloroplast gene in mitochondria. Curr Genet 13: 247–254 (1988).

77. Wissinger B, Hiesel R, Schuster W, Brennicke A: The NADH-dehydrogenase subunit 5 gene in *Oenothera* mitochondria contains two introns and is co-transcribed with the 5S rRNA gene. Mol Gen Genet 212: 56–65 (1988).

78. Wissinger B, Schuster W, Brennicke A: *Trans*-splicing in *Oenothera* mitochondria: nad1 mRNAs are edited in exon and *trans*-splicing group II intron sequences. Cell 65: 473–482 (1992).

Plant Molecular Biology **32:** 315–326, 1996.
© 1996 *Kluwer Academic Publishers.*

Regulation of gene expression in chloroplasts of higher plants

Mamoru Sugita* & Masahiro Sugiura
*Center for Gene Research, Nagoya University, Nagoya 464–01, Japan (*author for correspondence)*

Key words: chloroplast gene expression, posttranscriptional regulation, polycistronic transcript, cotranscription, RNA-binding protein, splicing, rna processing, mRNA stability

Contents

Abstract	315	Chloroplast small RNAs involved in	
Introduction	315	splicing	320
Features of gene expression in		Posttranscriptional control of chloroplast	
chloroplasts of higher plants	316	gene expression	320
Transcription by multiple RNA		Processing and stability of mRNAs	321
polymerases	316	Processing enzymes	321
Polycistronic and multiple transcripts	317	RNA-binding proteins	322
Introns of chloroplast genes and splicing	318	Conclusion	323
Group I and group II introns	318	Acknowledgements	324
Intron-encoded proteins	319	References	324

Abstract

Chloroplasts contain their own genetic system which has a number of prokaryotic as well as some eukaryotic features. Most chloroplast genes of higher plants are organized in clusters and are cotranscribed as polycistronic pre-RNAs which are generally processes into many shorter overlapping RNA species, each of which accumulates of steady-state RNA levels. This indicates that posttranscriptional RNA processing of primary transcripts is an important step in the control of chloroplast gene expression. Chloroplast RNA processing steps include RNA cleavage/trimming, RNA splicing, ENA editing and RNA stabilization. Several chloroplast genes are interrupted by introns and therefore require processing for gene function. In tobacco chrloroplasts, 18 genes contain introns, six for tRNA genes and 12 for protein-encoding genes. A number of specific proteins and RNA factors are believed to be involved in splicing and maturation of pre-RNAs in chrloroplasts. Processing enzymes and RNA-binding proteins which could be involved in posttranscriptional steps have been identified in the last several years. Our current knowledge of the regulation of gene expression in chloroplasts of higher plants is overviewed and further studies on this matter are also considered.

Introduction

Chloroplasts have their own genome and a transcription-translation machinery which is distinct from that of the nucleo-cytoplasm. Chloroplast DNAs of higher plants are circular molecules of 120–160 kb and contain about 130 genes ([73], Table 1). Unlike chloroplast genes of algal species (paper by Rochaix in this issue), most chloroplast genes of higher plants are organized in clusters and are cotranscribed as polycistronic pre-RNAs which are then extensively processed into shorter RNA species [4, 84]. Transcription rates and steady-state RNA levels of chloroplast genes are generally not coincident, and many chloro-

316

plast genes are known to be constitutively transcribed, suggesting that post-transcriptional RNA processing of primary transcripts represents an important step in the control of chloroplast gene expression [21, 52]. Chloroplast RNA processing steps include endonucleolytic cleavage, 3'-end trimming, cis/trans-splicing and RNA editing ([27, 74], papers by Maier et al. and by Rochaix in this issue). All of these steps are believed to be accomplished by specific proteins and RNA factors in higher-plant chloroplasts. This paper describes our current knowledge of transcription and RNA processing in chloroplasts of higher plants, with emphasis on post-transcriptional regulation in chloroplast gene expression. RNA editing in chloroplasts is presented by Maier et al. in this issue. Regulation of gene expression in chloroplasts has been discussed in several previous reviews [27, 29, 33, 51, 74].

Features of gene expression in chloroplasts of higher plants

The transcription-translation apparatus of chloroplasts has a number of prokaryote-like features [27, 29, 33, 73]. For example, chloroplast ribosomes are 70S in size, chloroplast rRNAs and tRNAs are very similar in sequence to their *Escherichia coli* counterparts, and chloroplast mRNAs contain triphosphates at their 5' ends and are believed to lack poly(A) tails. Unlike *E. coli*, however, chloroplast tRNA genes do not code for 3'-CCA ends. Most components of the chloroplast genetic system are not coded for by chloroplast DNA but instead are nuclear-encoded. Therefore, formation of the chloroplast genetic system requires the coordinate expression of nuclear and chloroplast genes.

Transcription by multiple RNA polymerases

The upstream regions of many transcription initiation sites on chloroplast DNA contain DNA sequences similar to the *E. coli* '−10' and '−35' consensus sequences [27, 73]. The *psbA* gene and several other photosynthesis-related genes in higher-plant chloroplasts contain both the prokaryotic-type promoter sequences and between them a sequence motif similar to the TATA box of nuclear genes [71]. Prokaryotic-like promoter sequences are required for proper transcription [27] and the TATA box-like region is also critical for correct *psbA* transcription *in vitro* [25]. On the other hand, several genes including some tRNA genes lack the prokaryotic promoter sequences in the

Table 1. Chloroplast genes described in this chapter

Genes	Products
Genes for the photosynthesis system	
rbcL	Ribulose-1,5-bisphosphate carboxylase, large subunit
psaA, B	Photosytem I, P700 apoproteins A1, A2
psaC	9 kDa protein
psbA	Photosystem II, D1 protein
psbB	47 kDa chlorophyll a-binding protein
psbC	43 kDa chlorophyll a-binding protein
psbD	D2 protein
psbE	cytochrome b559 (8 kDa protein)
psbF	cytochrome b559 (4 kDa protein)
psbH	10 kDa phosphoprotein
psbI, J, K, L, M, N	I-, J-, K-, L-, M-, N-proteins
atpA, B, E	H^+-ATPase, CF_1 subunits α, β, ε
atpF, H, I	CF_0 subunits I, III, IV
petB, D	Cytochrome b_6/f complex, subunits b_6, IV
ndhA − K	NADH dehydrogenase, subunits ND1 − ND11
Genes for the genetic system	
16S rDNA	16S rRNA
23S rDNA	23S rRNA
trnA-UGC	Alanine tRNA (UGC)
trnG-UCC	Glycine tRNA (UCC)
rnH-GUG	Histidine tRNA (GUG)
trnI-GAU	Isoleucine tRNA (GAU)
trnK-UUU	Lysine tRNA (UUU)
trnL-UAA	Leucine tRNA (UAA)
rps2, 7, 12, 16	30S ribosomal proteins CS2, CS7, CS12, CS16
rpl2, 20, 32	50S ribosomal proteins CL2, CL20, CL32
rpoA, B, C1, C2	RNA polymerase, subunits α, β, β', β''
matK	maturase-like protein
sprA	small plastid RNA
Others	
clpP	ATP-dependent protease, proteolytic subunit
irf168 (ycf3)	intron-containing reading frame (168 codons)

regions upstream of their transcription initiation sites [27], while the *rps16* gene possesses the '−10' motif but is missing the '−35' motif [54]. Multiple transcription initiation sites are known especially for the chloroplast genes *psbD* [10], *atpB* [36], and *16S rDNA* [34, 80]. At least one of the several transcription initiation sites of *psbD* is directed by a light-responsive promoter, whose activity is regulated differentially by chloroplast-specific protein factors and/or distinct RNA polymerase species [9, 82]. One of the multiple promoters in *atpB* or *16S rDNA* lacks any conserved sequence motifs and is active irrespective of the developmental stage of chloroplasts, suggesting that it is responsible for the constitutive transcription of these genes [36, 80].

Multiple RNA polymerase activities have been found in chloroplasts [27]. For example, in mustard chloroplasts, two distinct RNA polymerase activities (peaks A and B) have been identified, both of which correspond to large multi-subunit complexes [59]. The peak A enzyme is resistant to rifampicin (an inhibitor of prokaryotic RNA polymerase) and is composed of ca. 13 polypeptides, while the rifampicin-sensitive peak B enzyme consists of only four polypeptides. Interestingly, the peak B enzyme is predominant in etioplasts of dark-grown mustard plants while the peak A enzyme is predominant in the chloroplasts of plants grown in the light. The relative activities of the two identified RNA polymerases appear to change during development from etioplasts to chloroplasts. Spinach chloroplasts also contain at least two types of RNA polymerases: one is multimeric and *E. coli*-like, while the other is a monomeric enzyme of 110 kDa with all the characteristics of single-subunit RNA polymerases of the T7 bacteriophage type [43]. It has not been determined which RNA polymerase(s) correspond to the subunits derived from the chloroplast *rpoA, B, C1* and *C2* genes. Transcript accumulation in ribosome-deficient plastids of a barley mutant, *albostrians*, provides additional evidence for one or more functional non-chloroplast (nuclear)-encoded RNA polymerases [31]. Thus, either chloroplast-encoded or nuclear-encoded RNA polymerases or both contribute to the transcription of individual chloroplast genes. In general, it appears that genes for the chloroplast genetic system are transcribed by the nuclear-encoded RNA polymerase during the early stage of plastid development and subsequently transcription of the genes for photosynthesis-related components are directed by the chloroplast-encoded RNA polymerase [36, 51].

For precise initiation of transcription or differential promoter selection, it is expected that the core RNA polymerase should require additional factors. Bacterial σ-like factors and polypeptides containing known DNA-binding motifs are not known to be encoded in the chloroplast genome and hence such factors must be of nuclear origin. Sigma-like factors have been isolated from chloroplasts of several higher-plant species [42, 77, 79], and phosphorylation/dephosphorylation of σ-like polypeptides has been reported to affect transcription initiation [78]. DNA-binding proteins interacting

Table 2. The representative transcription units of higher plant chloroplast genes.

---

1. Monocistronic transcription unit

ndhF, psbA*, psbM, psbN, rbcL, most of the tRNA genes

2. Dicistronic or polycistronic transcription unit

a. Related functions
atpB-atpE
3' *rps12-rps7*
psbE-psbF-psbL-psbJ
psbD-psbC**-orf62 (in dicots)
psbK-psbI-psbD**-psbC**-orf62-trnG (in monocots)
ndhC-ndhK-ndhJ
16S rDNA-trnI-trnA-23S rDNA-4.5S rDNA-5S rDNA
rpoB-rpoC1-rpoC2
rpl23-rpl2-rps19-rpl22-rps3-rpl16-rpl14-rps8-infA-rpl36-rps11-rpoA
trnE-trnY-trnD

b. Unrelated functions
clpP-5' rps12-rpl20
orf31-petG-psaJ-rpl33-rps18
psaA-psaB-rps14
psaC-ndhD
psbB-psbH-petB-petD
psbK-psbI-trnG
ndhA-ndhI-ndhG-ndhE-psaC
rpl32-sprA**
rps2-atpI-atpH**-atpF-atpA

---

* A dicistronic mRNA of *trnK-psbA* was observed in mustard [56].
** The genes are cotranscribed with the preceding gene but also are transcribed by own promoter(s) [9, 10, 68, 81, 82, 87].

with a light-responsive element have recently been identified in tobacco chloroplast extracts [2]. Using a similar approach, a sequence-specific DNA-binding factor, CDF2, has been identified as a repressor for transcription initiation from one of the three transcription initiation sites of *16S rDNA* that is directed by an *E. coli*-like RNA polymerase activity in spinach [34].

Polycistronic and multiple transcripts

Though several chloroplast genes are monocistronically transcribed, most of the chloroplast genes of higher plants are cotranscribed. Polycistronic primary transcripts thus obtained consist of messages for proteins or RNAs with related functions, such as photosynthesis, transcription and translation, or other functions (Table 2). A transcription unit is defined as a primarily transcribed region and can be determined by the combination of northern analysis, primer extension, S1 nuclease or ribonuclease protection, and *in vitro* capping experiments. Chloroplast primary transcripts harbor 5′-triphosphates that can be specifically labeled with $[\alpha\text{-}^{32}\text{P}]$GTP and guanylyltransferase. The so-called *in vitro* capping analysis is, therefore, a critical experimental tool to identify the *bona fide* initiation site of transcription. In tobacco, about 60 transcription units have thus far been identified in this manner (unpublished).

Chloroplast polycistronic primary transcripts are generally processed into many overlapping shorter RNA species, each of which accumulates at steady-state levels. For example, the *psbB* operon is transcribed from a bacterial-type promoter as a tetracistronic precursor (consisting of *psbB – psbH – petB – petD*) of ca. 6 kb [4, 84]. Two *cis*-splicing events and multiple cutting steps are necessary to form the functional shorter RNA species from the primary transcript. Many intermediate transcripts are produced during these processes, and 17 different RNA species in spinach [84] and at least 20 RNA species in maize [4] have been observed. Processing of the spinach primary transcript results ultimately in the formation of monocistronic mRNAs for *psbB* and *psbH* and a dicistronic mRNA for *petB* and *petD* [84]. In maize, however, the respective dicistronic mRNAs for *petB* or *petD* are further processed to monocistronic units or messages [4, 7]. Almost all of the transcripts derived from the maize *psbB* operon cosediment with polysomes, suggesting that they are all translated, and that 5′ end and intercistronic cleavages are not always required for translation of chloroplast mRNAs [4].

A set of chloroplast transcripts contain multiple 5′ ends, which result from the endonucleolytic cutting of precursor RNAs (e.g. [9, 54]). 5′-end cleavage may provide a mechanism for translational control or differential stabilization of the mRNA. One suggestive example for this regulatory mechanism was obtained from barley, in which translational inhibition but not alteration in levels of the *rbcL* mRNA was observed upon treatment with jasmonate [62]. In this case, an alternative form of *rbcL* mRNAs accumulated and failed to be incorporated into the polysomal fraction. The alternative transcript contains a 35 nucleotide extra sequence which is highly complementary to the 3′ end of chloroplast 16S rRNA. It is possible that formation of an intermolecular complex between the *rbcL* mRNA and 16S rRNA could hinder the translation of this message [62]. Thus, multiple steps of chloroplast mRNA processing are involved in the formation of

mature mRNAs and appear to play a prominent role in the control of chloroplast gene expression.

Introns of chloroplast genes and splicing

Group I and group II introns

Several chloroplast genes are interrupted by introns. In the tobacco chloroplast genome, 18 genes contain introns of 503–2526 bp in length, six for tRNA genes and 12 for protein-encoding genes. Two of the latter genes, *clpP* and *irf168* (*ycf3*), contain two introns each, indicating the presence of a total of 20 *cis* introns [73]. Most introns can be folded into a characteristic, evolutionarily conserved secondary structure similar to that of the group II intron class of fungal mitochondrial genes. As shown in Fig. 1, chloroplast group II-type introns (class III) have conserved boundary sequences (5′-GTGYGRY···RYCNAYYyYRAY-3′), which are similar in part to those of nuclear pre-mRNA introns [26]. The putative branch points of the spinach *atpF* and *petD* introns were found to be eight nucleotides upstream from their respective 3′ intron/exon boundaries, which lie in a bulged A residue in the domain VI of group II intron [39]. The three introns from *trnI* (GAU), *trnA* (UGC) and *clpP* (intron 2) can form secondary structures similar to those of fungal mitochondrial group II introns but their boundary sequences are not conserved (class II, in Fig. 1). The intron of the gene encoding tRNA^{Leu} (UAA) can be classified into the group I-type intron of fungal mitochondrial genes and the *Tetrahymena* nuclear 21S rRNA gene. A new class of short introns of ca. 100 bp has been found in the *Euglena gracilis* chloroplast genome and appears to be unique to this alga [19].

Efficient self-splicing of pre-RNAs from algal chloroplast genes containing group I-type introns has been demonstrated *in vitro* (paper by Rochaix in this issue). In higher-plant chloroplasts, *trnL* (UUA) is the only gene harboring a group I-type intron [73], but attempts to demonstrate self-splicing of its RNA *in vitro* have so far been unsuccessful. No self-splicing has been reported for chloroplast group II-type introns. It is well known that splicing of nuclear pre-mRNAs and maturation of nuclear pre-rRNAs are catalyzed by protein-RNA complexes. Therefore, it is possible that protein-RNA complexes are also involved in splicing and maturation of pre-RNAs in chloroplasts. Surprisingly, unspliced pre-RNAs accumulate at substantial levels even in polysomal fraction in chloroplasts,

A.

Class	Gene	5'Exon	Intron		3'Exon	Length (bp)
I	trnL-UAA	-GACTT	AATTGGATTG ----------	TATCGTAAGAGG	AAAAT-	503
II	trnI-GAU	-GATAA	TTGCGTCGTT ----------	GATTTACTTCAC	GGGCG-	707
	trnA-UGC	-TGCAA	TTGGGTCGTT ----------	GGTTTACCCTGC	GGCGG-	709
	clpP(I-2)	-ACGCT	TGGCGCCAAT ----------	TGTTATATCATC	AGGGT-	637
III	trnV-UAC	-TACAC	GCGCGCCAAT ----------	ACCTGTTTT AC	CGAGA-	571
	trnG-UCC	-TAAAA	GTGTGATTCG ----------	GTCGACTATAAC	CCCTA-	691
	trnK-UUU	-TTTAA	GTGCGGCTAG ----------	ATCTACTCC AT	CCGAC-	2526
	rpl2	-TTTGA	GTGCGGTTTG ----------	ATCTACTTC AA	CCGAT-	666
	rpl16	-TTAGT	GTGTGACTCC ----------	ATCAACTATAAC	CCCAA-	1020
	rps12(trans)	-TGTAT	GTGCGTTGTA ----------	GTCAACTTTTCC	ACTAT-	
	rps12(cis)	-TTCTA	GTGCGTTGTA ----------	ATCCACCCT AC	AATAT-	536
	rps16	-GCAAC	GTGCGACTTG ----------	ATCAATCCCAAT	GAGCC-	860
	rpoC1	-CCCAT	GTGTGATTTG ----------	TCCTATCCCAAT	TTTTC-	825
	petB	-TGAGT	GTGTGACTTG ----------	GCCTATCTCAAT	AAAGT-	753
	petD	-GGAGT	GTGTGACTTG ----------	ACCTATCCCAAT	AACAA-	742
	ndhA	-TCTAC	GTGTGATTCG ----------	ATCGACTATGAT	TATCT-	1148
	ndhB	-AAGGA	GTGCGGTTCG ----------	TTCGACTCTGAC	TCTCC-	679
	clpP(I-1)	-GTATA	GTGCGACTTG ----------	TTTTACCCTAAT	CAACC-	807
	atpF	-AGTGT	GTGCGAGTTG ----------	ATCTACTTTCAT	TAAGT-	695
	irf168(I-1)	-AGATG	GTGCGATTTG ----------	GACGACCGTAAC	GGATG-	738
	irf168(I-2)	-ATTAC	GTGCGACTAT ----------	ACCTATTCCGGC	CGGGG-	783
	Conserved:		GTGYGRYtyg ----------	RYCNAYYyYRAY		

B.

		Intron	Branch point	3'Exon
			*	
Spinach	atpF	..UCG GGAAGGGAU....30......AUCU	A CUUUCAU	UAAGU-
	petD	..UUC GGGGGAUGAAUU.11.AAUUCACCU	A UCCUAAU	AACAA-
	petB	..UUA GGGGGGUUUAAU----AGUUUACCU	A UCUCAAU	AAAGU-
	rpoC1	..GGG GGGAGAUCCUAUAGGAUAGGAUCCU	A UCCCAAU	UUUUC-
Tobacco	rps12(cis)	..UUG GAGGGAGAUCUU..11..GAGAUCC	A CCCU AC	AAUAU-

Figure 1. Comparison of exon-intron junctions and branch points of chloroplast split genes. A. Exon-intron boundary sequences of 18 tobacco chloroplast split genes [75]. clpP and irf168 (ycf3) contain two introns each (I-1 and I-2). B. Sequences surrounding the branch point. Lariat intron RNAs detected are presented, from spinach [39] and tobacco (unpublished). Asterisks indicate branch points determined experimentally. Domain VI is shown by divergent horizontal arrows.

and the efficiency of splicing of pre-RNA varies during plastid development and/or in different tissues [3, 5], thereby implying that posttranscriptional modification of pre-RNAs is complex and apparently plays an important role in chloroplast gene expression.

Intron-encoded proteins

The intron of *Chlamydomonas 23S rDNA* encodes a double-stranded DNA endonuclease (paper by Rochaix in this issue). However, in higher plant chloroplasts, no intron is present in *23S rDNA* and neither introns nor genes encoding a double-stranded DNA endonuclease have been found. A potential protein coding frame of 509 codons was first described in the longest intron (2529 bp) of *trnK* coding for tRNA^{Lys} (UUU) from tobacco chloroplasts [70]. The sequence of the ca. 500 amino acid protein encoded within *trnK* is well conserved in land plants [65], and has been shown to be structurally related to mitochondrial intron-encoded maturases from fungi, and hence the gene encoding it has been designated *matK* [53]. The antibody raised against a synthetic peptide of the C-terminal nine amino acids of the predicted potato *matK* product cross-reacted with 43 and 41.7 kDa proteins present in potato chloroplast extracts [43]. UV cross-linking and gel-shift assays with the mustard *matK-β-*galactosidase fusion protein made in *E. coli* indicated that of the three RNA probes containing chloroplast group II introns the *matK* protein can recognize pre-tRNA^{Gly} and pre-tRNA^{Lys} but not pre-mRNA from *rps16* [46]. This suggests that the *matK* protein may be a factor involved in splicing of pre-*trnK* and pre-*trnG* RNAs. The product of *matK* has also been implicated in the gene-specific splicing of a second gene, *rpl2*, because the pre-RNA from this gene is completely unspliced in plastid ribosome-deficient mutants of barley whereas the pre-RNAs from *rpl16*, *rps16*, *ndhB*, *petD*, *irf170* (*ycf3*) and *trnL* (UAA) are spliced [30]. It is proposed that lack of *rpl2* processing results from the loss of translation of chloroplast-encoded *matK*.

Direct biochemical analyses are needed to verify the above hypothesis as was done with the binding of *matK* protein to the intron of the pre-*rps16* mRNA [46]. However, another interesting correlation is that both *rpl2* and *matK* are retained in the plastid genome of the non-photosynthetic parasitic plant *Epifagus virginiana*, which is much smaller (70 kb) than typical chloroplast genomes mainly due to the lack of all genes for photosynthesis [85]. There are still dozens of ORFs on the chloroplast genome and we cannot exclude the possibility that some of their products are required for splicing.

Chloroplast small RNAs involved in splicing

The most striking feature among split chloroplast genes is found in *rps12* in land plants [73] and *psaA* in *C. reinhardtii* (paper by Rochaix in this issue). The *rps12* gene is split into three separate parts which are distantly located from each other on the chloroplast genome. The exon 1 is cotranscribed with the upstream *clpP* and the downstream *rpl20* genes, and the exons 2 and 3 are separated by a *cis* intron, located on the large inverted repeat (hence two copies per genome), and are transcribed with the downstream *rps7*. The two separate transcripts are assembled by a *trans*-splicing process to form the functional mRNA. The chloroplast-encoded *tscA* RNA of *C. reinhardtii* (430 nt) has been shown to be necessary for *trans*-splicing of exon 1 and exon 2 of *psaA* pre-mRNAs (paper by Rochaix). No such small RNAs for *trans*-splicing have so far been found in higher plants. The spinach *atpF* group II-type intron is not spliced in transgenic *C. reinhardtii* chloroplasts [22], suggesting that splicing of chloroplast introns may be species-specific or that its mechanism differs between algae and higher plants. It has been shown that, in ribosome-deficient plastids of barley,

the *rps12 cis* intron separating exons 2 and 3 is completely unspliced whereas the splicing of the bipartite *rps12 trans*-splicing introns between exons 1 and 2 occurs but at a reduced level [32]. This observation suggests that *cis* splicing requires a factor(s) encoded by the chloroplast genome. The putative factor may be either a protein or a small RNA transcribed specifically by the chloroplast-encoded RNA polymerase. No *in vitro* RNA splicing systems from chloroplasts are currently available, which makes it difficult to analyze individual steps in RNA splicing in chloroplasts and to detect the various factors involved. Recently an *in vivo* system using a chloroplast transformation technique was developed to study splicing in higher-plant chloroplasts [11]. This system can be useful for the identification of *cis* elements required for pre-mRNA splicing in chloroplasts.

A small plastid RNA (spRNA, 218 nt) is encoded by *sprA* of the tobacco chloroplast genome [81]. Part of the spRNA sequence exhibits the potential for base-pairing with a 10 nucleotide region in the leader sequence of pre-16S rRNA, suggesting a role for spRNA in chloroplast ribosome biogenesis, i.e. 16S rRNA maturation [81]. As the tobacco chloroplast genome still includes about ten spacer regions of ca. 1 kb which do not harbor any significant ORFs, additional small RNA species may be encoded in these spacers.

Posttranscriptional control of chloroplast gene expression

The steady-state level of chloroplast RNAs varies substantially for different genes and for developmental stages and environmental stimuli, and fluctuates during circadian rhythms [51, 60]. In general, mRNAs encoding photosynthesis-related proteins accumulate at high levels in chloroplasts from green tissues or during chloroplast development while their levels are very low in plastids of non-photosynthetic tissues, such as amyloplasts in roots and chromoplasts in fruits. The steady-state RNA level can be determined in principle by two factors, transcription activity of individual genes and stability of their transcripts. All mRNAs for 10 spinach genes accumulate at different but generally high levels during light-dependent chloroplast development and during leaf maturation, while there are no major changes in their relative transcription activities [21]. Similarly, the changes in transcription rates and steady-state RNA levels of various transcripts were not parallel during the growth of dark-grown or illuminated barley seedlings [52, 61]. These observations suggest that transcriptional regulation is limited in chloroplast gene expression and differential changes in steady-state RNA levels are due mainly to the stability of RNAs rather than to the transcription rate of their respective genes.

Processing and stability of mRNAs

The stability of chloroplast RNAs vary from RNA species to RNA species, depending on chloroplast development and environmental conditions [21, 27, 52]. Most chloroplast transcription units contain at their 3'-untranslated regions (UTR), a unique short inverted repeat (IR) which can potentially fold a stem-loop structure. These IRs were originally thought to function as transcription terminators as is known for prokaryotes. However, it has been demonstrated that chloroplast IR structures at 3'-UTRs probably act as RNA processing signals rather than termination signals [66]. The IRs present in *psbA*, *rbcL*, *petD* and *atpE* are ineffective as transcription terminators but serve as accurate and efficient RNA-processing signals as examined using an *in vitro* transcription system. IR-containing pre-mRNAs are processed in a 3' → 5' direction by 3' exonucleases present in the transcription system, generating nearly homogeneous 3' ends distal to the IR. When IR structures are removed from the pre-mRNAs, the mutant RNAs are rapidly degraded *in vitro* and *in organello* [1, 15]. These observations indicate that the mRNA 3' IR structures act as *cis*-acting elements critical for the stability of RNAs in chloroplasts from higher plants. However, we cannot exclude the possibility that the ineffective termination observed *in vitro* might be an artifact of extract preparation causing, for example, the loss of factors required for efficient termination.

Processing enzymes

The chloroplast RNase P-like enzyme, one of the tRNA processing enzymes, has been characterized biochemically in spinach and shown to differ from the *E. coli* RNase P [83]. The spinach chloroplast activity has a buoyant density of 1.28 g/ml (vs. 1.73 g/ml for *E. coli* RNase P) and is resistant to treatment with micrococcal nuclease, suggesting that the chloroplast RNase P does not contain an RNA subunit. However, the RNase P activity in tobacco chloroplasts was reported to be sensitive to nuclease treatment [86] and several algal chloroplast DNAs have been shown to encode RNase

P RNA homologues [41, 63, 67]. In contradiction to the earlier reports, these observations therefore suggest that the chloroplast enzyme contains an RNA component.

Endonucleolytic cleavage sites have been mapped in the primary transcript from the spinach *psbB* operon and a hexanucleotide motif, YGGA(A/T)↓Y was identified at these sites [84]. Three endonuclease activities were identified while studying processing of *petD* 3'-UTR in a spinach chloroplast extract. An endonuclease, termed Endo C1 (chloroplast endonuclease 1) cleaves within the loop of *petD* IR structure and disrupts the stabilizing hairpin structure [15]. The second activity, Endo C2, cleaves *petD* RNA both at the termination codon and at the mature RNA 3' end [15]. This cleavage could be involved in an early step in mRNA decay. Interestingly, the latter cleavage site lies within the binding site of a 57 kDa RNA-binding protein. Cleavage of *petD* mRNA at the termination codon leads to rapid degradation of its upstream RNA. The third enzyme, p67, which has *E. coli* RNAase E-like activity, is a site-specific endonuclease and cleaves the site just upstream of IR structure of *petD* mRNA [28]. The p67 activity is associated with a large complex containing a 100 kDa protein (100RNP) which has polyribonucleotide phosphorylase (PNPase) activity, and might be regulated by other RNA-binding proteins which mask the cleavage site *in vivo*. Recently, *psbA* mRNA decay has been studied using an *in vitro* mRNA decay system based on lysed chloroplasts from spinach leaves [40]. The early cleavage events which may induce *psbA* mRNA degradation occur within the 5' region of the message and are followed by further endonucleolytic cleavages within the downstream part of the RNA. Most of the cleavages occur between an adenosine and an uridine. Finally, the endonucleolytic fragments are rapidly degraded by processive exonucleolytic cleavage.

A mustard 54 kDa protein was first identified as a protein interacting specifically with U-rich sequence elements, UUU(A/C)U(C/A)U followed by another U-rich stretch, present in the 3'-flanking regions of *trnK*, *rps16* and *trnH* [55, 57]. The gel-purified 54 kDa protein was then shown to have an endonucleolytic activity very similar to that of spinach Endo C1. The spinach 57 kDa protein recognizes specifically a conserved AU-rich sequence motif, AUUYNAUU, located before and after the IR [16] and appears to be a homologue of the mustard 54 kDa protein. Four additional proteins of 28, 32, 33 and 100 kDa, which interact *in vitro* with the spinach *petD* 3'-UTR, were identified by UV-crosslinking with partially purified chloroplast extracts [16]. Of these, the 33 kDa protein binds specifically to the stem region of the IR structure and its binding may decrease activity of 3'-end formation, probably by impeding the progress of the processive 3' → 5' exonuclease activity. An RNase T1-treated complex of petD 5'-UTR and proteins contains the 55 kDa (resized as 57 kDa), 29 kDa (as 33 kDa) proteins, and an additional 41 kDa protein [14]. The 41 kDa protein, which binds to the specific region, called box II (AUUCAAUU), located immediate downstream of the IR of petD 3'-UTR, also has a nuclease activity ([14]; D. Stern, personal communication). The identified RNA protein complex may direct correct 3'-end processing and/or influence the stability of petD mRNA in chloroplasts.

RNA-binding proteins

A defined class of RNA-binding proteins, suggested to be involved in splicing and/or processing of pre-RNAs, was described for tobacco [44, 88] and spinach [64]. In tobacco, five chloroplast proteins ranging from 28 kDa to 33 kDa (named as cp28 to cp33), were isolated by a one-step procedure of ssDNA column chromatography. These proteins are nuclear-encoded and each of the precursor polypeptides is composed of four domains: a transit peptide, an acidic amino terminal domain and two conserved RNA-binding domains. The RNA-binding domains of ca. 80 amino acids belong to a superfamily, called consensus sequence-type RNA-binding domains (CS-RBD) or RNA recognition motifs (RRM), and are responsible for RNA binding. Similar chloroplast proteins have been isolated from spinach (28RNP) [64], *Mesembryanthemum crystallium* (ice plant, cRBP) [12] and pea [69]. cDNAs encoding similar proteins have also been isolated from maize (NBP or nucleic acid binding protein) [18], *N. plumbaginifolia* (CP-RNP30 and 31) [50], *Arabidopsis* [8, 17, 20, 23, 58], and *Phaseolus vulgaris* [37] (Table 3). Based on amino acid identities, these chloroplast RNA-binding proteins can be categorized into three groups: I (cp28, cp31, 28RNP and NBP), II (cp29A and B, cp29, CP-RNP30 and 31) and III (cp33). The five tobacco chloroplast RNA-binding proteins have higher affinities for poly(G) and poly(U) *in vitro*, suggesting that they bind preferentially to U- or G-rich stretches in chloroplast RNA molecules [45, 89]. A similar result was obtained with spinach 28RNP [48]. The nucleic acid-binding properties of these proteins resemble those of HeLa hnRNP proteins. Interest-ingly, RNA-binding proteins similar to those in chloroplasts have been found in cyanobacteria, suggesting that chloroplast RNA-binding proteins derived from an ancient cyanobacterium, and thus this finding provides additional evidence for the endosymbiotic theory of chloroplast origin [72].

The half-lives of most bacterial mRNAs are known to increase upon their association with ribosomes in the translation complex. However, a substantial portion of some of the chloroplast mRNAs including *psbA* mRNA is not associated with polysomes [38], and hence, unlike in *E. coli*, lack of ribosome association does not necessarily imply transcript instability in chloroplasts. It is unlikely that nascent, i.e. unspliced or unedited, chloroplast mRNAs bind ribosomes and form a translation complex and therefore transcription appears not to be coupled with translation in chloroplasts. Relevant to this, the spinach 28RNP which is required for mRNA 3'-end processing [64] is neither associated with thylakoid-bound nor with soluble polysomes in spinach [28, 49]. Our preliminary data also indicate that non-polysomal *psbA* is immunoprecipitated by antibodies against tobacco chloroplast RNA-binding proteins and hence these proteins probably stabilize non-polysomal RNAs (unpublished). These observations raise the possibility that, in analogy to hnRNP particles, chloroplast RNA-binding proteins are associated with non-polysomal pre-RNAs as stable RNA-protein complexes before being replaced by ribosomes for translation or being assembled with ribosomal proteins for ribosome formation. RNA-binding properties of the recombinant form of 28RNP (from *E. coli*) has been shown to differ from that of the native form in chloroplasts, suggesting that posttranslational modifications can modulate RNA-binding specificity of the 28RNP [49]. In fact, phosphorylation of the 28RNP has been demonstrated to modulate its affinity to several RNA species [35, 47]. Therefore, it seems that phosphorylation and dephosphorylation also play a significant role in the post-transcriptional regulation.

A large number of nuclear mutants of higher plants have been isolated that show reduced amounts of photosynthetic complexes [76]. Of those, several maize mutants exhibit alterations in chloroplast mRNA levels. For example, a maize nuclear mutant, *hcf-38* (high chlorophyll fluorescence-38), has been shown to accumulate aberrant amounts and sizes of transcripts of the *psbB* cluster [6] and other mutants *hcf7*, *cps1-1*, *cps1-2* (chloroplast protein synthesis 1) have a defect in the maturation of the 16S and 23S rRNAs, resulting in accumulation of pre-rRNAs [5]. These obser-

Table 3. RNA-binding proteins found in chloroplasts from higher plants.

Proteins	Plant species	Features or putative functions	References
Proteins containing two CS-RBDs			
cp28, cp29A, cp29B, cp31, cp33	*Nicotiana sylvestris*	preferential binding to poly(G), poly(U) general RNA-binding protein present as complexes of 20–30S	44, 88
CP-RNP30, CP-RNP31	*N. plumbaginifolia*	correspondence to cp29A and cp29B	50
cp29, cp31, cp33	*Arabidopsis thaliana*	homologues of *N. sylvestris*	58
RNP-T	*Arabidopsis thaliana*		8
ATRBP31, 33	*Arabidopsis thaliana*		17, 20, 23
cRBP	*Mesembryanthemum crystallinum*		12
28RNP	*Spinacia oleracea*	processing of 3'-UTR of mRNA	64
NBP	*Zea mays*	isolated as a DNA-binding protein	18
Bean-RNP1	*Phaseolus vulgaris*	specific binding to T-rich ssDNA identified as a AG–1-binding protein	37
Others			
100RNP	*S. oleracea*	polyribonucleotide phosphorylase-like activity	28
55RNP	*S. oleracea*	specific binding to U-rich region after and before *petD* 3'-IR	16
33RNP	*S. oleracea* (29 kDa protein)	*petD* 3'-UTR-binding associated with 100RNP	16, 28
41 kDa protein	*S. oleracea*	specific binding to a box-II of *petD* 3'-UTR nuclease activity	14
p67	*S. oleracea*	RNAase E-like activity specific cleavage of a site before *petD* IR	28
Endo C1	*S. oleracea*	site-specific endonucleolytic cleavage	15
Endo C2	*S. oleracea*	site-specific endonucleolytic cleavage	15
54 kDa protein	*Sinapis alba*	binding to *trnK*, *psbA*, *rps 16* 3'-UTRs endonuclease activity	55, 57
RNase P	*N. tabacum*	5' end formation of pre-tRNAs	86
	S. oleracea		83

vations suggest that these loci affect chloroplast transcription or RNA processing. The pre-16S rRNA thus accumulated in *hcf7* chloroplasts has been shown to be excluded from the polysomal fraction, therefore suggesting that maturation of pre-rRNA is necessary for efficient function in translation. The maize mutant *crp1* (chloroplast *R*NA *p*rocessing-*1*) lacks monocistronic forms of the *petB* and *petD* mRNAs and instead accumulates the dicistronic form, resulting in loss of cytochrome b_6/f complex [7]. In barley, several nuclear mutants have been shown to exhibit impaired splicing of *rps12* and *rpl2* mRNA precursors [30, 32]. It is possible that at least some of these loci encode RNA-binding proteins present in chloroplasts. Biological analyses using the mutants will be a promising way to elucidate the molecular mechanism of chloroplast RNA processing.

Conclusion

The mode of chloroplast gene expression exhibits several unique features not found in prokaryotic and eukaryotic nuclear systems, and hence chloroplast gene expression and its dynamic regulation are very interesting subjects. The control of chloroplast gene expression is affected by environmental factors and a developmental program, and operates at several steps, transcription, posttranscription, translation and posttranslation. Chloroplast gene expression is largely controlled at the post-transcriptional level. This includes: RNA cutting/trimming, RNA splicing, RNA editing and RNA stabilization/storage. Chloroplast mRNA stability is currently being studied in detail and factors involved in this process are being identified. Because very little is known about splicing of chloroplast pre-

RNA, this area of study presents one of the most interesting and challenging targets for further study. However, for analysis of this process it is essential to develop an *in vitro* splicing system from chloroplasts. Chloroplast transformation systems recently established for *Chlamydomonas* and tobacco also hold promise for the analysis of splicing pathways. Most of the regulatory factors for the chloroplast genetic system are believed to be nuclear-encoded as is the case for the chloroplast DNA- and RNA-binding proteins identified so far. Therefore, extensive biochemical approaches using nuclear mutants affecting chloroplast development and photosynthetic activity could facilitate our understanding of the regulatory components for RNA cleavage and splicing. The involvement of RNA components in chloroplast RNA cleavage and splicing should also be considered. Candidates for genes encoding such RNA species may be present in the spacer regions between known chloroplast genes. Further, more RNA components involved in post-transcriptional control, if they exist, might also be imported from the cytoplasm.

Acknowledgements

We thank Drs T. Wakasugi, S. Kapoor, T. Hirose and J.Y. Suzuki for useful comments and critical reading of the manuscript.

References

1. Adams CC, Stern DB: Control of mRNA stability in chloroplasts by 3′ inverted repeats: effects of stem and loop mutations on degradation of *psb*A mRNA *in vitro*. Nucl Acids Res 18: 6003–6010 (1990).
2. Allison LA, Maliga P: Light-responsive and transcription-enhancing elements regulate the plastid *psb*D core promoter. EMBO J 14: 3721–3730 (1995).
3. Barkan A: Tissue-dependent plastid RNA splicing in maize: transcripts from four plastid genes are predominantly unspliced in leaf meristems and roots. Plant Cell 1: 437–445 (1989).
4. Barkan A: Proteins encoded by a complex chloroplast transcription unit are each translated from both monocistronic and polycistronic mRNAs. EMBO J 7: 2637–2644 (1988).
5. Barkan A: Nuclear mutants of maize with defects in chloroplast polysome assembly have altered chloroplast RNA metabolism. Plant Cell 5: 389–402 (1993).
6. Barkan A, Miles D, Taylor WC: Chloroplast gene expression in nuclear, photosynthetic mutants of maize. EMBO J 5: 1421–1427 (1986).
7. Barkan A, Walker M, Nolasco M, Johnson D: A nuclear mutation in maize blocks the processing and translation of several chloroplast mRNA and provides evidence for the differential translation of alternative mRNA forms. EMBO J 13: 3170–3181 (1994).
8. Bar-Zvi D, Shagan T, Schindler U, Cashmore AR: RNP-T, a ribonucleoprotein from *Arabidopsis thaliana*, contains two RNP–80 motifs and a novel acidic repeat arranged in an α-helix conformation. Plant Mol Biol 20: 833–838 (1992).
9. Berends Sexton T, Christopher DA, Mullet JE: Light-induced switch in barley *psb*D-*psb*C promoter utilization: a novel mechanism regulating chloroplast gene expression. EMBO J 9: 4484–4494 (1990).
10. Berends Sexton T, Jones JT, Mullet JE: Sequence and transcriptional analysis of the barley ctDNA region upstream of *psb*D-*psb*C encoding *trn*K(UUU), *rps*16, *trn*Q(UUG), *psb*K, *psb*I, and *trn*S(GCU). Curr Genet 17: 445–454 (1990).
11. Bock R, Maliga P: Correct splicing of a group II intron from a chimeric reporter gene transcript in tobacco plastids. Nucl Acids Res 23: 2544–2547 (1995).
12. Breiteneder H, Michalowski CB, Bohnert HJ: Environmental stress-mediated differential 3′ end formation of chloroplast RNA-binding protein transcripts. Plant Mol Biol 26: 833–849 (1994).
13. Bülow S, Link G: Sigma-like activity from mustard (*Sinapis alba* L.) chloroplasts conferring DNA-binding and transcription specificity to *E. coli* core RNA polymerase. Plant Mol Biol 10: 349–357 (1988).
14. Chen Q, Adams CC, Usack L, Yang J, Monde R-A, Stern D: An AU-rich element in the 3′ untranslated region of the spinach chloroplast *pet*D gene participates in sequence-specific RNA-protein complex formation. Mol Cell Biol 15: 2010–2018 (1995).
15. Chen H-C, Stern D: Specific ribonuclease activities in spinach chloroplasts promote mRNA maturation and degradation. J Biol Chem 266: 24205–24211 (1991).
16. Chen H-C, Stern D: Specific binding of chloroplast proteins in vitro to the 3′ untranslated region of spinach chloroplast *pet*D mRNA. Mol Cell Biol 11: 4380–4388 (1991).
17. Cheng S-H, Cline K, DeLisle AJ: An *Arabidopsis* chloroplast RNA-binding protein gene encodes multiple mRNAs with different 5′ ends. Plant Physiol 106: 303–311 (1994).
18. Cook WB, Walker JC: Identification of a maize nucleic acid-binding protein (NBP) belonging to a family of nuclear-encoded chloroplast proteins. Nucl Acids Res 20: 359–364 (1992).
19. Copertino DW, Hallick RB: Group II and group III introns of twintrons: potential relationships with nuclear pre-mRNA introns. Trends Biochem Sci 18: 467–471 (1993).
20. Delisle AJ: RNA-binding protein from *Arabidopsis*. Plant Physiol 102: 313–314 (1993).
21. Deng X-W, Gruissem W: Control of plastid gene expression during development: the limited role of transcriptional regulation. Cell 49: 379–387 (1987).
22. Deshpande NN, Hollingsworth M, Herrin DL: The *atp*F group-II intron-containing gene from spinach chloroplasts is not spliced in transgenic *Chlamydomonas* chloroplasts. Curr Genet 28: 122–127 (1995).
23. Didier DK, Klee HJ: Identification of an *Arabidopsis* DNA-binding protein with homology to nucleolin. Plant Mol Biol 18: 977–979 (1992).
24. Du Jardin P, Portetelle D, Harvengt L, Dumont M, Wathelet B: Expression of intron-encoded maturase-like polypeptides in potato chloroplasts. Curr Genet 25: 158–163 (1994).
25. Eisermann A, Tiller K, Link G: *In vitro* transcription and DNA binding characteristics of chloroplast and etioplast extracts from mustard (*Sinapis alba*) indicate differential usage of the *psb*A promoter. EMBO J 9: 3981–3987 (1990).

26. Filipowicz W, Gniadkowski M, Klahre U, Liu HX: Pre-mRNA splicing in plants. In: Lamond AI (ed) Pre-mRNA Processing, pp. 65–77. RG Landes Co. (1995).

27. Gruissem W, Tonkyn JC: Control mechanisms of plastid gene expression. Crit Rev Plant Sci 12: 19–55 (1993).

28. Hayes R., Kudla J, Schuster G, Gabay L, Maliga P, Gruissem W: Chloroplast mRNA 3' end processing by a high molecular weight protein complex is regulated by nuclear-encoded RNA-binding proteins. EMBO J 15: 1132–1141 (1996).

29. Herrmann RG, Westhoff P, Link G: Biogenesis of plastids in higher plants. In: Herrmann RG (ed) Cell Organelles, pp. 276–349. Springer-Verlag, Wien/New York (1992).

30. Hess WR, Hoch B, Zeltz P, Hübschmann T, Kössel H, Börner T: Inefficient rpl2 splicing in barley mutants with ribosome-deficient plastids. Plant Cell 6: 1455–1465 (1994).

31. Hess WR, Prombona A, Fieder B, Subramanian AR, Borner T: Chloroplast rps15 and the rpoB/C1/C2 gene cluster are strongly transcribed in ribosome-deficient plastids: evidence for a functioning non-chloroplast-encoded RNA polymerase. EMBO J 12: 563–571 (1993).

32. Hübschmann T, Hess WR, Börner T: Impaired splicing of the rps12 transcript in ribosome-deficient plastids. Plant Mol Biol 30: 109–123 (1996).

33. Igloi GL, Kössel H: The transcriptional apparatus of chloroplasts. Crit Rev Plant Sci 10: 525–558 (1992).

34. Iratni R, Baeza L, Andreeva A, Mache R, Lerbs-Mache S: Regulation of rDNA transcription in chloroplasts: promoter exclusion by constitutive repression. Genes Devel 8: 2928–2938 (1994).

35. Kanekatsu M, Ezumi A, Nakamura T, Ohtsuki K: Chloroplast ribonucleoproteins (RNPs) as phosphate acceptors for casein kinase. II. Purification by ssDNA-cellulose column chromatography. Plant Cell Physiol 36: 1649–1656 (1995).

36. Kapoor S, Suzuki JY, Sugiura M: Functional significance of a novel (basal) type promoter in plastids: evidence for differential promoter utilization by plastid- and non-plastid-encoded RNA polymerases. (Submitted.)

37. Kawagoe Y, Achberger EC, Bartlett SG, Murai N: A nuclear-encoded chloroplast RNP–80 protein from bean binds to a thymine-rich sequence of single-stranded DNA. Plant Sci 111: 199–207 (1995).

38. Kim J, Mullet JE: Ribosome-binding sites on chloroplast rbcL and psbA mRNAs and light-induced initiation of D1 translation. Plant Mol Biol 25: 437–448 (1994).

39. Kim J-K, Hollingsworth J: Splicing of group II introns in spinach chloroplasts (in vivo): analysis of lariat formation. Curr Genet 23: 175–180 (1993).

40. Klaff P: mRNA decay in spinach chloroplasts: psbA mRNA degradation is initiated by endonucleolytic cleavages within the coding region. Nucl Acids Res 23: 4885–4892 (1995).

41. Kowallik KV, Stoebe B, Schaffran I, Kroth-Pancic P, Freier U: The chloroplast genome of a chlorophyll a+c-containing alga, Odontella sinensis. Plant Mol Biol Rep 13: 336–342 (1995).

42. Lerbs S, Brautigam E, Mache R: DNA-dependent RNA polymerase of spinach chloroplasts: characterization of α-like and σ-like polypeptides. Mol Gen Genet 211: 459–464 (1988).

43. Lerbs-Mache S: The 110-kDa polypeptide of spinach plastid DNA-dependent RNA polymerase: single-subunit enzyme or catalytic core of multimeric enzyme complexes? Proc Natl Acad Sci USA 90: 5509–5513 (1993).

44. Li Y, Sugiura M: Three distinct ribonucleoproteins from tobacco chloroplasts: each contains a unique amino terminal acidic domain and two ribonucleoprotein consensus motifs. EMBO J 9: 3059–3066 (1990).

45. Li Y, Sugiura M: Nucleic acid-binding specificities of tobacco chloroplast ribonucleoproteins. Nucl Acids Res 19: 2893–2896 (1991).

46. Liere K, Link G: RNA-binding activity of the matK protein encoded by the chloroplast trnK intron from mustard (Sinapis alba L.). Nucl Acids Res 23: 917–921 (1995).

47. Lisitsky I, Schuster G: Phosphorylation of a chloroplast RNA-binding protein changes its affinity to RNA. Nucl Acids Res 23: 2506–2511 (1995).

48. Lisitsky I, Liveanu V, Schuster G: RNA-binding activities of the different domains of a spinach chloroplast ribonucleoprotein. Nucl Acids Res 22: 4719–4724 (1994).

49. Lisitsky I, Liveanu V, Schuster G: RNA-binding characteristics of a ribonucleoprotein from spinach chloroplast. Plant Physiol 107: 933–941 (1995).

50. Mieszczak M, Klahre U, Levy JH, Goodall GJ, Filipowicz W: Multiple plant RNA binding proteins identified by PCR: expression of cDNAs encoding RNA binding proteins targeted to chloroplasts in Nicotiana plumbaginifolia. Mol Gen Genet 234: 390–400 (1992).

51. Mullet JE: Dynamic regulation of chloroplast transcription. Plant Physiol 103: 309–313 (1993).

52. Mullet JE, Klein RR: Transcription and RNA stability are important determinants of higher plant chloroplast RNA levels. EMBO J 6: 1571–1579 (1987).

53. Neuhaus H, Link G: The chloroplast tRNA^{Lys} (UUU) gene from mustard (Sinapis alba) contains a class II intron potentially coding for a maturase-related polypeptide. Curr Genet 11: 251–257 (1987).

54. Neuhaus H, Scholz A, Link G: Structure and expression of a split chloroplast gene from mustard (Sinapis alba): ribosomal protein gene rps16 reveals unusual transcriptional features and complex RNA maturation. Curr Genet 15: 63–70 (1989).

55. Nickelsen J, Link G: Interaction of a 3' RNA region of the mustard trnK gene with chloroplast proteins. Nucl Acids Res 17: 9637–9648 (1989).

56. Nickelsen J, Link G: RNA-protein interactions at transcript 3' ends and evidence for trnK-psbA cotranscription in mustard chloroplasts. Mol Gen Genet 228: 89–96 (1991).

57. Nickelsen J, Link G: The 54 kDa RNA-binding protein from mustard chloroplasts mediates endonucleolytic transcript 3' end formation in vitro. Plant J 3: 537–544 (1993).

58. Ohta M, Sugita M, Sugiura M: Three types of nuclear genes encoding chloroplast RNA-binding proteins (cp29, cp31 and cp33) are present in Arabidopsis thaliana: presence of cp31 in chloroplasts and its homologue in nuclei/cytoplasms. Plant Mol Biol 27: 529–539 (1995).

59. Pfannschmidt T, Link G: Separation of two classes of plastid DNA-dependent RNA polymerases that are differentially expressed in mustard (Sinapis alba L.) seedlings. Plant Mol Biol 25: 69–81 (1994).

60. Piechulla B, Gruissem W: Diurnal mRNA fluctuations of nuclear and plastid genes in developing tomato fruits. EMBO J 6: 3593–3599 (1987).

61. Rapp JC, Baumgartner BJ, Mullet J: Quantitative analysis of transcription and RNA levels of 15 barley chloroplast genes, transcription rates and mRNA levels vary over 300-fold; predicted mRNA stabilities vary 30-fold. J Biol Chem 267: 21404–21411 (1992).

62. Reinbothe S, Reinbothe C, Heintzen C, Seidenbecher C, Parthier B: A methyl jasmonate-induced shift in the length of the 5' untranslated region impairs translation of the plastid rbcL transcript in barley. EMBO J 12: 1505–1512 (1993).

326

63. Reith M, Munholland J: Complete nucleotide sequence of the *Porphyra purpurea* chloroplast genome. Plant Mol Biol Rep 13: 333–335 (1995).

64. Schuster G, Gruissem W: Chloroplast mRNA 3′ end processing requires a nuclear-encoded RNA-binding protein. EMBO J 10: 1493–1502 (1991).

65. Shimada H, Sugiura M: Fine structural features of the chloroplast genome: comparison of the sequenced chloroplast genomes. Nucl Acids Res 19: 983–995 (1991).

66. Stern DB, Gruissem W: Control of plastid gene expression: 3′ inverted repeats act as mRNA processing and stabilizing elements, but do not terminate transcription. Cell 51: 1145–1157 (1987).

67. Stirewalt VL, Michalowski CB, Löffelhardt W, Bohnert HJ, Bryant D: Nucleotide sequence of the cyanelle genome from *Cyanophora paradoxa*. Plant Mol Biol Rep 13: 327–332 (1995).

68. Stollar NE, Hollingsworth MJ: Expression of the large ATP synthase gene cluster from spinach chloroplasts. J Plant Physiol 144: 141–149 (1994).

69. Subbaiah CC, Tewari KK: Purification and characterization of ribonucleoproteins from pea chloroplasts. Eur J Biochem 211: 171–179 (1993).

70. Sugita M, Shinozaki K, Sugiura M: Tobacco chloroplast tRNA^{Lys} (UUU) gene contains a 2.5-kilobasepair intron: an open reading frame and a conserved boundary sequence in the intron. Proc Natl Acad Sci USA 82: 3557–3561 (1985).

71. Sugita M, Sugiura M: Nucleotide sequence and transcription of the gene for the 32,000 dalton thylakoid membrane protein from *Nicotiana tabacum*. Mol Gen Genet 195: 308–313 (1984).

72. Sugita M, Sugiura M: The existence of eukaryotic ribonucleoprotein consensus sequence-type RNA-binding proteins in a prokaryote, *Synechococcus* 6301. Nucl Acids Res 22: 25–31 (1994).

73. Sugiura M: The chloroplast genome. Plant Mol Biol 19: 149–168 (1992).

74. Sugiura M: Transcript processing in plastids: trimming, cutting, splicing. In: Bogorad L, Vasdil IK (eds) The Molecular Biology of Plastids. Cell Culture and Somatic Cell Genetics of Plants, vol. 7A, pp. 125–137. Academic Press, San Diego, FL (1991).

75. Sugiura M, Sugita M, Wakasugi T, Li Y, Ye L, Torazawa K: Organization and expression of the chloroplast genome from higher plants. In: Ishikawa H (ed) Endocytobiology V, pp. 307–312. Tübingen University Press, Tübingen (1993).

76. Taylor WC: Regulatory interactions between nuclear and plastid genomes. Annu Rev Plant Physiol Plant Mol Biol 40:211–233 (1989).

77. Tiller K, Link G: Sigma-like transcription factors from mustard (*Sinapis alba* L.) etioplast are similar in size to, but functionally distinct from, their chloroplast counterparts. Plant Mol Biol 21: 503–513 (1993).

78. Tiller K, Link G: Phosphorylation and dephosphorylation affect functional characteristis of chloroplast transcription systems from mustard (*Sinapis alba* L.). EMBO J 12: 1745–1753 (1993).

79. Troxler RF, Zhang F, Hu J, Bogorad L: Evidence that s factors are components of chloroplast RNA polymerase. Plant Physiol 104: 753–759 (1994).

80. Vera A, Sugiura M: Chloroplast rRNA transcription from structurally different tandem promoters: an additional of transcriptional novel type promoter. Curr Genet 27: 280–284 (1995).

81. Vera A, Sugiura M: A novel RNA gene in the tobacco plastid genome: its possible role in the maturation of 16S rRNA. EMBO J 13: 2211–2217 (1994).

82. Wada T, Tunoyama Y, Shiina T, Toyoshima Y: In vitro analysis of light-induced transcription in the wheat *psbD/C* gene cluster using plastid extracts from dark-grown and short-term-illuminated seedlings. Plant Physiol 104: 1259–1267 (1994).

83. Wang MJ, Davis NW, Gegenheimer P: Novel mechanisms for maturation of chloroplast transfer RNA precursors. EMBO J 7: 1567–1574 (1988).

84. Westhoff P, Herrmann RG: Complex RNA maturation in chloroplasts: the *psbB* operon from spinach. Eur J Biochem 171: 551–564 (1988).

85. Wolfe KH, Morden CW, Palmer JD: Function and evolution of a minimal plastid genome from a nonphotosynthetic parasitic plant. Proc Natl Acad Sci USA 89: 10648–10652 (1992).

86. Yamaguchi-Shinozaki K, Shinozaki K, Sugiura M: Processing of precursor tRNAs in a chloroplast lysate. Processing of the 5′-end involves endonucleolytic cleavage by an RNase P-like enzyme and precedes 3′-end maturation. FEBS Lett 215: 132–136 (1987).

87. Yao WB, Meng BY, Tanaka M, Sugiura M: An additional promoter within the protein-coding region of the *psbD-psbC* gene cluster in tobacco chloroplast DNA. Nucl Acids Res 17: 9583–9591 (1989).

88. Ye L, Li Y, Fukami-Kobayashi K, Go M, Konishi T, Watanabe A, Sugiura M: Diversity of a ribonucleoprotein family in tobacco chloroplasts: two new chloroplast ribonucleoproteins and a phylogenetic tree of ten chloroplast RNA-binding domains. Nucl Acids Res 19: 6485–6490 (1991).

89. Ye L, Sugiura M: Domains required for nucleic acid binding activities in chloroplast ribonucleoproteins. Nucl Acids Res 20: 6275–6279 (1992).

Plant Molecular Biology **32**: 327–341, 1996.
© 1996 *Kluwer Academic Publishers.*

Post-transcriptional regulation of chloroplast gene expression in *Chlamydomonas reinhardtii*

J.-D. Rochaix
Departments of Molecular Biology and Plant Biology, University of Geneva, Geneva, Switzerland

Key words: Chlamydomonas, chloroplast gene expression, RNA stability, RNA processing, splicing, translation

Contents

Abstract   327
Introduction   327
Chloroplast RNA stability and processing   328
   Role of 3′-untranslated region   329
   Role of 5′-untranslated region   329
   Maturation of polycistronic transcripts 330
Chloroplast RNA splicing   331
Translation   333
   Several nuclear mutants of *C.*
   reinhardtii* are affected at the level of
   translation initiation   333

Cis elements in the 5′-untranslated
   regions of chloroplast mRNAs   334
Initiation codons   334
Relationship between mRNA stability
   and translational capacity   334
Trans-acting factors involved in
   translation   335
Assembly of the photosynthetic
   complexes   337
Conclusions and perspectives   338
Acknowledgements   338
References   338

Abstract

The biosynthesis of the photosynthetic apparatus depends on the concerted action of the nuclear and chloroplast genetic systems. Numerous nuclear and chloroplast mutants of *Chlamydomonas* deficient in photosynthetic activity have been isolated and characterized. While several of these mutations alter the genes of components of the photosynthetic complexes, a large number of the mutations affect the expression of chloroplast genes involved in photosynthesis. Most of these mutations are nuclear and only affect the expression of a single chloroplast gene. The mutations examined appear to act principally at post-transcriptional steps such as RNA stability, RNA processing, *cis*- and *trans*-splicing and translation. Directed chloroplast DNA surgery through biolistic transformation has provided a powerful tool for identifying important *cis* elements involved in chloroplast gene expression. Insertion of chimeric genes consisting of chloroplast regulatory regions fused to reporter genes into the chloroplast genome has led to the identification of target sites of the nuclear-encoded functions affected in some of the mutants. Biochemical studies have identified a set of RNA-binding proteins that interact with the 5′-untranslated regions of plastid mRNAs. The binding activity of some of these factors appears to be modulated by light and by the growth conditions.

Introduction

The green unicellular alga *Chlamydomonas reinhardtii* has emerged as a powerful model system for study-

ing chloroplast gene expression for several reasons. First, this photosynthetic eukaryotic organism can be grown under controlled laboratory conditions in large amounts and is thus suitable for biochemical analysis. Second, *C. reinhardtii* cells, which exist either as mating type + or −, undergo a well-defined sexual cycle and are amenable to extensive genetic analysis. Cells can be propagated as well in the haploid as in the diploid state (for details, see [30]). Third, photosynthetic function in *C. reinhardtii* is dispensable provided a carbon source such as acetate is included in the growth medium. It has therefore been relatively easy to isolate a large number of mutants deficient in photosynthetic activity. Fourth, methods and tools have been developed recently for efficient nuclear and organellar transformation of *C. reinhardtii* [3, 14, 44, 45, 58]. The efficiency of nuclear transformation is sufficiently high to make gene isolation through genomic complementation of mutants with cosmid libraries feasible [68, 94]. Chloroplast transformation can be achieved by bombardment of cells with DNA-coated tungsten or gold particles. The transforming DNA is integrated efficiently into the chloroplast genome by homologous recombination. This technology together with the availability of selectable markers allows one to perform directed gene disruptions and site-directed mutagenesis of any chloroplast gene [26, 59, 76, 86].

It is well established that the biosynthesis of the photosynthetic apparatus depends on the concerted action of the nuclear and chloroplast genetic systems. Both chloroplast and nuclear mutants deficient in photosynthetic activity have been isolated which can be readily distinguished based on the segregation patterns during crosses. While nuclear mutations follow Mendelian inheritance, chloroplast mutations are uniparentally inherited from the mating-type+parent. Nuclear mutations affecting photosynthesis fall into two groups. The first includes mutations located within the genes of structural components of the photosynthetic apparatus whereas the second comprises mutations that affect the expression of chloroplast genes. A striking feature is that these mutations act specifically on the expression of individual genes, i.e. in these mutants the expression of a single chloroplast gene is affected and other chloroplast genes are expressed normally. A large number of mutations of this type corresponding to distinct nuclear loci have been identified. The mutations examined to date appear to act mostly at post-transcriptional steps such as RNA stability, RNA processing, *cis-* and *trans-*splicing and translation (Fig. 1). The analysis of these mutants has provided new insights into the molecular mechanisms underlying chloroplast gene expression, in particular its dependence on nuclear gene products. Since several reviews covering various aspects of chloroplast gene

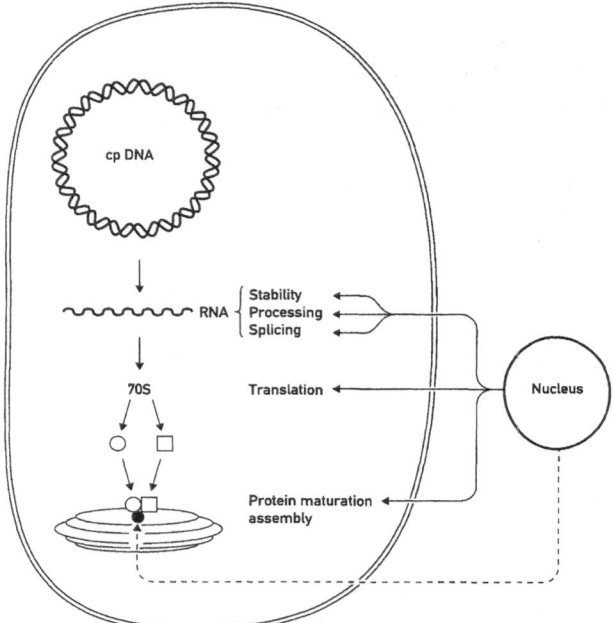

Figure 1. Scheme of biosynthesis of photosynthetic complexes. Photosynthetic complexes consist of chloroplast (open symbols) and nuclear-encoded subunits (black symbols). Numerous nuclear-encoded functions are required for the expression of chloroplast genes at the level of RNA stability, processing, splicing, translation, protein maturation and assembly of complexes. cp, chloroplast DNA; 70S, chloroplast ribosome.

expression in *C. reinhardtii* have appeared recently [23, 57, 60, 70], this article will focus mostly on recent developments in this area.

Chloroplast RNA stability and processing

It is well documented that the levels of chloroplast transcripts vary considerably during plastid development and differentiation in higher plants (see [29]). Although it has been recognized that this differential accumulation of chloroplast mRNAs is partly controlled at the transcriptional level, a major part of this regulation occurs through stabilization or destabilization of transcripts. Further, chloroplast RNA stability is closely linked to RNA processing and translation. *C. reinhardtii* offers interesting possibilities to study this problem using genetic, molecular and biochemical approaches.

Role of 3'-untranslated region

A characteristic feature of most chloroplast mRNAs is the presence of inverted repeats within the 3'-untranslated region that can fold into stem-loop structures. Similar structures are also found in prokaryotes where they appear to play an important role in rho-independent transcription termination. The role of the 3' inverted repeat of the *atpB* mRNA of *C. reinhardtii* was studied using both chloroplast transformation and *in vitro* assays [83]. Deletion of this structure reduces the amount of the *atpB* mRNA to 20–35% of wild-type levels without affecting the transcription rate and leads to the appearance of *atpB* RNA of heterogeneous size [83]. These data indicate that the 3' inverted repeat of *atpB* plays a role both in RNA stabilization and in 3'-end formation.

Although run-on transcription analysis has shown that the *atpB* 3' inverted repeat is involved in transcription termination, this only occurs with a frequency of 50%. Hence, the 3' inverted repeat is also required for 3'-end processing (see below). *In vitro* studies revealed that the 3'-end maturation of the *atpB* mRNA is a two-step process [84]. Endonucleolytic cleavage at a site downstream of the 3' inverted repeat is followed by exonucleolytic trimming to the stem of the 3' inverted repeat.

Blowers *et al.* [2] examined the role of the 3'-untranslated region of *rbcL* and *psaB* and their associated inverted repeat elements. They introduced into the chloroplast of *C. reinhardtii* chimeric GUS genes containing the *rbcL* or *psaB* 3'-untranslated regions with one or more 3' inverted repeats or with the inverted repeat in opposite orientation. The analysis of these transformants revealed that these 3' stem-loop structures act principally as signals for RNA 3'-end formation by transcription termination or endonucleolytic cleavage. Since their removal does not affect RNA decay, it was concluded that they play no major role in RNA stabilization [2] in contrast to the results obtained with the *atpB* 3' end [83]. Further, efficient 3'-end formation occurred only with constructs containing the 3' inverted repeat in the normal forward orientation. It is interesting to contrast these results with those obtained with the spinach *petD* 3' inverted repeat which was shown to promote correct 3'-end formation both in the forward and reverse orientation when inserted downstream of *atpB* in *C. reinhardtii* [83].

The picture which emerges from these studies is that each chloroplast transcript 3' end is distinct and that a number of mechanisms underlie the formation of discrete ends and the stabilization of the transcripts. From the work performed in higher plants it is clear that the 3' inverted repeat alone is not sufficient to account for the stabilization and processing of mRNA [82]. Indeed several proteins that bind specifically to the 3'-untranslated region of mRNAs have been identified in higher plants and include polypeptides of 100, 54, 33, 28 and 24 kDa. The 28 kDa protein appears to be involved in 3'-end processing [79] and a 54 kDa protein from mustard was shown to copurify with an endoribonuclease activity [65]. Little is known on RNA-binding proteins of this sort in *C. reinhardtii*.

Role of 5'-untranslated region

Evidence that determinants for chloroplast mRNA stability reside within the 5'-untranslated region arose from the analysis of a nuclear mutant of *C. reinhardtii* deficient in PSII activity [46]. This mutant, nac2–26, was shown to lack *psbD* mRNA and appears to be specifically affected in the stabilization of this RNA since *psbD* RNA is transcribed and other chloroplast RNAs accumulate to wild-type levels [46]. To identify the determinants for degradation within the *psbD* message, chimeric genes consisting of the *psbD* 5'-untranslated region fused to the reporter gene *aadA* (conferring resistance to spectinomycin [26]) were introduced into the chloroplast genome through biolistic transformation [66]. The transformants of mating type + were subsequently crossed to the nac2–26 mutant strain. Tetrad analysis revealed cosegregation between photosystem II deficiency and spectinomycin sensitivity. Furthermore, it was shown that the chimeric message is destabilized in the nac2–26 nuclear background, but not in the wild-type background thus indicating that the 74 nucleotide *psbD* leader contains one of the major target sites for *psbD* RNA degradation in the absence of wild-type NAC 2 function. Similar experiments in which the *psbD* 3'-untranslated region was fused to *aadA* did not reveal cosegregation between the two phenotypes indicating that the 3'-untranslated region is not sufficient to promote RNA degradation in the nac2–26 nuclear background.

The selective instability of *psbD* 5' leader RNA could also be demonstrated *in vitro* [66]. Incubation of this RNA with lysates from purified chloroplasts of wild type and nac2–26 revealed that the RNA is degraded considerably faster with the mutant extract. This rapid degradation appears to be mediated by specific endonucleolytic cleavages within the *psbD* 5' leader. UV crosslinking analysis of proteins binding to the *psbD* leader revealed a 47 kDa protein in wild-type extracts whose binding activity was altered in the mutant. Binding activity could only be recovered from lysates treated with non-ionic detergents, suggesting that the 47 kDa protein may be associated with membranes.

The *psbD* RNA is produced as a precursor with a 74 nucleotide leader which is processed at residue −47 (relative to the initiation codon) to give rise to the mature RNA. The 47 kDa protein binds only to the −74 leader, but not to the mature −47 leader suggesting that the binding site is in the 5′ region of the leader and/or overlapping the processing site [66]. Processing at −47 could not be achieved *in vitro*, possibly because the activity and/or essential cofactors are lost during the preparation of the lysate. Surprisingly, deletion of the 5′-terminal region of the *psbD* leader from −74 to −47 leads to destabilization of *psbD* message *in vivo* although *psbD* is still transcribed at wild-type levels under these conditions (J. Nickelsen and J.-D. Rochaix, unpublished results). We are thus left with a paradox since in wild-type the *psbD* precursor RNA is processed to produce the mature RNA which is stable. One way to reconcile these results is to assume that there is an obligate maturation pathway for *psbD* mRNA in which *psbD* RNA processing is closely coupled to some early event in the initiation of translation, possibly binding of a factor or protein complex to the 5′ end of the mature *psbD* mRNA which may protect the RNA against degradation and make it competent for translation.

The nuclear gene deficient in nac2–26 has recently been isolated through genomic complementation using a wild-type cosmid library (J. Nickelsen and J.-D. Rochaix, unpublished results). Preliminary data indicate that the encoded polypeptide does not encode known RNA-binding motifs.

Other nuclear mutants have been characterized which fail to accumulate the chloroplast transcripts of either *psbB* [42, 63], *psbC* [81] or *atpA* [17]. Preliminary results indicate that in the case of mutant 222E the target site of the nuclear-encoded function defined by the mutation lies within the 5′-untranslated region of the *psbB* mRNA (F. Vaistij, M. Goldschmidt-Clermont and J. D. Rochaix, unpublished results). These genetic data suggest the existence of nuclear encoded factors which interact specifically with the 5′-untranslated regions of chloroplast mRNAs and play a major role in transcript stabilization.

Additional evidence that 5′ sequences of chloroplast transcripts contain important positive and negat-

ive determinants for RNA stability has emerged from studies of chimeric plastid genes in transgenic *C. reinhardtii* [77]. It was found that fusion of the 5′ *rbcL* leader to various reporter genes strongly destabilizes the chimeric transcript upon illumination of cells previously grown in the dark. However, addition of 5′ sequences from the coding part of *rbcL* prevented this transcript destabilization.

Recently, a nuclear photosynthetic deficient mutant was identified that accumulates minute amounts of *rbcL* mRNA [38]. Pulse-labelling experiments indicated that the rate of *rbcL* mRNA synthesis is apparently reduced in this mutant. Whether this phenotype is indeed due to a specific alteration of the rate of transcription of *rbcL* (which has never been observed before for any chloroplast gene) or whether the *rbcL* mRNA is highly unstable in this mutant strain remains to be explored.

Considerable changes in the levels of chloroplast mRNAs of *C. reinhardtii* cells grown in a 12 h light/12 h dark regime have been found to be due to changes in both transcription and decay rates [78]. A striking drop in stability in the light was observed for the transcripts of *rbcL*, *atpB*, *tufA*, *psaB*, *psbA* and 16S rDNA. In contrast to the genetically defined factors involved in the stability of specific transcripts, this fast light-induced acceleration of RNA degradation does not appear to be transcript specific as all transcripts tested were affected.

Maturation of polycistronic transcripts

In contrast to higher plants only few chloroplast genes of *C. reinhardtii* have been shown to be organized as operons that are transcribed into polycistronic primary transcripts. Besides the ribosomal operon, polycistronic transcription units include *psbB-ycf8* [43, 68], *psbD-* exon 2 of *psaA* [7], *psaC*-ORF58-*ycf7* [86], *rps7*-atpE [69], *tscA-chlN* (M. Goldschmidt-Clermont, unpublished results) and *petA-petD* [33, 75, 85] (Fig. 2). In the latter case, *petD* can be transcribed from its own promoter, or cotranscribed with *petA*. Use of specific deletions of the 5′ region of *petD* allowed the identification of an RNA processing site which is used to generate monocistronic *petD* mRNA from the *petA-petD* dicistronic RNA and most likely also from the transcript initiated from the *petD* promoter [75]. The nature of the processing site was further investigated by inserting various chimeric genes, consisting of the 5′-untranslated region of *petD* fused to *uidA*, downstream of *petA* into the chloroplast genome. Trans-

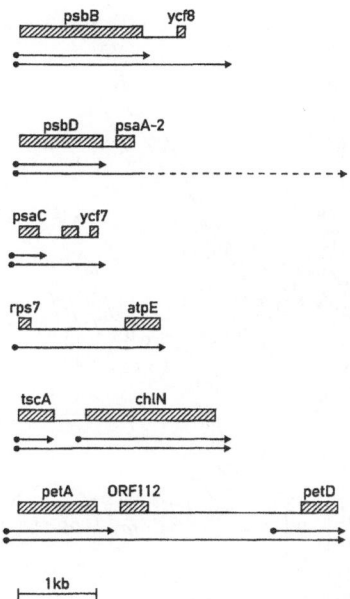

Figure 2. Di-and tri-cistronic chloroplast transcription units in *C. reinhardtii.* The 5' ends are indicated by dots and the 3' ends by arrows. In the case of *psbD-psaA–2,* several large transcripts with sizes ranging between 2.6 and 7 kb have been detected [7]. However, their ends have not been mapped precisely.

formants with the deletion of the *petD* promoter still accumulated monocistronic *uidA* transcript whereas in transformants with a larger deletion removing in addition the first 25 bases downstream of the mature *petD* 5' end only dicistronic *petA-uidA* transcript could be detected. These results suggest that the recognition site for processing is localized at least partially within the first 25 bases of the 5'-untranslated region of *petD.* It is likely that additional polycistronic transcription units will be found in the chloroplast of *C. reinhardtii.*

Chloroplast RNA splicing

Chloroplast introns of higher plants and algae fall into two distinct classes, group I and group II introns, each of which has characteristic secondary structure features [61]. In the chloroplast of *C. reinhardtii* only two genes, the 23S rRNA gene [71] and *psbA* [21] contain group I introns and one gene, *psaA,* contains two unusual group II introns that are split [48] (see below). It is not known whether other group II introns are present in the chloroplast genome of *C. reinhardtii.*

Efficient *in vitro* self-splicing was demonstrated for the ribosomal intron [19, 35] and for three *psbA*

introns [34]. These findings do not necessarily imply that self-splicing also occurs *in vivo.* The existence of nuclear mutations that prevent splicing of mitochondrial introns known to self-splice *in vitro* is well-documented in fungi [9]. Also, a nuclear mutant of *C. reinhardtii* partially deficient in chloroplast ribosomes was shown to accumulate unspliced 23SrRNA precursor [35].

The *psaA* gene of *C. reinhardtii,* which encodes one of the major reaction center polypeptides of photosystem I, consists of three exons that are widely separated on the chloroplast genome [48]. The observation that exons 1 and 2 are directed in opposite orientation on the chloroplast genome implies that transcription of these exons is discontinuous [7]. Each of the *psaA* exons is surrounded by typical 5'- and/or 3'-terminal sequences of group II introns [48]. Hence, maturation of the *psaA* mRNA depends on two *trans*-splicing reactions.

Surprisingly, amongst nuclear mutants deficient in photosystem I activity more than one fourth are affected in the maturation of *psaA* mRNA [27]. These mutants fall into three classes based on their *psaA* RNA phenotype. Class A mutants are defective in exon 2-exon 3 splicing, class B mutants are unable to perform either of the two *trans*-splicing reactions and class C mutants are deficient in exon 1-exon 2 splicing. These mutations fall into a large number of complementation groups: at least 5 for class A, 2 for class B and 7 for class C [27, 28]. Since several of these complementation groups include a single allele, it is likely that many more nuclear loci exist that are involved, directly or indirectly, in *psaA trans*-splicing. Besides these nuclear loci, a chloroplast locus, *tscA,* was identified and shown to be involved in *trans*-splicing [27, 28]. This locus is distinct from either *psaA* exon. Mutants with deletions in *tscA* are unable to splice exons 1 and 2. Analysis of the *tscA* mutants revealed that this locus encodes a 430 nucleotide RNA required for the first *trans*-splicing reaction [28].

A model was proposed in which the *tscA* RNA interacts with the 5' and 3' intron regions flanking the first and second *psaA* exons so that a composite structure with features of group II introns is reconstituted [28]. This secondary structure consists of a characteristic wheel and six protruding stem-loop domains. In this model the *tscA* RNA includes domains II and III of the intron and its terminal regions reconstitute the disrupted helices corresponding to domains I and IV by base-pairing with the 5' and 3' parts of the intron (Fig. 3). The model is based on the presence of short conserved sequence motifs of group II introns with-

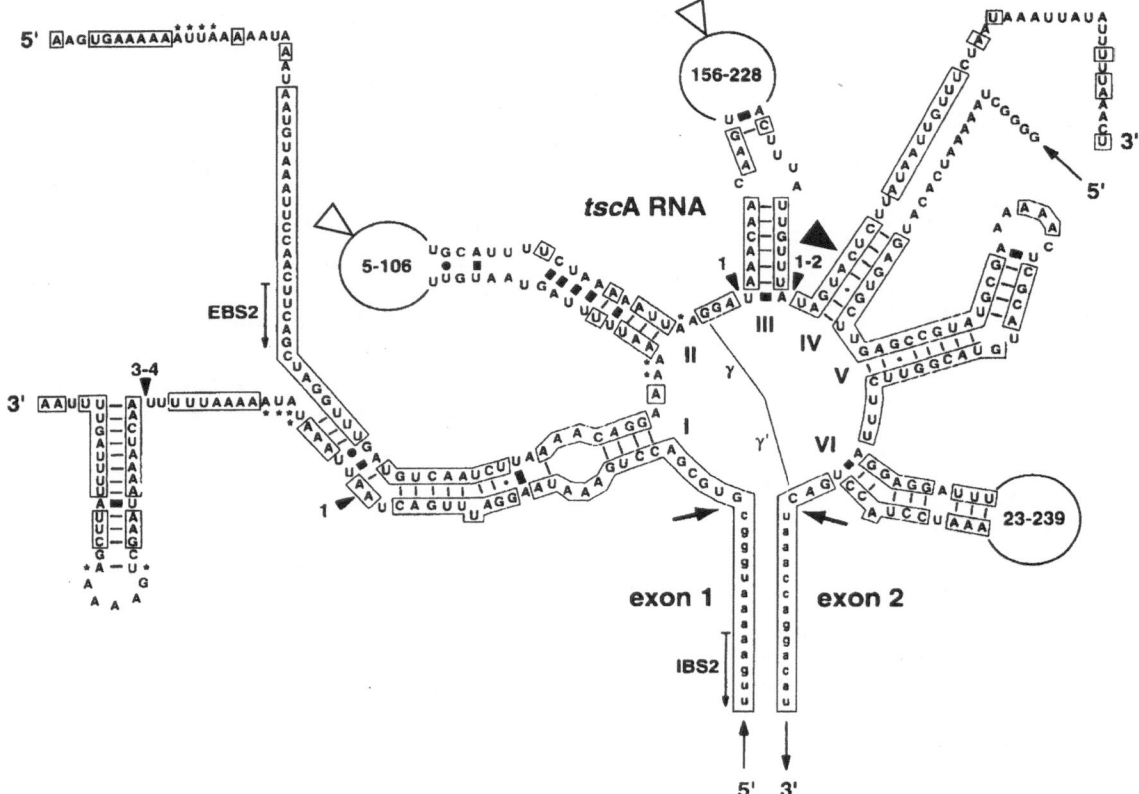

Figure 3. Secondary structure model of the tripartite intron 1 of *psaA* in *C. reinhardtii, C. gelatinosa* and *C. zebra.* Roman numerals indicate the six major secondary structure domains of group II introns. Conserved bases in the three *Chlamydomonas* are boxed. Residues that are missing in *C. gelatinosa* and/or *C. zebra* are marked by asterisks, whereas those representing additions in the latter two algae relative to *C. reinhardtii* are indicated by small solid triangles with their accompanying numbers indicating the size of the insertions in nucleotides. Numbers in the loops indicate the minimum and maximum size of these regions. Blocked arrows denote a potential pairing between EBS2 (exon-binding site) and IBS3 (intron-binding site). The long-range tertiary interaction γ-γ' is indicated. The thick arrows point to the intron-exon junctions. Thin arrows indicate the 5'- and 3'-strand polarity of the separate transcripts containing exon 1 and exon 2. Closed and open wedges indicate the sites of insertion of the *aadA* cassette which block and which do not inhibit *trans*-splicing, respectively (adapted from Turmel *et al.* [89], with permission).

in *tscA* and on the observation that disruption of *tscA* by the *aadA* expression cassette within the external loops of domains II and III does not interfere with *trans*-splicing, whereas disruption of *tscA* within the stem of domain IV blocks *trans*-splicing. It is known from studies of mitochondrial group II introns that the external loops are variable in size and that they do not appear to play an active role in splicing, in contrast to the more conserved intron core [62]. Further support for this model has arisen from the sequence comparison of the equivalent introns in the *psaA* genes from *C. gelatinosa* and *C. zebra* [89] which are also tripartite and which can be folded in a similar way (Fig. 3). It is noticeable that high sequence conservation within *tscA* is found in the intron core region and also in its

5' and 3' parts suggesting that these regions may have some important functional role perhaps by interacting with specific (*trans*)splicing factors.

A striking feature of the split intron 1 of *psaA* is that it lacks any recognizable domain I structure which is usually well conserved within group II introns. This raises the possibility that this domain has diverged considerably in *C. reinhardtii* or that another *trans*-acting RNA specifies domain I. The latter possibility is of special interest in the context of the idea that group II intron domains may have evolved into snRNAs during the evolution of nuclear pre-mRNA introns (80).

The *tscA* gene of *C. reinhardtii* appears to be cotranscribed with the downstream *chlN* gene. This is suggested by the existence of a nuclear mutant deficient in splicing exon 1 and exon 2 of *psaA* which accumulates *tscA-chlN* cotranscripts (M. Goldschmidt-Clermont, unpublished results). Whether the splicing defect is directly linked to the failure to process *tscA* RNA or whether the accumulation of the *tscA* precursor RNA results from a secondary effect of the mutation is not yet known.

Besides the two discontinuous group II introns of *psaA*, no other group II intron has been detected in the chloroplast genome of *C. reinhardtii*. To determine whether the group II *trans*-splicing machinery of *C. reinhardtii* is capable of splicing a heterologous group II *cis* intron, the *atpF* gene of spinach was introduced into the chloroplast genome of *C. reinhardtii* [16]. This gene contains a 764 bp group II intron which does not self-splice *in vitro*. Although the precursor RNA was expressed, no spliced product could be detected suggesting that group II *cis*-intron splicing is either species specific or distinct from *trans*-splicing.

In contrast, the mitochondrial group II intron rI1 from *Scenedesmus obliquus* was shown to splice efficiently when introduced into the *C. reinhardtii* chloroplast [32]. This intron is capable of self-splicing *in vitro* [49] and the question remains whether splicing of this intron is dependent on specific factors in *C. reinhardtii*.

Translation

Several nuclear mutants of C. reinhardtii *are affected at the level of translation initiation*

Translation is a key step in the control of expression of chloroplast genes (cf. [22, 60]). Several mutants of *C. reinhardtii* affected in translation have been characterized and have provided valuable insights into the underlying molecular mechanisms.

At least two nuclear mutations, F34 and F64, and one chloroplast mutation, Fud34, were shown to block synthesis of the photosystem II core subunit P6 (CP43 in higher plants) which is encoded by the chloroplast *psbC* gene [1, 8, 15]. The chloroplast mutation has been mapped in a stem-loop structure within the 550 base *psbC* 5′ leader region [73]. Moreover, a point mutation within the same stem-loop structure partially suppresses the F34 mutation [73]. To confirm that these mutations interfere with translation rather than turnover of the protein, chimeric genes consisting of the *psbC* 5′ leader region fused to the reporter *aadA* were introduced into the chloroplast via biolistic transformation. Expression of the reporter gene was shown

to occur only with the wild type, but not with the mutant F34 and F64 alleles thus indicating that the *psbC* 5′ leader contains *cis*-acting target(s) for these F34- and F64-dependent *trans*-acting factors [93].

To gain further insight into these target regions various deletions were introduced in the *psbC* 5′-untranslated region of the chimeric *aadA* construct (W. Zerges and J.-D. Rochaix, unpublished results). Removal of the stem-loop considerably reduced expression of *aadA* suggesting that this structure is required for efficient translation. This deletion partially relieved the dependence on F34 function. Deletion of the 3′ part of the leader severely diminished this dependence indicating that the F34 factor interacts with this region including the stem loop. In contrast none of the deletions tested relieved the dependence of translation on the F64 function suggesting that the F64 factors acts on several sites of the *psbC* leader. UV crosslinking experiments with chloroplast extracts and the *psbC* leader revealed a binding activity at 46 kDa in the F64 extract that was absent in wild type [93]. One possibility is that F64 encodes an anti-repressor which prevents tight binding of the 46 kDa protein to the *psbC* leader in wild type, but not in the mutant. Alternatively the F64 factor, together with the 46 kDa protein, may be part of a protein complex required for translation which is non-functional in the F64 mutant and releases the 46 kDa RNA-binding protein.

Another nuclear mutant, F15, is unable to synthesize the PsaB protein, one of the reaction center subunits of photosystem I. Since the *psaB* mRNA accumulates to wild-type levels in F15, the mutation must act at a post-transcriptional step. Similar experiments as described above for *psbC* with chimeric genes consisting of the *psaB* leader fused to *aadA* revealed that in this case, too, the *psaB* 5′-untranslated region mediates the translational requirement for the nuclear-encoded F15 function (O. Stampacchia, unpublished results). A chloroplast suppressor mutation of F15 was found within the 5′-untranslated region of *psaB* near a potential Shine-Dalgarno sequence (J.L. Zanasco, P. Bennoun, J.-D. Rochaix, unpublished results). The latter appears to be occluded in the wild type because of base-pairing with the upstream region. The suppressor mutation could destabilize this pairing and may thereby allow interactions between the Shine-Dalgarno sequence and the small ribosomal subunit. The function affected in F15 may be required for promoting this interaction.

At least two nuclear loci NAC1 and AC115 were shown to be required for the synthesis of the *psbD* product D2 [47]. Pulse-labelling experiments revealed that synthesis of theD2 protein is severely reduced in both mutants [47, 91]. Mutations at the NAC1 locus appear to act at a step following the initiation of translation since chimeric genes consisting of the *psbD* 5'-untranslated region fused to *aadA* are still expressed in the presence of the mutant *nac1* allele (J.-D. Rochaix and J. van Dillewijn, unpublished results). A nuclear suppressor was isolated which is able to overcome the effects of two different allelic mutations at the NAC1 locus as well as mutations in AC-115. The behavior of this suppressor, which is neither allele- nor gene-specific, is consistent with a bypass suppressor [91]. The molecular basis of this interesting suppression remains to be elucidated.

Cis *elements in the 5'-untranslated regions of chloroplast mRNAs*

The finding of potential Shine-Dalgarno (SD) sequences in close vicinity upstream of the start codons of many, but not all, chloroplast genes of *C. reinhardtii* suggests that the initiation of translation in the chloroplast may occur in a similar way as in prokaryotes where pairing between the 3' end of 16S rRNA with the SD sequence facilitates the positioning of the 30S ribosomal subunit on mRNA [25]. The 3' end of 16S rRNA of *C. reinhardtii* and *E. coli* are highly homologous and pairings with putative chloroplast SD sequences are possible [18]. The functional role of SD sequences has been tested only recently in *C. reinhardtii*. Mutations in the potential SD sequences of *petD* (GGA to TTA) did not affect translation of *petD* [74]. Similarly, a change in the *psbD* SD sequence from GGAG to AAAG did not affect phototrophic growth and photosystem II activity (J.-D. Rochaix, unpublished results). In contrast, deletion of the *psbA* SD sequence abolished *psbA* translation [59]. However, since *psbA* mRNA was strongly reduced in this mutant, the significance of the result is not clear.

The 362 nucleotide *petD* 5' leader was shown to be both necessary and sufficient to drive expression of GUS in transgenic chloroplasts [76]. However, removal of the first two thirds of the *petD* leader in the chimeric gene considerably reduced the expression of GUS [76]. Mutational studies of other regions of the *petD* 5' leader revealed that two regions are required for translation, one located between 150 and 200 nucleotides, the other ca. 40 nucleotides upstream of the initiation codon. Site-directed mutations of the *psbA* [59] and *psbD* 5'-untranslated regions (J. Nickelsen

and J.-D. Rochaix, unpublished results) also revealed short sequence motifs that are important for translation.

Initiation codons

Although AUG is used in most cases as initiation codon, GUG can also be utilized efficiently, for example for *psbC* translation in *C. reinhardtii* [73]. Initiation codon recognition in *C. reinhardtii* was examined by changing the AUG initiation codon of *petD* to either AUU or AUC [5]. These mutations reduce the rate of translation initiation to 10 to 20% of wild-type level. Phototrophic growth of these mutants is reduced at room temperature and completely blocked at 35 °C. In the case of *psbD*, replacement of the AUG initiation codon by AUA severely reduced expression of *psbD* whereas the AUC replacement had only a minor effect (J.-D. Rochaix, unpublished results). In a more thorough study, five mutations were created in the initiation codon of *petA*, which encodes cytochrome *f* of the cytochrome *b6/f* complex [6]. Mutant strains with single-basepair substitutions, AUU and ACG, accumulated cytochrome *f* to 20 and 2–5% of wild-type levels. In mutant strains with double-basepair changes, ACC and ACU, cytochrome *f* levels were 1–2% and 0.8% relative to wild-type. The mutant strain with the UUC codon did not accumulate detectable levels of cytochrome *f*. Of these mutants only the AUU codon mutant was able to grow phototrophically. The use of alternative initiation codons in these mutants could be ruled out by introducing stop codons either immediately upstream or downstream of the initiation codon: cytochrome *f* could be detected with the upstream, but not with the downstream stop codon [6]. These results suggest that the initiation codon in *C. reinhardtii* is important for determining the efficiency of translation initiation. However, its role in establishing the location of the initiation site seems to be minimal. The latter appears to be determined by the sequence context around the initiation site, such as the Shine Dalgarno sequence. For those mRNAs lacking an SD sequence other interactions between rRNA and mRNA may occur [6].

Relationship between mRNA stability and translational capacity

In prokaryotes mRNA stability is usually closely correlated to the translatability of the mRNA with strains deficient in translation accumulating less of the corresponding RNA (cf. [67]). However, the mechanism by which a given mRNA is degraded can affect this correlation. Amongst the mutants of *C. reinhardtii* deficient in translation of specific chloroplast mRNAs, three categories have been described. The first contains mutants in which the corresponding mRNA level is decreased. These include mutants with altered 5'-untranslated region of *psbA* [59] and *petD* [74] and strain Fud47 harboring a frame-shift mutation within *psbD* [20]. In the second category, the mRNA levels in the mutants are unaltered relative to wild type or only slightly reduced. Here one finds a mutant deficient in *psbA* translation [24] and the nuclear mutants F34 and F64 affected in the initiation of translation of *psbC* [73]. However, chimeric transcripts consisting of the 5'-untranslated region of *psbC* fused to *aadA* accumulate to higher levels in the F34 and F64 mutants than in wild type [93]. This clearly shows that translatability is not the only determinant of mRNA stability. Other mutants in which the mRNA levels are increased include mutants probably affected at the level of translation elongation of *psbA* RNA [47], mutants deficient in *atpA* translation [17] and a strain with a frameshift mutation within *psaB* [92]. Taken together, these data indicate that there is no simple relationship between mRNA stability on the one hand, and ribosome association and translation on the other.

Trans-*acting factors involved in translation*

Although *C. reinhardtii* cells are able to synthesize and assemble photosynthetic complexes in the dark, there are considerable differences in the rate of synthesis of several chloroplast encoded subunits of these complexes between dark- and light-grown cells [55]. In particular, synthesis of the photosystem II reaction center polypeptides D1 and D2 increases considerably during dark-light transitions in both wild type and the y-1 mutant strain which is unable to synthesize chlorophyll in the dark [55]. Since the corresponding mRNA levels remain unchanged in the light, the observed increase of D1 and D2 synthesis must occur at a post-transcriptional level. It has indeed been shown that translation is a key step in this process [10, 55].

A protein complex has been isolated and shown to bind to the 5'-untranslated region of *psbA* mRNA [10]. This complex includes polypeptides of 60 and 47 kDa. The latter binds to a 36 base stem-loop structure located upstream of a potential SD sequence of the *psbA* mRNA. A good correlation was found between the RNA binding activity of the complex and the level of translation of the *psbA* mRNA under light and dark

growth conditions [10]. In contrast, only minor differences in the amount of 47 and 60 kDa protein could be detected in light- and dark-grown wild-type cells. These observations suggest that modulation of the binding of the protein complex to the 5'-untranslated region of *psbA* mRNA controls its translation.

To test the function of the stem-loop structure of *psbA* RNA in translation, several mutations were generated in this region and tested *in vivo* [59]. Deletion of the terminal loop or disruption of base pairing in the stem significantly reduced D1 synthesis to levels less that 5% of wild type. These mutants were however still able to accumulate D1 to 20% of wild-type levels when grown under non-saturating light. Translation of *psbA* mRNA in dark-grown cells was found to be the same for these mutants and wild type. Surprisingly, deletion of the entire stem-loop structure did not affect appreciably D1 synthesis and accumulation. Whether light activation of D1 translation occurs in this mutant was not determined. Taken together, these data suggest that the *psbA* stem loop region could act as a translational attenuator that is overcome in wild-type cells to induce translation [59].

Recent work has identified some of the molecular mechanisms controlling the binding of the protein complex to the 5'-untranslated region of *psbA* mRNA. A serine/threonine protein phosphotransferase associated with the *psbA* mRNA-binding complex has been identified that utilizes the β-phosphate of ADP to phosphorylate the 60 kDa protein and to inactivate *psbA* mRNA binding *in vitro* [11]. These data raise the possibility that translation of *psbA* mRNA is decreased by phosphorylation of the *psbA* mRNA complex in response to an increase in the stromal ADP levels upon transfer of cells to the dark.

It has been proposed recently that light-regulated translation of chloroplast *psbA* mRNA may occur through photosynthesis-generated redox potential [12]. The evidence is based mostly on *in vitro* experiments showing that oxidized *psbA* protein complex no longer binds to *psbA* RNA and that binding can be restored upon incubation of the protein complex with dithiol-containing reducing agents. These results raise the interesting possibility that thioredoxin, which is known to activate the enzymes of the reductive pentose phosphate cycle [4], could also modulate binding of the *psbA* protein complex to *psbA* RNA and, hence, D1 translation in response to changes in the reducing power generated by photosynthesis. The *in vivo* effect of chloroplast redox potential on D1 expression was examined in a photosystem I-deficient mutant strain

336

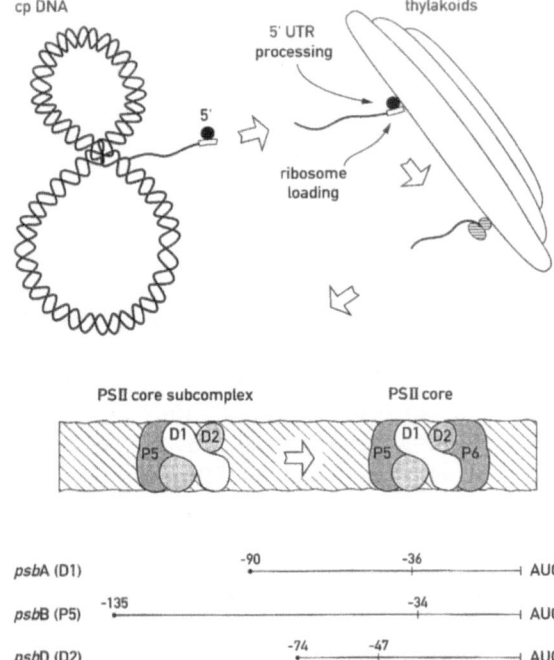

Figure 4. Model for synthesis of the photosystem II core polypeptides. The mRNAs of D1 (*psbA*), D2 (*psbD*) and P5 (*psbB*) are synthesized as precursors. A protein complex binds to the 5' end of the leader, thereby targeting and/or docking the RNA to the thylakoid membrane. Cleavage of the leader is required to induce a conformational change allowing ribosome binding and initiation of translation. Synthesis of D1, D2 and P5 which form an intermediate photosystem II core subcomplex may proceed by a similar mechanism, including the same maturation pathway for the 5' leader. The *psbC* product P6 is integrated in a subsequent step to form the photosystem II core complex. The size of the 5'-untranslated regions of *psbA*, *psbB* and *psbD* mRNA with the cleavage site are indicated in the lower part of the figure. Similar models have been proposed by others [22, 60].

in which thioredoxin is no longer reduced [4]. The observed decreased binding of the complex to *psbA* RNA and the reduced synthesis of D1 and the other photosystem II core polypeptides in light-grown cells are difficult to interpret because photosystem I-deficient mutants are highly sensitive to light. In contrast to this report, earlier pulse-labeling studies with photosystem I deficient mutants did not reveal a diminished synthesis of the core photosystem II polypeptides [23, 86].

A puzzling observation is that both S1 nuclease mapping and primer extension studies have revealed that the *psbA* mRNA is made as a precursor with its 5' end at position −90 relative to the initiation codon and that this RNA is cleaved at −36 to give rise to the more

abundant mature *psbA* RNA [21, 66]. The presumed processing site is located within the stem loop structure shown to constitute the binding site of the *psbA* protein complex [10]. Cleavage at this site would most probably abolish the RNA binding activity of the protein complex. This case resembles the processing of the *psbD* precursor RNA described above which abolishes binding of the 47 kDa protein [66]. It is possible that binding of the protein complex to the *psbA* mRNA is required for its targeting or docking to the thylakoid membrane, the site of translation of *psbA* mRNA [34]. Proper docking would be followed by cleavage of the precursor RNA which would make the *psbA* RNA competent for ribosome loading and translation initiation (Fig. 4).

Besides the mRNAs of *psbA* and *psbD*, the *psbB* RNA is also synthesized as a precursor and cleaved within its 5'-untranslated region to generate the mature mRNA (J. Nickelsen, unpublished results). Previous work has revealed the concerted expression of the *psbA*, *psbD* and *psbB* genes [13, 20, 42]. Their corresponding products, D1, D2 and P5, form a protein complex intermediate during the biosynthesis of the photosystem core [13]. Possibly, synthesis of these three hydrophobic polypeptides might follow a common pathway as outlined above for *psbA* mRNA translation (Fig. 4).

The analysis of proteins binding to the 5'-untranslated regions of several chloroplast mRNAs (*psbA*, *atpB*, *rbcL*, *rps7*, *rps12*) by UV crosslinking revealed at least seven polypeptides of 81, 62, 56, 47, 38, 36 and 15 kDa [31]. These RNA-binding proteins were detectable in cells grown under mixotrophic, heterotrophic and phototrophic conditions. Competition experiments demonstrated the binding specificity of some of these *trans*-acting proteins for the chloroplast 5'-untranslated regions [31]. These regions are highly AU-rich (70–84%) in *C. reinhardtii* with no significant sequence homology to one another. However, most of these regions appear to have a secondary structure, in particular stem-loops, which might be required for the binding. The RNA binding activity of the 36 kDa protein was undetectable in mutant cells with reduced chloroplast protein synthesis capacity. Such cells are known to preferentially translate mRNAs for chloroplast-encoded ribosomal proteins and to severely reduce the translation of photosynthetic proteins [54]. Whether the 36 kDa protein is synthesized on chloroplast ribosomes or is required for translation of mRNAs of photosynthetic proteins remains to be determined.

The observation that six of the RNA-binding proteins bind to five distinct chloroplast leaders under a wide variety of environmental conditions suggests that at least some of them, in particular the 81 and 47 kDa proteins, are general leader binding proteins. However, it is likely that there is a whole class of RNA-binding proteins in the 47 kDa range, some of which may be message-specific while others may bind ubiquitously to all chloroplast leaders.

Assembly of the photosynthetic complexes

The numerous mutants of *C. reinhardtii* deficient in photosynthesis have provided unique opportunities to dissect the assembly process of photosynthetic complexes and, in particular, to study how the absence of certain polypeptides alters the assembly of the other subunits. These studies have revealed that impaired synthesis of one subunit usually leads to the destabilization of the corresponding complex through increased turnover of the other subunits [13, 20, 42, 72, 81a]. It is striking that the complexes of *C. reinhardtii* are considerably more sensitive to small alterations than their homologues in cyanobacteria. In *C. reinhardtii*, loss of the *psbK* subunit of photosystem II leads to the destabilization of the complex which accumulates at vastly reduced levels relative to wild type [87]. In cyanobacteria, however, a mutant lacking *psbK* is still able to grow photoautotrophically [41]. Similarly, disruption of the *psaC* gene prevents stable accumulation of the photosystem II complex in *C. reinhardtii* [86], but not in cyanobacteria [56]. These observations point to the existence of a clearing system in *C. reinhardtii* which recognizes and degrades misfolded polypeptides and protein complexes, and that is more efficient than its cyanobacterial counterpart. The molecular components of this clearing system remain to be identified.

Assembly of the cytochrome *b6f* compex of *C. reinhardtii* has been examined by using mutants deficient in cytochrome *b6f* activity which were unable to synthesize the chloroplast-encoded subunits cytochrome *f* or cytochrome *b6*. In these mutants, the other subunits of the complex do not accumulate [53]. In contrast, in the absence of the nuclear-encoded Rieske protein, the chloroplast-encoded subunits accumulate to over 50% of wild-type levels [53]. This concerted accumulation of the chloroplast-encoded subunits has been studied further by constructing mutants with deletions of *petA*, *petB* or *petD*, which do not synthesize cytochrome *f*, cytochrome *b6* and subunit IV,

respectively [51]. Analysis of these mutants revealed that the rate of synthesis of cytochrome *b6* and subunit IV are unaffected by the absence of the other subunits of the complex, although the accumulation of either of these two subunits is considerably reduced. In marked contrast, the rate of cytochrome *f* synthesis was severely decreased in the absence of either subunit IV or cytochrome *b6*, but cytochrome *f* was stable under these conditions [51]. Hence, two distinct mechanisms operate to achieve stoichiometric accumulation of the chloroplast-encoded subunits of the cytochrome *b6f* complex: proteolytic degradation of unassembled subunits and a regulatory process in which the production of cytochrome *f* depends on its immediate interaction with cytochrome *b6* and subunit IV.

The stability of cytochrome *f* observed in the absence of the other subunits suggests that it is the core subunit through which the other subunits of the complex are stabilized. One would therefore expect that the stoichiometric accumulation of cytochrome *f* with the other subunits is controlled to prevent overproduction of the protein. The C-terminal region of cytochrome *f* which includes a transmembrane anchoring domain appears to be involved in the control based on the observation that its removal leads to a 3-fold increase in the rate of synthesis of the truncated soluble form of cytochrome *f*, which is redox-active and accumulates in the thylakoid lumen [52]. It has been proposed that the C-terminal region of cytochrome *f* down-regulates synthesis of the protein when it is not properly assembled within the complex. Thus, mutants lacking the C-terminal region or expressing unstable cytochrome *f* lack the feedback control [90]. The nature of this feedback contol remains to be determined.

Synthesis of cytochrome *f* involves translation of the *petA* mRNA, N-terminal processing of the precursor protein and heme attachment. To test whether the latter two events are interdependent, chloroplast cytochrome *f* mutants were constructed whose heme-binding site or the processing site was inactivated [50]. The mutants affected in heme binding were non-phototrophic, accumulated reduced levels of cytochrome *f* and revealed that heme binding is not required for cytochrome *f* processing. Similar conclusions were reached from the analysis of mutants defective in heme attachment to apocytochromes *c6* and *f* [39, 40]. The analysis of the processing site mutants showed that preapocytochrome *f* can bind heme and that it is competent for assembly in cytochrome *b6f* complexes [50]. Incorporation of these mutations in the truncated form of cytochrome *f* lacking the C-terminal membrane anchor led in all cases to a stimulation of synthesis of cytochrome *f*. Hence, the two maturation events of cytochrome *f* do not appear to be rate-limiting for its synthesis [50].

Other cases are known in *C. reinhardtii* in which some chloroplast-encoded subunits display decreased rates of synthesis when another subunit of the same complex is no longer synthesized. These epistatic effects in protein synthesis lead to the stoichiometric accumulation of the various subunits of photosynthetic complexes and ensure that some subunits, such as cytochrome *f*, are produced at a rate which is governed by the assembly process.

In mutants unable to synthesize the PsaB reaction center subunit of photosystem I, the PsaA subunit is no longer produced [23]. However, the opposite is not true: mutants unable to synthesize the PsaA subunit still synthesize the PsaB subunit [23]. Similarly, analysis of several photosystem II-deficient mutants has revealed a dependence of D1 synthesis on the presence of D2 and of P5 synthesis on D1 [13, 20, 42]. Further, for the chloroplast ATP synthase, the β-subunit appears to control the rate of synthesis of the α-subunit [17]. As the chloroplast genes of these subunits are transcribed independently, the observed coordinated synthesis cannot be explained by translational coupling as described for bacterial polycistronic RNAs, in which translation of an upstream open reading frame facilitates translation of a downstream sequence (cf. [25]).

Conclusions and perspectives

Analysis of numerous nuclear mutants of *C. reinhardtii* deficient in photosynthetic activity has revealed that the interactions between the nuclear and chloroplast genetic systems are highly complex. A surprisingly large number of nuclear loci are involved in the expression of chloroplast genes. These loci define factors that are most likely imported into the chloroplast, where they act in a gene-specific manner, mostly at posttranscriptional steps such as RNA stabilization, RNA processing and splicing and translation. Several recent studies based on the use of the powerful chloroplast transformation technology and molecular-biochemical methods have identified the targets on the chloroplast messages of several of these *trans*-acting factors. The picture which emerges from this work is that the 5′- and 3′-untranslated regions of the plastid mRNA play a key role in RNA metabolism and several *cis* elements, in particular the stem-loop structures of the 5′ leaders of *psbC* and *psbA*, appear to be especially important. Although gel mobility shift assays and UV crosslinking experiments have revealed the existence of several RNA-binding reactions, little is known on how these proteins interact with their targets and how they modulate the expression of the chloroplast genes. It is likely that several of these RNA-binding proteins form protein complexes with other factors that could be involved in various functions such as RNA targeting or docking to the thylakoid membrane, RNA stability and activation of translation, and whose action may be environmentally controlled. With the improved nuclear transformation efficiency in *C. reinhardtii* and the availability of cosmid libraries, genomic complementation of specific nuclear photosynthetic mutants has been achieved. Gene tagging has also been used successfully to isolate nuclear genes [88]. It is therefore likely that several mutant loci will be cloned in the next years and the corresponding products identified and characterized.

An important task will be to complement these studies with *in vitro* systems. While RNA stability can be studied *in vitro* with suitable chloroplast extracts, processing of RNA precursors has met with limited success and no functional *in vitro* splicing system has been reported, except for the self-splicing of the group I introns. The recent development of a faithful plastid *in vitro* translation system in tobacco [37] opens promising possibilities for a biochemical dissection of the mechanisms of plastid translation and its relation to RNA stability and processing.

Acknowledgements

I thank N. Gillham, J. Boynton and F.A. Wollmann for providing unpublished results, M. Goldschmidt-Clermont, J. Nickelsen and W. Zerges for helpful comments, and N. Roggli for drawings. The work described from my laboratory was supported by grant 31.34014.92 from the Swiss National Foundation.

References

1. Bennoun P, Masson A, Delosme N: A method for complementation analysis of nuclear and chloroplast mutants of photosynthesis in *Chlamydomonas*. Genetics 95: 39–47 (1980).
2. Blowers A, Klein U, Ellmore GS, Bogorad L: Functional *in vivo* analysis of the 3′ flanking sequences of the *Chlamydo-*

monas chloroplast *rbcL* and *psaB* genes. Mol Gen Genet 238: 339–349 (1993).

3. Boynton JE, Gillham NW, Harris EH, Hosler JP, Johnson AR, Jones BL, Randolph-Anderson D, Robertson TM, Klein KB, Shark B, Sanford JC: Chloroplast transformation in *Chlamydomonas* with high velocity microprojectiles. Science 240: 1534–1538 (1988).

4. Buchanan BB: Regulation of CO_2 assimilation in oxygenic photosynthesis: the ferredoxin thioredoxin system. Perspective on its discovery, present status, and future development. Arch Biochem Biophys 288: 1–9 (1991).

5. Chen X, Kindle KL, Stern DB: Initiation codon mutations in the *Chlamydomonas* chloroplast *petD* gene result in temperature-sensitive photosynthetic growth. EMBO J 12: 3627–3635 (1993).

6. Chen X, Kindle KL, Stern DB: The initiation codon determines the efficiency but not the site of translation initiation in *Chlamydomonas* chloroplasts. Plant Cell 7: 1295–1315 (1995).

7. Choquet Y, Goldschmidt-Clermont M, Girard-Bascou H, Kuck U, Bennoun P *et al.*: Mutant phenotypes support a *trans*-splicing mechanism for the expression of the tripartite *psaA* gene in the *C. reinhardtii* chloroplast. Cell 52: 903–913 (1988).

8. Chua NH, Bennoun P: Thylakoid membranes polypeptides of *Chlamydomonas reinhardtii*: wild-type and mutant strains deficient in photosystem II reaction center. Proc Natl Acad Sci USA 72: 2175–2179 (1975).

9. Collins RA, Lambowitz AM: RNA splicing in *Neurospora mitochondria*. Defective splicing of mitochondrial mRNA precursors in the nuclear mutant cyt 18–1. J Mol Biol 184: 413–428 (1985).

10. Danon A, Mayfield SPY: Light-regulated translational activators: identification of chloroplast gene-specific mRNA binding proteins. EMBO J 10: 3993–4002 (1991).

11. Danon A, Mayfield SP: ADP-dependent phosphorylation regulates RNA-binding in vitro: implications in light-modulated translation. EMBO J 13: 2227–2235 (1994).

12. Danon A, Mayfield SP: Light-regulated translation of chloroplast mRNAs through redox potential. Science 266: 1717–1719 (1995).

13. De Vitry C, Olive J, Drapier D, Recouvreur M, Wollman F-A: Posttranslational events leading to the assembly of photosystem II protein complex: a study using photosynthesis mutants from *Chlamydomonas reinhardtii*. J Cell Biol 109: 991–1006 (1989).

14. Debuchy R, Purton S, Rochaix J-D: The argininosuccinate lyase gene of *Chlamydomonas reinhardtii*: an important tool for nuclear transformation and for correlating the genetic and molecular maps of the ARG7 locus. EMBO J 8: 2803–2809 (1989).

15. Delepelaire P: Partial characterization of the biosynthesis and integration of the photosystem II reaction centers in the thylakoid membrane of *Chlamydomonas reinhardtii*. EMBO J 3: 701–706 (1984).

16. Deshpande NN, Hollingsworth M, Herrin DL: The *atpF* group II intron-containing gene from spinach chloroplasts is not spliced in transgenic *Chlamydomonas* chloroplasts. Curr Genet 28: 122–127 (1995).

17. Drapier D, Girard-Bascou J, Wollman FA: Evidence for a nuclear control on the expression of the *atpA* and *atpB* chloroplast genes in *Chlamydomonas reinhardtii*. Plant Cell 4: 283–295 (1992).

18. Dron M, Rahire M, Rochaix J-D: Sequence of the 16SrRNA gene and its surrounding regions of *Chlamydomonas reinhardtii*. Nucl Acids Res 10: 7609–7620 (1982).

19. Dürrenberger F, Rochaix J-D: Chloroplast ribosomal intron of *Chlamydomonas reinhardtii*: in vitro self-splicing, DNA endonuclease activity and in vivo mobility. EMBO J 10: 3495–3501 (1991).

20. Erickson JM, Rahire M, Malnoe P, Girard-Bascou J, Pierre Y *et al.*: Lack of the D2 protein in a *Chlamydomonas reinhardtii psbD* mutant affects photosystem II stability and D1 expression. EMBO J 5: 1745–1754 (1986).

21. Erickson JM, Rahire M, Rochaix J-D: *Chlamydomonas reinhardtii* gene for the 32000 mol. wt. protein of photosystem II contains four large introns and is located entirely within the chloroplast inverted repeat. EMBO J 3: 2753–2762 (1984).

22. Gillham NW, Boynton JE, Hauser CR: Translational regulation of gene expression in chloroplasts and mitochondria. Annu Rev Genet 28: 71–93 (1994).

23. Girard-Bascou J, Choquet Y, Schneider M, Delosme M, Dron M: Characterization of a chloroplast mutation in the *psaA2* gene of *Chlamydomonas reinhardtii*. Curr Genet 12: 489–495 (1987).

24. Girard-Bascou J, Pierre Y, Drapier D: A nuclear mutation affects the synthesis of the chloroplast *psbA* gene product in *Chlamydomonas reinhardtii*. Curr Genet 22: 47–52 (1992).

25. Gold L: Posttranscriptional regulatory mechanism in *Escherichia coli*. Annu Rev Biochem 57: 199–233 (1968).

26. Goldschmidt-Clermont M: Transgenic expression of aminoglycoside adenine transferase in the chloroplast: a selectable marker for site-directed transformation of *Chlamydomonas*. Nucl Acids Res 19: 4083–4090 (1991).

27. Goldschmidt-Clermont M, Girard-Bascou J, Choquet Y, Rochaix J-D: *Trans*-splicing mutants of *Chlamydomonas reinhardtii*. Mol Gen Genet 223: 417–425 (1990).

28. Goldschmidt-Clermont M, Choquet Y, Girard-Bascou H, Michel F, Schirmer-Rahire M, Rochaix J-D: A small chloroplast RNA may be required for *trans*-splicing in *Chlamydomonas reinhardtii*. Cell 65: 135–143 (1991).

29. Gruissem W, Schuster G: Control of mRNA degradation in organelles. In: Brawerman G, Belasco J (eds) Control of Messenger RNA Stability, pp. 329–365. Academic Press, Orlando, FL (1993).

30. Harris EH: The *Chlamydomonas* Sourcebook: A Comprehensive Guide to Biology and Laboratory Use. Academic Press, San Diego, CA (1989).

31. Hauser CR, Gillham NW, Boynton JE: Translational regulation of chloroplast mRNAs in *Chlamydomonas reinhardtii*. J Biol Chem, in press (1996).

32. Herdenberger F, Hollander V, Kück U: Correct in vivo RNA splicing of a mitochondrial intron in algal chloroplasts. Nucl Acids Res 22: 2869–2875 (1994).

33. Herrin D, Michaels A: The chloroplast 32kDa protein is synthesized on thylakoid-bound ribosomes in *Chlamydomonas reinhardtii*. FEBS Lett 184: 90–94 (1985).

34. Herrin DL, Bao Y, Thompson AJ, Chen Y-F: Self-splicing of the *Chlamydomonas* chloroplast *psbA* introns. Plant Cell 3: 1095–1107 (1991).

35. Herrin DL, Chen Y-F, Schmidt GW: RNA splicing in *Chlamydomonas* chloroplasts: self-splicing of 23S preRNA. J Biol Chem 265: 21134–21140 (1990).

36. Herrin DL, Schmidt GW: *Trans*-splicing of transcripts for the chloroplast *psaA1* gene. In vivo requirement for nuclear gene products. J Biol Chem 263: 14601–14604 (1988).

37. Hirose T, Sugita M, Sugiura M: *In vitro* analysis of *cis*- and *trans*-acting regulatory elements for translation of photosynthetic gene transcription in tobacco chloroplasts. Photosynth Res Suppl 1: 153 (1995).

38. Hong S, Spreitzer RJ: Nuclear mutation inhibits expression of the chloroplast gene that encodes the large subunit of ribulose–1,5-bisphosphate carboxylase/oxygenase. Plant Physiol 106: 673–678 (1994).

39. Howe G, Merchant S: The biosynthesis of membrane and soluble plastic c-type cytochromes of *Chlamydomonas reinhardtii* is dependent on multiple common gene products. EMBO J 11: 2789–2801 (1992).

40. Howe G, Mets L, Merchant S: Biosynthesis of cytochrome f in *Chlamydomonas reinhardtii*: analysis of the pathway in gabaculine treated cells and in the heme attachment mutant B6. Mol Gen Genet 246: 156–165 (1995).

41. Ikeuchi M, Eggers B, Shen G, Webber A, Yu J, Hirano A, Inoue Y, Vermaas W: Cloning of the *psbK* gene from *Synechocystis* sp. PCC 6803 and characterization of photosystem II in mutants lacking PSII-K. J Biol Chem 266: 11111–11115 (1991).

42. Jensen KH, Herrin DL, Plumley FG, Schmidt GW: Biogenesis of photosystem II complexes: transcriptional translational and posttranslational regulation. J Cell Biol 103: 701–706 (1986).

43. Johnson CH and Schmidt GW: The *psbB* gene cluster of the *Chlamydomonas reinhardtii* chloroplast: sequence and transcriptional analyses of *psbN* and *psbH*. Plant Mol Biol 22: 645–658.

44. Kindle KL: High frequency nuclear transformation of *Chlamydomonas reinhardtii*. Proc Natl Acad Sci USA 87: 1228–1232 (1990).

45. Kindle KL, Schnell RA, Fernandez E, Lefebvre PA: Stable transformation of *Chlamydomonas* using the *Chlamydomonas* gene for nitrate reductase. J Cell Biol 109: 2589–2601 (1989).

46. Kuchka M, Goldschmidt-Clermont M, van Dillewijn J. Rochaix JD: Mutation at the *Chlamydomonas* nuclear NAC2 locus specifically affects stability of the chroloplast *psbD* transcript encoding polypeptide D2 of photosystem II in *Chlamydomonas reinhardtii*. Cell 58: 869–876 (1989).

47. Kuchka MR, Mayfield SP, Rochaix J.-D: Nuclear mutations specifically affect the synthesis and/or degradation of the chloroplast-encoded D2 polypeptide of photosystem II in *Chlamydomonas reinhardtii*. EMBO J 7: 319–324 (1988).

48. Kück U, Choquet Y, Schneider M, Dron M, Bennoun P. Structural and transcription analysis of two homologous genes for the P700 chlorophyll a apoproteins *Chlamydomonas reinhardtii*: evidence for in vivo *trans*-splicing. EMBO J 6: 2185–2195 (1987).

49. Kück U, Godehart I, Schmidt U: A self-splicing group-II intron in the mitochondrial large subunit rRNA (LSUrRNA) gene of the eucaryotic alga *Scenedesmus obliquus*. Nucl Acids Res 18: 2691–2697 (1990).

50. Kuras R, Büschlen S, Wollman FA: Maturation of pre-apocytochrome f in vivo. J Biol Chem 270, in press (1995).

51. Kuras R, Wollman F-A: The assembly of cytochrome b$_6$/f complexes: An approach using genetic transformation of the green alga *Chlamydomonas reinhardtii*. EMBO J 13: 1019–1027 (1994).

52. Kuras R, Wollman FA, Joliot P: Conversion of cytochrome f to a soluble form in vivo in *Chlamydomonas reinhardtii*. Biochemistry 34: 7468–7475 (1995).

53. Lemaire C, Girard-Bascou J, Wollman F-A, Bennoun P: Studies on the cytochrome b$_6$/f complex. I. Characterization of the complex subunits in *Chlamydomonas reinhardtii*. Biochim Biophys Acta 851: 229–238 (1986).

54. Liu X-Q, Hosler JP, Boynton JE, Gillham NW: mRNAs for two ribosomal proteins are preferentially translated in the chloroplast of *Chlamydomonas reinhardtii* under conditions of reduced protein synthesis. Plant Mol Biol 12: 385–394 (1989).

55. Malnoe P, Mayfield SP, Rochaix JD: Comparative analysis of the biogenesis of photosystem II in the wild-type and Y-1 mutant of *Chlamydomonas reinhardtii*. J Cell Biol 106: 609–616 (1988).

56. Mannan RM, Whitmarsh J, Nyman P, Pakrasi H: Directed mutagenesis of an iron-sulfur protein of the photosystem I complex in the filamentous cyanobacterium *Anabaena variabilis* ATCC 29413. Proc Natl Acad Sci USA 88: 10168–10172 (1991).

57. Mayfield SP: Chloroplast gene regulation: interaction of the nuclear and chloroplast genomes in the expression of photosynthetic proteins. Curr Opin Cell Biol 2: 509–513 (1990).

58. Mayfield SP, Kindle KL: Stable nuclear transformation of *Chlamydomonas reinhardtii* using a *C. reinhardtii* gene as the selectable marker. Proc Natl Acad Sci USA 87: 2087–2091 (1990).

59. Mayfield SP, Cohen A, Danon A, Yohn CB: Translation of the *psbA* mRNA of *Chlamydomonas reinhardtii* requires a structured RNA element contained within the 5′ untranslated region. J Cell Biol 127: 1537–1545 (1994).

60. Mayfield SP, Yohn CB, Cohen A, Danon A: Regulation of chloroplast gene expression. Annu Rev Plant Physiol Plant Mol Biol 46: 147–166 (1994).

61. Michel F, Jacquier A, Dujon B: Comparison of fungal mitochondrial introns reveals extensive homologous in RNA secondary structure. Biochimie 64: 867–881 (1982).

62. Michel F, Umesono K, and Ozeki H: Comparative and functional analysis of group II catalytic introns: a review. Gene 82: 5–30 (1989).

63. Monod C, Goldschmidt-Clermont M, Rochaix J.-D: Accumulation of chloroplast *psbB* RNA requires a nuclear factor in *Chlamydomonas reinhardtii*. Mol Gen Genet 231: 449–459 (1992).

64. Monod C, Takahashi Y, Goldschmidt-Clermont M, Rochaix J-D: The chloroplast ycf8 open reading frame encodes a photosystem II polypeptide which maintains photosynthetic activity under adverse growth conditions. EMBO J 13: 2747–2754 (1994).

65. Nickelsen J, Link G: The 54kDa RNA-binding protein from mustard chloroplasts mediates endonucleolytic transcript 3′ end formation in vitro. Plant J 3: 537–544 (1993).

66. Nickelsen J, Van Dillewijn J, Rahire M, Rochaix J-D: Determinants for stability of the chloroplast psbD RNA are located within its short leader region in *Chlamydomonas reinhardtii*. EMBO J 13: 3182–3191 (1994).

67. Petersen C: Translation and mRNA stability in bacteria: a complex relationship. In: Belasco JG, Brawerman G (eds) Control of Messenger RNA Stability, pp. 117–145. Academic Press, San Diego, CA (1993).

68. Purton S, Rochaix J-D: Complementation of a *Chlamydomonas reinhardtii* mutant using a genomic cosmid library. Plant Mol Biol 24: 533–537 (1994).

69. Robertson D, Boynton JE, Gillham NW: Cotranscription of the wild-type chloroplast *atpE* gene encoding the CF$_1$/CF$_0$ epsilon subunit with the 3′ half of the *rps7* gene in *Chlamydomonas reinhardtii* and characterization of frameshift mutations in *atpE*. Mol Gen Genet 221: 155–163 (1990).

70. Rochaix J-D: Post-transcriptional steps in the expression of chloroplast genes. Annu Rev Cell Biol 8: 1–28 (1992).

71. Rochaix J-D, Rahire M, Michel F: The chloroplast ribosomal intron of *Chlamydomonas reinhardtii* codes for a polypeptide related to mitochondrial maturases. Nucl Acids Res 13: 975–984 (1985).

72. Rochaix JD, Erickson JM: Function and assembly of photosystem II: genetic and molecular analysis. Trends Biochem Sci 13: 56–59 (1988).

73. Rochaix J-D, Kuchka M, Mayfield S, Schirmer-Rahire M, Girard-Bascou J, Bennoun P: Nuclear and chloroplast mutations affect the synthesis or stability of the chloroplast *psbC* gene product in *Chlamydomonas reinhardtii*. EMBO J 8: 1013–1022 (1989).

74. Sakamoto W, Chen X, Kindle KL, Stern DB: Function of the *Chlamydomonas reinhardtii petD* 5' untranslated region in regulating the accumulation of subunit IV of the cytochrome b6f complex. Plant J 6: 503–512 (1994).

75. Sakamoto W, Sturm NR, Kindle KL, Stern DB: *petD* mRNA maturation in *Chlamydomonas reinhardtii* chloroplasts: role of 5' endonucleolytic processing. Mol Cell Biol 14: 6180–6186 (1994).

76. Sakamoto WK, Kindle KL, Stern DB: In vivo analysis of *Chlamydomonas* chloroplast *petD* gene expression using stable transformation of *β*-glucuronidase translational fusions. Proc Natl Acad Sci. USA 90: 497–501 (1993).

77. Salvador ML, Klein U, Bogard L: 5' sequences are important positive and negative determinants of the longevity of *Chlamydomonas* chloroplast gene transcripts. Proc Natl Acad Sci USA 90: 1556–1560 (1993).

78. Salvador ML, Klein U, Bogorad L: Light-regulated and endogenous fluctuations of chloroplast transcript levels in *Chlamydomonas*. Regulation by transcription and RNA degradation. Plant J 3: 213–219 (1993).

79. Schuster G, Gruissem W: Chloroplast mRNA 3' end processing requires a nuclear-encoded RNA binding protein. EMBO J 1: 1493–1502 (1982).

80. Sharp P: Five easy pieces. Science 254: 663 (1991).

81. Sieburth LE, Berry-Lowe S, Schmidt GW: Chloroplast RNA stability in *Chlamydomonas*: rapid degradation of *psbB* and *psbC* transcripts in two nuclear mutants. Plant Cell 3: 175–189 (1991).

81a. Spreitzer RJ, Goldschmidt-Clermont M, Rahire M, Rochaix J-D: Nonsense mutations in the *Chlamydomonas* chloroplast gene that codes for the large subunit of ribulose carboxylase/oxygenase. Proc Natl Acad Sci USA 82: 5460–5464 (1985).

82. Stern DB, Gruissem W: Control of plastid gene expression: 3' inverted repeats acts as mRNA processing and stabilizing elements, but do not terminate transciption. Cell 51: 1145–1157 (1987).

83. Stern DB, Radwanski ER, Kindle KL: A 3' stem/loop structure of the *Chlamydomonas* chloroplast *atpB* gene regulates mRNA accumulation in vivo. Plant Cell 3: 285–297 (1991).

84. Stern DB, Kindle KL: 3' end maturation of the *Chlamydomonas reinhardtii* chloroplast atp mRNA is a two-step process Mol Cell Biol 13: 2277–2285 (1993).

85. Sturm NR, Kuras R, Büschlen S, Sakamoto W, Kindle KL, Stern DB, Wollman F-A: The *petD* gene is transcribed by functionally redundant promoters in *Chlamydomonas reinhardtii* chloroplasts. Mol Cell Biol 14: 6180–6186 (1994).

86. Takahashi Y, Goldschmidt-Clermont M, Soen S-Y, Franzen LG, Rochaix J-D: Directed chloroplast transformation in *Chlamydomonas reinhardtii*: insertional inactivation of the *psaC* gene encoding the iron sulfur protein destabilizes photosystem I. EMBO J 10: 2033–2040 (1991).

87. Takahashi Y, Matsumoto H, Goldschmidt-Clermont M, Rochaix J-D: Directed disruption of the *Chlamydomonas* chloroplast *psbK* gene destabilizes the photosystem II reaction center complex. Plant Mol Biol 24: 779–788 (1994).

88. Tam LW, Lefevbre PA: Cloning of flagellar genes in *Chlamydomonas reinhardtii* by insertional mutagenesis. Genetics 135: 375–394 (1993).

89. Turmel M, Choquet Y, Goldschmidt-Clermont M, Rochaix J-D, Otis C, Lemieux C: The *trans*-spliced intron 1 in the *psaA* gene of the *Chlamydomonas* chloroplast: a comparative analysis. Curr Genet 27: 270–279 (1995).

90. Wollman FA, Kuras R, Choquet Y: Epistatic effects in thylakoid protein synthesis: the example of cytochrome f. In: Proceedings 10th International Congress of Photosynthesis. Kluwer Academic Publishers, Dordrecht, in press (1995).

91. Wu HY, Kuchka M: A nuclear suppressor overcomes defects in the synthesis of the chloroplast *psbD* gene product caused by mutations in two distinct nuclear genes of *Chlamydomonas*. Curr Genet 27: 263–269 (1995).

92. Xu R, Bingham SE, Webber AN: Increased mRNA accumulation in *psaB* frame-shift mutant of *Chlamydomonas reinhardtii* suggests a role for translation in *psaB* mRNA stability. Plant Mol Biol 22: 465–474 (1993).

93. Zerges W, Rochaix J-D: The 5' leader of a chloroplast mRNA mediates the translational requirements for two nucleus-encoded functions in *Chlamydomonas reinhardtii*. Mol Cell Biol 14: 5268–5277 (1994).

94. Zhang H, Herman PL, Weeks DP: Gene isolation through genomic complementation using an indexed library of *Chlamydomonas reinhardtii* DNA. Plant Mol Biol 24: 663–672 (1994).

Plant Molecular Biology **32**: 343–365, 1996.
© 1996 *Kluwer Academic Publishers.*

RNA editing in plant mitochondria and chloroplasts

Rainer M. Maier[1], Patric Zeltz[1], Hans Kössel[1], Géraldine Bonnard[2], José M. Gualberto[2] & Jean Michel Grienenberger[2],*
[1]*Institut für Biologie III, Universität Freiburg, Schänzlestrasse 1, D–79104 Freiburg, Germany ;* [2]*Institut de Biologie Moléculaire des Plantes du CNRS, Université Louis Pasteur, 12 rue du Général Zimmer, 67084 Strasbourg Cedex, France (*author for correspondence)*

Key words: RNA editing, tRNA editing, chloroplast, mitochondrion, post-transcriptional modification, initiation codon, stop codon, deamination, evolution, guide RNA, transgenic plants, plastid transformation

Contents

Abstract	343
Introduction	344
mRNA editing in higher-plant mitochondria	344
Occurrence and extent of mRNA editing in plant mitochondria	344
Editing is a post-transcriptional event	348
Influence of editing at the protein level	349
Biochemistry of RNA editing in plant mitochondria	349
Evolution of mt RNA editing	350
RNA editing in mitochondrial structural RNA	351
RNA editing in chloroplasts of higher plants	351
Creation of start codons by editing	351
Editing of internal codons	352
Estimate of the editing frequency in the transcripts of the maize plastome	353
General occurrence and characteristics of chloroplast editing sites	353
Phylogenetic aspects of editing in chloroplasts	355
Editing of untranslated regions	357
The chloroplast translational apparatus and plastome encoded proteins are not involved in editing	358
Editing in chloroplasts is an early RNA processing step independent of splicing and cleavage of polycistronic transcripts	359
Use of the plastid transformation technique as a tool for studying the editing process	359
Comparative aspects of plant mitochondrial and chloroplast RNA editing	361
Acknowledgements	362
References	362

Abstract

In the mitochondria and chloroplasts of higher plants there is an RNA editing activity responsible for specific C-to-U conversions and for a few U-to-C conversions leading to RNA sequences different from the corresponding DNA sequences. RNA editing is a post- transcriptional process which essentially affects the transcripts of protein coding genes, but has also been found to modify non-coding transcribed regions, structural RNAs and intron sequences. RNA editing is essential for correct gene expression: proteins translated from edited transcripts are different from the ones deduced from the genes sequences and usually present higher similarity to the corresponding non-plant homologues. Initiation and stop codons can also be created by RNA editing. RNA editing has also been shown to be required for the stabilization of the secondary structure of introns and tRNAs.

The biochemistry of RNA editing in plant organelles is still largely unknown. In mitochondria, recent experiments indicate that RNA editing may be a deamination process. A plastid transformation technique showed to be a powerful tool for the study of RNA editing. The biochemistry as well as the evolutionary features of RNA editing in both organelles are compared in order to identify common as well as organelle-specific components.

Introduction

Since the discovery of RNA editing in a mitochondrially encoded mRNA of trypanosomes [5], post-transcriptional modification of transcripts has been detected in a number of different genetic systems, affecting nuclear, mitochondrial as well as chloroplast RNAs. Up to now, RNA editing has been found only in eukaryotes. While RNA editing affects primarily messengers, it has been shown to be involved also in the post-transcriptional maturation of structural RNAs, essentially in tRNAs.

RNA editing was initially described in trypanosomes as the deletion and insertion of uridines. It now appears that RNA editing can also produce transcripts that contain adenosines, guanosines, cytidines and possibly inosines that are not present in the genomic coding sequence. These modifications can be obtained either by insertion-deletion as well as by conversion mechanisms. In paramyxoviruses, a co-transcriptional event that can be considered as RNA editing involves the insertion of guanosines in the P-gene mRNA of this virus due possibly to a stuttering of the polymerase. Table 1 shows the different systems in which RNA editing has been found. Excellent reviews describing all cases of RNA editing can be found in a book edited by R. Benne [4].

This review will focus on the occurrence of RNA editing in plants. Up to now, RNA editing has been discovered only in chloroplast and mitochondrial transcripts. In all cases but a few, it involves the post-transcriptional conversions of cytidine into uridine. By far, RNA editing mostly affects mRNA, only a few examples have been described that modify the sequence of structural RNA. This review will describe RNA editing successively in plant mitochondria and in plant chloroplasts. At the end of this review, we will try to establish whether RNA editing in both organelles can be compared.

Excellent reviews have already been published on this subject [16, 37, 61, 104]; the reader is referred to them.

mRNA editing in higher-plant mitochondria

Occurrence and extent of mRNA editing in plant mitochondria

The discovery of RNA editing in plant mitochondrial transcripts was provoked by the apparent confusion induced by the study of mitochondria-encoded protein genes. Analysis of these genes indicates that the CGG codon which was first believed to specify tryptophan in a plant mitochondrial specific genetic code [30] can also by aligned with arginine, as specified by the universal genetic code [8, 40, 43]. It was also intriguing that a number of sequence differences between homologous plant mitochondrial genes would be eliminated by C-to-U substitutions. Analysis of cDNA sequences for the corresponding genes led to the finding that RNA editing by C-to-U conversion solves these contradictions. This phenomenon was independently discovered the same year in three laboratories [23, 41, 52].

It then became evident that the real genetic information should be found on the cDNA sequence of mitochondrial genes. Numerous cDNA sequences have now been published and it is clear that, in plant mitochondria, RNA editing involves the conversion of a cytidine encoded by the DNA into a uridine on the RNA. Nearly all editing sites in all plant mitochondrial cDNAs reported fit with this definition. A few exceptions have been reported which correspond to the reverse conversion, namely a uridine encoded by the DNA into a cytidine on the cDNA (U-to-C conversion). Up to now there are only four occurrences of this reverse conversion in higher plants, in wheat cox3 [42], Oenothera cob and cox2 [51, 88]´and pea cox2 [24]. In lower plants, several U-to-C conversions have been described mainly in the pteridophyta Asplenium nidus [50]. It is not clear whether the U-to-C conversions are real editing events or due to some kind of mis-editing. Information on this matter will probably be very useful for the analysis of the biochemical basis of RNA editing as the biochemical requirements for the two reactions are quite different.

RNA editing is essential for the correct expression of plant mitochondrial genes. It is found to affect all mitochondrial transcripts whose cDNA sequence

Table 1. Compilation of editing events in different systems.

Organism	Type of editing	Cell compartment	Ref.
Trypanosomes	U insertion/deletion	mitochondrion	[5]
Physarum	Nucleotide insertion	mitochondrion	[72]
Paramyxoviruses	G insertion	cytoplasm	[100]
Higher plants	C-to-U conversion	mitchondrion	[23, 41, 52]
Higher plants	C-to-U conversion	mitchondrion	[42]
Higher plants	C-to-U conversion	mitchondrion	[55]
Mammals	C-to-U conversion	nucleus	[22, 95]
Hepatitis delta virus	C-to-U conversion	nucleus	[112]

Extent of editing in *nad* mRNAS

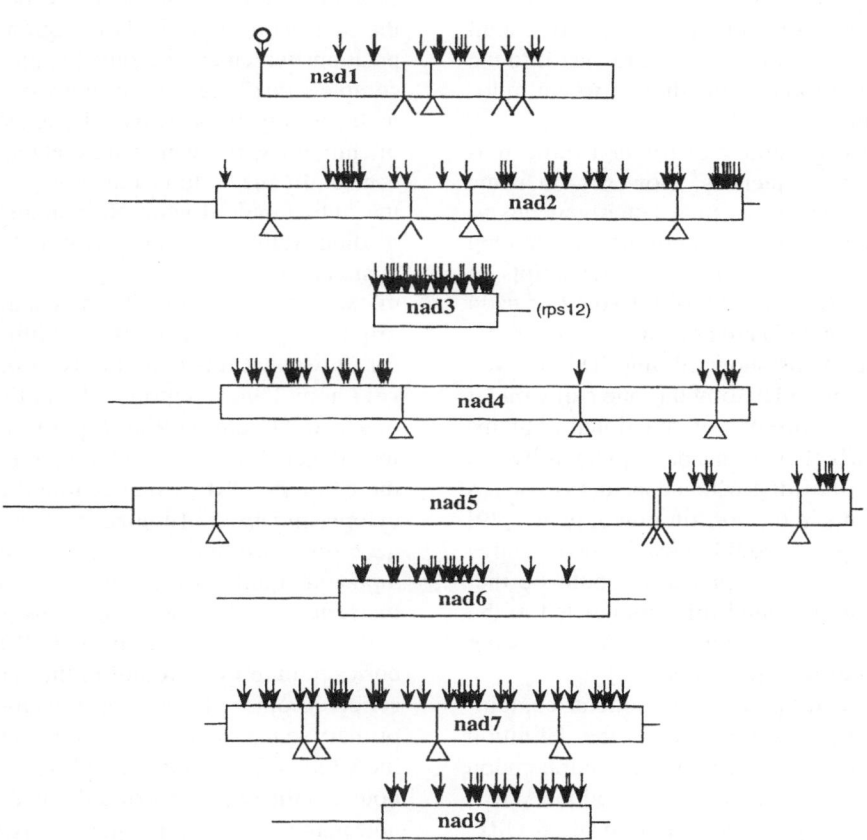

Figure 1. Distribution of all editing sites along the transcripts coding for the mitochondrially encoded subunits of Complex I of wheat. Each arrow corresponds to one editing site. The arrow with an open circle indicates an editing site creating an initiation codon. The coding region of the genes are indicated by open boxes which is divided when the gene is composed of exons. Triangles indicate a *cis*-spliced whereas ∧ indicate *trans*-spliced introns. The data used in this figure are adapted from the following references: *nad1* [20], *nad2* [84], *nad3* [39], *nad4* [67], *nad5* [27], *nad6* [47], *nad7* [14] and *nad9* [66].

has been compared with the corresponding genomic sequence. Frequency of RNA editing is gene-specific as its influence on different transcripts vary deeply can be seen in Fig. 1 which shows the position of all editing sites on the transcripts for the mitochondrially encoded subunits of Complex I (nad). The distribution of the editing sites seems to be at random. The number of editing sites reported in wheat are summarized in Table 2. Based on these data, one can estimate the number of editing sites present in the transcripts corresponding to the complete genetic information of the mitochondrial DNA. In wheat mitochondria, 19 mRNAs have been shown to be edited at 364 positions. The least edited transcript to date is nad5 (5.5 editing sites per 1000 bp) whereas the most edited is an open reading frame (ORF) involved in the biogenesis of c-type cytochromes, orf206 (67.9 editing sites per 1000 bp). If the mitochondrial genome of Marchantia polymorpha [82] is taken as a model for higher plants with a total of 62 expressed genes, a rough calculation leads to the prediction of about 1200 editing sites in wheat mitochondrial transcripts.

Editing sites are generally distributed quite uniformly along coding sequences. However, in some instances, this distribution is quite heterogeneous as exemplified by the distribution of editing sites among the different exons of nad4 and nad5 transcripts of wheat mitochondria (Fig. 1) [27, 67]. Exon 2 of nad4 and exons 1,2 and 3 of nad5 are not edited at all whereas the surrounding exons show editing. This uneven distribution is not yet understood but one can wonder if it is related to the presence of functional domains in the protein or whether it can be explained by the hypothesis of exon shuffling which considers exons as independent elements in the constitution of genes [29] and therefore each exon could show its own degree of RNA editing. Another explanation could be that reverse transcription of edited mRNAs has led to the incorporation of the corresponding cDNA at the same place by homologous recombination [33].

Because RNA editing modifies codons containing a C, it has a profound influence in the definition of open reading frames. A genomic threonine codon (ACG) can be converted to a methionine codon (AUG), thereby leading to the possible initiation of an ORF. Conversely, the editing of genomic glutamine codons (CAA and CAG) and arginine codon (CGA) will introduce a premature termination of the ORF. All these cases have in fact be found in different genes and species, an extreme case being editing in rps10 of potato mitochondria which creates a correct start codon and a

new stop codon [109]. In wheat, start codons are created by editing for the transcript of nad1 [20] (Fig. 1), whereas stop codons are created for atp9 [3] and rps1 [35]. It is therefore absolutely mandatory to determine the effect of RNA editing on a transcript before investigating the functional significance of a putative ORF. Conversely, identification of editing sites in an ORF is a good indication that it constitutes a functional gene.

It is now well known that plant mitochondrial genomes are subject to multiple recombination events which often induce the close association of partial gene sequences with genuine mitochondrial genes. Analysis of the editing status of these partial gene sequences, when they are part of transcripts, shows that RNA editing is sequence-specific. In wheat mitochondria, a large transcript of 2.9 kb allows the co-expression of 3 genes: nad3, rps12 and orf156 (orfB) [39] . This transcript contains a partial sequence corresponding to 193 bp of the first exon of cox2. This region is edited at three positions which are the same as the ones found on the complete cox2 mRNA. In many cases of cytoplasmic male-sterility, the sterility is linked with the expression of chimeric genes which are created by recombination events. In the Polima male-sterile line of rapeseed, the mitochondrial genome contains a chimeric open reading frame (orf224) constituted of three parts, the upstream and 5' coding regions of orfB, part (43 bp) of exon 1 of rps3 and 456 bp of sequence of unknown origin [46]. Analysis of RNA editing of this chimeric orf shows that it is edited only at one position, in the rps3 homologous region and that this position corresponds to the same editing event found in the real rps3 messenger (Fig. 2). Another example is the editing of the chimeric ORF which is found in the male-sterile cytoplasm cms-C of maize [63]. This ORF is a triple gene fusion comprising sequences derived from atp9, atp6 and of unknown origin. The transcript of this chimeric gene is edited at 19 positions, 1 in the atp9 region and 18 in the atp6 region, and all these editing sites correspond to those found in the same complete transcripts in other plants. These examples show that the primary sequence of a transcript is sufficient to determine which Cs are to be edited. It is also interesting to note that the sequences of unknown origin (not detectable in the mitochondrial genomes of the fertile plants) are not edited.

If RNA editing is found in all normal mRNAs as stated above, some transcripts appears to be not edited. This is the case for T-urf13, which has been demonstrated to be the cause of the Texas male sterility of maize [28]. T-urf13 is a chimeric gene composed of 3'

Table 2. Editing sites in the protein coding regions of wheat mitochondrial transcripts.

Gene	C to U	Editing sites per 1000 bp	Modified amino acids	%	Ref.
nad1	17	17.4	14	4.3	[20]
nad2	36	24.5	28	6.5	[84]
nad3	21	59.3	22	13.5	[39]
nad4	23	15.5	22	4.5	[67]
nad5	11	5.5	10	1.5	[27]
nad6	15	23.4	9	3.6	[47]
nad7	32	27.1	27	6.9	[14]
nad9	14	24.3	12	6.2	[66]
atp9	8	33.3	5	6.8	[3]
cob	18	41.3	17	4.3	[108]
cox2	17	21.8	15	5.8	[24]
cox3	12	15.1	12	4.5	[42]
rps1	4	7.8	4	2.3	[35]
rps2	7	6.4	7	1.9	*
rps7*	1	–	–	–	[113]
rps12	6	16	6	4.8	[42]
orf156	4	8.5	3	1.9	[42]
orf206	42	67.9	32	15.5	*
orf240	43	59.7	33	13.8	[15]
orf575	34	19.7	30	5.2	[35]

* Incomplete cDNA sequence, * Unpublished results from the Strasbourg laboratory.

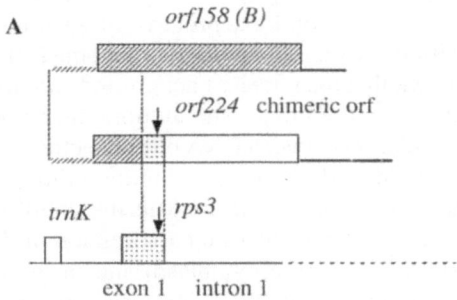

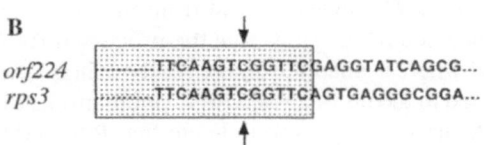

Figure 2. A. Organization of the chimeric open reading frame found in the 'Polima' male-sterile cytoplasm of rapeseed. Regions corresponding to *orf158* (*orfB*) sequences are slanting hatched, to *rps3* are dotted. The open box corresponds to the sequence of unknown origin. B. The position of the editing site in the rps3 homologous regions is shown by arrows. The box indicates the end of the homologous regions.

regions of the 26S rRNA gene, an unknown sequence and a part of the 26S rRNA gene. Transcripts of T-*urf13* are not edited [103] although the co-transcribed *orf221*, a common gene in plant mitochondria is edited. This is probably due to the fact that about 92% of T-*urf13* is derived from coding and 3'-flanking regions of the 26S rRNA gene which are largely unedited (see section RNA editing in mitochondrial structural RNA). This is up to now the only plant mitochondrial mRNA which is not edited.

A small number of editing sites have also been found in transcribed non-coding sequences both upstream or downstream of the coding region [15, 24, 90] and also in intergenic region when two or several genes are co-transcribed [35, 39]. The effect of these editing sites is still poorly understood. One editing site located upstream of *Oenothera rps14* [90] can improve a putative ribosome-binding site but the role of these sequences has not been firmly established in plant mitochondria [13].

Editing sites have been found in the intronic sequences of some transcripts, either in *trans*-spliced [7, 60, 105] or in *cis*-spliced introns[109]. Most of these editing sites are located in stems of the well conserved secondary structure of these group II introns [77] and improve a C/A mismatch to an U-A pairing. These editing sites may therefore be very important for the correct folding of the corresponding intron and play a major role during the maturation of the transcript. To test this assumption, domain 6 of the *trans*-spliced intron c/d of *Oenothera* mitochondrial *nad1* gene [105], which contains a C/A mismatch was transplanted into the autocatalytic yeast intron aI5c, creating chimeras with either the genomic C or the edited U at the nucleotide 5′ to the branchpoint A [17]. After incubation under the conditions which allow the autocatalytic excision, only the edited version of the chimera shows the creation of the lariat while the unedited version remains inactive. These experiments strengthen the assumption that editing is a prerequisite for splicing in plant mitochondrial introns.

Editing is a post-transcriptional event

Analysis of RNA editing by direct determination of the cDNA sequence of wheat *nad3* [41] has shown that transcripts can be partially edited. In this case, the concomitant presence of a C and a U signal in the sequencing ladder led to the conclusion that edited and unedited transcripts are simultaneously present in the RNA fraction. This situation has been proved to be very frequent and was described in almost all species analysed. Further analysis of cloned cDNAs (representing the copy of individual mRNA molecules) has shown that one can find molecules having a C or a U at a particular editing site, with only a fraction of the transcripts being fully edited at all known sites [39, 68, 91, 98, 106]. The distribution of edited sites along the sequence of numerous transcripts of the co- transcribed wheat *nad3-rps12* [39] has shown that there is no polarity of RNA editing. The distribution of edited sites is essentially random, in contrast with the evident 3′ to 5′ polarity found in trypanosome kinetoplasts [94]. The presence of these partially edited transcripts has given the opportunity to study the timing of RNA editing as compared with transcription, splicing and translation.

In wheat seedling mitochondria, most of the genes are expressed by a single mature transcript which is fully edited. A notable exception are the co-transcribed *nad3* and *rps12* genes which are expressed through a 2900 nt precursor and maturated in a 900 nt product. Both forms are partially edited [39]. The analysis of the frequency of partial editing in the precursor and mature forms in total RNA and in a polysomal enriched fraction allowed the study of the timing of RNA editing. It is possible to find precursor transcripts completely un-edited whereas transcripts engaged in translation tend to be fully edited (see next section). The correlation between RNA editing and maturation of transcripts is further strengthened by comparison between species with rapid maturation (e.g. wheat seedlings with single transcript) and species which show complex pattern of expression with multiple transcript forms. When the transcription pattern is complex, the presence of partially edited transcripts is always observed.

The frequency of partial editing has also been studied in spliced and unspliced transcripts [68, 98, 106]. In each case, the comparison clearly shows that unspliced molecules are less edited than spliced ones and that any site can be edited before the splicing event. Unspliced and partially edited transcripts clearly represent intermediates in the RNA maturation process. Here also, the trend is that mature transcripts are more edited than precursor RNAs. However, there is no influence of editing of the coding region on the splicing, RNA editing occurring independently and in parallel of splicing. This is probably not the case for editing sites located in the intronic sequences [17] (see above).

Studies of systems in which the abundance as well as the processing of transcripts have an influence on the frequency of editing again point out that RNA editing is a post-transcriptional process. In several *Petunia* lines with the same mitochondrial genome [70], it has been shown that the nuclear background can influence the efficiency of editing. The authors have noticed a good correlation between RNA editing activity and the abundance of *nad3* transcript, linking editing with a nuclear-encoded factor that controls the stability of the transcripts. In rice, study of the expression of the *atp6* transcript in the fertile cytoplasm and in the *cms-bo* male-sterile cytoplasm shows that RNA processing is different between the two lines [58]. Sequence analysis of cDNA clones derived from the processed and unprocessed RNAs show that the processed RNAs are edited whereas unedited and partially edited RNAs are detected in the unprocessed RNA population. All these results are convergent to indicate that RNA editing is a post-transcriptional process, that the amount of partial editing is directly linked to the presence of non-mature transcripts and that the stability and structure of the RNA can influence the accumulation of partially edited transcripts in the mitochondria. In tissues where mainly mature transcripts accumulate (i.e. wheat seedlings) the number of partially edited transcripts is very low and is undetectable unless thousands of cDNA clones would be sequenced. In species where the different transcript intermediates are readily detected by northern hybridization experiments, partial editing is common.

Influence of editing at the protein level

Apart from the creation of start and stop codons (see above), RNA editing has a profound influence on protein sequence after translation. Most of the editing events (86%) induce a change in the encoded amino acid whereas 14% of editing sites affecting the wobble position and are neutral with respect to the encoded amino acids. Table 2 lists the 364 editing sites found in 19 wheat mitochondrial transcripts to alter the signification of 308 codons. The difference between the two numbers is due to the existence of silent editing sites, editing at the wobble base as already mentioned and a few editing sites that modify the CCR proline codons into leucine codons (either CUR or UUR). Editing of one or both of the adjacent cytidines leads to the same amino acid and one can consider the second editing event as a silent editing site.

RNA editing seems essential for the synthesis of a functional protein. The effect of editing at the protein level has been well documented in the case of wheat mitochondrial ATP9 [3, 36]. Sequencing of this protein has shown that every editing site identified at the cDNA level is reflected at the protein level, both in the coding region and for the creation of a stop codon. Other circumstantial evidences lead to the same conclusion. Editing of a threonine codon in a methionine codon at position 235 of the wheat *cox2* transcript is essential considering the role of this methionine in the structure of one of the copper binding domain of COX2 [25]. In two genes which are proposed to be involved in c-type cytochromes biogenesis, *orf575* and *orf240* [15, 35], editing of three arginine codons (2 in *orf575* and 1 in *orf240*) to thryptophane codons restores a consensus sequence present in both proteins that is most probably required for heme binding.

Another indication that RNA editing is essential for the correct structure and assembly of the mitochondrial protein is given by the creation of transgenic tobacco lines producing a protein synthesized from an unedited messenger sequence [48]. The unedited and edited forms of *atp9* (coding for a subunit of the mitochondrial ATPase) have been cloned downstream of the signal peptide sequence from COXIV yeast protein in order to target the fusion protein intomitochondria. These constructions have been used to transform tobacco. After regeneration, it has been verified that the nuclear form of ATP9 is present in the mitochondria. A significant part of the plants recovered have a cytoplasmic male sterility (CMS) phenotype. This disease is provoked by nucleo-cytoplasmic incompatibility inducing a mitochondrial dysfunction. The translation of unedited *atp9* mRNA has resulted in a protein which is different at 7 positions of a protein of 74 amino acids. The authors postulate that the competition between the genuine and the modified protein induces the synthesis of chimeric ATPases and therefore a loss of ATP production that is causative of CMS.

As described above, partially edited transcripts can accumulate in mitochondria and correspond to a significant proportion of the mRNA population which therefore contain different more or less edited messages. Translation of these messages into protein would probably be detrimental to the biogenesis of the mitochondrial complexes which implies very specific protein-protein interactions. This point has been analysed in two cases at the protein level. The potato NAD9 protein was purified from isolated complex I and partially sequenced. It was found that one amino acid corresponding to a partially edited position in the mRNA is specified by the edited codon, with no indication of heterogeneity at this position [38]. In *Petunia*, *atp6* transcripts are partially edited at a position that create a stop codon and shortens the putative protein by 12 amino acids. Extraction of ATP6 using organic solvents allows the determination of its mass by laser desorption mass spectroscopy, a mass compatible with the product of a fully edited transcript [71].

All these data, together with the preferential association of fully edited mRNA with polysomes [39], indicate that most probably only the fully edited transcripts are translated giving rise to a unique protein sequence. This raises the problem of the mechanism which is active to prevent partially edited transcripts to access the ribosome. One can postulate the existence of a complex structure sequestering partially edited transcripts until all editing sites have been modified. Isolation and analysis of such a complex would be crucial to obtain information on the biochemical mechanisms involved as well as on the factors giving specificity to the process.

Biochemistry of RNA editing in plant mitochondria

There are three possible biochemical mechanisms for the conversion of a cytidine into a uridine: base modification, base exchange and nucleotide exchange. The simplest mechanism is a deamination at position 4 of cytosine leading to a uracil residue, but the reverse conversion, U-to-C, found in a few cases would be obtained by a different mechanism, using a CTP synthase in an ATP-dependent reaction. There are now very few reports that give clear cut evidence for the existence of one of these mechanisms. Two reports investigate the fate of the α-phosphate of cytidines after their incorporation into the mRNA. After *in organello* incorporation of [α-^{32}P]CTP into maize or *Petunia* mitochondria, mitochondrial RNAs were extracted, digested with P1 ribonuclease and the resulting nucleotide monophosphates analysed by two-dimensional thin-layer chromatography (TLC) [85]. A small proportion of the radioactivity co-migrates with UMP, and the amount of UMP increases with the time of labelling. These experiments exclude that editing can be obtained by a nucleotide deletion-addition mechanism because the phosphate backbone appears not to be cut. In another *in vitro* approach, the same result was obtained using a pea mitochondrial crude lysate [107]. Comparison of unedited and edited transcripts in these experiments leads to the conclusion that only unedited transcripts allow the apparition of labelled UMP in TLC, indicating that labelled UMP is present at editing sites. In the same study [107], ^{3}H-labelled CTP, with the label present on the cytosine, was incorporated into mitochondrial RNAs and treated the same way, which resulted in a small incorporation of ^{3}H into UMP. Therefore exchange of bases (*trans*-glycosylation) seems also to be excluded as editing mechanism. Altogether, these results indicate that the mechanism of RNA editing probably involves a deamination reaction.

Further progress in the identification of the factors that act for editing will be obtained by analysis of the editing process using an *in vitro* system. Several attempts have been done to prepare such a system by preparing mitochondrial lysates from different species (wheat, pea, potato) and different organs of these plants (embryos, etiolated leaves or shoots, tubers). Up to now, editing activity could be obtained from a wheat embryo mitochondrial extract prepared by lysis of the mitochondria with detergent and high salt and fractionation of the S100 fraction by chromatography on DEAE-cellulose [2]. Using unedited *atp9* RNA as substrate, it has been possible to show the *in vitro* conversion of C into U. This editing activity is dependent on Mg^{2+} ions, is sensitive to protease digestion and to high temperature. Pre-treatment of the extract by micrococcal nuclease decreases RNA editing activity,

indicating the presence of a nucleic acids important for the reaction. More experimental work is however necessary to extend these observations and to have a fully reliable *in vitro* system. A similar *in vitro* editing system from pea mitochondria has also recently been described [107].

One of the most important question raised by RNA editing is to know what is the basis for the specificity, in other words, how the edited cytidine is selected. As shown in Table 2, for an extensively edited transcript as for the wheat mitochondrial *orf206*, up to 68 editing sites can be found for 1000 bp, corresponding to more than 25% of all cytidines. What is the mechanism which allows to choose one cytidine and not its neighbour? Analysis of the sequence around an editing site delivers very little information. 5' to the editing sites, one can find most frequently a pyrimidine (90%) with often a uridine (more than 60%) and rarely a guanidine (less than 2%). Aside from this observation, no sequence consensus or secondary structure can be identified. When the sequences surrounding numerous editing sites are aligned, it is possible to group these sequences in families that share some homology [42] but that does not provide a predictive tool. Secondary structures around the editing sites or obtained after allowing for long-range interactions on the same mRNA molecule do not give evidence for a conserved structural motif. Because our estimation that about 1200 editing sites will be present on all the mRNAs of wheat mitochondria, it is highly unlikely that site selection would be achieved by specific protein factors like for the apolipoprotein B editing mechanism [80]. It is necessary to postulate the existence of nucleic acids *trans*-factors such as guide RNAs like those found in trypanosome mitochondria [9]. Up to now, no evidence has been published for the existence of such guide RNAs in plant mitochondria. Development of an efficient *in vitro* editing system should allow to identify such guide RNAs, associated protein factors and other enzymatic activity involved.

Evolution of mt RNA editing

RNA editing has been found in every angiosperm species investigated. It has also been found in the gymnosperm *Thuya* [34]. A survey of evolutionary distant plants has been published, based on the good conservation of a sequence coding for a subunit of complex IV [50]. Part of this gene and of the corresponding cDNA was amplified from total tissue of representative species of Spermatophyta, Pteridophyta, Bryophita and Chlorophyta and their sequences compared. RNA editing was found in all species except for the Bryophyta and for the Chlorophyta as representative of green algae. In all ferns, RNA editing was found mostly as C-to-U conversions but also as U-to-C conversions in a significantly higher frequency than in higher plants. No evidence for RNA editing has been found in pluricellular red algae like *Chondrus crispus* [19]. This study indicates that RNA editing occurs in all vascular plants. The fact that it does not exist in bryophytes and in green and red algae reinforces the hypothesis of the apparition of RNA editing in plants after the divergence between bryophytes and vascular plants against that of specific loss of this mechanism in the branches leading to these organisms. In the section 'Comparative aspects of plant mitochondrial and chloroplast RNA editing', the comparative aspects of RNA editing in plant mitochondria and chloroplasts are described. There are clearly common features between editing in chloroplasts and mitochondria. Whether the two mechanisms evolved from a common ancestor is unclear.

RNA editing in mitochondrial structural RNA

The majority of editing sites has been found in precursor and mature mRNAs implying the importance of RNA editing for the expression of correct protein sequences. Analysis of RNA editing of structural RNAs has shown that there are only a few editing sites on these RNA. For ribosomal RNA, two editing sites have been found in *Oenothera* 26S RNA, a C-to-U and a U-to-C conversion, 4 bases apart, both situated in a conserved loop of the structural model of 26S RNA [89]. Their vicinity may be an indication of the existence of a transamination reaction between the two bases. However, these editing sites have been found in one out of five cDNAs sequenced and this low frequency raises questions about their biological significance. The sequences of *Oenothera* 5S and 18S rRNA are strictly identical to their genes. In wheat mitochondria, the same situation was observed for the last 1800 nucleotides of the 18S rRNA (G. Bonnard, J.M. Gualberto and J.M. Grienenberger, unpublished results).

Up to now, three editing sites have been found in tRNAs. In bean and potato mitochondria, the tRNA^{Phe}(GAA) is edited at position 4 by conversion of C-to-U, correcting a C.A mismatch in the acceptor stem [76]. This editing site was determined by the comparison of the sequence of the gene with that of its product obtained by direct sequencing of the mature tRNA. In *Oenothera* mitochondria [6], the same editing event also correct the C4-A69 mismatch of tRNA^{Phe}(GAA). Study of the frequency of this editing event by cDNA analysis showed that the tRNA is partially edited (46.7%), most probably because sequences corresponding to precursor RNA were also analysed. Another editing event was discovered in tRNA^{Cys}(GCA) which change a C.U mispairing in the anticodon stem into a U.U, a mismatch currently found in tRNAs. Two other C-to-U transitions were found in only one cDNA leading to the modification of G-C pairs to G-U pairs. The importance of these editing sites is unclear.

The significance of editing in tRNA^{Phe}(GAA) is clear as it enhances the stability of the acceptor stem of the tRNA, regenerating a correct double-stranded structure of the tRNA. This may be very important for the recognition of the tRNA by its cognate aminoacyl-tRNA synthetase if the edited pair is part of the identity element and/or for maturation of the precursor molecule by RNase P.

RNA editing in chloroplasts of higher plants

Creation of start codons by editing

Detection of RNA editing events in chloroplast transcripts was provoked by the apparent absence of the translational start codon ATG in the *rpl2* gene of maize [55]. ACG codons had been reported in rice [53, 78] and maize [59] at positions homologous to the ATG start codons present in plastomes of other species such as tobacco [92] and the liverwort *Marchantia polymorpha* [83]. Originally these ACG codons were taken as an indication that the translational apparatus of chloroplasts might in rare cases use them as functional initiation codons instead of AUG codons [53, 78]. A similar situation had been reported for the *psbL* genes from tobacco and spinach where ACG codons replace the ATG start codons present in plastomes from other species. However, using the RT-PCR technique for the production of cDNAs followed by sequence analysis of the respective cDNA amplification products, the conversion of the ACG codons to AUG codons by C-to-U editing events could be demonstrated for the *rpl2* transcript of maize [55] as well as for the *psbL* transcript of tobacco [62]. It is reasonable to assume, but not demonstrated experimentally, that creation of

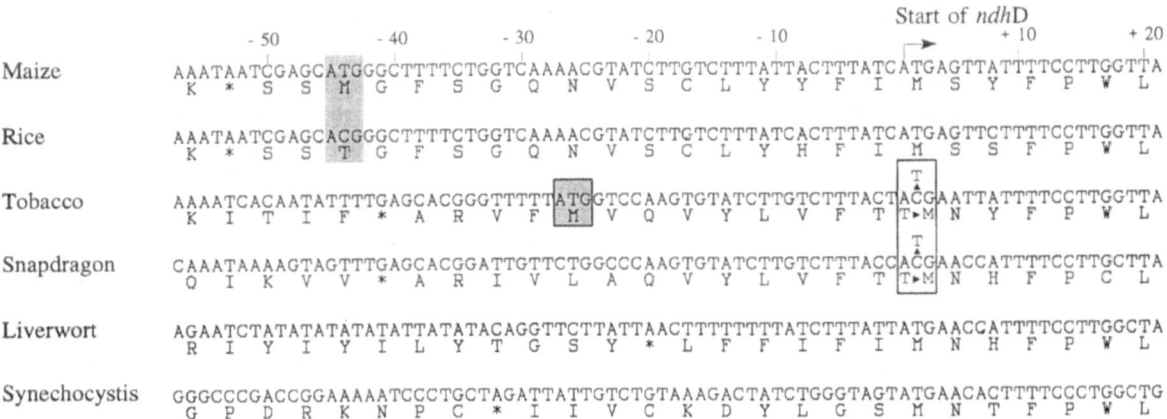

Figure 3. Alignment of the nucleotide sequences flanking the translational start of *ndhD* from maize, rice, tobacco, snapdragon, the liverwort *Marchantia polymorpha* and the cyanobacterium *Synechocystis* sp. PCC 6803. The amino acid sequence in phase with the *ndhD* reading frame is shown below each nucleotide sequence. The ACG codons in tobacco and spinach (nucleotide position +1 to +3) converted to AUG translational start codons as well as the originally proposed ATG start codon (nucleotide position −27 to −25) of tobacco *ndhD* are framed. The originally proposed *ndhD* ATG start codon of maize and the corresponding ACG codon of rice (nt −45 to −43) are shaded. For references of the aligned sequences, see [81].

the AUG codon by editing converts the two transcripts from a untranslatable form into a translatable message.

As a more general consequence, cryptic initiation codons 'activated' by editing have to be taken into consideration for the definition of reading frames encoded in chloroplast DNAs. This is exemplified in Fig. 3 for the translational start of *ndhD*-encoded transcripts for which at a first glance no common AUG codon at a conserved homologous position could be identified in the various plastome species. The apparently divergent N-terminal region of the *ndhD*-encoded peptide could, however, be unified to a single N-terminus by editing of the cryptic ACG initiation codons encoded in the plastomes of tobacco, spinach and snapdragon to a conserved AUG initiation codon. cDNA sequence analysis for these three species indeed showed that about half the transcripts contain AUG codons in the expected position [81]. This partial editing is in marked contrast to the virtually complete editing observed for the above mentioned *rpl2*- and *psbL*-encoded initiation codons. Whether the partial editing reflects simply exhaustion of the editing capacity (see also 'Use of the plastid transformation technique...') or whether it is of some physiological significance, remains to be determined. It is noteworthy in this connection that editing of the *psbL* initiation codon (together with editing of an internal codon of the cotranscribed *psbF* gene) is also reduced to a partial state in certain non-green tissues of spinach [10].

Editing of internal codons

After the detection of editing in the chloroplast genetic system for the *rpl2* and *psbL* start codons, the question arose whether editing is restricted to the rare cases in which ACG codons are converted to initiator codons or whether it also occurs at higher frequency within internal codons of chloroplast mRNAs. As a first step towards answering this question, alignments of several plastome-encoded peptides were screened for deviating positions within highly conserved regions. Subsequently, those positions were selected for which identity could be restored by C-to-U transitions within the respective codons. Several such candidate positions could be identified in the *ndhA*-encoded peptide of maize chloroplast and subsequent cDNA analyses allowed the experimental verification of four editing sites [73]. An alignment of amino acid sequences encoded by the regions surrounding the four maize editing sites with homologous sequences of other *ndhA* gene products and of mitochondrial *nad1* gene products is depicted in Fig. 4. At site I, for which homologous sequences are only present in the chloroplast peptides, editing creates a UUG leucine codon that corresponds to the leucine residue encoded in the plastome DNAs of the three other plant species. Thus, editing results in the conservation of this apparently functionally important residue. Similarly, the UUA leucine codons produced at editing sites II and III restore highly conserved leucine residues observed in other *ndhA* genes and in most of the homologous mitochondrial *nad1* genes. Editing of the UCA serine codons found at sites II and III in rice to UUA codons could be verified (unpublished work from the Freiburg laboratory). A similar situation was observed for the UCC serine codon at site IV which is present not only in rice but even in the more distantly related tobacco *ndhA* gene. Editing at this site to a UUC phenylalanine codon could be demonstrated for all of the three species maize, rice and tobacco.

In summary, the codon changes caused by the four editing events restore conserved amino acid residues which are probably essential for the structure/function of the *ndhA*-encoded peptides. A similar restoration of codons for conserved amino acid residues is evident for virtually all the editing events observed in other chloroplast transcripts. Therefore, editing may be regarded as a genetic repair mechanism acting at the RNA level against missense mutations introduced at the gene level and maintained during evolution for reasons which are not yet apparent.

The alignment in Fig. 4 demonstrates that in the liverwort *Marchantia*, the four amino acid residues restored by editing of the *ndhA* transcript in the other species are determined by the 'correct' codon already at the DNA level. Again the same situation is observed for virtually all the other editing events in chloroplast transcripts. It appears therefore likely that editing in the chloroplasts of the phylogenetically distant liverwort *Marchantia polymorpha* is minimal or perhaps even non-existent. This situation is paralleled by the apparent absence of editing in mitochondrial transcripts of this species [82].

Estimate of the editing frequency in the transcripts of the maize plastome

The availability of the complete nucleotide sequence of the maize plastome allowed a systematic computer-aided screening for putative editing sites using the *Marchantia* plastome as a reference system. This analysis resulted in the identification of more than 200 candidate sites. Whereas the majority of these putative sites could be experimentally disproved, altogether 25 functional C-to-U editing sites (including the ones in the *rpl2* and the *ndhA* transcripts already mentioned) within the transcripts of 13 different genes could be confirmed (Fig. 5) [75]. A more detailed analysis of maize *ndhG* transcripts leads to the identification of two additional, unexpected editing sites, one leading to a triplet not coding for an amino acid conserved in the

Marchantia plastome, and one which is located in the untranslated leader sequence (K. Neckermann, R.M. Maier and H. Kössel, manuscript in preparation; see section 'Editing of untranslated regions' below). Altogether, more than 50% of the maize plastome-encoded mRNA sequences, including all of the most probable editing sites, were analysed in the form of RT-PCR products. Complete cDNA sequence analysis would, of course, be necessary in order to identify all the editing events occurring in the transcripts of a given plastome. However, a rough estimate based on the alignments of all the encoded peptides (similar to the alignment shown in Fig. 4 for the *ndhA*-encoded peptides), taking into account also the bias for certain codon transitions (see below), supports the conclusion that the 27 sites identified in maize must be very close to the total number of sites encoded in this plastome. So far no successful attempts have been reported which identify editing events within chloroplast rRNAs or tRNAs, even though editable mismatches can be recognized in some of the secondary structures. For instance, the tRNA^Cys from spinach and tobacco chloroplasts contains an AC base pair in the T-stem [56, 101] which, by C-to-U editing, would be converted to an AU pair present in the tRNA^Cys of other chloroplast species. cDNA sequence analysis failed, however, to confirm a C-to-U transition (P. Zeltz and H. Kössel, unpublished).

The 14 different genes whose transcripts are subject to editing, also show that a large proportion of the altogether 70 different protein-coding genes and several additional conserved open reading frames of the maize plastome do not require editing events for maturation of their mRNAs. The comparatively low number of both editing sites and edited transcripts is in marked contrast to the much higher editing frequencies of plant mitochondrial transcripts and to the estimate that probably all plant mitochondrial mRNAs – albeit to a varying degree – are subject to editing (see section 'Occurrence and extent of mRNA editing in plant mitochondria').

General occurrence and characteristics of chloroplast editing sites

Without any exception, only C-to-U editing events have so far been observed in chloroplast transcripts. The total number of editing sites encoded in the plastomes of rice [53], tobacco [92] and black pine [102] appears to be within the same order of magnitude as in maize. This inference remains, however,

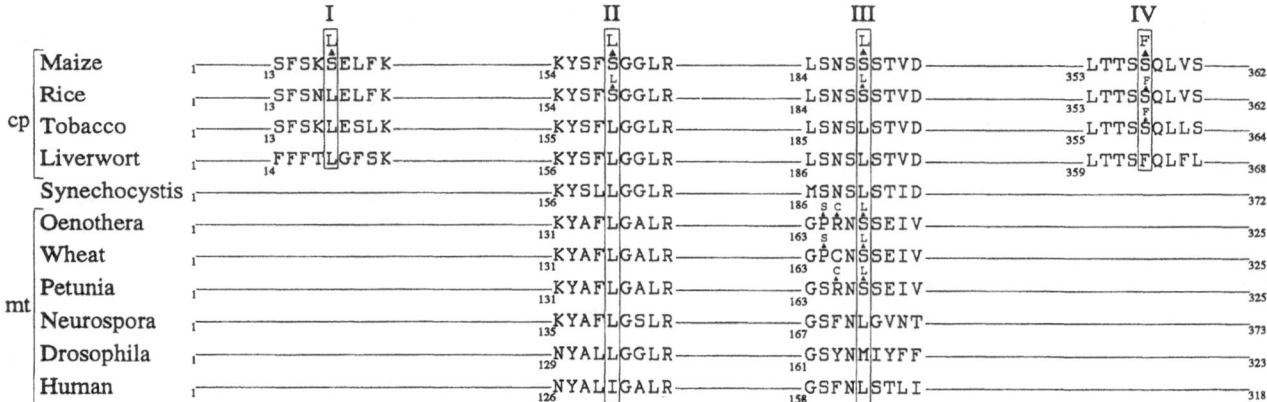

Figure 4. Alignment of the amino acid sequences encoded by the maize *ndhA* regions containing the four editing positions (I to IV) with homologous sequences from rice, tobacco, the liverwort *Marchantia polymorpha* and the cyanobacterium *Synechocystis* sp. PCC 6803. Amino acid sequences encoded by the homologous regions of the mitochondrial *nad1* genes from *Oenothera*, wheat, *Petunia*, Neurospora, Drosophila and humans are also included. Amino acid substitutions resulting from editing are marked by arrowheads. For references of the aligned sequences see [73].

to be verified by more exhaustive cDNA sequence analyses, as only a limited number of editing sites have been identified in various transcripts from barley (*rpoB* [110], *rpl2* [49], *ndhB* [32]), rice (*ndhB* [32]), bell pepper (*psbL* [64]), tobacco (*psbL* [62], *petB* [54], *ndhB* [32] *ndhD* [81]) and spinach (*psbF* [10], *psbL* [10] *ndhD* [81]). Some of these sites, such as the ones encoded in the *psbF*, *psbL* and *ndhD* genes, are not conserved in maize and other monocots. More than 20 editing events have been identified within 8 different transcripts of the black pine plastome (M. Sugiura, personal communication). Most of these sites are 'new' ones not previously found in homologous transcripts of angiosperms. This adds to the impression that, with the exception of the liverwort *Marchantia*, editing events occur in a wide variety of land plant species.

An overview of the genes encoding transcripts requiring editing and of the number of editing sites contained in these transcripts is presented in Fig. 5. It shows that the transcripts of at least 16 genes (14 genes in maize, 2 additional genes in dicots) are subject to editing and that the number of editing sites within the individual mRNAs varies between a single site to as many as 9 sites in the *ndhB*-encoded transcript of barley and tobacco chloroplasts [32]. Except for rRNAs and tRNAs for which no editing sites have been reported to date, editing is not restricted to the transcripts of certain protein-coding gene classes and occurs in transcripts of both intron-containing and intronless genes. Compared to the number of altogether 18 introns contained in 10 mRNA and 6 tRNA coding genes, the

number of editing sites encoded in the maize plastome is at least 27. In view of the overall number of codons in the maize plastome the number of codons containing editing sites is low (0.13%). Thus, in quantitative terms, editing causes only fine-tuning of the genetic information defined by a chloroplast DNA sequence. However, in qualitative terms, editing still has to be regarded as an essential prerequisite for chloroplast gene expression. This is evident not only from the few cases in which initiation codons have to be created by editing [55, 62, 81] but also from the fact that a lack of editing at a single position can result in a mutant phenotype [11] (see also section 'Use of the plastid transformation technique...' above).

The codon transitions observed for 33 independent editing events (in several species but mostly from maize transcripts) are summarized in Table 3. Given the one transition occurring in first codon positions and none in third positions, a strong bias exists for second codon positions. Furthermore, a strong preference for certain codon transitions, with UCA (Ser) to UUA (Leu) being by far the most frequent, can be recognized, while other possible transitions such as AC(U,C,A) to AU(U,C,A) and GCN to GUN are completely absent (or remain to be detected at very low frequency). It is not clear yet whether these preferences reflect mechanistic constraints, such as inhibitory effects of purines at 5'-neighbouring positions and/or stimulatory effects at 3'-neighbouring positions. Alternatively, they may be connected to the evolutionary pathways which by conferring selective advantages allow gain, maintenance

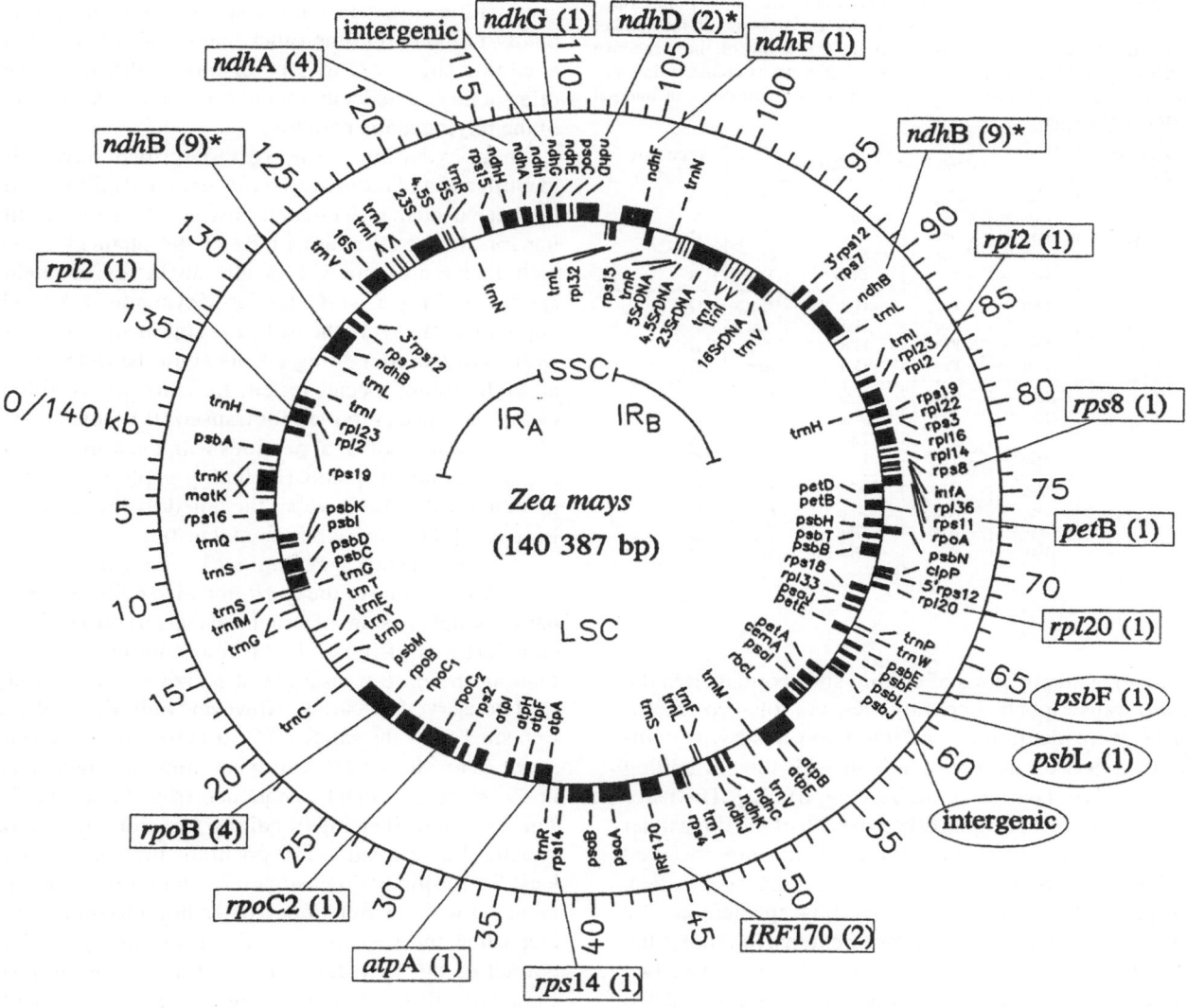

Figure 5. Summary of editing sites identified in transcripts encoded by higher-plant plastomes using the gene organization of the maize chloroplast genome as a reference system. The inverted repeat regions IR$_A$ and IR$_B$, respectively, divide the rest of the circular genome into a large (LSC) and a small (SSC) single-copy region. Genes outside the circle are transcribed clockwise. Genes and intergenic regions in which editing sites have been detected in maize and other species are denoted by enlarged gene symbols and by framing. Those detected only in dicot species are marked by circling. The numbers within the brackets following the gene symbols indicate the number of editing sites observed for the respective gene. Numbers marked by asterisks refer to the maximum number of editing sites in those cases where interspecific variation was detected.

or loss of only a subfraction of possible editing events. However, as outlined below, the codon preferences do not reflect a dependence of the editing process on the translation machinery.

Phylogenetic aspects of editing in chloroplasts

The relative abundance of editing events in the *ndhB* encoded transcript for which six sites have been reported in maize [74] renders this transcript as a suitable candidate for studying the structural and functional conservation of chloroplast editing sites within the plant kingdom. As a first step in this direction, the

Table 3. Codon transitions caused by editing of chloroplast transcripts. The arrows show the direction of the codon transitions with the number above the arrow indicating the observed frequency of the respective transition. Three threonine to methionine transitions create an initiator codon (Start), the fourth (Int.) leads to the formation of an internal methionine codon.

First Base of Codon	Second Base of Codon				Third Base of Codon
	U	C	A	G	
U	Phe ←1x Ser		→Tyr	Cys	U
	Phe ←3x Ser		Tyr	Cys	C
	Leu ←13x Ser		1x Stop	Stop	A
	Leu ←4x Ser		Stop	Trp	G
C	Leu ←1x Pro		—His	Arg	U
	Leu Pro		His	Arg	C
	Leu ←5x Pro		Gln	Arg	A
	Leu ←1x Pro		Gln	Arg	G
A	Ile Thr		Asn	Ser	U
	Ile Thr		Asn	Ser	C
	Ile Thr		Lys	Arg	A
	Met Start←3x/Int.←1x Thr		Lys	Arg	G
G	Val Ala		Asp	Gly	U
	Val Ala		Asp	Gly	C
	Val Ala		Glu	Gly	A
	Val Ala		Glu	Gly	G

numbers and positions of editing sites contained in the *ndhB* transcripts from barley, rice and tobacco chloroplasts were identified in order to allow a pairwise comparison as well as a comparison with the six editing sites analysed earlier in maize (Fig. 6) [32]. The latter sites are found to be conserved in each of the three other species, whereas several additional sites are observed in barley (3 sites), rice (2 sites) and tobacco (3 sites). Surprisingly, the additional sites show species-specific divergence, which is even more extensive among the closely related graminean species. For instance, two of the sites are shared between the distantly related species barley and tobacco but, as already pointed out, none of them with the more closely related maize. A similar situation has been reported for the single editing site of the *petB* transcript which is present in maize [31] and tobacco [54] but not in rice. As a consequence of these observations, the presence or absence of individual editing sites per se cannot be utilized as a criterion for evaluating phylogenetic relationships.

The apparent phylogenetic paradox of editing sites shared between more distantly related species but showing divergence between more closely related species can, however, be reconciled by postulating that evolutionary divergence of editing sites is caused by loss of pre-existing sites rather than by acquisition of new sites. Such acquisition would require convergent evolution by independent formation of identical new sites within distantly related species, which appears highly unlikely. On the other hand, loss of individual editing sites by C-to-T reversions at the gene level offers a mechanistically much more plausible solution of the phylogenetic paradox.

In an extension of the work described above, the identification of *ndhB*- and *ndhA*-encoded editing sites has been used in order to demonstrate chloroplast editing for almost all major phyla of the plant kingdom including representatives of all subclasses of angiosperms (R. Freyer, M.-C. Kiefer-Meyer and H. Kössel, unpublished). With the sole exception of the liverwort *Marchantia*, editing events could be detected in all higher-plant species tested. In contrast, no editing events have been reported for transcripts of algal plastids. Alignments of algal peptides with the homologous peptides of higher plants (including peptides encoded by edited mRNAs) have so far failed to reveal promising candidates for experimental testing.

An interesting evolutionary intermediate is observed with one of the *rpoB*-encoded editing sites in barley which, in spite of structural conservation, shows complete loss of function. All four editing sites identified in the *rpoB* transcripts of maize are structurally well conserved in barley. However, only three of the four sites show the expected C-to-U transitions whereas the C residue of the fourth site remains unchanged in the respective cDNA sequence from barley [110]. This apparent silencing of editing function, in spite of structural conservation (19 positions upstream and at least 24 positions downstream of the editing site are identical with the functional maize homologue), could mean that the sequences flanking an editing site do not act as the only determinants for the editing process. This interpretation is, however, at variance with the observation of certain sets of consensus sequences in the immediate vicinity of chloroplast editing sites. Some of these consensus sequences are even shared with mitochondrial editing sites (Fig. 7) [74]. A more likely interpretation of the functional loss is, therefore, that the editing information is indeed contributed by the sequences flanking individual sites, but that the still unidentified substrate recognition device (e.g. a guide RNA) of the editing machinery is lost or functionally impaired whenever a structurally conserved editing site loses its function. A similar functional silencing of two structurally conserved editing sites has been observed more recently in the *ndhB* transcript from two gymnosperm species (M.-C. Kiefer-Meyer, R. Freyer and H. Kössel, unpublished).

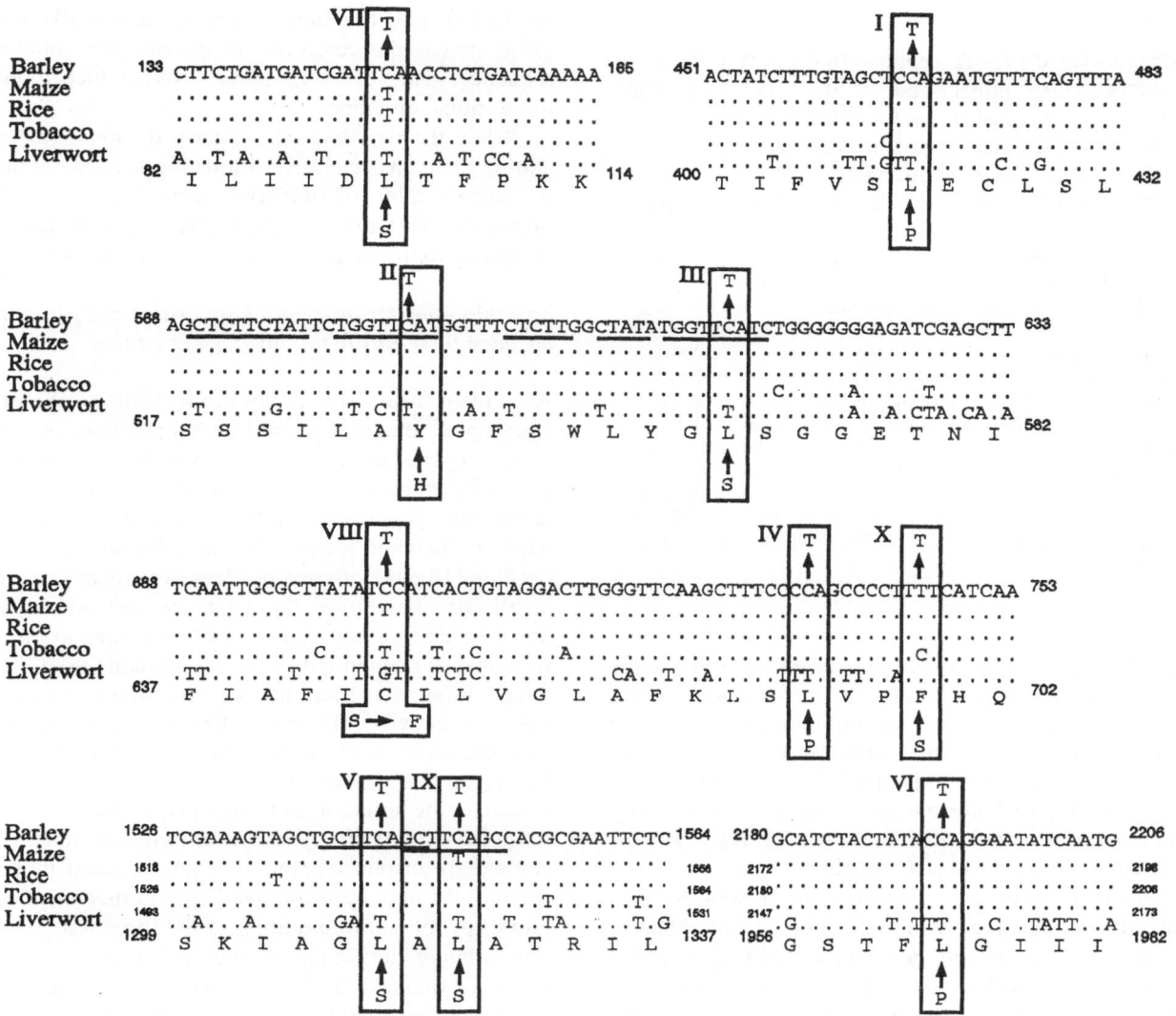

Figure 6. Alignment of the amino acid sequences encoded by the *ndhB* regions containing editing positions from barley, maize, rice, tobacco and the liverwort *Marchantia polymorpha*. The amino acid sequence below the aligned nucleotide sequences is derived from the liverwort *ndhB* gene. Editing positions functional in at least one of the listed species are marked by framing. Repetitive sequences surrounding editing sites II/III and V/IX, respectively, are marked by underlining.

The existence of at least two non-functional but structurally highly conserved editing sites shows also that the assignment of editing sites for a different species cannot rely solely on DNA sequence alignments and on the presence of sites already identified in other species. Instead, the experimental verification by cDNA sequence analysis is an absolute necessity even for homologous sites.

Editing of untranslated regions

As the two criteria used for screening of editing sites are amino acid alignments and codon transition frequencies, the estimate of the total number of editing sites encoded in the maize plastome neglects the possible occurrence of editing events within untranslated regions. In the absence of other criteria, the observation of editing events within untranslated regions depends on their fortuitous detection. No editing could

358

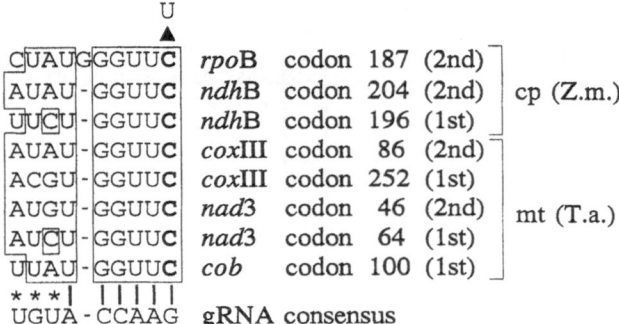

C̲U̲A̲U̲G GGUUC	*rpo*B codon 187 (2nd)	⎤
A̲U̲A̲U̲-GGUUC	*ndh*B codon 204 (2nd)	cp (Z.m.)
U̲U̲C̲U̲-GGUUC	*ndh*B codon 196 (1st)	⎦
A̲U̲A̲U̲-GGUUC	*cox*III codon 86 (2nd)	⎤
A̲C̲G̲U̲-GGUUC	*cox*III codon 252 (1st)	
A̲U̲G̲U̲-GGUUC	*nad*3 codon 46 (2nd)	mt (T.a.)
A̲U̲C̲U̲-GGUUC	*nad*3 codon 64 (1st)	
U̲U̲A̲U̲-GGUUC	*cob* codon 100 (1st)	⎦

```
* * * |   | | | | |
UGUA - CCAAG   gRNA consensus
```

Figure 7. Sequence similarities around editing sites of chloroplasts and mitochondria. The position where C residues are edited to U residues is marked by an arrowhead. The position of the edited nucleotide (1st or 2nd) within the modified codon is also given. Nucleotide sequences which could be involved in base pairing with a putative guide RNA are boxed. The occurrence of G-U base pairs is indicated by asterisks. The codons 196 and 204 of the maize (Z.m.) *ndhB* transcript coincides with the editing positions II and III shown in Fig. 6. For references of the wheat (T.a.) mitochondrial sequences see [74].

be detected in the intron sequences tested so far, such as the introns contained in the *rpl2*, *rps16*, *ndhA* transcripts and in the intron-containing reading frame 170 (IRF170, *ycf3*) from maize (unpublished work of the Freiburg laboratory). This analysis, being limited to only five out of 18 intron sequences does, however, not yet permit the conclusion that intron editing is generally absent in chloroplast transcripts.

RT-PCR-mediated amplification products frequently include part of the 5'- and 3'-untranslated regions or complete intergenic regions of polycistronic transcripts. In such a case, an editing site positioned 10 nucleotides upstream of the start codon of the *ndhG* transcript from maize and rice was detected (K. Neckermann, R.M. Maier and H. Kössel, manuscript in preparation). It appears that the C-to-U transition at this site destabilizes a secondary structure which in turn may facilitate ribosome binding during the translation process. It remains to be seen whether this inferred effect on translational efficiency can be confirmed experimentally.

Likewise, in the *psbL/psbJ* intergenic region from tobacco and *Ginkgo biloba*, an editing site could be identified (J. Kudla and H. Kössel, unpublished). The site (located 32 nt upstream of the *psbJ* start codon) in its unedited form would confer stability on a six base pair hairpin structure which comprises the *psbJ* ribosome-binding sequence within its unpaired loop. Transition to the edited form leads to a destabiliz-

ing GU pair in the stem which in turn is likely to cause increased accessibility of the ribosome-binding sequence. Thus, editing again could induce higher translational efficiency.

While the two examples clearly demonstrate that editing sites can occur also within untranslated regions, the identification of only two such sites leaves the impression that their frequency is relatively low as compared to the frequency within coding regions.

The chloroplast translational apparatus and plastome encoded proteins are not involved in editing

As evident from Table 3, chloroplast editing shows a strong preference for second codon positions and for certain types of codon transitions which appeared suggestive of an involvement of the translational apparatus in the editing process. The barley mutant *albostrians* which in its white sectors contains ribosome-deficient plastids [18, 44] offered in ideal system to test this possibility. Therefore, the existence and functionality of several editing sites identified earlier in maize were investigated in wild-type and mutant barley. The three editing sites identified in wild type *rpoB* were, however, equally effective in the transcripts isolated from the white sectors of the *albostrians* mutant [110]. Editing in the *albostrians* plastids could also be confirmed for the initiation codon of the *rpl2*-encoded transcript [49]. Interestingly, the latter transcript remained completely unspliced in the ribosome-deficient mutant whereas the major portion (60–70%) of this unspliced transcript still showed editing of the initiation codon. The splicing deficiency of the *rpl2* transcript does not allow translation into the encoded ribosomal protein L2 which in turn is probably the reason (or at least one reason) why functional ribosomes cannot be formed in the plastids of the *albostrians* mutant. This also excludes the possibility that a small number of ribosomes undetectable with standard techniques suffices for a ribosome-dependent editing process. It appears therefore safe to conclude that plastid editing, despite its preference for certain codon positions and codon transitions, does not depend on plastid ribosomes and, therefore, is not linked to codon recognition during translation. This conclusion is confirmed by the more recent identification of editing sites within untranslated regions (see above) although the proximity of the two sites to translational initiation sites would not entirely exclude ribosomal contacts.

As a further consequence of unimpaired or, in the *rpl2* transcript, slightly reduced editing in the *albostri*-

ans plastids, the involvement of plastome-encoded peptides in the editing machinery can be excluded. It appears therefore that all the peptide components which are probably necessary for plastid editing are encoded by nuclear genes and, after translation in the cytoplasm, are imported into the organelle.

Editing in chloroplasts is an early RNA processing step independent of splicing and cleavage of polycistronic transcripts

As described in the preceding sections codon changes caused by editing events restore conserved amino acid residues which are probably essential for the function of the respective peptides. Consequently, in case of delayed events translation of unedited mRNAs would lead to aberrant peptides.

Therefore, processing intermediates of several chloroplast transcripts with respect to their editing status were investigated.

First, an editing site within the *petB* transcript of maize was studied. cDNAs still containing the intron of the *petB* message and/or still tethered to the downstream *petD* sequences (also harbouring an intron in its primary form) were tested. Complete C-to-T transitions were observed in all cDNAs, irrespective of their origin from unspliced, partially or fully spliced and/or dicistronic or monocistronic states. Therefore, editing of the *petB* message must be regarded as an early step in mRNA processing which precedes both splicing and cleavage to monocistronic mRNAs and which must, therefore, be independent of the latter two steps [31].

This conclusion could be further substantiated by a similar study with transcripts encoded by *ycf3* from maize which is split by two introns. Again editing could be shown to precede both splicing and cleavage to monocistronic mRNA [87]. Thus editing as an early processing step demonstrated for the *petB* and *ycf3* transcripts from maize is likely to act as a safeguard against the production of aberrant peptides by premature translation of unedited transcripts.

The primary transcript of the *rpl2* gene contains an intron and is part of a large polycistronic mRNA. A cDNA population derived from an unspliced maize *rpl2* transcript still linked to the proximal *rpl23* sequences shows only incomplete (40–60%) editing of the *rpl2* initiation codon [31]. This observation contrasts with the complete editing observed for the unspliced dicistronic transcripts of the *petB/petD* and the *rps4/ycf3* genes. It appears that in this case the three processing steps necessary to obtain mature monocis-

tronic *rpl2* message are more balanced with respect to their efficiency and sequential occurrence. On the other hand, it should be noted that, contrary to the editing of the internal sites of the *petB*- and *ycf3*-encoded transcripts, a rapid editing of the *rpl2* transcript is not necessary to avoid translation of unedited mRNA into an aberrant protein, as the unedited *rpl2* transcript is likely to be untranslatable.

A high molecular weight transcriptional active complex (TAC) consisting of DNA, nascent RNA and protein can be isolated from various plastids [45, 65, 86]. The TAC was initially defined through its ability to elongate *in vivo* initiated transcripts *in vitro* (for review see [57]). By using various TAC preparations from barley the splicing and editing states of TAC-associated mRNAs were tested. Most of the investigated mRNAs were found to be already fully spliced and edited (unpublished work from the Freiburg laboratory). This implicates that editing and splicing activities are associated or even integral parts of the TAC. At the same time this provides further evidence that editing and also splicing are very early processes during plastid mRNA maturation which appear to take place in a concerted manner at the site of transcription.

Use of the plastid transformation technique as a tool for studying the editing process

The development of a biolistic transformation system for tobacco chloroplasts provides a powerful tool for studying chloroplast gene expression *in vivo* [1, 96]. The usefulness of this system for the analysis of chloroplast editing events has been documented by several recent reports [11, 12, 21].

In one study, an editing site contained in the *psbF* transcript of spinach but not of tobacco was introduced into the tobacco plastome by replacement of the endogenous *psbF* gene [11]. The *psbF* genes of these two dicot species are highly conserved showing 100% identity of the 39 encoded amino acids and deviating at the nucleotide level at only 8 silent positions. The generation of transplastomic tobacco containing the heterologous spinach *psbF* gene should therefore yield information on whether the editing capacity present in spinach would also exist in tobacco in spite of the absence of an editable site in its own *psbF* transcript. It could be demonstrated that the heterologous site remains unmodified in transplastomic tobacco plants thus indicating that the editing capacity for individual sites is not necessarily maintained even among two dicot species. The lack of editing which causes in the mutant a phenylalanine-to-serine substitution within a transmembrane helix of the *psbF*-encoded peptide (*β*-subunit of cytochrome b559) is associated with a phenotype (slower growth, lowered chlorophyll content, high chlorophyll fluorescence) characteristic of photosynthetic mutants. This finding confirms that editing of the *psbF* transcript is an essential processing step for protein function and thus provides the first direct proof for the biological significance of plant organellar RNA editing [11].

In a similar study, a test system for the analysis of *cis*-acting determinants required for creation of the initiation codon of the tobacco *psbL* transcript was developed [21]. Transformants were produced by using constructs in which a 98 nt fragment spanning the *psbL* editing site (63 positions upstream including 40 nt of the *psbF* 3′ terminus, and 34 positions downstream belonging to the *psbL*-coding region) was translationally fused to bacterial reporter genes encoding spectinomycin (*aadA*) or kananycin (*kan*) resistances. Expression of the reporter genes thus becomes dependent on the *psbL* initiation codon the formation of which in turn is dependent on editing. It could be demonstrated that both the reporter genes are expressed in the plastome transformants due to creation of the *psbL* start codon by editing. cDNA sequence analysis showed that the editing efficiency is only about 70% but obviously allows sufficient translation of both the reporter gene transcripts. Editing of the chimeric gene transcripts, though not complete, indicates, however, that the 98 nt segment comprising the *psbL* editing site contains most of the *cis* information necessary for editing. It should now be possible to define the *cis*-acting sequence determinants more precisely by systematic alterations of the constructs, such as size reductions (from both sides of the editing site), internal deletions and point mutations within the 98 nt fragment.

Expression of the chimeric gene leads to a small but significant decrease (ca. 10%) in the editing efficiency of the endogenous *psbL* mRNA which is virtually fully edited in wild-type tobacco. As a control, several editing sites of other transcripts (*rpoB* and *ndhB*) were tested in the transplastomic plants. Their editing efficiencies remained, however, unchanged as compared with wild-type tobacco. It is therefore concluded that the reduced efficiency of both the endogenous *psbL* transcript and the chimeric *psbL* reporter gene transcript reflects limiting concentrations of *psbL*-specific factor(s) which apparently are titrated out by the increased levels of editable *psbL* sequences [21]. The existence of structurally conserved but function-

ally silent editing sites occurring naturally (see section 'Phylogenetic aspects of editing in chloroplasts') or by introduction of heterologous sites (see first part of this section) was already suggestive of the function of site-specific factors in the editing process. The observations from the experiments with the chimeric *psbL* transcripts further substantiate the existence of such *trans*-acting factors, which, however, remain to be identified and characterized by *in vitro* studies.

In an attempt to identify sequences complementary to editing sites which in the form of small RNAs would guide the editing of individual sites (as is well established for the guide RNAs of the trypanosomal editing system; for review see [93]) a screening of the tobacco chloroplast DNA sequence was performed [12]. This resulted in the detection of a 14 nucleotide sequence motif that is complementary to the *psbL* editing site. This 14-mer sequence (located on the antisense strand of the *psaB*-coding region) contains an A nucleotide at the position opposite to the C nucleotide to be edited and thus could be a possible guide sequence for the conversion to a U nucleotide. In order to test the putative function of this sequence as a guide RNA *in vivo*, the critical A nucleotide was changed to a G by plastid transformation which should abolish *psbL* editing if the 14-mer sequence would function as a guiding sequence. Editing of the *psbL* transcript turned out, however, not to be impaired in the transplastomic mutant. From this it was concluded that the 14-mer sequence does not provide guiding information for editing of the *psbL* transcript [12]. Computer-aided screening for guide sequences complementary to editing sites was performed for the plastome of maize (K. Neckermann, R.M. Maier and H. Kössel, unpublished) which, however, did not give hints as to the existence of promising candidate sequences. In view of this, the concept of plastome-encoded guide RNAs complementary to the sequences surrounding an editing site is difficult to maintain. On the other hand, given the sequence variability around editing sites, it is also difficult to imagine that nuclear-encoded proteins alone without participation of guiding RNA (or DNA) templates could provide the basis for the specificity of the editing process. Therefore, more remote possibilities such as scrambled guide RNAs (either composed of several RNAs, or *trans*-spliced from several RNA precursors), or guide RNAs functioning with only limited complementarities to their targets, or even nuclear-encoded guide RNAs imported into chloroplasts have to be considered for further investigations.

Comparative aspects of plant mitochondrial and chloroplast RNA editing

The editing processes of the two plant organelles share several characteristics. First of all, only C-to-U transitions have been identified in chloroplast transcripts and the same transitions are also observed in the vast majority of mitochondrial editing events with only a few U-to-C transitions being the exceptions. Furthermore, the strong preference for second codon positions and for certain codon transitions (as depicted in Table 3 for chloroplast editing) is valid for both organelles. However, with more than 1000 independent editing sites identified in mitochondrial transcripts, the occurrence of rare transitions so far not observed for the 33 chloroplast editing sites (Table 3) could also be documented. It remains to be seen from the analyses of additional chloroplast transcripts whether the transitions still missing do in fact occur with low frequency or whether they are completely avoided by the chloroplast editing apparatus. Several new transition types such as the GCA (Ala)-to-GUA (Val) transition, the CGG (Arg)-to-UGG (Trp) transition and the CAA (Gln)-to-UAA (stop) transition could be observed in black pine chloroplasts (M. Sugiura, personal communication). This adds to the impression that the list of actually occurring transitions in chloroplasts may finally match the mitochondrial list if a sufficient number of chloroplast editing events becomes analysed during further work. As already mentioned, a difference between the two organellar editing systems still exists with respect to editing of tRNAs and rRNAs, which has so far been reported only for plant mitochondria but not for chloroplasts.

The homology between the chloroplast *ndh* genes and the mitochondrial *nad* genes provided the basis for the identification of editing events occurring at homologous positions of the respective transcripts from the organelles. As depicted in Fig. 4, one such example of an editing site shared between the two organelles is the editing site III of the chloroplast *ndhA* transcript. The same transition from a serine to a leucine codon is observed in the homologous position of the *nad1*-encoded messages from mitochondria of at least three plant species [73]. Similar situations have been found for two of the *ndhB*-encoded sites [74], although in these cases – due to editing of different synonymous serine codons – phenylalanine codons are produced in the mitochondrial transcript and leucine codons in the chloroplast transcript.

Certain sets of consensus sequences shared between editing sites encoded in either mitochondrial or chloroplast transcripts had been noted soon after the detection of the two editing systems [42, 74, 110] (Fig. 6). In addition to intraorganellar sets of consensus sequences even interorganellar consensus sequences could be deduced [74]. As depicted in Fig. 7, such consensus sequences do not only occur within transcripts of homologous genes (such as the *ndhA/nad1* and the *ndhB/nad2* genes of the two organelles) but also between transcripts of non-homologous genes, as for instance the mitochondrial *cox3*, *nad3* and *cob* genes and the chloroplast *rpoB* and *ndhB* genes. As also outlined in Fig. 7, the sequence similarities are suggestive of the existence of putative guide sequences which would be complementary to the respective sets of editing sites. The existence of A/G and U/C substitutions observed at several positions of the shared sequences, which would imply UG base pairing between the editing sites and the hypothetical guide RNAs lends additional support to this suggestion [42, 74]. However, as already pointed out above, attempts to isolate and characterise guide RNAs or to screen for putative guide sequences on the tobacco and maize plastomes have so far not been successful. Nevertheless, the existence of common sequence motifs shared between homologous and even non-homologous transcripts of the two plant organelles are suggestive of common components and/or mechanistic steps in the two editing systems. In this connection, it is interesting that a nuclear-encoded peptide synthesized in the cytoplasm can be targeted with comparable efficiency to either of the two plant organelles [26].

Finally, the apparent absence of editing in both organelles from the liverwort *Marchantia* is also suggestive of a linkage between the chloroplast and mitochondrial editing systems in all the other plants by common components which may have been lost in an ancestor of *Marchantia*.

From all these similarities it appeared not unlikely that mitochondrial sequences could even be targets for the editing machinery of chloroplasts and *vice versa*. That this is not the case could be demonstrated by two recent *in vivo* studies [99, 111]. In order to test the activities of mitochondrial editing sites in chloroplasts, transplastomic tobacco plants were produced that contained chimeric constructs in which the second exon of a *Petunia* mitochondrial *cox2* gene was under control of a strong or a weak chloroplast promoter. When editing of the *cox2* transcripts in the transgenic chloroplasts was examined, no editing of any of the seven sites active in *Petunia* mitochondria could be detected nor was there any novel editing at other sites [99]. Absence of editing was also observed in the low level transcripts produced by a weak promoter. This excludes the possibility that the failure to observe editing of the high-level transcript is merely caused by an exhaustive overproduction of the editing substrate. However, in this study no evidence has been provided that the editing sites active in *Petunia* mitochondria would maintain their functionality in tobacco mitochondria. Therefore, the failure to observe editing of the *Petunia* mitochondrial transcript in tobacco chloroplasts might also be due to a species barrier similar to the heterologous situation observed with the *psbF* transcript which is edited in spinach chloroplasts but not edited in tobacco chloroplasts [11].

No transformation system for plant mitochondria is available in order to test the activities of chloroplast editing sites in mitochondria. But, fortunately, Nature has already performed the appropriate transformation steps during evolution by transferring a variety of chloroplast DNA sequences into the mitochondrial genome [69, 79, 97]. The *rpoB* sequences from rice chloroplasts were selected for an investigation as they encode within a short region three putative editing sites and as this region after its transfer to the rice mitochondrial genome has diverged by only very few (but diagnostic) nucleotide substitutions. Examination of cDNAs derived from either chloroplast or mitochondrial *rpoB* transcripts did, however, not reveal any editing of the three sites (or of novel sites) in mitochondria, whereas complete editing was evident in the chloroplast-derived cDNA sequence [111]. The results of the two reciprocal studies indicate that the RNA editing mechanisms of chloroplasts and mitochondria in spite of the similarities described above are not identical but must have at least some organelle-specific components.

Acknowledgements

This work was supported by grants from the Deutsche Forschungsgemeinschaft (SFB 206 and 388) and the Fonds der Chemischen Industrie for the Freiburg laboratory, by the Centre National de la Recherche Scientifique for the Strasbourg laboratory. Both laboratories acknowledge the support of the Human Frontier Science Program Organisation for grant RG-437/94 M. We thank Ralph Bock and Dr Gabor Igloi for critical reading of the manuscript.

References

1. Allison LA, Maliga P: Light-responsive and transcription-enhancing elements regulate the plastid *psbD* core promoter. EMBO J 14: 3721–3730 (1995).
2. Araya A, Domec C, Bégu D, Litvak S: An *in vitro* system for the editing of ATP synthase subunit-9 messenger RNA using wheat mitochondrial extracts. Proc Natl Acad Sci USA 89: 1040–1044 (1992).
3. Bégu D, Graves PV, Domec C, Arselin G, Litvak S, Araya A: RNA editing of wheat mitochondrial ATP synthase subunit 9: direct protein and cDNA sequencing. Plant Cell 2: 1283–1290 (1990).
4. Benne R: RNA Editing. The Alteration of Protein Coding Sequences of RNA. Ellis Horwood, Chichester (1993).
5. Benne R, Van den Burg J, Brakenhoff J, Sloof P, Van Boom JH, Tromp MC: Major transcript of the frameshift *coxII* from trypanosome mitochondria contains four nucleotides that are not encoded in the DNA. Cell 46: 819–826 (1986).
6. Binder S, Marchfelder A, Brennicke A: RNA editing of tRNA(Phe) and tRNA(Cys) in mitochondria of *Oenothera berteriana* is initiated in precursor molecules. Mol Gen Genet 244: 67–74 (1994).
7. Binder S, Marchfelder A, Brennicke A, Wissinger B: RNA editing in trans-splicing intron sequences of *nad2* messenger RNAs in *Oenothera* mitochondria. J Biol Chem 267: 7615–7623 (1992).
8. Bland MM, Levings CS III, Matzinger DF: The tobacco mitochondrial ATPase subunit 9 gene is closely linked to an open reading frame for a ribosomal protein. Mol Gen Genet 204: 8–16 (1986).
9. Blum B, Balakara N, Simpson L: A model for RNA editing in kinetoplastid mitochondria: small 'guide RNA' molecules transcribed from maxicircle DNA provide the edited sequence information. Cell 60: 189–198 (1990).
10. Bock R, Hagemann R, Kössel H, Kudla J: Tissue-specific and stage-specific modulation of RNA editing of the *psbF* and *psbL* transcript from spinach plastids – a new regulatory mechanism? Mol Gen Genet 240: 238–244 (1993).
11. Bock R, Kössel H, Maliga P: Introduction of a heterologous editing site into the tobacco plastid genome: the lack of RNA editing leads to a mutant phenotype. EMBO J 13: 4623–4628 (1994).
12. Bock R, Maliga P: *In vivo* testing of a tobacco plastid DNA segment for guide RNA function in *psbL* editing. Mol Gen Genet 247: 439–443 (1995).
13. Boer PH, McIntosh JE, Gray MW, Bonen L: The wheat mitochondrial gene for apocytochrome b: absence of a prokaryotic ribosome binding site. Nucl Acids Res 13: 2281–2292 (1985).
14. Bonen L, Williams K, Bird S, Wood C: The NADH dehydrogenase subunit 7 gene is interrupted by four group II introns in the wheat mitochondrial genome. Mol Gen Genet 244: 81–89 (1994).
15. Bonnard G, Grienenberger JM: A gene proposed to encode a transmembrane domain of an ABC transporter is expressed in wheat mitochondria. Mol Gen Genet 246: 81–99 (1995).
16. Bonnard G, Gualberto JM, Lamattina L, Grienenberger JM: RNA editing in plant mitochondria. CRC Crit. Rev. Plant Sci 10: 503–524 (1992).
17. Börner GV, Morl M, Wissinger B, Brennicke A, Schmelzer C: RNA editing of a group II intron in *Oenothera* as a prerequisite for splicing. Mol Gen Genet 246: 739–744 (1995).

18. Börner T, Schumann B, Hageman R: Biochemical studies on a plastid ribosome-deficient mutant of *Hordeum vulgare*. In Bücher T, Neupert W, Sebald S , Werner S (eds), Genetics and biogenesis of chloroplasts and mitochondria, pp. 41–48. Elsevier, Amsterdam (1976).

19. Boyen C, Leblanc C, Bonnard G, Grienenberger JM, Kloareg B: Nucleotide sequence of the *cox3* gene from *Chondrus crispus*: evidence that UGA encodes tryptophan and evolutionary implications. Nucl Acids Res 22: 1400–1403 (1994).

20. Chapdelaine Y, Bonen L: The wheat mitochondrial gene for subunit-I of the NADH dehydrogenase complex – A transsplicing model for this gene-in-pieces. Cell 65: 465–472 (1991).

21. Chaudhuri S, Carrer H, Maliga P: Site-specific factor involved in the editing of the *psbL* mRNA in tobacco plastids. EMBO J 14: 2951–2957 (1995).

22. Chen SH, Habib G, Yang CY, Gu ZW, Lee BR, Weng SA, Silberman SR, Cai SJ, Deslypere JP, Rosseneu M, Gotto AM, Li WH, Chan L: Apolipoprotein B-48 is the product of a messenger RNA with an organ-specific in-frame stop codon. Science 238: 363–366 (1987).

23. Covello PS, Gray MW: RNA editing in plant mitochondria. Nature 341: 662–666 (1989).

24. Covello PS, Gray MW: Differences in editing at homologous sites in messenger RNAs from angiosperm mitochondria. Nucl Acids Res 18: 5189–5196 (1990).

25. Covello PS, Gray MW: RNA sequence and the nature of the Cu_A-binding site in cytochrome c oxidase. FEBS Lett 268: 5–7 (1990).

26. Creissen G, Reynolds H, Xue Y, Mullineaux P: Simultaneous targeting of pea glutathione reductase and of a bacterial fusion protein to chloroplasts and mitochondria in transgenic tobacco. Plant J 8: 167– 175 (1995).

27. De Souza AP, Jubier MF, Delcher E, Lancelin D, Lejeune B: A *trans*-splicing model for the expression of the tripartite *nad5* gene in wheat and maize mitochondria. Plant Cell 3: 1363–1378 (1991).

28. Dewey RE, Levings CS III, Timothy DH: Novel recombinations in the maize mitochondrial genome produce a unique transcriptional unit in the Texas male sterile cytoplasm. Cell 44: 439– 449 (1986).

29. Dorit LR, Schoenbach L, Gilbert W: How big is the universe of exons? Science 250: 1377–1382 (1990).

30. Fox TD, Leaver CJ: The *Zea mays* mitochondrial gene coding cytochrome oxidase subunit II has an intervening sequence and does not contain TGA codons. Cell 26: 315–323 (1981).

31. Freyer R, Hoch B, Neckermann K, Maier RM, Kössel H: RNA editing in maize chloroplasts is a processing step independent of splicing and cleavage to monocistronic mRNAs. Plant J 4: 621–629 (1993).

32. Freyer R, López C, Maier RM, Martin M, Sabater B, Kössel H: Editing of the chloroplast *ndhB* encoded transcripts shows divergence between closely related members of the grass family (poaceae). Plant Mol Biol, in press: (1995).

33. Geiss KT, Abbas GM, Makaroff CA: Intron loss from the NADH dehydrogenase subunit 4 gene of lettuce mitochondrial DNA: evidence for homologous recombination of a cDNA intermediate. Mol Gen Genet 243: 97–105 (1994).

34. Glaubitz JC, Carlson JE: RNA editing in the mitochondria of a conifer. Curr Genet 22: 163–165 (1992).

35. Gonzalez DH, Bonnard G, Grienenberger JM: A gene involved in the biogenesis of c-type cytochromes is cotranscribed with a ribosomal protein gene in wheat mitochondria. Curr Genet 21: 248– 255 (1993).

36. Graves PV, Bégu D, Velours J, Neau E, Belloc F, Litvak S, Araya A: Direct protein sequencing of wheat mitochondrial ATP synthase subunit 9 confirms RNA editing in plants. J Mol Biol 214: 1–6 (1990).

37. Gray MW, Covello PS: RNA editing in plant mitochondria and chloroplasts. FASEB J 7: 64–71 (1993).

38. Grohmann L, Thieck O, Herz U, Schroder W, Brennicke A: Translation of *nad9* mRNAs in mitochondria from *Solanum tuberosum* is restricted to completely edited transcripts. Nucl Acids Res 22: 3304–3311 (1994).

39. Gualberto JM, Bonnard G, Lamattina L, Grienenberger JM: Expression of the wheat mitochondrial *nad3-rps12* transcription unit: correlation between editing and mRNA maturation. Plant Cell 3: 1109–1120 (1991).

40. Gualberto JM, Domon C, Weil JH, Grienenberger JM: Structure and transcription of the gene coding for subunit 3 of cytochrome oxidase in wheat mitochondria. Curr Genet 17: 41–47 (1990).

41. Gualberto JM, Lamattina L, Bonnard G, Weil JH, Grienenberger JM: RNA editing in wheat mitochondria results in the conservation of protein sequences. Nature 341: 660–662 (1989).

42. Gualberto JM, Weil JH, Grienenberger JM: Editing of the wheat *coxIII* transcript : evidence for twelve C-to-U and one U-to-C conversions and for sequence similarities around editing sites. Nucl Acids Res 18: 3771–3776 (1990).

43. Gualberto JM, Wintz H, Weil JH, Grienenberger JM: The genes coding for subunit 3 of NADH dehydrogenase and for ribosomal protein S12 are present in the wheat and maize mitochondrial genomes and are co-transcribed. Mol Gen Genet 215: 118–127 (1988).

44. Hagemann R, Scholz F: Ein Fall Gen-induzierter Mutationen des Plasmotyps bei Gerste. Züchter 32: 50–59 (1962).

45. Hallick RB, Lipper C, Richards OC, Rutter WJ: isolation of a transcriptionally active chromosome from chloroplasts of *Euglena gracilis*. Biochemistry 15: 3039–3045 (1976).

46. Handa H, Gualberto JM, Grienenberger JM: Characterization of the mitochondrial *orfB* gene and its derivative, *orf224*, a chimeric open reading frame specific to the mitochondrial genome of 'Polima' male-sterile cytoplasm in rapeseed (*Brassica napus* L.). Curr Genet 28: in press (1995).

47. Haouazine N, De Souza AP, Jubier MF, Lancelin D, Delcher E, Lejeune B: The wheat mitochondrial genome contains an ORF showing sequence homology to the gene encoding the subunit 6 of the NADH ubiquinone oxidoreductase. Plant Mol Biol. 20: 395–404 (1992).

48. Hernould M, Mouras A, Litvak S, Araya A: RNA editing of the mitochondrial *atp9* transcript from tobacco. Nucl Acids Res 20: 1809 (1992).

49. Hess WR, Hoch B, Zeltz P, Hübschmann T, Kössel H, Börner T: Inefficient *rpl2* splicing in barley mutants with ribosome-deficient plastids. Plant Cell 6: 1455–1465 (1994).

50. Hiesel R, Combettes B, Brennicke A: Evidence for RNA editing in mitochondria of all major groups of land plants except the bryophyta. Proc Natl Acad Sci USA 91: 629–633 (1994).

51. Hiesel R, Wissinger B, Brennicke A: Cytochrome oxidase subunit II mRNAs in *Oenothera* mitochondria are edited at 24 sites. Curr Genet 18: 371–375 (1990).

52. Hiesel R, Wissinger B, Schuster W, Brennicke A: RNA editing in plant mitochondria. Science 246: 1632–1634 (1989).

53. Hiratsuka J, Shimada H, Whittier R, Ishibashi T, Sakamoto M, Mori M, Kondo C, Honji Y, Sun CR, Meng BY, Li YQ, Kanno A, Nishikawa Y, Hirai A, Shinozaki K, Sugiura M: The

364

complete sequence of the rice (*Oryza sativa*) chloroplast genome: intermolecular recombination between distinct tRNA genes accounts for a major plastid DNA inversion during the evolution of cereals. Mol Gen Genet 217: 185–194 (1989).

54. Hirose T, Wakasugi T, Sugiura M, Kössel H: RNA editing of tobacco *petB* mRNAs occurs both in chloroplasts and non-photosynthetic proplastids. Plant Mol Biol. 26: 509–513 (1994).

55. Hoch B, Maier RM, Appel K, Igloi GL, Kössel H: Editing of a chloroplast mRNA by creation of an initiation codon. Nature 353: 178–180 (1991).

56. Holschuh K, Bottomley W, Whitfeld PR: Sequence of the genes for tRNA^{Cys} and tRNA^{Asp} from spinach chloroplasts. Nucl Acids Res 11: 8547–8554 (1983).

57. Igloi GL, Kössel H: The transcriptional apparatus of chloroplasts. CRC Crit. Rev. Plant Sci 10: 525–558 (1992).

58. Iwabuchi M, Kyozuka J, Shimamoto K: Processing followed by complete editing of an altered mitochondrial *atp6* RNA restores fertility of cytoplasmic male sterile rice. EMBO J 12: 1437–1446 (1993).

59. Kavousi M, Giese K, Larrinua IM, McLaughlin WE, Subramanian AR: Nucleotide sequence and map positions of the duplicated gene for maize (*Zea mays*) chloroplast ribosomal protein L2. Nucl Acids Res 18: 4244–4244 (1990).

60. Knoop V, Schuster W, Wissinger B, Brennicke A: *Trans*-splicing integrates an exon of 22 nucleotides into the *nad5* messenger RNA in higher plant mitochondria. EMBO J 10: 3483–3493 (1991).

61. Kössel H, Hoch B, Maier RM, Igloi GL, Kudla J, Zeltz P, Freyer R, Neckermann K, Ruf S: RNA editing in chloroplasts of higher plants. In Brennicke A , Kück U (eds), Plant mitochondria, pp. 93– 102. VCH, Weinheim (1993).

62. Kudla J, Igloi G, Metzlaff M, Hagemann R, Kössel H: RNA editing in tobacco chloroplasts leads to the formation of a translatable *psbL* messenger RNA by a C-to-U substitution within the initiation codon. EMBO J 11: 1099–1103 (1992).

63. Kumar R, Levings CS III: RNA editing of a chimeric maize mitochondrial gene transcript is sequence specific. Curr Genet 23: 154–159 (1993).

64. Kuntz M, Camara B, Weil JH, Schantz R: The *psbL* gene from bell pepper (*Capsicum annuum*) : plastid RNA editing also occurs in non-photosynthetic chromoplasts. Plant Mol Biol. 20: 1185– 1188 (1992).

65. Kuprinska K, Falk J: Changes in RNA-polymerase activity during development and senescence of barley chloroplasts. Comparative analysis of transcripts synthesized either in run-on assays or by transcriptionally active chromosomes. J Plant Physiol. 143: 298–305 (1994).

66. Lamattina L, Gonzalez D, Gualberto JM, Grienenberger JM: Higher plant mitochondria encode an homologue of the nuclear-coded 30 kDa subunit of bovine mitochondrial complex I. Eur J Biochem 217: 831–838 (1993).

67. Lamattina L, Grienenberger JM: RNA editing of the transcript coding for subunit-4 of NADH dehydrogenase in wheat mitochondria – Uneven distribution of the editing sites among the 4 exons. Nucl Acids Res 19: 3275–3282 (1991).

68. Lippok B, Brennicke A, Wissinger B: Differential RNA editing in closely related introns in *Oenothera* mitochondria. Mol Gen Genet 243: 39–46 (1994).

69. Lonsdale DM, Hodge TP, Howe CJ, Stern DB: Maize mitochondrial DNA contains a sequence homologous to the ribulose-1,5-bisphosphate carboxylase large subunit gene of chloroplast DNA. Cell 34: 1007–1014 (1983).

70. Lu BW, Hanson MR: A single nuclear gene specifies the abundance and extent of RNA editing of a plant mitochondrial transcript. Nucl Acids Res 20: 5699–5703 (1992).

71. Lu BW, Hanson MR: A single homogeneous form of ATP6 protein accumulates in *petunia* mitochondria despite the presence of differentially edited atp6 transcripts. Plant Cell 6: 1955–1968 (1994).

72. Mahendran R, Spottswood MR, Miller DL: RNA editing by cytidine insertion in mitochondria of *Physarum polycephalum*. Nature 349: 434–438 (1991).

73. Maier RM, Hoch B, Zeltz P, Kössel H: Internal editing of the maize chloroplast *ndhA* transcript restores codons for conserved amino acids. Plant Cell 4: 609–616 (1992).

74. Maier RM, Neckermann K, Hoch B, Akhmedov NB, Kössel H: Identification of editing positions in the *ndhB* transcript from maize chloroplasts reveals sequence similarities between editing sites of chloroplasts and plant mitochondria. Nucl Acids Res 20: 6189–6194 (1992).

75. Maier RM, Neckermann K, Igloi GL, Kössel H: Complete sequence of the maize chloroplast genome: gene content, hotspots of divergence and fine tuning of genetic information by transcript editing. J Mol Biol. 251: 614–628 (1995).

76. Maréchal-Drouard L, Ramamonjisoa D, Cosset A, Weil JH, Dietrich A: Editing corrects mispairing in the acceptor stem of bean and potato mitochondrial phenylalanine transfer RNAs. Nucl Acids Res 21: 4909–4914 (1993).

77. Michel F, Umesono K, Oseki H: Comparative and functional anatomy of group II catalytic introns: a review. Gene 82: 5–30 (1989).

78. Moon E, Wu R: Organization and nucleotide sequence of genes at both junctions between the two inverted repeats and the large single- copy region in the rice chloroplast genome. Gene 70: 1–12 (1988).

79. Nakazono M, Hirai A: Identification of the entire set of transferred chloroplast DNA sequences in the mitochondrial genome of rice. Mol Gen Genet 236: 341–346 (1993).

80. Navaratnam N, Shah R, Patel D, Fay V, Scott J: Apolipoprotein-B messenger RNA editing is associated with UV crosslinking of proteins to the editing site. Proc. Natl. Acad. Sci USA 90: 222–226 (1993).

81. Neckermann K, Zeltz P, Igloi GL, Kössel H, Maier RM: The role of RNA editing in conservation of start codons in chloroplast genomes. Gene 146: 177–182 (1994).

82. Oda K, Yamato K, Ohta E, Nakamura Y, Takemura M, Nozato N, Akashi K, Kanegae T, Ogura Y, Kohchi T, Ohyama K: Gene organization deduced from the complete sequence of liverwort *Marchantia polymorpha* mitochondrial DNA: a primitive form of plant mitochondrial genome. J Mol Biol. 223: 1–7 (1992).

83. Ohyama K, Fukuzawa H, Kohchi T, Shirai H, Sano T, Sano S, Umesono K, Shiki Y, Takeuchi M, Chang Z, Aota S, Inokuchi H, Ozeki H: Chloroplast gene organization deduced from complete sequence of liverwort *Marchantia polymorpha* chloroplast DNA. Nature 322: 572–574 (1986).

84. Patell V, Bonnard G, Lamattina L, Gualberto JM, Grienenberger JM: *Trans-*, *cis*-splicing and RNA editing for *nad2* gene expression in wheat mitochondria. In: Brennicke A (eds), Proceedings of the HFSP-workshop RNA Editing in Plant Mitochondria, pp. 69. Berlin (1992).

85. Rajasekhar VK, Mulligan RM: RNA editing in plant mitochondria: alpha-phosphate is retained during C-to-U conversion in mRNAs. Plant Cell 5: 1843–1852 (1993).

86. Reiss T, Link G: Characterization of transcriptionally active DNA-protein complexes from chloroplasts and etioplasts of

mustard (*Sinapis alba* L.). Eur. J Biochem. 148: 207–212 (1985).

87. Ruf S, Zeltz P, Kössel H: Complete RNA editing of unspliced and dicistronic transcripts of the intron-containing reading frame *IRF170* from maize chloroplasts. Proc. Natl. Acad. Sci USA 91: 2295–2299 (1994).

88. Schuster W, Hiesel R, Wissinger B, Brennicke A: RNA editing in the cytochrome *b* locus of the higher plant *Oenothera berteriana* includes a U-to-C transition. Mol Cell. Biol. 10: 2428–2431 (1990).

89. Schuster W, Ternes R, Knoop V, Hiesel R, Wissinger B, Brennicke A: Distribution of RNA editing sites in *Oenothera* mitochondrial mRNAs and rRNAs. Curr Genet 20: 397–404 (1991).

90. Schuster W, Unseld M, Wissinger B, Brennicke A: Ribosomal protein S14 transcripts are edited in *Oenothera* mitochondria. Nucl Acids Res 18: 229–233 (1990).

91. Schuster W, Wissinger B, Unseld M, Brennicke A: Transcripts of the NADH-dehydrogenase subunit 3 gene are differentially edited in *Oenothera* mitochondria. EMBO J 9: 263–269 (1990).

92. Shinozaki K, Ohme M, Tanaka M, Wakasugi T, Hayashida N, Matsubayashi T, Zaita N, Chunwongse J, Obokata J, Yamaguchi- Shinozaki K, Ohto C, Torazawa K, Meng BY, Sugita M, Deno H, Kamogashira T, Yamada K, Kusuda J, Takaiwa F, Kato A, Tohdoh N, Shimida H, Sugiura M: The complete nucleotide sequence of the tobacco chloroplast genome: its gene organization and expression. EMBO J 5: 2043–2049 (1986).

93. Simpson L, Maslov DA, Blum B: RNA editing in *Leishmania* mitochondria. In: Benne R (eds), RNA editing. The Alteration of Protein Coding Sequences of RNA. Ellis Horwood, Chichester (1993).

94. Simpson L, Shaw J: RNA editing and the mitochondrial cryptogenes of kinetoplastid protozoa. Cell 57: 355–366 (1989).

95. Sommer B, Kohler M, Sprengel R, Seeburg PH: RNA editing in brain controls a determinant of ion flow in glutamate-gated channels. Cell 67: 11–19 (1991).

96. Staub JM, Maliga P: Accumulation of D1 polypeptide in tobacco plastids is regulated via the untranslated region of the *psbA* mRNA. EMBO J 12: 601–606 (1993).

97. Stern DB, Lonsdale DM: Mitochondrial and chloroplast genomes of maize have a 12-kilobase DNA sequence in common. Nature 299: 698–702 (1982).

98. Sutton CA, Conklin PL, Pruitt KD, Hanson MR: Editing of pre-mRNAs can occur before cis- and trans-splicing in *Petunia* mitochondria. Mol Cell. Biol. 11: 4274–4277 (1991).

99. Sutton CA, Zoubenko OV, Hanson MR, Maliga P: A plant mitochondrial sequence transcribed in transgenic tobacco chloroplasts in not edited. Mol Cell Biol. 15: 1377–1381 (1995).

100. Thomas SM, Lamb RA, Paterson RG: Two mRNAs that differ by two nontemplated nucleotides encode the amino coterminal proteins P and V of the paramyxovirus SV5. Cell 54: 891–902 (1988).

101. Wakasugi T, Ohme M, Shinozaki K, Suguira M: Structures of tobacco chloroplast genes for tRNA(Ile,CAU), tRNA(Leu,CAA), tRNA(Cys,GCA), tRNA(Ser,UGA) and tRNA(Thr,GGU): a compilation of tRNA genes from tobacco chloroplasts. Plant Mol Biol. 7: 385–392 (1986).

102. Wakasugi T, Tsudzuki J, Ito S, Nakashima K, Tsudzuki T, Suguira M: Loss of all *ndh* genes as determined by sequencing the entire chloroplast genome of the black pine *Pinus thunbergii*. Proc. Natl. Acad. Sci USA 91: 9794–9798 (1994).

103. Ward GC, Levings CS III: The protein-encoding gene*T- urf13* is not edited in maize mitochondria. Plant Mol Biol. 17: 1083–1088 (1991).

104. Wissinger B, Brennicke A, Schuster W: Regenerating good sense: RNA editing and *trans*-splicing in plant mitochondria. Trends Genet 8: 322–328 (1992).

105. Wissinger B, Schuster W, Brennicke A: Trans splicing in *Oenothera* mitochondria: *nad1* mRNAs are edited in exon and *trans*-splicing group-II intron sequences. Cell 65: 473–482 (1991).

106. Yang AJ, Mulligan RM: RNA editing intermediates of *cox2* transcripts in maize mitochondria. Mol Cell Biol 11: 4278–4281 (1991).

107. Yu W, Schuster W: Evidence for a site-specific cytidine deamination reaction involved in C-to-U RNA editing of plant mitochondria. J Biol. Chem. 270: 18227–18233 (1995).

108. Zanlungo S, Bégu D, Quinones V, Araya A, Jordana X: RNA editing of apocytochrome-b (cob) transcripts in mitochondria from two genera of plants. Curr Genet 24: 344–348 (1993).

109. Zanlungo S, Quinones V, Moenne A, Holuigue L, Jordana X: Splicing and editing of *rps10* transcripts in potato mitochondria. Curr Genet 27: 565–571 (1995).

110. Zeltz P, Hess WR, Neckermann K, Börner T, Kössel H: Editing of the chloroplast*rpoB* transcript is independent of chloroplast translation and shows different patterns in barley and maize. EMBO J 12: 4291–4296 (1993).

111. Zeltz P, Kadowaki K, Kubo N, Maier RM, Hirai A, Kössel H: A promiscuous chloroplast DNA fragment is transcribed in plant mitochondria but the encoded RNA is not edited. Submitted

112. Zheng H, Fu TB, Lazinski D, Taylor J: Editing on the genomic DNA of human hepatitis delta virus. J Virol 66: 4693–4697 (1992).

113. Zhuo DG, Bonen L: Characterization of the S7 ribosomal protein gene in wheat mitochondria. Mol Gen Genet 236: 395–401 (1993).

Plant Molecular Biology **32:** 367–391, 1996.
© 1996 *Kluwer Academic Publishers.*

Gene expression from viral RNA genomes

Ivan G. Maia[1], Karin Séron, Anne-Lise Haenni & Françoise Bernardi*
Institut Jacques Monod, 2 Place Jussieu - Tour 43, 75251 Paris Cedex 05, France (*author for correspondence);
[1]*Present address: Universidade Estadual de Campinas, Cidade Universitária 'Zeferino Vaz', Cx. Postal 6109, Campinas, Brazil*

Key words: plant RNA viruses, translation, subgenomic RNA synthesis, leaky scanning, frameshift, readthrough, proteolytic maturation

Contents

Abstract	367
Introduction	368
Subgenomic RNAs	368
Subgenomic promoter	369
Regulation of synthesis of sgRNAs	373
Mechanism of sgRNA synthesis	373
5′ and 3′ end structures of sgRNAs	373
Encapsidation	374
Expression of sgRNAs	374
Viruses using the ambisense strategy	375
Initiation of translation: leaky scanning	375
Overlap in different reading frames	376
In-frame initiation	376
Regulation at the level of elongation:	
frameshift	377
−1 Frameshift	377
+1 Frameshift	380

Regulation at the level of termination:	
readthrough	380
mRNAs	380
Suppressor tRNAs	382
Proteolytic processing	382
Virus-encoded proteinases	383
Plant RNA viruses and proteolytic	
processing	383
Potyviruses	383
Comoviruses, nepoviruses	384
Sobemoviruses, luteoviruses	385
Tymoviruses	385
Other viruses with papain-like	
proteinases	385
Closteroviruses	385
Discussion	385
Acknowledgements	385
Addendum	386
References	386

Abstract

This review is centered on the major strategies used by plant RNA viruses to produce the proteins required for virus multiplication. The strategies at the level of transcription presented here are synthesis of mRNA or subgenomic RNAs from viral RNA templates, and 'cap-snatching'. At the level of translation, several strategies have been evolved by viruses at the steps of initiation, elongation and termination. At the initiation step, the classical scanning mode is the most frequent strategy employed by viruses; however in a vast number of cases, leaky scanning of the initiation complex allows expression of more than one protein from the same RNA sequence. During elongation, frameshift allows the formation of two proteins differing in their carboxy terminus. At the termination step, suppression of termination produces a protein with an elongated carboxy terminus. The last strategy that will be described is co- and/or post-translational cleavage of a polyprotein precursor by virally encoded proteinases. Most (+)-stranded RNA viruses utilize a combination of various strategies.

Abbreviations. Viruses: (genera of viruses are based on [198]): ACLSV, apple chlorotic leaf spot closterovirus; AlMV, alfalfa mosaic virus; AMCV, artichoke mottle crinkle tombusvirus; ASGV, apple stem grooving capillovirus; ASPV, apple stem pitting virus; BaYMV, barley yellow mosaic bymovirus; BBScV, blueberry scorch carlavirus; BMV, brome mosaic bromovirus; BNYVV, beet necrotic yellow vein furovirus; BSMV, barley stripe mosaic hordeivirus; BWYV, beet western yellows luteovirus; BYDV, barley yellow dwarf luteovirus; BYV, beet yellows closterovirus; CaMV, cauliflower mosaic caulimovirus; CarMV, carnation mottle carmovirus; CCFV, cardamine chlorotic fleck carmovirus; CCSV, cucumber chlorotic spot closterovirus; CfMV, cockfoot mottle sobemovirus; CLRV, cherry leafroll (nepo?)virus; CMV, cucumber mosaic cucumovirus; CNV, cucumber necrosis tombusvirus; CPMV, cowpea mosaic comovirus; CRSV, carnation ringspot dianthovirus; CTV, citrus tristeza closterovirus; CyRSV, cymbidium ringspot tombusvirus; FMV, foxtail mosaic potexvirus; GFMV, grapevine fanleaf nepovirus; LIYV, lettuce infectious yellows closterovirus; MCMV, maize chlorotic mottle machlomovirus; MNSV, melon necrotic spot carmovirus; OCSV, oat chlorotic stunt virus; PCV, peanut clump furovirus; PEBV, pea early browning tobravirus; PEMV, pea enation mosaic enamovirus; PLRV, potato leafroll luteovirus; PMTV, potato mop-top potexvirus; PPV, plum pox potyvirus; PSbMV, pea seedborne mosaic potyvirus; PVM, potato carlavirus M; PVX, potato potexvirus X; PYFV, parsnip yellow fleck sequivirus; RBDV, raspberry bushy dwarf ilarvirus; RCNMV, red clover necrotic mosaic dianthovirus; RTSV, rice tungro spherical waikavirus; SBWMV, soil-borne wheat mosaic furovirus; SCNMV, sweet clover necrotic mosaic dianthovirus; SDV, soybean dwarf luteovirus; STMV, satellite of tobacco mosaic virus; TBSV, tomato bushy stunt tombusvirus; TCV, turnip crinkle carmovirus; TEV, tobacco etch potyvirus; TMV, tobacco mosaic tobamovirus; TNV, tobacco necrosis necrovirus; TRV, tobacco rattle tobravirus; TSV, tobacco streak ilarvirus; TYMV, turnip yellow mosaic tymovirus; WClMV, white clover mosaic potexvirus. *Others*: CP, coat protein; ds, double-strand; g, genomic; GUS, β-glucuronidase; ICR, internal control region; IR, intercistronic (or intergenic) region; IRES, internal ribosome entry site; kDa, kilodalton; nt, nucleotide; ORF, open reading frame; RdRp, RNA-dependent RNA polymerase; sg, subgenomic; SGP, subgenomic promoter; ss, single strand; TGB, triple gene block; UTR, untranslated region; v, viral; vc, viral complementary; VPg, virus protein genomic; wt, wild type.

Introduction

Positive-stranded RNA viruses express their genes through a large variety of strategies at the level of transcription as well as at the level of translation. There are several reasons for this complexity, the finite size of RNA genomes which leads to the compactness of their genetic information, the dependence on the host machinery for translation and by whose rules the viral RNAs must abide, and regulation of synthesis of the viral products which must be achieved at the level of RNA transcription and translation.

The genes encoded by (+)-stranded RNA viruses are therefore those required for viral replication, particle formation, movement of the virus from cell to cell, interaction of the virus with its vector, and maturation of polyproteins.

Plant viruses differ from animal viruses in several respects. Plant viruses do not enter the plant cell as do animal viruses that require cell receptors; they enter through wounding of the cell wall by viral vectors, or by mechanical injuries. Once in a plant cell, the viruses require specific proteins designated movement proteins, to migrate to other cells and to spread throughout the plant. The RNA segments that compose the genome of animal viruses are encapsidated within the same particle; this is not the case of plant RNA viruses whose RNA segments frequently reside in distinct particles, so that a plant cell must be infected by distinct particles for the virus to multiply. Plant RNA viruses are frequently accompanied by satellites, a rather rare phenomenon among animal viruses.

Here, we have attempted to give an overall view of the numerous strategies used by viruses to produce and regulate the production of their proteins.

Subgenomic RNAs

The first report on the existence of a subgenomic (sg) RNA concerned RNA4 of BMV [163]; later, the TMV sgRNA was postulated to be the mRNA for the expression of the viral coat protein (CP [84]). Since then, among the various expression strategies used by (+)-stranded RNA viruses, the production of sgRNAs has been the most frequently encountered. Viruses from three supergroups, the alpha-, carmo-, and sobemo-like supergroups [64], utilize this type of strategy, albeit their genomic organization and other expression strategies differ widely. In addition, tospoviruses (Bunyaviridae) and the tenuiviruses, which are ambisense, express their genes through the production of sgRNAs.

In general, sgRNAs are the mRNAs for the 3' proximal genes of polycistronic viral RNAs, and are identical in sequence to the 3' end of the genomic (g) RNA. When several genes are present at the 3' end of the gRNA, a family of 3' collinear sgRNAs is usually produced such that each gene to be expressed is located at the 5' end of one sgRNA, a situation required for proper expression in eukaryotes.

The mechanism of sgRNA synthesis supported by *in vitro* and *in vivo* data [119, 113, 57] involves transcription from a template RNA [the (−) strand] starting at an internal site, the subgenomic promoter (SGP), by the viral RNA-dependent RNA polymerase (RdRp).

Subgenomic promoter

The viral RdRps are difficult to obtain in an active form and in large quantities, hindering analyses by physical methods of protein-RNA interactions involved in initiation of transcription. Identification of the SGP has therefore relied on mutational analyses of sequences of about 100 to 200 nucleotides (nt) overlapping the start site of the sgRNA. The 5' end of the sgRNA can be easily determined by direct RNA sequencing, primer extension or RNAse protection experiments and numerous sequences have thus been determined. Examples, when available, for each viral group are presented in Table 1 with information in particular about the 5' sequence of the corresponding gRNA. As in most publications on this subject, we will refer to the SGP sequence as the sequence present on the (+) strand, even though the sequence recognized by the RdRp is on the (−) strand, and sequence numbering will start at the +1 nt of the sgRNA. The 5' end of the sgRNA can be localized in an intercistronic region (IR) or in the coding sequence of the upstream gene (Table 1). The IR may vary in length from one virus to another and the corresponding SGP which encompasses about 150 nt may overlap the coding region even in the case of an intercistronic localization of the 5' end of the sgRNA. Therefore additional constraints are added to the mutational analysis of the promoter, particularly when the upstream gene encodes the RdRp or is otherwise required for viral amplification.

Perusal of the sequences corresponding to the SGP regions reveals common features for all groups of viruses. The first nt of the gRNAs and sgRNAs are often the same (Table 1). Conservation of the first nt is particularly important: a mutation at this position completely abolishes production of the sgRNA *in vivo* in the case of AlMV [182]. In most cases, homologies between gRNAs and sgRNAs extend further downstream for up to 20 nt from the 5' end, and these regions are often AU- or U-rich [113, 127, 117]. Deletion of these sequences abolishes correct initiation of the sgRNA [113, 182].

For a given virus group, sequence homologies are found in the vicinity − mostly upstream − of the 5' sgRNA start site, as for potexviruses [166], tobraviruses [68] and tymoviruses [45]. These sequences are also candidate domains of the corresponding SGP.

Two SGP have been extensively characterized *in vitro* and *in vivo*, those for AlMV [181, 182], and for BMV [53, 113, 168]. A common organization of the SGP emerges from these studies. A 'core promoter' can be defined which corresponds to the smallest region capable of promoting sgRNA synthesis at a basal level and with low accuracy. In addition, several 'enhancer' regions with appropriate spacing provide accuracy of replication initiation and yields of sgRNAs comparable to those obtained *in vivo*. The fully functional SGP encompasses about 150 nt, and when inserted into a new site on the viral RNA it is sufficient for the production of a new sgRNA species.

The SGP of BMV is located in an IR of 250 nt between the 3a (movement protein) and the CP genes. *In vitro* studies [113] have identified four functional domains, the core region (−20 to +1), and three 'enhancer' regions, the 16 nt downstream of the start site (+1 to +16) which provide accurate initiation, and two upstream domains, the polyA stretch present in all the bromoviruses (−20 to −37) and a repeat of UUA (three times, between positions −38 and −48). Characterization of the SGP *in vivo* [53] revealed that the core region had to be extended (−20 to +16), and that the polyA tract was absolutely required as were three repeats of the sequence AUCUAUGUU, extending the complete SGP to a site between −74 and −95 upstream of the +1 start site. The number of these latter repeats affected the level of sgRNA synthesis; increasing their number led to levels of sgRNA above normal. The role of the polyA stretch, present in all the bromoviruses and essential for sgRNA synthesis has been investigated in detail [168]. After deletion of the polyA tract, three revertants were found for which sgRNA synthesis was restored; for two of these revertants the mutations

Table 1. Plant RNA viruses that produce sgRNAs.

Virus	5' sequence		sgRNA designation	Locus	5' end	Encoded gene	Virion	References and comments
Alpha-like								
AlMV	g3	GUUUUA			cap			[8]
(3)	3sg1	GUUUUU	RNA4	IR	+	CP	+	[23]
bromo								
BMV	g3	GUAAAA			cap			[4]
(3)	3sg1	GUAUUA	RNA4	IR	+	CP	+	[4]
carla								
PVM	g	NNUAAA						[199]
(1)	sg1	post		IR		TGB	+/−	[156]
	sg2	GAAAU		CS		CP, 11K	+/−	
clostero								
BYV					cap			[2]; 7 ORFs deduced from
(1)								the partial sequence;
								5 sg RNAs detected in plants
cucumo								
Q-CMV	g2	GUUUAU			cap			[146]
(3)	2sg1	GUUUUG	RNA4A	CS	+	2b	+	[44]
	g3	GUAAUC			cap			[67], reviewed in [135]
	3sg1	GUUUAG	RNA4	IR	+	CP	+	
furo								
SBWMV	g1	GUAUUU			cap			[164]
(2)	1sg	post		IR		37K	−	[164]
	g2	GUAUUU			cap			[164]
	2sg	post		IR		19K	−	
BNYVV	g2	AAAUUC			cap			[18]
(4)	2sg1	nd		CS		TGB: 42K	+/−	
	2sg2	nd		CS		TGB: 13K	+/−	
	2sg3	AAUGUC	2subc	CS	+	14K		[62]
	g3	AAAAUU			cap			[17]
	3sg	AAUCG	3sub	IR		4.5K	−	[7]
hordei								
BSMV	g2	GUAAAA	β		cap			[74]
(3)	2sg1	post		IR		TGB	−	
	2sg2	post		CS			−	
	g3	GUAUAG	γ		cap			[75]
	3sg	GUUUAA	γsg	IR	+	17K	+/−	
ilar								
TSV	g3	GUAUUC			cap			[34]
(3)	3sg	nd	RNA4	IR		CP	+/−	
RBDV	g2	AUAUAU						[127]
(2)	2sg	UAUUUC	RNA3	IR		CP		[114]
potex								
PVX	g	GAAAAC			cap			[83, 125, 166]
(1)	sg1	GAAUA		IR		TGB: 25K	−	
	sg2	GAAAU		CS		TGB: 12K+8K	−	
	sg3	GAAAG		CS		CP	−	
tobamo								
TMV	g	GUAUUU			cap			[63]
(1)	sg1	CUCCAG		CS	−	30K	−	[107]
	sg2	GUUUUA		CS	+	CP	−	

Table 1. Continued.

Virus	5' sequence		sgRNA designation	Locus	5' end	Encoded gene	Virion	References and comments
tobra								
TRV-PSG	g1	AUAAAA			cap			strain PSG: [35],
(2)	1sg1	AUAUUA	RNA3	CS		29K		strain TCM: [5];
	1sg2	AUAAAG	RNA5	CS		16K		this latter strain has two
	g2	AUAAAA			cap			additional sgRNA from RNA2
	2sg1	AUAAAU	RNA4	IR	+	CP (PSG)		
tymo								
TYMV	g	GUAAUC			cap			[123]
(1)	sg	AAUAGC		CS	+	CP	+	[72]
Carmo-like								
carmo								
CarMV	g	GGGUAA						[73]
(1)	sg1	GGUAAC		CS		7K	+	[33]
	sg2	GUGAAG		CS		CP	+/−	reviewed in [157]
diantho								
RCNMV	g1	ACAAAC			cap			[197]
(2)	1sg	nd		CS		CP		
luteo								
BYDV-PAV	g	AGUGAA						[94], sequence of
(1)	sg1	GUGAAG				CP/Rt, 17K[N]	−	Australian isolate
	sg2	AGUGAA				6K	−	
	sg3	GACGAC				non coding	+/−	
BYDV-PAV	g	XGUGAA			VPg			[120]
(1)	sg1	GAUAG	sgRNA1			CP/Rt, 17K[N]		[42]
	sg2	GGCAG	sgRNA2			6.7K		
MCMV	g	AGGUAA			cap			[108, 130]
(1)	sg	AUCAGA		CS			+/−	
TNV	g	AGUAUU			ppA			[116]
(1)	sg1	nd		CS		7.9K		
	sg2	nd		CS		CP		
tombus								
CNV	g	NNAAAU						[150, 151],
(1)	sg1	ACCAA		IR		CP	+	reviewed in [157]
	sg2	GAAUCU		IR		22K, 19K[N]	+	
CyRSV	g	NGAAAUC						[71], only the 22K
(1)	sg1	GACCAA		IR		CP	+	protein found *in vivo*
	sg2	GAACCU		IR		22K, 19K[N]	+	
Sobemo-like								
sobemo								
SBMV-B	g	CACAAA			VPg			[61, 133]
(1)	sg	nd		IR	+	CP	+/−	
luteo								
PLRV	g	CAAAAG			VPg			[183]
(1)	sg	ACAAAA	sgRNA1	CS		CP/Rt, 17K[N]	−	[117]

Table 1. Continued.

Virus	5' sequence		sgRNA designation	Locus	5' end	Encoded gene	Virion	References and comments
Others								
PEMV	g1	NGUGAA						unique member of enamovirus
(2)	1sg			IR	VPg	CP/Rt	+/−	genus [38]
RTSV	g	UGAAAA						member of MCDV family [162];
(1)	sg1	GAUGAC		CS		Mvt or trans-		the 2 sgRNAs differ only by the
	sg2	GCUGGG		IR		mission		length of the leader sequence
CLRV	g1	GAAAA			VPg			presumed nepovirus [25]
(2)	1sg1	AAAAGG	RNA 1A	IR		non-coding		

Data concerning sgRNAs are presented for the viruses of the alpha-, carmo- and sobemo-like supergroups based on the classification proposed [64]. The name of the supergroup is in bold, and of the genus in italic. The number in parentheses below the abbreviation of the virus name indicates the total number of viral RNA genome segments. A general nomenclature has been used for the g and sgRNAs, the gRNAs are numbered according to decreasing size and the corresponding sgRNA is designated based on its cognate gRNA, again in order of decreasing size when multiple sgRNAs are produced. The 5' sequences of the gRNAs and sgRNAs are presented, the underlined sequences indicating that the sequence is postulated on the basis of homologies; for some viruses the sgRNA is postulated (post), in other cases the sgRNA has been detected but not sequenced (nd). The usual name of the sgRNA is given in column sgRNA designation. Loc is the location of the 5' end of the sgRNA, either in the intergenic region (IR) or in a coding sequence (CS); when the precise position of the 5' end is unknown, IR is indicated. 5' end is the structure given for the gRNA and its presence on the sgRNA is indicated by + or −. Encoded gene: the nature of the proteins encoded by the sgRNA is given; TGB is the triple gene block; [N] indicates a nested gene; CP/Rt indicates CP/readthrough; Mvt indicates movement protein. Virion is the extent of encapsidation, denoted by +, − or +/−.K, Kilodalton.

were in the SGP and they also showed increased levels of gRNA3. The third mutation was located upstream of the SGP in a sequence designated box B which led to decreased levels of RNA3. Box B also designated ICR2-like (internal control region 2) is present in the promoter of the cellular RNA polymerase III, in the tRNA TΨU loop as well as at the 5' ends of gRNA1 and gRNA2 of CMV and BMV [14].

As compared to BMV, characterization of the AlMV SGP is more complex particularly *in vivo*. First, the sgRNA (RNA4) is the mRNA for the CP which is required for AlMV (−)-strand replication. Second, the IR region in which the SGP is located consists of 49 nt, too short to accomodate the SGP which overlaps into the coding region of the 3a protein necessary for the spread of the virus in plants; as a further complication, it is the carboxy terminus of the 3a protein which is essential for activity [182], hampering the possibilities of mutational analyses of the SGP. *In vitro* studies of AlMV have shown that a core promoter is located between −55 and −8 [181]. Characterization of the SGP *in vivo* (in protoplasts and plants) was achieved after inserting a fragment large enough to contain the SGP at the 5' end of the gRNA so as to allow production of active 3a protein and CP. The core promoter was located between −26 and +1, comparable to the BMV core region (−20 to +1). Two enhancer regions, one upstream (−136 to −94) and a second one downstream

(+1 to +12), allowed near maximum levels of sgRNA synthesis *in vivo*. Point mutations showed the importance of +1 and of a small conserved sequence (AAU) present in BMV and AlMV [182]. These results show a common spatial organization for BMV and AlMV, and the presence of conserved sequences which bear homologies with a consensus sequence derived from the RNA of all the alphaviruses of plant and animal origin [134].

The SGP of CMV has been characterized to a lesser extent [14]. A region from −70 to +20 carrying the ICR2-like sequence is sufficient to provide wild-type (wt) levels of sgRNA *in vivo*. As for BMV, common regions are involved in the accumulation of RNA3 and its corresponding sgRNA. The organization of the SGP of the BNYVV sgRNA '3sub' is significantly different from that of BMV, AlMV and CMV. The core promoter defined *in vivo* is located between −16 and +100/208, thus mostly downstream of +1, and the U-rich sequence is located upstream of the start site [7].

CNV represents the first case of a carmo-like virus whose SGP has been analyzed [88]. Two sgRNAs are produced, the core region for sgRNA2 is located between −20 and +6 and is sufficient for the production of wt levels of sgRNA. As opposed to what is observed for the alpha-like viruses, no enhancer regions seem to be involved in the production of this sgRNA. The precise role of the different domains, core and flanking regions has not yet been established. Based on the sequence homology and specificity of the 5' ends of the gRNAs and sgRNAs it seems likely that this sequence is the recognition site of the RdRp, whereas the other domains could interact with viral or host factors.

Regulation of synthesis of sgRNAs

Synthesis of a sgRNA often exceeds the amount necessary for the production of the corresponding gene; in the case of BMV lower levels of sgRNA as a result of mutations, did not affect the amount of CP produced [168].

When an appropriate SGP region is inserted into a new location on the viral RNA, a new sgRNA is produced (BMV [53], TMV [107], CMV [14], AlMV [182]) demonstrating the functionality of the SGP. Another outcome of these experiments was to show that expression of the sgRNA was dependent on the position of the corresponding SGP on the gRNA. Higher levels of the sgRNA lying closer to the 3' end of the gRNA are observed for TMV, BMV and CNV, whereas the opposite situation occurs for AlMV [182].

When multiple SGPs are present on a given gRNA, the same type of regulation dependent on the position of the SGP is observed: the smallest sgRNA produced is the most abundant, and it often corresponds to the CP mRNA. In several cases expression of the sgRNA is modulated in time and amount. This effect was observed *in vivo* in the case of TMV: expression of the 30 kDa movement protein from sgRNA1 is low and time-dependent, whereas synthesis of the CP from sgRNA2 is delayed but thereafter expressed at a constant and very high rate. This pattern of expression has been attributed to the positions of the genes relative to the 3' end of the viral RNA rather than to the difference in sequence of the putative SGPs, which have no sequence homologies (reviewed in [36]). Different types of regulation for two colinear sgRNAs have been reported for BYDV [42], CLRV [25] and CNV [88]. Therefore sgRNAs may be expressed for short periods and can be missed if time course experiments are not performed. Moreover comparison of *in vitro* and *in vivo* sgRNA expression may also help to detect this type of time-dependent regulation.

Mechanism of sgRNA synthesis

The mechanism of sgRNA synthesis is still poorly understood. It awaits characterization of the viral RdRp and identification of other factors that may be involved in this process, either from the host or of viral origin. A model for AlMV sgRNA synthesis has been proposed by van der Vossen *et al.* [182] based on experimental data and on a model first presented for the replication of Qβ [9]. These authors postulate that the RdRp is initially bound to two internal sites on (−)-strand AlMV RNA3, the SGP and a cryptic promoter site close to the 3' end of the template strand. This binding would in turn induce a conformational change in the RNA followed by correct initiation at the 3' end and/or at the SGP depending on the presence of other factors. In agreement with the proposed model for AlMV, binding of factors to two internal sites had already been postulated for BMV RNA replication [53]. This mechanism involves looping out of the intervening RNA sequence facilitating interaction between the different factors. *Trans*-acting factors bound to distinct nucleic acid sites might interact by protein-protein contacts to stimulate gene expression. The above model proposed for sgRNA synthesis takes into account several features that have been observed for gRNA and sgRNA synthesis: involvement of internal sequences such as box B in both gRNA and sgRNA synthesis (BMV, CMV, AlMV), differential expression of multiple sgRNAs present on the same gRNA by competition between the SGPs, and enhancement of sgRNA levels by increasing the number of repeated sequences upstream of +1 [14, 182].

5' and 3' end structures of sgRNAs

In the case of (+)-stranded RNA viruses and as mentioned earlier, the 3' ends of the sgRNA have the same structure as the corresponding gRNA (polyA, tRNA-like or OH) due to the lack of termination signals for the transcription machinery. The only exception recorded to date concerns BSMV; the sgRNA produced from RNAγ has a polyA tail whereas the corresponding gRNA has a tRNA-like structure. The 3' terminal polyA tail of the sgRNA corresponds to a polyA region of the gRNA located about 200 nt upstream of the 3' end [169]. On the other hand, and in spite of the presence of the polyA tract in all three gRNAs of BSMV, the sgRNA derived from RNAβ terminates with the tRNA-like structure [74].

Since sgRNAs share the same 3' sequence as, and are more abundant than the gRNAs, they may interfere with virus replication. Therefore double-stranded (ds) RNAs which represent replicative forms corresponding to sgRNAs have been sought. In the early days of sgRNA studies, the lack of dsRNA forms was used as an argument against the replicative model for sgRNA synthesis. However since then dsRNAs derived from sgRNAs have been found in several cases. High levels of dsRNA have been described for the sgRNA of TNV [116], BYDV [94] and RCNMV [132]. A detailed study of the dsRNA forms of the PVX and BSMV sgRNAs has been carried out [47, 46]; the 3' end of the (−) strand of the gRNA has an extra unpaired G residue as compared to the (+) strand, this extra nucleotide being necessary for new synthesis of (+) strand gRNA. On the contrary the 3' end of the (−) strand of the sgRNA in the ds form lacks this extra nucleotide and both strands are perfectly matched at this end, abolishing production of sgRNA through *bona fide* replication from a sgRNA template.

The 5' end of the gRNA can be constituted by a cap, a viral-encoded protein designated VPg, or ppX (Table 1); the first nt of a sgRNA is often the same as for the gRNA and this may reflect the necessity for the sgRNA to have the same 5'-end structure as the gRNA. In the case of TMV where two sgRNAs are produced, the highly expressed sgRNA corresponding to the CP is capped whereas the less abundant sgRNA of the 30 kDa protein is probably uncapped [89, 107].

Encapsidation

sgRNAs are often found in the virion particles, albeit in varying amounts. The BMV, AlMV and RBDV sgRNAs are encapsidated as efficiently as the gRNA whereas no encapsidation of the potexvirus sgRNAs is observed since the encapsidation site is located at the 5' end of the gRNA [83]. In the case of TMV and TRV, two sgRNAs are produced from the same gRNA, but only the larger sgRNA is encapsidated. In the case of TMV, this results from the position of the encapsidation site which in certain strains is present in sgRNA1 but not in sgRNA2. By analogy, the same situation was postulated for TRV [13]. However in many cases low amounts of the sgRNA are present in the virions. They are sometimes only detected by hybridization with radioactive probes or by their capacity to support reasonable levels of protein synthesis.

Expression of sgRNAs

The translation strategies for the expression of the genes encoded by sgRNAs are the same as those reported for gRNAs. sgRNAs are very efficient mRNAs, such as those expressing the CP gene, but several examples exist in which multiple genes are expressed from a single sgRNA. Nested genes are expressed by leaky scanning from the bifunctional sgRNAs of CNV [150]. Expression of three proteins from sgRNA1 of BYDV, the CP, the CP-readthrough protein, and the nested 17 kDa protein, occurs by readthrough and leaky scanning [42].

As previously mentioned, sgRNAs can code for the movement proteins. In the case of the carla-, furo-, hordei- and potexviruses, this function is encoded by the triple gene block (TGB) constituted by three overlapping genes which have been shown by mutational analyses to be involved in movement. The first and largest gene encompasses a conserved nucleotide binding motif (PMTV [161], FMV [154], BSMV [65]. In the case of FMV an apparent ATPase activity has been demonstrated as well as an RNA binding activity [154]. This first open reading frame (ORF) is expressed via a sgRNA in the case of BNYVV [62], WClMV [10], PVX [125] and BSMV [202]. The two smaller ORFs are putative membrane associated proteins [126]; sgRNAs corresponding to these genes have been detected in the case of BNYVV [62] and PVX [125].

In the case of the TRV PSG strain, even though gRNA2 only encodes the CP gene, its expression seems to occur from a sgRNA [5].

Expression of a sgRNA-encoded gene may be regulated in time as for the γb gene of BSMV which is the prevalent product obtained by *in vitro* translation but is barely detectable *in vivo* as a result of either restricted expression in time or of the localization of the protein [75].

Two reports concern sgRNAs that are expressed at high levels but possess no coding capacity: CLRV [25] and the BYDV-PAV isolate [94]. These sgRNA have been postulated to be involved in as yet unidentified regulatory functions.

In conclusion, sgRNAs are synthesized by a large number of viruses, and serve many different purposes: from high-level gene expression of CP genes to time- and amount-limited protein synthesis. These features make sgRNAs very interesting tools to study the expression of genes under the control of virus RdRps; they could be used to encode resistance genes that would be expressed only upon infection as already suggested [45]. This type of strategy requires in *trans* action of the RdRp, and all the studies carried out so far on sgRNAs have always been achieved for SGPs carried by gRNAs. It would be interesting to determine whether sgRNA synthesis can occur *in trans*, and what the minimal requirements are, as for example the 3' and 5' sequences necessary for gRNA replication.

Viruses using the ambisense strategy

The segments of the RNA genome of the tospoviruses and the tenuiviruses are either of (−) polarity or ambisense. In the ambisense strategy, the viral (v) RNA contains an ORF in its 5' region, whereas another ORF is located in the 5' region of the viral complementary (vc) RNA. The ORFs do not overlap, but are separated by an IR that in most cases can potentially adopt a stable hairpin structure (reviewed in [37, 51, 143]).

Viruses using the ambisense strategy as well as all animal viruses possessing a segmented (−)-single-stranded (ss) RNA genome have evolved a unique mechanism to produce mRNAs for the synthesis of their viral proteins. Initiation of synthesis of the viral mRNAs is primer-dependent, the primers deriving from the 5' end of host mRNAs, a mechanism known as 'cap-snatching'. As a consequence, the viral mRNAs are capped ([97, 142], reviewed in [16]) as opposed to the vRNAs and the vcRNAs that are uncapped. In the case of the plant viruses resorting to this strategy, the exquisite mechanisms of cleavage of host mRNAs to generate primers that initiate viral RNA transcription are unknown. Termination of transcription is believed to occur within the IR.

In spite of the fact that the viral RNAs are presumably uncapped, some of them can serve directly as template in translation experiments *in vitro*. This is the case of certain tenuivirus RNAs (reviewed in [143]). Translation of these RNAs follows the classical mode of initiation.

The ambisense coding strategy could provide independent regulation for the synthesis of proteins whose ORFs are located on opposite strands of a given RNA segment. Such regulation could be important for virus development, as for instance if one assumes that in certain cells synthesis of one mRNA is specifically reduced or even inhibited as opposed to synthesis of the complementary mRNA. Selective inhibition could occur at the level of the mRNAs, if the vcRNA is only produced in very low amounts in these cells, or if certain RNA species are not used for the production of

mRNAs, but rather preferentially used for viral RNA replication.

Initiation of translation: leaky scanning

Initiation of protein synthesis in eukaryotes corresponds to a series of events that ultimately lead to the formation of an 80S ribosomal complex at the level of the initiation codon on the mRNA. Both *cis*- and *trans*-acting factors affect initiation. The following *cis*-acting factors influence initiation.

1. Nature of the 5' end of the RNA. This is a cap structure, a VPg, or in the case of necroviruses and of certain satellites, it is ppX.
2. Nature of the initiator. In plant RNA viruses, the only example of a viral protein that has been shown to produce a protein whose initiation triplet is not AUG, is in the N-terminal extension (28 kDa) of the SBW-MV capsid (19 kDa). Both proteins are detected upon *in vitro* translation. A valine GUG codon upstream of the capsid ORF has tentatively been designated as the possible initiator codon for the extended form. Evidence for a function of this protein in the virus life cycle stems from the observation that equivalent proteins have been detected in two SBWMV-related viruses [164]. In addition, in certain chimeric systems, non-canonical initiation at AUU codons in the TMV 5'-untranslated region (UTR) can be recognized as initiation codons [160].
3. Nucleotide context surrounding the initiator AUG codon. Whereas the general most favorable context among eukaryotes is CA/GCCAUGG [100], in plant systems, it is AACAAUGG [107] with a purine at position −3 and/or a guanine at position +4 [position numbering relative to the A (position +1) of the AUG triplet] playing essential roles [55].
4. Nature of the leader sequence. If this sequence is poorly structured (i.e. AU-rich) it is less cap-dependent, as is the case of AlMV and TMV. Highly structured leader sequences can resort to internal initiation, as do the picornaviruses. The position of the AUG with respect to a hairpin structure also affects initiation of translation [99, 101, 118].
5. Possible interaction between the 5' and the 3' UTRs. This aspect of regulation of translation is discussed in another contribution to this voulume (Gallie).

The trans-acting factors are either cell factors distinct from the initiation factors such as those described for the picornaviruses, or viral trans-activators such as the product of gene VI of CaMV. To date, no trans-acting factor has been described for plant RNA viruses.

As with eukaryotic cellular systems, the scanning mode for translation initiation is the most common mechanism among plant RNA viruses. Multipartite RNA viruses correspond to the most straightforward adaptation of RNA viruses to the expression strategy encountered in eukaryotes and in which one gene is expressed from one mRNA, each gRNA coding for a unique protein.

The scanning mode is also the most common mode of translation initiation among ssRNA satellite viruses that code for their CP, and it is the mode used by the large ssRNA satellites that encode a non-structural protein (reviewed in [153, 54]). In addition, 'unconventional' modes of initiation are also frequently observed; they have been reviewed recently [59, 152, 104]. Three main types of unconventional or leaky scanning can be distinguished: (1) autonomous upstream ORF that leads to termination-reinitiation common among prokaryotes, (2) upstream ORF overlapping the downstream ORF, and (3) upstream ORF in frame with a downstream ORF. The first situation is encountered in CaMV, but has not been described among plant RNA viruses. On the other hand, overlapping and in-frame ORFs are extremely common, and allow viruses to maximize their translation capacity and to translation-ally control the expression of their genes. Finally, in addition to these three modes of leaky scanning, certain viral and cellular systems also initiate protein synthesis by internal initiation.

In leaky scanning, the 'first-AUG' rule that generally modulates initiation of translation in eukaryotic mRNAs is broken. In this mechanism, a portion of the 43S subunits bypasses the 5'-proximal AUG and initiates at a downstream AUG if this one is placed in a better context for initiation of translation. Two independently initiated proteins are thus produced from the same mRNA [98, 100]. Although in most cases leaky scanning was shown to occur as a result of a more favorable context around the second AUG triplet, other cis-acting elements such as the presence of secondary structures, the length of the leader sequence, and the distance separating the two initiation AUGs also contribute to the strength of initiation at a given start codon [102, 103].

Overlap in different reading frames

An interesting and well-documented example of overlap was recently shown to occur during translation of the second ORF of the PCV RNA2 [80]. RNA2 directs the synthesis of the CP (5'-proximal ORF) and of a 39 kDa protein which is probably translated from an AUG triplet that overlaps the terminator codon of the CP ORF. The authors have shown that the second AUG serves as an efficient start codon for translation of the 39 kDa protein, which is initiated in vitro by a context-dependent leaky scanning mechanism in spite of the unusually long distance (620 nt) that separates the two initiation codons.

Overlap also explains the translational control of two ORFs of sgRNA1 of BYDV and PLRV. In sgRNA1, the ORF of the 17 K protein is nested within the virus CP (ca. 23 kDa) ORF but in a distinct frame [170, 42]. The 5'-proximal CP AUG in its naturally occurring unfavorable sequence context (U at position −3 and A at +4 for BYDV; [43]) has no inhibitory effect on initiation of the 5'-distal 17 kDa protein AUG, and synthesis of the latter protein in vivo is two- to sevenfold higher than the former [170, 43]. Surprisingly, in BYDV, initiation at the CP AUG was severely impaired by mutations that decreased the initiation at the 17 kDa protein AUG [43]. To explain this intriguing observation, the authors proposed a model in which the formation of the 80S complex or pausing of the 43S complex at the 17 kDa protein AUG allows unwinding of an upstream stem loop that contains the CP AUG, and consequently initiation at the 17 kDa protein AUG codon.

In STMV, a ssRNA satellite virus, two overlapping ORFs are present that both lead to the synthesis of proteins of the expected size in vitro [121]. The CP encoded by the downstream ORF could be initiated by leaky scanning: the context surrounding its AUG is more favorable than the one surrounding the AUG of the upstream ORF.

In-frame initiation

The second ORF of RNAβ of BSMV codes for a pair of proteins, named βb and βb', that are translated in vivo from two alternative in-frame initiator AUGs. Mutagenesis analyses have demonstrated that translation of the βb' protein is initiated in vivo by ribosomes that bypass the βb start codon as a result of context-dependent leaky scanning [140].

Leaky scanning has usually been ascribed to a less favorable context around the first AUG codon as compared to the second AUG codon. Therefore in all cases in which mutagenesis rendered the sequence context around the first AUG more favorable, the frequency of translation initiation at the second initiator AUG was reduced. However at least in one case, in the ATCC66 strain of BMV, a correlation between the length of the 5' UTR of the viral RNA and occurrence of in-frame initiation was proposed. In this strain, synthesis of a full-length and of an N-terminally truncated form of the CP, was attributed not only to a codon context effect but possibly also to the reduction in length of the 5' UTR resulting from the absence of two adjacent adenine residues immediately upstream of the first AUG codon [122]. This truncated CP form was shown to be functional *in vivo*.

A conventional leaky scanning mechanism was proposed to explain how translation of PPV genomic RNA initiates at the second in-frame AUG at position 147 despite the presence of a first AUG codon in a poor context at position 36 [147]. However the 5' UTR contains an AU-rich region that was shown in TEV and in PSbMV to enhance translation *in vitro* and *in vivo* [30, 128] and has led these authors to suggest that potyviruses could use the same mechanism of internal initiation of translation as described for picornaviruses (reviewed in [87]), even though the 5' UTRs of picornaviruses (that vary in length from 610 to 1200 nt) contain several silent AUGs preceding the authentic initiation codon and are significantly more structured and longer than the ones present in potyviruses. Internal initiation involves binding of ribosomes and of *trans*-acting factors to a complex secondary/tertiary structure within the 5' UTR referred to as internal ribosome entry site (IRES [1]). A possible potyvirus IRES was proposed to lie within the last 63 nt of the UTR of TEV [30] but definitive evidence for this is still lacking.

Another controversial example is observed in CPMV. The middle component RNA (M-RNA) codes for two colinear polyproteins of 105 and 95 kDa that are initiated at two in-frame AUG codons located at positions 161 and 512. *In vitro* translation of a dicistronic mRNA in a wheat germ or reticulocyte system has demonstrated that M-RNA allows internal binding of ribosomes as do picornaviruses [174, 190]. Surprisingly, the synthesis but not the ratio of both proteins was enhanced by the addition of eIF-4F to the *in vitro* assays. However, *in vivo* experiments using an animal system failed to confirm the internal initiation observed *in vitro* and led to the proposal that access to the AUG codon is instead modulated by a leaky scanning mechanism [12] that seems to be largely independent of the presence of a cap structure. The fact that the AUG of the 95 kDa protein lies in a more favorable context (G at positions −3 and +4) for initiation than the AUG of

the 105 kDa protein (A at −3 and U at +4) argues in favor of this possibility.

Regulation at the level of elongation: frameshift

Various types of regulation can occur at the level of elongation of protein synthesis. One of them, frameshifting occurs on a single mRNA. On the other hand, editing and splicing that both result from transcriptional or post-transcriptional modifications, lead to the production of two mRNAs. In all cases, these regulation mechanisms lead to the synthesis of two proteins that are identical in their N-terminal region, but differ in their C-terminal region. Transcriptional editing is encountered among the Paramyxoviridae, but has not been reported among plant viruses. Splicing has been reported to occur among plant DNA viruses. As for frameshift, it is a very frequent strategy used by RNA viruses, and it will be discussed here.

A number of review articles have recently appeared on frameshifting in viruses [152, 24]. Frameshift results from movement of ribosomes either in the 5' direction (−1 frameshift) or in the 3' direction (+1 frameshift) on the mRNA. Two proteins (the frame and the transframe proteins) are produced that are identical from the N terminus to the frameshift point, but differ beyond that point; the frame protein is always more abundant than the transframe protein.

This strategy is frequently encountered for viral RNAs of (+) polarity and is also found in certain cellular systems. In animal and plant viruses, frameshift usually permits the expression of the viral replicase; in most cases, the frame protein harbors the RdRp. The occurrence of −1 frameshift is far more common than +1 frameshift; the structural elements and the mechanisms involved are distinct for these two modes of frameshifting.

−1 Frameshift

The following signals are required in the RNA to promote -1 frameshifting.

1. A heptanucleotide sequence also known as 'slippery' sequence. In most cases, it contains two homopolymeric triplets of the type XXXYYYZ (where X = A, G, U; Y = A, U; Z = A, C, U). Upon reaching the shifty heptanucleotide sequence, the two ribosome-bound tRNAs that are in one reading frame (X.XXY.YYZ) shift by one nucleotide towards the

5′ direction (XXX.YYY.Z) retaining two out of three base-paired nucleotides with the mRNA [86].

2. Nucleotides downstream of the point of frameshift. In virtually all cases, this comprises a hairpin structure that in many instances can additionally form a pseudoknot with downstream RNA stretches.

3. A spacer region between the slippery sequence and the hairpin structure. The length of the spacer varies between 4 and 9 nt.

An interesting feature of viruses resorting to frameshifting is that the extent of frameshifting in animal viruses is always higher than in plant viruses.

Table 2 lists the plant genera and viruses in whose genome frameshift has been demonstrated either on the basis of *in vitro* (and sometimes *in vivo*) translation studies accompanied or not by mutations in the signals required for frameshifting, or postulated on theoretical grounds. Table 2 also indicates the nature of the slippery sequence, and whether the presence of a pseudoknot structure is (or could potentially be) required for efficient frameshift. It also designates the proteins participating in the frameshift event.

It is among the luteoviruses that the requirements for frameshifting have been examined the most thoroughly. The overlap region between the two ORFs is several hundred nt long in all cases except in the BYDV-PAV isolate where it is 13 nt long. The slippery sequence is bordered in BYDV-PAV by a UAG codon terminating the 39 kDa ORF, and in BWYV by a UAA terminating the 66 kDa ORF. Frameshift was monitored in plant protoplasts by introducing the β-galacturonidase (GUS) coding region in the −1 reading frame relative to the AUG of the 39 kDa protein [21]. A complex pseudoknot structure or a large stem-loop can be constructed downstream of the slippery sequence. In the case of the BYDV-RPV isolate, a possible slippery sequence followed by a simple pseudoknot has been proposed to be implicated in frameshifting [172]. Studies on the frameshift region in a German isolate of PLRV (PLRV-G) *in vitro* and in protoplasts [141] indicate that the slippery sequence is followed by a stable hairpin, but without the requirement for a pseudoknot. These results contrast with those obtained using a Polish isolate of PLRV (PLRV-P); in this latter case, frameshift is pseudoknot-dependent [105]. Finally, *in vitro* translation studies with BWYV RNA have located the site of frameshift; the requirement for a pseudoknot very similar to the one described for PLRV-P, has been proposed [56].

PEMV harbors a bipartite RNA genome. Both RNA1 and RNA2 of PEMV possess ORFs whose proteins potentially contain RdRp-related motifs that appear to be produced by −1 frameshifting. The fact that RNA1 bears strong sequence similarity with the luteovirus genome (PLRV and BWYV) [38] has led to the suggestion [39] that PEMV resulted from the association of two distinct viral RNA species, one of which would be derived from the luteovirus group. It has been suggested [172] that in RNA1 the slippery sequence might be followed by a pseudoknot structure. In the case of RNA2, a protein of the size expected of the frameshift product was detected by *in vitro* translation. The transframe protein in RNA2 contains the helicase and the RdRp-related motifs.

Dianthoviruses contain a bipartite RNA genome. Sequence data are available for three viruses of this genus, CRSV, RCNMV and SCNMV. RNA1 contains the elements for RNA replication. Translation studies performed *in vitro* with CRSV [158] and RCNMV [196, 95] have demonstrated that a hairpin structure downstream of the slippery sequence can be formed; no potential pseudoknot has been detected. In all three dianthoviruses, the slippery sequence is identical. It is followed immediately by the triplet terminating the frame protein. However, in the case of SCNMV, the frameshift sequence, as well as the possibility of forming a hairpin downstream of the slippery sequence have been proposed solely by comparison with CRSV and RCNMV [158].

PVM presents a rather unique situation. The 34 kDa CP lies just upstream of a 12 kDa nucleic acid-binding protein [69]. Experimental evidence suggests that two mechanisms probably lead to the formation of the 12 kDa protein, either internal initiation, or by −1 frameshift which produces the CP/12 kDa transframe protein. An unorthodox slippery sequence has been identified on the basis of site-directed mutagenesis. It contains the UGA for the CP which overlaps the initiation codon for the 12 kDa protein. A 'P-site slippage' model has been proposed for this −1 frameshift event, in which peptide bond formation would take place after slippage at the frameshift site. There is no evidence for the presence of a stem loop structure or a pseudoknot downstream of the slippery sequence.

+1 Frameshift

This strategy requires a slippery run of bases, and an A-site with a rare or 'hungry' codon or a termination codon. It is rarely encountered among viruses, and among plant viruses, only the closteroviruses probably resort to this strategy. The members of this virus group

Table 2. Plant RNA viruses that resort to frameshifting.

Virus	RNA	Shifty sequence	Proteins	Mutations	Assay	DS sequence	References and comments
Alpha-like							
carla							
PVM (1)	1	AAAAUGA	CP/12K	+	a		[70] reviewed in [152]; Overlap of AUG and UGA
clostero							
BYV (1)	1	GGGUUUA	295K/48K	−		PK	[3]; +1; UA is followed by G, terminating 295K protein
CCSV (1)	1	GUUUGAC	ORF1a/b[1]	+	a	PK	[172]; +1; UGA terminates ORF1a
CTV (1)	1	GCGUUCG	349K/57K	−			[91]; +1; no 'slippery' sequence
LIYV (2)	1	AAAG	217K/55K	−			[96]; +1; no 'slippery' sequence
Carmo-like							
diantho							
CRSV (2)	1	GGAUUUU	27K/54K	−	a	HP	[158]
RCNMV (2)	1	GGAUUUU	27K/57K	+	a	HP	[95, 196, 197]
SCNMV	1	GGAUUUU	27K/57K	−			[58]
luteo							
BYDV-PAV (1)	1	GGGUUUU	39K/60K	−	a,b	PK	[21, 41]
BYDV-RPV (1)	1	GGGAAAC	71K/72K	−		PK	[172, 191]
Sobemo-like							
sobemo							
BWYV (1)	1	GGGAAAC	66K/67K	+	a	PK	[56, 191]
CfMV (1)	1	UUUAAAC	64K/56K	−	a	HP	[111]
luteo							
PLRV (1)	1	UUUAAAU	70K/67K	+	a,b	HP	[141]
					a	PK	[105]
Others							
enamo							
PEMV (2)	1	GGGAAAC	84K/67K	−		PK	[36]
	2	GAUUUUU	33K/65K	−	a		[39]

The presentation of the viruses is as in Table 1. RNA indicates the RNA whose proteins undergo frameshifting. Shifty sequence is the heptanucleotide sequence in which frameshift occurs. Proteins designates the CP or the size of the two ORFs. [1] indicates that the size of the corresponding ORFs is unknown. Mutations indicates whether mutation analyses have been (+) or have not (−) been performed. Assay indicates if *in vitro* (a) or *in vivo* (b) assays have been performed; DS Seq is the nature of the downsteam sequence involved in frameshifting, either a hairpin (HP) or a pseudoknot (PK). Frameshift is −1 except for the closteroviruses where +1 is specified. K, Kilodalton.

are very heterogeneous, and this is reflected in the size of their monopartite genome, the presence in certain closteroviruses of a bipartite instead of a monopartite genome, and the elements presumably required for frameshift.

In BYV genome, a shifty heptanucleotide sequence encompasses the first 2 nt of the codon terminating the frame ORF, and is followed by a hairpin structure that could potentially form a pseudoknot [3].

A similar situation is encountered in CCSV. In CCSV RNA it is as yet unclear whether the rather weak pseudoknot and/or other elements downstream of the frameshift signal are required for frameshift [172].

The proposed shifty sequence in CTV RNA terminates with the makings of the CGG codon, possibly a rare arginine codon [91]. No hairpin structure is detected downstream of the shifty region.

LIYV possesses a bipartite RNA genome [96]. RNA1 bears the domains related to genome replication. No shifty sequence could be detected, although the amino acid sequence in the region of overlap shows that the N-terminal region of the 54 kDa ORF aligns with the C-terminal region of the BYV 295 kDa ORF. +1 frameshift in LIYV could involve the sequence AAAG with slippage of a tRNA^{Lys}. No downstream structure that might suggest frameshift could be detected.

Finally, the situation in ACLSV is different still. This virus possesses a monopartite genome that is much shorter than that of the other closteroviruses with a monopartite genome discussed above. The ORFs for a 216 kDa followed by a 50 kDa protein overlap; however, frameshift is highly unlikely to occur in this virus, since the helicase and the RdRp domains are located in the 216 kDa protein, and since the 50 kDa protein ORF begins with an AUG codon [60].

It is interesting to examine the role played by the codon terminating the frame protein, when it is located just downstream of the slippery sequence. This is the case of BYDV-PAV, BWYV, PVM and the dianthoviruses. Changing the termination codon to a sense codon dramatically reduces frameshifting in the case of BYDV-PAV [21]. Likewise, in PVM where the termination codon is part of the slippery sequence, mutation of this codon to a sense codon demonstrated that the termination codon is indispensable for frameshift [70]. On the other hand, the same type of experiment performed with RCNMV RNA1 has no significant effect on the level of frameshift [95].

Regulation at the level of termination: readthrough

Under 'normal' circumstances, the presence of an in-frame termination codon in an mRNA dictates termination of translation. In recent years, there has been increased interest in the study of recognition of termination codons by tRNAs as a means of regulating the synthesis of given proteins at the level of termination. This phenomenon known as suppression of termination, is frequently encountered among plant viruses (reviewed in [173, 180, 152]).

In the presence of the appropriate nonsense suppressor aminoacyl-tRNA which recognizes the termination (or nonsense) codon at the end of a given cistron, two proteins are synthesized, a stopped protein and a readthrough protein that are identical over the total length of the stopped protein. The stopped protein is always more abundant than the readthrough protein.

Readthrough has been investigated from two points of view, the elements in the mRNAs that favor readthrough in *cis*, and the nature of the suppressor tRNAs.

mRNAs

Table 3 lists the genera and plant viruses whose genome has been shown or postulated to resort to readthrough, the nature of the suppressible termination codon, the designation of the stopped and readthrough proteins, whether *in vitro* or *in vivo* assays were performed to verify readthrough and the function that the readthrough protein provides for the virus wherever this has been established. As opposed to frameshift which among plant viruses always provides the polymerase, readthrough can in addition to providing the polymerase, also lead to fusion of the CP to the protein produced by the readthrough domain. In certain cases, it has been demonstrated that the latter protein is required for aphid transmission and/or viral RNA encapsidation.

The furoviruses in particular, are a heterogeneous group of viruses, and this is reflected among others by their use of suppressible termination codons. SBW-MV uses two suppressible UGA codons, one in RNA1 (polymerase) and one in RNA2 (capsid/fusion protein) respectively [164], PCV uses a UGA codon in RNA1 to produce the RdRp [81], whereas BNYVV uses a UAG codon in RNA2 to produce the fusion protein [129].

The monopartite RNA genome of carmo-, necro-, and tombusviruses, as well as of MCMV, and of OCSV

Table 3. Plant RNA viruses that resort to readthrough.

Virus	RNA	Term codon	Proteins	Assay	References and comments
Alpha-like					
furo					
BNYVV (4)	2	UAG	CP/75K	a,b	[129, 159, 203]
SBWMV	1	UGA	150K/209K		[82, 164]
(2)	2	UGA	CP/84K	a	
PCV (2)	1	UGA	130K/191K	a	[81, 164]
tobamo					
TMV (1)	1	UAG	126K/183K	a,b	[85, 137, 167]
tobra					
TRV (2)	1	UGA	134K/194K	a	[76, 138]
PEBV	1	UGA	141K/201K		[110]
Carmo-like					
carmo					
CarMV (1)	1	UAG, UAG	27K/86K/98K	a	[73, 77]
TCV (1)	1	UAG	28K/88K	a	[194]
CCFV (1)	1	UAG	28K/87K		[165]
MCMV (1)	1	UAG	50K/111K		[130]
		UGA	9K/33K		
luteo					
BYDV-PAV (1)	1	UAG	CP/72K	a,b	[42, 52, 120, 193]
necro					
TNV (1)	1	UAG	23K/82K		[116]; 82K readthrough protein not observed *in vitro*
tombus					
AMCV (1)	1	UAG	33K/92K		[171]
CNV (1)	1	UAG	33K/92K		[151]
CyRSV (1)	1	UAG	33K/92K		[71]
TBSV	1	UAG	33K/92K	a	[78, 79]
Sobemo-like					
luteo					
BWYV (1)	1	UAG	CP/74K	b	[22, 185, 186]
PLRV (1)	1	UAG	CP/80K	a,b	[6, 152]
SDV (1)	1	UAG	CP/80K		[144]

Table 3. Continued.

Virus	RNA	Term codon	Proteins	Assay	References and comments
Others					
OCSV	1	UAG	23K/84K		[15]
(1)					
MNSV	1	UAG	29K/89K		[149]
(1)		UAG	7K/14K		
enamo					
PEMV	1	UGA	CP/55K		[38]
(2)					

The presentation of the viruses is as in Table 1. RNA indicates the RNAs whose proteins undergo readthrough. Term codon indicates the nature of the suppressible codon. Proteins designates the stopped/readthrough protein; the stopped protein is indicated by CP or by the size when it does not correspond to the CP, and the readthrough protein is indicated by the total size of the resulting protein. The readthrough protein always encompasses the polymerase when the stopped protein is not CP. Assay is as in Table 2. K, Kilodalton.

which is carmo- and tombus-like, contains a long ORF of 87 kDa to 111 kDa interrupted by a UAG codon. The readthrough domain contains the conserved GDD motif that characterizes RdRps. Translation studies performed *in vitro* using the RNA of CarMV and TCV have demonstrated that the protein corresponding to the readthrough domain is indeed synthesized. CarMV is unique in so far as its 98 kDa protein results from a double readthrough event, as suggested by *in vitro* and *in vivo* translation studies [77] and indicated by RNA sequencing studies [73].

In addition to the termination codon, features in the RNA are also required *in cis* for efficient readthrough. With the exception of TMV RNA, relatively few studies have centered on the nature of these *cis* elements. The two downstream codons appear to be crucial for efficient readthrough *in vivo* and *in vitro* [167, 179]. No evidence for the requirement of a downstream hairpin structure exists, as has been demonstrated for certain retroviruses (reviewed in [172]).

Suppressor tRNAs

To date, suppressor tRNAs have been isolated from tobacco leaves, wheat leaves and lupin. Two tRNA^{Tyr} bearing a 5'-GΨA-3' anticodon have been isolated from tobacco leaves and shown to suppress the TMV UAG codon *in vitro*. Since insertion of a tyrosine codon in place of the UAG codon in the TMV genome does not abolish infection, and since the stopped and the readthrough proteins are found in infected tobacco protoplasts, it is reasonable to assume that the two tRNA^{Tyr} are responsible for suppression *in vivo* [11, 200].

tRNA^{Trp} and tRNA^{Cys} have been isolated from tobacco plants. They are capable of overcoming the suppressible UGA codon in TRV RNA1 [201, 178]. A suitable context around the suppressible UGA codon is important for efficient readthrough, since none of these suppressor tRNAs can overcome the UGA codon in β-globin mRNA.

It is interesting to note that to date, only UAG and UGA suppressible codons have been described. Nevertheless, mature TMV virions are produced when the naturally occurring suppressible UAG codon in the genome is replaced by a UAA codon [85], suggesting that the host contains a tRNA capable of suppressing the UAA codon.

Proteolytic processing

Proteolytic processing of a precursor polyprotein is a translation strategy employed by many viruses to produce more than one protein from a single mRNA. To date, all known RNA viruses that express a polyprotein which is then cleaved by a virus-encoded proteinase are (+)-stranded. Cleavage of the polyprotein precursor can occur co- or post-translationally, and it can occur in *cis* and/or in *trans*. In addition, the activity of a processed protein may be different from its activity when it is in the precursor form. Since the structural and functional features of viral-encoded proteinases have been discussed in detail elsewhere [50], only a brief overview of the different classes of proteinases employed by plant viruses during proteolytic processing will be presented here.

Virus-encoded proteinases

The virus-encoded proteinases are related in sequence and activity to known cellular proteinases. Plant RNA viruses encode chymotrypsin-like proteinases, or papain-like proteinases.

Chymotrypsin-like proteinases. They are structurally related to chymotrypsin. Their catalytic site is composed of a triad (H-D/E-S or H-D/E-C). A serine is present in the active site of chymotrypsin, whereas a serine or a cysteine is present in the active site of chymotrypsin-like proteinases.

Papain-like cysteine proteinases. They are structurally related to papain. Their active site is a dyad composed of a cysteine and a histidine (C-H) residue. Here, the histidine is C-terminal to the cysteine residue, whereas the opposite applies to the corresponding residues in the chymotrypsin-like proteinases.

Plant RNA viruses and proteolytic processing

A compilation of the results reported to date concerning all known plant RNA viruses (and their genera) making use of proteolytic processing during their life cycle is shown in Table 4. Additionally, for each virus, the size of the protein precursor, the final cleavage products and the proteinase responsible for cleavage are indicated.

Potyviruses

The potyviruses, a well-studied group of viruses belonging to the picorna-like superfamily, make extensive use of proteinases during expression of their genome [148]. They constitute a remarkable example since all of their functional proteins derive from proteolytic processing of a single polyprotein precursor.

Processing of the viral polyprotein precursor into functional proteins is ensured by three virus-encoded proteinases: the P1 proteinase and the helper component-proteinase (HC-Pro) that are responsible for two autoproteolytic cleavage events within the N-terminal region of the polyprotein, and the nuclear inclusion–a (NIa) proteinase that is responsible for the remaining cleavage events within the C-terminal two-thirds of the polyprotein. Additionally, all these proteinases are multifunctional proteins that also play other important roles during the life cycle of the virus.

P1 proteinase. P1 is derived from the N-terminal end of the full-length potyvirus polyprotein and is one of the most variable proteins both in length and in base composition between potyviruses [175]. In TEV, the P1 proteolytic domain is located within the C-terminal 147 amino acid residues that contain the functional catalytic triad [188]. This proteolytic domain is structurally related to the chymotrypsin class of serine proteinases and is responsible for autoproteolytic cleavage between itself and HC-Pro [189, 188]. Proteolytic separation of P1 from HC-Pro, but not P1 proteolytic activity per se, is essential for TEV viability in plants [187].

HC-Pro. This protein is adjacent to the C-terminus of the P1 proteinase. The C-terminal region of HC-Pro is a papain-type proteinase that releases HC-Pro from the P3 protein [26, 32]. In TEV, the catalytic dyad was shown by mutational analysis to be composed of cysteine and histidine [131].

NIa proteinase. The N-terminal region of NIa harbors the VPg, whereas the C-terminal region is associated with a trans- and autoproteolytic activity responsible for most of the processing that occurs in the potyvirus polyprotein [28, 27]. The N-terminus of NIa also contains a nuclear localization signal that directs the protein to the cell nucleus [31]. In infected plants, proteolytic activity is initially associated with a 49 kDa polyprotein precursor. This precursor can be internally processed at a suboptimal cleavage site to liberate the VPg and a functional 27 kDa proteinase domain [49] that is normally present later during infection. NIa is a chymotrypsin-like proteinase that is structurally related to the picornavirus 3C proteinase, and to the comovirus 24 kDa proteinase described below. The proteinase cleaves at specific sites characterized by highly conserved heptapeptide sequences that present different cleavage rates [29, 48]. Recently, a self-cleavage site located within the C-terminus of the processed 27 kDa proteinase has also been described [136]. Regulation of processing efficiency has been suggested to occur not only by the amino acid sequence at the cleavage site but also by a differential compartmentalization of the processed forms of NIa within the cell. The fact that absence of proteolytic separation between NIa and the downstream 6 kDa peptide, described as a membrane-associated protein, hindered translocation of the resulting 55 kDa protein to the nucleus [145], is an example of such a possible regulation.

Based on sequence comparisons, it has been proposed that in BaYMV, an HC-Pro-type proteinase is encoded by RNA1, and an NIa-type proteinase by RNA2 [93, 92].

Table 4. Plant RNA viruses encoding proteinases.

Virus	RNA	Precursor	Cleavage products	Proteinases	References and comments
Picorna-like					
como					
CPMV	B	200K	32K/58K/VPg/24K/87K	24K[a] and 32K	[40, 139]
(2)	M	105K (95K)	58K (48K)/VP37/VP23		
nepo					
GFLV	1	253K	~133K/VPg/24K/RdRp	24K[a]	[112]
(2)	2	122K	28K/38K/56K		
poty					
TEV	1	351K	P1/HC-Pro/P3/CI/6K/NIa/NIb/CP	P1[a], HC-Pro[b] and NIa[a]	[148]
(1)					
Alpha-like					
carla					
BBScV	1	223K	166K/57K	166K[b]	[106]
(1)					
clostero					
BYV	1	295K (348K)	66K/26K	26K[b]	[3]
(1)					
tymo					
TYMV	1	206K	140K/66K	140K[b]	[20, 90, 124]
(1)					

The presentation of the viruses is as in Table 1. RNA designates the RNAs whose proteins undergo proteolytic maturation. Precursor is size of the polyprotein; the protein in brackets results from in-frame initiation at a second AUG. Cleavage products are the final proteins; underlined are the proteinases. Proteinases are either chymotrypsin-like proteinase (a) or papain-like proteinase (b). In CPMV, the 32K (32 kDa) protein is an accessory protein. K, Kilodalton.

Comoviruses, nepoviruses

Como- and nepoviruses are bipartite RNA viruses that are very similar in genome organization to potyviruses. They synthesize polyproteins that are cleaved in *cis* and in *trans* by a virus-encoded chymotrypsin-like proteinase.

The best studied comovirus is CPMV. The B-RNA (which encodes the replication functions), codes for a 24 kDa chymotrypsin-like proteinase [40]. This proteinase cleaves the polyproteins encoded by B-RNA and M-RNA (which encodes the movement and structural proteins). An interesting feature of the CPMV proteinase is that its activity is regulated by a two-component system, since it requires an accessory viral-encoded protein of 32 kDa in addition to the 24 kDa proteinase [192, 139]. The 32 kDa protein is essential for cleavage of the polyprotein encoded by the M-RNA. This processing leads to the 58 and 48 kDa movement proteins and to the CP precursor of 60 kDa. The 60 kDa precursor is then cleaved by the 24 kDa proteinase but in this case the 32 kDa is not essential. The 32 kDa protein plays an important role in the processing of the 170 kDa polyprotein encoded by the B-RNA, since the rate of cleavage is decreased in the presence of the 32 kDa protein. It seems that the 32 kDa protein is essential for the expression of the CPMV protein and is the main regulator of the CPMV life cycle during infection.

The chymotrypsin-like 24 kDa proteinase of GFMV is encoded by RNA-1 [112]. The main difference between this proteinase and the other chymotrypsin-like proteinases of plant RNA viruses resides in their substrate specificity which is dependent on a conserved leucine residue. The nepovirus proteinase does not resort to an accessory protein.

Based on sequence comparisons, a chymotrypsin-like proteinase has been postulated for two other groups of viruses, PYFV [176] and RTSV [162].

Sobemoviruses, luteoviruses

In these two groups of RNA viruses, sequence alignments suggest that proteinases are encoded by the viral genome, but to date no biochemical evidence for their activity has appeared. Their proteinases are closer in sequence to cellular serine proteinases than any of the viral chymotrypsin-like proteinases known to date [66, 186, 115].

It has been suggested that RNA1 of PEMV codes for a proteinase related to sobemo- and luteovirus proteinases [38].

Tymoviruses

TYMV belongs to the alpha-like superfamily (sindbis-like). Its papain-like proteinase is situated in the central part of the large nonstructural polyprotein expressed by the virus and presents features in common with papain [19, 155, 90]. Processing of the TYMV polyprotein by the proteinase is essential for replication of the virus in plant protoplasts. When the catalytic site of the proteinase is mutated, cleavage is abolished and the virus is unable to replicate ([19]; K. Séron and G. Kadaré, unpublished data).

Other viruses with papain-like cysteine proteinases

In additions to the HC-Pro of potyviruses which is a cysteine proteinase, evidence also exists for a papain-like proteinase activity in BBScV [106]. The presence of cysteine-like proteinases has been proposed on the basis of computer analyses, but without biochemical evidence [155] for ASGV, ACLSV, PVM, BNYVV and ASPV.

Closteroviruses

BYV contains a papain-like proteinase activity [3]. In addition, the possible existence of an aspartic proteinase has been postulated for BYV by computer sequence analysis [3]; however, there is as yet no experimental evidence in support for such an activity.

Discussion

Probably one of the most striking features that emerges from this review, is the extraordinary variety of strategies used by viruses for the expression and regulation of their proteins. It is moreover noteworthy that most of the expression strategies known to date were first demonstrated among virus systems of animal or plant origin, and only subsequently shown to also occur in eukaryotic cells. Consequently, detailed characterization of these mechanisms among viruses is mandatory to better understand the fundamental reactions under-

lying these processes, and it is often a stepping stone that allows us to unravel similar processess in eukaryotic cells.

It should be possible to subvert some of these strategies of expression to develop methods of resistance that would interfere only little with the host cell. Such an approach would be particularly important to counteract animal viruses. Knowledge of the organization and expression of viruses also makes viruses very useful tools for the introduction of novel genes, either fused to the viral genes, or encoded by a new sgRNA. This approach has for instance been successfully used to produce antigens exposed on the surface of TMV particles [177]. It would be pertinent to extend such an approach to other viral systems.

It is surprising that among all the strategies used by eukaryotic RNA viruses, no post-translational regulation mechanism (putting aside proteinase activity) by a virally-encoded protein has been detected to date. Indeed, this has been well documented for prokaryotic RNA viruses of the R17 family [195].

Finally, it should be emphasized that many viruses resort to several strategies to express all their proteins. Having reached the end of this review, we would like to award a prize to the virus group that combines the largest number of strategies. In this respect, the luteoviruses most certainly deserve the prize of highest complexity: they combine sgRNA synthesis, leaky scanning, readthrough and frameshift. Moreover, we have not given up hope that data will soon emerge demonstrating that they also use post-translational cleavage.

Acknowledgements

I.G.M. is grateful to CNPq-Brazil for a fellowship, and K.S. was recipiant of a fellowship from the 'Ministère de la Recherche et de l'Enseignement Supérieur'. The Institut Jacques Monod is an 'Institut Mixte, CNRS-Université Paris 7'.

Addendum

Two publications, concerning the characterization of two sgRNAs, have appeared since completion of this manuscript. The first one concerns the mapping of the RCNMV sgRNA and the expression of the encoded CP gene (Zavriev SK, Hickey CM & Lommel SA. 1996 Mapping of the red clover necrotic mosaic subgenomic RNA. Vicology **216**: 407–410). The second publication deals with the analysis of the *cis*-acting elements required for gRNA and sgRNA synthesis of PVX (Kim K-H & Hemenway C. 1996. The 5' non-translated region of potato virus X RNA affects both genomic and subgenomic RNA synthesis. J. Virol. **70**: 5533–5540).

References

1. Agol VI: The 5'-untranslated region of picornaviral genomes. Adv Virus Res 40: 103–180 (1991).
2. Agranovsky AA, Bokyo VP, Karasev AV, Lunina NA, Koonin EV, Dolja VV: Nucleotide sequence of the 3'-terminal half of beet yellows closterovirus RNA genome: unique arrangement of eight virus genes. J Gen Virol 72: 15–23 (1991).
3. Agranovsky AA, Koonin EV, Boyko VP, Maiss E, Frötschl R, Lunina NA, Atabekov JG: Beet yellows closterovirus: complete genome structure and identification of a leader papain-like thiol protease. Virology 198: 311–324 (1994).
4. Ahlquist P, Luckow, V, Kaesberg P: Complete nucleotide sequence of brome mosaic virus RNA 3. J Mol Biol 153: 23–38 (1981).
5. Angenent GC, Linthorst HJM, van Belkum AF, Cornelissen BJC, Bol JF: RNA 2 of tobacco rattle virus strain TCM encodes an unexpected gene. Nucl Acids Res 14: 4673–4682 (1986).
6. Bahner I, Lamb J, Mayo MA, Hay RT: Expression of the genome of potato leafroll virus: readthrough of the coat protein termination codon *in vivo*. J Gen Virol 71: 2251–2256 (1990).
7. Balmori E, Gilmer D, Richards K, Guilley H, Jonard G: Mapping the promoter for subgenomic RNA synthesis on beet necrotic yellow vein virus RNA3. Biochimie 75: 517–521 (1993).
8. Barker RF, Jarvis NP, Thompson DV, Loesch-Fries LS, Hall TC: Complete nucleotide sequence of alfalfa mosaic virus RNA3. Nucl Acids Res 11: 2881–2891 (1983).
9. Barrera I, Schuppli D, Sogo JM, Weber H: Different mechanisms of recognition of bacteriophage Qβ plus and minus strand RNAs by Qβ replicase. J Mol Biol 232: 512–521 (1993).
10. Beck DL, Guilford PJ, Voot DM, Andersen MT, Forster RLS: Triple gene block proteins of white clover mosaic potexvirus are required for transport. Virology 183: 695-702 (1991).
11. Beier H, Barciszewska M, Krupp G, Mittnacht R, Gross HJ: UAG readthrough during TMV RNA translation: isolation and sequence of two tRNAs^Tyr with suppressor activity from tobacco plants. EMBO J 3: 351–356 (1984).
12. Belsham GJ, Lomonossoff GP: The mechanism of translation of cowpea mosaic virus middle component RNA: no evidence for internal initiation from experiments in an animal cell transient expression system. J Gen Virol 72: 3109–3113 (1991).
13. Boccara M, Hamilton WDO, Baulcombe DC: The organisation and interviral homologies of genes at the 3' end of tobacco rattle virus RNA1. EMBO J 5: 223–229 (1986).
14. Boccard F, Baulcombe D: Mutational analysis of cis-acting sequences and gene function in RNA3 of cucumber mosaic virus. Virology 193: 563–578 (1993).
15. Boonham N, Henry CM, Wood KR: The nucleotide sequence and proposed genome organization of oat chlorotic stunt virus, a new soil-borne virus of cereals. J Gen Virol 76: 2025–2034 (1995).
16. Bouloy M: Bunyaviridae: genome organization and replication strategies. Adv Virus Res 40: 235–275 (1991).
17. Bouzoubaa S, Niesbach-Klösgen U, Jupin I, Guilley H, Richards K, Jonard G: Shortened forms of beet necrotic yellow vein virus RNA-3 and -4: internal deletions and a subgenomic RNA. J Gen Virol 72: 259–266 (1991).
18. Bouzoubaa S, Ziegler V, Beck D, Guilley H, Richards K, Jonard G: Nucleotide sequence of beet necrotic yellow vein virus RNA-2. J Gen Virol 67: 1689–1700 (1986).
19. Bransom KL, Dreher TW: Identification of the essential cysteine and histidine residues of the turnip yellow mosaic virus protease. Virology 198: 148–154 (1994).
20. Bransom KL, Weiland JJ, Dreher TW: Proteolytic maturation of the 206-kDa nonstructural protein encoded by turnip yellow mosaic virus RNA. Virology 184: 351–358 (1991).
21. Brault V, Miller WA: Translational frameshifting mediated by a viral sequence in plant cells. Proc Natl Acad Sci USA 89: 2262–2266 (1992).
22. Brault V, van den Heuvel JFJM, Verbeek M, Ziegler-Graff V, Reutenauer A, Herrbach E, Garaud J-C, Guilley H, Richards K, Jonard G: Aphid transmission of beet western yellows luteovirus requires the minor capsid read-through protein P74. EMBO J 14: 650–659 (1995).
23. Brederode FT, Koper-Zwarthoff EC, Bol JF: Complete nucleotide sequence of alfalfa mosaic virus RNA 4. Nucl Acids Res 8: 2213–2223 (1980).
24. Brierley I: Ribosomal frameshifting on viral RNAs. J Gen Virol 76: 1885–1892 (1995).
25. Brooks M, Bruening G: A subgenomic RNA associated with cherry leafroll virus infections. Virology 211: 33–41 (1995).
26. Carrington JC, Cary SM, Parks TD, Dougherty WG: A second proteinase encoded by a plant potyvirus genome. EMBO J 8: 365–370 (1989).
27. Carrington JC, Dougherty WG: Small nuclear inclusion protein encoded by a plant potyvirus genome is a protease. J Virol 61: 2540–2548 (1987).
28. Carrington JC, Dougherty WG: Processing of the tobacco etch virus 49K protease requires autoproteolysis. Virology 160: 355–362 (1987).
29. Carrington JC, Dougherty WG: A viral cleavage site cassette: identification of amino acid sequences required for tobacco etch virus polyprotein processing. Proc Natl Acad Sci USA 85: 3391–3395 (1988).
30. Carrington JC, Freed DD: Cap-independent enhancement of translation by a plant potyvirus 5' nontranslated region. J Virol 64: 1590–1597 (1990).
31. Carrington JC, Freed DD, Leinicke AJ: Bipartite signal sequence mediates nuclear translocation of the plant potyviral NIa protein. Plant Cell 3: 953–962 (1991).
32. Carrington JC, Herndon KL: Characterization of the potyviral HC-Pro autoproteolytic cleavage site. Virology 187: 308–315 (1992).
33. Carrington JC, Morris TJ: High resolution mapping of carnation mottle virus-associated RNAs. Virology 150: 196–206 (1986).
34. Cornelissen BJC, Janssen H, Zuidema D, Bol JF: Complete nucleotide sequence of tobacco streak virus RNA3. Nucl Acids Res 12: 2427–2437 (1984).

35. Cornelissen BJC, Linthorst HJM, Brederode FT, Bol JF: Analysis of the genome structure of tobacco rattle virus strain PSG. Nucl Acids Res 14: 2157–2169 (1986).

36. Dawson WO, Lehto KM: Regulation of tobamovirus gene expression. Adv Virus Res 38: 307–342 (1990).

37. de Haan P, Wagemakers L, Goldbach R, Peters D: Tomato spotted wilt virus, a new member of the Bunyaviridae? In: Kolakofsky, D, Mahy, BWJ (eds) Genetics and Pathogenicity of Negative Strand Viruses, pp. 287–290. Elsevier Science Publishers, Amsterdam (1989).

38. Demler SA, de Zoeten GA: The nucleotide sequence and luteovirus-like nature of RNA 1 of an aphid non-transmissible strain of pea enation mosaic virus. J Gen Virol 72: 1819–1834 (1991).

39. Demler SA, Rucker DG, de Zoeten GA: The chimeric nature of the genome of pea enation mosaic virus: the independent replication of RNA 2. J Gen Virol 74: 1–14 (1993).

40. Dessens JT, Lomonossoff GP: Mutational analysis of the putative catalytic triad of the cowpea mosaic virus 24K protease. Virology 184: 738–746 (1991).

41. Di R, Dinesh-Kumar SP, Miller WA: Translational frameshifting by barley yellow dwarf virus RNA (PAV serotype) in *Escherichia coli* and in eukaryotic cell-free extracts. Mol Plant-Microbe Interact 6: 444–452 (1993).

42. Dinesh-Kumar SP, Brault V, Miller WA: Precise mapping and *in vitro* translation of a trifunctional subgenomic RNA of barley yellow dwarf virus. Virology 187: 711–722 (1992).

43. Dinesh-Kumar SP, Miller WA: Control of start codon choice on a plant viral RNA encoding overlapping genes. Plant Cell 5: 679–692 (1993).

44. Ding S-W, Anderson BJ, Haase HR, Symons RH: New overlapping gene encoded by the cucumber mosaic virus genome. Virology 198: 593–601 (1994).

45. Ding S-W, Howe J, Keese P, Mackenzie A, Meek D, Osorio-Keese M, Skotnicki M, Srifah P, Torronen M, Gibbs A: The tymobox, a sequence shared by most tymoviruses: its use in molecular studies of tymoviruses. Nucl Acids Res 18: 1181–1187 (1990).

46. Dolja VV, Atabekov JG: The structure of barley stripe mosaic virus double-stranded RNAs. FEBS Lett 214: 313–316 (1987).

47. Dolja VV, Grama DP, Morozov SY, Atabekov JG: Potato virus X-related single- and double-stranded RNAs. FEBS Lett 214: 308–312 (1987).

48. Dougherty WG, Parks TD: Molecular genetic and biochemical evidence for the involvement of the heptapeptide cleavage sequence in determining the reaction profile at two tobacco etch virus cleavage sites in cell-free assays. Virology 172: 145–155 (1989).

49. Dougherty WG, Parks TD: Post-translational processing of the tobacco etch virus 49-kDa small nuclear inclusion polyprotein: identification of an internal cleavage site and delimitation of VPg and proteinase domains. Virology 183: 449–456 (1991).

50. Dougherty WG, Semler BL: Expression of virus-encoded proteinases: functional and structural similarities with cellular enzymes. Microbiol Rev 57: 781–822 (1993).

51. Elliott RM: Molecular biology of the *Bunyaviridae*. J Gen Virol 73: 501–522 (1990).

52. Filichkin SA, Lister RM, McGrath PF, Young MJ: *In vivo* expression and mutational analysis of the barley yellow dwarf virus readthrough gene. Virology 205: 290–299 (1994).

53. French R, Ahlquist P: Characterization and engineering of sequences controlling *in vivo* synthesis of brome mosaic virus subgenomic RNA. J Virol 62: 2411–2420 (1988).

54. Fritsch C, Mayo M, Hemmer O: Properties of the satellite RNA of nepoviruses. Biochimie 75: 561–567 (1993).

55. Gallie DR: Posttranscriptional regulation of gene expression in plants. Annu Rev Plant Physiol Plant Mol Biol 44: 77–105 (1993).

56. Garcia A, van Duin J, Pleij CWA: Differential response to frameshift signals in eukaryotic and prokaryotic translational systems. Nucl Acids Res 21: 401–406 (1993).

57. Gargouri R, Joshi RL, Bol JF, Astier-Manifacier S, Haenni A-L: Mechanism of synthesis of turnip yellow mosaic virus coat protein subgenomic RNA *in vivo*. Virology 171: 386–393 (1989).

58. Ge Z, Hiruki C, Roy KL: Nucleotide sequence of sweet clover necrotic mosaic virus RNA–1. Virus Res 28: 113–124 (1993).

59. Geballe AP, Morris DR: Initiation codons within 5′-leaders of mRNAs as regulators of translation. Trends Biochem Sci 19: 159–164 (1994).

60. German S, Candresse T, Lanneau M, Huet JC, Pernollet JC, Dunez J: Nucleotide sequence and genomic organization of apple chlorotic leaf spot closterovirus. Virology 179: 104–112 (1990).

61. Ghosh A, Rutgers T, Ke-Qiang M, Kaesberg P: Characterization of the coat protein mRNA of southern bean mosaic virus and its relationship to the genomic RNA. J Virol 39: 87–92 (1981).

62. Gilmer D, Bouzoubaa S, Hehn A, Guilley H, Richards K, Jonard G: Efficient cell-to-cell movement of beet necrotic yellow vein virus requires 3′ proximal genes located on RNA 2. Virology 189: 40–47 (1992).

63. Goelet P, Lomonossoff GP, Butler PJG, Akam ME, Gait MJ, Karn J: Nucleotide sequence of tobacco mosaic virus RNA. Proc Natl Acad Sci USA 79: 5818–5822 (1982).

64. Goldbach R, Le Gall O, Wellink J: Alpha-like viruses in plants. Semin Virol 2: 19–25 (1991).

65. Gorbalenya AE, Koonin EV: Viral proteins containing the purine NTP-binding sequence pattern. Nucl Acids Res 17: 8413–8440 (1989).

66. Gorbalenya AE, Koonin EV, Blinov VM, Donchenko AP: Sobemovirus genome appears to encode a serine protease related to cysteine proteases of picornaviruses. FEBS Lett 236: 287–290 (1988).

67. Gould AR, Symons RH: Cucumber mosaic virus RNA 3. Eur J Biochem 126: 217–226 (1982).

68. Goulden MG, Lomonossoff GP, Davies JW, Wood KR: The complete nucleotide sequence of PEBV RNA2 reveals the presence of a novel open reading frame and provides insights into the structure of tobraviral subgenomic promoters. Nucl Acids Res 18: 4507–4512 (1990).

69. Gramstat A, Courtpozanis A, Rohde W: The 12 kDa protein of potato virus M displays properties of a nucleic acid-binding regulatory protein. FEBS Lett 276: 34–38 (1990).

70. Gramstat A, Prüfer D, Rohde W: The nucleotide acid-binding zinc finger protein of potato virus M is translated by internal initiation as well as by ribosomal frameshifting involving a shifty stop codon and a novel mechanism of P-site slippage. Nucl Acids Res 22: 3911–3917 (1994).

71. Grieco F, Burgyan J, Russo M: The nucleotide sequence of cymbidium ringspot virus RNA. Nucl Acids Res 17: 6383 (1989).

72. Guilley H, Briand JP: Nucleotide sequence of turnip yellow mosaic virus coat protein mRNA. Cell 15: 113–122 (1978).

388

73. Guilley H, Carrington JC, Balàzs E, Jonard G, Richards K, Morris TJ: Nucleotide sequence and genome organization of carnation mottle virus RNA. Nucl Acids Res 13: 6663–6677 (1985).

74. Gustafson G, Armour SL: The complete nucleotide sequence of RNAβ from the type strain of barley stripe mosaic virus. Nucl Acids Res 14: 3895–3909 (1986).

75. Gustafson G, Hunter B, Hanau R, Armour SL, Jackson AO: Nucleotide sequence and genomic organization of barley stripe mosaic virus RNAγ. Virology 158: 394–406 (1987).

76. Hamilton WDO, Boccara M, Robinson DJ, Baulcombe DC: The complete nucleotide sequence of tobacco rattle virus RNA–1. J Gen Virol 68: 2563–2575 (1987).

77. Harbison S-A, Davies JW, Wilson TMA: Expression of high molecular weight polypeptides by carnation mottle virus RNA. J Gen Virol 66: 2597–2604 (1985).

78. Hayes RJ, Brunt AA, Buck KW: Gene mapping and expression of tomato bushy stunt virus. J Gen Virol 69: 3047–3057 (1988).

79. Hearne PQ, Knorr DA, Hillman BI, Morris TJ: The complete genome structure and synthesis of infectious RNA from clones of tomato bushy stunt virus. Virology 177: 141–151 (1990).

80. Herzog E, Guilley H, Fritsch C: Translation of the second gene of peanut clump virus RNA 2 occurs by leaky scanning in vitro. Virology 208: 215–225 (1995).

81. Herzog E, Guilley H, Manohar SK, Dollet M, Richards K, Fritsch C, Jonard G: Complete nucleotide sequence of peanut clump virus RNA 1 and relationships with other fungus-transmitted rod-shaped viruses. J Gen Virol 75: 3147–3155 (1994).

82. Hsu YH, Brakke MK: Cell-free translation of soil-borne wheat mosaic virus RNAs. Virology 143: 272–279 (1985).

83. Huisman MJ, Linthorst HJM, Bol JF, Cornelissen BJC: The complete nucleotide sequence of potato virus X and its homologies at the amino acid level with various plus-stranded RNA viruses. J Gen Virol 69: 1789–1798 (1988).

84. Hunter TR, Hunt T, Knowland J, Zimmern D: Messenger RNA for the coat protein of tobacco mosaic virus. Nature 260: 759–764 (1976).

85. Ishikawa M, Meshi T, Motoyoshi F, Takamatsu N, Okada Y: In vitro mutagenesis of the putative replicase genes of tobacco mosaic virus. Nucl Acids Res 14: 8291–8305 (1986).

86. Jacks T, Madhani HD, Masiarz FR, Varmus HE: Signals for ribosomal frameshifting in the Rous sarcoma virus gag-pol region. Cell 55: 447–458 (1988).

87. Jackson RJ, Howell MT, Kaminski A: The novel mechanism of initiation of picornavirus RNA translation. Trends Biochem Sci 15: 477–483 (1990).

88. Johnston JC, Rochon DM: Deletion analysis of the promoter for the cucumber necrosis virus 0.9-kb subgenomic RNA. Virology 214: 100–109 (1995).

89. Joshi S, Pleij CWA, Haenni A-L, Chapeville F, Bosch L: Properties of the tobacco mosaic virus intermediate length RNA–2 and its translation. Virology 127: 100–111 (1983).

90. Kadaré G, Rozanov M, Haenni A-L: Expression of the turnip yellow mosaic virus proteinase in Escherichia coli and determination of the cleavage site within the 206 kDa protein. J Gen Virol 76: 2853–2857 (1995).

91. Karasev AV, Boyko VP, Gowda S, Nikolaeva OV, Hilf ME, Koonin EV, Niblett CL, Cline K, Gumpf DJ, Lee RF, Garnsey SM, Lewandowski DJ, Dawson WO: Complete sequence of the citrus tristeza virus RNA genome. Virology 208: 511–520 (1995).

92. Kashiwazaki S, Minobe Y, Hibino H: Nucleotide sequence of barley yellow mosaic virus RNA 2. J Gen Virol 72: 995–999 (1991).

93. Kashiwazaki S, Minobe Y, Omura T, Hibino H: Nucleotide sequence of barley yellow mosaic virus RNA 1: a close evolutionary relationship with potyviruses. J Gen Virol 71: 2781–2790 (1990).

94. Kelly L, Gerlach WL, Waterhouse PM: Characterization of the subgenomic RNAs of an Australian isolate of barley yellow dwarf luteovirus. Virology 202: 565–573 (1994).

95. Kim KH, Lommel SA: Identification and analysis of the site of −1 ribosomal frameshifting in red clover necrotic mosaic virus. Virology 200: 574–582 (1994).

96. Klaassen VA, Boeshore ML, Koonin EV, Tian T, Falk BW: Genome structure and phylogenetic analysis of lettuce infectious yellows virus, a whitefly-transmitted, bipartite closterovirus. Virology 208: 99–110 (1995).

97. Kormelink R, van Poelwijk F, Peters D, Goldbach R: Non-viral heterogeneous sequences at the 5′ ends of tomato spotted wilt mRNAs. J Gen Virol 73: 2125–2128 (1992).

98. Kozak M: Bifunctional messenger RNAs in eukaryotes. Cell 47: 481–483 (1986).

99. Kozak M: Influences of mRNA secondary structure on initiation by eukaryotic ribosomes. Proc Natl Acad Sci USA 83: 2850–2854 (1986).

100. Kozak M: The scanning model for translation: an update. J Cell Biol 108: 229–241 (1989).

101. Kozak M: Downstream secondary structure facilitates recognition of initiator codons by eukaryotic ribosomes. Proc Natl Acad Sci USA 87: 8301–8305 (1990).

102. Kozak M: Structural features in eukaryotic mRNAs that modulate the initiation of translation. J Biol Chem 266: 19867–19870 (1991).

103. Kozak M: Determinants of translational fidelity and efficiency in vertebrate mRNAs. Biochimie 76: 815–821 (1994).

104. Kozak M: Adherence to the first-AUG rule when a second AUG codon follows closely upon the first. Proc Natl Acad Sci USA 92: 2662–2666 (1995).

105. Kujawa AB, Drugeon G, Hulanicka D, Haenni A-L: Structural requirements for efficient translational frameshifting in the synthesis of the putative viral RNA-dependent RNA polymerase of potato leafroll virus. Nucl Acids Res 21: 2165–2171 (1993).

106. Lawrence DM, Rozanov MN, Hillman BI: Autocatalytic processing of the 223-kDa protein of blueberry scorch carlavirus by papain-like proteinase. Virology 207: 127–135 (1995).

107. Lehto K, Grantham GL, Dawson WO: Insertion of sequences containing the coat protein subgenomic RNA promoter and leader in front of the tobacco mosaic virus 30K ORF delays its expression and causes defective cell-to-cell movement. Virology 174: 145–157 (1990).

108. Lommel SA, Kendall TL, Xiong Z, Nutter RC: Identification of the maize chlorotic mottle virus capsid protein cistron and characterization of its subgenomic messenger RNA. Virology 181: 382–385 (1991).

109. Lütcke HA, Chow KC, Mickel FS, Moss KA, Kern HF, Scheele GA: Selection of AUG initiation codons differs in plants and animals. EMBO J 6: 43–48 (1987).

110. MacFarlane SA, Taylor SC, King DI, Hughes G, Davies JW: Pea early browning virus RNA1 encodes four polypeptides including a putative zinc-finger protein. Nucl Acids Res 17: 2245–2260 (1989).

111. Mäkinen K, Nöss V, Tamm T, Truve E, Aaspællu A, Saarma M: The putative replicase of the cocksfoot mottle sobemovir-

us is translated as a part of the polyprotein by −1 ribosomal frameshift. Virology 207: 566–571 (1995).

112. Margis R, Pinck L: Effects of site-directed mutagenesis on the presumed catalytic triad and substrate-binding pocket of grapevine fanleaf nepovirus 24-kDa proteinase. Virology 190: 884–888 (1992).

113. Marsh LE, Dreher TW, Hall TC: Mutational analysis of the core and modulator sequences of the BMV RNA3 subgenomic promoter. Nucl Acids Res 16: 981–995 (1988).

114. Mayo MA, Joly CA, Murant AF, Raschke JH: Nucleotide sequence of raspberry bushy dwarf virus RNA-3. J Gen Virol 72: 469–472 (1991).

115. Mayo MA, Robinson DJ, Jolly CA, Hyman L: Nucleotide sequence of potato leafroll luteovirus RNA. J Gen Virol 70: 1037–1051 (1989).

116. Meulewaeter F, Seurinck J, van Emmelo J: Genome structure of tobacco necrosis virus strain A. Virology 177: 699–709 (1990).

117. Miller JS, Mayo MA: The location of the 5′ end of the potato leafroll luteovirus subgenomic coat protein mRNA. J Gen Virol 72: 2633–2638 (1991).

118. Miller WA, Dinesh-Kumar SP, Paul CP: Luteovirus gene expression. Crit Rev Plant Sci 14: 179–211 (1995).

119. Miller WA, Dreher TW, Hall TC: Synthesis of brome mosaic virus subgenomic RNA in vitro by internal initiation on (−)-sense genomic RNA. Nature 313: 68–70 (1985).

120. Miller WA, Waterhouse PM, Gerlach WL: Sequence and organization of barley yellow dwarf virus genomic RNA. Nucl Acids Res 16: 6097–6111 (1988).

121. Mirkov TE, Mathews DM, Du Plessis DH, Dodds JA: Nucleotide sequence and translation of satellite tobacco mosaic virus RNA. Virology 170: 139–146 (1989).

122. Mise K, Tsuge S, Nagao K, Okuno T, Furusawa I: Nucleotide sequence responsible for the synthesis of a truncated coat protein of brome mosaic virus strain ATCC66. J Gen Virol 73: 2543–2551 (1992).

123. Morch M-D, Boyer J-C, Haenni A-L: Overlapping open reading frames revealed by complete nucleotide sequencing of turnip yellow mosaic virus genomic RNA. Nucl Acids Res 16: 6157–6173 (1988).

124. Morch M-D, Drugeon G, Szafranski P, Haenni A-L: Proteolytic origin of the 150-kilodalton protein encoded by turnip yellow mosaic virus genomic RNA. J Virol 63: 5153–5158 (1989).

125. Morozov SY, Miroshnichenko NA, Solovyev AG, Fedorkin ON, Zelenina DA, Lukasheva LI, Karasev AV, Dolja VV, Atabekov JG: Expression strategy of the potato virus X triple gene block. J Gen Virol 72: 2039–2042 (1991).

126. Mushegian AR, Koonin EV: Cell-to-cell movement of plant viruses: insights from amino acid sequence comparisons of movement proteins and from analogies with cellular transport systems. Arch Virol 133: 239–257 (1993).

127. Natsuaki T, Mayo MA, Joly CA, Murant AF: Nucleotide sequence of raspberry bushy dwarf virus RNA-2: a bicistronic component of a bipartite genome. J Gen Virol 72: 2183–2189 (1991).

128. Nicolaisen M, Johansen E, Poulsen GB, Borkhardt B: The 5′ untranslated region from pea seedborne mosaic potyvirus RNA as a translational enhancer in pea and tobacco protoplasts. FEBS Lett 303: 169–172 (1992).

129. Niesbach-Klösgen U, Guilley H, Jonard G, Richards K: Immunodetection in vivo of beet necrotic yellow vein virus-encoded proteins. Virology 178: 52–61 (1990).

130. Nutter RC, Scheets K, Panganiban LC, Lommel SA: The complete nucleotide sequence of the maize chlorotic mottle virus genome. Nucl Acids Res 17: 3163–3177 (1989).

131. Oh C-S, Carrington JC: Identification of essential residues in potyvirus proteinase HC-Pro by site-directed mutagenesis. Virology 173: 692–699 (1989).

132. Osman TAM, Buck KW: Double-stranded RNAs isolated from plant tissue infected with red clover necrotic mosaic virus correspond to genomic and subgenomic single-stranded RNAs. J Gen Virol 71: 945–948 (1990).

133. Othman Y, Hull R: Nucleotide sequence of the bean strain of southern bean mosaic virus. Virology 206: 287–297 (1995).

134. Ou J-H, Rice CM, Dalgarno L, Strauss EG, Strauss JH: Sequence studies of several alphavirus genomic RNAs in the region containing the start of the subgenomic RNA. Proc Natl Acad Sci USA 79: 5253–5239 (1982).

135. Palukaitis P, Roossinck MJ, Dietzgen RG, Francki RIB: Cucumber mosaic virus. Adv Virus Res 41: 281–348 (1992).

136. Parks TD, Howard ED, Wolpert TJ, Arp DJ, Dougherty WG: Expression and purification of a recombinant tobacco etch virus NIa proteinase: biochemical analyses of the full-length and a naturally occurring truncated proteinase form. Virology 210: 194–201 (1995).

137. Pelham HRB: Leaky UAG termination codon in tobacco mosaic virus RNA. Nature 272: 469–471 (1978).

138. Pelham HRB: Translation of tobacco rattle virus RNAs in vitro: four proteins from three RNAs. Virology 97: 256–265 (1979).

139. Peters SA, Voorhorst WGB, Wery J, Wellink J, van Kammen A: A regulatory role for the 32K protein in proteolytic processing of cowpea mosaic virus polyproteins. Virology 191: 81–89 (1992).

140. Petty ITD, Jackson AO: Two forms of the major barley stripe mosaic virus nonstructural protein are synthesized in vivo from alternative initiation codons. Virology 177: 829–832 (1990).

141. Prüfer D, Tacke E, Schmitz J, Kull B, Kaufmann A, Rohde W: Ribosomal frameshifting in plants: a novel signal directs the −1 frameshift in the synthesis of the putative viral replicase of potato leafroll luteovirus. EMBO J 11: 1111–1117 (1992).

142. Ramírez B-C, Garcin D, Calvert LA, Kolakofsky D, Haenni A-L: Capped nonviral sequences at the 5′ end of the mRNAs of rice hoja blanca virus RNA4. J Virol 69: 1951–1954 (1995).

143. Ramírez B-C, Haenni A-L: Molecular biology of tenuiviruses, a remarkable group of plant viruses. J Gen Virol 75: 467–475 (1994).

144. Rathjen JP, Karageorgos LE, Habili N, Waterhouse PM, Symons RH: Soybean dwarf luteovirus contains the third variant genome type in the luteovirus group. Virology 198: 671–679 (1994).

145. Restrepo-Hartwig MA, Carrington JC: Regulation of nuclear transport of a plant potyvirus protein by autoproteolysis. J Virol 66: 5662–5666 (1992).

146. Rezaian MA, Williams RHV, Gordon KHJ, Gould AR, Symons RH: Nucleotide sequence of cucumber-mosaic-virus RNA 2 reveals a translation product significantly homologous to corresponding proteins of other viruses. Eur J Biochem 143: 277–284 (1984).

147. Riechmann JL, Laín S, García JA: Identification of the initiation codon of plum pox potyvirus genomic RNA. Virology 185: 544–552 (1991).

148. Riechmann JL, Laín S, García JA: Highlights and prospects of potyvirus molecular biology. J Gen Virol 73: 1–16 (1992).

149. Riviere CJ, Rochon DM: Nucleotide sequence and genomic organization of melon necrotic spot virus. J Gen Virol 71: 1887–1896 (1990).

150. Rochon DM, Johnston JC: Infectious transcripts from cloned cucumber necrosis virus cDNA: evidence for a bifunctional subgenomic mRNA. Virology 181: 656–665 (1991).

151. Rochon DM, Tremaine JH: Complete nucleotide sequence of the cucumber necrosis virus genome. Virology 169: 251–259 (1989).

152. Rohde W, Gramstat A, Schmitz J, Tacke E, Prüfer D: Plant viruses as model systems for the study of non-canonical translation mechanisms in higher plants. J Gen Virol 75: 2141–2149 (1994).

153. Roossinck MJ, Sleat D, Palukaitis P: Satellite RNAs of plant viruses: structures and biological effects. Microbiol Rev 56: 265–279 (1992).

154. Rouleau M, Smith RJ, Bancroft JB, Mackie GA: Purification, properties, and subcellular localization of foxtail mosaic potexvirus 26-kDa protein. Virology 204: 254–265 (1994).

155. Rozanov MN, Drugeon G, Haenni A-L: Papain-like proteinase of turnip yellow mosaic virus: a prototype of a new viral proteinase group. Arch Virol 140: 273–288 (1995).

156. Rupasov VV, Morozov SY, Kanuyka KV, Zavriev SK: Partial nucleotide sequence of potato virus M RNA shows similarities to potexviruses in gene arrangement and the encoded amino acid sequences. J Gen Virol 70: 1861–1869 (1989).

157. Russo M, Burgyan J, Martelli GP: Molecular biology of Tombusviridae. Adv Virus Res 44: 381–428 (1994).

158. Ryabov EV, Generozov EV, Kendall TL, Lommel SA, Zavriev SK: Nucleotide sequence of carnation ringspot dianthovirus RNA–1. J Gen Virol 75: 243–247 (1994).

159. Schmitt C, Balmori E, Jonard G, Richards KE, Guilley H: *In vitro* mutagenesis of biologically active transcripts of beet necrotic yellow vein virus RNA 2: evidence that a domain of the 75-kDa readthrough protein is important for efficient virus assembly. Proc Natl Acad Sci USA 89: 5715–5719 (1992).

160. Schmitz J, Prüfer D, Rohde W, Tacke E: Non-canonical translation mechanisms in plants: efficient *in vitro* and *in planta* initiation at AUU codons of the tobacco mosaic virus enhancer sequence. Nucl Acids Res 24: 257–263 (1996).

161. Scott KP, Kashiwazaki S, Reavy B, Harrison BD: The nucleotide sequence of potato mop-top virus RNA 2: a novel type of genome organization for a furovirus. J Gen Virol 75: 3561–3568 (1994).

162. Shen P, Kaniewska M, Smith C, Beachy RN: Nucleotide sequence and genomic organisation of rice tungro spherical virus. Virology 193: 621–630 (1993).

163. Shih D-S, Lane LC, Kaesberg P: Origin of the small component of brome mosaic virus RNA. J Mol Biol 64: 353–362 (1972).

164. Shirako Y, Wilson TMA: Complete nucleotide sequence and organization of the bipartite RNA genome of soil-borne wheat mosaic virus. Virology 195: 16–32 (1993).

165. Skotnicki ML, Mackenzie AM, Torronen M, Gibbs AJ: The genomic sequence of cardamine chlorotic fleck carmovirus. J Gen Virol 74: 1933–1937 (1993).

166. Skryabin KG, Morozov SY, Kraev AS, Rozanov MN, Chernov BK, Lukasheva LI, Atabekov JG: Conserved and variable elements in RNA genomes of potexviruses. FEBS Lett 240: 33–40 (1988).

167. Skuzeski JM, Nichols LM, Gesteland RF, Atkins JF: The signal for a leaky UAG stop codon in several plant viruses includes the two downstream codons. J Mol Biol 218: 365–373 (1991).

168. Smirnyagina E, Hsu Y-H, Chua N, Ahlquist P: Second-site mutations in the brome mosaic virus RNA3 intercistronic region partially suppress a defect in coat protein mRNA transcription. Virology 198: 427–436 (1994).

169. Stanley J, Hanau R, Jackson AO: Sequence comparison of the 3' ends of a subgenomic RNA and the genomic RNAs of barley stripe mosaic virus. Virology 139: 375–383 (1984).

170. Tacke E, Prüfer D, Salamini F, Rohde W: Characterization of a potato leafroll luteovirus subgenomic RNA: differential expression by internal translation initiation and UAG suppression. J Gen Virol 71: 2265–2272 (1990).

171. Tavazza M, Lucioli A, Calogero A, Pay A, Tavazza R: Nucleotide sequence, genomic organization and synthesis of infectious transcripts from a full-length clone of artichoke mottle crinkle virus. J Gen Virol 75: 1515–1524 (1994).

172. ten Dam EB: Pseudoknot-dependent ribosomal frameshifting. Ph. D. thesis, University of Leiden (1995).

173. ten Dam EB, Pleij CWA, Bosch L: RNA pseudoknots: translational frameshifting and readthrough on viral RNAs. Virus Genes 4: 121–136 (1990).

174. Thomas AAM, ter Haar E, Wellink J, Voorma HO: Cowpea mosaic virus middle component RNA contains a sequence that allows internal binding of ribosomes and that requires eukaryotic initiation factor 4F for optimal translation. J Virol 65: 2953–2959 (1991).

175. Tordo VM-J, Chachulska AM, Fakhfakh H, Le Romancer M, Robaglia C, Astier-Manifacier S: Sequence polymorphism in the 5' NTR and in the P1 coding region of potato virus Y genomic RNA. J Gen Virol 76: 939–949 (1995).

176. Turnbull-Ross AD, Mayo MA, Reavy B, Murant AF: Sequence analysis of the parsnip yellow fleck virus polyprotein: evidence of affinities with picornaviruses. J Gen Virol 74: 555–561 (1993).

177. Turpen TH, Reinl SJ, Charoenvit Y, Hoffman SL, Fallarme V, Grill LK: Malarial epitopes expressed on the surface of recombinant tobacco mosaic virus. Bio/Technology 13: 53–57 (1995).

178. Urban C, Beier H: Cysteine tRNAs of plant origin as novel UGA suppressors. Nucl Acids Res 1995: 4591–4597 (1995).

179. Valle RPC, Drugeon G, Devignes-Morch M-D, Legocki AB, Haenni A-L: Codon context effect in virus translational readthrough. A study in vitro of the determinants of TMV and Mo-MuLV amber suppression. FEBS Lett 306: 133–139 (1992).

180. Valle RPC, Haenni A-L: Peptide chain termination. In: Trachsel H (ed) Translation in Eukaryotes, pp. 177–189. CRC Press, Boca Raton, FL (1991).

181. van der Kuyl AC, Langereis K, Houwing CJ, Jaspars EMJ, Bol JF: *cis*-Acting elements involved in replication of alfalfa virus RNAs *in vitro*. Virology 176: 346–354 (1990).

182. van der Vossen EAG, Notenboom T, Bol JF: Characterization of sequences controlling the synthesis of alfalfa mosaic virus subgenomic RNA *in vivo*. Virology 212: 663–672 (1995).

183. van der Wilk F, Huisman MJ, Cornelissen BJC, Huttinga H, Goldbach R: Nucleotide sequence and organization of potato leafroll virus genomic RNA. FEBS Lett 245: 51–56 (1989).

184. van Lent J, Storms M, van der Meer F, Wellink J, Goldbach R: Tubular structures involved in movement of cowpea mosaic virus are also formed in infected cowpea protoplasts. J Gen Virol 72: 2615–2623 (1991).

185. Veidt I, Bouzoubaa SE, Leiser R-M, Ziegler-Graff V, Guilley H, Richards K, Jonard G: Synthesis of full-length transcripts of beet western yellows virus RNA: messenger properties and biological activity in protoplasts. Virology 186: 192–200 (1992).

186. Veidt I, Lot H, Leiser M, Scheidecker D, Guilley H, Richards K, Jonard G: Nucleotide sequence of beet western yellows virus RNA. Nucl Acids Res 16: 9917–9932 (1988).

187. Verchot J, Carrington JC: Debilitation of plant potyvirus infectivity by P1 proteinase-inactivating mutations and restoration by second-site modifications. J Virol 69: 1582–1590 (1995).

188. Verchot J, Herndon KL, Carrington JC: Mutational analysis of the tobacco etch potyviral 35-kDa proteinase: identification of essential residues and requirements for autoproteolysis. Virology 190: 298–306 (1992).

189. Verchot J, Koonin EV, Carrington JC: The 35-kDa protein from the N-terminus of the potyviral polyprotein functions as a third virus-encoded proteinase. Virology 185: 527–535 (1991).

190. Verver J, Le Gall O, van Kammen A, Wellink J: The sequence between nucleotides 161 and 512 of cowpea mosaic virus M RNA is able to support internal initiation of translation *in vitro*. J Gen Virol 72: 2339–2345 (1991).

191. Vincent JR, Lister RM, Larkins BA: Nucleotide sequence analysis and genomic organization of the NY-RPV isolate of barley yellow dwarf virus. J Gen Virol 72: 2347–2355 (1991).

192. Vos P, Verver J, Jaegle M, Wellink J, van Kammen A, Goldbach R: Two viral proteins involved in the proteolytic processing of the cowpea mosaic virus polyproteins. Nucl Acids Res 16: 1967–1985 (1988).

193. Wang JY, Chay C, Gildow FE, Gray SM: Readthrough protein associated with virions of barley yellow dwarf luteovirus and its potential role in regulating the efficiency of aphid transmission. Virology 206: 954–962 (1995).

194. White KA, Skuzeski JM, Li W, Wei N, Morris TJ: Immunodetection, expression strategy and complementation of turnip crinkle virus p28 and p88 replication components. Virology 211: 525–534 (1995).

195. Witherell GW, Gott JM, Uhlenbeck OC: Specific interaction between RNA phage coat proteins and RNA. Prog Nucl Acid Res Mol Biol 40: 185–220 (1991).

196. Xiong Z, Kim KH, Kendall TL, Lommel SA: Synthesis of the putative red clover necrotic mosaic virus RNA polymerase by ribosomal frameshifting *in vitro*. Virology 193: 213–221 (1993).

197. Xiong Z, Lommel SA: The complete nucleotide sequence and genome organization of red clover necrotic mosaic virus RNA-1. Virology 171: 543–554 (1989).

198. Zaccomer B, Haenni A-L, Macaya G: The remarkable variety of plant RNA virus genomes. J Gen Virol 76: 231–247 (1995).

199. Zavriev SK, Kanyuka KV, Levay KE: The genome organization of potato virus M RNA. J Gen Virol 72: 9–14 (1991).

200. Zerfass K, Beier H: Pseudouridine in the anticodon GψA of plant cytoplasmic tRNA^{Tyr} is required for UAG and UAA suppression in the TMV-specific context. Nucl Acids Res 22: 5911–5918 (1992).

201. Zerfass K, Beier H: The leaky UGA termination codon of tobacco rattle virus RNA is suppressed by tobacco chloroplast and cytoplasmic tRNAs^{Trp} with CmCA anticodon. EMBO J 11: 4167–4173 (1992).

202. Zhou H, Jackson AO: Expression of the barley stripe mosaic virus RNAβ 'triple gene block'. Virology 216: 367–379 (1996).

203. Ziegler V, Richards K, Guilley H, Jonard G, Putz C: Cell-free translation of beet necrotic yellow vein virus: readthrough of the coat protein cistron. J Gen Virol 66: 2079–2087 (1985).

Plant Molecular Biology **32**: 393–405, 1996.
© 1996 *Kluwer Academic Publishers.*

Optimizing expression of transgenes with an emphasis on post-transcriptional events

Michael G. Koziel*, Nadine B. Carozzi & Nalini Desai
Ciba Agricultural Biotechnology Unit, 3054 Cornwallis Road, Research Triangle Park, NC 27709, USA (*author for correspondence)

Key words: gene expression, endotoxin, untranslated leader, intron, splicing, synthetic genes

Contents

Abstract   393
Introduction   393
5′-Untranslated leaders   394
Bacterial genes expressed in plants   395
The use of introns to increase gene
   expression   396
Synthetic genes   398
Chloroplast sequestering and targeting   401
3′ regions   402
Effects of protein folding   402
Conclusions   402
References   403

Abstract

Introducing a foreign gene into a new plant host does not always result in a high level of expression of the incoming gene. Numerous promoters have been used to express foreign genes in different plant tissues, but there are sometime various features of the new gene which are deleterious to expression in the new host. There are a number of post-transcriptional steps in the expression of a gene and sometimes sequences present in a particular coding region can resemble the signals which initiate these processing steps. When aberrantly carried out, these steps diminish the level of expression. By removing such fortuitous signals, one can dramatically increase expression of a transgene in plants. Ensuring proper protein folding and/or targeting the protein product to a particular cellular compartment can also be used to increase the level of protein obtained. The various methods used to optimize expression of a foreign gene in plants by concentrating on post-transcriptional events are discussed.

Introduction

The introduction of foreign genes into a variety of plant species is becoming increasingly routine. However, consistent levels of expression of the transgenes is not a foregone conclusion of a successful transformation process. Certain of the 'rules' for obtaining good expression are becoming clear. For instance choice of a proper 5′-untranslated leader or use of an intron in a gene that lacks its own intron, such as a bacterial gene or a cDNA can increase the level of expression obtained from a given gene. Nevertheless, in addi-

tion to large variations in expression levels due to the positional insertion site, a large number of potential problems with a deleterious impact on gene expression exist, especially when using a gene from a heterologous source. Improper splicing, improper polyadenylation, improper nuclear transport, or instability of the resulting cytosolic mRNA can result in accumulation of only a low level of both mRNA and the resulting protein. Additionally, the form of a particular protein chosen to be expressed may play an important role in the steady state level of the protein which is obtained. For instance, if a protein requires a stepwise folding with subsequent processing of the properly folded form to produce a mature active protein, attempting to express only the mature protein may result in a protein which is not properly folded and is either inactive, unstable and rapidly degraded, or both. In this chapter we will review what is known about optimizing expression of transgenes by focusing on post-transcriptional events. Typically, optimization of expression means obtaining a high level of expression. Inducible, developmental, and tissue specific patterns of expression are obtained by using specific promoters with the desired pattern of expression and will not be dealt with here.

5'-Untranslated leaders

The optimization of transgene expression in plants must include signals for efficient initiation of protein translation. The 5'-untranslated leader (5'-UTL) sequence of eukaryotic mRNA plays a major role in translational efficiency. Its role is presumed to influence the efficiency with which the bound 40S ribosomal subunits migrate and recognize the translational start site. The design of chimeric promoter gene fusions, particularly those which involve a fusion between a plant promoter and a bacterial coding region, allow for great latitude in the composition of 5'-UTL sequences. Many early chimeric transgenes expressed using the cauliflower mosaic virus (CaMV) 35S promoter used an arbitrarily chosen length of viral sequence after the 35S start of transcription and fused this artificially defined 5'-UTL to the AUG of the coding region. Several studies have now shown that the 5'-UTL sequence and sequences directly surrounding the AUG can have a large effect in translational efficiency in plants and that this effect can be different depending on the plant. Although a great deal remains to be understood in the role of the 5'-UTL it is helpful to be aware of the importance of this region.

In most eukaryotic mRNAs, translational initiation occurs at the AUG codon closest to the 5' cap of the mRNA transcript. Mutagenesis studies of specific mRNAs have demonstrated that the sequences surrounding the AUG of the 5'-untranslated leader are involved in translational efficiency. Comparison of vertebrate mRNA sequences and the results of site-directed mutagenesis experiments have demonstrated the existence of a preferred nucleotide context surrounding the initiation codon [42, 43]. Kozak has defined an optimal AUG context for vertebrates as GCC(A/G)CCAUGG. The most highly conserved position of the consensus is position −3 (the A of the AUG codon being +1) where 97% of vertebrate mRNAs have a purine in that position. The −3 position appears to mediate the efficiency of initiation of translation in animal systems, allowing about a five-fold higher rate of translation with an A or purine than with a pyrimidine [42]. A consensus sequence can also be found among plant genes, UAAACAAUGGCU [40, 47]. In position +4 the preference for a guanine is significantly greater in plants (85%) compared with animals (38%). The preference for G in position +4 (85%) and C at +5 (77%) in plants resembles the preference for A in position −3 (80%) in animals [47]. In plant systems, the −3 position does not appear to effect translational efficiency and the sequence requirements do not appear to be as stringent as in vertebrates. A compilation of sequences surrounding the AUG from 85 maize genes yields a consensus of (C/G)AUGGCG [46]. In contrast to mammalian translation, there is no apparent advantage to having an A at −3 in maize. *β-glucuronidase* (GUS) gene expression was enhanced 4-fold in tobacco protoplasts when the start codon was ACCAUGG rather than UCCUAUGG [22]. An 8-fold increase in expression of bacterial chitinase was observed when ACCAUGG was used in place of the native UUUAUGG [39]. In these instances the U at −1 may be the problem since this is rare in both vertebrates and plants.

The construction of chimeric genes, particularly bacterial genes, has included the use of 5'-UTL sequences from plant viruses. Chimeric mRNAs containing plant viral RNA leaders have been shown to act as efficient enhancers of translation in several systems. Plant viral mRNAs, in particular those encoding the coat protein, are efficiently translated and have UTLs which must compete effectively with plant cellular mRNAs for available ribosomes. The alfalfa mosaic virus (AMV) coat protein and brome mosaic virus (BMV) coat protein UTLs have been shown to enhance mRNA translation 8-fold in electroporated tobacco protoplasts [22]. A 67 nucleotide derivative (Ω) of the 5'-UTL of tobacco mosaic virus RNA (TMV) fused to the chloramphenicol acetyltransferase (CAT) gene and GUS gene has been shown to enhance translation of the reporter genes *in vitro* [22, 69, 70]. Electroporation of tobacco mesophyll protoplasts with transcripts containing the TMV leader fused to reporter genes CAT, GUS and LUC produced a 33-, 21-, and 36-fold level of enhancement, respectively [21, 25]. At least in the case of the TMXV Ω 5'-UTL sequences, the effect was independent of the coding region. An 83 nt 5'-UTL of potato virus X RNA was shown to enhance expression of the neomycin phosphotransferase II (NptII) gene in *Nicotiana tabacum* protoplasts 4-fold [65]. The 5'-UTL sequences from other highly efficient viral mRNAs do not always enhance translational efficiency of a chimeric mRNA as shown in studies with the turnip yellow mosaic virus coat protein mRNA and black beetle virus RNA 2 [22].

The effect of a 5'-UTL may be different depending on the plant, particularly between dicots and monocots. The TMV 5'-UTL has been shown to be less effective in maize protoplasts, producing ca. 2.6-fold enhancement [26] as opposed to a 30-fold enhancement in tobacco protoplasts [24]. In maize protoplasts, we have demonstrated that 5'-UTLs from the CaMV 35S transcript, both a 73 nucleotide leader [67] and a longer hybrid between the CaMV 35S and luciferase native 5'-UTL, function to strongly enhance expression of the luciferase reporter gene 35-fold and 18-fold respectively compared to the native luciferase 5'-UTL. The 5'-untranslated leaders from the maize genes glutelin [6], PEP-carboxylase [35] and ribulose bisphosphate carboxylase increased expression in maize protoplasts 12-fold, 3.7-fold, and 3.4-fold respectively. The 5'-UTLs from TMV-Ω [23] and three plant viral coat protein genes AMV-coat [27, 38], TMV-coat [30], and BMV-coat [18] worked poorly in maize and inhibited expression of the luciferase gene in maize relative to its native leader.

In tobacco, the effects of these 5'-UTLs is strikingly different, suggesting distinct differences in mechanisms of translational initiation between tobacco and maize. In contrast to maize, the TMV Ω 5'-UTL and the AMV coat protein 5'-UTL enhanced expression in tobacco 5.4-fold and 3.0-fold respectively, whereas the glutelin, maize PEP-carboxylase, and maize ribulose-1,5-bisphosphate carboxylase 5'-UTLs showed no enhancement relative to the native luciferase 5'-UTL. Only the CaMV 35S/Luc hybrid and the CaMV 35S 5'-UTLs enhanced luciferase expression in both maize and tobacco. The TMV and BMV coat protein 5'-UTLs were inhibitory in both maize and tobacco protoplasts. The 5'-UTLs derived from the TMV Ω and AMV coat protein genes produced translational enhancement in tobacco whereas the 5'-UTLs from highly expressed maize cellular genes worked poorly in tobacco. Significant differences exist between maize and tobacco translational systems in their response to the various leaders. The differences in leader activity appear to be a function of the dicot or monocot nature of the cells, and may represent differences in translational protein factors and binding regions.

Many plant 5'-UTLs are considered to be A-rich but this alone does not appear to be significant. Plant genes generally have a high frequency of A and T nucleotides in their 5'-UTL region which may create a relatively unstructured region accessible to the scanning ribosomal 40S subunit [33]. With respect to percent A+T, although TMV omega at 72% and AMV at 70% are the highest, clearly this alone is not critical particularly for maize. A long 5'-UTL does not necessarily confer good activity, as evidenced by the low activity of the TMV Ω 5'-UTL (81 nucleotides) in maize and the poor activities of 5'-UTLs glutelin [6] (87 nucleotides), maize PEP-carboxylase [35] (103 nucleotides), and ribulose-1,5-bisphosphate carboxylase (82 nucleotides) in tobacco.

Bacterial genes expressed in plants

Antibiotic resistance genes of bacterial origin such as neomycin phosphotransferase (NptII), hygromycin phosphotransferase (Hpt) and phosphinothricin acetyltransferase (*bar*) have been used successfully for the selection of transgenic plants. These bacterial genes have been introduced into the monocots rice [4, 12, 68], maize [19, 32, 44, 66], and wheat [76], under control of the CaMV 35S promoter. Improvements in bacterial gene expression have been achieved in monocots through the use of strong constitutive promoters such as the rice actin and maize ubiquitin promoters [8, 72, 77]. The G+C content of *bar*, Hpt, and NptII are 68%, 58%, and 59% respectively.

Bias towards high G+C has made these genes good candidates for expression in plants. The β-glucuronidase (GUS) gene encoded by the *uidA* locus of *Escherichia coli* and the luciferase gene encoded by the firefly, *Photinus pyralis*, are two of the most widely used reporter genes in both dicots and monocots. Expression of luciferase (44% G+C) has been demonstrated in maize [19, 41] whereas expression of the GUS gene (52% G+C) has been reported to be lower and more sporadic [32]. Higher levels of GUS expression have been achieved in the monocots rice [80] and wheat [77] using either the rice actin promoter or the maize ubiquitin promoter.

The use of introns to increase gene expression

Including introns in the transcribed portion of a gene has been found to increase heterologous gene expression in both animal [5] and plant systems [7, 48, 49, 51, 75]. The mechanism of enhancement produced by introns in improving gene expression is not clear. Not all introns produce a stimulatory effect and the degree of stimulation varies, most probably because the effect depends on many factors. However, the effect appears to be accompanied by an increase in the steady-state levels of mRNA lending support to the hypothesis that introns somehow improve efficiency of mRNA processing. In plants the enhancing effect of introns is more apparent in monocots than in dicots. Tanaka *et al.* [71] reported that the presence of the catalase intron 1 isolated from castor beans resulted in an increase in gene expression in rice but not in tobacco when using GUS as a marker gene. Working with a transient expression system in *Arabidopsis* leaf Norris *et al.* [55] found the polyubiquitin intron to have little effect on expression of luciferase using either a CaMV 35S or ubiquitin promoter. Genschick *et al.* [29] found that in a transient expression system derived from tobacco the presence of an intron in the 5'-UTR of the ubiquitin promoter had no effect on expression of the GUS gene.

The first report that the presence of introns could enhance gene expression in transformed monocot cells used the maize alcohol dehydrogenase1 (*Adh1*) gene [7, 13]. When a genomic clone of the *Adh1* gene with its endogenous promoter was used to transform maize cells, the expression of the Adh-s protein was about 100-fold higher than if the coding region came from cDNA clones. To study how the various introns in the *Adh* gene affected expression, regions of the genomic clone were replaced with corresponding regions from the cDNA clone. The level of expression obtained with the genomic clone was restored fully when the sequence downstream of the first intron was replaced by the cDNA sequence. Addition of more genomic sequences did not produce a further increase in expression. In the absence of intron 1, inclusion of introns 8 and 9 restored only 1/3 of the original activity of the genomic clone. This led the authors to conclude that the first intron of *Adh1* was capable of producing up to a 100-fold increase in gene expression. Expression of the *NptII* gene driven by the CaMV 35S promoter was also enhanced 14.6-fold when the region containing the first exon and the first intron of the *Adh1* gene was included in the construct, demonstrating that the stim-

ulatory effect could be observed with a heterologous gene.

The degree to which gene expression is affected in a heterologous system is variable. An important factor appears to be the strength of the promoter being used in the expression cassette [7]. A chimeric gene consisting of an enhanced version of the 35S promoter driving expression of the genomic *Adh1* coding sequences resulted in a 6-fold higher expression compared to that obtained with the native Adh promoter. With the *Adh* cDNA and the CaMV 35S promoter expression was about 500 fold lower then when using the genomic clone [7]. Addition of the first intron from the *Adh1* gene to this construct increased the expression to 1/5 of the level obtained when the complete genomic clone was used. When a shorter 5' leader was used in the CaMV 35S promoter cassette, there was no apparent loss of expression using the genomic clone but replacing the region between the end of intron 1 and the beginning of intron 8 with cDNA resulted in a two fold increase in expression over that obtained with the complete genomic clone.

Luehrsen and Walbot [45] found that addition of the maize *Adh1* intron 1 at the 5' end of luciferase or GUS reporter genes driven by Adh1 promoter stimulated expression 2–4-fold. In the presence of a longer leader than that used in the above experiments and the CaMV 35S promoter, expression was reduced to 0.6× of the control construct without introns, mainly due to a significantly higher level of expression from the CaMV 35S promoter and leader combination. Mascarenhas *et al.* [49] found that when using an 'improved' version of the CaMV 35S promoter to drive CAT expression in BMS cells, adding the *Adh1* intron 1 increased expression only 1–2-fold. They reported that under the same conditions *Adh1* introns 2 and 6 produced a 12- and 20-fold increase in expression, respectively. They concluded from their work and results of others [7, 45] that the degree of enhancement by a particular intron depends on the strength of the promoter. This is primarily reflected in the effect of the leader and presence of enhancer elements. Indeed, this appears to be supported by the results of Tanaka *et al.* [71]. Using the catalase intron in an expression cassette containing the CaMV 35S promoter found in pBI221 [37] to drive GUS expression in transformed rice protoplasts, they observed a 80–90-fold increase in expression compared with a construct lacking introns.

Not all introns appear to be capable of increasing gene expression. However, introns other than the first intron of Adh1 have been shown to have a favorable effect. Callis *et al.* [7] demonstrated that intron 1 of the maize bronze1 gene had a stimulatory effect on CAT expression in maize BMS cells using a 35S promoter and a nos terminator. Enhanced expression has also been reported in protoplasts of bread wheat using the CaMV 35S promoter and the *Adh1* intron 6 [56] and in maize using the *Adh1* intron 2 and intron 6 driven by an enhanced 35S promoter [49]. Vasil *et al.* [75] reported that the presence of the maize sucrose synthase intron 1 provides a 10-fold enhancement of CAT expression over that obtained by the *Adh1* intron 1 in a maize transient expression assay. Intron 3 of the rice actin gene can stimulate reporter gene expression 2- to 6-fold although the overall level of expression was much lower than under the same conditions using the *Adh1* 5' region [45]. Plant transformation vectors utilizing the rice actin promoter [51] and the maize ubiquitin promoter [11] have retained the first intron in the 5'-UTR when optimizing for expression on monocots. The presence of these introns is required for expression.

Monocots and dicots have been shown to have different requirements for intron recognition and efficient splicing [31]. According to Goodall and Filipowicz, maize protoplasts are more efficient at processing introns with secondary structure and therefore have less of a dependence on AU-rich sequences which, in addition to other effects, are presumed to lower intron secondary structure (see also chapter by Simpson and Filipowicz in this volume).

Gallie and Young [26], working with aleurone and endosperm protoplasts found that the presence of the Adh1 intron1 within a transcript enhances the expression of GUS but the degree of stimulation may vary due to cell type. Mascarenhas *et al.* [49] found that the expression enhancing activity of an intron was affected by the length of the flanking exon sequences. With *Adh1* intron 6 as the stimulatory element, inclusion of 56 nucleotides of the 5'- and 6 nucleotides of the 3'-flanking region in the construct, gave the greatest enhancement. In the case of the *Adh1* intron 6 the largest effect was seen when the construct contained 76 nucleotides of the 5' and 53 nt of the 3' sequences flanking the intron. The combination of the maize sucrose synthase exon 1 and intron 1 gave a 1000-fold increase in CAT activity using the CaMV 35S promoter for expression in maize and rice protoplasts [48]. Taken together, these data indicate that the degree of enhancement of gene expression by a particular intron depends on the strength of the promoter, cell type, and flanking exon sequences.

Although the quantitative effect of enhancement due to the presence of introns depends on many factors, there appears to be a requirement for the placement of the intron in the 5' transcriptional unit in the correct orientation with respect to the splice junction sequences [7, 48, 49, 56, 71, 75]. Introns placed at the 5' end of the promoter are not active suggesting that the introns do not contain sequences that act as transcriptional enhancers [7, 48, 49, 56, 71, 75]. A consensus has emerged that splicing per se is important to increase stability of the mRNA resulting in a subsequent enhancement of gene expression. [45, 49] This view does not explain the observation that not all introns increase expression in heterologous systems and even the ones that do seem to increase expression do not show uniformity in effect. Results of experiments to correlate splicing efficiency with expression enhancing activity are ambiguous as complete splicing of the transgene is not observed in transient assays [45]. However, the requirement for the presence of sequences flanking the intron for activity suggest that splicing is probably necessary for the enhancement of gene expression [49]. The placement of a particular intron may also be important. For example, when the *Adh1* intron 9 was placed at the 5' end of a GUS gene expressed using the CaMV 35S promoter, no increase in expression was observed [49]. However, its inclusion in at the 3' end of the GUS gene increased expression about 3-fold [7].

Of the introns found to positively influence expression four, rice actin intron1, maize ubiquitin intron1, maize bronze1 intron1 and maize sucrose synthase intron1, are present naturally in the 5'-untranslated regions. Perhaps the presence of the intron in the 5'-untranslated region increases the length and strength of the untranslated leader in the resulting processed mRNA. This speculation is supported by the observation that in studies with these introns the highest enhancement has been found when using the native promoter and the first exon [7, 48]. When stronger promoters or those with a strong 5'-untranslated leader were used the degree of enhancement of the *Adh1* intron 1 drops from 100-fold to 2–6-fold [45, 49, 75]. The stimulatory effect resulting from inclusion of maize *Adh1* intron 2, maize *Adh1* intron 6 and the rice actin intron 3 do not fit this speculation unless one assumes that these introns have sequence motifs that enhance gene expression in some unknown manner which has not yet been supported by other evidence.

When optimizing a gene for expression, the role of introns is best evaluated on a case-by-case basis since the size of increase in expression depends on many factors. In most cases the use of a promoter with a strong 5'-untranslated leader would appear to be more effective in increasing expression than simply including an intron with a weaker promoter. This phenomenon should be studied further in transgenic plants before definitive conclusions can be made about the impact of an intron on increasing levels of transgene expression. Other factors such as the nucleotide composition of the coding region and the 3'-untranslated region can have significant effects on gene expression. For example, the effect of the nucleotide composition of the coding region can far outweigh any enhancement achieved by the inclusion of an intron and/or a strong promoter if one is dealing with a A+T-rich coding region.

Synthetic genes

Heterologous genes may contain fortuitous processing and/or instability signals that have a deleterious impact on gene expression. This is especially true when introducing a prokaryotic gene into a eukaryotic host. The large differences in the gene expression mechanisms between prokaryotes and eukaryotes provides the opportunity for prokaryotic coding regions to contain eukaryotic processing signals which block or diminish expression in a eukaryotic host. Precise removal or alteration of processing signals can be difficult due to the imprecise nature of these signals and the affects of the surrounding nucleotides. It can be difficult to make a few minor, but precise, changes which result in a large increase in expression of the desired gene. The minor changes may each contribute only a minor improvement in expression. Rather than carrying out multiple rounds of mutagenesis to change several scattered sites, it may be easier to construct a synthetic gene of a desired sequence. Synthesis of a gene provides the ability to remove all potentially deleterious processing signals and produce a gene which may more nearly resemble genes of the new host than does the heterologous coding sequence. The greatest amount of experience using synthetic genes to increase expression of a desired protein is derived from expressing the insecticidal proteins, δ-endotoxins, from *Bacillus thuringiensis* (Bt). Attempts to express these genes in plants over several years has provided a basis for comparing levels of expression obtained from modified and unmodified coding regions in a number of plant hosts.

The first attempts to express δ-endotoxins in plants used the native coding sequences derived from Bt.

Vaeck *et al.* [73] used the mannopine synthetase promoter and 3' polyadenylation region of T-DNA gene 7 to express the *cryIA(b)* δ-endotoxin gene. Constructs examined in tobacco included a NH_2-terminal fragment encoding 610 amino acids and two constructs encoding translational fusions between the NH_2-terminal fragment of CryIA(b) and the *NptII* gene. Levels of Bt δ-endotoxin expressed in the plants containing the truncated proteins or the fusion proteins ranged from 2.6–190 ng CryIA(b)/mg soluble protein, or 0.0002–0.02% of total soluble protein. Barton *et al.* [2] analyzed expression of the native *cryIA(a)* gene in tobacco using both a full-length gene and a truncated gene encoding a 644 amino acid protein. Both versions used the CaMV 35S promoter, the 5'-UTL of alfalfa mosaic virus RNA 4, and the nopaline synthase 3' polyadenylation region. No plants transformed with the full-length gene produced detectable levels of Bt protein or mRNA. Northern blot analysis of these plants showed mRNA species shorter than the expected full-length transcript with distinct shorter fragments. The existence of incomplete mRNA in transgenic plants was attributed to inefficient post-transcriptional processing or rapid turnover of the full-length transcript.

Fischhoff *et al.* [15] transformed tomato with a native *cryIA(b)* gene. Two truncated versions of the gene were used, one encoding a protein of 646 amino acids and the other a protein of 725 amino acids. Each version was driven by the CaMV 35S promoter and the 3' polyadenylation region from the nopaline synthetase gene. Expression of only the NH_2-terminal fragment of a δ-endotoxin gene produced plants with insecticidal activity. Likewise a native *cryIA(c)-NptII* fusion in potato showed little insecticidal activity [10], indicating a low level of expression [14]. An active *cryIIIA* gene starting at amino acid 48 was introduced into tomato and potato plants [58]. The expression levels in these plants were very low, less than 0.001% of total soluble protein. In 1989, Vaeck *et al.* published results of studies on the transformation of tomato and potato plants with the NH_2-terminal Bt *cryIA(b)* gene [74]. Transgenic tomato plants produced 60–80 ng CryIA(b) per gram leaf tissue and potato plants produced 90–150 ng CryIA(b) per gram leaf tissue. Expression of a native *cryIA(b)* endotoxin gene in field grown tobacco was characterized by Carozzi *et al.* [9]. Six transgenic tobacco lines, both homozygous and hemizygous, expressing a 645 amino acid CryIA(b) protein from a truncated native gene present under control of the CaMV 35S promoter and 35S 3' polyadenylation sequences were studied. Bt δ-endotoxin levels increased throughout the course of plant development, with a substantial increase at the time of flowering. CryIA(b) levels at flowering ranged from 400 to 1000 ng per gram fresh weight or up to 0.01% of the total soluble protein. Bt δ-endotoxin mRNA of the expected size was readily detected, but there were also distinct truncated RNA forms, perhaps resulting from incomplete transcripts or cleavage products.

In all the above reports, native truncated Bt genes could be expressed in dicots but only at low levels. Analysis of mRNA from the native genes showed transcripts of less than the expected size. There are no reports of expression of native Bt genes in monocots. While it is difficult to precisely compare the expression levels reported by the various groups because different promoters and 3' ends were used, the observation of poor expression is consistent. Various explanations were put forth to explain this observation, but none was readily proven. To study mRNA stability, Murray et al. [53] examined expression of cryIA(b), cryIA(c), and cryIIIA genes in both transgenic tobacco and electroporated carrot protoplasts. The study examined full-length and truncated cryIA(b) and cryIA(c) genes and a full-length cryIIIA gene. The cryIA(b) gene was under control of the CaMV 35S promoter and 3' polyadenylation sequences from ORF 26 of the TR-DNA. The cryIA(c) gene was under control of the mannopine synthetase promoter. Northern analysis of mRNA isolated from transgenic tobacco plants transformed with either full-length or truncated genes from both cryIA(b) and cryIA(c) showed only truncated transcripts. Northern analysis of cryIA(b) and cryIA(c) constructs electroporated into carrot protoplasts showed full-length undegraded forms during the first 8 h after electroporation; but by 18 h the cryIA mRNA was degraded.

These results suggested that the truncated transcripts observed in plants were the result of message instability rather than truncation due to improper processing. A series of 3' deletion constructs in electroporated carrot cells showed that deletion of sequences in the 3' end of the cryIA(b) gene did not increase message stability. mRNA instability was retained in the first 570 bases of the gene. The cryIIIA gene was likewise poorly expressed in electroporated carrot cells.

Codon usage in the native δ-endotoxin genes is considerably different from that found in typical plant genes, which have a higher G+C content. Native endotoxin genes tend to have a very low G+C content, around 37%. Plants genes in general tend to have a higher G+C content, with maize showing a strong preference for G+C-rich coding regions [52]. Trun-

cated δ-endotoxin gene transcripts in transgenic plants could result from a number of events relating to their high A+T content. These include premature transcriptional termination or polyadenlyation in regions of high A+T content or inappropriate splicing or cleavage. Instability of the mRNA could be the result of endonucleolytic or exonucleolytic degradation at specific sequences that destabilize the message during transcription or create pausing due to the formation of secondary structures. Instability of the mRNA could also be the result of inefficient translation due to poor codon usage. To solve these problems and increase expression of δ-endotoxins in plants, synthetic genes were constructed using different strategies which emphasized eliminating different problems. These strategies all alter the overall G+C content of the genes. Several laboratories have now made partially or completely modified Bt δ-endotoxin genes that have resulted in significant improvements in expression of endotoxins in cotton [59, 78], tomato [60], and potato [61]. Synthetic Bt genes are a requirement to obtain expression of δ-endotoxins in monocots such as maize [44].

Transgenic cotton expressing truncated forms of two Bt endotoxin genes including native and modified sequences showed the efficacy of modified sequences. In 1990, Perlak et al. [59] published reports on the performance of transgenic cotton plants, Gossypium hirsutum cv. Coker 312, expressing truncated forms of two Bt δ-endotoxin genes. Constructs included genes expressing amino acids 1–612 of the cryIA(b) and amino acids 1–640 or 1–615 of the cryIA(c) gene. Both genes used a CaMV 35S promoter containing a duplicated enhancer region. Regions targeted for sequence modification included those with potential roles as regulatory sequences, or sequences with predicted mRNA secondary structure. These modifications increased the levels of both CryIA(b) and CryIA(c) δ-endotoxins to 0.05–0.1% of the total soluble protein, or what the authors estimated to be a 100-fold increase in expression compared to the truncated wild-type gene.

Perlak et al. [60] examined several versions of modified cryIA(b) and cryIA(c) genes in both transgenic tobacco and tomato to further analyze the increased expression associated with various sequence modifications. All genes were under control of a CaMV 35S promoter with a duplicated enhancer region. Two types of modified genes were used. One was partially modified, while the second was fully modified. The partially modified cryIA(b) gene had 62 of 1743 bases changed to eliminate regions with potential polyadenylation signals and A+T-rich regions. The partially modified *cryIA(b)* gene had 97% homology with the wild-type gene and a G+C content of 41%, compared to 37% in the native *cryIA(b)* gene. The fully modified *cryIA(b)* gene had 390 of 1845 bases changed to remove all ATTTA sequences and regions of potential mRNA secondary structure and also replaced bacterial codons with plant-preferred codons. The fully modified gene had 79% homology with the wild type gene with a G+C content increased to 49%. The majority of the transgenic tomato and tobacco plants expressing the partially modified gene produced δ-endotoxin at levels of 1–200 ng CryIA(b) per mg total protein. Over 10% of the fully modified *cryIA(c)* and *cryIA(b)* transgenic tomato and tobacco plants expressed between 600–2000 ng CryIA per mg. Compared with the native truncated gene, the most highly expressing transgenic plants containing the partially modified gene and the fully modified gene increased expression 10- and 100-fold, respectively. Levels of expression with the modified *cryIA* genes were now up to 0.2% of the total protein in the best plants. Constructs with different combinations of changes were tested in tobacco to study the effects of particular sequences. Modifications in the 5′ one-third of the *cryIA(b)* gene were sufficient to produce expression levels comparable to the partially modified gene whereas modifications in the 3′ half of the gene had no effect. The 5′ one-third of the *cryIA(b)* gene and nucleotides 246–283, which contain three potential polyadenylation signals, were identified as important, but no single region could be identified as being critical to expression levels. Northern analysis of the modified *cryIA(b)* genes in transgenic plants indicated that, although the levels of mRNA were increased, the mRNA increase was not proportional to the level of protein increase. This observation led the authors. to conclude that the low gene expression originated at the level of protein translation and not at the level of transcription.

Expression levels of CryIIIA were improved by using synthetic *cryIIIA* genes. A modified version of the *cryIIIA* gene was engineered to remove potential polyadenylation sites and A+T-rich regions, resulting in a final G+C content of 49% as compared to 37% in the native gene [61]. Expression of the *cryIIIA* gene in transgenic potatoes was under control of the CaMV 35S promoter with a duplicated enhancer region. Expression levels attained were between 0.002–0.3% of total soluble protein. Adang *et al.* [1] made a synthetic *cryIIIA* gene and examined its expression in carrot and maize protoplasts as well as transgenic potato plants. The codon pattern of this synthetic gene was altered to match the codon pattern of a dicot. The G+C content of the modified *cryIIIA* gene in this study was 45%. Only the synthetic gene, not the native gene, produced the expected size mRNA in electroporated carrot or maize protoplasts. CryIIIA protein was detected in these protoplasts only with the synthetic gene. The levels of CryIIIA obtained in carrot and maize protoplasts was similar, 0.001 to 0.005%, or 10–50 ng/mg total protein. The native gene did not produce CryIIIA. Transgenic potatoes containing the synthetic *cryIIIA* gene were estimated to produce CryIIIA up to 0.025% of total protein.

Barton *et al.* [3] constructed a synthetic *cryIA(a)* gene which resembled plant genes more closely, using a series of constructs with blocks of synthetic DNA representing 50 codons. Each of the constructs was studied to examine the effect of the modified block. Large increases (100-fold) in gene expression were seen with as little as 10% of the peptide coding region modified in the NH$_2$-terminal region. A modified truncated *cryIA(c)* gene under the control of the *Arabidopsis thaliana* ribulose-1,5-bisphosphate carboxylase small subunit promoter was tested in tobacco [79]. The small subunit promoter with its own 5′-untranslated leader and chloroplast transit peptide provided a 10–20-fold increase in CryIA(c) expression levels compared to the CaMV 35S promoter with a double enhancer. The increase in expression was the result of a combined effect due to both the 5′-untranslated leader and the transit peptide. A similar increase in expression of the CryIA(c) protein was observed when the 5′-untranslated leader and transit peptide were fused behind the CaMV 35S promoter. In this case, optimizing the translational leader improved expression and so did sequestering the final protein to an intracellular compartment.

The first example of a cereal plant expressing a δ-endotoxin used a synthetic gene [44]. A synthetic gene encoding the first 648 amino acids of the 1155 amino acid CryIA(b) protein produced by Bt var. *kurstaki* HD–1 [28] was constructed using the most preferred codon from maize for each amino acid [44, 52]. The synthetic gene had 65% homology with the native gene and a G+C content of 65%, compared to 37% for the native gene. When compared with the native gene, the synthetic gene produced significantly higher levels of CryIA(b) protein in both tobacco and maize. In transgenic tobacco plants, the synthetic gene produced about five times the level of CryIA(b) protein as did the native truncated gene using the CaMV 35S promoter and terminator regions. Transgenic maize lines expressing this gene were characterized. Plants containing the synthetic gene under control of the maize PEP-carboxylase [35] and pollen-specific promoters produced up to 4000 ng CryIA(b) per mg soluble protein in certain plants. Since native Bt genes are not expressed in maize, the magnitude of increase cannot be calculated, but 1–5 ng of CryIA(b) per mg total protein can be detected so this level of expression could be taken as the baseline. Transformation of rice with a native *cryIA(b)* gene produced no detectable protein so a highly modified *cryIA(b)* gene was designed based on the codon usage of rice genes [20]. The modified *cryIA(b)* gene had 66.6% of the codons changed to produce an overall G+C content of 59.2%. Bioassays of transgenic rice showed 10–50% mortality against the striped stemborer (*Chilo suppresalis*) and 45–55% mortality against the leaffolder (*Cnaphalocrosis medinalis*) indicating expression of the synthetic gene. The success of synthetic *cry* genes in improving gene expression facilitates studies on transcriptional and translational regulation of the *cry* genes, as well as other genes, in plants.

Synthetic genes encoding proteins other than δ-endotoxins have been made and tested for expression in plants. Hightower *et al.* [34] introduced two versions of a gene encoding a cecropin, a protein with anti-bacterial activity, into tobacco. The first synthetic gene used the same nucleotide sequence present in the source organism, the *Cecropia* moth. Cecropin protein was barely detectable in the transgenic plants. A second cecropin gene was synthesized using plant preferred codons, but transgenic tobacco plants showed no increased level of cecropin from this gene. The authors showed that synthetic cecropin protein is rapidly degraded in plant extracts and suggest that the low levels of protein observed with both versions of the cecropin genes was due to rapid protein degradation. In this case, it would appear to be more important to be concerned with the final stability or perhaps cellular location of the protein product than to use different coding regions to optimize expression. Florack *et al.* [17] also tested expression of a cecropin gene in tobacco. They used different versions of cDNA clones. The short version, lacking the amino terminal signal peptide, showed the poorest accumulation of mRNA. The cDNA containing the amino terminal signal peptide showed higher mRNA levels, while a version of the cecropin cDNA fused with a plant signal peptide had the highest levels of mRNA accumulation. In all cases, the cecropin protein could not be detected, apparently due to rapid degradation. It is interesting to note that the protein being produced appears to have an affect on the level of steady-state mRNA, even when the same final protein is being expressed. Addition of protein processing signals to proteins normally processed appears to have an affect on mRNA levels. The reason for this is unclear.

A gene encoding an insect specific scorpion toxin, I_5A, was synthesized and expressed in bacteria, yeast, and plants by Pang *et al.* [57]. Expression in tobacco was driven by the CaMV 35S promoter and the nopaline synthase 3' end was used for the polyadenylation site. mRNA derived from the synthetic gene was observed in transgenic plants and appeared to be the expected size. The I_5A peptide was purified from the transgenic plants, but it lacked biological activity, presumably due to improper folding.

Chloroplast sequestering and targeting

Another approach for optimizing expression of A+T-rich Bt genes in plants has been to transform the chloroplast of tobacco. McBride *et al.* [50] have reported very high levels of expression of an unmodified Bt *cryIA(c)* gene in tobacco. The tobacco chloroplast genome has a relatively high A+T content. The introduced Bt gene, present at up to about 10 000 copies per cell in the chloroplast genome, is not improperly processed as it is in the nucleus but is expressed at a high level, yielding up to 3–5% of total leaf protein as CryIA(c). The limitation at present for this approach is the availability of a broadly applicable chloroplast transformation system and the result of having the trait transmissible from only the female parent.

Chloroplast targeting has also been used to increase the expression of polyhydroxybutyrate (PHB) in plants. PHB is the product of a three enzyme pathway starting with acetyl-CoA. When these three proteins were expressed in the cytoplasm of *Arabidopsis*, PHB accumulated at a low level, 20–100 μg per gram fresh weight, and the plants were stunted [62–64]. Nawrath *et al.* [54] targeted these enzymes to the chloroplast using a pea chloroplast transit peptide. The resulting transgenic *Arabidopsis* plants accumulated PHB in plastids up to a level of 14% of the dry weight without any apparent deleterious effects on the plant. The targeting of the enzymes to an intracellular compartment with a high flux of acetyl-CoA increased accumulation of PHB about 100-fold. Choice of the proper intracellular compartment was critical in providing the required substrates for the introduced enzymes. Optimization of product yield in this case did not require modification of the coding region other than to add an appropriate targeting sequence. This underscores the importance of understanding the biochemistry underlying the mechanisms being introduced into a plant to achieve optimal expression of the desired product.

3' regions

Ingelbrecht et al. [36] examined 3'-end regions of transgenes in plants and found a large difference in expression could be obtained in stable plants. Interestingly, this difference was not detected in transient expression assays. Neomycin phosphotransferase II (NptII) was used as the reporter gene for this study. Chimeric gene expression was driven by the CaMV 35S promoter. A NptII gene containing no plant 3'-end sequences could be expressed in transgenic tobacco, but at a level about 12-fold lower, and much more variable, than a comparable construct containing the 3' end of the octopine synthase (OCS) gene. In transient expression assays, the construct containing the OCS 3' sequences expressed about 20-fold higher than the construct with no 3'-end sequences. Different 3' ends were compared in both transient and stable transformants. In transient expression assays, the CaMV 35S promoter/NptII gene was expressed at comparable levels regardless of the source of the 3' end. The different 3' ends used were obtained form the octopine synthase gene, the 2S seed protein from Arabidopsis, the small subunit of rbcS from Arabidopsis, extensin from carrot, and chalcone synthase from Antirrhinum. However, in stable tobacco transformants, there was about a 60-fold difference between the best-expressing construct (small subunit rbcS 3' end) and the lowest-expressing construct (chalcone synthase 3' end), with the other 3' ends producing different expression levels between these two extremes. Levels of mRNA corresponded well with the observed differences in protein levels. The authors suggested that the higher level of expression obtained with the small subunit rbcS 3' end was a reflection of the need of the plant to express high levels of the small subunit rbcS protein and therefore the mRNA was likely to be more stable than the mRNA encoding proteins present at lower levels. The precise role of 3'-end sequences and how to optimize them for maximal expression in transgenic plants is an area where much remains to be learned.

The 3'-untranslated region has been implicated in determining stability or instability of a mRNA. For a detailed discussion on determinants of mRNA instability and their effect on gene expression in plants, readers should see the review by Green et al. in this volume.

Effects of protein folding

The effects of pre- and pro-protein sequences on the level of mature protein obtained in transgenic plants was studied by Florack et al. [16]. Small anti-bacterial proteins called hordothionins (HDH) are found in barley. The mature proteins are processed from larger precursors which contain the mature protein, an amino terminal signal peptide (SP), and a carboxy-terminal acidic peptide (AP). Different versions of the mature protein were expressed in transgenic tobacco. The chimeric genes were derived from barley cDNA clones as well as synthetic sequences encoding the mature protein and the signal peptide. Plants containing the synthetic gene encoding only the mature peptide produced the lowest level of mRNA. Addition of sequences encoding the SP or adding both the SP and AP increased the level of mRNA detected. Plants containing the gene encoding only the mature protein did not produce detectable amounts of mature protein even though there was mRNA. Addition of the SP to this synthetic gene yielded plants expressing HDH up to about 0.1% of total protein. Addition of both SP and AP sequences increased maximum expression levels almost ten fold to about 0.7% of total protein. Using either the synthetic or barley cDNA to encode the mature protein did not affect levels of mature protein obtained from the constructs containing the SP and AP sequences. In this case, the coding region itself was not responsible for altering the level of the final protein product. Rather, the initial protein produced had a much more dramatic affect. Attempting to produce only the mature protein produced the poorest results and expressing the entire precursor protein produced the best results. The SP was required to produce detectable HDH, likely because this sequence targets the protein to the endoplasmic reticulum where the needed disulfide bonds can be properly formed. Optimization of HDH peptides requires use of the proper pro- and pre-sequences of the protein rather than changing of the coding region.

Conclusions

There are several methods for increasing expression of a particular gene in a plant host. While several opportunities exist to increase expression, which ones are likely to produce the best effect need to be determined on a case by case basis. Addition of introns to the transcript, use of a good leader for a particular host plant, removing RNA instability signals, and synthesis of optimized coding regions have all been shown to increase levels of expression. Use of a stable 3' end can help increase expression. While changes in steady-state RNA levels have proven successful in changing levels of protein obtained, the fate of the encoded protein should also be kept in mind. Using truncated versions containing only the desired activity may omit certain folding events necessary for proper folding, stability and activity. Further, the desired protein might not be stable in the plant cell and be rapidly degraded after expression. In such cases, targeting to intracellular compartments or secretion outside the cell might improve levels of the desired protein. For a given protein, not all of the problems are likely to occur, but at present the best method for determining how to optimize expression lies with empirical observation and trial and error. There are likely further refinements used by cells which remain to be elucidated and put to use in optimizing transgene expression.

References

1. Adang MJ, Brody MS, Cardineau G, Eagan N, Roush RT, Shewmaker CK, Jones A, Oakes JV, McBride KE: The reconstruction and expression of a *Bacillus thuringiensis cryIIIA* gene in protoplasts and potato plants. Plant Mol Biol 21: 1131–1145 (1993).
2. Barton KA, Whiteley HR, Yang N: *Bacillus thuringiensis* delta-endotoxin expressed in transgenic *Nicotiana tabacum* provides resistance to lepidopteran insects. Plant Physiol 85: 1103–1109 (1987).
3. Barton KA, Miller MJ: Production of *Bacillus thuringiensis* insecticidal proteins in plants. In: Kung SD, Wu R (eds) Transgenic Plants, vol 1: Engineering and Utilization, pp. 297–315. Academic Press, New York (1993).
4. Battraw M, Hall TC: Expression of a chimeric neomycin phosphotransferase II gene in first and second generation transgenic rice plants. Plant Sci 86: 191–202 (1992).
5. Borenstein P, McKay J, Morishima Jk, Devarayalu S, Gelinas RE: Regulatory elements in the first intron contribute to transcriptional control of the human a1(I) collagen gene. Proc Natl Acad Sci USA 84: 8869–8873 (1987).
6. Boronat A, Martinez MC, Reina M, Puigdomenech P, Palau J: Isolation and sequencing of a 28 kd glutelin–2 gene from maize: Common elements in the 5' flanking regions among zein and glutelin genes. Plant Sci 47: 95–102 (1986).
7. Callis J, Fromm M, Walbot V: Introns increase gene expression in cultured maize cells. Genes Devel 1: 1183–1200 (1987).
8. Cao J, Duan X, McElroy D, Wu R: Regeneration of herbicide resistant rice plants following microprojectile-mediated transformation of suspension culture cells. Plant Cell Rep 11: 586–591 (1992).
9. Carozzi NB, Warren GW, Desai N, Jayne SM, Lotstein R, Rice DA, Evola S, Koziel MG: Expression of a chimeric CaMV 35S *Bacillus thuringiensis* insecticidal protein gene in transgenic tobacco. Plant Mol Biol 20: 539–548 (1992).
10. Cheng J, Bolyard MG, Saxena RC, Sticklen MB: Production of insect resistant potato by genetic transformation with a delta-endotoxin gene from *Bacillus thuringiensis* var. *kurstaki*. Plant Sci 81: 83–91 (1992).
11. Christensen AH, Sharrock RA, Quail PH: Maize polyubiquitin genes: Structure, thermal perturbation of expression and transcript splicing, and promoter activity following transfer to protoplasts by electroporation. Plant Mol Biol 18: 675–689 (1992).
12. Christou P, Ford T, Kofron M: Production of transgenic rice (*Oryza sativa* L.) plants from agronomically important indica and japonica varieties via electric discharge particle acceleration of exogenous DNA into immature zygotic embryos. Bio/technology 9: 957–962 (1991).
13. Dennis ES, Gerlach WL, Pryor AJ, Bennetzen JL, Inglis A, Llewellyn D, Sachs MM, Ferl RJ, Peackocock WJ: Molecular analysis of the alcohol dehydrogenase (*Adh1*) gene of maize. Nucl Acids Res 12: 3983–4000 (1984).
14. Ebora RV, Ebora MM, Sticklen MB: Transgenic potato expressing the *Bacillus thuringiensis cryIA(c)* gene: effects on the survival and food consumption of *Phthorimea operculella* (Lepidoptera: Gelechiidae) and *Ostrinia nubilalis* (Lepidoptera: Noctuidae). J Econ Entomol 87: 1122–1127 (1994).
15. Fischhoff DA, Bowdish KS, Perlak FJ, Marrone PG, McCormick SM, Niedermeyer JG, Dean DA, Kusano-Kretzmer K, Mayer EJ, Rochester DE, Rogers SG, Fraley RT: Insect tolerant transgenic tomato plants. Bio/technology 5: 807–813 (1987).
16. Florack D, Dirkse WG, Visser B, Heidekamp F, Stiekema W: Expression of biologically active hordothionins in tobacco. Effects of pre- and pro-sequences at the amino and carboxyl termini of the hordothionin precursor on the mature protein expression and sorting. Plant Mol Biol 24: 83–96 (1994).
17. Florack D, Allefs S, Bollen R, Bosch D, Visser B, Stiekema W: Expression of giant silkmoth cecropin B genes in tobacco. Transgenic Res 4: 132–141 (1995).
18. French R, Janda M, Ahlquist P: Bacterial gene inserted in an engineered RNA virus: efficient expression in monocotyledonous plant cells. Science 231: 1294–1297 (1986).
19. Fromm ME, Morrish F, Armstrong C, Williams R, Thomas J, Klein TM: Inheritance and expression of chimeric genes in the progeny of transgenic plants. Bio/technology 8: 833–844 (1990).
20. Fujimoto H, Itoh K, Yamamoto M, Kyozuka J, Shimamoto K: Insect resistant rice generated by introduction of a modified delta-endotoxin gene of *Bacillus thuringiensis*. Bio/technology 11: 1151–1155 (1993).
21. Gallie DR, Sleat DE, Watts JW, Turner PC, Wilson TMA: The 5'-leader sequence of tobacco mosaic virus RNA enhances the expression of foreign gene transcripts *in vitro* and *in vivo*. Nucl Acids Res 15: 3257–3273 (1987).
22. Gallie DR, Sleat DE, Watts JW, Turner PC, Wilson TMA: A comparison of eukaryotic viral 5'-leader sequences as enhancers of mRNA expression *in vivo*. Nucl Acids Res 15: 8693–8711 (1987).
23. Gallie DR, Sleat DE, Turner PC, Wilson TMA: Mutational analysis of the tobacco mosaic virus 5'-leader for altered ability to enhance translation. Nucl Acids Res 16: 883–893 (1988).

404

24. Gallie DR, Lucas WJ, Walbot V: Visualizing mRNA expression in plant protoplasts: factors influencing efficient mRNA uptake and translation. Plant Cell 1: 301–311 (1989).

25. Gallie DR, Feder JN, Schimke, RT, Walbot V: Post-transcriptional regulation in higher eukaryotes: the role of the reporter gene in controlling expression. Mol Gen Genet 228: 258–264 (1991).

26. Gallie DR, Young TE: The regulation of expression in transformed maize aleurone and endosperm protoplasts. Plant Physiol 106: 929–939 (1994).

27. Gehrke L, Auron PE, Quigley GQ, Rich A, Sonenberg N: 5′-Conformation of capped alfalfa mosaic virus ribonucleic acid 4 may reflect its independence of the cap structure or of cap-binding protein for efficient translation. Biochemistry 22: 5157–5164 (1983).

28. Geiser M, Schweitzer S, Grimm C: The hypervariable region in the genes coding for entomopathogenic crystal proteins of *Bacillus thuringiensis*: nucleotide sequence of the kurhd1 gene of subsp. *kurstaki* HD1. Gene 48: 109–118 (1986).

29. Genschick P, Marbach M, Uze M, Feuerman B, Plesse B, Fleck J: Structure and promoter activity of a stress and developmentally regulated polyubiquitin-encoding gene of *Nicotiana tabacum*. Gene 148: 195–202 (1994).

30. Goelet P, Lomonossoff GP, Butler PJG, Akam ME, Gait MJ, Karn J: Nucleotide sequence of tobacco mosaic virus RNA. Proc Natl Acad Sci USA 79: 5818–5822 (1982).

31. Goodall GJ, Filipowicz W: Different effects of intron nucleotide composition and secondary structure on pre-mRNA splicing in monocot and dicot plants. EMBO J 10: 2635–2644 (1991).

32. Gordon-Kamm WJ, Spencer M, Mangano ML, Adams TR, Daines RJ, Start WG, O'Brien JV, Chambers SA, Adams WR, Willetts NG, Rice TB, Mackey CJ, Krueger RW, Kausch AP, Lemaux PG: Transformation of maize cells and regeneration of fertile transgenic plants. Plant Cell 2: 603–618 (1990).

33. Heidecker G, Messing J: Structural analysis of plant genes. Annu Rev Plant Physiol 37: 439–466 (1986).

34. Hightower R, Baden C, Penzes E, Dunsmuir P: The expression of cecropin peptide in transgenic tobacco does not confer resistance to *Pseudomonas syringae* pv. *tabaci*. Plant Cell Rep 13: 295–299 (1994).

35. Hudspeth RL, Grula JW: Structure and expression of the maize gene encoding the phosphoenolpyruvate carboxylase isozyme involved in C4 photosynthesis. Plant Mol Biol 12: 579–589 (1989).

36. Ingelbrecht LW, Herman LMF, Dekeyser RA, Van Montagu MC, Depicker AG: Different 3′ end regions strongly influence the level of gene expression in plant cells. Plant Cell 1: 671–680 (1989).

37. Jefferson RA, Kavanaugh TA, Bevan MW: GUS Fusions: β-glucuronodase as a sensitive and versatile gene fusion marker in higher plants. EMBO J 6: 3901–3907 (1987).

38. Jobling SA, Gehrke L: Enhanced translation of chimaeric messenger RNAs containing a plant viral untranslated leader sequence. Nature 325: 622–625 (1987).

39. Jones JDG, Dean C, Gidoni D, Gilbert D, Bond-Nutter D, Lee R, Bedbrook J, Dunsmuir P: Expression of bacterial chitinase protein in tobacco leaves using two photosynthetic gene promoters. Mol Gen Genet 212: 536–542 (1988).

40. Joshi CP: An inspection of the domain between putative TATA box and translation start site in 79 plant genes. Nucl Acids Res 15: 6643–6653 (1987).

41. Klein TM, Roth BA, Fromm ME: Regulation of anthocyanin biosynthetic genes introduced into intact maize tissues by microprojectiles. Proc Natl Acad Sci USA 86: 6681–6685 (1989).

42. Kozak M: Point mutations define a sequence flanking the AUG initiator codon that modulates translation by eukaryotic ribosomes. Cell 44: 283–292 (1986).

43. Kozak M: An analysis of 5′-noncoding sequences from 699 vertebrate messenger RNAs. Nucl Acids Res 15: 8125–8132 (1987).

44. Koziel MG, Beland GL, Bowman C, Carozzi NB, Crenshaw R, Crossland L, Dawson J, Desai N, Hill M, Kadwell S, Launis K, Lewis K, Maddox D, McPherson K, Meghji MR, Merlin E, Rhodes R, Warren GW, Wright M, Evola SV: Field performance of elite transgenic maize plants expressing an insecticidal protein derived from *Bacillus thuringiensis*. Bio/technology 11: 194–200 (1993).

45. Luehrsen KR, Walbot V: Intron enhancement of gene expression and the splicing efficiency of introns in maize cells. Mol Gen Genet 225: 81–93 (1991).

46. Luehrsen KR, Walbot V: The impact of AUG start codon context on maize gene expression *in vivo*. Plant Cell Rep 13: 454–458 (1994).

47. Lutcke HA, Chow KC, Mickel FS, Moss KA, Kern HF, Scheele GA: Selection of AUG initiation codons differs in plants and animals. EMBO J 6: 43–48 (1987).

48. Maas C, Laufs J, Grant S, Korfhage C, Werr W: The combination of a novel stimulatory element in the first exon of the maize shrunke-1 gene with the following intron enhances reporter gene expression 1000-fold. Plant Mol Biol 16: 199–207 (1991).

49. Mascerenhas D, Mettler IJ, Pierce DA, Lowe HW: Intron mediated enhancement of heterologous gene expression in maize. Plant Mol Biol 15: 913–920 (1990).

50. McBride KE, Svab Z, Schaaf DJ, Hogan PS, Stalker DM, Maliga P: Amplification of a chimeric *Bacillus* gene in chloroplasts leads to an extraordinary level of an insecticidal protein in tobacco. Bio/technology 13: 362–365 (1995).

51. McElroy D, Zhang W, Wu R: Isolation of an efficient promoter for use in rice transformation. Plant Cell 2: 163–171 (1990).

52. Murray EE, Lotzer J, Eberle M: Codon usage in plants. Nucl Acids Res 17: 477–498 (1989).

53. Murray EE, Rocheleau T, Eberle M, Stock C, Sekar V, Adang M: Analysis of unstable RNA transcripts of insecticidal crystal protein genes of *Bacillus thuringiensis* in transgenic plants and electroporated protoplasts. Plant Mol Biol 16: 1035–1050 (1991).

54. Nawrath C, Poirier Y, Somerville C: Targeting of the poly-hydroxybutyrate biosynthetic pathway to the plastids of *Arabidopsis thaliana* results in high levels of polymer accumulation. Proc Natl Acad Sci USA 91: 12760–12764 (1994).

55. Norris SR, Meyer SE, Callis J: The intron of *Arabidopsis thaliana* polyubiquitin genes is conserved in location and is a quantitative determinant of chimeric gene expression. Plant Mol Biol 21: 895–906 (1993).

56. Oard JH, Paige D, Dvorak J: Chimeric gene expression using maize intron in cultured cells of breadwheat. Plant Cell Rep 8: 156–160 (1989).

57. Pang SZ, Oberhaus SM, Rasmussen JL, Knipple DC, Bloomquist JR, Dean DH, Bowman KD, Sanford JC: Expression of a gene encoding a scorpion insectotoxin in yeast, bacteria, and plants. Gene 116: 165–172 (1992).

58. Perlak FJ, McPherson SA, Fuchs RL, Macintosh SC, Dean DA, Fischhoff DA: Expression of *Bacillus thuringiensis* protein in transgenic plants. In: Roberts DW, Granados RR (eds) Proceedings of a Conference on Biotechnology, Biological Pesti-

cides and Novel Plant-Pest Resistance for Insect Pest Management, pp. 77–81. Boyce Thompson Institute for Plant Research, Ithaca, NY (1988).

59. Perlak FJ, Deaton RW, Armstrong TA, Fuchs RL, Sims SR, Greenplate JT, Fischhoff DA: Insect resistant cotton plants. Bio/technology 8: 939–943 (1990).

60. Perlak FJ, Fuchs RL, Dean DA, McPherson SL, Fischhoff DA: Modification of the coding sequence enhances plant expression of insect control protein genes. Proc Natl Acad Sci USA 88: 3324–3328 (1991).

61. Perlak FJ, Stone TB, Muskopf YM, Peterson LJ, Parker GB, McPherson SA, Wyman J, Love S, Reed G, Biever D, Fischhoff DA: Genetically improved potatoes: protection from damage by Colorado potato beetles. Plant Mol Biol 22: 313–321 (1993).

62. Poirier Y, Dennis DE, Klomparens K, Sommerville CR: Polyhydroxybutyrate, a biodegradable thermoplastic, produced in transgenic plants. Science 256: 520–523 (1992).

63. Poirier Y, Dennis DE, Nawrath C, Somerville CR: Progress toward biologically produced biodegradable thermoplastics. Adv Mat 5: 30–37 (1993).

64. Poirier Y, Schechtman LA, Satkowski MM, Noda I, Somerville CR: Synthesis of high-molecular-weight poly([R]-(-)-3-hydroxybutyrate) in transgenic Arabidopsis thaliana plant cells. Int J Biol Macromol 17: 7–12 (1995).

65. Poogin MM, Skryabin KG: The 5' untranslated leader sequence of potato virus X RNA enhances the expression of the heterologous gene in vivo. Mol Gen Genet 234: 329–331 (1992).

66. Rhodes CA, Pierce DA, Mettler IJ, Mascarenhas D, Detmer JJ: Genetically transformed maize plants from protoplasts. Science 240: 204–207 (1988).

67. Rothstein SJ, Lahners KN, Lotstein RJ, Carozzi NB, Jayne SM, Rice DA: Promoter cassettes, antibiotic-resistance genes, and vectors for plant transformation. Gene 53: 153–161 (1987).

68. Shimamoto K, Terada R, Izawa T, Fujimoto H: Fertile transgenic rice plants regenerated from transformed protoplasts. Nature 338: 274–276 (1989).

69. Sleat DE, Gallie, DR, Jefferson RA, Bevan MW, Turner PC, Wilson TMA: Characterization of the 5'-leader sequence of tobacco mosaic virus RNA as a general enhancer of translation in vitro. Gene 217: 217–225 (1987).

70. Sleat DE, Hull R, Turner PC, Wilson TMA: Studies on the mechaism of translational enhancement by the 5'-leader sequence of tobacco moasaic virus RNA. Eur J Biochem 175: 75–86 (1988).

71. Tanaka A, Mita S, Ohta S, Kyozuka J, Shimamoto K, Nakamura K: Enhancement of foreign gene expression by a dicot intron in rice but not in tobacco is correlated with an increased level of mRNA and an efficient splicing of the intron. Nucl Acids Res 18: 6767–6770 (1990).

72. Toki S, Takamatsu S, Nojiri C, Ooba S, Anzai H, Iwata M, Christensen AH, Quail PH, Uchimiya H: Expression of a maize ubiquitin gene promoter-bar chimeric gene in transgenic rice plants. Plant Physiol 100: 1503–1507 (1992).

73. Vaeck M, Reynaerts A, Hofte H, Jansens S, De Beuckeleer M, Dean C, Zabeau M, Van Montagu M, Leemans J: Transgenic plants protected from insect attack. Nature 328: 33–37 (1987).

74. Vaeck M, Reynaerts A, Hofte H: Protein engineering in plants: expression of Bacillus thuringiensis insecticidal protein genes. In: Schell J, Vasil IK (eds) Cell Culture and Somatic Cell Genetics of Plants, vol 6. Academic Press, Troy, MO (1989).

75. Vasil V, Clancy M, Ferl RJ, Vasil IK, Hannah LC: Increased gene expression by the first intron of maize shrunken–1 locus in grass species. Plant Physiol 91: 1575–1579 (1989).

76. Vasil V, Castillo AM, Fromm ME, Vasil IK: Herbicide resistant fertile transgenic wheat plants obtained by microprojectile bombardment of regenerable embryogenic callus. Bio/technology 10: 667–674 (1992).

77. Weeks JT, Anderson OD, Blechl A: Rapid production of multiple independent lines of fertile transgenic wheat (Triticum aestivum). Plant Physiol 102: 1077–1084 (1992).

78. Wilson FD, Flint HM, Deaton WR, Fischhoff DA, Perlak FJ, Armstrong TA, Fuchs RL, Berberich SA, Parks NJ, Stapp BR: Resistance of cotton lines containing a Bacillus thuringiensis toxin to pink bollworm (Lepidopteran: Gelechiidae) and other insects. J Econ Entomol 4: 1516–1521 (1992).

79. Wong EY, Hironaka CM, Fischoff DA: Arabidopsis thaliana small subunit leader and transit peptide enhance the expression of Bacillus thuringiensis proteins in transgenic plants. Plant Mol Biol 20: 81–93 (1992).

80. Zhang W, McElroy D, Wu R: Analysis of rice act1 5' region activity in transgenic rice plants. Plant Cell 3: 1155–1165 (1991).

Subject index

A complex, spliceosome 6
AAPT1, 5′ leader sequence 166
Abscisic acid effects
   HSP100 197
   mRNA stability 67
   poly (A) polymerase inhibition 56
ACC synthase, proteolytic regulation 290–291
Adh1, intron enhancement 24–25, 396–397
ADP-ribosylation 128
albostrians 358–359
Alfalfa mosaic virus
   5′ leader sequence 394–395
   subgenomic promoter 372
Algae, RNA editing 356
Alternative splicing 21–24
Ambisense strategy, RNA virus 375
Amino acids, proteolytic supply 290
α-Amylase
   mRNA stability 68
   translational enhancement 152
ANJ1 200
Antirrhinum 263 ff.
Antisense RNA 90 ff.
   gene silencing 81–83, 86
   stability 72
AP 265 ff.
Arabidopsis
   flowering genes 265 ff.
   gene silencing 81
   KNAT1 260–261
   SAUR transcripts 69–71
Ascobolus immersus, methylation 83
ATH1, 5′ leader sequence 166
ATPase
   BiP – see under Chaperones
   HSP70 – see under Chaperones
   S complex 281–282
atp genes 95, 347, 349
AUG codon – see Start codons
AU-rich sequences
   chloroplast mRNA 336
   intron 13, 16, 17–19
   5′ leader sequence 151
   mRNA stability 70
   poly(A) tail 52–55
   U-islands 18
Auxin effects
   eEFIA expression 124
   poly(A) polymerase inhibition 56
   proteolysis 292
   ribosome proteins 132

B complex, spliceosome 6

Bacterial gene expression in plants 395
Bacterial mRNA, folding 100
Barley, RNA editing 356
Barley yellow dwarf virus, translation 150–151
Bean dwarf mosaic virus, movement protein 255
BiP – see under Chaperones
BP-80 239
Branch site – see under Intron
Broad bean, mtRNA processing 309
Brome mosaic bromevirus
   coat protein 5′ leader sequence 394–395
   subgenomic promoter 369
Bulge loop, RNA 91

C complex, spliceosome 6
Calnexin – see under Chaperones
Calpain 285
Calreticulin 204
Cap (see also Initiation of translation) 160, 304, 306
   mRNA stability 69
   regulation of translation 146 ff.
   RNA virus 375
Cap-binding protein 118–120, 121, 133
Cap-snatching 375
Cauliflower mosaic virus
   gene silencing 80 ff.
   5′ leader sequence 167–168
   poly(A) signals 47–52, 55
Cecropin 401
Cell cycle
   effect on spliceosome 29
   nuclear bodies 31–32
   proteolytic control 293
Chalcone synthase, gene silencing 83 ff.
Chaperones 191–212
   BiP 193–194, 201–202, 233, 287
   calnexin 193, 203–204
   Clp proteins 196, 210–211, 286, 288–289, 294
   GRP94 202–203
   HSP70 193 ff.
   HSP90 195 ff.
   HSP100 196 ff.
   Hsp104 196–197
   nucleoplasmin 193
   small HSPs 198 ff.
Chaperonins 193–212
   proteolysis 288–289
Chlamydomonas reinhardtii, chloroplast gene expression 327–338
Chloroplast
   chaperones 204–211
   chaperonins 204–205
   cotranscription 317
   endonuclease 321

evolution 322
foldases 204–211
gene expression in *Chlamydomonas reinhardtii* 327–338
introns 318–319
mRNA processing 318–323, 328–331
mutants 328 ff.
phosphorylation 322, 335
polycistronic transcripts 317, 330–331
polyhydroxybutyrate transgenese 401–402
postranscriptional control 320
promoters 316–318
protein transport 224–225
proteolysis 286–287
regulation of gene expression 315–323
RNA-binding proteins 322–323, 336
RNA degradation 329
RNA editing 351–362
RNA polymerase 316–318
RNA-processing mutants 322
RNA splicing 318, 320, 331–333
RNA stability 321–323, 328–330
self-splicing 331
transformation 328, 401–402
transcription 316–317, 334
trans-splicing 95, 331–333
Clathrin-coated vesicles 194
Cleavage and polyadenylation specificity factor 53–54, 56
Cleavage factors, poly(A) tail 53–54
Cleavage stimulation factor 53–54
Closterovirus
frameshift 380
proteinase 385
Clp – see under Chaperones
Coatomers 235–236
Co-chaperones 193, 199, 205–211
cpn10 205 ff.
DnaJ 199–201
GrpE 199–201
Coiled body, see also Dense body 26–28
Cold tolerance 125
Comovirus, proteinase 384
Companion cell 253, 268
β-Conglycinin 202
Cotranscription
chloroplast 317
mitochondria 304
cox genes 344 ff.
cry transgenes 398–401
C-terminal propeptides 238
Cucumber mosaic virus
movement protein 255
subgenomic promoter 372
Cyclin 293–294
Cyclin-dependent kinase 293
Cyclophilin – see under Foldases
Cytochrome b6 337
Cytochrome f 334, 337
Cytokinin, effect on mRNA stability 67
Cytoplasm, proteolysis 277–284
Cytoskeleton 133–134

DEAD box 5
Dense body 30–31
Desiccation, effect on translation 154

Dianthovirus, frameshifting 378
Dicot vs. monocot
AU-rich sequences 13
eIF4 gene structure 114
intron enhancement 396
intron processing 8
5′ leader sequence 395
poly(A) tail 53
transgene expression 399
Diphthamide 128
DNA
ectopic pairing 85
methylation 83
DnaJ – see under Co-chaperone
Downstream element, poly(A) 49, 53–54
dsRNA 90–95
subgenomic forms 374

E complex, spliceosome 5
Ectopic pairing 85
Editing – see RNA editing
Efficiency element, poly(A) signal 49, 53–54
Elongation, translation 121–129, 162–163
Elongation factors 121–129, 162
Encapsidation, subgenomic RNA 374
Endonuclease, chloroplast 321
Endoplasmic reticulum
budding 240
calnexin 203–204
calreticulin 204
chaperones 200–204
foldases 200–204
heat shock effect 67–68
plasmodesmata 252
protein folding 200
protein secretion 229–235
proteolysis 287
small heat shock proteins 204
Endosperm mutants 202
Endotoxin, transgene expression 398–401
Enzymes, proteolytic regulation 290–291
Erd2p receptor 232
Ethylene, effect on mRNA stability 67
Euglena gracilis, chloroplast introns 318
Evolution
chloroplasts 322
eEF1B 127
histone mRNA 149–150
leucine zipper 127
RNA editing 350, 355–357
snRNA 332
Exon
intron enhancement 25
RNA editing 346
role in splicing 17
Exon enhancer 17
External transcribed spacer 98

Far-upstream element, poly(A) tail 45, 50–54, 56
Ferrodoxin (Fed) genes 68, 70–71, 74
Fibrillarin 29
Figwort mosaic virus, poly(A) signals 47–52
FK506-binding protein – see under Foldases
floricaula 263–264

Flowering
  genes 266
  gene silencing 87
  signal 268–270
  supracellular control proteins 262–270
Foldases 191–212
  cyclophilin 199, 211
  FK506-binding protein 199, 302, 211
  immunophilin 199, 211
Frameshifting 174–176
  RNA virus 377–380
  structural basis 97
Frameshift mutations 74
Fruit ripening, gene silencing 84
Fungal elicitors, effect on mRNA stability 69, 72
Fungal mitochondria 306, 308
Furin 236

Gene silencing 79–87
Gene targeting 401–402
Gibberellic acid
  poly(A) polymerase stimulation 56
  protein disulfide isomerase induction 203
Golgi complex
  protein transport 232, 235–237
GRP94 – see under Chaperones
GrpE – see under Co-chaperones
GTPase superfamily 124, 127
Guide protein 117
Guide RNA 350

Hairpin, RNA 91 ff.
  editing 358
  translation inhibition 151, 161
  viral frameshifting 378 ff.
Hairpin ribozyme 7
Half-life
  mRNA 65
  protein 277, 290–291
Hammerhead ribozyme 98
Heat shock effects
  mRNA stability 67–68
  ribosomes 132
  spliceosome 29–30
  splicing 26
  translation 153–154, 178
Heat shock granules 153
Heat shock proteins (see also Chaperones) 153–154
  p102 148
  PRP73 286
Helicase 5, 113
Helix, spliceosome 6–8
Heme, cytochrome f attachment 337
Heterogenous nuclear ribonucleoprotein particle (hnRNP) 5, 30–32
Histone mRNA 149–150
  poly(A) tail 45
HnRNPs 5, 30–32
Hordothionins 402
hymns 117
Hypoxia effects
  eEF1A expression 125
  phosphorylation 132
  ribosomes 132
  translation 154

Hypusine 116–117

I5A (scorpion toxin) transgene expression 401
Immunization 295
Immunophilin – see Foldases
Importins 228
Initiation, translation (see also Cap, 5′ leader sequence, Start codons) 108–121, 146 ff., 160–162
  cap independent 150, 152, 171–173
  efficiency 394
  5′ end effect 160
  internal ribosome entry 171
  leaky scanning 168–170, 376–377
  multiple sites 164–165
  mutants 333–334
  RNA structure 94
  RNA virus 375–377
  scanning 16, 161–162, 164, 167–170, 376
  site selection 162
  trans-acting factors 335–337
  3′-UTR effect 173–174
Initiation codon – see Start codons
Initiation factors 108–121, 146, 160, 162–164, 172
Interchromatin granule clusters 5, 26–28
Internal ribosome entry site 171–172
Internal loop, RNA 91
Internal transcribed spacer 98
Intron – see also Splicing
  Adh1 396–397
  AT-AC class 32
  AU-rich sequence – see principal heading
  branch site 6, 12–13, 16–17
  chloroplast genes 318
  enhancement of gene expression 24–26
  functional RNA 32
  group I 99, 331
  group II 309, 331
  group III 95
  RNA editing 347, 357–359
  size 9
  structure 9–14
  use in transgenes 396
Intron-encoded protein 318–320
Inverted repeat
  chloroplast RNA 321, 329
  stem-loop structure 95

Jasmonate-induced protein 133
Junction, RNA 95

K+ channels 199
Karyopherins 228
KDEL sequence 232
Kin recognition, Golgi 235 ff.
Kinase, eIF2 phosphorylation 110–111
KKXX signal 235
KNAT1 260–261
KNOTTED1 257–261
Kunitz trypsin inhibitor (KTi3) transcript 74

Lariat – see Stem-loop
Lc protein 166
5′ leader sequence, mRNA 161–162
  chloroplast 329–330

dicot vs. monocot 395
light response 70–71
pseudoknot 167
regulation of translation 151–152
RNA editing 358
RNA virus 172, 375
secondary structure 151, 167
short open reading frames 152–153, 165–168
stem-loop structures 94, 167
sucrose synthase gene 24
transgenes 394–395
translation enhancers 152
Leaky scanning – see under Initiation of translation
Leucin zipper 127
Light effects
chloroplast transcription 330
eEF1A expression 124
mRNA stability 68, 70, 74, 335
protein transport 228
translation 335–336
transcription 330
Long-distance interaction, phage RNA 104
Luteovirus
frameshifting 378
proteinase 385

Macropain – see Proteasome
MADS box 21, 265
Maize
chloroplast transcription 317–318
endotoxin transgene expression 400–401
gene expression in Petunia 80
KNOTTED1 257–261
mitochondrial RNA processing 304 ff.
RNA editing 353 ff.
kDa zein poly(A) tail 45
Male sterility – see under Mitochondrion
Marchantia, RNA editing 351, 361
matK 319–320
Membrane proteins 230
Meristem, flowering signals 268–270
Metastable RNA 93, 98–102
Methylation
gene silencing 83 ff.
Microbody, proteolysis 287
Microtubules
eEF1A 133
eIFiso4F 118, 133
Mitochondrion
animals 306, 308
chaperones 211
fungi 306, 308
male sterility 311–312, 346
protein transport 226–227
protein turnover 310–311
proteolysis 287, 311
RNA editing 309–310, 344–351, 361–362.
RNA processing 303 ff.
regulation of gene expression 303–323
Monocot vs. dicot – see Dicot vs. monocot
Movement protein
maize 257 ff.
RNA virus 254 ff., 374
mRNA

3′ end [see also Cap, Poly(A) tail, 3′ untranslated region] 43–57
histones 149–150
5′ leader sequence – see principal heading
steady-state levels 66
5′ untranslated region – see 5′ leader sequence
mRNA decay 72–75
half-life 65
kinetics 65
nonsense mediated 74–75
mRNA stability 63–75, 89 ff.
cap 69
chloroplasts 320–323, 328–330
cis-acting factors 69–70
heat shock effects 67–68
hormonal regulation 67
light effects 68, 70, 74, 330
measurement 64–67
mitochondria 307
poly(A) tail 69
rbcS 68–69, 72
PRP-BP 72
RNAse 72
sucrose starvation effect 68
trans-acting factors 71–72
transgenes 399
translation coupling 74–75
Multibranched loop, RNA 95
Multicatalytic protease – see Proteasome
Multiubiquitin binding protein (MBP1) 282, 284
Mustard, chloroplast RNA polymerase 317
Myrosinase genes 9

nad genes 345 ff.
Nascent chain-associated complex 230
ndh 352 ff.
Near upstream element 44, 49–54
N-end rule pathway 289
Nepovirus proteinase 384
Nitrate reductase, proteolytic regulation 290–291
Nodulation 237
Nodulin-24 237
N-terminal propeptides 238
Nuclear bodies 30–32
Nuclear photosynthetic-deficient mutant 330
Nuclear pore complex 227–228
Nuclear run-on transcription 64–65
Nucleic acid transport 251–270
Nucleolar vacuole 29
Nucleolus-associated body 30–31
Nucleoplasmin – see under Chaperones
Nucleus
protein targeting 261
protein transport 227–228
proteolysis 277–284

Oenothera, mtRNA processing 304
Ω leader 152, 172, 394–395
opaque2
eEF1A overexpression 125
5′ leader sequence 166
Open reading frame
5′ leader sequence 152–153, 165–168
overlapping 168–173, 376
orf genes 346 ff.

OSH1 260

p102 148
Paperclip ribozyme 98
Pararetrovirus, poly(A) signal 47–49
Pea, mtRNA processing 305 ff.
Peptidyl-prolyl *cis/trans* isomerase (PPIase) 199, 211
Peribacteroid membrane 237
Perichromatin fibrils 26
Peroxisome, protein transport 227
pet genes
    A 337
    B 337, 359
    D 95, 334 ff.
Petunia
   gene silencing 80 ff.
   mitochondrial RNA processing 309
   RNA editing 348, 361
PFL locus 132
Phage RNA 104
Phaseolin 202
Phloem
   proteins 267 ff.
   viral transport 257
Phosphorylation 295
   cap-binding protein 118–119
   chloroplast 323, 335
   eEF1 124
   eEF1B 127
   eEF2 128
   eIF2 109–111
   hypoxia effect 132
   poly(A) polymerase 56
   ribosome proteins 132
Photosynthetic complexes 337
Photosystem mutants 335, 338
PHYA, mRNA decay 72–75
Phytochrome A 292
Plasma membrane, protein transport 237
Plasmid copy number 100
Plasmodesmata 251–270
   KN1 transport 260–261
   size exclusion limit 252 ff.
   structure 252–254
   supracellular control proteins 261 ff.
   viral RNA/DNA movement 255
Plastid transformation 359 ff.
Pollen development 311–312
Polyadenylation – see Poly(A) tail
Polyadenylation factor 54
Poly(A) binding protein 53, 146–147, 173
   mRNA stability 71
Poly(A) polymerase 53, 55–56
Poly(A) signal/poly(A) tail
   Agrobacterium 45–46
   cleavage factors 53–54
   far upstream element 44, 49–54, 56
   histone mRNA 45–46
   maize 27 kDa zein 45
   molecular architecture 49–53
   mRNA stability 69
   near upstream element 44, 49–54
   processing 44–57
   regulation of translation 146 ff.

T-DNA genes 45–46
Polycistronic transcripts in chloroplasts 317, 330–331
Polyhydroxybutyrate 401–402
Porphyra pupurea 308
Post-transcriptional gene silencing 79–87
Potato, mitochondria 308
Potexvirus, gene silencing 82
Potyvirus
   gene silencing 82
   proteinase 383
Pre-mRNA splicing – see Splicing
Pre-protein sequences 402
Prenuclear body – see Dense body
Prion 196
Programmed cell death 293–294
Prolamine 202, 240
Promoter
   chloroplast 316–317
   gene silencing 80 ff.
   intron enhancement 25, 396
   maize tubulin gene 395
   mitochondrial 304–306, 308
   mRNA half-life measurement 66–67
   rice actin gene 395
   Top10 67
Pro-protein sequences 402
Protease – see also Proteolysis
   calpain 285
   KEX2 287
   La 287
   NIa 383
   P1 383
   Tsp 289
   virus-encoded 383 ff.
Proteasome
   S complex 280–283
   S 280 ff
   S 277 ff.
Protein accumulation 294–295
Protein disulfide isomerase 203
Protein folding 191–212
   Rubisco 205, 208
Protein half-life 277, 290–291
Protein kinase, dsRNA dependent 93
Protein secretion 228–235
Protein transport 223–241, 251–270
   chloroplast 224–225
   endoplasmic reticulum to Golgi 233–235
   Golgi complex 235–237
   HSP70 194, 210–211
   mitochondria 226–227
   nucleus 227–228
   peroxisome 227
   plasma membrane 237
   thylakoid membrane 225–226
   vacuole 237–240
Protein turnover in mitochondria 310–311
Proteolysis (see also Protease) 275–296
   biotechnology applications 294–295
   carnivorous plants 290
   chloroplast 286–287
   cytoplasmic 277–284
   endoplasmic reticulum 287
   enzymes 290–291

functions 287–294
heat schock proteins 196
immunization 295
microbody 287
mitochondria 287, 311
nuclear 277–284
plant defense 286
RNA virus 382 ff.
storage proteins 290
ubiquitin independent 285
vacuolar 285–286
Proximal pyrimidine tract, 3′ splice site 13–14
PRP proteins 5
psa genes
  A 95, 331–333
  B 333
psb genes
  B 336
  C 333 ff.
  D 333 ff.
Pseudoknot – see under Tertiary structure
Pulse-chase 66
PvPRP1, mRNA stability 69, 71–72

rbcS
  mRNA stability 68–69, 72
  poly(A) signal 44 ff.
Readthrough – see Termination suppression
Red clover necrotic mosaic virus, movement protein 254–258
Redox potential, translation control 335
Release factor 129–130, 163
Restorer gene 311–312
Ribosome 130–133
  mitochondrial regulation 310
  pausing 151
  proteins 130–133
  RNA editing 358
  shunt 167–168
Ribosome-inactivating proteins 133
Ribozyme
  paperclip 98
  structure 97–98
Rice
  endotoxin transgene expression 401
  mitochondria 307
  OSH1 260
  RNA editing 348, 356
Rice tungro bacilliform virus
  AUU start codon 164
  5′ leader sequence 167–168
  poly(A) signal 47–52
Rieske FeS protein 209
RNA – see principal headings dsRNA, mRNA, rRNA, snoRNA, tRNA
RNA-binding protein, chloroplast 321–323, 336
RNA degradation, chloroplast 329
RNA-dependent ATPase – see Helicase
RNA:DNA interaction 85
RNA editing 343–362
  biochemistry 349–350
  chloroplasts 351–362
  evolution 350, 355–357
  5′ leader 358
  mitochondria 309–310, 344–351, 361–362

protein effects 349
  sequence guide 360
  specificity 350
  timing 348
  transformants 360
RNA electroporation, mRNA half-life measurement 65–66
RNA polymerase
  chloroplast 316–317
  mitochondria 306
  RNA-dependent 369
RNA processing 43–67, 89 ff.
  chloroplasts 318, 321–323, 328–331
  mitochondria 306–308
  mutants 323
RNA structure – see also Secondary structure, Tertiary structure
  kinetics 99
  prediction 99–102
  regulation of gene expression 89–102
  structural motifs 90–98
RNA transport 133, 260
RNA virus
  frameshifting 377–380
  gene expression 367–386
  movement proteins 254 ff.
  proteolytic processing 382–385
  readthrough 380–382
  scanning 376
  start codons 169, 375–379
  subgenomic RNA 368 ff.
  translation 146–155, 159–179, 375–377
RNAse
  dsRNA specific 93
  mRNA stability 72
  RNAse P 98, 321
Rotomase 199
rRNA 304–308
  editing 351, 353, 361
  leader sequence structure 98
rpl2 359–360
rpoB 355 ff.
rps 320, 346 ff.
Rubisco 205, 208
  proteolytic control 291
Run-on transcription 64–65

Satellite tobacco necrosis virus, translation 150–151
Scanning – see under Initiation of translation
Secondary structure, RNA (see also Hairpin, Stem-loop) 90–95
  effect on splicing 20–21
  inhibition of translation 151
  5′ leader sequence 151, 167
  poly(A) signal 52
  ribosome translocation 162
  transcription termination 307
  translation efficiency 94·
Secretin-defective mutants 236
Seed development 197
Seed germination 285–286, 289
Seed storage protein folding 201
Selenocysteine 178
Self-incompatibility alleles 21
Self-splicing
  chloroplast RNA 318, 331
  role of base triples 96

Senescence, proteolysis 286, 294
Sense-antisense dsRNA 90
Shine–Dalgarno sequence 94, 100
    chloroplast genes 334
Short-stop RNA 49
Sigma-like factors 317
Signal anchor 229
Signal peptide 229
Signal recognition particle 230–231
Signal transduction
    dsRNA 93
    HSP90 195–196
Size exclusion limit, plasmodesmata 252 ff.
Slippery sequence 97, 377
Small Auxin-Up RNA (SAUR) genes 69–71
Small plastid RNA 320
SNARE proteins 234, 236
snoRNA 27, 32
Snurposome 29
Sobemovirus proteinase 385
Soybean
    mitochondria 305
    SAUR transcripts 69–71
Spinach, chloroplast transcription 317
Spliceosome 3 ff.
    A complex 6
    B complex 6
    C complex 6
    E complex 5
Splicing (see also Self-splicing) 1–32
    alternative 21–24
    chloroplast RNA 318, 320, 331–333
    intron effect 397
    mechanism 3–8
    mitochondrial RNA 309–310
    polycistronic mRNAs 170–171
    RNA editing 348, 359
    species differences 8–9, 22–24
    3′ splice site 7, 10–12, 16–17, 22
    5′ splice site 6, 9–10, 15–16, 22
    trans-splicing in chloroplast 95, 331–333
    unspliced RNA 21
SR protein 5–6, 10, 15–17, 27–28
SRK genes 23
SRS4, mRNA decay 72–74
Start codons (see also Initiation of translation) 108, 162–164, 168–170
    Chlamydomonas 334
    chloroplast 334
    editing 346, 352
    mitochondria 310
    structure 94
    RNA virus 169, 375–379
Stem-loop 93–95
    chloroplast RNA 329 ff.
    frameshifting 378
    histone mRNA 149–150
    5′ leader sequence 167
    mitochondrial RNA 307
    splicing 7–9, 12
    transcription termination 95
    3′-UTR 95
Stop codon – see under Termination of translation
Storage protein proteolysis 290

Stress (see also Heat shock)
    eEF1A 124–125
    HSP70 194
    HSP90 196
    proteolysis 286, 288
    ribosome proteins 132–133
    splicing 26
    translation 153–154, 178
Subgenomic promoter 369 ff.
Subgenomic RNA 368 ff.
Sucrose-binding protein 237
Sucrose starvation
    mRNA stability 68
Sucrose synthase genes 24–25
Sunflower,
    mitochondria 312
SUP 266–267
Suppression – see under Termination of translation
Supracellular control protein 261 ff.
    cofactor 261 ff.
Symplasm concept 252
Syntaxin 239
Synthetic genes 398–401

Tail anchor 230
TCP-1 197
T-DNA, poly(A) tail 45–46
Termination, translation 129–130, 163
    stop codon 163
    stop codon editing 346
    suppression 130, 176, 380–382
    suppressor tRNA 176–178, 382
Tertiary structure, RNA 95–98
    bifurcations 92, 95
    pseudoknot 92, 96, 148–149, 167, 378
    triple helix 92, 95–96
    tRNA 96
Thermotolerance 196, 198
Thioredoxin 335
Thylakoid membrane
    protein transport 225–226
    mRNA docking 336
α-TIP 240
Tobacco
    gene silencing 80 ff.
    RNA editing 356
    transgenic 80 ff.
Tobacco etch virus, gene silencing 82, 84
Tobacco mosaic virus, movement proteins 254, 258
Tobamovirus, pseudoknots 148
Tomato
    floral morphogenesis 262
    transgenic 82
Tonoplast, protein transport 240
Top10 67
Transactivation 170–171
Transcription
    chloroplast 316, 317, 334
    gene silencing 79–87
    inhibitors for mRNA half-life measurement 65
    initiation in chloroplasts 316, 317
    light effect 330
    mitochondria 304–306
    regulation by supracellular control proteins 261 ff.

414

termination 57, 95
Transcriptional active complex 359
Transgene
    gene silencing 79–87
    3′ end 402
    5′ leader sequence 394–395
    optimization of expression 393–403
Transgenic plants, gene silencing 79–87
Transit peptide 224
Translation – see Elongation, Initiation, Splicing, Termination
Translocation chain-associated membrane protein 231
Transposable element 8, 23–24
TriC 197
Triple helix, RNA 92, 95–96
tRNA
    editing 351, 353 361
    mitochondrial processing 308–309
    suppressor 176–178
    tertiary structure 96
tRNA-like element, viral 96
tscA 95, 331–333
T-urf 346
Tymovirus, proteinase 383

UA-rich sequence – see AU-rich sequence
Ubiquitin 277–294
    activating enzyme (E1) 279 ff.
    activation 279
    C-terminal hydrolase 280
    conjugating enzyme (E2) 279 ff.
    functions 288–294
    genes 278–279
    multichains 280
    protein accumulation 294–295
    protein ligase (E3) 279 ff.
    structure 277–278
UFO 266–267
3′ untranslated region (3′-UTR) – see also Poly(A) tail, Termination
        of translation
    AUUUA motif 70–71
    chloroplast mRNA 321, 329

effect on initiation of translation 173–174
    poly(A) site 51
    regulation of translation 148–150
    SAUR transcripts 70–71
    stem-loop structures 95
    sucrose synthase gene 24
    viral 148–149
5′ untranslated region – see 5′ leader sequence
U snRNAs 4 ff.
U snRNPs 4 ff., 54–55

Vacuole
    protein transport 237–240
    proteolysis 285–286
Viral resistance 80 ff.
Viral transport 257
Viroid
    replication 100–102
    self-cleavage 98
Virusoids
    self-cleavage 98

Water stress, effect on translation 154
Wheat
    histone H3 mRNA poly(A) tail 45
    mitochondrial editing 346 ff.
    mitochondrial gene expression 305
Wounding
    eEF1A expression 125
    plasmodesmal control 254
    proteolysis 292

Xylogenesis 294

ycf3 359
Yeast
    secretion-defective mutants 233, 236
    vacuolar protein transport 238

Zea mays – see Maize
Zein 202